THE

INTERNATIONAL SERIES OF MONOGRAPHS ON PHYSICS

GENERAL EDITORS

W. MARSHALL   D. H. WILKINSON

THEORY OF THERMAL NEUTRON SCATTERING

The Use of Neutrons for the Investigation of Condensed Matter

W. Marshall and S. W. Lovesey

Atomic Energy Research Establishment
Harwell

OXFORD
AT THE CLARENDON PRESS
1971

Oxford University Press, Ely House, London W. 1

GLASGOW NEW YORK TORONTO MELBOURNE WELLINGTON
CAPE TOWN SALISBURY IBADAN NAIROBI DAR ES SALAAM LUSAKA ADDIS ABABA
BOMBAY CALCUTTA MADRAS KARACHI LAHORE DACCA
KUALA LUMPUR SINGAPORE HONG KONG TOKYO

PRINTED IN GREAT BRITAIN
AT THE UNIVERSITY PRESS, OXFORD
BY VIVIAN RIDLER
PRINTER TO THE UNIVERSITY

PREFACE

In the last decade there has been a rapid expansion in the use of neutron-beam techniques to investigate the properties of solids, liquids, and gases. Nevertheless there is no easy access to the theory of neutron scattering from these materials, and workers in the subject are obliged to work through the original papers to get the necessary theoretical background. The intention of this book is therefore to give a comprehensive review of all the theory that is relevant to the interpretation of experiment.

Inevitably, the limitations of time and space have obliged us to omit many points of interest. We have given almost nothing on experimental techniques, on extinction, and on the refraction of long wavelength neutrons. Liquids are given a long chapter although they deserve a book to themselves. The discussion of critical phenomena is handicapped by the fact that many ideas and results are being published as the book goes to press; we have therefore been content to describe simple theories of critical phenomena and merely indicate some of the more recent work. Except in a few special cases we have not aimed to quote original references for the various ideas or observations which are described in the book. Parts of Chapter 6, on the theory of magnetic scattering from salts, and of Chapter 8, on the use of correlation functions in magnetic scattering, have not previously been published, as far as I know.

This is a book that has been started many times but finished only once. The first attempt was made in collaboration with my colleague Roger Elliott more than a decade ago. But we both found research too time-demanding for the more leisurely exercise of book-writing to be given any attention. A second attempt with Marty Blume as partner suffered the same fate. A third, solo, attempt was smothered by growing responsibilities and their demands for time.

The modest idea of these three early attempts was simply to revise, update, and extend the lecture notes that I had prepared for a course at Harvard in 1959. The fourth attempt started inauspiciously: Stephen Lovesey rejected the modest idea and argued for a fresh start altogether, not based on material written a decade earlier supplemented with miscellaneous lectures and notes I had prepared since then. His argument seemed logical but likely to set back the completion date for another decade. This pessimism was unfounded, because fifteen months later he produced a manuscript which, with little alteration, has now become this book.

W. M.

AERE, Harwell
May 1970

ACKNOWLEDGEMENTS

WE are pleased to acknowledge the help we have received from various discussions with our colleagues, particularly members of the Materials Physics and Theoretical Physics Divisions of the Atomic Energy Research Establishment, Harwell. Particular thanks are due to Dr. J. Hubbard and Dr. R. D. Lowde. One of us (S. W. L.) wishes to thank Dr. E. Balcar for acting as the devil's advocate during the writing of the manuscript, and Margaret Lovesey because without her support and tolerance it could not have been accomplished.

Sid Marlow made calculations for some of the figures in Chapters 1 and 15. We are indebted to the following for permission to use diagrams as a basis for figures in the text: G. E. Bacon, S. Brimberg, R. J. Elliot, R. A. Erickson, J. M. Hastings, Yu. Izyumov, W. C, Koehler, J. U. Koppel, G. G. E. Low, M. Medvedev, H. A. Mook, R. Moon, R. Nathans, G. F. Nardelli, M. P. Schulhof, G. Shirane, C. G. Shull, A. Sjölander, H. Smith, M. W. Stringfellow, K. C. Turberfield, P. Vashishta, F. J. Webb, W. Whittemore, J. Young, *Acta Crystallographica*, The Royal Society, Pergamon Press, Academic Press, American Institute of Physics, North-Holland Publishing Company, *The Physical Review* and *Physical Review Letters*, Institute of Physics and The Physical Society (London). D. E. Rimmer and N. W. Dalton kindly supplied the tables of spin traces in Chapter 13.

The manuscript was completed in a concentrated work period at the Research Establishment at Risø, Denmark. We thank the Director of Risø for his hospitality and are grateful to our Danish colleagues, Hans Bjerrum-Møller, Ove Dietrich, Jans Als-Nielson, Per-Ander Lindgard, and Bente Lebech, each of whom read part of the manuscript during our stay with them. Finally we thank Ann and Margaret, because without their encouragement and patience we could not have stolen time from our families to complete this book.

CONTENTS

LIST OF IMPORTANT SYMBOLS

$A = \{(i+1)b^{(+)}+ib^{(-)}\}/(2i+1)$ (1.27 a)

$\quad = \bar{b}$ coherent scattering amplitude (1.28 a)

$\mathbf{a}_1, \mathbf{a}_2, \mathbf{a}_3$ basic vectors of unit cell (2.1)

$\hat{A}$ Hermitian operator (Appendix B)

$\hat{a}^+, \hat{a}$ Bose creation and annihilation operators (Chapters 3 and 4)

$A_{\alpha\beta}(l, l')$ second-order derivatives of crystal potential (4.3) and (4.26)

$\mathscr{A}_{\alpha\beta}(\mathbf{q})$ dynamical matrix (4.11 b) and (4.28)

$a_\lambda^\Lambda(\mathbf{d}; \mathbf{k})$ tight binding transformation coefficients (5.72)

$B = 2\{b^{(+)}-b^{(-)}\}/(2i+1)$ (1.27 b)

$\hat{B}$ Hermitian operator (Appendix B)

$b^{(+)}$ scattering length for $i+\frac{1}{2}$ state } Chapter 1

$b^{(-)}$,, ,, $i-\frac{1}{2}$ state }

$\hat{b}$ scattering amplitude operator (1.26)

$\bar{b}$ coherent scattering length (1.24)

$\mathscr{b}_l = \bar{b}_l \exp[-\frac{1}{2}\langle\{\boldsymbol{\kappa}.\hat{\mathbf{u}}(l,0)\}^2\rangle]$ (15.7)

$B_s[x]$ Brillouin function (13.22)

$\hat{c}, \hat{c}^+$ Fermion operators for electrons (§ 5.5 and Chapter 9)

c velocity of light (Chapter 5), c concentration of impurities (Chapters 14 and 15)

C_P, C_V specific heats at constant pressure, C_P, and constant volume, C_V (Chapter 11)

c_ξ concentration of ξ isotope (1.21)

c_0 velocity of sound (11.138)

$$\operatorname{curl}\mathbf{A} \equiv \boldsymbol{\nabla}\times\mathbf{A} = \tilde{\mathbf{x}}\left(\frac{\partial A_z}{\partial y}-\frac{\partial A_y}{\partial z}\right)+\tilde{\mathbf{y}}\left(\frac{\partial A_x}{\partial z}-\frac{\partial A_z}{\partial x}\right)+\tilde{\mathbf{z}}\left(\frac{\partial A_y}{\partial x}-\frac{\partial A_x}{\partial y}\right)$$

$\mathbf{d}$ position vector of atom within unit cell (2.2)

$\mathbf{D}(\mathbf{m}-\mathbf{n})$ anisotropic exchange parameter (13.55)

$\mathscr{D}(\mathbf{q})$ Fourier transform of $\mathbf{D}(\mathbf{m}-\mathbf{n})$ (13.56)

D spin-wave stiffness (Chapter 9)

$D(T)$ temperature-dependent spin-wave stiffness (Chapter 9)

$\mathrm{d}\tilde{\mathbf{k}}' = \mathrm{d}\Omega$ (Chapters 1 and 2)

$\mathrm{d}\sigma/\mathrm{d}\Omega$ differential cross-section (§ 1.2 and Appendix A)

$\mathrm{d}^2\sigma/\mathrm{d}\Omega\mathrm{d}E'$ partial differential cross-section ((1.10) and Appendix A)

$\mathscr{D}^{(J)}_{MK}(\alpha\beta\gamma)$ rotation matrix (12.14)

$$\operatorname{div}\mathbf{A} = \boldsymbol{\nabla}.\mathbf{A} = \frac{\partial A_x}{\partial x}+\frac{\partial A_y}{\partial y}+\frac{\partial A_z}{\partial z}$$

E, E' initial and final energy of neutron, respectively (§ 1.2)
$\mathscr{E}_\lambda(\mathbf{k})$ band energy (5.46)
$\mathscr{E}$, $\mathscr{E}_\mathbf{q}$ reduced spin-wave energy (§ 9.6 (*b*) and § 15.2)
e charge of single electron
$\mathbf{E}$ electric field due to atomic electrons and nuclei in target ((5.1) and § 10.3)
E_λ, $E_{\lambda'}$ initial and final target energies (§ 1.2)

$f(\mathbf{k},\mathbf{k}')$ scattering amplitude (1.8 b)
$F(\boldsymbol{\varkappa})$ atomic form factor ((5.15), § 5.3, (6.120), and (6.121))
$\mathscr{F}(\boldsymbol{\tau})$ magnetic unit-cell vector structure factor (7.16)
$\mathscr{F}_\mathrm{d}(\boldsymbol{\varkappa})$ vector form factor (7.21)
$F_\mathrm{N}(\boldsymbol{\tau})$ nuclear unit-cell structure factor (2.12)
$F_\mathrm{M}(\boldsymbol{\tau})$ magnetic unit-cell structure factor (7.27 a)
$F_\mathrm{E}(\boldsymbol{\tau})$ electrostatic unit-cell structure factor (10.56)
$F^{\alpha\beta}(\boldsymbol{\varkappa},\omega)$ spin spectral weight function (Chapter 8 and (13.15))
$f(r)$ radial part of one-electron wave function (§ 5.2 and (6.19))

g gyromagnetic ratio, Landé splitting factor (5.24)
$g_{OP}(\omega)$ spectral weight function (§ B.5)
$g_{OP}(t) = \langle\hat{O}(t)\hat{P}(0)\pm\hat{P}(0)\hat{O}(t)\rangle$ Fourier transform of spectral weight function (Appendix B, (B.36))
$G^{<}_{OP}(\omega)$ spectral function of arbitrary operators $\hat{O}$ and $\hat{P}$ (Appendix B, (B.33 a))
$G^{>}_{OP}(\omega) = \exp(\hbar\omega\beta)G^{<}_{OP}(\omega)$ spectral function (Appendix B, (B.33 b) and (B.35))
$G^{(\pm)}_{OP}$ advanced ($-$) and retarded ($+$) thermal two-time Green functions (§§ 8.3 and B.8)
$G(\mathbf{r},t)$ pair correlation function (3.9)
$G_\mathrm{s}(\mathbf{r},t)$ self pair correlation function (3.19 a)
$G^\mathrm{cl}(\mathbf{r},t)$ classical pair correlation function (3.21)
$G^\mathrm{cl}_\mathrm{s}(\mathbf{r},t)$ classical self pair correlation function (3.20)
$\tilde{G}(\mathbf{r},t) = G(\mathbf{r},t+\tfrac{1}{2}\mathrm{i}\hbar\beta)$ (3.46)
$\bar{G}_\mathrm{s}(\mathbf{r},t)$ ((3.59 a) and § 3.5)
$\bar{G}_\mathrm{d}(\mathbf{r},t)$ ((3.59 b) and § 3.5)
$G'(\mathbf{r},t)$ (3.26 a)
$G'_\mathrm{s}(\mathbf{r},t)$ (3.27 a)

$g(\mathbf{r})$ static pair-distribution function (3.23)

$\operatorname{grad}\phi = \nabla\phi = \tilde{\mathbf{x}}\frac{\partial\phi}{\partial x}+\tilde{\mathbf{y}}\frac{\partial\phi}{\partial y}+\tilde{\mathbf{z}}\frac{\partial\phi}{\partial z}$

$G_l'(\mathbf{r},t)$ (8.9)

$G_l(\mathbf{r},t)$ (8.9)

$\mathscr{H}$ quantum mechanical Hamiltonian

$\mathscr{H}$ classical Hamiltonian

H, $\mathbf{H}$ magnetic field (Chapters 8, 9, and 13, and Appendix B)

H_{A} single-ion anisotropy field (§ 9.6 (*a*))

h_{A} reduced single-ion anisotropy field (9.245)

$\hbar$ Planck's constant/2π

$\hat{\mathbf{i}}$ nuclear spin operator of magnitude i

I Coulomb matrix element (9.95)

$I(\boldsymbol{\kappa},t)$ spatial Fourier transform of pair correlation function (3.15 b)

$I_{\mathrm{s}}(\boldsymbol{\kappa},t)$ spatial Fourier transform of self pair correlation function (11.2 b)

Im imaginary part

$\hat{\mathbf{J}} = \hat{\mathbf{L}}+\hat{\mathbf{S}}$ total angular momentum operator of magnitude J (§ 5.3 and Chapter 6)

$J(\mathbf{m}-\mathbf{n})$ exchange parameter in Heisenberg Hamiltonian between ions at lattice sites **m** and **n** (Chapter 9)

$\mathscr{J}(\mathbf{q})$ spatial Fourier transform of exchange parameter J (9.10 a)

$J^{(n)}$ moments of exchange parameter (9.240), see also $\overline{l^n}$

$j_K(x)$ spherical Bessel function of order K

$\bar{j}_K(\kappa)$ radial integral ((5.20) and (6.38))

$\mathbf{k}$ wave vector of incident neutron, electron momentum (Chapters 5 and 9)

$\mathbf{k}'$ wave vector of scattered neutron

k_{B} Boltzmann's constant

k_f electron wave vector at Fermi surface (§ 9.2)

$\mathscr{k}$, $\mathscr{k}'$ initial and final neutron wave vectors in centre of mass frame of reference (§ 12.2)

l cell indices, $\mathbf{l}$ lattice vector — use also m, n, and k (2.1)

$\overline{l^n}$ moments of exchange parameter (9.22), see also $J^{(n}$

l orbital quantum number (Chapters 5 and 6)

$\hat{\mathbf{l}}$ electron orbital angular momentum operator (Chapter 6)

$\hat{\mathbf{L}}$ total orbital angular momentum operator of magnitude L (Chapter 5)

m mass of neutron
m_e mass of electron
M mass of nucleus
$M^\alpha = -g\mu_B\langle\hat{S}^\alpha\rangle$ magnetic moment
M_L, M_S total orbital and spin magnetic quantum numbers

$$-L \leqslant M_L \leqslant L, \quad -S \leqslant M_S \leqslant S$$

(Chapter 6)
m_λ^* effective mass (§ 5.5)

N number of unit cells in crystal (or nuclei of same type)
$n_j(\mathbf{q}) = [\exp\{\beta\hbar\omega_j(\mathbf{q})\}-1]^{-1}$ (4.18 a)
$n(\omega) = [\exp(\hbar\omega\beta)-1]^{-1}$ Bose factor
$\tilde{\mathbf{n}} = \mathbf{k}'\times\mathbf{k}/(k^2\sin\theta)$ (10.25)
n number of electrons in incomplete atomic shell (§ 6.3)
$\hat{n}_\nu$ particle number operator (5.50)
$N(k)$, $N(\lambda)$ incident flux of neutrons (§§ 1.1 and 2.4)

$\hat{O}$ arbitrary operator (Appendix B)
$\mathscr{O}$ = average over nuclear spin orientations (§ 10.4)

$\mathbf{P}$ polarization vector of incident neutrons (§ 10.1)
$\mathbf{P}'$ polarization vector of scattered neutrons (§ 10.2)
$\hat{\mathbf{p}} = -i\hbar\boldsymbol{\nabla}$ linear momentum operator (3.37)
$\hat{\not{p}}$ linear momentum density operator (5.91), see also (6.3)
$\hat{\mathbf{p}}(l)$ nuclear momentum operator (4.16 b)
p_σ incident neutron spin probability (§§ 1.2 and 10.2)
p_λ probability distribution for initial target states (§ 1.2)
$\mathrm{P}\int$ principal part integral
$\mathbf{P}_\perp = \tilde{\boldsymbol{\varkappa}}\times(\mathbf{P}\times\tilde{\boldsymbol{\varkappa}})$ (10.79)
$P_{\alpha\beta}(l, l'; \omega)$ unperturbed phonon Green function (15.15)
$\mathscr{P}^j(\mathbf{q})$ (15.24)
$\mathscr{P}(\omega)$ reduced unperturbed phonon Green function (15.34)
$P(m, n; \mathscr{E})$ unperturbed ferromagnetic spin-wave Green function (15.87) and (15.89)
$\hat{P}$ arbitrary operator (Appendix B)

$\mathbf{q}$ reciprocal vector
$\hat{\mathbf{Q}}_\perp$ total magnetic interaction operator ((5.10) and § 6.4)
$\hat{\mathbf{Q}}$, where $\hat{\mathbf{Q}}_\perp = [\tilde{\boldsymbol{\varkappa}}\times(\hat{\mathbf{Q}}\times\tilde{\boldsymbol{\varkappa}})]$ ((5.12 a), (6.57), and § 6.4)
$\hat{\mathbf{Q}}^{(\mathrm{D})}$ dipole approximation to $\hat{\mathbf{Q}}$ ((5.19) and § 6.5)

$\mathbf{R}_{ld} = \mathbf{l}+\mathbf{d}$ position vector of nucleus (2.3)
$R_{AB}(t)$ relaxation function (Appendix B, § B.4)
$\mathscr{R}_{\mathbf{q}}(t)$ spin relaxation function ((8.31) and Chapter 13); see also (8.89)
$\mathscr{R}_{\mathbf{q}}(\omega)$ Fourier transform of spin relaxation function
r number of atoms in a unit cell (Chapter 2) and number of nearest-neighbour ions to given ion (Chapter 9)
Re real part

$\hat{S}$ spin operator of magnitude S for magnetic ion (Chapter 5)
$\hat{s}$ spin operator for a single electron (Chapter 5)
$\hat{\mathfrak{s}}(\mathbf{r})$ electron spin density operator (5.59)
$\hat{\mathfrak{s}}(\mathbf{q})$ Fourier transform of electron spin density operator (5.60)
$S(\boldsymbol{\varkappa}, \omega)$ scattering law (Chapter 3, nuclear, and Chapter 8, magnetic)
$\hat{S}^{\pm} = \hat{S}^{x} \pm i\hat{S}^{y}$ spin angular momentum raising and lowering operators
$\tilde{S}(\boldsymbol{\varkappa}, \omega) = \exp(-\frac{1}{2}\hbar\omega\beta)S(\boldsymbol{\varkappa}, \omega)$ (3.45)
$S_{\mathrm{i}}(\boldsymbol{\varkappa}, \omega)$ 'incoherent' nuclear scattering law (3.18)
$\hat{S}^{\alpha}_{\mathbf{q}} = \sum \exp(-i\mathbf{q}.\mathbf{l})\hat{S}^{\alpha}_{l}$ (Appendix B, (B.81 b), Chapters 5, 8, 9, and 13)
$\hat{S}^{\pm}_{\mathbf{q}} = \sum \exp(\mp i\mathbf{q}.\mathbf{l})\hat{S}^{\pm}_{l} = \hat{S}^{x}_{\pm\mathbf{q}} \pm i\hat{S}^{y}_{\pm\mathbf{q}}$ (Appendix B, (B.81), Chapters 9 and 13)

T absolute temperature $\beta = (k_{\mathrm{B}}T)^{-1}$
t time variable
t_1, t_2, t_3 reciprocal lattice indices (2.5)
Tr ≡ trace
$T^{dd'}_{\Lambda\Lambda'}(\mathbf{l}-\mathbf{l}')$ (5.77)

$\hat{u}_{\bar{\alpha}}(l, t)$ displacement operator (4.16 a) and (4.30)
$u_{\mathbf{k}\lambda}(\mathbf{r}) = \exp(-i\mathbf{k}.\mathbf{r})\psi_{\mathbf{k}\lambda}(\mathbf{r})$ periodic part of Bloch function (5.45 a)
$u_{\mathbf{q}}$, $\bar{u}_{\mathbf{q}}$ transformation coefficients (§§ 9.4, 9.6, and 9.7)

V volume of target
v_0 volume of unit cell (2.7)
$\hat{V}(\boldsymbol{\varkappa})$ Fourier transform of total neutron–target interaction potential (§ 1.2 and Appendix A)
$\hat{V}_{\mathrm{E}}(\boldsymbol{\varkappa})$ Fourier transform of total electrostatic interaction potential (10.24)
$\hat{V}_{\mathrm{N}}(\boldsymbol{\varkappa})$ Fourier transform of total nuclear interaction potential (§§ 1.3 and 1.4, and (10.20))
$\hat{V}_{\mathrm{M}}(\boldsymbol{\varkappa})$ Fourier transform of total magnetic interaction potential (§ 5.2 and (10.22))
$v_{\mathbf{q}}$, $\bar{v}_{\mathbf{q}}$ transformation coefficients (§§ 9.4, 9.6, and 9.7)

v quantum numbers needed when the set $SM_{\mathrm{S}}\,LM_{\mathrm{L}}$ fail to define atomic state uniquely (6.49)

$\langle V(\mathbf{r})\rangle$ mean potential seen by neutron (2.14)

$\delta\hat{V}(\mathbf{r})$ deviation in potential from mean value (2.15)

$V(\boldsymbol{\tau}) = \int \mathrm{d}\mathbf{r}\,\exp(\mathrm{i}\boldsymbol{\tau}.\mathbf{r})\langle V(\mathbf{r})\rangle$ (2.20 b)

v_{ld} potential due to atom at site $\mathbf{R}_{ld}$ (2.26)

$W(\boldsymbol{\varkappa})$ Debye–Waller factor (4.35), (4.38) and (4.51)

$W_{k\to k'}$ Golden rule transition probability ((1.3) and Appendix B, § B.6)

$w(t), \bar{w}(t)$ width function (11.14)

$Y_Q^K(\theta, \phi)$ spherical harmonic of rank K and order Q: definition in accord with A. R. Edmonds, *Angular momentum in quantum mechanics*, Princeton University Press, 1960.

$Z = \mathrm{Tr}\exp(-\beta\mathscr{H})$ partition function

$Z(\omega)$ normalized vibrational density of states ((4.39) and Chapter 15)

Z number of electrons associated with ion (§ 10.3)

$z(\tau)$ number of reciprocal lattice vectors with magnitude $|\boldsymbol{\tau}| = \tau$ ((2.41) and Chapter 4, Table 4.3)

α (with β and γ) Euler angle (§ 12.1)

α (with β) Cartesian component index

α coefficient of thermal expansion (11.135)

$\hat{\boldsymbol{\alpha}}$ vector part of general interaction potential between neutron and target (10.26)

β (with α and γ) Euler angle (§ 12.1)

β (with α) Cartesian component index

$\beta = (k_{\mathrm{B}}\,T)^{-1}$

$\hat{\beta}$ scalar part of general interaction potential between neutron and target (10.26)

γ (with α and β) Euler angle (§ 12.1)

$\Gamma(\mathbf{q}, \omega)$ width in energy (§ 4.4, (9.79), (15.64), and Appendix C)

$r\gamma_{\mathbf{q}} = \sum \exp(\mathrm{i}\mathbf{q}.\boldsymbol{\rho})$ (9.15)

$\gamma(\boldsymbol{\varkappa})$ Fourier transform of static pair distribution function (11.4 a)

$\gamma(t)$ (3.57) and (4.96)

$\hat{\Gamma}_{jj'}(t)$ (3.52)

$\Gamma_{\boldsymbol{\varkappa}}^{\alpha}$ width in energy (§ 13.3 (*d*), (*e*))

$\gamma = -1{\cdot}91$ gyromagnetic ratio of neutron (5.2)

Γ_n Bethe's notation for irreducible representations of cubic group (§§ 6.6 (*b*), 15.2)
$\Gamma_{\alpha\beta}(\mathbf{r},t)$ (8.5)
$\gamma_{\alpha\beta}(l,t)$ (8.8)
$\gamma'_{\alpha\beta}(l,t)$ (8.8)

$\delta(x)$ Dirac delta function
$\delta_{l,m}$ Kronecker delta function
Δ band splitting (9.110 b)
$\Delta(\mathbf{q},\omega)$ real part of self-energy (shift) (§ 4.4, (9.79), (15.64), and Appendix C)

$\epsilon^{\alpha\beta\gamma}$ antisymmetric unit tensor with three indices (10.35)
ϵ perturbation parameter (15.96)

$\zeta(n)$ Riemann's zeta function of order n (9.32)

$\tilde{\boldsymbol{\eta}}_{ld}(\boldsymbol{\kappa})$ unit vector in direction of spin at site $\mathbf{R}_{ld}$ (7.23)
η, η' first, η, and second, η', viscosity coefficients (§ 11.4 (*b*))
η_{B} bulk viscosity coefficient (§ 11.4 (*b*))

$\theta(t)$ unit step function ((8.48) and Appendix B, (B.12))
θ scattering angle (§ 1.2)
θ, θ' shorthand for vSL and $v'S'L'$ defining terms of l^n (6.49)
$\bar{\theta}$ shorthand for $\bar{v}\bar{S}\bar{L}$ defining terms of l^{n-1} (6.49)

$\boldsymbol{\kappa} = \mathbf{k}-\mathbf{k}'$ scattering vector (§ 1.3)
κ_S adiabatic compressibility (11.154)
κ_T isothermal compressibility (11.135)

λ perturbation parameter (15.3)
λ, Λ band indices (§ 5.5)
λ thermal conductivity (§ 11.4 (*b*))
λ, λ' quantum numbers specifying initial and final target states (§ 1.2 and Appendix A)

$\mu_{\mathrm{B}} = e\hbar/2m_{\mathrm{e}}c$ Bohr magneton (Chapter 5)
$\mu_{\mathrm{N}} = e\hbar/2m_{\mathrm{p}}c$ nuclear magneton (Chapter 5)
$\hat{\mu}$ magnetic moment operator (Chapter 5)
μ (together with ν) spherical coordinate index (Chapter 8 and Appendix B, § B.9)
μ reduced mass (§ 12.2)

ν (together with μ) spherical coordinate index (Chapter 8 and Appendix B, § B.9)

ξ isotope label (1.21)

$\Xi(\boldsymbol{\varkappa})$ Fourier transform of mean square moment fluctuation in random mixed system (14.6)

ρ perturbation parameter (15.96)

$\hat{\rho}$ density matrix (§ 10.2, Appendix B, § B.3)

ρ mean density (§§ 3.2, 11.1)

$\hat{\rho}(\mathbf{r},t) = \sum \delta\{\mathbf{r}-\hat{\mathbf{R}}_j(t)\}$ particle density operator (3.8)

$\hat{\rho}_{\mathbf{q}}(t) = \sum \exp\{-i\mathbf{q}\,.\,\hat{\mathbf{R}}_j(t)\}$ Fourier component of particle density operator (3.13)

$\boldsymbol{\rho}$ vector to nearest-neighbour ions (Chapter 9)

$\frac{1}{2}\hat{\boldsymbol{\sigma}}$ neutron spin operator (§ 1.4)

$\boldsymbol{\sigma}^j(\mathbf{q})$ phonon polarization vector for jth branch (4.11 a)

$\hat{\sigma}^\alpha$ Pauli matrices (§ 5.5)

σ, σ' initial and final neutron spin quantum numbers in cross-section; also, electron spin index (in §§ 5.5 and 9.3)

$\Sigma_{\mathbf{q}}(\omega)$ self-energy function (9.78) and Appendix C

$\sigma = \langle \hat{S}^z \rangle / S$ reduced moment (13.21)

$\sigma'_{\alpha\beta}$ stress tensor (11.130)

σ_c single (bound) nucleus coherent cross-section (§ 1.5)

$\sigma_i = \sigma - \sigma_c$ (Chapters 4 and 11)

σ_{ab} single (bound) nucleus absorption cross-section (§ 1.5 and Appendix A, § A.1)

σ total single (bound) nucleus cross-section

$\boldsymbol{\tau} = t_1\boldsymbol{\tau}_1 + t_2\boldsymbol{\tau}_2 + t_3\boldsymbol{\tau}_3$ reciprocal lattice vectors (2.5)

$\varphi_{\mathbf{k}\lambda\sigma}(\mathbf{r})$ Bloch function: λ band index, σ spin index (5.46)

$\phi_\Lambda(\mathbf{r}-\mathbf{R}_{ld})$ localized electron wave function (5.72 and § 9.3)

$\phi_{AB}(t)$ response function (Appendix B, § B.1 and (B.24 b))

$\phi_{\mathbf{q}}^{(\pm)}(t)$ wave vector dependent response function (Appendix B, § B.9 and Chapter 8)

$\phi_{\mathbf{q}}^{\alpha\beta}(t)$ wave vector dependent response function (8.29 b); see also (8.87)

χ_σ electron spin spinor (§ 5.5)

χ_0 susceptibility of an isolated ion (13.13)

$\chi_{AB}[\omega] = \chi'_{AB}[\omega] + i\chi''_{AB}[\omega]$ generalized susceptibility (Appendix B, § B.1)

$\chi_{\mathbf{q}}^{\alpha\beta}$ isothermal wave vector dependent susceptibility (Appendix B, (B.97))

$\chi_{\mathbf{q}}^{(\pm)}[\omega]$ generalized wave vector dependent susceptibility (Appendix B, (B.91) and (8.30 b)); see also (8.88)

$\chi^{\alpha\beta}[l,\omega]$ generalized susceptibility tensor (8.28 b)

$\Psi(\mathbf{q},\omega)$ wave vector dependent susceptibility for liquid in hydrodynamic region (11.144)

$\hat{\psi}$, $\hat{\psi}^+$ particle field operators (5.56)

$\hbar\omega = \hbar^2(k^2-k'^2)/2m$ energy lost by neutron (§ 1.2)

$\omega_j(\mathbf{q})$ frequency function of jth phonon branch (§ 4.1)

$\omega_{\mathbf{q}}$ frequency function for spin waves (Chapter 9)

$\Omega_{\mathbf{k}}$ frequency function for spin waves in itinerant electron system (§ 9.3 (*d*))

Ω solid angle

$\Omega(\mathbf{q})$ (§ 9.6 (*a*), (*b*))

$\overline{\omega^n}$ moments, of $F^{\alpha}(\boldsymbol{\varkappa},\omega)$ (§ 13.3), of scattering law (§ 11.6)

1

ELEMENTARY THEORY

1.1. Introduction

THE application of neutron scattering to the understanding of the physical properties of matter can be said to have started in earnest from about 1950. Before that date the neutron fluxes available for experimental research were small, and subsequently total cross-sections were measured. The theory was much in advance of the experimental situation at this time, the papers of Fermi (1936), Schwinger (1937 and 1948), and Halpern and Johnson (1939) being fundamental in the development of the subject. When higher fluxes became available the theoretical work of Placzek (1952) and Van Hove (1954) did much to stimulate the whole field of thermal neutron scattering into becoming what is now one of the major subjects of experimental research.

The utility of thermal neutrons in the study of solids, liquids, and gases arises because of their following basic properties:

(*a*) the wavelength of thermal neutrons is comparable to the separation of nuclei in crystals,

(*b*) the change in the energy of thermal neutrons due to inelastic scattering processes involving the creation and annihilation of excitations in solids, liquids, and gases is frequently of the same order of magnitude as their initial energy,

(*c*) the neutron is a neutral particle, and

(*d*) the neutron has a magnetic moment.

From the property (*a*) it is evident that thermal neutrons will display pronounced interference effects when scattered from condensed atomic systems, and from (*b*) that they afford a very sensitive means of studying the energy spectra of elementary excitations and the dynamics of the target system on an atomic scale. Because of (*c*) the neutron is able to penetrate very deeply into solids, in contrast, for example, with electrons, while (*d*) its magnetic moment means that it is very sensitive to the magnetic structure and dynamics of the target.

Because of the properties of the neutron its interaction with the scattering system is, in general, the sum of two parts, namely

(1) the interaction of the neutron with the nuclei of the target via purely nuclear forces, and

(2) the interaction between the magnetic moment of the neutron and the magnetic moment of the electrons of the target.

A third interaction between the neutron magnetic moment and the Coulomb field of the electrons and nuclei can, in almost all cases, be completely neglected in comparison to the contributions from the purely nuclear interactions and the purely magnetic interactions mentioned above (cf. § 10.3).

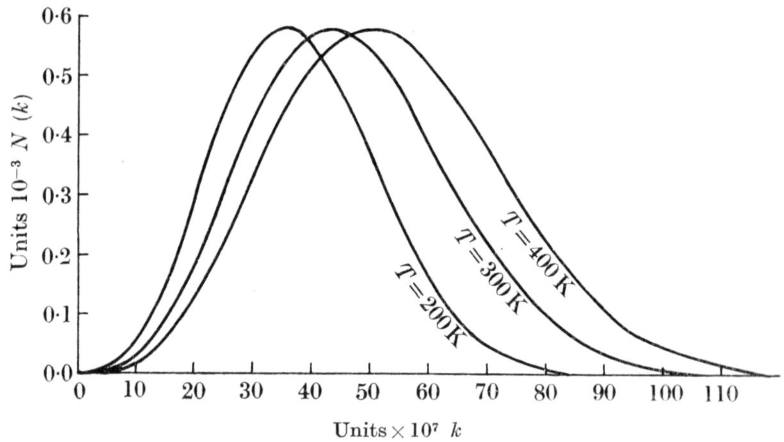

FIG. 1.1. Maxwell distribution $N(k)$ defined by eqn (1.1) as a function of wave vector k for various temperatures T (unit neutron density).

Though we are not concerned here with the techniques and physics of thermal neutron production, it will be profitable for future discussions to consider briefly the energy spectrum and flux that reactors make available for experimental research. Thermal neutrons are obtained by slowing down fission neutrons with light materials (moderators), e.g. H_2O, D_2O, and graphite. Very high flux reactors produce a maximum thermal flux of about 10^{19} neutrons m^{-2} s^{-1}.

If we denote the flux of neutrons, irrespective of direction but in the wave-vector range k to $k+dk$, by $N(k)\,dk$, then, for k in the thermal range, this can be approximately represented by a Maxwell distribution

$$N(k) = n\frac{4\pi\hbar^4}{m}\left(\frac{\beta}{2\pi m}\right)k^3\exp\left(-\frac{\beta\hbar^2k^2}{2m}\right), \tag{1.1}$$

corresponding to a moderator temperature $T = 1/k_B\beta$, typically about 100 °C, and a neutron density n. Fig. 1.1 shows the neutron flux as given by eqn (1.1) for various moderator temperatures.

Actually (1.1) does not represent the neutron flux over the entire energy range, because the formula describes the flux distribution for a neutron gas in thermal equilibrium. The fission events that are constantly taking place in a reactor act as a source of neutrons in the MeV range, and the slowing of these neutrons down to thermal energies gives an additional flux distribution, roughly like $1/E$, between thermal energies and fission energies. This higher energy flux is not important for our purposes, though it is sometimes troublesome in actual experiments.

The maximum of $N(E)$ occurs typically at a neutron energy $E \sim 0{\cdot}03$ eV, which is very small on the usual energy range considered in nuclear structure theory (several MeV). The wavelength λ of thermal neutrons is given by the de Broglie relationship

$$\lambda = \frac{2\pi}{k} = \frac{2\pi\hbar}{m}(\text{velocity})^{-1} = \frac{0{\cdot}286}{\sqrt{E}}\,\text{Å} \tag{1.2}$$

with the neutron energy E in electron volts. Thus for $E \sim 0{\cdot}03$,

$$\lambda \sim 1{\cdot}65\ \text{Å},$$

which, as stated above, is on the same scale as the spacing of nuclei in crystals.

To illustrate the energy changes to be expected in an inelastic scattering process involving the creation or annihilation of an elementary excitation we consider acoustic phonons. Typically these have an energy $\sim 10^{-2}$ eV, which is a large fraction of the energy of a thermal neutron and, therefore, results in energy changes that are easily observable. We also cite as further examples the rotational and vibrational energy states of molecules, which can, again, have level spacings that are the same order of magnitude as the energy of thermal neutrons.

Experimental results are presented in a number of units and for this reason we list below a useful set of conversion data.

1 eV (electron volt) is equivalent to:

neutron velocity	$1{\cdot}38 \times 10^{4}$ m s^{-1}
neutron wavelength	$0{\cdot}286$ Å
neutron wave number	$21{\cdot}95$ Å^{-1}
temperature	$1{\cdot}16 \times 10^{4}$ K
frequency	$1{\cdot}52 \times 10^{15}$ rad s^{-1}

Our first task must be to define the neutron cross-section and formulate an expression that will enable us to calculate it for the various types of scattering process. This we do in the next section.

1.2. General expression for the neutron cross-section

The geometry of the scattering problem is shown in Fig. 1.2. The incident neutron has the state $\psi_{\mathbf{k}}$ and the scattered neutron the state $\psi_{\mathbf{k}'}$. The direction of propagation of the latter with respect to the incident

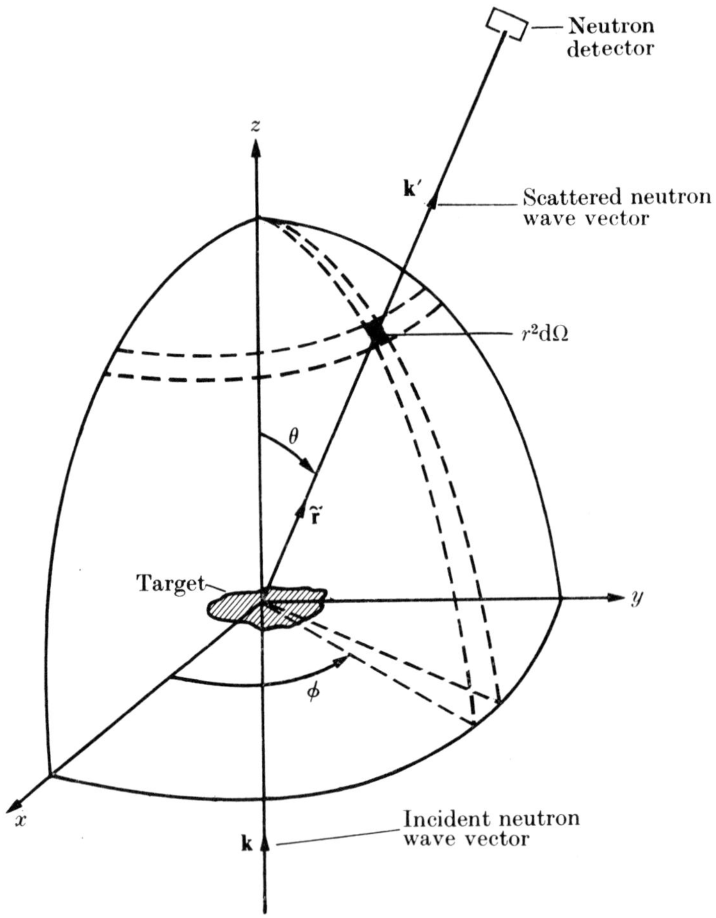

FIG. 1.2. Geometry of scattering problem.

neutron is defined by the polar angle θ and the azimuthal angle ϕ. If the flux of incident neutrons, defined as the number per unit area per unit time, is N, then the number scattered per unit time into the element of solid angle $\mathrm{d}\Omega = \sin\theta \,\mathrm{d}\theta\mathrm{d}\phi$ is

$$N\left(\frac{\mathrm{d}\sigma}{\mathrm{d}\Omega}\right)\mathrm{d}\Omega,$$

where $\mathrm{d}\sigma/\mathrm{d}\Omega$ is the differential cross-section. It follows that $\mathrm{d}\sigma/\mathrm{d}\Omega$ and the total cross-section σ have the dimensions of area.

If the target system has no internal structure then clearly the energy of the scattered neutron is identical to that of the incident neutron, i.e. the scattering is elastic. To calculate the differential cross-section for this case we need to know the probability of a transition from the plane-wave state defined by the wave vector $\mathbf{k}$ to the plane-wave state defined by the wave vector $\mathbf{k}'$, both having the same energy $E = \hbar^2k^2/2m$. This probability is given by Fermi's Golden rule, namely†

$$W_{k\to k'} = \frac{2\pi}{\hbar}\left|\int \mathrm{d}\mathbf{r}\, \psi^*_{\mathbf{k}'}\, \hat{V}\, \psi_{\mathbf{k}}\right|^2 \rho_{\mathbf{k}'}(E). \tag{1.3}$$

Here $\hat{V}$ is the interaction potential that causes the transition, in our case the interaction between the incident neutron and the target system, and $\rho_{\mathbf{k}'}(E)$ is the density of final scattering states per unit energy range.

To evaluate the latter we must first decide upon the normalization of our wave functions. For a large box of volume L^3 the states $\psi_{\mathbf{k}}$ and $\psi_{\mathbf{k}'}$ are, respectively,

$$\psi_{\mathbf{k}} = \frac{1}{L^{\frac{3}{2}}}\exp(i\mathbf{k}.\mathbf{r})$$

and

$$\psi_{\mathbf{k}'} = \frac{1}{L^{\frac{3}{2}}}\exp(i\mathbf{k}'.\mathbf{r}).$$

It follows that our density of final states is

$$\rho_{\mathbf{k}'}(E) = \left(\frac{L}{2\pi}\right)^3 \frac{\mathrm{d}\mathbf{k}'}{\mathrm{d}E}.$$

Clearly,

$$\mathrm{d}\mathbf{k}' = k'^2\,\mathrm{d}\Omega\mathrm{d}k' = k^2\,\mathrm{d}\Omega\mathrm{d}k$$

and

$$\mathrm{d}E = \frac{\hbar^2 k}{m}\,\mathrm{d}k,$$

so that

$$\rho_{\mathbf{k}'}(E) = \left(\frac{L}{2\pi}\right)^3 \frac{mk}{\hbar^2}\,\mathrm{d}\Omega. \tag{1.4}$$

To obtain the cross-section we have now only to determine the incident flux of neutrons. This is

$$(\text{velocity of incident neutrons})/L^3$$

$$= \frac{\hbar k}{mL^3}. \tag{1.5}$$

† See, for example, L. I. Schiff, *Quantum mechanics*, § 35 (McGraw-Hill, third edition, 1968).

Hence, from (1.3), (1.4), and (1.5),

$$\mathrm{d}\sigma = W_{k\to k'}/\text{incident flux}$$
$$= L^6\left(\frac{m}{2\pi\hbar^2}\right)^2 \left|\int \mathrm{d}\mathbf{r}\, \psi^*_{\mathbf{k}'}\hat{V}\psi_{\mathbf{k}}\right|^2 \mathrm{d}\Omega.$$

With the notation

$$\int \mathrm{d}\mathbf{r}\, \psi^*_{\mathbf{k}'}\hat{V}\psi_{\mathbf{k}} = \frac{1}{L^3}\langle\mathbf{k}'|\hat{V}|\mathbf{k}\rangle \tag{1.6}$$

we have

$$\frac{\mathrm{d}\sigma}{\mathrm{d}\Omega} = \left(\frac{m}{2\pi\hbar^2}\right)^2 |\langle\mathbf{k}'|\hat{V}|\mathbf{k}\rangle|^2, \tag{1.7}$$

which is the desired result.

A useful concept is that of the scattering amplitude $f(\mathbf{k}, \mathbf{k}')$. This is defined so that

$$\frac{\mathrm{d}\sigma}{\mathrm{d}\Omega} = |f(\mathbf{k}, \mathbf{k}')|^2. \tag{1.8 a}$$

(The phase of the scattering amplitude is not defined by this relationship, but has been chosen to agree with that obtained from a more rigorous analysis. See Appendix A.) We can therefore take

$$f(\mathbf{k}, \mathbf{k}') = -\frac{m}{2\pi\hbar^2}\langle\mathbf{k}'|\hat{V}|\mathbf{k}\rangle. \tag{1.8 b}$$

We need now to calculate the cross-section for inelastic scattering, i.e. when the energy of the scattered neutron differs from that of the incident neutron. This means, of course, that the state of the target must change. We denote the initial and final states of the target by $|\lambda\rangle$ and $|\lambda'\rangle$ respectively, with corresponding energies E_λ and $E_{\lambda'}$. The cross-section (1.7) is then

$$\left(\frac{\mathrm{d}\sigma}{\mathrm{d}\Omega}\right)_{\substack{\mathbf{k},\lambda\\ \mathbf{k}',\lambda'}} = \frac{k'}{k}\left(\frac{m}{2\pi\hbar^2}\right)^2 |\langle\mathbf{k}'\lambda'|\hat{V}|\mathbf{k}\lambda\rangle|^2,$$

with the condition of conservation of energy

$$\frac{\hbar^2k^2}{2m}+E_\lambda = \frac{\hbar^2k'^2}{2m}+E_{\lambda'}.$$

It is convenient to build this condition into the cross-section and define a *partial differential cross-section* for the scattering process in which the incident neutron state $|\mathbf{k}\rangle$ changes to $|\mathbf{k}'\rangle$, the state of the target changing from $|\lambda\rangle$ to $|\lambda'\rangle$:

$$\left(\frac{\mathrm{d}^2\sigma}{\mathrm{d}\Omega\mathrm{d}E'}\right)_{\substack{\mathbf{k},\lambda\\ \mathbf{k}',\lambda'}} = \frac{k'}{k}\left(\frac{m}{2\pi\hbar^2}\right)^2 |\langle\mathbf{k}'\lambda'|\hat{V}|\mathbf{k}\lambda\rangle|^2\delta(\hbar\omega+E_\lambda-E_{\lambda'}), \tag{1.9 a}$$

where

$$\hbar\omega = \frac{\hbar^2}{2m}(k^2-k'^2) = E-E'. \tag{1.9 b}$$

To obtain the total partial differential cross-section for all possible scattering processes, we must sum (1.9 a) over all the final states of the target λ' and average it over all the initial states, which occur with probabilities p_λ. In this context the quantum numbers λ and λ' denote all quantum numbers required to specify completely the initial and final states of the target system: the distribution of isotopes over the target, the nuclear spin orientations, the precise positions of the nuclei, and, in magnetic materials, the electron spin orientations. Finally we must take proper account of the spin states of the neutron. We denote these by σ and σ' and the probability distribution of the polarization of the incident neutrons by p_σ. Thus our master formula for the partial differential cross-section is

$$\frac{\mathrm{d}^2\sigma}{\mathrm{d}\Omega\mathrm{d}E'} = \frac{k'}{k}\left(\frac{m}{2\pi\hbar^2}\right)^2 \sum_{\lambda,\sigma} p_\lambda p_\sigma \sum_{\lambda',\sigma'} |\langle\mathbf{k}'\sigma'\lambda'|\hat{V}|\mathbf{k}\sigma\lambda\rangle|^2 \,\delta(\hbar\omega + E_\lambda - E_{\lambda'}). \tag{1.10}$$

Since expression (1.3) for the transition probability $W_{k\to k'}$ is derived from perturbation theory our expression for the neutron cross section is itself only approximate. The expression (1.10) is the first Born approximation (or more colloquially, the Born approximation) to the cross-section. This point is discussed in more detail in Appendix A.

1.3. Scattering from bound nuclei: coherent and incoherent scattering

A relatively simple, yet fundamental problem is the scattering of neutrons by independent, bound nuclei. In reality, of course, the nuclei in a solid are not strictly bound nor are they completely free. Whether or not they can be regarded as being in either of these two extreme states, or intermediate to them, depends, however, on the energy of the incident neutrons. We set aside a discussion of these points for the present and concentrate on the concepts of coherent and incoherent nuclear scattering of unpolarized thermal neutrons.

There is at present no complete theory of the nucleon–nucleon interaction, but we know from experimental results that it has a very short range (of the order $1{\cdot}5\times10^{-15}$ m). Because this is much less than the wavelength of thermal neutrons and the nuclear radius is only about an order of magnitude greater, the neutron–nucleus scattering can contain only s-wave components, i.e. the scattering is isotropic, and, as is well known, can therefore be characterized by a single parameter b, called the scattering length.† b can be complex, and the real part may be either

† L. D. Landau and E. M. Lifshitz, *Quantum mechanics*. See also Appendix A.

positive or negative depending on the energy of the incident neutron and the particular nucleus involved in the scattering. The imaginary part of b represents absorption (principally radiative capture for thermal neutrons) and in most cases it is small.

(The concept of an imaginary scattering length and the absorption of neutrons by nuclei is discussed in Appendix A. One example where the imaginary part of the scattering length is not small is $_{64}$Gd. This has a total cross-section for 0·01 eV neutrons of $\sim 7\times10^4$ b, which is due mainly to a radiative capture resonance at 0·028 eV for $^{157}_{64}$Gd with a cross-section of about 12×10^4 b. Gd also serves as an example of how neutron cross-sections can vary quite dramatically between different isotopes of the same element, since $^{160}_{64}$Gd has a radiative capture cross-section of only $\sim$ 0·15 b.)

In general we can have different scattering lengths not only for each atomic type but also for each isotope, and furthermore the scattering length depends on the relative orientation of the neutron spin and nuclear spin (if the latter exists). For the moment we set aside these complications by assuming the target is a single spinless isotope with scattering length b. In the following b is taken as complex but independent of energy.

The only form of $\hat{V}(\mathbf{r})$ that, using the Born approximation, gives isotropic scattering is a delta function (Fermi 1936). If the nucleus is at the position defined by $\mathbf{R}$, the Fermi pseudo-potential is defined as

$$\hat{V}(\mathbf{r}) = \frac{2\pi\hbar^2}{m}\, b\, \delta(\mathbf{r}-\mathbf{R}). \tag{1.11}$$

(The sign of the scattering length is defined such that b is positive for the scattering by an impenetrable sphere of radius b, cf. eqn (1.12) and Appendix A.)

We emphasize that (1.11) is not the true potential.† Indeed, perturbation theory is inapplicable to the scattering of a neutron by a nucleus because, although the interaction potential has a very short range, it is very strong. The Fermi pseudo-potential as given by eqn (1.11) is a formal artifice defined to give, in the Born approximation, what we know to be correct for simple s-wave scattering.†

If we substitute (1.11) in (1.8 b), the scattering amplitude f is (we take $\mathbf{R} = 0$)

$$f = -\left(\frac{m}{2\pi\hbar^2}\right)\frac{2\pi\hbar^2}{m}\, b \int \mathrm{d}\mathbf{r}\, \exp(-i\mathbf{k}'.\mathbf{r})\,\delta(\mathbf{r})\exp(i\mathbf{k}.\mathbf{r}) = -b. \tag{1.12}$$

† For a discussion of the validity of the Fermi pseudo-potential see Blatt and Weisskopf, *Theoretical nuclear physics*, Chapter 2, and references therein.

From the definition of f,
$$\frac{d\sigma}{d\Omega} = |b|^2 \tag{1.13 a}$$

and hence the total cross-section is
$$\sigma_b = 4\pi|b|^2. \tag{1.13 b}$$

Next consider the scattering from a rigid array of N nuclei, the position vector of the lth nucleus being denoted by $\mathbf{R}_l$. We also remove the restriction to a single spinless isotope and describe the scattering length of the lth nucleus as b_l. The interaction potential in this instance is
$$\hat{V}(\mathbf{r}) = \frac{2\pi\hbar^2}{m}\sum_l b_l\,\delta(\mathbf{r}-\mathbf{R}_l). \tag{1.14}$$

Thus
$$\langle\mathbf{k}'|\hat{V}|\mathbf{k}\rangle = \frac{2\pi\hbar^2}{m}\sum_l b_l \int d\mathbf{r}\, \exp(-i\mathbf{k}'.\mathbf{r})\,\delta(\mathbf{r}-\mathbf{R}_l)\exp(i\mathbf{k}.\mathbf{r})$$
$$= \frac{2\pi\hbar^2}{m}\sum_l b_l \exp(i\boldsymbol{\varkappa}.\mathbf{R}_l),$$

with $\boldsymbol{\varkappa} = \mathbf{k}-\mathbf{k}'$ and the cross-section (1.10) becomes
$$\frac{d^2\sigma}{d\Omega dE'} = \frac{k'}{k}\sum_{\lambda,\sigma} p_\lambda p_\sigma \sum_{\lambda',\sigma'} \left|\left\langle\sigma'\lambda'\right|\sum_l b_l \exp(i\boldsymbol{\varkappa}.\mathbf{R}_l)\left|\sigma\lambda\right\rangle\right|^2 \delta(\hbar\omega+E_\lambda-E_{\lambda'}). \tag{1.15}$$

Since the nuclei are rigidly bound at the sites $\mathbf{R}_l$, the quantum numbers λ and λ' refer only to the isotope distribution and nuclear spin orientations. Now to an excellent approximation we can assume that the energy of the crystal is independent both of the isotope distribution and the nuclear spin orientations. The scattering is therefore elastic and the sum over $\lambda'\sigma'$ is done by closure. Thus
$$\frac{d\sigma}{d\Omega} = \sum_{\lambda,\sigma} p_\lambda p_\sigma \sum_{l,l'} \exp\{i\boldsymbol{\varkappa}.(\mathbf{R}_l-\mathbf{R}_{l'})\}\langle\sigma\lambda|b_{l'}^* b_l|\sigma\lambda\rangle. \tag{1.16 a}$$

We find it convenient to denote by $\overline{b_{l'}^* b_l}$ the value of $b_{l'}^* b_l$ averaged over random nuclear spin orientations and random isotope distributions. Formally therefore
$$\overline{b_{l'}^* b_l} = \sum_\lambda p_\lambda \langle\lambda|b_{l'}^* b_l|\lambda\rangle.$$

Having performed this averaging, all dependence on σ must disappear, and since $\sum_\sigma p_\sigma$ is unity by definition, (1.16 a) becomes
$$\frac{d\sigma}{d\Omega} = \sum_{ll'} \exp\{i\boldsymbol{\varkappa}.(\mathbf{R}_l-\mathbf{R}_{l'})\}\overline{b_{l'}^* b_l}. \tag{1.16 b}$$

In general b_l will depend on which isotope is at the site $\mathbf{R}_l$ and what nuclear spin is associated with the isotope. Clearly there is no correlation between values of b_l and $b_{l'}$ if l and l' refer to different sites, i.e.

$$\overline{b_{l'}^* b_l} = \overline{b_{l'}^*}\,\overline{b_l} = |\bar{b}|^2 \quad \text{if } l \neq l'.$$

But if $l = l'$,

$$\overline{b_{l'}^* b_l} = \overline{|b_l|^2} = \overline{|b|^2},$$

so that in general

$$\overline{b_{l'}^* b_l} = |\bar{b}|^2 + \delta_{l,l'}(\overline{|b|^2} - |\bar{b}|^2). \tag{1.17}$$

Substituting (1.17) into (1.16 b), the cross-section can be written as the sum of two parts,

$$\frac{\mathrm{d}\sigma}{\mathrm{d}\Omega} = \left(\frac{\mathrm{d}\sigma}{\mathrm{d}\Omega}\right)_{\mathrm{coh}} + \left(\frac{\mathrm{d}\sigma}{\mathrm{d}\Omega}\right)_{\mathrm{incoh}}, \tag{1.18}$$

where the *coherent* cross-section is

$$\left(\frac{\mathrm{d}\sigma}{\mathrm{d}\Omega}\right)_{\mathrm{coh}} = |\bar{b}|^2 \left| \sum_l \exp(\mathrm{i}\boldsymbol{\kappa}.\mathbf{R}_l) \right|^2 \tag{1.19}$$

and the *incoherent* cross-section

$$\left(\frac{\mathrm{d}\sigma}{\mathrm{d}\Omega}\right)_{\mathrm{incoh}} = N\{\overline{|b|^2} - |\bar{b}|^2\} = N\overline{|b-\bar{b}|^2}. \tag{1.20}$$

It is obvious from these formulae that the coherent and the incoherent scattering are profoundly different. In the coherent scattering there is a strong interference between the waves scattered from each nucleus; so strong that we find coherent scattering from a crystal only if strict geometrical conditions are satisfied. In the incoherent scattering there is no interference at all and the cross-section is completely isotropic. (This is not strictly true for a real lattice, i.e. a lattice in which the nuclei can move. See Chapter 4.) Physically we can think of the coherent and incoherent scattering as follows: because the scattering lengths vary from one isotope to another and also depend on the nuclear spin orientation (relative to the neutron) the neutron does not see a crystal of uniform scattering potential but one in which the scattering varies from one point to the next. It is only the mean scattering potential that can give interference effects and thus coherent scattering; this mean scattering potential is proportional to $\bar{b}$ and hence the coherent scattering cross-section is proportional to $|\bar{b}|^2$. The deviations from the mean potential are randomly distributed and therefore cannot give interference effects; they therefore give incoherent scattering proportional to the mean-square deviation, i.e. to $\overline{|b-\bar{b}|^2} = \overline{|b|^2} - |\bar{b}|^2$.

We now seek the values of $\bar{b}$ and $\overline{|b|^2}$. As a first example we assume the scattering lengths to be independent of spin, so that only the variation with isotope need be considered. We label the isotopes with the subscript ξ, so that b_ξ is the scattering length and c_ξ the concentration of the ξ isotope. Then

$$\bar{b} = \sum_\xi c_\xi b_\xi \tag{1.21 a}$$

and

$$\overline{|b|^2} = \sum_\xi c_\xi |b_\xi|^2. \tag{1.21 b}$$

As a second example we assume a single isotope but recall that the interaction between two nucleons depends on their spins (Wigner 1936). Thus, if the nuclei have a spin i we must make allowance for this in our scattering cross-section. Physically, the basis of this dependence on spin orientation can be seen as follows: when the neutron, with spin $\frac{1}{2}$, interacts with a nucleus of spin i it can do so in states of total spin $i+\frac{1}{2}$ or $i-\frac{1}{2}$, and, accordingly, there are scattering lengths $b^{(+)}$ and $b^{(-)}$ associated with the two states. (Recall that only s-wave components of the neutron wave function are involved.) We remember that there are $2t+1$ states of total angular momentum t, there are therefore $2(i+\frac{1}{2})+1$ states of spin $i+\frac{1}{2}$ and $2(i-\frac{1}{2})+1$ states of spin $i-\frac{1}{2}$. Each of these has the same intrinsic probability, so we conclude that the probability of interaction in the $i+\frac{1}{2}$ state is

$$\frac{2i+2}{(2i+2)+2i} = \frac{i+1}{2i+1},$$

while the probability of interaction in the $i-\frac{1}{2}$ state is

$$\frac{2i}{(2i+2)+2i} = \frac{i}{2i+1}.$$

Hence

$$\bar{b} = \left(\frac{i+1}{2i+1}\right)b^{(+)}+\left(\frac{i}{2i+1}\right)b^{(-)} \tag{1.22}$$

and

$$\overline{|b|^2} = \left(\frac{i+1}{2i+1}\right)|b^{(+)}|^2+\left(\frac{i}{2i+1}\right)|b^{(-)}|^2. \tag{1.23}$$

From these two examples we can now combine (1.21), (1.22), and (1.23) to give the general expressions for the case where there is more than one isotope present and some of these have non-zero spins:

$$\bar{b} = \sum_\xi c_\xi \frac{1}{2i_\xi+1}\{(i_\xi+1)b_\xi^{(+)}+i_\xi b_\xi^{(-)}\} \tag{1.24}$$

and

$$\overline{|b|^2} = \sum_\xi c_\xi \frac{1}{2i_\xi+1}\{(i_\xi+1)|b_\xi^{(+)}|^2+i_\xi|b_\xi^{(-)}|^2\}. \tag{1.25}$$

These expressions have to be inserted into (1.19) and (1.20) to give the coherent and incoherent cross-sections.

1.4. Scattering amplitude operator

In some circumstances (for example in Chapters 10 and 12) it is more convenient to allow for the spin dependence of the neutron–nucleus interaction by introducing the formal concept of a scattering amplitude operator $\hat{b}$ (Schwinger and Teller 1937). It is convenient to discuss this concept here and reproduce the above results by this alternative technique. For a single nucleus we define

$$\hat{b} = A + \tfrac{1}{2}B\hat{\boldsymbol{\sigma}}.\hat{\mathbf{i}}, \tag{1.26}$$

where $\frac{1}{2}\hat{\boldsymbol{\sigma}}$ is the spin of the neutron and $\hat{\mathbf{i}}$ that of the nucleus. The coefficients A and B are determined from the requirement that $\hat{b}$ should have eigenvalues $b^{(\pm)}$ for the two possible states of total spin, $t = i \pm \frac{1}{2}$. Since

$$\hat{\mathbf{t}}^2 = (\tfrac{1}{2}\hat{\boldsymbol{\sigma}} + \hat{\mathbf{i}})^2 = \tfrac{1}{4}\hat{\boldsymbol{\sigma}}^2 + \hat{\mathbf{i}}^2 + \hat{\boldsymbol{\sigma}}.\hat{\mathbf{i}}$$

and $\hat{\mathbf{t}}^2 = t(t+1)$, $\hat{\boldsymbol{\sigma}}^2 = 3$, $\hat{\mathbf{i}}^2 = i(i+1)$, then

$$\hat{\boldsymbol{\sigma}}.\hat{\mathbf{i}} = t(t+1) - \tfrac{3}{4} - i(i+1),$$

and hence we have, for total spin $t = i + \frac{1}{2}$,

$$b^{(+)} = A + B\tfrac{1}{2}\{(i+\tfrac{1}{2})(i+\tfrac{3}{2}) - i(i+1) - \tfrac{3}{4}\},$$

and for $t = i - \frac{1}{2}$,

$$b^{(-)} = A + B\tfrac{1}{2}\{(i-\tfrac{1}{2})(i+\tfrac{1}{2}) - i(i+1) - \tfrac{3}{4}\}.$$

These two equations give

$$A = \frac{1}{2i+1}\{(i+1)b^{(+)} + ib^{(-)}\} \tag{1.27 a}$$

and

$$B = \frac{2}{2i+1}\{b^{(+)} - b^{(-)}\}. \tag{1.27 b}$$

We have now to calculate the average values of the operators $\hat{b}$ and $|\hat{b}|^2$ for unpolarized incident neutrons. The directions of the spins of the neutrons and the nuclei are completely random, and the averaging of the spins of the neutrons and of the nuclei are independent, each spin averaging to zero. We therefore have

$$\overline{\hat{b}} = A. \tag{1.28 a}$$

To calculate $\overline{|\hat{b}|^2}$ we use the identity†

$$(\hat{\mathbf{i}}.\hat{\boldsymbol{\sigma}})^2 = \hat{\mathbf{i}}^2 + \mathrm{i}\hat{\boldsymbol{\sigma}}.(\hat{\mathbf{i}}\times\hat{\mathbf{i}}) = \hat{\mathbf{i}}^2 + \mathrm{i}\hat{\boldsymbol{\sigma}}.(\mathrm{i}\hat{\mathbf{i}}) = \hat{\mathbf{i}}^2 - \hat{\boldsymbol{\sigma}}.\hat{\mathbf{i}},$$

† If $\hat{\mathbf{A}}$ and $\hat{\mathbf{B}}$ commute with $\hat{\boldsymbol{\sigma}}$ then $[\mathrm{i} = \sqrt{(-1)}]$

$$(\hat{\mathbf{A}}.\hat{\boldsymbol{\sigma}})(\hat{\mathbf{B}}.\hat{\boldsymbol{\sigma}}) = \hat{\mathbf{A}}.\hat{\mathbf{B}} + \mathrm{i}\hat{\boldsymbol{\sigma}}.\hat{\mathbf{A}}\times\hat{\mathbf{B}}.$$

This identity is proved, for example, by A. Messiah in Chapter 13 § 20 of *Quantum mechanics* (North-Holland).

which leads to the result

$$\overline{|b|^2} = |A|^2+|B|^2\tfrac{1}{4}i(i+1) = \frac{1}{2i+1}\{(i+1)|b^{(+)}|^2+i|b^{(-)}|^2\}. \tag{1.28b}$$

Eqns (1.28) are in agreement with (1.22) and (1.23).

1.5. Examples of total single-atom cross-sections

To illustrate how the above formulae work out in practice, and to indicate the orders of magnitude involved, we give a few examples in Table 1.1. The data presented are intended only for demonstration purposes.

The various cross-sections tabulated are the coherent cross-section σ_c, the total scattering cross-section σ, and the absorption cross-section σ_{ab}, each being the single-atom value (bound nuclei). Thus σ_c and σ are

$$\sigma_c = 4\pi|\bar{b}|^2 \tag{1.29}$$

and

$$\sigma = 4\pi\overline{|b|^2}. \tag{1.30}$$

The absorption cross-section is that for incident neutrons with an energy 0·025 eV. The elements Eu and Gd are included in the table to indicate the dominance of the radiative capture cross-sections over their total scattering cross-sections (cf. Appendix A).

The scattering of neutrons by protons affords a simple example of the use of the above formulae. The relevant scattering lengths are

$$b^{(+)} = 1{\cdot}04\ 10^{-14}\ \text{m (triplet)}$$

and

$$b^{(-)} = -4{\cdot}74\ 10^{-14}\ \text{m (singlet)},$$

and hence

$$\bar{b} = \tfrac{3}{4}b^{(+)}+\tfrac{1}{4}b^{(-)} = -0{\cdot}375\ 10^{-14}\ \text{m}$$

and

$$\overline{|b|^2} = \tfrac{3}{4}|b^{(+)}|^2+\tfrac{1}{4}|b^{(-)}|^2 = 6{\cdot}49\ 10^{-28}\ \text{m}^2,$$

which give

$$\sigma_c = 1{\cdot}77\ \text{b}$$

$$\sigma = 81{\cdot}6\ \text{b},$$

as given in the table.

The incoherent scattering, $\sigma-\sigma_c$, is in the case of hydrogen purely spin dependent. Vanadium is another example of an element that scatters almost entirely incoherently, since $\sigma_c = 0{\cdot}03$ b and $\sigma = 5{\cdot}1$ b. This comes about because, although there is only one stable V isotope, the scattering lengths (+) and (−) have opposite signs and are almost exactly in the ratio $(i+1):i$, i.e. almost $9/2:7/2$, so that $\bar{b}$ is very close to zero. In contrast to H and V, iron has a very small incoherent cross-section, due to the high abundance of an isotope with zero spin. On the other hand nickel has a relatively high incoherent cross-section, even though its three principal isotopes have zero spin, because one of these has a large negative scattering length.

TABLE 1.1

Representative single-atom cross-sections

Element	Isotope(s) (abundance %)	Nuclear spin	$\bar{b}$ (10^{-14} m)	σ_c (b)	σ (b)	σ_{ab} (at 0·025 eV) (b)
H		$\frac{1}{2}$	−0·375	1·77	81·6	0·330
V	^{51}V	7/2	(—)	0·032	5·1	5·1
Fe			0·946	11·3	11·6	2·5
	^{54}Fe (5·84)	0	0·42	2·2	2·5	2·2
	^{56}Fe (91·68)	0	1·01	12·8	12·8	2·6
	^{57}Fe (2·17)	$\frac{1}{2}$	0·23	0·64	2·0	2·4
Ni			1·024	13·2	18·4	4·6
	^{58}Ni (67·76)	0	1·47	26	—	4·2
	^{60}Ni (26·16)	0	0·28	1·1	—	2·5
	^{62}Ni (3·66)	0	−0·85	9·5	—	15
Eu	153, 151				29·7	4600
Gd	158, 160, 156 157, 155				26	$\sim 10^5$

1.6. Scattering by a single free nucleus

In § 1.3 we calculated the cross-section for the scattering by a single bound nucleus of zero spin. We now consider the problem in the opposite limit, namely when the nucleus is completely free. The state of the latter is completely determined by specifying its momentum $\mathbf{q}$. The wave function is simply

$$|\mathbf{q}\rangle = V^{-\frac{1}{2}}\exp i\mathbf{q}.\mathbf{R},$$

where V is the volume of the box in which the motion of the nucleus is quantized and, as before, $\mathbf{R}$ is its position vector. (1.10) now becomes, with the potential (1.11),

$$\frac{d^2\sigma}{d\Omega dE'} = \overline{|b|^2}\frac{k'}{k}\sum_{\mathbf{q}',\mathbf{q}} p_{\mathbf{q}}\left|\int d\mathbf{R}\,\frac{1}{V}\exp\{i\mathbf{R}.(\boldsymbol{\varkappa}+\mathbf{q}-\mathbf{q}')\}\right|^2 \times$$

$$\times\delta\left\{\hbar\omega+\frac{\hbar^2}{2M}(q^2-q'^2)\right\}, \quad (1.31)$$

where $\boldsymbol{\varkappa} = \mathbf{k}-\mathbf{k}'$ and M is the mass of the nucleus. The integral in (1.31) is zero unless $\boldsymbol{\varkappa}+\mathbf{q}-\mathbf{q}' = 0$; hence

$$\left|\int d\mathbf{R}\exp\{i\mathbf{R}.(\boldsymbol{\varkappa}+\mathbf{q}-\mathbf{q}')\}\right|^2 = \delta(\boldsymbol{\varkappa}+\mathbf{q}-\mathbf{q}')\times(\text{constant}),$$

and the constant of proportionality is obtained by integrating over $\boldsymbol{\varkappa}$. This gives

$$\text{constant} = (2\pi)^3 V.$$

(The proof is similar to that given in § 2.2.)

If in conjunction with this result we use the identity

$$\sum_{\mathbf{q}'} \Rightarrow \frac{V}{(2\pi)^3} \int \mathrm{d}\mathbf{q}'$$

and suppose that initially the target nucleus is at rest, so that

$$p_{\mathbf{q}} = \delta_{\mathbf{q},0},$$

the cross-section (1.31) becomes

$$\frac{\mathrm{d}\sigma^2}{\mathrm{d}\Omega \mathrm{d}E'} = \overline{|b|^2} \frac{k'}{k} \delta\left(\hbar\omega - \frac{\hbar^2\kappa^2}{2M}\right)$$

$$= \overline{|b|^2} \frac{k'}{k} \frac{2m}{\hbar^2} \delta\{k^2(1-\gamma) - k'^2(1+\gamma) + \gamma 2\mathbf{k}.\mathbf{k}'\}, \tag{1.32}$$

where $$\gamma = m/M.$$

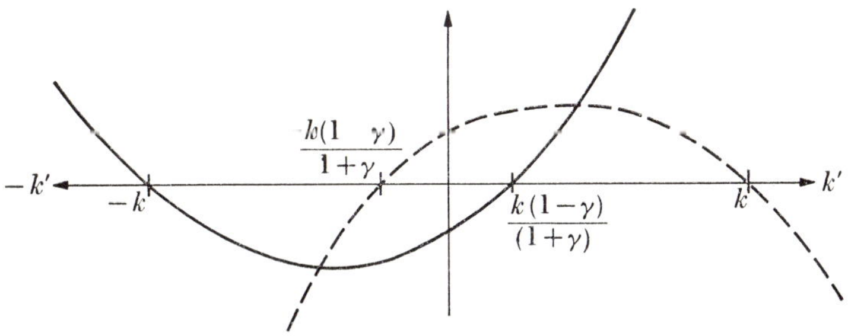

FIG. 1.3. The broken line is the l.h.s. of eqn (1.34 a) and the solid line is the l.h.s. of eqn (1.34 b) as functions of k'.

We denote the cosine of the angle between $\mathbf{k}$ and $\mathbf{k}'$ by μ; then

$$\frac{\mathrm{d}\sigma}{\mathrm{d}E'} = \overline{|b|^2} \frac{4\pi m}{\hbar^2} \int_{-1}^{1} \mathrm{d}\mu \frac{k'}{k} \delta\{k^2(1-\gamma) - k'^2(1+\gamma) + 2\gamma kk'\mu\}$$

$$= \frac{4\pi \overline{|b|^2} m}{\hbar^2} \frac{k'}{k} \frac{1}{2kk'\gamma} = \frac{\pi \overline{|b|^2}}{\gamma E} \tag{1.33}$$

for all values of k' that satisfy

$$-2kk'\gamma \leqslant k^2(1-\gamma) - k'^2(1+\gamma) \leqslant 2kk'\gamma,$$

or, alternatively,

$$-k'^2(1+\gamma) + 2\gamma kk' + k^2(1-\gamma) > 0 \tag{1.34 a}$$

and

$$k'^2(1+\gamma) + 2\gamma kk' - k^2(1-\gamma) > 0. \tag{1.34 b}$$

From Fig. 1.3 it is obvious that both these conditions are fulfilled if k' satisfies

$$\frac{k(1-\gamma)}{1+\gamma} \leqslant k' \leqslant k, \tag{1.35 a}$$

which is equivalent to

$$E(1-\gamma)^2/(1+\gamma)^2 \leqslant E' \leqslant E. \tag{1.35 b}$$

From (1.33) we see that the differential cross-section $d\sigma/dE'$ is a constant for all values of E' from E to $E(1-\gamma)^2/(1+\gamma)^2$. Thus the total cross-section is

$$\sigma = \frac{\pi\overline{|b|^2}}{\gamma E}\int\limits_{\frac{E(1-\gamma)^2}{(1+\gamma)^2}}^{E} dE' = \frac{\pi\overline{|b|^2}}{\gamma E}\frac{4\gamma E}{(1+\gamma)^2} = \frac{\sigma_b}{(1+\gamma)^2}. \tag{1.36}$$

The cross-section for the scattering by a free nucleus, eqn (1.36), is smaller than that for the scattering by a bound nucleus, σ_b, because the target nucleus is able to move aside in the collision process.

2

NUCLEAR BRAGG SCATTERING

THE central topic of this chapter is a detailed discussion of the coherent cross-section given by eqn (1.19) in the case of a rigid perfect crystal. First, however, we digress from this task to consider a general description of crystal lattices and their associated reciprocal lattices. This description will serve to define a notation that we use throughout this book.

2.1. Crystal lattices and reciprocal lattices

A perfect crystal lattice may be constructed by repeating a unit cell periodically in space. The unit cell is defined in terms of three non-coplanar *basic vectors* $\mathbf{a}_1$, $\mathbf{a}_2$, and $\mathbf{a}_3$ and has a volume $v_0 = \mathbf{a}_1 . (\mathbf{a}_2 \times \mathbf{a}_3)$. In general the unit cell will contain more than one lattice site. The vector leading from a point in one unit cell to the corresponding point in another cell is called a *lattice vector*. We denote the lattice vectors by $\mathbf{l}$. By definition

$$\mathbf{l} = l_1 \mathbf{a}_1 + l_2 \mathbf{a}_2 + l_3 \mathbf{a}_3, \tag{2.1}$$

where the integers l_1, l_2, l_3, known as cell indices and frequently denoted for brevity by the single letter l, take all values.

If the unit cell contains just one lattice site, so that the lattice vectors $\mathbf{l}$ give every lattice site in the crystal ($\mathbf{R}_l = \mathbf{l}$), then we have a Bravais lattice. We note that in a Bravais lattice each lattice point is a centre of symmetry, as is evident from its definition in terms of the lattice vectors $\mathbf{l}$. There are in total fourteen Bravais lattices,† but we shall be concerned in the main with the three cubic and the hexagonal Bravais lattices. One simple choice of the basic vectors of these four Bravais lattices is given in Table 2.1 in Cartesian components in terms of the lattice constant a. The basic vectors for the body-centred and face-centred cubic and the hexagonal lattices are illustrated in Fig. 2.1.

In the fourth example in Table 2.1, the hexagonal Bravais lattice, the vectors $\mathbf{a}_1$ and $\mathbf{a}_2$ are seen to generate a plane of hexagonal network, and $\mathbf{a}_3$ displaces sites in a direction perpendicular to this plane, with a separation γa between neighbouring planes.

In general, crystal lattices have more than one atom per unit cell and are constructed from several (r) interpenetrating *identical* Bravais lattices

† All the Bravais lattices and their fundamental symmetry properties are given by L. D. Landau and E. M. Lifshitz, *Statistical physics*, § 129 (Pergamon Press, 1959).

TABLE 2.1

Basic vectors of four Bravais lattices

	$\mathbf{a}_1$	$\mathbf{a}_2$	$\mathbf{a}_3$	v_0
Simple cubic (s.c.)	$a(1, 0, 0)$	$a(0, 1, 0)$	$a(0, 0, 1)$	a^3
Body-centred cubic (b.c.c.)	$\frac{a}{2}(-1, 1, 1)$	$\frac{a}{2}(1, -1, 1)$	$\frac{a}{2}(1, 1, -1)$	$\frac{1}{2}a^3$
Face-centred cubic (f.c.c.)	$\frac{a}{2}(0, 1, 1)$	$\frac{a}{2}(1, 0, 1)$	$\frac{a}{2}(1, 1, 0)$	$\frac{1}{4}a^3$
Hexagonal	$a(1, 0, 0)$	$a(\frac{1}{2}, \frac{1}{2}\sqrt{3}, 0)$	$a(0, 0, \gamma)$	$\gamma\frac{\sqrt{3}}{2}a^3$

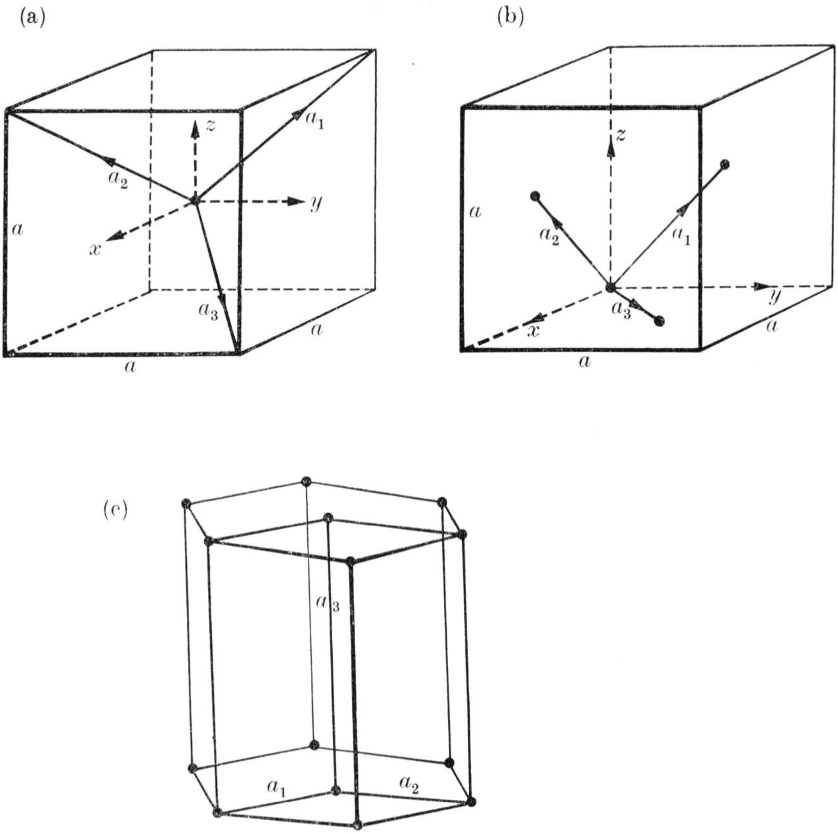

FIG. 2.1. The basic vectors of three Bravais lattices: (a) body-centred cubic, (b) face-centred cubic, and (c) hexagonal. In (a) and (b) a is the lattice constant.

(i.e. Bravais lattices with identical basic vectors), each lattice not necessarily containing atoms of the same type.

The positions of the r atoms within a unit cell are denoted by the vectors $\mathbf{d}$, and these can be expressed in terms of the basic vectors $\mathbf{a}_1$, $\mathbf{a}_2$, and $\mathbf{a}_3$ that define the unit cell. The $r-1$ non-null vectors $\mathbf{d}$ are

$$\mathbf{d} = d_1\,\mathbf{a}_1 + d_2\,\mathbf{a}_2 + d_3\,\mathbf{a}_3, \tag{2.2}$$

with $0 \leqslant d_i \leqslant 1$, $i = 1, 2, 3$, the site $\mathbf{d} = 0$ coinciding with the corner of the unit cell. The position vector $\mathbf{R}_{ld}$ of an atom in the crystal is now given by

$$\mathbf{R}_{ld} = \mathbf{l} + \mathbf{d}. \tag{2.3}$$

As a first example of a lattice with more than one atom per unit cell, consider a body-centred crystal with different atoms at the two sets of sites

$$(n_1\,a,\ n_2\,a,\ n_3\,a)$$

and

$$\{(n_1+\tfrac{1}{2})a,\ (n_2+\tfrac{1}{2})a,\ (n_3+\tfrac{1}{2})a\},$$

where the integers n_1, n_2, and n_3 take all possible values. Each set generates a simple cubic lattice, so that this crystal structure consists of two interpenetrating simple cubic lattices with a lattice constant a, the unit cell being identical with the unit cell of one of the single cubic lattices and the $r = 2$ atoms within the cell located at the sites

$$\mathbf{d} = (0, 0, 0) \quad \text{and} \quad \mathbf{d} = (\tfrac{1}{2}a,\ \tfrac{1}{2}a,\ \tfrac{1}{2}a).$$

Another example of a crystal with two atoms per unit cell is the hexagonal close-packed lattice (h.c.p.). This type of lattice is constructed from a hexagonal Bravais lattice, with basic vectors $\mathbf{a}_1$, $\mathbf{a}_2$, $\mathbf{a}_3$, by adding to each lattice site another at a distance

$$\boldsymbol{\rho} = \tfrac{1}{3}\mathbf{a}_1 + \tfrac{1}{3}\mathbf{a}_2 + \tfrac{1}{2}\mathbf{a}_3.$$

Thus the lattice consists of two identical hexagonal Bravais lattices, the six nearest-neighbour sites on the other lattice to a given site being at a distance $|\boldsymbol{\rho}| = a(\tfrac{1}{3}+\tfrac{1}{4}\gamma^2)^{\frac{1}{2}}$. There are also six nearest-neighbour sites on the same lattice. These are at a distance a and lie in the basal plane through the atom (perpendicular to the hexagonal axis). The close-packed structure is formed when all these twelve neighbouring sites, six on the second lattice and six on the same lattice, are at the same distance away, i.e. $|\boldsymbol{\rho}| = a$, which requires $\gamma = (8/3)^{\frac{1}{2}}$. This structure is illustrated in Fig. 2.2.

It is to be noted that the vector $-\boldsymbol{\rho}$ does not coincide with the position of a lattice site in the h.c.p. crystal structure, i.e. there is not inversion symmetry between neighbouring sites on different sublattices though, of course, there is within each sublattice because they are Bravais. This simply means that the h.c.p. structure is not a Bravais lattice, in contrast to our first example of a crystal lattice with two atoms per unit cell, in which a b.c.c. lattice was generated from two s.c. sublattices. It is worth while to note in this connection that the f.c.c. lattice can be generated from four s.c. sublattices, and a s.c. lattice from two f.c.c. lattices. The most general lattice point for a f.c.c. lattice can be expressed as

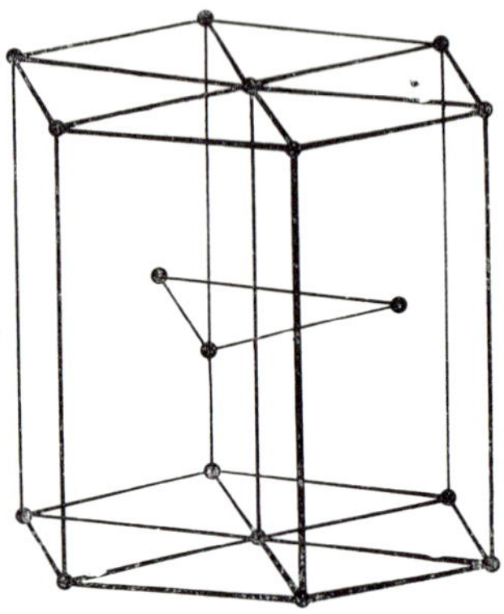

FIG. 2.2. The ideal hexagonal close-packed lattice, i.e. with $\gamma = (8/3)^{\frac{1}{2}}$.

$$\tfrac{1}{2}a(n_1, n_2, n_3),$$

with $n_1+n_2+n_3$ an even integer. Thus the two sets of vectors

$$\mathbf{l} = \tfrac{1}{2}a(n_1, n_2, n_3), \quad n_1+n_2+n_3 \text{ even},$$

and

$$\mathbf{l} = \tfrac{1}{2}a(n_1, n_2, n_3), \quad n_1+n_2+n_3 \text{ odd},$$

together generate a simple cubic lattice of side $\frac{1}{2}a$.

For every crystal lattice defined by vectors $\mathbf{l}$ as in (2.1), we can also define a reciprocal lattice with vectors $\boldsymbol{\tau}$ such that

$$\exp(i\boldsymbol{\tau}.\mathbf{l}) = 1 \quad \text{for all } \mathbf{l}. \tag{2.4}$$

If we write, by analogy with (2.1),

$$\boldsymbol{\tau} = t_1\boldsymbol{\tau}_1+t_2\boldsymbol{\tau}_2+t_3\boldsymbol{\tau}_3, \tag{2.5}$$

where the integers t_1, t_2, and t_3 take all possible values, the basic vectors of the reciprocal lattice are given by

$$\boldsymbol{\tau}_1 = \frac{2\pi}{v_0}\mathbf{a}_2\times\mathbf{a}_3, \qquad \boldsymbol{\tau}_2 = \frac{2\pi}{v_0}\mathbf{a}_3\times\mathbf{a}_1, \tag{2.6}$$

and

$$\boldsymbol{\tau}_3 = \frac{2\pi}{v_0}\mathbf{a}_1\times\mathbf{a}_2,$$

as is easily verified. The cartesian components of $\boldsymbol{\tau}_1$, $\boldsymbol{\tau}_2$, and $\boldsymbol{\tau}_3$ for the s.c., b.c.c., f.c.c., and hexagonal lattices are given in Table 2.2.

On comparing the results given in Table 2.1 with those given in Table 2.2 we note that the reciprocal lattice of a s.c. crystal corresponds to a s.c. lattice, while the reciprocal lattice of a b.c.c. crystal corresponds to a f.c.c. lattice and vice versa.

From (2.6) we notice that the volume of the unit cell of the reciprocal lattice

$$\boldsymbol{\tau}_1.(\boldsymbol{\tau}_2\times\boldsymbol{\tau}_3) = (2\pi)^3/v_0. \tag{2.7}$$

TABLE 2.2

Reciprocal lattices

	$\boldsymbol{\tau}_1$	$\boldsymbol{\tau}_2$	$\boldsymbol{\tau}_3$
s.c.	$\frac{2\pi}{a}(1, 0, 0)$	$\frac{2\pi}{a}(0, 1, 0)$	$\frac{2\pi}{a}(0, 0, 1)$
b.c.c.	$\frac{2\pi}{a}(0, 1, 1)$	$\frac{2\pi}{a}(1, 0, 1)$	$\frac{2\pi}{a}(1, 1, 0)$
f.c.c.	$\frac{2\pi}{a}(-1, 1, 1)$	$\frac{2\pi}{a}(1, -1, 1)$	$\frac{2\pi}{a}(1, 1, -1)$
hexagonal	$\frac{2\pi}{a}\left(1, -\frac{1}{\sqrt{3}}, 0\right)$	$\frac{2\pi}{a}\left(0, \frac{2}{\sqrt{3}}, 0\right)$	$\frac{2\pi}{a}\left(0, 0, \frac{1}{\gamma}\right)$

2.2. Coherent nuclear cross-sections

We have now defined the notation we need immediately and therefore return to an evaluation of (1.19). We suppose first we have a rigid Bravais lattice of N atoms so that (1.19) becomes simply

$$\left(\frac{d\sigma}{d\Omega}\right)_{\text{coh}} = |\bar{b}|^2 \left| \sum_l \exp(i\boldsymbol{\varkappa}.\mathbf{l}) \right|^2. \tag{2.8}$$

Obviously when the scattering vector $\boldsymbol{\varkappa}$ is equal to zero, or indeed to any reciprocal lattice vector, the right-hand side is large because all the terms in the sum over $\mathbf{l}$ add up in phase. However, as $\boldsymbol{\varkappa}$ moves away from a reciprocal lattice vector the terms rapidly come out of phase and the right-hand side drops to a negligible value. For a large crystal we can therefore see immediately that

$$\left| \sum_l \exp(i\boldsymbol{\varkappa}.\mathbf{l}) \right|^2 = \sum_{\boldsymbol{\tau}} \delta(\boldsymbol{\varkappa}-\boldsymbol{\tau}) \times (\text{constant}). \tag{2.9}$$

To determine the constant we integrate both sides of the equation over an arbitrary unit cell of the reciprocal lattice. Then we get

$$(\text{constant}) = \int_{\text{cell}} d\boldsymbol{\varkappa} \left| \sum_l \exp(i\boldsymbol{\varkappa}.\mathbf{l}) \right|^2 = \sum_{l,l'} \int_{\text{cell}} d\boldsymbol{\varkappa} \exp\{i\boldsymbol{\varkappa}.(\mathbf{l}-\mathbf{l}')\}.$$

In this last expression the integral vanishes if $\mathbf{l} \neq \mathbf{l}'$, and is the reciprocal cell volume, $(2\pi)^3/v_0$, if $\mathbf{l} = \mathbf{l}'$. Hence

$$(\text{constant}) = \sum_{l,l'} \delta_{l,l'}(2\pi)^3/v_0 = N(2\pi)^3/v_0,$$

and (2.8) becomes

$$\left(\frac{d\sigma}{d\Omega}\right)_{\text{coh}} = \frac{N(2\pi)^3}{v_0} |\bar{b}|^2 \sum_{\boldsymbol{\tau}} \delta(\boldsymbol{\varkappa}-\boldsymbol{\tau}). \tag{2.10}$$

The generalization of this formula to the case with more than one atom per unit cell is immediately clear as

$$\left(\frac{d\sigma}{d\Omega}\right)_{\text{coh}} = N\frac{(2\pi)^3}{v_0}\sum_{\boldsymbol{\tau}} \delta(\boldsymbol{\varkappa}-\boldsymbol{\tau})|F_N(\boldsymbol{\tau})|^2, \tag{2.11}$$

where the nuclear *unit-cell structure factor* $F_N(\boldsymbol{\tau})$ is defined as

$$F_N(\boldsymbol{\tau}) = \sum_{d} \exp(i\boldsymbol{\tau}.\mathbf{d})\,\bar{b}_d. \tag{2.12}$$

Similarly, the generalization of (1.20) is simply

$$\left(\frac{d\sigma}{d\Omega}\right)_{\text{incoh}} = N\sum_{d}\{\overline{|b_d|^2}-|\bar{b}_d|^2\}. \tag{2.13}$$

2.3. Crystal symmetry effects

The form of the results (2.10) and (2.11) is noteworthy because they demonstrate that no coherent elastic scattering occurs from a perfect crystal unless $\boldsymbol{\varkappa}$ coincides with a reciprocal lattice vector. This result is a direct consequence of the symmetry of the crystal lattice and does not, for example, depend upon the nature of the neutron–lattice interaction; we later derive similar formulae for the magnetic scattering of neutrons. To illustrate this point, it is convenient in this section to consider in a very general way the scattering from a periodic structure. We shall then verify that the general formulae agree with those of § 2.2 for the special case of coherent nuclear scattering.

The interaction potential between the neutron and the crystal, $\hat{V}(\mathbf{r})$, does not, in general, have the periodicity of the lattice, because the potential will depend upon the nuclear spin orientation, electron spin orientation, particular isotope, etc., in the neighbourhood of $\mathbf{r}$. Nevertheless, the major property of a solid is that it has, on average, a periodic structure, and this property must show up in the properties of $\hat{V}(\mathbf{r})$. We therefore define the mean potential seen by the neutron as

$$\langle V(\mathbf{r})\rangle = \sum_{\lambda} p_\lambda \langle\lambda|\hat{V}(\mathbf{r})|\lambda\rangle, \tag{2.14}$$

where $\langle V(\mathbf{r})\rangle$ is now a c-number so far as the target is concerned. (It could itself be an operator in so far as $\langle V(\mathbf{r})\rangle$ could depend on the neutron spin, but we set this aside for ease of notation.) The deviation in the potential is then $\delta\hat{V}(\mathbf{r})$, where

$$\hat{V}(\mathbf{r}) = \langle V(\mathbf{r})\rangle + \delta\hat{V}(\mathbf{r}). \tag{2.15}$$

Note that $\langle V(\mathbf{r})\rangle$ can, in general, be temperature dependent and that

$$\sum_{\lambda} p_\lambda \langle\lambda|\delta\hat{V}(\mathbf{r})|\lambda\rangle = 0 \tag{2.16}$$

by definition. Now from the master formula (1.10) we need to calculate

$$\sum_{\lambda\lambda'} p_\lambda \langle\lambda|\{\hat{V}(\boldsymbol{\varkappa})\}^+|\lambda'\rangle\langle\lambda'|\hat{V}(\boldsymbol{\varkappa})|\lambda\rangle\delta(\hbar\omega+E_\lambda-E_{\lambda'}), \tag{2.17}$$

where $\hat{V}(\boldsymbol{\varkappa})$ is

$$\hat{V}(\boldsymbol{\varkappa}) = \int d\mathbf{r}\, \exp(-i\mathbf{k}'.\mathbf{r})\hat{V}(\mathbf{r})\exp(i\mathbf{k}.\mathbf{r}). \tag{2.18}$$

Because $\langle V(\mathbf{r})\rangle$ is a c-number

$$\langle\lambda|\langle V(\mathbf{r})\rangle|\lambda'\rangle = \langle V(\mathbf{r})\rangle\delta_{\lambda,\lambda'},$$

and substituting (2.15) into (2.17) then gives

$$\langle V(\boldsymbol{\varkappa})\rangle^2\delta(\hbar\omega)+\sum_{\lambda\lambda'} p_\lambda\langle\lambda|\{\delta\hat{V}(\boldsymbol{\varkappa})\}^+|\lambda'\rangle\langle\lambda'|\delta\hat{V}(\boldsymbol{\varkappa})|\lambda\rangle\delta(\hbar\omega+E_\lambda-E_{\lambda'}). \tag{2.19}$$

The first term of (2.19) gives rise to Bragg scattering. The second term gives either inelastic scattering or elastic diffuse or incoherent scattering. In this chapter we concentrate on Bragg scattering.

The mean potential seen by the neutron must have the periodicity of the lattice. Hence we define

$$\langle V(\mathbf{r})\rangle = \frac{1}{V}\sum_{\boldsymbol{\tau}} \exp(-i\boldsymbol{\tau}.\mathbf{r})V(\boldsymbol{\tau}) \tag{2.20 a}$$

and

$$V(\boldsymbol{\tau}) = \int d\mathbf{r}\, \exp(i\boldsymbol{\tau}.\mathbf{r})\langle V(\mathbf{r})\rangle. \tag{2.20 b}$$

Then

$$\langle V(\boldsymbol{\varkappa})\rangle = \langle\mathbf{k}'|\langle V(\mathbf{r})\rangle|\mathbf{k}\rangle = \int d\mathbf{r}\, \exp(i\boldsymbol{\varkappa}.\mathbf{r})\langle V(\mathbf{r})\rangle$$

$$= \frac{1}{V}\sum_{\boldsymbol{\tau}} V(\boldsymbol{\tau})\int d\mathbf{r}\, \exp\{i(\boldsymbol{\varkappa}-\boldsymbol{\tau}).\mathbf{r}\} = \frac{(2\pi)^3}{V}\sum_{\boldsymbol{\tau}} V(\boldsymbol{\tau})\delta(\boldsymbol{\varkappa}-\boldsymbol{\tau}). \tag{2.21}$$

To obtain $\langle V(\boldsymbol{\varkappa})\rangle^2$ we cannot simply square (2.21), because this would give the square of a Dirac δ-function, but instead we notice that it must have the form

$$\langle V(\boldsymbol{\varkappa})\rangle^2 = \sum_{\boldsymbol{\tau}} x_{\boldsymbol{\tau}}\,\delta(\boldsymbol{\varkappa}-\boldsymbol{\tau}),$$

and to determine $x_{\boldsymbol{\tau}}$ we multiply by $\exp(i\boldsymbol{\varkappa}.\mathbf{y})$ and integrate over all $\boldsymbol{\varkappa}$ to give

$$\sum_{\boldsymbol{\tau}} x_{\boldsymbol{\tau}} \exp(i\boldsymbol{\tau}.\mathbf{y}) = \int d\boldsymbol{\varkappa}\int d\mathbf{r}\int d\mathbf{r}'\, \exp\{i\boldsymbol{\varkappa}.(\mathbf{r}-\mathbf{r}'+\mathbf{y})\}\times$$

$$\times\frac{1}{V^2}\sum_{\boldsymbol{\tau},\boldsymbol{\tau}'} \exp(-i\boldsymbol{\tau}.\mathbf{r}+i\boldsymbol{\tau}'.\mathbf{r}')\,V(\boldsymbol{\tau})V(\boldsymbol{\tau}')^*$$

$$= \frac{(2\pi)^3}{V^2}\int d\mathbf{r}\sum_{\boldsymbol{\tau},\boldsymbol{\tau}'} \exp\{i(\boldsymbol{\tau}'-\boldsymbol{\tau}).\mathbf{r}+i\boldsymbol{\tau}'.\mathbf{y}\}V(\boldsymbol{\tau})V(\boldsymbol{\tau}')^*$$

$$= \frac{(2\pi)^3}{V}\sum_{\boldsymbol{\tau},\boldsymbol{\tau}'} \delta_{\boldsymbol{\tau}',\boldsymbol{\tau}}\exp(i\boldsymbol{\tau}'.\mathbf{y})V(\boldsymbol{\tau})V(\boldsymbol{\tau}')^*$$

$$= \frac{(2\pi)^3}{V}\sum_{\boldsymbol{\tau}} \exp(i\boldsymbol{\tau}.\mathbf{y})|V(\boldsymbol{\tau})|^2.$$

Hence
$$x_{\tau} = \frac{(2\pi)^3}{V}|V(\boldsymbol{\tau})|^2,$$
and we obtain
$$|\langle\mathbf{k}'|\langle V(\mathbf{r})\rangle|\mathbf{k}\rangle|^2 = \left|\int \mathrm{d}\mathbf{r}\, \exp(\mathrm{i}\boldsymbol{\varkappa}.\mathbf{r})\langle V(\mathbf{r})\rangle\right|^2 = \frac{(2\pi)^3}{V}\sum_{\boldsymbol{\tau}}\delta(\boldsymbol{\varkappa}-\boldsymbol{\tau})|V(\boldsymbol{\tau})|^2. \tag{2.22}$$

This last formula shows that, irrespective of the scattering potential, the coherent elastic cross-section must contain the characteristic Bragg peaks at the reciprocal lattice positions.

Returning for the moment to (2.19), it is worth while to give a brief discussion of the second term, which involves the deviations $\delta\hat{V}(\mathbf{r})$ from the average scattering potential. The deviations are of two types; static variations from the overall crystal periodicity, and dynamic fluctuations due to the movement of atoms or electrons. We shall find that the static variations give rise to scattering that is strictly elastic, i.e. the partial differential cross-section contains $\delta(\hbar\omega)$, whereas the dynamic fluctuations give rise to inelastic scattering. One good example of the effect of static variations is given by the isotope disorder scattering we have considered earlier. Another example, which is discussed in Chapter 14, is the scattering from a magnetic alloy where the time-averaged magnetic moment, and therefore the interaction potential, varies from one type of atom to the next. The scattering from crystal vibrations (Chapter 4) or from spin waves (Chapter 9) are examples of the effects of dynamic fluctuations. Sometimes the distinction between static variations and dynamic fluctuations is, to some extent, arbitrary. An example of this is the nuclear spin orientation effects we discussed in Chapter 1. Clearly the orientation of a nuclear spin in the target must change with time, and therefore, formally, the spin-dependent scattering potential can be classified as a dynamic fluctuation and the corresponding scattering as 'inelastic'. In a formal sense this is true, but the energies connected with different nuclear spin orientations are typically 10^{-4} K, and therefore the frequencies at which they change orientation are extremely low compared to those the neutron can observe. Therefore, for all practical purposes, the nuclear spin orientation effects can be treated as static and the corresponding scattering as strictly elastic.

Throughout this book we use the following nomenclature. The scattering due to random isotope distributions and random nuclear spin orientations we call *incoherent*. Scattering due to other random effects in the crystal we call *diffuse*. Both give rise to elastic and inelastic scattering; the elastic scattering comes from the static variations directly and the inelastic scattering from the dynamic fluctuations superimposed on the random static variations. Similarly we get coherent elastic scattering from the average potential $\langle V(\mathbf{r})\rangle$ and coherent inelastic scattering from the dynamic fluctuations not associated with some intrinsic inhomogeneity in the sample.

We now examine the form of $\hat{V}(\mathbf{r})$ in more detail. In the simple case examined in Chapter 1, $\hat{V}(\mathbf{r})$ was the sum of interaction potentials with each atom, see (1.14) for example. In general this simple superposition of potentials is a good, but not rigorous, approximation. (It is rigorous to express $\hat{V}(\mathbf{r})$ as a superposition of potentials from each nucleus and each electron; the problem is that, when we assemble this into a sum over atoms, we cannot rigorously identify any part of the electron density with a specific atom, without introducing a model of the target system.) We shall return to a discussion of this point in Chapter 6 in connection with magnetic scattering. But, whether or not the superposition of potentials is a good enough approximation, we know by definition that the average potential is periodic. Hence (2.20 b) becomes

$$V(\boldsymbol{\tau}) = N\int_{\text{cell}} d\mathbf{r}\,\exp(i\boldsymbol{\tau}.\mathbf{r})\langle V(\mathbf{r})\rangle = N\left(\frac{2\pi\hbar^2}{m}\right)F(\boldsymbol{\tau}), \tag{2.23}$$

where in general we define the *unit-cell structure factor* $F(\boldsymbol{\tau})$ as

$$F(\boldsymbol{\tau}) = \left(\frac{m}{2\pi\hbar^2}\right)\int_{\text{cell}} d\mathbf{r}\,\exp(i\boldsymbol{\tau}.\mathbf{r})\langle V(\mathbf{r})\rangle, \tag{2.24}$$

and then (2.22) becomes ($V = Nv_0$)

$$|\langle\mathbf{k}'|\langle V(\mathbf{r})\rangle|\mathbf{k}\rangle|^2 = N\frac{(2\pi)^3}{v_0}\left(\frac{2\pi\hbar^2}{m}\right)^2\sum_{\boldsymbol{\tau}}\delta(\boldsymbol{\varkappa}-\boldsymbol{\tau})|F(\boldsymbol{\tau})|^2. \tag{2.25}$$

When it is satisfactory to regard $\hat{V}(\mathbf{r})$ as a superposition of potentials from each atom, we put

$$\hat{V}(\mathbf{r}) = \sum_{l,d}\hat{v}_{l,d}(\mathbf{r}-\mathbf{l}-\mathbf{d}) \tag{2.26}$$

and

$$\langle V(\mathbf{r})\rangle = \sum_{l,d} v_d(\mathbf{r}-\mathbf{l}-\mathbf{d}), \tag{2.27}$$

where v_d is now the average potential given by atoms of type d. Then (2.20 b) gives

$$V(\boldsymbol{\tau}) = N\sum_d\int d\mathbf{r}\,\exp(i\boldsymbol{\tau}.\mathbf{r})\,v_d(\mathbf{r}), \tag{2.28}$$

and the unit-cell structure factor is written

$$F(\boldsymbol{\tau}) = \left(\frac{m}{2\pi\hbar^2}\right)\sum_d\int d\mathbf{r}\,\exp(i\boldsymbol{\tau}.\mathbf{r})\,v_d(\mathbf{r}). \tag{2.29}$$

For nuclear scattering it is rigorous to express $\hat{V}(\mathbf{r})$ as a superposition of potentials, and $v_d(\mathbf{r})$ is simply $(2\pi\hbar^2/m)\bar{b}_d\,\delta(\mathbf{r}-\mathbf{d})$. The unit-cell structure factor is then

$$F_N(\boldsymbol{\tau}) = \sum_d \exp(\mathrm{i}\boldsymbol{\tau}.\mathbf{d})\,\bar{b}_d, \tag{2.30}$$

and this reproduces the results of (2.11) and (2.12).

2.4. Bragg scattering

From (2.11) we see that there is no *Bragg* scattering unless the condition

$$\boldsymbol{\varkappa} = \mathbf{k}-\mathbf{k}' = \boldsymbol{\tau} \tag{2.31}$$

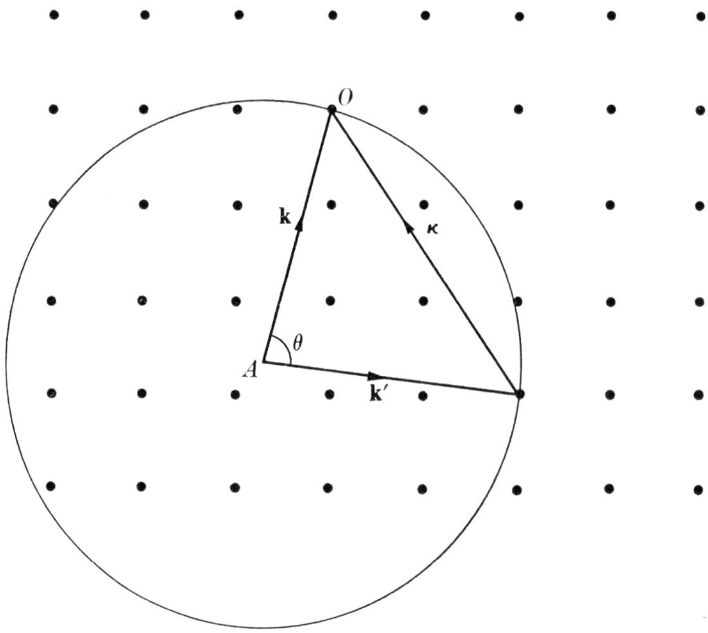

FIG. 2.3. Condition on the scattering vector $\boldsymbol{\varkappa} = \mathbf{k}-\mathbf{k}'$ for Bragg scattering. $|\boldsymbol{\varkappa}| = 2|\mathbf{k}|\sin\frac{1}{2}\theta$.

is satisfied. We can discuss this condition with the aid of Fig. 2.3. We first mark the reciprocal lattice points, and then, to represent the incident neutron, we draw a vector $\mathbf{k}$ from some point A chosen so that $\mathbf{k}$ ends at the origin O. Now because the scattering is elastic the final wave vector $\mathbf{k}'$ must have the same magnitude as $\mathbf{k}$ and therefore if drawn from A must end on the surface of a sphere drawn with centre A and radius $|\mathbf{k}| = |\mathbf{k}'| = k$. But in general the surface of this sphere does not pass through any reciprocal lattice points other than the origin, so in general (2.31) is not satisfied except for $\boldsymbol{\tau} = 0$, which corresponds to no scattering. Therefore in general a single crystal gives no elastic coherent scattering and, in fact, if the incoherent scattering happens to be zero then it gives no elastic scattering at all.

But for certain special orientations of the crystal, i.e. of the reciprocal lattice, relative to the incident beam, or with fixed orientation for certain special values of the wave vector **k**, the point A is such that the sphere drawn around it does pass through a reciprocal lattice point, so that (2.31) can be satisfied. Then we get scattering in a direction given by **k′**. If θ is the angle between **k** and **k′**

$$|\boldsymbol{\tau}| = |\boldsymbol{\varkappa}| = 2k\sin\tfrac{1}{2}\theta.$$

Hence scattering occurs only at angles θ given by ($|\boldsymbol{\tau}| = \tau$)

$$\sin\tfrac{1}{2}\theta = \frac{\tau}{2k}, \tag{2.32}$$

and then only if the crystal is correctly orientated. To show that (2.32) is the familiar Bragg law we remember that each vector $\boldsymbol{\tau}$ is perpendicular to a set of planes in the original lattice and has a magnitude that is an integral multiple of 2π times the inverse of the plane spacing. Thus

$$\tau = \frac{2\pi}{d}n \quad \text{and} \quad k = \frac{2\pi}{\lambda},$$

so

$$2d\sin\tfrac{1}{2}\theta = n\lambda, \tag{2.33}$$

which is the more familiar form of Bragg's law.

Notice that (2.32) cannot possibly be satisfied if

$$k < \tfrac{1}{2}\tau_{\min}, \tag{2.34}$$

where $\tau_{\min}$ is the smallest τ (other than zero). Hence if (2.34) is satisfied, i.e. if the neutron wavelength is long enough, there is no Bragg scattering, whatever the crystal orientation.

We obtain the total cross-section from (2.11) by integrating over $\mathrm{d}\Omega$, i.e. over $\mathrm{d}\tilde{\mathbf{k}}'$, where the tilde $\sim$ denotes a unit vector. Hence

$$\sigma_{\mathrm{c}} = N\frac{(2\pi)^3}{v_0}\sum_{\boldsymbol{\tau}}|F_N(\boldsymbol{\tau})|^2\int \mathrm{d}\tilde{\mathbf{k}}'\,\delta(\mathbf{k}-\mathbf{k}'-\boldsymbol{\tau}).$$

To perform this integral we write $\tilde{\mathbf{k}}' = \tilde{\mathbf{s}}$ and multiply by

$$(2/k')\int_0^\infty \mathrm{d}s\, s^2\,\delta(s^2-k'^2),$$

which is unity; we then rearrange the order of integration, make use of the vector $\mathbf{s} = s\tilde{\mathbf{s}}$, so $\mathrm{d}\mathbf{s} = s^2\,\mathrm{d}s\mathrm{d}\tilde{\mathbf{s}}$, and finally recall that $k' = k$ for elastic scattering. Thus

$$\int d\tilde{\mathbf{k}}'\,\delta(\mathbf{k}-\mathbf{k}'-\boldsymbol{\tau}) = (2/k')\int d\tilde{\mathbf{s}}\,\delta(\mathbf{k}-k'\tilde{\mathbf{s}}-\boldsymbol{\tau})\int_0^\infty ds\,s^2\,\delta(s^2-k'^2)$$

$$= (2/k')\int d\mathbf{s}\,\delta(\mathbf{k}-\mathbf{s}-\boldsymbol{\tau})\delta(s^2-k'^2)$$

$$= (2/k')\delta\{(\mathbf{k}-\boldsymbol{\tau})^2-k'^2\} = (2/k)\delta(\tau^2-2k\tau\sin\tfrac{1}{2}\theta).$$

Hence
$$\sigma_c = N\frac{(2\pi)^3}{v_0}\frac{2}{k}\sum_{\boldsymbol{\tau}}|F_N(\boldsymbol{\tau})|^2\,\delta(\tau^2-2k\tau\sin\tfrac{1}{2}\theta). \tag{2.35}$$

The fact that the total cross-section involves a δ-function tells us that there is no scattering unless very special conditions are satisfied.

There are three methods of observing Bragg scattering and we will discuss these in turn. Of course they correspond exactly to the three methods of observing the Bragg scattering of X-rays.

Method 1: The Laue method utilizes a fixed single crystal and a 'white beam', i.e. an incident beam which contains all wavelengths. In Fig. 2.3 this corresponds to keeping the reciprocal lattice fixed and the orientation of $\mathbf{k}$ fixed but varying the length k. For some length k the sphere drawn about A will intersect a reciprocal lattice point and these neutrons are those suitable for scattering. If the incident flux is such that $N(\lambda)\,d\lambda$ is the flux of neutrons between λ and $\lambda+d\lambda$, then the scattered intensity in one of the scattered peaks is

$$P_s = \int d\lambda\,N(\lambda)\sigma_c = VQ^\lambda N(\lambda), \tag{2.36}$$

where
$$Q^\lambda = \frac{2\pi|F_N(\boldsymbol{\tau})|^2\lambda^3}{v_0^2\tau\sin\frac{1}{2}\theta} = \frac{\lambda^4|F_N(\boldsymbol{\tau})|^2}{2v_0^2\sin^2\frac{1}{2}\theta}. \tag{2.37}$$

This formula is obtained by inserting (2.35) into (2.36), writing $k = 2\pi/\lambda$, and integrating. Here λ is the particular wavelength that is suitable for scattering and V is, as usual, the volume of the crystal.

Notice that in this method every reciprocal lattice vector $\boldsymbol{\tau}$ gives rise to some scattering. Each one selects out from the incident beam those neutrons that have suitable wavelengths and scatters them out in a direction characteristic of that particular $\boldsymbol{\tau}$ vector. This Laue method is clearly a very convenient way of producing a monochromatic beam from reactor beams, because if we look at those neutrons scattered in a particular Bragg peak $\boldsymbol{\tau}$ they will all have the same wavelength. Unfortunately there is a complication. We notice that if the crystal is scattering neutrons of wavelength λ in a certain direction via the reciprocal lattice vector $\boldsymbol{\tau}$ then it is also scattering neutrons of wavelength $\frac{1}{2}\lambda$

in the same direction via the reciprocal lattice vector $2\boldsymbol{\tau}$, and those of $\frac{1}{3}\lambda$ via $3\boldsymbol{\tau}$, etc. The scattered beam is therefore contaminated by these higher-order reflections. From (2.36) and (2.37) the ratio of the second-order contaminant to the primary scattered beam is

$$\frac{P_{\mathrm{s}}(\frac{1}{2}\lambda)}{P_{\mathrm{s}}(\lambda)} = \frac{N(\frac{1}{2}\lambda)}{N(\lambda)} \cdot \frac{1}{2^4} \frac{|F_N(2\boldsymbol{\tau})|^2}{|F_N(\boldsymbol{\tau})|^2}.$$

The ratio of the structure factors is fixed by the structure of the crystal: for simple crystals the ratio is unity. Except when working at long wave-

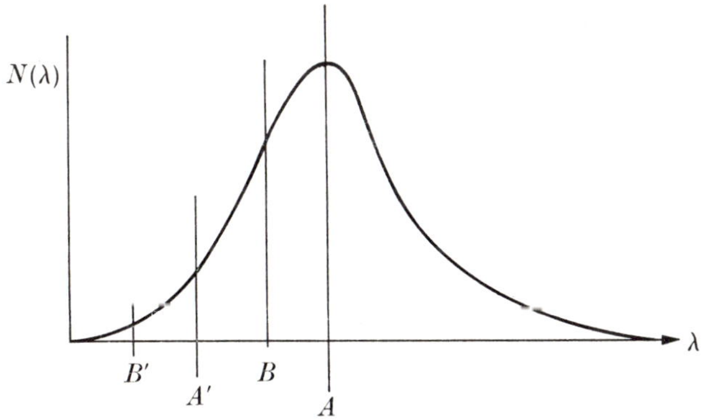

FIG. 2.4. Selecting Bragg conditions to reduce second-order contaminant.

lengths the factor 2^{-4} usually ensures that the second-order contaminant is small compared to the primary beam, but sometimes (e.g. when examining antiferromagnets) it is particularly important to reduce this second-order contaminant to a very small quantity. A convenient way of doing this is to reduce the ratio $N(\frac{1}{2}\lambda)/N(\lambda)$ by choosing λ suitably. We notice from (2.35) that by varying the orientation of the crystal relative to the incident beam we can change the wavelength that a vector $\boldsymbol{\tau}$ will scatter. Now if $N(\lambda)$ is a function roughly of the form shown in Fig. 2.4, one would first guess it is best to orientate the crystal so that neutrons with wavelengths corresponding to the peak, point A, are those that are scattered. But then we get a second-order contaminant with an intensity corresponding to the height of the curve at the point A', i.e. $\frac{1}{2}\lambda$. It is preferable to reorient the crystal so as to scatter neutrons at wavelengths B, thus sacrificing some intensity, so that the second-order contaminant, represented by the height at B', is smaller proportionally.

If this way is not convenient to reduce the second-order contaminant it is possible to place a filter in the beam that absorbs neutrons of wavelength $\frac{1}{2}\lambda$ much more strongly than those of λ. Of course this also reduces the primary intensity, but the second-order contaminant is reduced even more so. Yet another idea is to use as monochromator a crystal such as germanium, which has a structure such that, working with the (1, 1, 1) reflection, the second-order contaminant is zero.

The analysis we have presented so far has presupposed a perfect crystal and a perfectly collimated incident beam. Of course neither condition is realized in practice, and it is of importance to inquire how monochromatic a beam is produced by the Laue method in practice. We will not have time to go into this question in great detail but we will give it a short discussion.

But first of all we notice that (2.35) and (2.36) are certainly wrong in one respect, they are both linear in the size of the crystal (i.e. in N and $V = Nv_0$). This suggests that the scattered flux can be increased by increasing V indefinitely. This of course is not true; in using the Born approximation we assumed the crystal was uniformly bathed in the incident beam. This is only a good approximation if the scattering is small, i.e. the formulae we have are the 'small crystal' limit. Just how small a crystal has to be to satisfy the 'small crystal' criterion varies considerably. For scattering from a weak reflection the critical volume may be quite large, so the same crystal may be 'large' for some reflections (the strong ones) and 'small' for others (the weak ones). The criterion is also very sensitive to the value of the mosaic spread in the crystal. All crystals, no matter how perfect they may seem, are made up of mosaic blocks of linear dimensions of the order of thousands of Ångströms, each separated from one another by dislocations. Each small block may be regarded as a truly perfect crystal, but they are slightly displaced and slightly misorientated relative to one another so that the scattering from them is not coherent, i.e. there is no interference between the scattered waves coming from different mosaic blocks. The degree of misorientation η is usually several minutes but may occasionally be as large as a degree.

An important consequence of the mosaic spread is that it is impossible to align the whole volume of the crystal in the same orientation. The crystal may nominally be aligned in a certain direction but in reality each mosaic block is slightly misaligned from this direction. This is an extremely fortunate circumstance, for if all the blocks were exactly aligned the first few layers of mosaic blocks would scatter all the 'suitable' neutrons, leaving none 'suitable' to be scattered by the more remote layers of the crystal. As it is, in a real crystal each block selects its own group of 'suitable neutrons' and there is only competition between them for the same neutrons if the crystal is very thick. When the crystal is 'thick'—and what this means depends very sensitively on the magnitude of the mosaic spread—we say 'secondary extinction' is present. ('Primary extinction' would occur in the less likely circumstance that the mosaic blocks themselves were so thick that the incident beam was appreciably attenuated in traversing each one.)

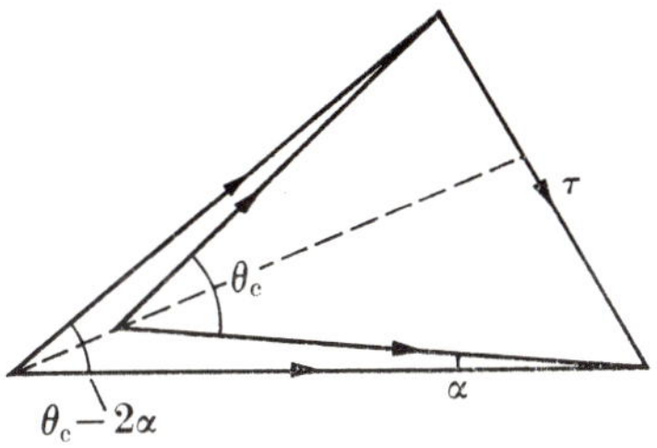

FIG. 2.5. The effect of collimation angle on Laue scattering with relative angle of emergence $+\alpha$. This is shown schematically in Fig. 2.6.

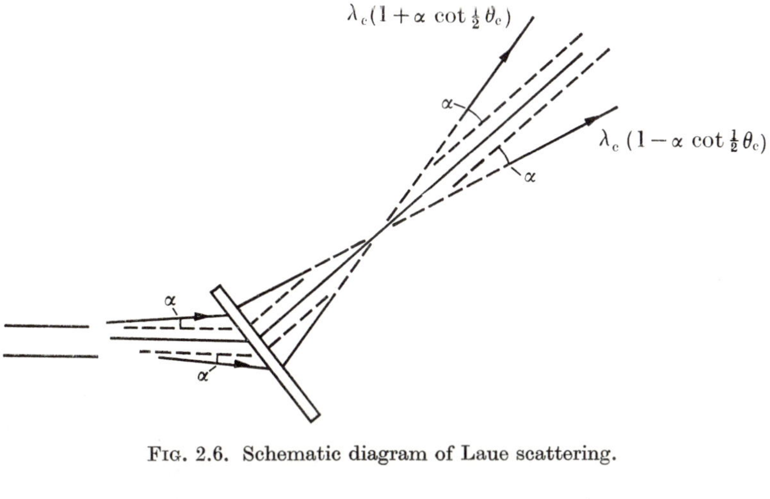

FIG. 2.6. Schematic diagram of Laue scattering.

Because of this mosaic structure it is clear that even if the incident beam were perfectly collimated the scattered beam would not be so (it would have an angular divergence of the order of the mosaic spread) and neither would it be monochromatic, because each mosaic block selects a slightly different wavelength neutron to scatter and does so through slightly different angles. But of course in practice the incident beam is not perfectly collimated and this leads to a further angular spread and wavelength variation in the scattered beam. These effects can be discussed with the aid of Figs. 2.5 and 2.6.

We suppose that the collimator is of such a length and width that it permits passage of neutrons at angles $\pm\alpha$ relative to the central path, which we suppose is scattered at an angle θ_c by the crystal. Now from Fig. 2.5 we notice that neutrons incident at an angle $+\alpha$ relative to the central path are scattered through an angle $\theta_c-2\alpha$ and therefore emerge at an angle $-\alpha$ relative to the central path. Similarly those incident at relative angle $-\alpha$ are scattered through $\theta_c+2\alpha$. If we ignore the effect of the mosaic spread for the moment the lines drawn in Fig. 2.6 represent the extreme paths of the neutrons. Now from (2.32) we have

$$\sin\tfrac{1}{2}\theta = \frac{\tau}{4\pi}\lambda,$$

where θ is the scattering angle. If scattering through θ_c corresponds to a wavelength λ_c, scattering through $\theta_c-2\alpha$ corresponds to

$$\lambda_c - \frac{4\pi}{\tau}\alpha\cos(\tfrac{1}{2}\theta_c) = \lambda_c - \lambda_c\,\alpha\cot(\tfrac{1}{2}\theta_c).$$

Hence the neutrons emerging at relative angle $-\alpha$ have wavelengths $\lambda_c(1-\alpha\cot\frac{1}{2}\theta_c)$, while those emerging at relative angle α have wavelengths $\lambda_c(1+\alpha\cot\frac{1}{2}\theta_c)$. This is indicated in Fig. 2.6. The total wavelength spread is

$$2\lambda_c\,\alpha\cot(\tfrac{1}{2}\theta_c).$$

Actually the mosaic spread of the crystal increases this wavelength spread somewhat but we will not discuss this. (For a more detailed discussion of beam widths and resolution problems we refer the reader to M. Nielsen and H. Bjerrum Møller, *Acta crystallogr.* A**25**, 547 (1969).)

Method 2: This method utilizes a monochromatic beam (produced for example by a Laue method) and a single crystal that is free to rotate. Referring to Fig. 2.3 we can think of this as keeping $\mathbf{k}$ fixed and rotating the reciprocal lattice so that points cross the surface of the sphere. It is customary to measure the orientation of the crystal by the value of the angle $\frac{1}{2}\theta$ and by rocking the crystal obtain a plot of scattered power against $\frac{1}{2}\theta$. According to (2.35) this plot is a δ-function, but of course in practice the peak has a non-zero width determined by the mosaic spread, the degree of collimation, and the wavelength spread in the incident beam, which, of course, is only nominally monochromatic. Whatever the width of the peak may be, the area is as determined from (2.35) and this is the significant parameter that is measured experimentally. The intensity integrated over the rocking curve is

$$\int d(\tfrac{1}{2}\theta)\, N(\lambda)\sigma_c = N(\lambda)VQ^\theta, \tag{2.38}$$

where $$Q^\theta = \frac{\lambda^3 |F_N(\boldsymbol{\tau})|^2}{v_0^2 \sin\theta}. \tag{2.39}$$

These formulae are obtained by substituting for σ_c and performing the integral.

It is rather interesting to consider the width of these rocking curves. In Fig. 2.6 we showed a schematic diagram of the production of a 'monochromatic' beam by a Laue type of method. In the figure the neutrons are shown as being scattered to the left. If the 'monochromatic' beam

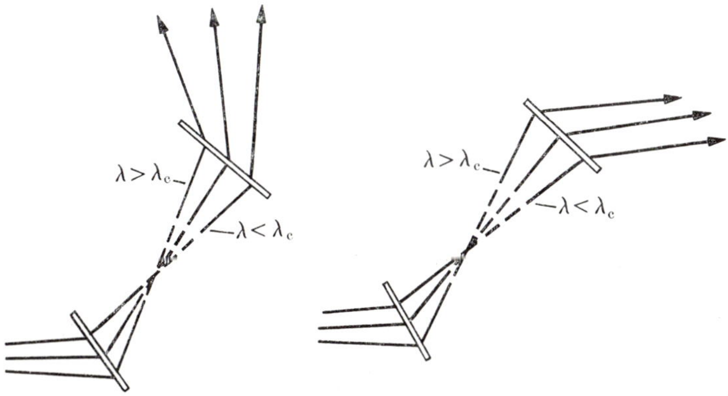

FIG. 2.7. Bragg scattering for 'left–left' and 'left–right' arrangements.

is now used for an experiment with a second crystal it is preferable to arrange this second crystal so that the second scattering takes place to the right, because the rocking curves are then narrower than they would be if the second scattering were also to go to the left. The reason for this asymmetry between left and right is that the beam produced by the first crystal is not really monochromatic nor perfectly collimated; those neutrons incident at relative angle $+\alpha$ (compared to the central path) have longer wavelengths than those incident at relative angle $-\alpha$. Now the longer wavelengths are scattered through wider angles than the shorter wavelengths, and hence if the neutrons are again scattered to the left the angular divergence between the extreme wavelengths will increase while if this second scattering takes place to the right the angular divergence will be decreased. Similarly the rocking curves for scattering to the right will be narrower than for scattering to the left. This is illustrated in Fig. 2.7. Of course, in either case the areas under the rocking curves are the same, but when the rocking curve is narrow it is easier to subtract off the background scattering and the peaks are easier to resolve.

Method 3: The powder method utilizes a monochromatic beam and a powder sample. This corresponds to keeping $\mathbf{k}$ fixed and averaging over all orientations of the reciprocal lattice. To obtain the cross-section for scattering from a powder sample we have to average (2.11) over all orientations of $\boldsymbol{\tau}$. This is equivalent to averaging over orientations of $\boldsymbol{\varkappa}$ and this is easier to do. The answer is:

$$\left(\frac{d\sigma}{d\Omega}\right)_{\text{coh}} = \frac{2\pi^2 N}{k^2 v_0} \sum_{\boldsymbol{\tau}} \frac{1}{\tau} |F_N(\boldsymbol{\tau})|^2 \delta\left(1 - \frac{\tau^2}{2k^2} - \cos\theta\right). \tag{2.40}$$

(The derivation is similar to that of (2.35).)

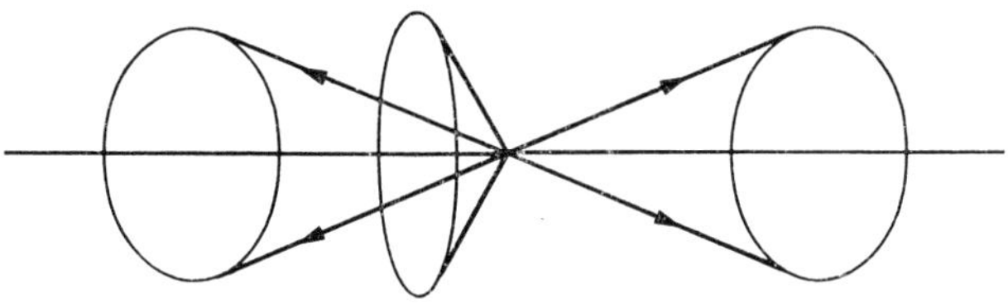

FIG. 2.8. Debye–Scherrer cones for Bragg scattering from a powder.

In this formula all reference to the azimuthal scattering angle has disappeared. Hence the scattering takes place in cones, called Debye–Scherrer cones, which have the direction $\mathbf{k}$ as axis and have semi-angles θ defined by

$$\cos\theta = 1 - \tau^2/2k^2.$$

We recognize this last condition to be the same as (2.32).

According to (2.40) each cone is infinitely thin; of course in practice each cone has a width determined by the degree of collimation and the wavelength spread in the incident beam. The total cross-section associated with each cone is

$$\sigma_c(\tau) = \frac{4\pi^3 N}{k^2 v_0} \frac{z(\tau)}{\tau} |F_\tau|^2, \tag{2.41}$$

where $z(\tau)$ is the number of $\boldsymbol{\tau}$ vectors with magnitude τ and $|F_\tau|^2$ is the mean value of $|F_N(\boldsymbol{\tau})|^2$ for just these vectors. When using this expression one must take care to allow for the fact that the counter can observe only a fraction of the complete cones. This fraction is $l/2\pi r \sin\theta$, where l is the height of the counter and r is the distance from the sample to the counter.

It is instructive to watch how the positions of these cones vary with increasing k. For very small k (2.32) cannot be satisfied and there is no scattering. When k is equal to the critical value $\frac{1}{2}\tau_{\min}$ scattering occurs for $\theta = \pi$, i.e. in the backward direction. As k slowly increases above this value this backward scattering opens out into a cone, which moves steadily towards the forward direction. Eventually, when k reaches a second critical value, backward scattering again appears and then in turn this opens out into a cone, and so on. When k is large there are many, many cones all closely spaced and, since they have an appreciable width in practice, they overlap and become indistinguishable. From (2.41) the total cross-section is

$$\sigma_c = \frac{4\pi^3 N}{k^2 v_0} \sum_{\tau}^{\tau<2k} \frac{1}{\tau} |F_\tau|^2. \tag{2.42}$$

Let us evaluate this for a simple lattice with one atom per unit cell, so that the form factor is simply $\bar{b}$, and in the limit when k is large. In this limit the sum can be replaced by an integral. We remember

$$\sum_{\boldsymbol{\tau}} \to \frac{v_0}{(2\pi)^3} \int d\boldsymbol{\tau}$$

to get
$$\sigma_c \to \frac{4\pi^3 N}{k^2 v_0} |\bar{b}|^2 \frac{v_0}{(2\pi)^3} \int\limits_0^{2k} d\boldsymbol{\tau}\, \frac{1}{\tau},$$

i.e.
$$\sigma_c \to 4\pi N |\bar{b}|^2 \quad \text{as } k \to \infty.$$

Similarly, for a more complex lattice it is possible to show that

$$\sigma_c \to 4\pi N \sum_{d=1}^{r} |\bar{b}_d|^2 \quad \text{as } k \to \infty. \tag{2.43}$$

Hence using (2.13) we get for the total cross-section

$$\sigma = \sigma_c + \sigma_i \to 4\pi N \sum_{d=1}^{r} \overline{|b_d|^2}, \tag{2.44}$$

i.e. all interference effects vanish for high incident momenta. This is just what we expect. Actually, it is not true that elastic cross-sections approach a non-zero limit as k increases. We have obtained answers that say so only because we have assumed our lattice is rigid. When we take account of lattice vibrations we will find all elastic cross-sections approach zero as k increases. It is true, however, that all interference effects vanish as k increases (cf. Chapter 4).

This powder method is often used because it is usually much easier to obtain a sample in powder form than as a single crystal and because there are no serious extinction problems with a powder. But, of course, powder intensities are much lower than single crystal intensities, so it is always preferable to use a single crystal if one is available.

In Fig. 2.9 we show a plot of the total coherent cross-section (2.42) for a body-centred cubic lattice of cube side a. Notice how this cross-section is zero for wavelengths beyond the Bragg cut-off and how the cross-section changes discontinuously as λ is decreased and new Debye–Sherrer cones appear. This curve is seriously modified for short wavelengths when proper account of lattice vibrations is made; instead of approaching a non-zero limit as $\lambda \to 0$, as it does in this graph, it should approach zero.

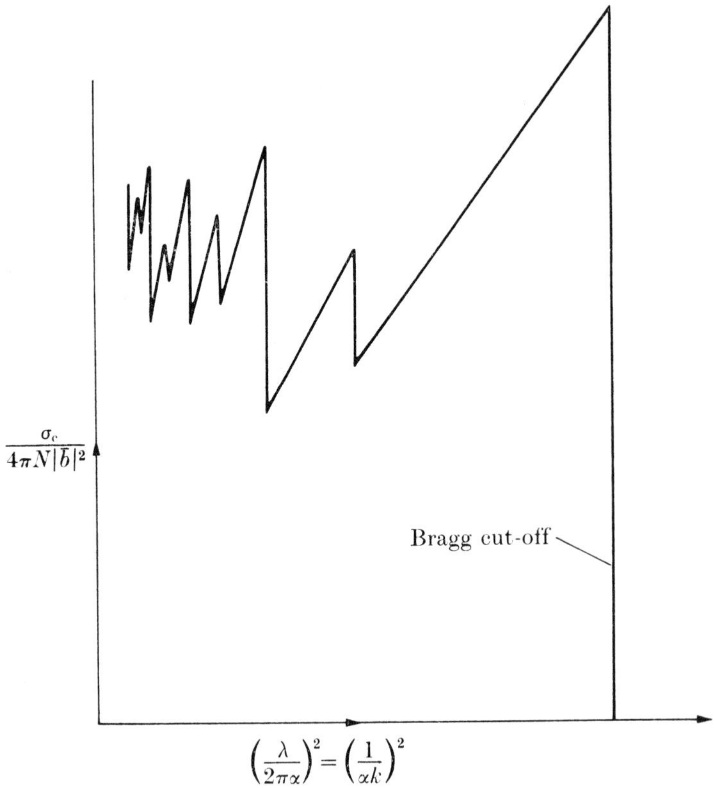

FIG. 2.9. The total coherent scattering from a body-centred cubic *rigid* lattice.

In practice nuclear Bragg scattering is used to investigate only those materials where it is impossible or inconvenient to use X-rays. For example, it is hard to locate light atoms in the presence of heavy atoms using X-rays because the X-ray scattering length is proportional to Z, the number of electrons in the atom. But for neutrons the scattering lengths for light and heavy nuclei are comparable and therefore the light nuclei can be located. Neutron Bragg scattering is specially valuable for the location of hydrogen atoms, i.e. protons. Also, using X-rays it is not easy to distinguish between atoms of similar Z values (though it can be done by using wavelengths close to an absorption edge of one of the atoms), and therefore it is often convenient to use neutrons because the neutron scattering lengths vary irregularly from one nucleus to another and therefore are likely to be quite different for the nuclei concerned. We shall not give explicit examples of these kinds of investigations, but a good reference is the review article by C. G. Shull and E. O. Wollan, *Solid state physics*, Volume 2 (Academic Press, 1956).

3

CORRELATION FUNCTIONS IN NUCLEAR SCATTERING

THE master formula we developed in Chapter 1 for the neutron cross-section can be expressed in terms of the thermal average of a combination of operators belonging to the target, so that the scattering problem is reduced essentially to a problem in statistical mechanics. The techniques for handling the latter have developed substantially in recent years and hence the reduction of the calculation of the neutron cross-section to a study of an appropriate correlation function is an exceedingly important one. In the following we pursue this problem, the results obtained being specifically concerned with purely nuclear scattering. The scattering of neutrons by magnetic systems can also be reduced to the study of an appropriate correlation function, but, because the interaction potential involved is more complicated than that for purely nuclear scattering, we postpone an analysis of the correlation functions involved till Chapter 8. For the present we demonstrate that for nuclear scattering (1.10) can be reduced to the study of a correlation function and indicate the manipulations involved to calculate these functions for a number of cases which, though very simple, we shall refer to frequently.

3.1. Scattering law†

The interaction potential between the neutrons and the target for purely nuclear scattering is of the general form

$$\hat{V}(\mathbf{r}) = \sum_j \hat{V}_j(\mathbf{r}-\mathbf{R}_j), \tag{3.1}$$

where $\mathbf{R}_j$ is the position vector of the jth scattering nucleus. If we substitute (3.1) into our master formula (1.10), then the cross-section for the scattering of unpolarized neutrons is easily shown to be

$$\frac{\mathrm{d}^2\sigma}{\mathrm{d}\Omega \mathrm{d}E'} = \frac{k'}{k}\left(\frac{m}{2\pi\hbar^2}\right)^2 \sum_{\lambda,\lambda'} p_\lambda \left|\left\langle \lambda' \left| \sum_j \hat{V}_j(\boldsymbol{\varkappa}) \exp(\mathrm{i}\boldsymbol{\varkappa}.\mathbf{R}_j) \right| \lambda \right\rangle\right|^2 \delta(\hbar\omega + E_\lambda - E_{\lambda'}), \tag{3.2}$$

where $$\hat{V}_j(\boldsymbol{\varkappa}) = \int \mathrm{d}\mathbf{r}\, \exp(\mathrm{i}\boldsymbol{\varkappa}.\mathbf{r})\, \hat{V}_j(\mathbf{r}).$$

† L. Van Hove, *Phys. Rev.* **95**, 249 (1954).

We can write the δ-function in (3.2) in the form:

$$\delta(\hbar\omega+E_\lambda-E_{\lambda'}) = \frac{1}{2\pi\hbar}\int_{-\infty}^{\infty} \mathrm{d}t \exp\{-\mathrm{i}t(\hbar\omega+E_\lambda-E_{\lambda'})/\hbar\},$$

and thus

$$\sum_{\lambda,\lambda'} p_\lambda \left|\left\langle\lambda'\right| \sum_j \hat{V}_j(\boldsymbol{\varkappa})\exp(\mathrm{i}\boldsymbol{\varkappa}.\mathbf{R}_j) \left|\lambda\right\rangle\right|^2 \delta(\hbar\omega+E_\lambda-E_{\lambda'})$$

$$= \frac{1}{2\pi\hbar}\int_{-\infty}^{\infty} \mathrm{d}t \exp(-\mathrm{i}\omega t) \sum_{\lambda,\lambda'} p_\lambda \left\langle\lambda\right| \sum_j \hat{V}_j^+(\boldsymbol{\varkappa})\exp(-\mathrm{i}\boldsymbol{\varkappa}.\mathbf{R}_j)\left|\lambda'\right\rangle \times$$

$$\times \left\langle\lambda'\right|\exp(\mathrm{i}tE_{\lambda'}/\hbar) \sum_{j'} \hat{V}_{j'}(\boldsymbol{\varkappa})\exp(\mathrm{i}\boldsymbol{\varkappa}.\mathbf{R}_{j'})\exp(-\mathrm{i}tE_\lambda/\hbar)\left|\lambda\right\rangle$$

$$= \frac{1}{2\pi\hbar}\int_{-\infty}^{\infty} \mathrm{d}t \exp(-\mathrm{i}\omega t) \sum_{\lambda,\lambda'} p_\lambda \left\langle\lambda\right| \sum_j \hat{V}_j^+(\boldsymbol{\varkappa})\exp(-\mathrm{i}\boldsymbol{\varkappa}.\mathbf{R}_j)\left|\lambda'\right\rangle \times$$

$$\times \left\langle\lambda'\right| \sum_{j'} \hat{V}_{j'}(\boldsymbol{\varkappa})\exp(\mathrm{i}t\mathscr{H}/\hbar)\exp(\mathrm{i}\boldsymbol{\varkappa}.\mathbf{R}_{j'})\exp(-\mathrm{i}t\mathscr{H}/\hbar)\left|\lambda\right\rangle, \quad (3.3)$$

where $\mathscr{H}$ is the Hamiltonian that describes the target system. The sum over the final states in (3.3) can be performed by closure, and in terms of Heisenberg operators

$$\exp\{\mathrm{i}\boldsymbol{\varkappa}.\hat{\mathbf{R}}_j(t)\} = \exp(\mathrm{i}t\mathscr{H}/\hbar)\exp(\mathrm{i}\boldsymbol{\varkappa}.\mathbf{R}_j)\exp(-\mathrm{i}t\mathscr{H}/\hbar),$$

the whole expression can be written

$$\frac{1}{2\pi\hbar}\int_{-\infty}^{\infty} \mathrm{d}t \exp(-\mathrm{i}\omega t) \sum_\lambda p_\lambda \sum_{j,j'} \times$$

$$\times\langle\lambda|\hat{V}_j^+(\boldsymbol{\varkappa})\exp\{-\mathrm{i}\boldsymbol{\varkappa}.\hat{\mathbf{R}}_j(0)\}\,\hat{V}_{j'}(\boldsymbol{\varkappa})\exp\{\mathrm{i}\boldsymbol{\varkappa}.\hat{\mathbf{R}}_{j'}(t)\}|\lambda\rangle. \quad (3.4)$$

We now recognize that the averaging over λ, with probability p_λ, involves first an average over nuclear spin orientations and isotope distributions and second, an average over other initial states of the target with these states distributed according to a Boltzmann law at temperature $T = 1/k_\mathrm{B}\beta$, i.e. with

$$p_\lambda = \exp(-\beta E_\lambda)/Z$$

and

$$\sum_\lambda p_\lambda = 1.$$

These two averaging processes can be carried out independently to give

$$\frac{1}{2\pi\hbar}\int_{-\infty}^{\infty} \mathrm{d}t \exp(-\mathrm{i}\omega t) \sum_{j,j'} \overline{\hat{V}_j^+(\boldsymbol{\varkappa})\hat{V}_{j'}(\boldsymbol{\varkappa})} \times$$

$$\times \sum_\lambda \frac{\exp(-\beta E_\lambda)}{Z} \langle\lambda|\exp\{-\mathrm{i}\boldsymbol{\varkappa}.\hat{\mathbf{R}}_j(0)\}\exp\{\mathrm{i}\boldsymbol{\varkappa}.\hat{\mathbf{R}}_{j'}(t)\}|\lambda\rangle.$$

We shall denote thermal averaging by angular brackets $\langle\ldots\rangle$, i.e.

$$\langle\ldots\rangle = \sum_{\lambda} \frac{\exp(-\beta E_{\lambda})}{Z} \langle\lambda|\ldots|\lambda\rangle. \tag{3.5}$$

With this notation the partial differential cross-section can be written in the general form

$$\frac{\mathrm{d}^2\sigma}{\mathrm{d}\Omega\mathrm{d}E'} = \frac{k'}{k}\left(\frac{m}{2\pi\hbar^2}\right)^2 \frac{1}{2\pi\hbar} \int_{-\infty}^{\infty} \mathrm{d}t \exp(-\mathrm{i}\omega t) \times$$
$$\times \sum_{j,j'} \overline{\hat{V}_j^+(\boldsymbol{\varkappa})\hat{V}_{j'}(\boldsymbol{\varkappa})} \langle \exp\{-\mathrm{i}\boldsymbol{\varkappa}.\hat{\mathbf{R}}_j(0)\}\exp\{\mathrm{i}\boldsymbol{\varkappa}.\hat{\mathbf{R}}_{j'}(t)\}\rangle. \tag{3.6}$$

When the two-body interaction potential $\hat{V}_j$ is independent of the type of nucleus, i.e. independent of j, (3.6) reduces to

$$\frac{\mathrm{d}^2\sigma}{\mathrm{d}\Omega\mathrm{d}E'} = N\frac{k'}{k}\left(\frac{m}{2\pi\hbar^2}\right)^2 \overline{|\hat{V}(\boldsymbol{\varkappa})|^2}\, S(\boldsymbol{\varkappa},\omega), \tag{3.7 a}$$

where

$$S(\boldsymbol{\varkappa},\omega) = \frac{1}{2\pi\hbar N} \int_{-\infty}^{\infty} \mathrm{d}t \exp(-\mathrm{i}\omega t) \sum_{j,j'} \langle \exp\{-\mathrm{i}\boldsymbol{\varkappa}.\hat{\mathbf{R}}_j(0)\}\exp\{\mathrm{i}\boldsymbol{\varkappa}.\hat{\mathbf{R}}_{j'}(t)\}\rangle \tag{3.7 b}$$

is usually called the *scattering law*. In (3.7), N is the total number of scattering nuclei.

An alternative expression to (3.7 b) for $S(\boldsymbol{\varkappa},\omega)$ can be obtained in terms of the microscopic particle density operator

$$\hat{\rho}(\mathbf{r},t) = \sum_j \delta\{\mathbf{r}-\hat{\mathbf{R}}_j(t)\}. \tag{3.8}$$

To see this, consider the *pair correlation function*

$$G(\mathbf{r},t) = \left(\frac{1}{2\pi}\right)^3 \int \mathrm{d}\boldsymbol{\varkappa} \exp(-\mathrm{i}\boldsymbol{\varkappa}.\mathbf{r}) \frac{1}{N} \sum_{j,j'} \langle \exp\{-\mathrm{i}\boldsymbol{\varkappa}.\hat{\mathbf{R}}_j(0)\}\exp\{\mathrm{i}\boldsymbol{\varkappa}.\hat{\mathbf{R}}_{j'}(t)\}\rangle. \tag{3.9}$$

We must be careful in evaluating the right-hand side of (3.9) because, except when $t = 0$, the two operators $\exp\{-\mathrm{i}\boldsymbol{\varkappa}.\hat{\mathbf{R}}_j(0)\}$ and $\exp\{\mathrm{i}\boldsymbol{\varkappa}.\hat{\mathbf{R}}_{j'}(t)\}$ do not commute. Hence we cannot replace their product by

$$\exp[-\mathrm{i}\boldsymbol{\varkappa}.\{\hat{\mathbf{R}}_j(0)-\hat{\mathbf{R}}_{j'}(t)\}]$$

and, indeed, must be careful in all algebraic manipulations to keep $\hat{\mathbf{R}}_{j'}(t)$ to the right of $\hat{\mathbf{R}}_j(0)$. Hence we write

$$G(\mathbf{r},t) = \left(\frac{1}{2\pi}\right)^3 \int \mathrm{d}\boldsymbol{\varkappa} \exp(-\mathrm{i}\boldsymbol{\varkappa}.\mathbf{r}) \frac{1}{N} \sum_{j,j'} \int \mathrm{d}\mathbf{r}' \times$$
$$\times \langle \exp\{-\mathrm{i}\boldsymbol{\varkappa}.\hat{\mathbf{R}}_j(0)+\mathrm{i}\boldsymbol{\varkappa}.\mathbf{r}'\}\,\delta\{\mathbf{r}'-\hat{\mathbf{R}}_{j'}(t)\}\rangle$$
$$= \frac{1}{N}\sum_{j,j'} \int \mathrm{d}\mathbf{r}' \langle \delta\{\mathbf{r}-\mathbf{r}'+\hat{\mathbf{R}}_j(0)\}\delta\{\mathbf{r}'-\hat{\mathbf{R}}_{j'}(t)\}\rangle, \tag{3.10}$$

and from (3.8) it follows that

$$G(\mathbf{r},t) = \frac{1}{N}\int d\mathbf{r}'\,\langle\hat{\rho}(\mathbf{r}'-\mathbf{r},0)\hat{\rho}(\mathbf{r}',t)\rangle, \tag{3.11 a}$$

with

$$\int d\mathbf{r}\; G(\mathbf{r},t) = N. \tag{3.11 b}$$

Thus

$$S(\boldsymbol{\varkappa},\omega) = \frac{1}{2\pi\hbar}\int\limits_{-\infty}^{\infty} dt\,\exp(-i\omega t)\int d\mathbf{r}\,\exp(i\boldsymbol{\varkappa}.\mathbf{r})\,G(\mathbf{r},t)$$

$$= \frac{1}{2\pi\hbar N}\int\limits_{-\infty}^{\infty} dt\,\exp(-i\omega t)\int d\mathbf{r}\int d\mathbf{r}'\,\exp(i\boldsymbol{\varkappa}.\mathbf{r})\times$$

$$\times\langle\hat{\rho}(\mathbf{r}'-\mathbf{r},0)\hat{\rho}(\mathbf{r}',t)\rangle. \tag{3.12}$$

In some instances it is convenient to introduce the Fourier components of $\hat{\rho}(\mathbf{r})$, i.e. we write

$$\hat{\rho}(\mathbf{r}) = \frac{1}{V}\sum_{\mathbf{q}}\hat{\rho}_{\mathbf{q}}\exp(i\mathbf{q}.\mathbf{r}). \tag{3.13}$$

The vectors $\mathbf{q}$ are defined by imposing cyclic boundary conditions on the target sample and have a density $V/(2\pi)^3$, where V is the volume of the sample. Therefore, from (3.13),

$$\sum_j \exp(-i\mathbf{q}'.\mathbf{R}_j) = \int d\mathbf{r}\,\exp(-i\mathbf{q}'.\mathbf{r})\frac{1}{V}\sum_{\mathbf{q}}\hat{\rho}_{\mathbf{q}}\exp(i\mathbf{q}.\mathbf{r})$$

$$= \frac{(2\pi)^3}{V}\sum_{\mathbf{q}}\hat{\rho}_{\mathbf{q}}\,\delta(\mathbf{q}-\mathbf{q}') = \frac{(2\pi)^3}{V}\frac{V}{(2\pi)^3}\int d\mathbf{q}\;\hat{\rho}_{\mathbf{q}}\,\delta(\mathbf{q}-\mathbf{q}') = \hat{\rho}_{\mathbf{q}'},$$

i.e.

$$\hat{\rho}_{\boldsymbol{\varkappa}}(t) = \sum_j \exp\{-i\boldsymbol{\varkappa}.\hat{\mathbf{R}}_j(t)\}. \tag{3.14}$$

From (3.12) (or directly from (3.7 b)) it follows that

$$S(\boldsymbol{\varkappa},\omega) = \frac{1}{2\pi\hbar N}\int\limits_{-\infty}^{\infty} dt\,\exp(-i\omega t)\langle\hat{\rho}_{\boldsymbol{\varkappa}}(0)\hat{\rho}_{-\boldsymbol{\varkappa}}(t)\rangle. \tag{3.15 a}$$

For many complex problems it is more convenient to deal with the spatial transform of $G(\mathbf{r},t)$ rather than $G(\mathbf{r},t)$ itself. The spatial transform is denoted as $I(\boldsymbol{\varkappa},t)$, where

$$I(\boldsymbol{\varkappa},t) = \int d\mathbf{r}\,\exp(i\boldsymbol{\varkappa}.\mathbf{r})G(\mathbf{r},t) = \frac{1}{N}\langle\hat{\rho}_{\boldsymbol{\varkappa}}(0)\hat{\rho}_{-\boldsymbol{\varkappa}}(t)\rangle, \tag{3.15 b}$$

and hence

$$S(\boldsymbol{\varkappa},\omega) = \frac{1}{2\pi\hbar}\int\limits_{-\infty}^{\infty} dt\,\exp(-i\omega t)I(\boldsymbol{\varkappa},t). \tag{3.15 c}$$

We draw attention to the form of $S(\varkappa, \omega)$ given by (3.12), which shows that $S(\varkappa, \omega)$ can be expressed in terms of a density–density correlation function. The reason for this is simply that, so far as coherent nuclear scattering is concerned, the interaction potential couples the neutron to the density of the target system. Indeed, using (3.8), the interaction potential (3.1), averaged over nuclear spins and isotopes, can be written in the form

$$\hat{V}(\mathbf{r}) = \int \mathrm{d}\mathbf{r}' \, \bar{V}(\mathbf{r}-\mathbf{r}')\hat{\rho}(\mathbf{r}', 0).$$

The cross-section as expressed by eqn (3.7 a) is the product of two factors, the first of which depends only on the properties of the individual scatterers in the target, while the second $S(\varkappa, \omega)$ is, for given energy and momentum transfers, determined solely by the structure and dynamical behaviour of the target system. Since it is $S(\varkappa, \omega)$ that is measured in an experiment, we obtain directly from an analysis of experimental data information about the structure and dynamics of the target.

3.2. Coherent and incoherent cross-sections

From Chapter 1 it follows that in (3.6)

$$\overline{\hat{V}_j^+(\varkappa)\hat{V}_{j'}(\varkappa)} = \left(\frac{2\pi\hbar^2}{m}\right)^2 \{|\bar{b}|^2 + \delta_{j,j'}[\overline{|b|^2} - |\bar{b}|^2]\} = \left(\frac{2\pi\hbar^2}{m}\right)^2 \left\{\frac{\sigma_c}{4\pi} + \delta_{j,j'} \frac{\sigma_i}{4\pi}\right\}.$$

Whence
$$\left(\frac{\mathrm{d}^2\sigma}{\mathrm{d}\Omega \mathrm{d}E'}\right)_{\mathrm{coh}} = N \frac{k'}{k} \frac{\sigma_c}{4\pi} S(\varkappa, \omega) \tag{3.16}$$

and
$$\left(\frac{\mathrm{d}^2\sigma}{\mathrm{d}\Omega \mathrm{d}E'}\right)_{\mathrm{incoh}} = N \frac{k'}{k} \frac{\sigma_i}{4\pi} S_i(\varkappa, \omega). \tag{3.17}$$

Here

$$S_i(\varkappa, \omega) = \frac{1}{2\pi\hbar N} \int_{-\infty}^{\infty} \mathrm{d}t \, \exp(-\mathrm{i}\omega t) \sum_j \langle \exp\{-\mathrm{i}\varkappa . \hat{\mathbf{R}}_j(0)\} \exp\{\mathrm{i}\varkappa . \hat{\mathbf{R}}_j(t)\}\rangle, \tag{3.18}$$

and, in analogy to the function $G(\mathbf{r}, t)$, of which, we recall, $S(\varkappa, \omega)$ is, apart from a constant, the space and time Fourier transform, we define the *self pair correlation function*,

$$G_s(\mathbf{r}, t) = \frac{1}{N} \sum_j \int \mathrm{d}\mathbf{r}' \, \langle \delta\{\mathbf{r}-\mathbf{r}'+\hat{\mathbf{R}}_j(0)\} \, \delta\{\mathbf{r}'-\hat{\mathbf{R}}_j(t)\}\rangle, \tag{3.19 a}$$

with
$$\int \mathrm{d}\mathbf{r} \, G_s(\mathbf{r}, t) = 1. \tag{3.19 b}$$

At this juncture in our discussion we ask what physical significance can be attached to the two functions $G_s(\mathbf{r}, t)$ and $G(\mathbf{r}, t)$ that have been introduced in this and the previous section. For this we examine the functions for special, simple conditions, since G_s and G are in general complicated functions on account of the fact that the operators involved in their definitions do not as a rule commute.

If the system under discussion is classical and all nuclei are equivalent (this enables us to drop the averaging $(1/N)\Sigma_j$ in both G_s and G), then, clearly,

$$G_s^{\mathrm{cl}}(\mathbf{r},t) = \langle\delta\{\mathbf{r}-\mathbf{R}_0(t)+\mathbf{R}_0(0)\}\rangle, \tag{3.20}$$

and this is the probability that, given a particle at the origin at time $t = 0$, the same particle is at $\mathbf{r}$ at time t. Also, from (3.10),

$$G^{\mathrm{cl}}(\mathbf{r},t) = \sum_j \langle\delta\{\mathbf{r}-\mathbf{R}_j(t)+\mathbf{R}_0(0)\}\rangle, \tag{3.21}$$

and this is just the probability that, given a particle at the origin at time $t = 0$, any particle (including the original particle) is to be found at the position $\mathbf{r}$ at time t.

Returning to the general discussion, the operators involved in the definitions of G_s and G commute at equal times. Thus,

$$G_s(\mathbf{r},0) = \delta(\mathbf{r}) \tag{3.22}$$

and

$$G(\mathbf{r},0) = \delta(\mathbf{r})+\sum_{j\neq 0}\langle\delta(\mathbf{r}-\mathbf{R}_j+\mathbf{R}_0)\rangle = \delta(\mathbf{r})+g(\mathbf{r}), \tag{3.23}$$

where $g(\mathbf{r})$ is the *static pair-distribution function*, and gives the average particle-density about a given particle.

Consider now the limit $t \to \infty$. In this instance there is, clearly, no correlation between the two terms on the right-hand sides of (3.10) and (3.19). From (3.11) it is clear that G in the limit $t \to \infty$ can be written

$$G(\mathbf{r},\infty) = \frac{1}{N}\int d\mathbf{r}'\,\langle\rho(\mathbf{r}'-\mathbf{r})\rangle\langle\rho(\mathbf{r}')\rangle. \tag{3.24}$$

In the case of a liquid the density is uniform, i.e. $\langle\rho(\mathbf{r})\rangle = \rho$, say. For this particular case (3.24) reduces to

$$G(\mathbf{r},\infty) = \frac{1}{N}\int d\mathbf{r}'\,\rho\,.\,\rho = \rho. \tag{3.25 a}$$

Also for this liquid example

$$G_s(\mathbf{r},\infty) = 0. \tag{3.25 b}$$

If we compare (3.22) and (3.25 b), and (3.23) and (3.25 a) then $G_s(\mathbf{r},t)$ and $G(\mathbf{r},t)$ are seen to possess a pronounced local structure for $t = 0$, but as t increases this dies away, and in the limit $t \to \infty$ the functions are independent of $\mathbf{r}$.

The results (3.25) are applicable only to the example of liquids. Thus, for example, for a rigid crystal $G_s(\mathbf{r},\infty)$ would be the same as $G_s(\mathbf{r},0)$, i.e. $\delta(\mathbf{r})$, and for a real crystal, as we shall show in Chapter 4, $G_s(\mathbf{r},\infty)$ is a Gaussian centred around the parent lattice site.

In general we can write

$$G(\mathbf{r},t) = G(\mathbf{r},\infty)+G'(\mathbf{r},t), \tag{3.26 a}$$

so that

$$\lim_{t\to\infty} G'(\mathbf{r},t) = 0. \tag{3.26 b}$$

Similarly, we define

$$G_s(\mathbf{r},t) = G_s(\mathbf{r},\infty)+G_s'(\mathbf{r},t), \tag{3.27 a}$$

where again

$$\lim_{t\to\infty} G_s'(\mathbf{r},t) = 0. \tag{3.27 b}$$

The cross-sections $(\mathrm{d}^2\sigma/\mathrm{d}\Omega\mathrm{d}E')_{\mathrm{coh}}$ and $(\mathrm{d}^2\sigma/\mathrm{d}\Omega\mathrm{d}E')_{\mathrm{incoh}}$ can now be written as the sum of elastic and inelastic parts:

$$\left(\frac{\mathrm{d}^2\sigma}{\mathrm{d}\Omega\mathrm{d}E'}\right)_{\mathrm{coh}} = \left(\frac{\mathrm{d}^2\sigma}{\mathrm{d}\Omega\mathrm{d}E'}\right)^{\mathrm{el}}_{\mathrm{coh}} + \left(\frac{\mathrm{d}^2\sigma}{\mathrm{d}\Omega\mathrm{d}E'}\right)^{\mathrm{inel}}_{\mathrm{coh}} \tag{3.28}$$

and

$$\left(\frac{\mathrm{d}^2\sigma}{\mathrm{d}\Omega\mathrm{d}E'}\right)_{\mathrm{incoh}} = \left(\frac{\mathrm{d}^2\sigma}{\mathrm{d}\Omega\mathrm{d}E'}\right)^{\mathrm{el}}_{\mathrm{incoh}} + \left(\frac{\mathrm{d}^2\sigma}{\mathrm{d}\Omega\mathrm{d}E'}\right)^{\mathrm{inel}}_{\mathrm{incoh}} \tag{3.29}$$

where

$$\left(\frac{\mathrm{d}^2\sigma}{\mathrm{d}\Omega\mathrm{d}E'}\right)^{\mathrm{el}}_{\mathrm{coh}} = N\frac{k'}{k}\frac{\sigma_{\mathrm{c}}}{4\pi}\,\delta(\hbar\omega)\int \mathrm{d}\mathbf{r}\,\exp(\mathrm{i}\boldsymbol{\kappa}.\mathbf{r})G(\mathbf{r},\infty)$$

$$= \frac{k'}{k}\frac{\sigma_{\mathrm{c}}}{4\pi}\,\delta(\hbar\omega)\left|\int \mathrm{d}\mathbf{r}\,\exp(\mathrm{i}\boldsymbol{\kappa}.\mathbf{r})\langle\rho(\mathbf{r})\rangle\right|^2,$$

where we have used (3.24). Thus

$$\left(\frac{\mathrm{d}\sigma}{\mathrm{d}\Omega}\right)^{\mathrm{el}}_{\mathrm{coh}} = \frac{\sigma_{\mathrm{c}}}{4\pi}\left|\int \mathrm{d}\mathbf{r}\,\exp(\mathrm{i}\boldsymbol{\kappa}.\mathbf{r})\langle\rho(\mathbf{r})\rangle\right|^2 \tag{3.30}$$

and

$$\left(\frac{\mathrm{d}^2\sigma}{\mathrm{d}\Omega\mathrm{d}E'}\right)^{\mathrm{inel}}_{\mathrm{coh}} = \frac{N\sigma_{\mathrm{c}}}{4\pi}\frac{k'}{k}\frac{1}{2\pi\hbar}\int_{-\infty}^{\infty} \mathrm{d}t\,\exp(-\mathrm{i}\omega t)\int \mathrm{d}\mathbf{r}\,\exp(\mathrm{i}\boldsymbol{\kappa}.\mathbf{r})G'(\mathbf{r},t). \tag{3.31}$$

Similarly

$$\left(\frac{\mathrm{d}^2\sigma}{\mathrm{d}\Omega\mathrm{d}E'}\right)^{\mathrm{el}}_{\mathrm{incoh}} = N\frac{k'}{k}\frac{\sigma_{\mathrm{i}}}{4\pi}\,\delta(\hbar\omega)\int \mathrm{d}\mathbf{r}\,\exp(\mathrm{i}\boldsymbol{\kappa}.\mathbf{r})G_s(\mathbf{r},\infty)$$

$$= \frac{k'}{k}\frac{\sigma_{\mathrm{i}}}{4\pi}\,\delta(\hbar\omega)\int \mathrm{d}\mathbf{r}\,\exp(\mathrm{i}\boldsymbol{\kappa}.\mathbf{r})\times$$

$$\times\sum_j\int \mathrm{d}\mathbf{r}'\,\langle\delta(\mathbf{r}-\mathbf{r}'+\mathbf{R}_j)\rangle\langle\delta(\mathbf{r}'-\mathbf{R}_j)\rangle.$$

Hence

$$\left(\frac{\mathrm{d}\sigma}{\mathrm{d}\Omega}\right)^{\mathrm{el}}_{\mathrm{incoh}} = \frac{\sigma_{\mathrm{i}}}{4\pi}\sum_j\langle\exp(-\mathrm{i}\boldsymbol{\kappa}.\mathbf{R}_j)\rangle\langle\exp(\mathrm{i}\boldsymbol{\kappa}.\mathbf{R}_j)\rangle. \tag{3.32}$$

Finally we have

$$\left(\frac{d^2\sigma}{d\Omega dE'}\right)^{\text{inel}}_{\text{incoh}} = N\frac{k'}{k}\frac{\sigma_i}{4\pi}\frac{1}{2\pi\hbar}\int_{-\infty}^{\infty} dt\,\exp(-i\omega t)\int d\mathbf{r}\,\exp(i\boldsymbol{\varkappa}.\mathbf{r})G'_s(\mathbf{r},t). \tag{3.33}$$

The formulae (3.30)–(3.33) are valid for any system with just one chemical species present. They are easily generalized to mixtures.

As a very simple example, consider the scattering from an array of bound nuclei, a problem already discussed in Chapters 1 and 2. In this instance $\mathbf{R}_j \equiv \mathbf{l}$ (we assume a Bravais lattice) and both $G'(\mathbf{r},t)$ and $G'_s(\mathbf{r},t)$ are zero. The number density is simply

$$\rho(\mathbf{r}) = \sum_l \delta(\mathbf{r}-\mathbf{l}).$$

Hence (3.30) becomes

$$\left(\frac{d\sigma}{d\Omega}\right)^{\text{el}}_{\text{coh}} = \frac{\sigma_c}{4\pi}\left|\int d\mathbf{r}\,\exp(i\boldsymbol{\varkappa}.\mathbf{r})\sum_l \delta(\mathbf{r}-\mathbf{l})\right|^2 = \frac{\sigma_c}{4\pi}\left|\sum_l \exp(i\boldsymbol{\varkappa}.\mathbf{l})\right|^2.$$

This is just the result derived in Chapter 1. (3.32) is simply

$$\left(\frac{d\sigma}{d\Omega}\right)^{\text{el}}_{\text{incoh}} = N\frac{\sigma_i}{4\pi}.$$

3.3. Analytic properties of $S(\varkappa, \omega)$ and $G(\mathbf{r}, t)$

In general it is not possible to evaluate the correlation functions occurring in the coherent and incoherent cross-sections for anything other than a much simplified model of the target system, some elementary examples of which we shall examine in a subsequent section. For this reason it is important to extract from the above formalism the analytic properties of the various functions we have introduced so that we can insure that approximations to these functions have the correct fundamental behaviour. The results of this section are of particular importance to give a ground work to the more detailed discussions of liquids (in Chapter 11) and gases (in Chapter 12).

First, the representation (3.15 a) of $S(\boldsymbol{\varkappa},\omega)$ yields

$$\int_{-\infty}^{\infty} d\omega\,\exp(i\omega t)S(\boldsymbol{\varkappa},\omega) = \frac{1}{\hbar N}\langle\hat{\rho}_{\boldsymbol{\varkappa}}(0)\hat{\rho}_{-\boldsymbol{\varkappa}}(t)\rangle;$$

so it immediately follows that

$$\int_{-\infty}^{\infty} d\omega\,S(\boldsymbol{\varkappa},\omega) = \frac{1}{\hbar N}\langle\hat{\rho}_{\boldsymbol{\varkappa}}(0)\hat{\rho}_{-\boldsymbol{\varkappa}}(0)\rangle. \tag{3.34 a}$$

Also $$i\hbar\frac{\partial}{\partial t}\int_{-\infty}^{\infty} d\omega \exp(i\omega t)S(\boldsymbol{\varkappa},\omega) = \frac{1}{\hbar N}\langle\hat{\rho}_{\boldsymbol{\varkappa}}(0)i\hbar\frac{\partial}{\partial t}\hat{\rho}_{-\boldsymbol{\varkappa}}(t)\rangle$$
$$= \frac{1}{\hbar N}\langle\hat{\rho}_{\boldsymbol{\varkappa}}(0)[\hat{\rho}_{-\boldsymbol{\varkappa}}(t),\mathscr{H}]\rangle,$$

i.e. $$\int_{-\infty}^{\infty} d\omega\; \omega S(\boldsymbol{\varkappa},\omega) = -\frac{1}{\hbar^2 N}\langle\hat{\rho}_{\boldsymbol{\varkappa}}(0)[\hat{\rho}_{-\boldsymbol{\varkappa}}(0),\mathscr{H}]\rangle. \qquad (3.34\text{ b})$$

The latter result has the following physical significance; the right-hand side is proportional to the average energy transferred to the target for a given momentum transfer $\hbar\boldsymbol{\varkappa}$. This is most clearly seen by writing the correlation function in (3.34 b) explicitly in terms of the eigenfunctions $|\lambda\rangle$ and eigenvalues E_λ of the target Hamiltonian $\mathscr{H}$. On introducing a complete set of eigenfunctions $|\lambda'\rangle$ we have

$$\langle\hat{\rho}_{\boldsymbol{\varkappa}}[\hat{\rho}_{-\boldsymbol{\varkappa}},\mathscr{H}]\rangle = \frac{1}{Z}\sum_{\lambda}\exp(-\beta E_\lambda)\langle\lambda|\hat{\rho}_{\boldsymbol{\varkappa}}[\hat{\rho}_{-\boldsymbol{\varkappa}},\mathscr{H}]|\lambda\rangle$$
$$= \frac{1}{Z}\sum_{\lambda}\exp(-\beta E_\lambda)\sum_{\lambda'}(E_\lambda - E_{\lambda'})(\hat{\rho}_{\boldsymbol{\varkappa}})_{\lambda\lambda'}(\hat{\rho}_{-\boldsymbol{\varkappa}})_{\lambda'\lambda}$$
$$= \frac{1}{Z}\sum_{\lambda}\exp(-\beta E_\lambda)\sum_{\lambda'}(E_\lambda - E_{\lambda'})|(\hat{\rho}_{\boldsymbol{\varkappa}})_{\lambda\lambda'}|^2.$$

Now, for a fixed momentum transfer to the target $\hbar\boldsymbol{\varkappa}$, the right-hand side of this expression represents the energy *lost* by the target medium on undergoing the transition from the state $|\lambda\rangle$ to the state $|\lambda'\rangle$, $|(\hat{\rho}_{\boldsymbol{\varkappa}})_{\lambda\lambda'}|^2$ being the weight attached to this transition, the whole being averaged over all possible initial states $|\lambda\rangle$.†

Alternatively, (3.34 b) may be regarded as the first moment of the scattering law. The higher moments are obtained in an analogous manner

$$\int_{-\infty}^{\infty} d\omega\; (\omega)^n S(\boldsymbol{\varkappa},\omega) = \frac{(-\hbar^{-1})^n}{\hbar N}\langle\hat{\rho}_{\boldsymbol{\varkappa}}(0)[[[\hat{\rho}_{-\boldsymbol{\varkappa}}(0),\mathscr{H}],\mathscr{H}],\mathscr{H}\ldots]\rangle, \qquad (3.34\text{ c})$$

there being an n-fold commutator of $\hat{\rho}_{-\boldsymbol{\varkappa}}(0)$ with $\mathscr{H}$ on the right-hand side. The relations (3.34) are often called *sum rules* for the scattering law $S(\boldsymbol{\varkappa},\omega)$.

The evaluation of (3.34 c) for n odd is greatly simplified on taking note of the following relationship. Because, as we show below,

$$\langle\lambda|\hat{\rho}_{\boldsymbol{\varkappa}}[\hat{\rho}_{-\boldsymbol{\varkappa}},\mathscr{H}]|\lambda\rangle = -\langle\lambda|[\hat{\rho}_{-\boldsymbol{\varkappa}},\mathscr{H}]\hat{\rho}_{\boldsymbol{\varkappa}}|\lambda\rangle,$$

then $$\langle\lambda|\hat{\rho}_{\boldsymbol{\varkappa}}[\hat{\rho}_{-\boldsymbol{\varkappa}},\mathscr{H}]|\lambda\rangle = \tfrac{1}{2}\langle\lambda|[\hat{\rho}_{\boldsymbol{\varkappa}},[\hat{\rho}_{-\boldsymbol{\varkappa}},\mathscr{H}]]|\lambda\rangle. \qquad (3.35\text{ a})$$

† G. Placzek, *Phys. Rev.* **86**, 377 (1952).

Similarly,

$$\langle\lambda|\hat{\rho}_{\varkappa}[[[\hat{\rho}_{-\varkappa},\mathscr{H}],\mathscr{H}],\mathscr{H}]\rangle = \tfrac{1}{2}\langle\lambda|[\hat{\rho}_{\varkappa},[[[\hat{\rho}_{-\varkappa},\mathscr{H}],\mathscr{H}],\mathscr{H}]]|\lambda\rangle, \tag{3.35 b}$$

and so on.

Because the right-hand sides of these equations involve expectation values of commutators, they are specially convenient for evaluation. The proof of these identities rests upon the fact that the matrix element $(\hat{\rho}_{\varkappa})_{\lambda\lambda'} = (\hat{\rho}_{\varkappa})_{\lambda'\lambda}$, which is a consequence of the fact that we can choose the eigenfunctions $|\lambda\rangle$ to be real because they represent stationary states and, in addition, that $\hat{\rho}_{\varkappa}$ acts only on the coordinates of the nuclei. We need only prove the first of the relations given by (3.35), since the second and higher-order ones follow by inspection. On introducing a complete set of states $|\lambda'\rangle$, we have

$$\begin{aligned}\langle\lambda|\hat{\rho}_{\varkappa}[\hat{\rho}_{-\varkappa},\mathscr{H}]|\lambda\rangle &= \sum_{\lambda'}(\hat{\rho}_{\varkappa})_{\lambda\lambda'}\langle\lambda'|[\hat{\rho}_{-\varkappa},\mathscr{H}]|\lambda\rangle \\ &= \sum_{\lambda'}(\hat{\rho}_{\varkappa})_{\lambda\lambda'}(\hat{\rho}_{-\varkappa})_{\lambda'\lambda}(E_{\lambda}-E_{\lambda'}) \\ &= -\sum_{\lambda'}(\hat{\rho}_{-\varkappa})_{\lambda\lambda'}(\hat{\rho}_{\varkappa})_{\lambda'\lambda}(E_{\lambda'}-E_{\lambda}) \\ &= -\langle\lambda|[\hat{\rho}_{-\varkappa},\mathscr{H}]\hat{\rho}_{\varkappa}|\lambda\rangle,\end{aligned}$$

which is the result we needed to derive (3.35 a).

One can similarly prove

$$\langle\lambda|\hat{\rho}_{\varkappa}[[\hat{\rho}_{-\varkappa},\mathscr{H}],\mathscr{H}]|\lambda\rangle = \langle\lambda|[[\hat{\rho}_{-\varkappa},\mathscr{H}],\mathscr{H}]\hat{\rho}_{\varkappa}|\lambda\rangle,$$

so that for n even we do not achieve any worthwhile simplification in the evaluation of the sum rules.

Using the first result of (3.35), we obtain for (3.34 b)

$$\int_{-\infty}^{\infty} \mathrm{d}\omega\, \omega S(\varkappa,\omega) = -\frac{1}{2\hbar^2 N}\langle[\hat{\rho}_{\varkappa},[\hat{\rho}_{-\varkappa},\mathscr{H}]]\rangle. \tag{3.36}$$

To discuss this it is convenient first to evaluate (3.36) for a free gas of nuclei, i.e. for

$$\mathscr{H} = \sum_{j=1}^{N}\frac{1}{2M}\hat{\mathbf{p}}_j^2.$$

First we need $\quad [\hat{\rho}_{-\varkappa},\mathscr{H}] = \sum_{j,j'}\frac{1}{2M}[\exp(\mathrm{i}\varkappa.\mathbf{R}_j),\hat{\mathbf{p}}_{j'}^2].$

This commutator is evaluated by writing it as

$$[\exp(\mathrm{i}\varkappa.\mathbf{R}_j),\hat{\mathbf{p}}_{j'}^2] = [\exp(\mathrm{i}\varkappa.\mathbf{R}_j),\hat{\mathbf{p}}_{j'}]\hat{\mathbf{p}}_{j'}+\hat{\mathbf{p}}_{j'}[\exp(\mathrm{i}\varkappa.\mathbf{R}_j),\hat{\mathbf{p}}_{j'}]$$

and making use of the result that if $f(\mathbf{r})$ is an arbitrary function of position

$$[f(\mathbf{r}),\hat{\mathbf{p}}] = \mathrm{i}\hbar\,\mathrm{grad}\,f(\mathbf{r}). \tag{3.37}$$

Hence

$$[\exp(i\boldsymbol{\varkappa}.\mathbf{R}_j),\, \hat{\mathbf{p}}_{j'}^2] = \delta_{j,j'}\{-\hbar\boldsymbol{\varkappa}\exp(i\boldsymbol{\varkappa}.\mathbf{R}_j)\hat{\mathbf{p}}_j - \hat{\mathbf{p}}_j\hbar\boldsymbol{\varkappa}\exp(i\boldsymbol{\varkappa}.\mathbf{R}_j)\}$$
$$= -\delta_{j,j'}\exp(i\boldsymbol{\varkappa}.\mathbf{R}_j)\{2\hbar\boldsymbol{\varkappa}.\hat{\mathbf{p}}_j + \hbar^2\kappa^2\}. \tag{3.38}$$

So
$$[\hat{\rho}_{-\varkappa}, \mathscr{H}] = -\frac{1}{2M}\sum_j \exp(i\boldsymbol{\varkappa}.\mathbf{R}_j)\{2\hbar\boldsymbol{\varkappa}.\hat{\mathbf{p}}_j + \hbar^2\kappa^2\}$$

and

$$[\hat{\rho}_{\varkappa}, [\hat{\rho}_{-\varkappa}, \mathscr{H}]] = -\frac{1}{2M}\sum_{j,j'} \delta_{j,j'}\exp(i\boldsymbol{\varkappa}.\mathbf{R}_j)2\hbar\boldsymbol{\varkappa}.[\exp(-i\boldsymbol{\varkappa}.\mathbf{R}_j),\, \hat{\mathbf{p}}_j]$$
$$= -\frac{1}{2M}\sum_j 2\hbar^2\kappa^2 = -N\frac{\hbar^2\kappa^2}{M}.$$

Hence
$$\langle \hat{\rho}_{\varkappa}[\hat{\rho}_{-\varkappa}, \mathscr{H}]\rangle = -N\frac{\hbar^2\kappa^2}{2M}, \tag{3.39 a}$$

and thus
$$\int_{-\infty}^{\infty} d\omega\, \omega S(\boldsymbol{\varkappa}, \omega) = \frac{\kappa^2}{2M}. \tag{3.39 b}$$

If we assume the nuclei of the target to be at rest prior to the scattering of a neutron, then the conservation of momentum demands for each nucleus

$$\hbar\mathbf{k} = \hbar\mathbf{k}' + M\mathbf{v}', \tag{3.40 a}$$

where $\mathbf{v}'$ is the velocity of the nucleus after the scattering process. The kinetic energy imparted to the nucleus, its recoil energy, is

$$\frac{M}{2}v'^2 = \frac{\hbar^2\kappa^2}{2M}. \tag{3.40 b}$$

From the discussion following (3.34 b), (3.39 a) is the energy lost by the target, which, in the light of (3.40), is equal to minus N times the recoil energy of a single nucleus. This is just what we should expect; the nuclei being free and initially at rest can only gain energy from the incident neutron, this accounts for the negative sign in (3.39 a), and since they are independent of one another the total energy gained is N times that for a single nucleus.

This simple argument gives a neat physical interpretation of (3.39 b), but two further points are worth noting. First, that (3.39 b) is a result independent of temperature; hence for scattering from a gas the first moment of the scattering law is always $\kappa^2/2M$. Second, we have only used the Hamiltonian once in the commutation leading to (3.39 b); hence, if we added terms representing the interactions between the nuclei in the target system, we would obtain an identical result provided those interactions were momentum independent and varied only with the nuclear positions $\mathbf{R}_j$. In practice this last point is very substantially true, and therefore we also conclude that the first moment of the scattering law is given to an excellent approximation by (3.39 b) even for a dense gas or liquid.

We now consider further basic properties of $S(\varkappa, \omega)$ and the corresponding properties of $G(\mathbf{r}, t)$.

Of course, the scattering law $S(\varkappa, \omega)$ is purely real because the cross-section is real. In detail, $S(\varkappa, \omega)$ is real because it is the time Fourier transform of a correlation function formed from two operators that are the Hermitian conjugate of each other, i.e. in (3.15 a) $\hat{\rho}_{\varkappa}^{+}(t) = \hat{\rho}_{-\varkappa}(t)$.

As a consequence of the reality of $S(\varkappa, \omega)$ it follows from (3.12) that $G(\mathbf{r}, t)$ satisfies

$$G(\mathbf{r}, t) = G^{*}(-\mathbf{r}, -t). \tag{3.41}$$

We also note at this point that the fact that $G(\mathbf{r}, t)$ is a complex function is directly related to the quantum mechanical aspect of the problem. This is made clear by (3.11 a), from which

$$\begin{aligned} \operatorname{Im} G(\mathbf{r}, t) &= \frac{1}{2\mathrm{i}N} \int \mathrm{d}\mathbf{r}' \, \langle \hat{\rho}(\mathbf{r}'-\mathbf{r}, 0)\hat{\rho}(\mathbf{r}', t) - \hat{\rho}(\mathbf{r}', t)\hat{\rho}(\mathbf{r}'-\mathbf{r}, 0) \rangle \\ &= \frac{\mathrm{i}}{2N} \int \mathrm{d}\mathbf{r}' \, \langle [\hat{\rho}(\mathbf{r}', t), \hat{\rho}(\mathbf{r}'-\mathbf{r}, 0)] \rangle. \end{aligned} \tag{3.42}$$

In the classical limit the commutator in (3.42) would vanish, and hence the classical correlation functions would be real, as is evident directly from eqns (3.20) and (3.21). Since $\hat{\rho}(\mathbf{r})$ is both real and Hermitian (cf. Appendix B, eqns (B.48), (B.49))

$$\langle \hat{\rho}(\mathbf{r}, t)\hat{\rho}(\mathbf{r}', 0) \rangle = \langle \hat{\rho}(\mathbf{r}', t)\hat{\rho}(\mathbf{r}, 0) \rangle,$$

and it therefore follows directly from (3.42) that $\operatorname{Im} G(\mathbf{r}, t)$ is an odd function of t. Similarly it can be shown that $\operatorname{Re} G(\mathbf{r}, t)$ is an even function of t.

We also note that because $S(\varkappa, \omega)$ is essentially the cross-section it must be positive definite. The analytical consequences of this show up most clearly in terms of the function $I(\varkappa, t)$ defined by (3.15 b). Inverting (3.15 c) gives

$$I(\varkappa, t) = \frac{1}{N} \langle \hat{\rho}_{\varkappa}(0)\hat{\rho}_{-\varkappa}(t) \rangle = \hbar \int \mathrm{d}\omega \; \mathrm{e}^{\mathrm{i}\omega t} S(\varkappa, \omega),$$

$$-\left(\frac{\partial}{\partial t}\right)^{2} I(\varkappa, t) = \hbar \int \mathrm{d}\omega \; \mathrm{e}^{\mathrm{i}\omega t} \omega^{2} S(\varkappa, \omega),$$

and

$$\left(\frac{\partial}{\partial t}\right)^{4} I(\varkappa, t) = \hbar \int \mathrm{d}\omega \; \mathrm{e}^{\mathrm{i}\omega t} \omega^{4} S(\varkappa, \omega), \quad \text{etc.}$$

Putting $t = 0$ in this series of equations we note that the integrals on the right-hand side are all positive definite and non-zero (unless the scattering is strictly elastic). It follows that either $I(\boldsymbol{\kappa}, t)$ is strictly independent of t, i.e. the target is rigidly fixed and the scattering is strictly elastic, or $I(\boldsymbol{\kappa}, t)$ at small times has a negative non-zero second derivative and a positive non-zero fourth derivative, with respect to t.

From the representation of $S(\boldsymbol{\kappa}, \omega)$ given by eqn (3.15a) it is a simple matter to extract the so-called *condition of detailed balance.*

To obtain the required relationship between $S(\boldsymbol{\kappa}, \omega)$ and $S(-\boldsymbol{\kappa}, -\omega)$ we use the identity (cf. Appendix B, eqn (B.34))

$$\langle\hat{\rho}_{\boldsymbol{\kappa}}(t)\hat{\rho}_{-\boldsymbol{\kappa}}(0)\rangle = \langle\hat{\rho}_{-\boldsymbol{\kappa}}(0)\hat{\rho}_{\boldsymbol{\kappa}}(t+\mathrm{i}\hbar\beta)\rangle \tag{3.43}$$

in eqn (3.15a). We find

$$S(\boldsymbol{\kappa}, \omega) = \exp(\hbar\omega\beta)S(-\boldsymbol{\kappa}, -\omega). \tag{3.44}$$

The physical significance of (3.44) is easily understood. When the incident neutrons come into thermal equilibrium with the target after many collisions with the nuclei, the number of neutrons that gain a given energy must equal the number that lose the same energy. Eqn (3.44) is a mathematical statement of this fact, i.e. the probability that a neutron loses an energy $\hbar\omega$ is equal to $\exp(\hbar\omega\beta)$ times the probability that a neutron gains an energy $\hbar\omega$.

Note that if we define

$$\tilde{S}(\boldsymbol{\kappa}, \omega) = \mathrm{e}^{-\frac{1}{2}\hbar\omega\beta}S(\boldsymbol{\kappa}, \omega) \tag{3.45}$$

then it follows from (3.44) that $\tilde{S}(\boldsymbol{\kappa}, \omega)$ is an even function in both $\boldsymbol{\kappa}$ and ω. Thus, if in analogy with (3.12) we define $\tilde{G}(\mathbf{r}, t)$ such that

$$\tilde{G}(\mathbf{r}, t) = \frac{\hbar}{(2\pi)^3}\int \mathrm{d}\boldsymbol{\kappa} \int \mathrm{d}\omega \exp(-\mathrm{i}\boldsymbol{\kappa}.\mathbf{r}+\mathrm{i}\omega t)\tilde{S}(\boldsymbol{\kappa}, \omega) = G(\mathbf{r}, t+\tfrac{1}{2}\mathrm{i}\hbar\beta), \tag{3.46}$$

then $\tilde{G}(\mathbf{r}, t)$ is purely real, and even in both $\mathbf{r}$ and t.

Finally we ask for the relation for $G(\mathbf{r}, t)$ that is equivalent to (3.44) for the scattering law. It is straightforward to show with the aid of the identity (3.43) (or directly from (3.44)) that

$$G(\mathbf{r}, t) = G(-\mathbf{r}, -t+\mathrm{i}\hbar\beta). \tag{3.47}$$

This result in conjunction with (3.41) can be used to establish a relationship between the real and imaginary parts of $G(\mathbf{r}, t)$. Consider

$$\int_{-\infty}^{\infty} \mathrm{d}t\, \mathrm{e}^{-\mathrm{i}\omega t}\,\mathrm{Im}\, G(\mathbf{r}, t) = \frac{1}{2\mathrm{i}}\int_{-\infty}^{\infty} \mathrm{d}t\, \mathrm{e}^{-\mathrm{i}\omega t}[G(\mathbf{r}, t)-G^*(\mathbf{r}, t)]. \tag{3.48}$$

From (3.41) and (3.47) we have

$$\int_{-\infty}^{\infty} dt\, e^{-i\omega t} G^*(\mathbf{r}, t) = \int_{-\infty}^{\infty} dt\, e^{-i\omega t} G(-\mathbf{r}, -t) = \int_{-\infty}^{\infty} dt\, e^{-i\omega t} G(\mathbf{r}, t+i\hbar\beta)$$

$$= e^{-\hbar\omega\beta} \int_{-\infty}^{\infty} dt\, e^{-i\omega t} G(\mathbf{r}, t),$$

so that the right-hand side of (3.48) is

$$\frac{1}{2i} \int_{-\infty}^{\infty} dt\, e^{-i\omega t}(1-e^{-\hbar\omega\beta}) G(\mathbf{r}, t). \tag{3.49}$$

Hence

$$\int_{-\infty}^{\infty} dt\, e^{-i\omega t}\, \mathrm{Im}\, G(\mathbf{r}, t) = -i \tanh(\tfrac{1}{2}\hbar\omega\beta) \int_{-\infty}^{\infty} dt\, e^{-i\omega t}\, \mathrm{Re}\, G(\mathbf{r}, t), \tag{3.50}$$

which is the desired relation. Alternatively we can write this in the form

$$\mathrm{Re}\, G(\mathbf{r}, t) = \frac{i}{2\pi} \int_{-\infty}^{\infty} d\omega\, e^{i\omega t} \coth(\tfrac{1}{2}\hbar\omega\beta) \int_{-\infty}^{\infty} dt'\, e^{-i\omega t'}\, \mathrm{Im}\, G(\mathbf{r}, t'). \tag{3.51}$$

This expression is formally identical to eqn (B.56) of Appendix B, which is itself a statement of the fluctuation-dissipation theorem. To see this we note that from the representation of $\mathrm{Im}\, G(\mathbf{r}, t)$ given by eqn (3.42) and the form of the linear response function ϕ given by (B.24 b) we may take

$$\phi \equiv \frac{2}{\hbar}\, \mathrm{Im}\, G(\mathbf{r}, t).$$

As a consequence of this formal relation we can regard $\mathrm{Im}\, G(\mathbf{r}, t)$ as measuring the response, or the disturbance, produced by the neutron in the density of the target medium.† Furthermore it follows that, by definition,

$$\frac{1}{i\hbar} \int_{-\infty}^{\infty} dt\, e^{i\omega t}\, \mathrm{Im}\, G(\mathbf{r}, t)$$

is the imaginary (or dissipative) part of the generalized susceptibility function (as a function of position), whose poles as a function of ω give the positions in energy of the collective excitations of the target (cf. § B.1 of Appendix B).

† L. Van Hove, *Physica 's Grav.* **24**, 404 (1958).

3.4. Classical approximation to the scattering cross-section

In some instances the quantum mechanical nature of $S(\boldsymbol{\varkappa},\omega)$ can justifiably be neglected, and this greatly simplifies the calculation of the cross-section. In this section we seek a rigorous prescription for effecting this approximation, it being clearly incorrect merely to replace $G(\mathbf{r},t)$ by the corresponding classical function.

In our discussion of the physical significance of $G(\mathbf{r},t)$ and $G_{\mathrm{s}}(\mathbf{r},t)$ we noted that if we treat the operators involved as c-numbers, the resulting expressions, eqns (3.20) and (3.21), admit a very simple interpretation, but that in general it is not possible to give G and G_{s} such simple meanings. Bearing this in mind we re-write the operators involved in the definition of $S(\boldsymbol{\varkappa},\omega)$ to bring out their non-commuting nature, namely we write†

$$\exp\{-\mathrm{i}\boldsymbol{\varkappa}.\hat{\mathbf{R}}_j(0)\}\exp\{\mathrm{i}\boldsymbol{\varkappa}.\hat{\mathbf{R}}_{j'}(t)\} = \exp\{-\mathrm{i}\boldsymbol{\varkappa}(\hat{\mathbf{R}}_j(0)-\hat{\mathbf{R}}_{j'}(t))+\mathrm{i}\hbar\hat{\Gamma}_{jj'}(t)\}. \tag{3.52}$$

To determine $\hat{\Gamma}$ we use the identity

$$\exp\hat{A}\exp\hat{B} = \exp(\hat{A}+\hat{B}+\hat{C}),$$

where‡

$$\hat{C} = \tfrac{1}{2}[\hat{A},\hat{B}]+\tfrac{1}{12}[[\hat{A},\hat{B}],\hat{B}]+\tfrac{1}{12}[[\hat{B},\hat{A}],\hat{A}]+\tfrac{1}{24}[[[\hat{B},\hat{A}],\hat{A}],\hat{B}]+\ldots, \tag{3.53}$$

and set $$\hat{A} = -\mathrm{i}\boldsymbol{\varkappa}.\hat{\mathbf{R}}_j(0)$$

and $$\hat{B} = \mathrm{i}\boldsymbol{\varkappa}.\hat{\mathbf{R}}_{j'}(t).$$

Hence
$$\begin{aligned}\mathrm{i}\hbar\hat{\Gamma}_{jj'}(t) = {} & \tfrac{1}{2}[\boldsymbol{\varkappa}.\hat{\mathbf{R}}_j(0),\boldsymbol{\varkappa}.\hat{\mathbf{R}}_{j'}(t)]+\\ & +\frac{\mathrm{i}}{12}[[\boldsymbol{\varkappa}.\hat{\mathbf{R}}_j(0),\boldsymbol{\varkappa}.\hat{\mathbf{R}}_{j'}(t)],\boldsymbol{\varkappa}.\hat{\mathbf{R}}_j(0)]+\\ & +\frac{\mathrm{i}}{12}[[\boldsymbol{\varkappa}.\hat{\mathbf{R}}_j(0),\boldsymbol{\varkappa}.\hat{\mathbf{R}}_{j'}(t)],\boldsymbol{\varkappa}.\hat{\mathbf{R}}_{j'}(t)]+\\ & +\ldots.\end{aligned} \tag{3.54}$$

Using (3.52) we obtain for $S(\boldsymbol{\varkappa},\omega)$

$$S(\boldsymbol{\varkappa},\omega) = \frac{1}{2\pi\hbar}\int\limits_{-\infty}^{\infty}\mathrm{d}t\;\mathrm{e}^{-\mathrm{i}\omega t}\frac{1}{N}\sum_{j,j'}\langle\exp\{-\mathrm{i}\boldsymbol{\varkappa}.(\hat{\mathbf{R}}_j(0)-\hat{\mathbf{R}}_{j'}(t))+\mathrm{i}\hbar\hat{\Gamma}_{jj'}(t)\}\rangle. \tag{3.55}$$

† Sunakawa *et al.*, *Prog. theor. Phys. Osaka*, **37**, 1051 (1967).

‡ See, for example, R. Englman and P. Levy, *J. math. Phys.* **4**, 105 § 4 (1963).

By writing $S(\varkappa, \omega)$ in this form we have in effect isolated the purely quantum mechanical features in $\hat{\Gamma}_{jj'}(t)$. The *classical approximation* to $S(\varkappa, \omega)$ is now obtained by (1) treating the operator

$$\exp[-\mathrm{i}\boldsymbol{\varkappa}.\{\hat{\mathbf{R}}_j(0)-\hat{\mathbf{R}}_{j'}(t)\}]$$

as a c-number and (2) replacing the commutators in the expansion for $\hat{\Gamma}$, eqn (3.54), by $\mathrm{i}\hbar\{\ ,\ \}$, where the Poisson bracket $\{\ ,\ \}$ is defined by

$$\{A, B\} = \frac{\partial A}{\partial x}\frac{\partial B}{\partial p} - \frac{\partial A}{\partial p}\frac{\partial B}{\partial x}, \tag{3.56}$$

p being the momentum variable conjugate to coordinate x.

If the commutator $[\boldsymbol{\varkappa}.\hat{\mathbf{R}}_j(0), \boldsymbol{\varkappa}.\hat{\mathbf{R}}_{j'}(t)]$ is a c-number, as is often the case, $\hat{\Gamma}$ is itself a c-number. If in addition the particles move independently of one another, $\hat{\Gamma}$ takes the form

$$\Gamma_{jj'}(t) = \delta_{j,j'}\gamma(t), \tag{3.57 a}$$

where, from (3.54),

$$\gamma(t) = \frac{1}{2\mathrm{i}\hbar}[\boldsymbol{\varkappa}.\hat{\mathbf{R}}_j(0), \boldsymbol{\varkappa}.\hat{\mathbf{R}}_j(t)]. \tag{3.57 b}$$

Inserting (3.57 a) into (3.55)

$$S(\varkappa, \omega) = \frac{1}{2\pi\hbar}\int_{-\infty}^{\infty} \mathrm{d}t\, \mathrm{e}^{-\mathrm{i}\omega t} \int \mathrm{d}\mathbf{r}\, \mathrm{e}^{\mathrm{i}\boldsymbol{\varkappa}.\mathbf{r}}[\mathrm{e}^{\mathrm{i}\hbar\gamma(t)}\bar{G}_\mathrm{s}(\mathbf{r}, t) + \bar{G}_\mathrm{d}(\mathbf{r}, t)], \tag{3.58}$$

where

$$\bar{G}_\mathrm{s}(\mathbf{r}, t) = \frac{1}{N}\sum_j \langle\delta\{\mathbf{r}+\hat{\mathbf{R}}_j(0)-\hat{\mathbf{R}}_j(t)\}\rangle \tag{3.59 a}$$

and

$$\bar{G}_\mathrm{d}(\mathbf{r}, t) = \frac{1}{N}\sum_{\substack{j,j'\\ j\neq j'}} \langle\delta\{\mathbf{r}+\hat{\mathbf{R}}_j(0)-\hat{\mathbf{R}}_{j'}(t)\}\rangle. \tag{3.59 b}$$

Note that $\bar{G}_\mathrm{s}$ and $\bar{G}_\mathrm{d}$ apparently have the same form as the classical functions G_s^cl and G^cl that could be defined from (3.20) and (3.21). Nevertheless the functions are not the same because (3.59 a) and (3.59 b) include quantum effects coming from the use of $\hat{\mathbf{R}}_j(0)$, etc., as operators. Also

$$S_\mathrm{i}(\varkappa, \omega) = \frac{1}{2\pi\hbar}\int_{-\infty}^{\infty} \mathrm{d}t\, \exp\{-\mathrm{i}\omega t+\mathrm{i}\gamma(t)\} \int \mathrm{d}\mathbf{r}\, \exp(\mathrm{i}\boldsymbol{\varkappa}.\mathbf{r})\bar{G}_\mathrm{s}(\mathbf{r}, t). \tag{3.60}$$

The two functions $\bar{G}_\mathrm{s}(\mathbf{r}, t)$ and $\bar{G}_\mathrm{d}(\mathbf{r}, t)$ are purely real and have a simple physical significance: $\bar{G}_\mathrm{s}(\mathbf{r}, t)$ can be interpreted in exactly the same way as $G_\mathrm{s}^\mathrm{cl}(\mathbf{r}, t)$, eqn (3.20), while $\bar{G}_\mathrm{d}(\mathbf{r}, t)$ is the probability of finding an atom at time $t = 0$ and another at time t separated by a distance $\mathbf{r}$.

In the following section we shall contrast the behaviour of $\bar{G}_\mathrm{s}$ and $\bar{G}_\mathrm{d}$ with the corresponding functions introduced in the previous section by evaluating them explicitly for simple models.

Before proceeding to this task it is worth while to note that $\tilde{G}(\mathbf{r},t)$, eqn (3.46), possesses the same analytical properties as $G^{\text{cl}}(\mathbf{r},t)$. This suggests that in some instances the cross-section may be well approximated by replacing $\tilde{G}$ by† G^{cl}. Indeed, for the two examples of a monatomic gas and an isotropic oscillator this procedure gives a cross-section that is formally correct to first order in $\hbar$. Also note that the resulting cross-sections automatically fulfil the condition of detailed balance.

3.5. Examples: simple scattering systems

We demonstrate how to use the formulae of the previous sections by examining in detail the scattering from two very simple systems: (a) a single free nucleus and (b) a single nucleus bound in an isotropic, harmonic potential. In the course of discussing these systems we shall have to introduce several results that are used throughout the remainder of the book.

(a) *Single free nucleus*

For a single free nucleus we can set $\mathbf{R}_j = \mathbf{r}$ and so from (3.7 b) the scattering law is

$$S(\boldsymbol{\varkappa},\omega) = \frac{1}{2\pi\hbar}\int\limits_{-\infty}^{\infty} \mathrm{d}t\, \mathrm{e}^{-\mathrm{i}\omega t}\langle\exp\{-\mathrm{i}\boldsymbol{\varkappa}.\hat{\mathbf{r}}(0)\}\exp\{\mathrm{i}\boldsymbol{\varkappa}.\hat{\mathbf{r}}(t)\}\rangle. \tag{3.61}$$

The Hamiltonian of the target is simply

$$\mathscr{H} = \frac{M}{2}\dot{\hat{\mathbf{r}}}^2 = \frac{1}{2M}\hat{\mathbf{p}}^2,$$

where M is the mass of the nucleus. The most direct way of determining the operator $\exp\{\mathrm{i}\boldsymbol{\varkappa}.\hat{\mathbf{r}}(t)\}$ is to form its equation of motion, namely

$$\mathrm{i}\hbar\frac{\partial}{\partial t}\exp\{\mathrm{i}\boldsymbol{\varkappa}.\hat{\mathbf{r}}(t)\} = [\exp\{\mathrm{i}\boldsymbol{\varkappa}.\hat{\mathbf{r}}(t)\},\mathscr{H}]. \tag{3.62}$$

The commutator on the right-hand side of this equation is

$$\begin{aligned}
\frac{1}{2M}&[\exp\{\mathrm{i}\boldsymbol{\varkappa}.\hat{\mathbf{r}}(t)\},\hat{\mathbf{p}}^2]\\
&= \frac{\exp(\mathrm{i}t\mathscr{H}/\hbar)}{2M}[\exp(\mathrm{i}\boldsymbol{\varkappa}.\mathbf{r}),\hat{\mathbf{p}}^2]\exp(-\mathrm{i}t\mathscr{H}/\hbar)\\
&= -\frac{\hbar}{2M}\exp(\mathrm{i}t\mathscr{H}/\hbar)\exp(\mathrm{i}\boldsymbol{\varkappa}.\mathbf{r})(2\boldsymbol{\varkappa}.\hat{\mathbf{p}}+\hbar\kappa^2)\exp(-\mathrm{i}t\mathscr{H}/\hbar)\\
&= -\frac{\hbar}{2M}\exp\{\mathrm{i}\boldsymbol{\varkappa}.\hat{\mathbf{r}}(t)\}(2\boldsymbol{\varkappa}.\hat{\mathbf{p}}+\hbar\kappa^2)
\end{aligned}$$

† P. Schofield, *Phys. Rev. Lett.* **4**, 239 (1960).

on using (3.38). If this result is inserted in (3.62) the equation of motion can be integrated straightforwardly. The result is

$$\exp\{i\boldsymbol{\varkappa}.\hat{\mathbf{r}}(t)\} = \exp\{i\boldsymbol{\varkappa}.\hat{\mathbf{r}}(0)\}\exp\left\{\frac{it}{2M}(2\boldsymbol{\varkappa}.\hat{\mathbf{p}}+\hbar\kappa^2)\right\}.$$

Hence

$$\langle\exp\{-i\boldsymbol{\varkappa}.\hat{\mathbf{r}}(0)\}\exp\{i\boldsymbol{\varkappa}.\hat{\mathbf{r}}(t)\}\rangle = \exp(it\hbar\kappa^2/2M)\langle\exp(it\boldsymbol{\varkappa}.\hat{\mathbf{p}}/M)\rangle. \quad (3.63)$$

By definition (eqn (3.5))

$$\begin{aligned}&\langle\exp(it\boldsymbol{\varkappa}.\hat{\mathbf{p}}/M)\rangle\\ &\quad = \left\{\int \mathrm{d}\mathbf{p}\, \exp(-\beta p^2/2M)\exp(it\boldsymbol{\varkappa}.\mathbf{p}/M)\right\}\Big/\left\{\int \mathrm{d}\mathbf{p}\, \exp(-\beta p^2/2M\right\}\\ &\quad = \exp(-t^2\kappa^2/2M\beta). \qquad (3.64)\end{aligned}$$

Here we have used the result

$$\int_{-\infty}^{\infty} \mathrm{d}x\, \exp(-ax^2+bx) = \sqrt{\left(\frac{\pi}{a}\right)}\exp(b^2/4a). \quad (3.65)$$

From (3.63) and (3.64) we have, finally,

$$\langle\exp\{-i\boldsymbol{\varkappa}.\hat{\mathbf{r}}(0)\}\exp\{i\boldsymbol{\varkappa}.\hat{\mathbf{r}}(t)\}\rangle = \exp\left[-\frac{\kappa^2}{2M}(t^2/\beta-i\hbar t)\right]. \quad (3.66)$$

Let us now calculate both $G(\mathbf{r},t)$ and $S(\boldsymbol{\varkappa},\omega)$. First

$$\begin{aligned}G(\mathbf{r},t) &= \left(\frac{1}{2\pi}\right)^3\int \mathrm{d}\boldsymbol{\varkappa}\, \exp(-i\boldsymbol{\varkappa}.\mathbf{r})\langle\exp\{-i\boldsymbol{\varkappa}.\hat{\mathbf{r}}(0)\}\exp\{i\boldsymbol{\varkappa}.\hat{\mathbf{r}}(t)\}\rangle\\ &= \left(\frac{\beta M}{2\pi t(t-i\hbar\beta)}\right)^{\frac{3}{2}}\exp\left[\frac{-\beta Mr^2}{2t(t-i\hbar\beta)}\right] \qquad (3.67)\end{aligned}$$

on using (3.65) and (3.66). Similarly we obtain for $S(\boldsymbol{\varkappa},\omega)$, eqn (3.61),

$$S(\boldsymbol{\varkappa},\omega) = \left(\frac{\beta M}{2\pi\hbar^2\kappa^2}\right)^{\frac{1}{2}}\exp\left[-\frac{\beta M}{2\hbar^2\kappa^2}\left(\hbar\omega-\frac{\hbar^2\kappa^2}{2M}\right)^2\right], \quad (3.68)$$

and thus the partial differential cross-section for the scattering by a single free nucleus at temperature $T = 1/k_B\beta$ is

$$\frac{\mathrm{d}^2\sigma}{\mathrm{d}\Omega\mathrm{d}E'} = \frac{k'}{k}\overline{|b|^2}\left(\frac{\beta M}{2\pi\hbar^2\kappa^2}\right)^{\frac{1}{2}}\exp\left[-\frac{\beta M}{2\hbar^2\kappa^2}\left(\hbar\omega-\frac{\hbar^2\kappa^2}{2M}\right)^2\right]. \quad (3.69)$$

If we take the limit $\beta\to\infty$ in (3.69) we regain the result derived in Chapter 1 for the scattering by a single free nucleus at zero temperature, since

$$\delta(x) = \lim_{\epsilon\to 0^+}\frac{1}{\sqrt{(\pi\epsilon)}}\exp(-x^2/\epsilon).$$

Also, if $M\to\infty$ then (3.69) reduces to the scattering from a single bound nucleus, as expected.

We notice that in (3.69) $\hbar^2\kappa^2/2M$ is the recoil energy of the nucleus and that the whole expression is, in fact, independent of $\hbar$ when written in terms of the momenta $\mathbf{p} = \hbar\mathbf{k}$ and $\mathbf{p}' = \hbar\mathbf{k}'$ and energies E and E' of the neutron. Thus we would expect to obtain the same result from the classical approximation to the cross-section given in § 3.4. To demonstrate that this is in fact the case, we proceed as follows.

Our classical approximation to the scattering law is

$$\bar{S}(\boldsymbol{\varkappa}, \omega) = \frac{1}{2\pi\hbar}\int\limits_{-\infty}^{\infty} \mathrm{d}t\, \exp(-\mathrm{i}\omega t)\exp\{\mathrm{i}\hbar\gamma(t)\}\langle\exp[-\mathrm{i}\boldsymbol{\varkappa}.\{\mathbf{r}(0)-\mathbf{r}(t)\}]\rangle, \tag{3.70}$$

where [cf. (3.56)] $\gamma(t) = \frac{1}{2}\{\boldsymbol{\varkappa}.\mathbf{r}(0), \boldsymbol{\varkappa}.\mathbf{r}(t)\}$.

Clearly, for free motion,

$$\mathbf{r}(t) = \mathbf{r}(0)+t\mathbf{p}/M,$$

so that

$$\langle\exp[-\mathrm{i}\boldsymbol{\varkappa}.\{\mathbf{r}(0)-\mathbf{r}(t)\}]\rangle = \langle\exp(\mathrm{i}t\boldsymbol{\varkappa}.\mathbf{p}/M)\rangle = \exp(-t^2\kappa^2/2M\beta) \tag{3.71}$$

by (3.64). Also it is quite easy to show that

$$\gamma(t) = \tfrac{1}{2}\sum_{\alpha,\beta}\kappa_\alpha\kappa_\beta\{r_\alpha(0), r_\beta(t)\} = \tfrac{1}{2}\sum_{\alpha,\beta}\kappa_\alpha\kappa_\beta\,\delta_{\alpha,\beta}(t/M) = \tfrac{1}{2}t\kappa^2/M. \tag{3.72}$$

If we insert (3.71) and (3.72) into (3.70) it is obvious that $\bar{S}(\boldsymbol{\varkappa}, \omega)$ is identical to $S(\boldsymbol{\varkappa}, \omega)$ given by (3.68), as expected.

We use the present example of the scattering by a single free nucleus to demonstrate the need for the discussion given in § 3.4 of how to obtain the classical approximation to $S(\boldsymbol{\varkappa}, \omega)$, which in this particular instance is, as we have just shown, identical to the exact quantum mechanical expression. For if we had simply said that in the exact expression for $S(\boldsymbol{\varkappa}, \omega)$, eqn (3.61), ignore the fact that $\exp\{\mathrm{i}\boldsymbol{\varkappa}.\hat{\mathbf{r}}(t)\}$ is an operator and replace it by its classical analogue, then (3.61) would read

$$\frac{1}{2\pi\hbar}\int\limits_{-\infty}^{\infty} \mathrm{d}t\, \mathrm{e}^{-\mathrm{i}\omega t}\langle\exp[-\mathrm{i}\boldsymbol{\varkappa}.\{\mathbf{r}(0)-\mathbf{r}(t)\}]\rangle,$$

which by (3.71) is equal to

$$\frac{1}{2\pi\hbar}\int\limits_{-\infty}^{\infty} \mathrm{d}t\, \mathrm{e}^{-\mathrm{i}\omega t}\exp(-t^2\kappa^2/2M\beta) = \left(\frac{\beta M}{2\pi\hbar^2\kappa^2}\right)^{\frac{1}{2}}\exp\left(-\frac{\beta M(\hbar\omega)^2}{2\hbar^2\kappa^2}\right).$$

This is clearly incorrect because the term giving the recoil energy of the nucleus is absent; the latter appears in $\bar{S}(\boldsymbol{\varkappa}, \omega)$, eqn (3.70), through the function $\gamma(t)$.

A final point is to evaluate the correlation function $\bar{G}(\mathbf{r},t)$ defined in § 3.4. For the present case, using (3.64), and $[\hat{\mathbf{r}}_\alpha(0),\, \hat{\mathbf{r}}_\beta(t)] = \delta_{\alpha,\beta}(\mathrm{i}\hbar t/M)$,

$$\bar{G}(\mathbf{r},t) = \left(\frac{1}{2\pi}\right)^3 \int \mathrm{d}\boldsymbol{\kappa}\, \exp(-\mathrm{i}\boldsymbol{\kappa}\,.\,\mathbf{r})\langle \exp[-\mathrm{i}\boldsymbol{\kappa}\,.\{\hat{\mathbf{r}}(0)-\hat{\mathbf{r}}(t)\}]\rangle$$

$$= \left(\frac{1}{2\pi}\right)^3 \int \mathrm{d}\boldsymbol{\kappa}\, \exp(-\mathrm{i}\boldsymbol{\kappa}\,.\,\mathbf{r})\exp(-t^2\kappa^2/2M\beta)$$

$$= \left(\frac{\beta M}{2\pi t^2}\right)^{\frac{3}{2}} \exp\left(-\frac{\beta M}{2t^2}\, r^2\right).$$

In contrast to $G(\mathbf{r},t)$, eqn (3.67), this expression for $\bar{G}(\mathbf{r},t)$ is purely real and independent of $\hbar$.

(b) *Nucleus in harmonic oscillator potential*

The Hamiltonian of a particle of mass M in an isotropic, harmonic oscillator potential is

$$\mathscr{H} = \frac{1}{2M}\,\hat{\mathbf{p}}^2 + \frac{M}{2}\,\omega_0^2\,\hat{\mathbf{r}}^2, \tag{3.73}$$

where ω_0 is the frequency of vibration. Because of the isotropic nature of the Hamiltonian we need only consider the motion of one cartesian component x, say.

As is well known the eigenvalues of the Hamiltonian

$$\mathscr{H}_1 = \frac{1}{2M}\,\hat{p}^2 + \frac{M\omega_0^2}{2}\,\hat{x}^2 \tag{3.74}$$

are
$$E_n = \hbar\omega_0(n+\tfrac{1}{2}), \quad \text{where } n = 0,\, 1,\, 2,\dots .$$

Furthermore, $\mathscr{H}_1$ can be expressed in terms of Bose operators $\hat{a}$ and $\hat{a}^+$, which satisfy

$$[\hat{a},\hat{a}^+] = 1, \tag{3.75}$$

all other commutators being zero.† Indeed, if we set

$$\hat{x} = \left(\frac{\hbar}{2M\omega_0}\right)^{\frac{1}{2}} (\hat{a}+\hat{a}^+) \tag{3.76 a}$$

and

$$\hat{p} = -\mathrm{i}\left(\frac{M\hbar\omega_0}{2}\right)^{\frac{1}{2}} (\hat{a}-\hat{a}^+), \tag{3.76 b}$$

then (3.74) reduces to

$$\mathscr{H}_1 = \hbar\omega_0(\hat{a}^+\hat{a}+\tfrac{1}{2}), \tag{3.77}$$

as is easily verified with the aid of (3.75). If the eigenkets of (3.74) are denoted by $|n\rangle$ then

$$\hat{a}^+\,|n\rangle = (n+1)^{\frac{1}{2}}\,|n+1\rangle \tag{3.78 a}$$

and

$$\hat{a}\,|n\rangle = (n)^{\frac{1}{2}}\,|n-1\rangle, \tag{3.78 b}$$

† A. Messiah, *Quantum mechanics*, Chapter 12 (North-Holland).

and hence
$$\hat{a}^+\hat{a}\,|n\rangle = n\,|n\rangle, \tag{3.78 c}$$
showing that the eigenvalues of (3.77) are as given above.

To calculate the scattering law for this model we need to evaluate the correlation function
$$\langle\exp\{-\mathrm{i}\kappa_x\hat{x}(0)\}\exp\{\mathrm{i}\kappa_x\hat{x}(t)\}\rangle, \tag{3.79}$$
where we have denoted the component of the scattering vector $\boldsymbol{\kappa}$ that is parallel to x by κ_x.

The equation of motion for $\hat{a}$ is
$$\mathrm{i}\hbar\frac{\partial}{\partial t}\hat{a} = [\hat{a},\mathscr{H}_1] = \hbar\omega_0[\hat{a},\hat{a}^+\hat{a}] = \hbar\omega_0\,\hat{a}.$$
Hence
$$\hat{a}(t) = \hat{a}\exp(-\mathrm{i}\omega_0 t). \tag{3.80 a}$$
Similarly
$$\hat{a}^+(t) = \hat{a}^+\exp(\mathrm{i}\omega_0 t). \tag{3.80 b}$$
(3.80) gives
$$\hat{x}(t) = \left(\frac{\hbar}{2M\omega_0}\right)^{\frac{1}{2}}\{\hat{a}\exp(-\mathrm{i}\omega_0 t)+\hat{a}^+\exp(\mathrm{i}\omega_0 t)\}. \tag{3.81}$$
We now make use of the identity (3.53) to write
$$\exp\{-\mathrm{i}\kappa_x\hat{x}(0)\}\exp\{\mathrm{i}\kappa_x\hat{x}(t)\} = \exp\{\tfrac{1}{2}\kappa_x^2[\hat{x}(0),\hat{x}(t)]\}\exp[-\mathrm{i}\kappa_x\{\hat{x}(0)-\hat{x}(t)\}], \tag{3.82}$$
since $[\hat{x}(0),\hat{x}(t)]$ is a c-number.

To proceed further we invoke the following identity (frequently referred to as Bloch's identity): if $\hat{Q}$ is a linear function of the operators $\hat{a}$ and $\hat{a}^+$, then
$$\langle\exp\hat{Q}\rangle = \exp(\tfrac{1}{2}\langle\hat{Q}^2\rangle). \tag{3.83}$$
A proof is given at the end of this chapter.

Applying this to (3.82) we find
$$\langle\exp\{-\mathrm{i}\kappa_x\hat{x}(0)\}\exp\{\mathrm{i}\kappa_x\hat{x}(t)\}\rangle = \exp\{\tfrac{1}{2}\kappa_x^2[\hat{x}(0),\hat{x}(t)]\}\times$$
$$\times\exp[-\tfrac{1}{2}\kappa_x^2\langle\{\hat{x}(0)-\hat{x}(t)\}^2\rangle] = \exp[-\kappa_x^2\{\langle(\hat{x}(0))^2\rangle-\langle\hat{x}(0)\hat{x}(t)\rangle\}]. \tag{3.84}$$
Now from (3.81) we have
$$\langle\hat{x}(0)\hat{x}(t)\rangle = \frac{\hbar}{2M\omega_0}\langle\hat{a}\hat{a}^+\exp(\mathrm{i}\omega_0 t)+\hat{a}^+\hat{a}\exp(-\mathrm{i}\omega_0 t)\rangle$$
$$= \frac{\hbar}{2M\omega_0}\{(1+\langle\hat{a}^+\hat{a}\rangle)\exp(\mathrm{i}\omega_0 t)+\langle\hat{a}^+\hat{a}\rangle\exp(-\mathrm{i}\omega_0 t)\} \tag{3.85}$$
and
$$\langle\hat{a}^+\hat{a}\rangle = \sum_{n=0}^{\infty}\exp(-\beta E_n)\langle n|\,\hat{a}^+\hat{a}\,|n\rangle\Big/\sum_{n=0}^{\infty}\exp(-\beta E_n)$$
$$= \sum_{n=0}^{\infty}\exp(-\beta\hbar\omega_0 n)n\Big/\sum_{n=0}^{\infty}\exp(-\beta\hbar\omega_0 n)$$
$$= -\left[\frac{\partial}{\partial\alpha}\left\{\ln\sum_{n=0}^{\infty}\mathrm{e}^{-\alpha n}\right\}\right]\Bigg|_{\alpha=\hbar\omega_0\beta},$$
which gives
$$\langle\hat{a}^+\hat{a}\rangle = \{\exp(\hbar\omega_0\beta)-1\}^{-1}. \tag{3.86}$$

If we use (3.85) in (3.84) the latter becomes

$$\exp\left\{-\frac{\hbar\kappa_x^2}{2M\omega_0}(2\langle\hat{a}^+\hat{a}\rangle+1)\right\}\exp\left[\frac{\hbar\kappa_x^2}{2M\omega_0}\{(\langle\hat{a}^+\hat{a}\rangle+1)\exp(\mathrm{i}\omega_0 t)+ +\langle\hat{a}^+\hat{a}\rangle\exp(-\mathrm{i}\omega_0 t)\}\right]. \quad (3.87)$$

The actual correlation function required to evaluate $S(\boldsymbol{\varkappa},\omega)$ is

$$\langle\exp\{-\mathrm{i}\boldsymbol{\varkappa}.\hat{\mathbf{r}}(0)\}\exp\{\mathrm{i}\boldsymbol{\varkappa}.\hat{\mathbf{r}}(t)\}\rangle = \prod_{\alpha=x,y,z}\langle\exp\{-\mathrm{i}\kappa_\alpha\hat{\mathbf{r}}_\alpha(0)\}\exp\{\mathrm{i}\kappa_\alpha\hat{\mathbf{r}}_\alpha(t)\}\rangle.$$

Thus the result (3.87), which is just one of the three identical correlation functions occurring in this expression, gives

$$\langle\exp\{-\mathrm{i}\boldsymbol{\varkappa}.\hat{\mathbf{r}}(0)\}\exp\{\mathrm{i}\boldsymbol{\varkappa}.\hat{\mathbf{r}}(t)\}\rangle = \exp\left\{-\frac{\hbar\kappa^2}{2M\omega_0}\coth(\tfrac{1}{2}\hbar\omega_0\beta)\right\}\times \times\exp\left[\frac{\hbar\kappa^2}{2M\omega_0}\operatorname{cosech}(\tfrac{1}{2}\hbar\omega_0\beta)\{\exp(\mathrm{i}\omega_0 t+\tfrac{1}{2}\hbar\omega_0\beta)+\exp(-\mathrm{i}\omega_0 t-\tfrac{1}{2}\hbar\omega_0\beta)\}\right]. \quad (3.88)$$

In this form it is not possible to perform the time Fourier integral of the correlation function as required in $S(\boldsymbol{\varkappa},\omega)$. However, if we make use of the identity†

$$\exp\{\tfrac{1}{2}y(t+t^{-1})\} = \sum_{n=-\infty}^{\infty} t^n I_n(y), \quad (3.89)$$

where $I_n(y)$ is the modified Bessel function of the first kind, we can write

$$\langle\exp\{-\mathrm{i}\boldsymbol{\varkappa}.\hat{\mathbf{r}}(0)\}\exp\{\mathrm{i}\boldsymbol{\varkappa}.\hat{\mathbf{r}}(t)\}\rangle = \exp\left\{-\frac{\hbar\kappa^2}{2M\omega_0}\coth(\tfrac{1}{2}\hbar\omega_0\beta)\right\}\times \times\sum_{n=-\infty}^{\infty}\exp(\tfrac{1}{2}\hbar\omega_0\beta n+\mathrm{i}\omega_0 tn)I_n(y), \quad (3.90)$$

where $$y = \frac{\hbar\kappa^2}{2M\omega_0}\operatorname{cosech}(\tfrac{1}{2}\hbar\omega_0\beta).$$

With this result we have, finally,

$$S(\boldsymbol{\varkappa},\omega) = \frac{1}{2\pi\hbar}\int_{-\infty}^{\infty}\mathrm{d}t\,\exp(-\mathrm{i}\omega t)\langle\exp\{-\mathrm{i}\boldsymbol{\varkappa}.\hat{\mathbf{r}}(0)\}\exp\{\mathrm{i}\boldsymbol{\varkappa}.\hat{\mathbf{r}}(t)\}\rangle$$
$$= \exp\left\{-\frac{\hbar\kappa^2}{2M\omega_0}\coth(\tfrac{1}{2}\hbar\omega_0\beta)\right\}\exp(\tfrac{1}{2}\hbar\omega\beta)\sum_{n=-\infty}^{\infty}\delta(\hbar\omega-n\hbar\omega_0)I_n(y). \quad (3.91)$$

The integer n in (3.91) measures the number of units of energy $\hbar\omega_0$ lost ($n>0$) or gained ($n<0$) by the neutron. The second factor in

† See, for example, *Table of integrals, series and products*, I. S. Gradshteyn and I. M. Ryzhik, Academic Press (1965).

(3.91) shows that the processes in which the neutron loses energy are more strongly weighted, at a given temperature, than those for which it gains energy; a not unexpected result. (N.B. $I_n(y) = I_{-n}(y)$.) In the limit $T \to 0$ the neutrons can only lose energy because the nucleus must be in its ground state; in (3.91) this means that at $T = 0$ only the values $n \geqslant 0$ occur.

Consider now the behaviour of $S(\boldsymbol{\varkappa}, \omega)$ for a given temperature T. From (3.76 a) we can easily show that

$$\langle n|\hat{x}^2|n\rangle = \frac{\hbar}{2M\omega_0}(2n+1).$$

Since

$$\langle n|\hat{\mathbf{r}}^2|n\rangle = 3\langle n|\hat{x}^2|n\rangle,$$

it follows that we can re-write the argument of I_n in the form

$$y = \tfrac{1}{3}\langle 0|\hat{\mathbf{r}}^2|0\rangle\kappa^2 \operatorname{cosech}(\tfrac{1}{2}\hbar\omega_0\beta).$$

If the wavelength of the neutrons is such that

$$\langle 0|\hat{\mathbf{r}}^2|0\rangle\kappa^2 \ll 1,$$

i.e. $|\boldsymbol{\varkappa}|$ is small compared to the characteristic dimension of the motion of the nucleus, then it follows from the small argument expansion of $I_n(y)$, namely

$$I_n(y) = (\tfrac{1}{2}y)^n/n!,$$

that only the first few terms in the summation over n need be retained.

3.6. Total cross-section for a gas of non-interacting nuclei

From example (a) of the previous section, the partial differential cross-section for a gas of N non-interacting nuclei, each of mass M, is clearly

$$\frac{d^2\sigma}{d\Omega dE'} = \overline{|b|^2}\frac{k'}{k}\frac{1}{2\pi\hbar}\int_{-\infty}^{\infty} dt\, e^{-i\omega t} \exp\left\{\frac{\hbar\kappa^2}{2M}\left(it - \frac{t^2}{\hbar\beta}\right)\right\}. \tag{3.92}$$

To calculate the total cross-section from this equation we first note that

$$\hbar\omega = E - E' = -\frac{\hbar^2}{2m}(\kappa^2 - 2\mathbf{k}.\boldsymbol{\varkappa}). \tag{3.93}$$

Hence (3.92) reads

$$\frac{d^2\sigma}{d\Omega dE'} = N\overline{|b|^2}\frac{k'}{k}\frac{1}{2\pi\hbar}\int_{-\infty}^{\infty} dt\, \exp\left\{-\frac{\hbar\kappa^2}{2m}\left(\frac{t^2E}{\lambda\hbar} - it\mu\right) - \frac{it\hbar}{m}\mathbf{k}.\boldsymbol{\varkappa}\right\}, \tag{3.94}$$

where for ease of notation we have defined

$$\mu = 1 + \frac{m}{M} \quad \text{and} \quad \lambda = \frac{EM\beta}{m}. \tag{3.95}$$

It follows from (3.94) that

$$\sigma = N\overline{|b|^2}\frac{\hbar}{2\pi km}\int_{-\infty}^{\infty} \mathrm{d}t \int \mathrm{d}\boldsymbol{\varkappa}\, \exp\left\{-\frac{\hbar\kappa^2}{2m}\left(\frac{t^2E}{\lambda\hbar}-\mathrm{i}t\mu\right)-\frac{\mathrm{i}t\hbar}{m}\mathbf{k}.\boldsymbol{\varkappa}\right\}$$

$$= N\overline{|b|^2}\left(\frac{\pi\hbar}{E}\right)^{\frac{1}{2}}\int_{-\infty}^{\infty} \mathrm{d}t\, \left(\frac{t^2E}{\lambda\hbar}-\mathrm{i}t\mu\right)^{-\frac{3}{2}} \exp\left[-\frac{t^2E}{\hbar\{(t^2E/\lambda\hbar)-\mathrm{i}t\mu\}}\right] \tag{3.96}$$

where in obtaining the second line we have made use of the integral (3.65).

The integral

$$\left(\frac{\pi\hbar}{E}\right)^{\frac{1}{2}}\int_{-\infty}^{\infty} \mathrm{d}t\, \left(\frac{t^2E}{\lambda\hbar}-\mathrm{i}t\mu\right)^{-\frac{3}{2}} \exp\left[-\frac{t^2E}{\hbar\{(t^2E/\lambda\hbar)-\mathrm{i}t\mu\}}\right] \tag{3.97}$$

can be rearranged into the form

$$\sqrt{\pi}\,\lambda^{\frac{3}{2}}\,\mathrm{e}^{-\lambda}\int_{-\infty}^{\infty} \frac{\mathrm{d}t}{(\delta-\mathrm{i}t)^{\frac{3}{2}}(\mu\lambda+\mathrm{i}t)^{\frac{3}{2}}}\exp\left(\frac{\lambda^2\mu}{\mu\lambda+\mathrm{i}t}\right),$$

where we have introduced a small positive constant δ to define the integral at $t = 0$, which we shall later let go to zero. Since

$$\int_{-\infty}^{\infty} \frac{\mathrm{d}t}{(\delta-\mathrm{i}t)^{\frac{3}{2}}(\lambda\mu+\mathrm{i}t)^{\frac{3}{2}}}\exp\left(\frac{\lambda^2\mu}{\lambda\mu+\mathrm{i}t}\right)$$

$$= \int_{-\infty}^{\infty} \mathrm{d}x\, \left\{\frac{1}{2\pi}\int_{-\infty}^{\infty} \mathrm{d}t\, (\lambda\mu+\mathrm{i}t)^{-\frac{3}{2}}\exp\left(\frac{\lambda^2\mu}{\lambda\mu+\mathrm{i}t}\right)\mathrm{e}^{\mathrm{i}xt}\right\}\times$$

$$\times\int_{-\infty}^{\infty} \mathrm{d}t'\, (\delta-\mathrm{i}t')^{-\frac{3}{2}}\,\mathrm{e}^{-\mathrm{i}xt'}, \tag{3.98}$$

and

$$\frac{1}{2\pi}\int_{-\infty}^{\infty} \mathrm{d}t\, (\lambda\mu+\mathrm{i}t)^{-\frac{3}{2}}\exp\left(\frac{\lambda^2\mu}{\lambda\mu+\mathrm{i}t}\right)\mathrm{e}^{\mathrm{i}xt}$$

$$= \begin{cases} \dfrac{1}{\lambda(\mu\pi)^{\frac{1}{2}}}\,\mathrm{e}^{-\mu\lambda x}\sinh\{2\lambda\sqrt{(\mu x)}\} & (x > 0), \\ 0 \quad (x < 0), \end{cases} \tag{3.99}$$

and

$$\int_{-\infty}^{\infty} \mathrm{d}t\, (\delta-\mathrm{i}t)^{-\frac{3}{2}}\,\mathrm{e}^{-\mathrm{i}xt} = \begin{cases} 4\sqrt{\pi}\,x^{\frac{1}{2}}\mathrm{e}^{-\delta x} & (x > 0), \\ 0 & (x < 0), \end{cases} \tag{3.100}$$

we finally obtain

$$\sigma = N\overline{|b|^2}\,4\sqrt{\left(\frac{\lambda\pi}{\mu}\right)}\,\mathrm{e}^{-\lambda}\int_{0}^{\infty} \mathrm{d}x\, x^{\frac{1}{2}}\,\mathrm{e}^{-\mu\lambda x}\sinh\{2\lambda\sqrt{(\mu x)}\}. \tag{3.101}$$

Alternatively this expression can be rewritten in the more usual form

$$\sigma = N\overline{|b|^2}\frac{4\pi}{\mu^2}\left\{\left(1+\frac{1}{2\lambda}\right)\Phi(\sqrt{\lambda})+\frac{e^{-\lambda}}{\sqrt{(\pi\lambda)}}\right\}, \tag{3.102}$$

where $\Phi(x)$ is the probability function†

$$\Phi(x) = \frac{2}{\sqrt{\pi}}\int_0^x dt\, e^{-t^2}. \tag{3.103}$$

When $\lambda \ll 1$,

$$\sigma \simeq N\overline{|b|^2}\frac{2}{\mu^2}(\pi\lambda)^{-\frac{1}{2}}. \tag{3.104}$$

This result is seen to diverge as $\lambda \to 0$. This is simply a consequence of the fact that the limit $\lambda = 0$ corresponds to zero-energy neutrons and these are struck by the free-moving nuclei and thus scattered in spite of the absence of the apparent incident flux.

In the opposite limit, $\lambda \gg 1$, we obtain from (3.102)

$$\sigma \simeq \frac{N\overline{|b|^2}}{\mu^2} \equiv \frac{N\overline{|b|^2}}{(1+m/M)^2}, \tag{3.105}$$

which is simply the scattering from N completely free and independent nuclei.

3.7. Proof of Bloch's identity

We give here a proof of the identity

$$\langle \exp \hat{Q}\rangle = \exp\{\tfrac{1}{2}\langle \hat{Q}^2\rangle\}, \tag{3.i}$$

where $\hat{Q}$ is a linear combination of the Bose operators $\hat{a}$ and $\hat{a}^+$.

Let us choose

$$\hat{Q} = A\hat{a}+B\hat{a}^+ \tag{3.ii}$$

where A and B are arbitrary constants, and use the identity (3.53) to write

$$\exp \hat{Q} = \exp A\hat{a} \exp B\hat{a}^+ \exp(-\tfrac{1}{2}AB). \tag{3.iii}$$

Thus we need to calculate, for a harmonic oscillator,

$$\langle \exp A\hat{a} \exp B\hat{a}^+\rangle = \frac{1}{Z}\sum_{n=0}^{\infty} \exp(-\beta E_n)\langle n|\exp A\hat{a} \exp B\hat{a}^+|n\rangle. \tag{3.iv}$$

Here

$$Z = \sum_{n=0}^{\infty} \exp(-\beta E_n) = \exp(-\tfrac{1}{2}\hbar\omega_0\beta)\sum_{n=0}^{\infty} \exp(-\hbar\omega_0\beta n)$$

$$= \frac{\exp(-\frac{1}{2}\hbar\omega_0\beta)}{1-\exp(-\hbar\omega_0\beta)}. \tag{3.v}$$

† loc cit., p. 59.

With the aid of the relation

$$|n\rangle = \left(\frac{1}{n!}\right)^{\frac{1}{2}} (\hat{a}^+)^n |0\rangle$$

it is straightforward to show that

$$\exp(B\hat{a}^+)|n\rangle = \sum_{m=0}^{\infty} \left\{\frac{(n+m)!}{n!}\right\}^{\frac{1}{2}} \frac{B^m}{m!} |n+m\rangle. \tag{3.vi}$$

Hence

$$\langle n|\exp A\hat{a} \exp B\hat{a}^+|n\rangle = \sum_{m,m'} \{(n+m')!\,(n+m)!\}^{\frac{1}{2}} \frac{A^{m'}B^m}{n!\,m'!\,m!} \langle n+m'|n+m\rangle$$

$$= \sum_{m=0}^{\infty} \frac{(n+m)!}{n!\,(m!)^2} (AB)^m,$$

and so $\sum_{n=0}^{\infty} \exp(-\beta E_n)\langle n|\exp A\hat{a} \exp B\hat{a}^+|n\rangle$

$$= \exp(-\tfrac{1}{2}\hbar\omega_0\beta) \sum_{n=0}^{\infty}\sum_{m=0}^{\infty} \exp(-\hbar\omega_0\beta n) \frac{(n+m)!}{(n!)(m!)^2} (AB)^m.$$

Now
$$\frac{1}{(1-y)^{m+1}} = \sum_{n=0}^{\infty} \frac{(n+m)!}{n!\,m!} y^n,$$

so if we choose $y = \exp(-\hbar\omega_0\beta)$,

$$\sum_{n=0}^{\infty} \exp(-\beta E_n)\langle n|\exp A\hat{a} \exp B\hat{a}^+|n\rangle$$

$$= \frac{\exp(-\frac{1}{2}\hbar\omega_0\beta)}{\{1-\exp(-\hbar\omega_0\beta)\}} \sum_{m=0}^{\infty} \left[\frac{AB}{\{1-\exp(-\hbar\omega_0\beta)\}}\right]^m \frac{1}{m!}$$

$$= Z\exp\left\{\frac{AB}{1-\exp(-\hbar\omega_0\beta)}\right\} = Z\exp\{AB(\langle\hat{a}^+\hat{a}\rangle+1)\}, \tag{3.vii}$$

where in the second line we have used (3.v).

If we now use (3.vii) in conjunction with (3.iii) and (3.iv) we have, finally,

$$\langle \exp \hat{Q}\rangle = \exp\{AB(\langle\hat{a}^+\hat{a}\rangle+\tfrac{1}{2})\}, \tag{3.viii}$$

which is identical to (3.i).

4

SCATTERING BY PHONONS

In Chapters 1 and 2 we calculated the cross-section for the scattering of neutrons from an array of bound nuclei. We consider now the calculation of the cross-section when the nuclei are allowed to undergo small displacements from an equilibrium configuration that is periodic in space.

The Hamiltonian describing the translational motion of the nuclei can be expanded about the equilibrium configuration in terms of their displacements and we shall concern ourselves initially with the case when the expansion is terminated at terms quadratic in the displacements (the harmonic approximation to give a non-interacting phonon system). The periodicity of the crystal requires the phonons to propagate with a wave vector $\mathbf{q}$. As is well known, creation and annihilation operators can be associated with these phonon states. Before we can calculate the cross-section for the scattering from the vibrating system of nuclei we must first develop the theory of phonons. This we do in the next section.

4.1. Phonon theory†

We need to calculate the displacements and vibrational frequencies of the nuclei in a crystal about their equilibrium configuration. In Chapter 2 we discussed how the latter can be defined in terms of basic vectors $\mathbf{a}_1$, $\mathbf{a}_2$, and $\mathbf{a}_3$. We shall first consider the problem of determining the vibrational states of N nuclei with an equilibrium configuration that coincides with a Bravais lattice, i.e. that is given by the lattice vectors

$$\mathbf{l} = l_1\,\mathbf{a}_1 + l_2\,\mathbf{a}_2 + l_3\,\mathbf{a}_3, \tag{4.1}$$

and then consider the general case of more than one atom per unit cell.

The displacement of the nucleus at the lth site is denoted by $\mathbf{u}(l)$. Thus the position vector of this nucleus is

$$\mathbf{R}_l = \mathbf{l} + \mathbf{u}(l). \tag{4.2}$$

† See, for example, R. E. Peierls, *Quantum theory of solids*, Clarendon Press, Oxford (1955); A. A. Maradudin, E. W. Montroll, and G. H. Weiss, 'Theory of lattice dynamics in the harmonic approximation', *Solid state physics*, Supp. 3, edited by F. Seitz and D. Turnbull, Academic Press (1963); W. Cochran and R. A. Cowley, *Handb. Phys.* **25**/2a, 59.

The potential energy of the nuclei U is assumed to be lowest when all **u**'s are zero. We can expand U in a Taylor series

$$U-U_0 = \tfrac{1}{2}\sum_{\alpha,l}\sum_{\beta,l'} u_\alpha(l)A_{\alpha\beta}(l,l')u_\beta(l')+$$
$$+\tfrac{1}{6}\sum_{\alpha,l}\sum_{\beta,l'}\sum_{\gamma,l''} u_\alpha(l)u_\beta(l')u_\gamma(l'')B_{\alpha\beta\gamma}(l,l',l'')+\ldots, \qquad (4.3)$$

where U_0 is the potential energy in equilibrium. We are concerned only with the terms quadratic in the displacements, i.e. the first term on the r.h.s. of (4.3). The $A_{\alpha\beta}(l,l')$, the second-order derivatives of the potential energy, clearly satisfy

$$A_{\alpha\beta}(l,l') = A_{\beta\alpha}(l',l) \qquad (4.4\,\text{a})$$

and can depend only on the relative positions of the relevant cells, i.e.

$$A_{\alpha\beta}(l,l') = A_{\alpha\beta}(l-l'). \qquad (4.4\,\text{b})$$

Also, since every atom is a centre of symmetry,

$$A_{\alpha\beta}(l,l') = A_{\alpha\beta}(l',l) = A_{\beta\alpha}(l,l'). \qquad (4.4\,\text{c})$$

The Hamiltonian of the classical motion of the nuclei in the harmonic approximation is, from (4.3),

$$\mathscr{H} = \sum_{\alpha,l}\frac{1}{2M}p_\alpha^2(l)+\tfrac{1}{2}\sum_{\alpha,l}\sum_{\beta,l'} u_\alpha(l)A_{\alpha\beta}(l,l')u_\beta(l'), \qquad (4.5)$$

where M is the mass of a nucleus and $p_\alpha(l)$ the momentum conjugate to $u_\alpha(l)$.

The equation of motion for the lth nucleus is

$$\dot{p}_\alpha(l) = -\frac{\partial\mathscr{H}}{\partial u_\alpha(l)} = M\ddot{u}_\alpha(l),$$

or

$$M\ddot{u}_\alpha(l) = -\sum_{\beta,l'} A_{\alpha\beta}(l,l')u_\beta(l'), \qquad (4.6)$$

with the aid of (4.4 a). Since the right-hand side of (4.6) gives the force acting on particle l, and this force must vanish if all nuclei are translated by an identical amount, it follows that $\sum_{l'} A_{\alpha\beta}(l,l') = 0$. The set of equations (4.6) can be reduced to a set of equations for $3N$ independent harmonic oscillators by introducing a set of normal coordinates, which we shall now calculate.

If we allow the displacements to have the time dependence $\exp(-i\omega_0 t)$, then (4.6) becomes

$$M\omega_0^2 u_\alpha(l) = \sum_{l',\beta} A_{\alpha\beta}(l,l')u_\beta(l'). \qquad (4.7)$$

The condition that the set of $3N$ linear equations (4.7) for the displacements should have a non-trivial solution is that the determinant of the coefficients vanishes,

$$\det|M\omega_0^2\,\delta_{\alpha,\beta}\,\delta_{l,l'}-A_{\alpha\beta}(l,l')| = 0. \tag{4.8}$$

The roots of this equation are the normal-mode frequencies of the crystal.

Because of the translational symmetry of the lattice, let us make the further substitution

$$u_\alpha(l) = \frac{\sigma_\alpha}{\sqrt{M}}\exp(\mathrm{i}\mathbf{q}\,.\,\mathbf{l}). \tag{4.9}$$

Then the eigenvalue equation for the normal-mode frequencies can be written

$$\begin{aligned} M\omega_0^2\sigma_\alpha &= \exp(-\mathrm{i}\mathbf{q}\,.\,\mathbf{l})\sum_{l',\beta} A_{\alpha\beta}(l,l')\exp(\mathrm{i}\mathbf{q}\,.\,\mathbf{l}')\sigma_\beta \\ &= \sum_{l,\beta} A_{\alpha\beta}(l)\exp(-\mathrm{i}\mathbf{q}\,.\,\mathbf{l})\sigma_\beta, \end{aligned} \tag{4.10}$$

where the last line follows because of (4.4 b).

For each value of $\mathbf{q}$ there are three solutions for ω_0^2 corresponding to the three values of α. We denote this fact by writing the frequencies as $\omega_j(\mathbf{q})$, the index j having three values. To each $\omega_j(\mathbf{q})$ there corresponds an eigenvector σ_α, and we shall henceforth write these as $\sigma_\alpha^j(\mathbf{q})$ (more usually referred to as a polarization vector). Thus the eigenvalue equation (4.10) now reads

$$\omega_j^2(\mathbf{q})\sigma_\alpha^j(\mathbf{q}) = \sum_\beta \mathscr{A}_{\alpha\beta}(\mathbf{q})\sigma_\beta^j(\mathbf{q}). \tag{4.11 a}$$

Here

$$\mathscr{A}_{\alpha\beta}(\mathbf{q}) = \frac{1}{M}\sum_l A_{\alpha\beta}(l)\exp(-\mathrm{i}\mathbf{q}\,.\,\mathbf{l}) \tag{4.11 b}$$

and is usually called the *dynamical matrix*.

From (4.11 b) we deduce that in general

$$\mathscr{A}^*_{\alpha\beta}(\mathbf{q}) = \mathscr{A}_{\alpha\beta}(-\mathbf{q}). \tag{4.11 c}$$

If every atom is a centre of symmetry (as it is for a Bravais crystal) so that (4.4 c) is satisfied, then

$$\mathscr{A}^*_{\alpha\beta}(\mathbf{q}) = \mathscr{A}_{\alpha\beta}(\mathbf{q}), \tag{4.11 d}$$

i.e. the dynamical matrix is purely real.

From (4.11 a) and (4.11 c) it follows that $\sigma_\alpha^j(\mathbf{q})$ and $\sigma_\alpha^{*j}(-\mathbf{q})$ satisfy the same equation (4.11 a) and are therefore identical except for an arbitrary phase factor. We adopt a convention that makes $\sigma_\alpha^j(\mathbf{q})$ real and even in $\mathbf{q}$ for Bravais crystals, i.e. in the general case we take

$$\sigma_\alpha^{*j}(\mathbf{q}) = \sigma_\alpha^j(-\mathbf{q}). \tag{4.11 e}$$

The eigenvectors $\sigma_\alpha^j(\mathbf{q})$ can be constructed so as to satisfy

$$\sum_j \sigma_\alpha^{*j}(\mathbf{q})\sigma_\beta^j(\mathbf{q}) = \delta_{\alpha,\beta} \tag{4.12 a}$$

and

$$\sum_\alpha \sigma_\alpha^{*j}(\mathbf{q})\sigma_\alpha^{j'}(\mathbf{q}) = \delta_{j,j'}. \tag{4.12 b}$$

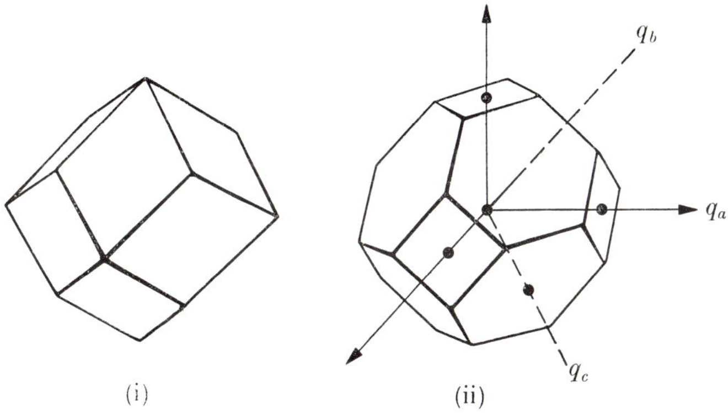

FIG. 4.1. The first Brillouin zone for (i) body-centred cubic and (ii) face-centred cubic lattices.

The number of vectors $\mathbf{q}$ is not unlimited. The vectors $\mathbf{q}$ and $\mathbf{q}+\boldsymbol{\tau}$ ($\boldsymbol{\tau}$ we recall is a reciprocal lattice vector, i.e. $\mathbf{l}.\boldsymbol{\tau} = 2\pi \times \text{integer}$) give the same displacement, as is seen by reference to (4.9). The matrix $\mathscr{A}_{\alpha\beta}(\mathbf{q})$, the dynamical matrix, remains unchanged under the transformation $\mathbf{q} \rightarrow \mathbf{q}+\boldsymbol{\tau}$, and hence its eigenvectors $\sigma_\alpha^j(\mathbf{q})$ must also be invariant under the same transformation. This means that we can restrict our values of $\mathbf{q}$ to lie within the first Brillouin zone. The first Brillouin zones for body-centred and face-centred cubic lattices are shown in Fig. 4.1. By applying periodic boundary conditions to the crystal we find that the density of vectors $\mathbf{q}$ is, for each value of j, $V/(2\pi)^3$, where V is the volume of the crystal. Hence

$$\sum_{\mathbf{q}} \rightarrow \frac{V}{(2\pi)^3}\int \mathrm{d}\mathbf{q}. \tag{4.13}$$

The eigenfrequencies $\omega_j(\mathbf{q})$ are shown in Fig. 4.2 for the particular case of a face-centred cubic lattice. The frequencies are shown for three directions of $\mathbf{q}$, each parallel to a symmetry direction. These are clearly shown in the diagram of the first Brillouin zone for a face-centred lattice given in Fig. 4.1. Because the directions of $\mathbf{q}$ are along symmetry directions, the eigenvectors or polarization vectors $\boldsymbol{\sigma}^j(\mathbf{q})$ have a simple relation to $\mathbf{q}$, namely two of them are perpendicular to $\mathbf{q}$ and the other is parallel to **q**. Hence, the corresponding eigenfrequencies have been labelled as transverse and longitudinal, the two transverse modes being degenerate in the examples chosen except for the case of **q** parallel to a face diagonal.

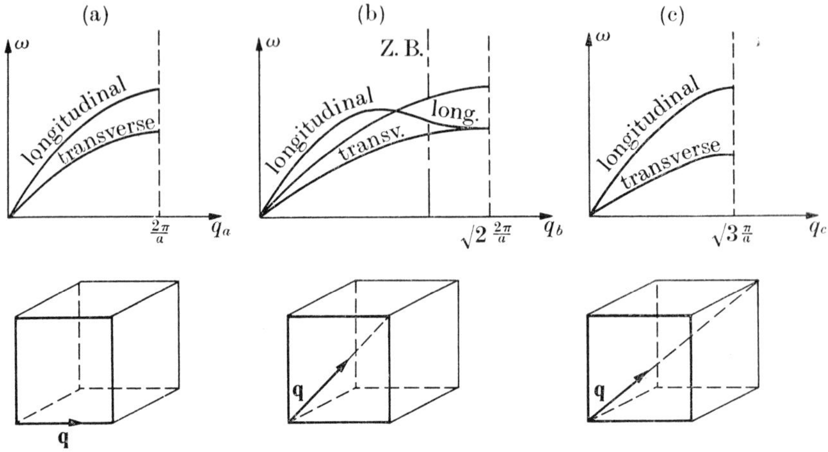

FIG. 4.2. The eigenfrequencies $\omega_j(\mathbf{q})$ are shown for **q** along special symmetry directions for a face-centred cubic structure. The three wave vectors q_a, q_b, and q_c are also shown in the first Brillouin zone of the lattice given in Fig. 4.1. Because the vectors **q** are taken along symmetry directions the eigenvectors $\boldsymbol{\sigma}^j(\mathbf{q})$ are either parallel or perpendicular to them, and the corresponding eigenfrequencies are denoted as longitudinal or transverse. In the particular example shown in the figure the two transverse modes are degenerate in cases (a) and (c). In (b) the curves have been extended into the next Brillouin zone to a total distance $\sqrt{2}(2\pi/a)$.

We now show that the quantum mechanical Hamiltonian that corresponds to (4.5) can be reduced to the sum of $3N$ independent harmonic oscillators. In Chapter 3 we noted that a component x of the displacement vector for a particle in a harmonic oscillator can be written in terms of Bose operators $\hat{a}$ and $\hat{a}^+$ in the form

$$\hat{x} = \left(\frac{\hbar}{2M\omega_0}\right)^{\frac{1}{2}}(\hat{a}+\hat{a}^+) \tag{4.14 a}$$

and, consequently, the momentum operator $\hat{p}$ conjugate to $\hat{x}$ as

$$\hat{p} = -\mathrm{i}\left(\frac{\hbar\omega_0 M}{2}\right)^{\frac{1}{2}}(\hat{a}-\hat{a}^+). \tag{4.14 b}$$

In the present case we can introduce Bose operators associated with a given lattice site, namely $\hat{\mathbf{a}}(l)$ and $\hat{\mathbf{a}}^+(l)$. We then define operators

$\hat{a}_j(\mathbf{q})$ and $\hat{a}_j^+(\mathbf{q})$ through the transformations

$$\hat{a}_\alpha(l) = \frac{1}{\sqrt{N}} \sum_{j,\mathbf{q}} \sigma_\alpha^j(\mathbf{q})\exp(\mathrm{i}\mathbf{q}.\mathbf{l})\hat{a}_j(\mathbf{q}) \tag{4.15 a}$$

and

$$\hat{a}_\alpha^+(l) = \frac{1}{\sqrt{N}} \sum_{j,\mathbf{q}} \sigma_\alpha^{*j}(\mathbf{q})\exp(-\mathrm{i}\mathbf{q}.\mathbf{l})\hat{a}_j^+(\mathbf{q}) \tag{4.15 b}$$

and thus, by analogy with (4.14),

$$\hat{u}_\alpha(l) = \sum_{j,\mathbf{q}} \left\{\frac{\hbar}{2NM\omega_j(\mathbf{q})}\right\}^{\frac{1}{2}} \{\sigma_\alpha^j(\mathbf{q})\exp(\mathrm{i}\mathbf{q}.\mathbf{l})\hat{a}_j(\mathbf{q}) + \sigma_\alpha^{*j}(\mathbf{q})\exp(-\mathrm{i}\mathbf{q}.\mathbf{l})\hat{a}_j^+(\mathbf{q})\} \tag{4.16 a}$$

and

$$\hat{p}_\alpha(l) = -\mathrm{i} \sum_{j,\mathbf{q}} \left\{\frac{\hbar\omega_j(\mathbf{q})M}{2N}\right\}^{\frac{1}{2}} \times$$

$$\times \{\sigma_\alpha^j(\mathbf{q})\exp(\mathrm{i}\mathbf{q}.\mathbf{l})\hat{a}_j(\mathbf{q}) - \sigma_\alpha^{*j}(\mathbf{q})\exp(-\mathrm{i}\mathbf{q}.\mathbf{l})\hat{a}_j^+(\mathbf{q})\}. \tag{4.16 b}$$

From

$$[\hat{u}_\alpha(l), \hat{p}_\beta(l')] = \mathrm{i}\hbar\, \delta_{\alpha,\beta}\, \delta_{l,l'} \tag{4.17}$$

it follows that

$$[\hat{a}_j(\mathbf{q}), \hat{a}_{j'}^+(\mathbf{q}')] = \delta_{j,j'}\, \delta_{\mathbf{q},\mathbf{q}'}. \tag{4.18}$$

The operators $\hat{a}_j(\mathbf{q})$ and $\hat{a}_j^+(\mathbf{q})$ commute with themselves, of course.

By analogy with the discussion of a simple harmonic oscillator in Chapter 3 it follows that $\hat{a}_j^+(\mathbf{q})\hat{a}_j(\mathbf{q})$ is the number operator for the state described by $\mathbf{q}$ and j. Hence†

$$\langle \hat{a}_j^+(\mathbf{q})\hat{a}_{j'}(\mathbf{q}')\rangle = \delta_{jj'}\,\delta_{\mathbf{q}\mathbf{q}'}\langle \hat{a}_j^+(\mathbf{q})\hat{a}_j(\mathbf{q})\rangle = \delta_{jj'}\,\delta_{\mathbf{q}\mathbf{q}'}\, n_j(\mathbf{q}), \tag{4.18 a}$$

where

$$n_j(\mathbf{q}) = [\exp\{\beta\hbar\omega_j(\mathbf{q})\} - 1]^{-1}.$$

With the transformations (4.16) we shall show that the Hamiltonian

$$\mathscr{H} = \sum_{\alpha,l} \frac{1}{2M} \{\hat{p}_\alpha(l)\}^2 + \tfrac{1}{2} \sum_{\alpha,l} \sum_{\beta,l'} \hat{u}_\alpha(l) A_{\alpha\beta}(l,l') \hat{u}_\beta(l') \tag{4.19}$$

becomes

$$\mathscr{H} = \sum_{j,\mathbf{q}} \hbar\omega_j(\mathbf{q})\{\hat{a}_j^+(\mathbf{q})\hat{a}_j(\mathbf{q}) + \tfrac{1}{2}\}. \tag{4.20}$$

† If $\hat{O}$ is any symmetry operator that commutes with the Hamiltonian and $\hat{P}$ an arbitrary operator then

$$\langle[\hat{O}, \hat{P}]\rangle = 0 \tag{a}$$

by the cyclic properties of the trace of two operators. Let us choose for $\hat{O}$

$$\sum_{j,\mathbf{q}} \mathbf{q}\hat{a}_j^+(\mathbf{q})\hat{a}_j(\mathbf{q}),$$

which clearly commutes with (4.20), and for P we take $\hat{a}_{j_1}^+(\mathbf{q}_1)\hat{a}_{j_2}(\mathbf{q}_2)$. Then

$$[\hat{O}, \hat{P}] = \hat{a}_{j_1}^+(\mathbf{q}_1)\hat{a}_{j_2}(\mathbf{q}_2)(\mathbf{q}_1 - \mathbf{q}_2), \tag{b}$$

so that from (a) it follows that we must have $\mathbf{q}_1 = \mathbf{q}_2$ in $\langle \hat{a}_j^+(\mathbf{q}_1)\hat{a}_j(\mathbf{q}_2)\rangle$. We can similarly prove that we must also have $j_1 = j_2$ by choosing $\hat{O} = \sum_{j,\mathbf{q}} j\hat{a}_j^+(\mathbf{q})\hat{a}_j(\mathbf{q})$.

That is to say (4.19), which is the quantum analogue of the classical Hamiltonian (4.5), can be represented as the sum of $3N$ independent harmonic oscillators; the operators $\hat{a}_j(\mathbf{q})$ and $\hat{a}_j^+(\mathbf{q})$ annihilate and create, respectively, phonons defined by the quantum numbers j and $\mathbf{q}$. It follows immediately from (4.20) that

$$i\hbar\frac{\partial}{\partial t}\hat{a}_j(\mathbf{q},t) = [\hat{a}_j(\mathbf{q},t),\mathscr{H}] = \hbar\omega_j(\mathbf{q})\hat{a}_j(\mathbf{q},t)$$

and hence
$$\hat{a}_j(\mathbf{q},t) = \hat{a}_j(\mathbf{q})\exp\{-i\omega_j(\mathbf{q})t\}. \tag{4.21 a}$$

Similarly we find
$$\hat{a}_j^+(\mathbf{q},t) = \hat{a}_j^+(\mathbf{q})\exp\{i\omega_j(\mathbf{q})t\}. \tag{4.21 b}$$

In verifying that the Hamiltonian (4.20) is equivalent to that given by (4.19) we find it convenient to use (4.11 e) and rewrite (4.16) in the alternative form

$$\hat{u}_\alpha(l) = \sum_{j,\mathbf{q}}\left\{\frac{\hbar}{2NM\omega_j(\mathbf{q})}\right\}^{\frac{1}{2}}\sigma_\alpha^j(\mathbf{q})\exp(i\mathbf{q}.\mathbf{l})\{\hat{a}_j(\mathbf{q})+\hat{a}_j^+(-\mathbf{q})\} \tag{4.22 a}$$

and

$$\hat{p}_\alpha(l) = -i\sum_{j,\mathbf{q}}\left\{\frac{\hbar\omega_j(\mathbf{q})M}{2N}\right\}^{\frac{1}{2}}\sigma_\alpha^j(\mathbf{q})\exp(i\mathbf{q}.\mathbf{l})\{\hat{a}_j(\mathbf{q})-\hat{a}_j^+(-\mathbf{q})\}. \tag{4.22 b}$$

The kinetic energy is then

$$\begin{aligned}\sum_{\alpha,l}\frac{1}{2M}\hat{p}_\alpha(l)\hat{p}_\alpha(l) &= \sum_{\alpha,l}\frac{1}{2M}\hat{p}_\alpha^+(l)\hat{p}_\alpha(l)\\ &= \sum_{j,\mathbf{q}}\sum_{j',\mathbf{q}'}\frac{\hbar}{4N}\{\omega_j(\mathbf{q})\omega_{j'}(\mathbf{q}')\}^{\frac{1}{2}}\sum_{\alpha,l}\sigma_\alpha^{*j}(\mathbf{q})\sigma_\alpha^{j'}(\mathbf{q}')\exp\{-i(\mathbf{q}-\mathbf{q}').\mathbf{l}\}\times\\ &\qquad\times\{\hat{a}_j^+(\mathbf{q})-\hat{a}_j(-\mathbf{q})\}\{\hat{a}_{j'}(\mathbf{q}')-\hat{a}_{j'}^+(-\mathbf{q}')\}\\ &= \tfrac{1}{4}\hbar\sum_{j,\mathbf{q}}\omega_j(\mathbf{q})\{2\hat{a}_j^+(\mathbf{q})\hat{a}_j(\mathbf{q})+1-\hat{a}_j(-\mathbf{q})\hat{a}_j(\mathbf{q})-\hat{a}_j^+(\mathbf{q})\hat{a}_j^+(-\mathbf{q})\},\end{aligned} \tag{4.23}$$

where we have used (4.12 b).

Similarly the potential energy is

$$\begin{aligned}\tfrac{1}{2}\sum_{\alpha,l}\sum_{\beta,l'}\hat{u}_\alpha^+(l)A_{\alpha\beta}(l,l')\hat{u}_\beta(l') &= \frac{\hbar}{4NM}\sum_{j,\mathbf{q}}\sum_{j',\mathbf{q}'}\left\{\frac{1}{\omega_j(\mathbf{q})\omega_{j'}(\mathbf{q}')}\right\}^{\frac{1}{2}}\times\\ &\quad\times\sum_{\alpha,l}\sum_{\beta,l'}\sigma_\alpha^{*j}(\mathbf{q})A_{\alpha\beta}(l,l')\exp(-i\mathbf{q}.\mathbf{l}+i\mathbf{q}'.\mathbf{l}')\sigma_\beta^{j'}(\mathbf{q}')\times\\ &\quad\times\{\hat{a}_j^+(\mathbf{q})+\hat{a}_j(-\mathbf{q})\}\{\hat{a}_{j'}(\mathbf{q}')+\hat{a}_{j'}^+(-\mathbf{q}')\}\\ &= \tfrac{1}{4}\hbar\sum_{j,\mathbf{q}}\omega_j(\mathbf{q})\{2\hat{a}_j^+(\mathbf{q})\hat{a}_j(\mathbf{q})+1+\hat{a}_j(-\mathbf{q})\hat{a}_j(\mathbf{q})+\hat{a}_j^+(\mathbf{q})\hat{a}_j^+(-\mathbf{q})\}.\end{aligned} \tag{4.24}$$

Here we have used (4.11 a) and (4.11 b). If we now add together the kinetic and potential energy terms we get the result (4.20) immediately.

Although (4.20) is the correct result, the derivation of the correct 'zero-point energy' (i.e. the factor $\frac{1}{2}$ within the bracket of (4.20)) is fortuitous. The classical Hamiltonian (4.5) could have been expressed in normal modes and then re-ordered in a way that, when replacing the classical variables by quantum mechanical operators, could give apparently any arbitrary 'zero-point energy'. Strictly speaking, therefore, the most the theory as described above can demonstrate is that the quantum Hamiltonian is

$$\mathscr{H} = \sum_{j,\mathbf{q}} \hbar\omega_j(\mathbf{q})\{\hat{a}_j^+(\mathbf{q})\hat{a}_j(\mathbf{q})+c\},$$

where c is to be determined.

To determine c we use the 'correspondence principle', which in brief states that at high temperatures no physical quantity can depend on $\hbar$.† We examine the thermal energy, which is simply

$$\mathscr{E}(T) = \sum_{j,\mathbf{q}} \hbar\omega_j(\mathbf{q})\{n_j(\mathbf{q})+c\},$$

where
$$n_j(\mathbf{q}) = [\exp\{\beta\hbar\omega_j(\mathbf{q})\}-1]^{-1}.$$

At high temperatures ($\beta \ll 1$) we then get

$$\mathscr{E}(T) \to \sum_{j,\mathbf{q}} \left\{k_\mathrm{B}\, T+\hbar\omega_j(\mathbf{q})(c-\tfrac{1}{2})+\frac{1}{12}\frac{\hbar^2\omega_j(\mathbf{q})^2}{k_\mathrm{B}\, T}+\ldots\right\}.$$

The first term is the classical result, the third and subsequent terms vanish as $T \to \infty$; the second term must vanish by the correspondence principle. Hence c is $\frac{1}{2}$.

The analysis we have given so far has been for the special case of a Bravais lattice, but it is quite easy to extend it to the general case of a lattice consisting of several interpenetrating Bravais lattices of identical structure each containing nuclei of different mass. The position vector of a nucleus is now

$$\mathbf{R}_{ld} = \mathbf{l}+\mathbf{d}+\mathbf{u}\binom{l}{d}. \tag{4.25}$$

Here $\mathbf{d}$ is the position vector to the dth nucleus within the lth cell, there being r nuclei in each unit cell.

The total number of degrees of freedom is now $3rN$ and hence the quantum number j takes $3r$ values. The eigenfrequencies $\omega_j(\mathbf{q})$ are shown schematically in Fig. 4.3 for a lattice with two atoms per cell ($r = 2$).

For ease of notation we shall denote α and d collectively by the single index $\bar{\alpha}$; there is then a correspondence between $\bar{\alpha}$ and j since the former,

† Cf. R. H. Fowler, *Statistical mechanics*, Cambridge University Press (1966).

of course, has also $3r$ values. Hence the displacement $u_\alpha\binom{l}{d}$ in (4.25) is now written

$$u_\alpha\binom{l}{d} \to u_{\bar{\alpha}}(l).$$

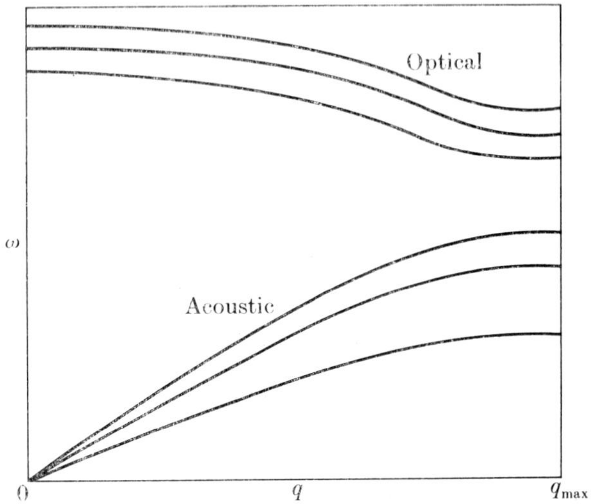

FIG. 4.3. The figure shows a schematic picture of the six eigenfrequencies for a crystal with two atoms per unit cell. The three lower curves are called the acoustic branches, the remaining three the optical branches.

The fundamental equation determining the displacements and eigenfrequencies, eqn (4.7), now reads

$$M_{\bar{\alpha}}\,\omega_0^2\,u_{\bar{\alpha}}(l) = \sum_{\bar{\beta},l'} A_{\bar{\alpha},\bar{\beta}}(l,l')u_{\bar{\beta}}(l'), \tag{4.26}$$

since the mass of the nucleus must now depend upon its position in the unit cell.

The substitution (4.9) can still be made

$$u_{\bar{\alpha}}(l) = \frac{\sigma_{\bar{\alpha}}}{\sqrt{M_{\bar{\alpha}}}}\exp(i\mathbf{q}\,.\,\mathbf{l}), \tag{4.27}$$

so that eqns (4.11 a) and (4.11 b) are replaced by

$$\omega_j^2(\mathbf{q})\sigma_{\bar{\alpha}}^j(\mathbf{q}) = \sum_{\bar{\beta}} \mathscr{A}_{\bar{\alpha},\bar{\beta}}(\mathbf{q})\sigma_{\bar{\beta}}^j(\mathbf{q}), \tag{4.28}$$

with $$\mathscr{A}_{\bar{\alpha},\bar{\beta}}(\mathbf{q}) = \sqrt{\left(\frac{1}{M_{\bar{\alpha}}\,M_{\bar{\beta}}}\right)}\sum_l A_{\bar{\alpha},\bar{\beta}}(l)\exp(-i\mathbf{q}\,.\,\mathbf{l}),$$

and (4.12 b), for instance, becomes

$$\sum_{\bar{\alpha}} \sigma_{\bar{\alpha}}^{*j}(\mathbf{q})\sigma_{\bar{\alpha}}^{j'}(\mathbf{q}) = \delta_{j,j'}. \tag{4.29}$$

Finally, the displacement operator is

$$\hat{u}_{\bar{\alpha}}(l,t) = \sum_{j,\mathbf{q}} \left\{\frac{\hbar}{2NM_{\bar{\alpha}}\,\omega_j(\mathbf{q})}\right\}^{\frac{1}{2}} [\sigma^j_{\bar{\alpha}}(\mathbf{q})\exp\{i\mathbf{q}\,.\,\mathbf{l}-i\omega_j(\mathbf{q})t\}\hat{a}_j(\mathbf{q})+$$
$$+\sigma^{*j}_{\bar{\alpha}}(\mathbf{q})\exp\{-i\mathbf{q}\,.\,\mathbf{l}+i\omega_j(\mathbf{q})t\}\hat{a}^+_j(\mathbf{q})]. \quad (4.30)$$

From this formula for the displacement operator we notice that if we can choose $\sigma^j_{\bar{\alpha}}(\mathbf{q}) = \sigma^{*j}_{\bar{\alpha}}(\mathbf{q})$, then the phonons are linearly polarized, otherwise the phonons are elliptically polarized. Hence if each lattice point is a centre of symmetry the phonons are linearly polarized; otherwise they are elliptically polarized, in general. Notice that this question of polarization is quite different from that for electromagnetic radiation. For the latter, in free space, the two polarization directions are always degenerate and the polarization is therefore undetermined; the radiation may be linearly polarized or elliptically polarized depending upon the way it is produced. For phonons a degeneracy between the two transverse modes can exist only for $\mathbf{q}$ along special symmetry directions. For all other directions of $\mathbf{q}$ there is no degeneracy and the polarization vectors are completely determined (apart from an unimportant phase factor) by the eigenvalue equation (4.28).

4.2. Elastic scattering

We now calculate the cross-sections for the elastic coherent and incoherent scattering of neutrons by nuclei undergoing harmonic vibrations. Initially we assume the equilibrium positions of the nuclei to form a Bravais lattice and then later extend the formulae to general lattice structures.

Consider first the elastic coherent cross-section. From the previous chapter

$$\left(\frac{d\sigma}{d\Omega}\right)^{\text{el}}_{\text{coh}} = \frac{\sigma_c}{4\pi}\left|\int d\mathbf{r}\,\exp(i\boldsymbol{\varkappa}\,.\,\mathbf{r})\langle\hat{\rho}(\mathbf{r})\rangle\right|^2, \quad (4.31)$$

where in the present case

$$\langle\hat{\rho}(\mathbf{r})\rangle = \sum_l \langle\delta\{\mathbf{r}-\hat{\mathbf{R}}_l(0)\}\rangle$$
$$= \left(\frac{1}{2\pi}\right)^3 \int d\mathbf{k} \sum_l \exp\{i\mathbf{k}\,.\,(\mathbf{r}-\mathbf{l})\}\langle\exp\{-i\mathbf{k}\,.\,\hat{\mathbf{u}}(l,0)\}\rangle. \quad (4.32)$$

If we invoke the identity (3.i),

$$\langle\exp\{-i\mathbf{k}\,.\,\hat{\mathbf{u}}(l,0)\}\rangle = \exp[-\tfrac{1}{2}\langle\{\mathbf{k}\,.\,\hat{\mathbf{u}}(l,0)\}^2\rangle]. \quad (4.33)$$

From the expansion of $\hat{\mathbf{u}}(l,0)$ in terms of the Bose operators $\hat{a}_j(\mathbf{q})$ and $\hat{a}_j^+(\mathbf{q})$, eqn (4.16 a), it follows that

$$\langle\{\mathbf{k}\,.\,\hat{\mathbf{u}}(l,0)\}^2\rangle = \frac{\hbar}{2NM}\sum_{j,\mathbf{q}}\sum_{j',\mathbf{q}'}\{\omega_j(\mathbf{q})\omega_{j'}(\mathbf{q}')\}^{-\frac{1}{2}}\sum_{\alpha,\beta}k_\alpha k_\beta\,\sigma_\alpha^j(\mathbf{q})\sigma_\beta^{j'}(\mathbf{q}')\times$$

$$\times[\exp\{-\mathrm{i}\mathbf{l}\,.\,(\mathbf{q}-\mathbf{q}')\}\langle\hat{a}_j^+(\mathbf{q})\hat{a}_{j'}(\mathbf{q}')\rangle+\exp\{\mathrm{i}\mathbf{l}\,.\,(\mathbf{q}-\mathbf{q}')\}\langle\hat{a}_j(\mathbf{q})\hat{a}_{j'}^+(\mathbf{q}')\rangle]$$

$$= \frac{\hbar}{2NM}\sum_{j,\mathbf{q}}\frac{|\mathbf{k}\,.\,\boldsymbol{\sigma}^j(\mathbf{q})|^2}{\omega_j(\mathbf{q})}\{2n_j(\mathbf{q})+1\}. \quad (4.34)$$

The second line of (4.34) follows from (4.18 a). Thus, on combining (4.33) and (4.34) it follows that

$$\langle\exp\{-\mathrm{i}\mathbf{k}\,.\,\hat{\mathbf{u}}(l,0)\}\rangle = \exp\left[-\frac{\hbar}{4NM}\sum_{j,\mathbf{q}}\frac{|\mathbf{k}\,.\,\boldsymbol{\sigma}^j(\mathbf{q})|^2}{\omega_j(\mathbf{q})}\{2n_j(\mathbf{q})+1\}\right]$$

$$= \exp\{-W(\mathbf{k})\}, \quad (4.35)$$

which is the well-known Debye–Waller factor.

It now follows that

$$\int \mathrm{d}\mathbf{r}\,\exp(\mathrm{i}\boldsymbol{\varkappa}\,.\,\mathbf{r})\langle\hat{\rho}(\mathbf{r})\rangle$$

$$= \sum_l\left(\frac{1}{2\pi}\right)^3\int \mathrm{d}\mathbf{k}\int \mathrm{d}\mathbf{r}\,\exp\{\mathrm{i}\mathbf{r}\,.\,(\boldsymbol{\varkappa}+\mathbf{k})-\mathrm{i}\mathbf{k}\,.\,\mathbf{l}\}\exp\{-W(\mathbf{k})\}$$

$$= \sum_l\exp(\mathrm{i}\boldsymbol{\varkappa}\,.\,\mathbf{l})\exp\{-W(\boldsymbol{\varkappa})\} \quad (4.36)$$

and thus, finally,

$$\left(\frac{\mathrm{d}\sigma}{\mathrm{d}\Omega}\right)^{\mathrm{el}}_{\mathrm{coh}} = \frac{N\sigma_{\mathrm{c}}}{4\pi}\frac{(2\pi)^3}{v_0}\sum_{\boldsymbol{\tau}}\delta(\boldsymbol{\varkappa}-\boldsymbol{\tau})\exp\{-2W(\boldsymbol{\varkappa})\}. \quad (4.37)$$

This result is simply the cross-section for coherent elastic scattering by a static lattice multiplied by the factor $\exp\{-2W(\boldsymbol{\varkappa})\}$. By definition,

$$2W(\boldsymbol{\varkappa}) = \langle\{\boldsymbol{\varkappa}\,.\,\hat{\mathbf{u}}(l,0)\}^2\rangle, \quad (4.38)$$

so that $2W$ is merely the mean square displacement of a nucleus multiplied by κ^2 (cf. § 3.5 (b)). The presence of the factor $\exp\{-2W(\boldsymbol{\varkappa})\}$ means that the intensity of the Bragg peaks decreases with increasing $|\boldsymbol{\varkappa}|$.

The formula for $W(\boldsymbol{\varkappa})$ can be greatly simplified if we introduce the normalized density of states

$$Z(\omega) = \frac{1}{3N}\sum_{j,\mathbf{q}}\delta\{\omega-\omega_j(\mathbf{q})\} \quad (4.39)$$

for then, from (4.35),

$$W(\boldsymbol{\varkappa}) = 3N\int_0^{\omega_{\mathrm{m}}}\mathrm{d}\omega\,Z(\omega)\frac{\hbar}{4NM\omega}\{2n(\omega)+1\}\{|\boldsymbol{\varkappa}\,.\,\boldsymbol{\sigma}|^2\}_{\mathrm{av}}. \quad (4.40)$$

In (4.40), ω_{m} denotes the maximum frequency of the phonons and in the last factor the subscript av means an average over a surface with a given ω must be taken. In cubic symmetry this average is $\frac{1}{3}\kappa^2$, and this result will be a fair approximation even for non-cubic crystals in most cases. Hence

$$W(\boldsymbol{\kappa}) = \frac{\hbar\kappa^2}{4M}\int\limits_0^{\omega_{\mathrm{m}}} \mathrm{d}\omega \frac{Z(\omega)}{\omega}\{2n(\omega)+1\} = \frac{\hbar\kappa^2}{4M}\int\limits_0^{\omega_{\mathrm{m}}} \mathrm{d}\omega \frac{Z(\omega)}{\omega}\coth(\tfrac{1}{2}\hbar\omega\beta). \tag{4.41}$$

For parts of the discussion on multi-phonon cross-sections (§ 4.5), it is convenient to have $W(\boldsymbol{\kappa})$ in an even more compact form. It is easy to show that, if we define negative frequencies by $Z(-\omega) = Z(\omega)$, we can rewrite (4.41) as

$$W(\boldsymbol{\kappa}) = \frac{\hbar\kappa^2}{4M}\int\limits_{\omega_{\mathrm{m}}}^{\omega_{\mathrm{m}}} \mathrm{d}\omega \frac{Z(\omega)}{\omega} n(\omega), \tag{4.41 a}$$

since $n(-\omega) = -\{n(\omega)+1\}$.

It is interesting to invert (4.36) to give

$$\langle\hat{\rho}(\mathbf{r})\rangle = \frac{1}{(2\pi)^3}\int \mathrm{d}\mathbf{k}\,\exp(-\mathrm{i}\mathbf{k}\,.\,\mathbf{r})\sum_l \exp(\mathrm{i}\mathbf{k}\,.\,\mathbf{l})\exp\{-W(\mathbf{k})\}.$$

If for simplicity we assume a cubic lattice, then, from (4.38),

$$W(\mathbf{k}) = \tfrac{1}{2}\langle(\mathbf{k}\,.\,\hat{\mathbf{u}})^2\rangle = \tfrac{1}{6}\mathbf{k}^2\langle\hat{\mathbf{u}}^2\rangle,$$

and we get

$$\langle\hat{\rho}(\mathbf{r})\rangle = \sum_l \left\{\frac{3}{2\pi\langle\hat{\mathbf{u}}^2\rangle}\right\}^{\frac{3}{2}}\exp\{-3(\mathbf{r}-\mathbf{l})^2/2\langle\hat{\mathbf{u}}^2\rangle\}.$$

This shows that $\langle\hat{\rho}(\mathbf{r})\rangle$ is a set of Gaussian functions centred on the lattice points $\mathbf{l}$. Each Gaussian has a mean square width

$$\langle\mathbf{r}^2\rangle = \langle\hat{\mathbf{u}}^2\rangle,$$

and by analogy with the magnetic scattering of neutrons, which we discuss in later chapters, we see that the Debye–Waller factor is effectively the 'form factor' of this nuclear density.

Anticipating (4.45), which shows that $\exp\{-2W(\mathbf{k})\}$ is the Fourier transform of the self-correlation function $G_{\mathrm{s}}(\mathbf{r},\infty)$ (cf. 3.32), we conclude that

$$G_{\mathrm{s}}(\mathbf{r},\infty) = \left\{\frac{3}{4\pi\langle\hat{\mathbf{u}}^2\rangle}\right\}^{\frac{3}{2}}\exp\{-3r^2/4\langle\hat{\mathbf{u}}^2\rangle\}.$$

We recall from Chapter 3 that $G_s(\mathbf{r}, 0)$ is simply $\delta(\mathbf{r})$, so as the time increases to infinity $G_s(\mathbf{r}, t)$ approaches the Gaussian distribution described above, whereas for a liquid $G_s(\mathbf{r}, t)$ approaches zero, because the motion of the particle is unrestrained.

To obtain some idea of the temperature dependence of $W(\boldsymbol{\varkappa})$ we can replace $Z(\omega)$ by the Debye spectrum, i.e.

$$Z(\omega) \to \frac{3\omega^2}{\omega_D^3}, \tag{4.42}$$

where ω_D is the Debye frequency, which is related to the Debye temperature θ_D through $\hbar\omega_D = k_B \theta_D$. If $T \gg \theta_D$, then in the integrand of (4.41) we can take

$$\coth(\tfrac{1}{2}\hbar\omega\beta) \sim \frac{2}{\hbar\omega\beta},$$

so that, with (4.42),

$$W(\boldsymbol{\varkappa}) \sim 3\left(\frac{\hbar^2\kappa^2}{2M}\right)\left(\frac{1}{\hbar\omega_D}\right)\left(\frac{T}{\theta_D}\right) \quad (T \gg \theta_D). \tag{4.43}$$

Note that $E_R = \hbar^2\kappa^2/2M$ is the energy of recoil of a nucleus of mass M, initially at rest.

In the limit $T \to 0$, $W(\boldsymbol{\varkappa})$ is not zero, as at first sight might be expected because of the zero-point motion of the nuclei. Indeed in this limit we obtain, with the use of (4.42),

$$W(\boldsymbol{\varkappa}) = \frac{3}{4}\left(\frac{\hbar^2\kappa^2}{2M}\right)\left(\frac{1}{\hbar\omega_D}\right) \quad (T = 0). \tag{4.44}$$

The evaluation of the incoherent elastic cross-section is quite straightforward. From Chapter 3,

$$\left(\frac{d\sigma}{d\Omega}\right)_{\text{incoh}}^{\text{el}} = \frac{\sigma_i}{4\pi} \sum_l \langle \exp\{-i\boldsymbol{\varkappa}.\hat{\mathbf{R}}_l(0)\}\rangle\langle \exp\{i\boldsymbol{\varkappa}.\hat{\mathbf{R}}_l(0)\}\rangle,$$

which from (4.35) reduces to

$$\left(\frac{d\sigma}{d\Omega}\right)_{\text{incoh}}^{\text{el}} = \frac{N\sigma_i}{4\pi} \exp\{-2W(\boldsymbol{\varkappa})\}. \tag{4.45}$$

Again this is the result for the static lattice multiplied by the factor $\exp\{-2W(\boldsymbol{\varkappa})\}$. Because of this factor, the incoherent elastic scattering is no longer isotropic but slightly larger in the forward direction than in the backward direction.

If we use (4.38) and assume a cubic crystal, then the total elastic incoherent cross-section is readily derived from (4.45). We have

$$\sigma_{\mathrm{i}}^{\mathrm{el}} = \frac{N\sigma_{\mathrm{i}}}{4\pi} 2\pi \int_{-1}^{1} \mathrm{d}\mu\, \exp\{-\tfrac{1}{3}\langle \hat{\mathbf{u}}^2\rangle(k^2+k'^2-2kk'\mu)\}$$

$$= \frac{N\sigma_{\mathrm{i}}}{4\pi} 2\pi \exp\{-\tfrac{2}{3}k^2\langle \hat{\mathbf{u}}^2\rangle\} \int_{-1}^{1} \mathrm{d}\mu\, \exp\{\tfrac{2}{3}k^2\langle \hat{\mathbf{u}}^2\rangle\mu\}$$

$$= N\sigma_{\mathrm{i}} \frac{3\hbar^2}{8mE\langle \hat{\mathbf{u}}^2\rangle}\left[1-\exp\left\{-\frac{8}{3}\frac{mE\langle \hat{\mathbf{u}}^2\rangle}{\hbar^2}\right\}\right]. \tag{4.46}$$

This formula replaces the simple result $N\sigma_{\mathrm{i}}$ quoted for a rigid lattice in Chapter 1. Note that when the atoms are rigidly fixed, i.e. $\langle \hat{\mathbf{u}}^2\rangle \to 0$, then the simple result is reproduced.

The total elastic coherent cross-section for the scattering from a powder sample is now, in place of (2.42),

$$\sigma_{\mathrm{c}}^{\mathrm{el}} = \frac{N\pi^2\hbar^2}{2mEv_0}\sigma_{\mathrm{c}} \sum_{\tau\neq 0}^{\tau<2k} \frac{1}{\tau}\exp\{-\tfrac{1}{3}\tau^2\langle \hat{\mathbf{u}}^2\rangle\}, \tag{4.47}$$

where we have again used (4.38) and assumed a cubic crystal structure. In the limit when the incident energy E is large, then, provided $\langle \hat{\mathbf{u}}^2\rangle$ is not too large, the summation can be replaced by an integral to give, as a reasonable approximation,

$$\sigma_{\mathrm{c}}^{\mathrm{el}} = \frac{N\pi^2\hbar^2\sigma_{\mathrm{c}}}{2mEv_0}\frac{v_0}{(2\pi)^3}4\pi \int_{0}^{(8mE/\hbar^2)^{\frac{1}{2}}} \mathrm{d}\tau\, \tau^2 \frac{1}{\tau}\times \exp\{-\tfrac{1}{3}\tau^2\langle \hat{\mathbf{u}}^2\rangle\}$$

$$= N\sigma_{\mathrm{c}} \frac{3\hbar^2}{8mE\langle \hat{\mathbf{u}}^2\rangle}\left[1-\exp\left\{-\frac{8}{3}\frac{mE\langle \hat{\mathbf{u}}^2\rangle}{\hbar^2}\right\}\right]. \tag{4.48}$$

We note that both (4.46) and (4.48) become zero as the incident energy E tends to infinity. This is just as we expect: the elastic scattering must approach zero as the energy of the incident neutrons increases, because all the scattering is becoming inelastic. At these high energies there are no interference effects in the elastic coherent cross-section, because the wavelength of the incident neutrons is very much smaller than the inter-nuclear spacing.

Let us now consider how eqns (4.37) and (4.45) for the elastic coherent and incoherent cross-sections become modified when we remove the restriction of a Bravais lattice structure.

Eqns (4.31) and (4.32) now give for the elastic coherent cross-section

$$\left(\frac{d\sigma}{d\Omega}\right)^{el}_{coh} = \left|\sum_{l,d} \bar{b}_d \exp\{i\boldsymbol{\kappa}.(\mathbf{l}+\mathbf{d})\}\left\langle \exp\left\{-i\boldsymbol{\kappa}.\hat{\mathbf{u}}\binom{l}{d}, 0\right\}\right\rangle\right|^2$$

$$= N\frac{(2\pi)^3}{v_0}\sum_{\boldsymbol{\tau}} |F_N(\boldsymbol{\kappa})|^2 \delta(\boldsymbol{\kappa}-\boldsymbol{\tau}), \quad (4.49)$$

where

$$F_N(\boldsymbol{\kappa}) = \sum_d \bar{b}_d \exp(i\boldsymbol{\kappa}.\mathbf{d})\exp\{-W_d(\boldsymbol{\kappa})\} \quad (4.50)$$

is the nuclear unit-cell structure factor. The Debye–Waller factor is given by

$$W_d(\boldsymbol{\kappa}) = \frac{1}{2}\left\langle\left\{\boldsymbol{\kappa}.\hat{\mathbf{u}}\binom{l}{d}, 0\right\}^2\right\rangle = \frac{\hbar}{4NM_d}\sum_{j,\mathbf{q}} \omega_j^{-1}(\mathbf{q})|\boldsymbol{\kappa}.\boldsymbol{\sigma}_d^j(\mathbf{q})|^2\{2n_j(\mathbf{q})+1\}. \quad (4.51)$$

We notice that $F_N(\boldsymbol{\kappa})$ depends upon the temperature through the Debye–Waller factor and, furthermore, that this does not appear as a simple scaling factor. The relative contributions to $F_N(\boldsymbol{\kappa})$ of different species of atoms can therefore change with temperature.

The elastic incoherent cross-section for a general lattice structure reads

$$\left(\frac{d\sigma}{d\Omega}\right)^{el}_{incoh} = N\sum_d (\overline{b_d^2}-\bar{b}_d^2)\exp\{-2W_d(\boldsymbol{\kappa})\}. \quad (4.52)$$

We now return to a more detailed discussion of the Debye–Waller factor. It is self-evident from (4.41) that $W(\boldsymbol{\kappa})$ is dependent on the phonon frequency spectrum; however, it is not immediately apparent that, for all practical purposes, W has only a single-parameter dependence on the spectrum. To demonstrate this we recall that at low temperatures W is small and unimportant and therefore we examine (4.41) in the high-temperature region. Expanding in powers of β we obtain

$$W(\boldsymbol{\kappa}) = \frac{\hbar\kappa^2}{4M}\int_0^{\omega_m} d\omega \frac{Z(\omega)}{\omega}\left\{\frac{2}{\hbar\omega\beta}+\frac{\hbar\omega\beta}{6}-\frac{(\hbar\omega\beta)^3}{360}+\ldots\right\}$$

$$= \frac{\hbar^2\kappa^2}{2M}k_B T\left\{\overline{(\hbar\omega)^{-2}}+\frac{1}{12k_B^2T^2}-\frac{\overline{(\hbar\omega)^2}}{720k_B^4T^4}+\ldots\right\}, \quad (4.53)$$

where we define the moments of the frequency distribution by

$$\overline{\omega^n} = \int_0^{\omega_m} d\omega\, \omega^n Z(\omega). \quad (4.54)$$

We notice that the second term of (4.53) is independent of the frequency spectrum and the third term has such a small coefficient that over a wide temperature range it can be neglected. Hence for all practical purposes $W(\varkappa)$ depends on the single parameter $\overline{\omega^{-2}}$. To get a quantitative estimate of the terms we can use the Debye spectrum (4.42) to give

$$\overline{(\hbar\omega)^{-2}} = 3/(k_B\,\theta_D)^2, \tag{4.55}$$

$$\overline{(\hbar\omega)^{2}} = \tfrac{3}{5}(k_B\,\theta_D)^2. \tag{4.56}$$

Hence
$$W(\varkappa) \sim \frac{\hbar^2\kappa^2}{2M}\frac{3k_B\,T}{(k_B\,\theta_D)^2}\left\{1+\frac{1}{36}\left(\frac{\theta_D}{T}\right)^2-\frac{1}{3600}\left(\frac{\theta_D}{T}\right)^4\right\}. \tag{4.57}$$

This shows that for all $T \gtrsim 0{\cdot}4\theta_D$ the third term is less than 1 per cent of the leading term.

This result is of the following value. Many thermodynamic quantities depend on $Z(\omega)$, the specific heat for example, but the dependence is not simple and depends upon the detailed shape of $Z(\omega)$. In contrast to this we see that the Debye–Waller factor is quite insensitive to the details of $Z(\omega)$; it depends only upon the single parameter $\overline{\omega^{-2}}$. We deduce that measurements of the Debye–Waller factor cannot be used to give the details of $Z(\omega)$, but they can be used to give $\overline{\omega^{-2}}$.

It is worth while giving a brief discussion of how the Debye–Waller factor is affected by anharmonic effects. The most important effect is that the spectrum $Z(\omega)$ and hence the parameter $\overline{\omega^{-2}}$ become themselves temperature dependent.

In addition there are explicit corrections to be considered because (4.33) is not valid in the presence of anharmonic terms. To display these corrections we notice from (4.32) that, in general,

$$e^{-W} = \langle \exp(-i\varkappa\,.\,\hat{\mathbf{u}})\rangle, \tag{4.58}$$

and we look for W as a power series in the vibration amplitudes, which by dimensional arguments must be identical to a series in powers of the vector $\varkappa$. To express this formally we write

$$\hat{A} = -i\varkappa\,.\,\hat{\mathbf{u}}$$

and
$$e^{-W(x)} = \langle e^{x\hat{A}}\rangle,$$

and look for power series in the parameter x, which we later set equal to unity. Hence

$$W(x) = -\ln\langle e^{x\hat{A}}\rangle. \tag{4.59}$$

The first derivative is

$$W'(x) = -\frac{\langle \hat{A}\, e^{x\hat{A}}\rangle}{\langle e^{x\hat{A}}\rangle}.$$

Similarly

$$W''(x) = -\frac{\langle \hat{A}^2\,\mathrm{e}^{x\hat{A}}\rangle}{\langle \mathrm{e}^{x\hat{A}}\rangle}+\frac{\langle \hat{A}\,\mathrm{e}^{x\hat{A}}\rangle^2}{\langle \mathrm{e}^{x\hat{A}}\rangle^2},$$

$$W'''(x) = -\frac{\langle \hat{A}^3\,\mathrm{e}^{x\hat{A}}\rangle}{\langle \mathrm{e}^{x\hat{A}}\rangle}+\frac{3\langle \hat{A}^2\,\mathrm{e}^{x\hat{A}}\rangle\langle \hat{A}\,\mathrm{e}^{x\hat{A}}\rangle}{\langle \mathrm{e}^{x\hat{A}}\rangle^2}-\frac{2\langle \hat{A}\,\mathrm{e}^{x\hat{A}}\rangle}{\langle \mathrm{e}^{x\hat{A}}\rangle^3}, \tag{4.60}$$

and

$$W^{\mathrm{iv}}(x) = -\frac{\langle \hat{A}^4\,\mathrm{e}^{x\hat{A}}\rangle}{\langle \mathrm{e}^{x\hat{A}}\rangle}+\frac{4\langle \hat{A}^3\,\mathrm{e}^{x\hat{A}}\rangle\langle \hat{A}\,\mathrm{e}^{x\hat{A}}\rangle}{\langle \mathrm{e}^{x\hat{A}}\rangle^2}+\frac{3\langle \hat{A}^2\,\mathrm{e}^{x\hat{A}}\rangle^2}{\langle \mathrm{e}^{x\hat{A}}\rangle^2}-$$

$$-\frac{12\langle \hat{A}^2\,\mathrm{e}^{x\hat{A}}\rangle\langle \hat{A}\,\mathrm{e}^{x\hat{A}}\rangle^2}{\langle \mathrm{e}^{x\hat{A}}\rangle^3}+\frac{6\langle \hat{A}\,\mathrm{e}^{x\hat{A}}\rangle^4}{\langle \mathrm{e}^{x\hat{A}}\rangle^4}.$$

Evaluating the derivatives at $x = 0$ and assuming $\langle \hat{A}\rangle = 0$ gives

$$\begin{aligned} W'(0) &= 0,\\ W''(0) &= -\langle \hat{A}^2\rangle,\\ W'''(0) &= -\langle \hat{A}^3\rangle,\\ W^{\mathrm{iv}}(0) &= -\langle \hat{A}^4\rangle+3\langle \hat{A}^2\rangle^2, \end{aligned} \tag{4.61}$$

and hence

$$W(x) = -x^2\frac{1}{2!}\langle \hat{A}^2\rangle - x^3\frac{1}{3!}\langle \hat{A}^3\rangle - x^4\frac{1}{4!}\{\langle \hat{A}^4\rangle - 3\langle \hat{A}^2\rangle^2\}+\dots.$$

Hence, finally,

$$W(\boldsymbol{\kappa}) = \tfrac{1}{2}\langle(\boldsymbol{\kappa}.\hat{\mathbf{u}})^2\rangle - \tfrac{1}{6}\mathrm{i}\langle(\boldsymbol{\kappa}.\hat{\mathbf{u}})^3\rangle - \tfrac{1}{24}\{\langle(\boldsymbol{\kappa}.\hat{\mathbf{u}})^4\rangle - 3\langle(\boldsymbol{\kappa}.\hat{\mathbf{u}})^2\rangle^2\}+\dots. \tag{4.62}$$

The leading term in this expression is the usual harmonic form for W. The second term vanishes if every atom is a centre of symmetry (if every atom is not at a centre of symmetry it can give contributions to the coherent cross-section). The third term vanishes if the vibration amplitudes have the Gaussian distribution due to harmonic forces. But in the presence of anharmonic forces this term does not vanish. According to Ambegaokar and Maradudin (1964)† these explicit anharmonic effects are of less importance than the corrections anharmonicity makes to $Z(\omega)$ and hence $\overline{\omega^{-2}}$. We shall therefore set aside any further discussion of the effects.

4.3. Inelastic one-phonon scattering

As in the previous section we shall first discuss the inelastic one-phonon cross-sections for Bravais lattices and then later generalize our results.

† V. Ambegaokar and A. A. Maradudin, *Westinghouse Research Laboratory Rept.* 64–929–100, P2 (1964).

We recall that the inelastic coherent cross-section can be written (cf. § 3.2)

$$\left(\frac{\mathrm{d}^2\sigma}{\mathrm{d}\Omega \mathrm{d}E'}\right)^{\mathrm{inel}}_{\mathrm{coh}} = \frac{N\sigma_{\mathrm{c}}}{4\pi}\frac{k'}{k}\frac{1}{2\pi\hbar}\int\limits_{-\infty}^{\infty}\mathrm{d}t\exp(-\mathrm{i}\omega t)\int \mathrm{d}\mathbf{r}\exp(\mathrm{i}\boldsymbol{\varkappa}.\mathbf{r})G'(\mathbf{r},t). \tag{4.63}$$

In (4.63),
$$G'(\mathbf{r},t) = G(\mathbf{r},t) - G(\mathbf{r},\infty),$$
where

$$G(\mathbf{r},t) = \frac{1}{(2\pi)^3 N}\sum_{l,l'}\int \mathrm{d}\mathbf{k}\exp(-\mathrm{i}\mathbf{k}.\mathbf{r})\langle\exp\{-\mathrm{i}\mathbf{k}.\hat{\mathbf{R}}_l(0)\}\exp\{\mathrm{i}\mathbf{k}.\hat{\mathbf{R}}_{l'}(t)\}\rangle. \tag{4.64}$$

For a Bravais lattice the correlation function in $G(\mathbf{r},t)$ reduces to

$$\exp\{\mathrm{i}\mathbf{k}.(\mathbf{l}'-\mathbf{l})\}\langle\exp\{-\mathrm{i}\mathbf{k}.\hat{\mathbf{u}}(l,0)\}\exp\{\mathrm{i}\mathbf{k}.\hat{\mathbf{u}}(l',t)\}\rangle. \tag{4.65}$$

If we set
$$\hat{A} = -\mathrm{i}\mathbf{k}.\hat{\mathbf{u}}(l,0)$$
and
$$\hat{B} = \mathrm{i}\mathbf{k}.\hat{\mathbf{u}}(l',t),$$
then, since $[\hat{A},\hat{B}]$ is a c-number,

$$\langle\exp\hat{A}\exp\hat{B}\rangle = \exp\{\tfrac{1}{2}[\hat{A},\hat{B}]\}\langle\exp(\hat{A}+\hat{B})\rangle$$

from (3.53), and further, using the identity (3.i), we obtain

$$\langle\exp\hat{A}\exp\hat{B}\rangle = \exp\{\tfrac{1}{2}[\hat{A},\hat{B}]\}\exp\{\tfrac{1}{2}\langle(\hat{A}+\hat{B})^2\rangle\}$$
$$= \exp\{\tfrac{1}{2}\langle\hat{A}^2+\hat{B}^2+2\hat{A}\hat{B}\rangle\}. \tag{4.66}$$

Thus the correlation function in (4.65) can be written

$$\begin{aligned}&\langle\exp\{-\mathrm{i}\mathbf{k}.\hat{\mathbf{u}}(l,0)\}\exp\{\mathrm{i}\mathbf{k}.\hat{\mathbf{u}}(l',t)\}\rangle\\ &\quad = \exp[\tfrac{1}{2}\langle-\{\mathbf{k}.\hat{\mathbf{u}}(l,0)\}^2-\{\mathbf{k}.\hat{\mathbf{u}}(l',0)\}^2+2\mathbf{k}.\hat{\mathbf{u}}(l,0)\mathbf{k}.\hat{\mathbf{u}}(l',t)\rangle]\\ &\quad = \exp\{-2W(\mathbf{k})\}\exp\{\langle\mathbf{k}.\hat{\mathbf{u}}(l,0)\mathbf{k}.\hat{\mathbf{u}}(l',t)\rangle\}.\end{aligned} \tag{4.67}$$

Clearly

$$\begin{aligned}&G(\mathbf{r},\infty)\\ &\quad = \frac{1}{(2\pi)^3 N}\sum_{l,l'}\int \mathrm{d}\mathbf{k}\exp(-\mathrm{i}\mathbf{k}.\mathbf{r})\langle\exp\{-\mathrm{i}\mathbf{k}.\hat{\mathbf{R}}_l(0)\}\rangle\langle\exp\{\mathrm{i}\mathbf{k}.\hat{\mathbf{R}}_{l'}(0)\}\rangle\\ &\quad = \frac{1}{(2\pi)^3 N}\sum_{l,l'}\int \mathrm{d}\mathbf{k}\exp(-\mathrm{i}\mathbf{k}.\mathbf{r})\exp\{-\mathrm{i}\mathbf{k}.(\mathbf{l}-\mathbf{l}')\}\exp\{-2W(\mathbf{k})\},\end{aligned}$$

so that the inelastic coherent cross-section is

$$\begin{aligned}\left(\frac{\mathrm{d}^2\sigma}{\mathrm{d}\Omega \mathrm{d}E'}\right)^{\mathrm{inel}}_{\mathrm{coh}} &= \frac{\sigma_{\mathrm{c}}}{4\pi}\frac{k'}{k}\frac{1}{2\pi\hbar}\int\limits_{-\infty}^{\infty}\mathrm{d}t\exp(-\mathrm{i}\omega t)\exp\{-2W(\boldsymbol{\varkappa})\}\times\\ &\quad\times\sum_{l,l'}\exp\{-\mathrm{i}\boldsymbol{\varkappa}.(\mathbf{l}-\mathbf{l}')\}[\exp\{\langle\boldsymbol{\varkappa}.\hat{\mathbf{u}}(l,0)\boldsymbol{\varkappa}.\hat{\mathbf{u}}(l',t)\rangle\}-1].\end{aligned} \tag{4.68}$$

Furthermore, it is readily seen from § 3.2 that the inelastic incoherent cross-section is

$$\left(\frac{\mathrm{d}^2\sigma}{\mathrm{d}\Omega\mathrm{d}E'}\right)^{\text{inel}}_{\text{incoh}} = \frac{\sigma_{\mathrm{i}}}{4\pi}\frac{k'}{k}\frac{1}{2\pi\hbar}\int\limits_{-\infty}^{\infty}\mathrm{d}t\exp(-\mathrm{i}\omega t)\exp\{-2W(\boldsymbol{\varkappa})\}\times$$
$$\times\sum_l[\exp\{\langle\boldsymbol{\varkappa}.\hat{\mathbf{u}}(l,0)\boldsymbol{\varkappa}.\hat{\mathbf{u}}(l,t)\rangle\}-1]. \quad (4.69)$$

These two equations for the inelastic coherent and incoherent partial differential cross-sections are exact within the harmonic approximation of the lattice vibrations. If we expand the exponentials in (4.68) and (4.69) containing the displacement correlation functions, the first non-zero terms give the one-phonon cross-sections, the second terms the two-phonon cross-sections, etc. For the present we shall consider in detail the one-phonon cross-sections. With regard to obtaining information on the dynamical motion of the nuclei these are by far the most important.

We need now to evaluate the correlation function

$$\langle\boldsymbol{\varkappa}.\hat{\mathbf{u}}(l,0)\boldsymbol{\varkappa}.\hat{\mathbf{u}}(l',t)\rangle.$$

From (4.22 a) this is

$$\sum_{\alpha,\beta}\kappa_\alpha\kappa_\beta\sum_{j,\mathbf{q}}\sum_{j',\mathbf{q}'}\left(\frac{\hbar}{2NM}\right)\sigma^j_\alpha(\mathbf{q})\sigma^{j'}_\beta(\mathbf{q}')\{\omega_j(\mathbf{q})\omega_{j'}(\mathbf{q}')\}^{-\frac{1}{2}}\times$$
$$\times[\exp(\mathrm{i}\mathbf{q}.\mathbf{l}-\mathrm{i}\mathbf{q}'.\mathbf{l}')\exp\{\mathrm{i}t\omega_{j'}(\mathbf{q}')\}\langle\hat{a}_j(\mathbf{q})\hat{a}^+_{j'}(\mathbf{q}')\rangle+$$
$$+\exp(-\mathrm{i}\mathbf{q}.\mathbf{l}+\mathrm{i}\mathbf{q}'.\mathbf{l}')\exp\{-\mathrm{i}t\omega_{j'}(\mathbf{q}')\}\langle\hat{a}^+_j(\mathbf{q})\hat{a}_{j'}(\mathbf{q}')\rangle]$$
$$=\sum_{\alpha,\beta}\kappa_\alpha\kappa_\beta\sum_{j,\mathbf{q}}\left(\frac{\hbar}{2NM}\right)\sigma^j_\alpha(\mathbf{q})\sigma^j_\beta(\mathbf{q})\omega_j^{-1}(\mathbf{q})\times$$
$$\times[\exp\{\mathrm{i}\mathbf{q}.(\mathbf{l}-\mathbf{l}')\}\exp\{\mathrm{i}t\omega_j(\mathbf{q})\}\{1+n_j(\mathbf{q})\}+$$
$$+\exp\{-\mathrm{i}\mathbf{q}.(\mathbf{l}-\mathbf{l}')\}\exp\{-\mathrm{i}t\omega_j(\mathbf{q})\}n_j(\mathbf{q})]$$

because of (4.18 a). Hence

$$\langle\boldsymbol{\varkappa}.\hat{\mathbf{u}}(l,0)\boldsymbol{\varkappa}.\hat{\mathbf{u}}(l',t)\rangle$$
$$=\frac{\hbar}{2NM}\sum_{j,\mathbf{q}}\frac{|\boldsymbol{\varkappa}.\boldsymbol{\sigma}^j(\mathbf{q})|^2}{\omega_j(\mathbf{q})}[\exp\{\mathrm{i}\mathbf{q}.(\mathbf{l}-\mathbf{l}')\}\exp\{\mathrm{i}t\omega_j(\mathbf{q})\}\{1+n_j(\mathbf{q})\}+$$
$$+\exp\{-\mathrm{i}\mathbf{q}.(\mathbf{l}-\mathbf{l}')\}\exp\{-\mathrm{i}t\omega_j(\mathbf{q})\}n_j(\mathbf{q})], \quad (4.70)$$

and it immediately follows that the one-phonon coherent partial differential cross-section

$$\left(\frac{d^2\sigma}{d\Omega dE'}\right)^{\text{inel}}_{\text{coh}} = \frac{\sigma_c}{4\pi}\frac{k'}{k}\frac{1}{2\pi\hbar}\int_{-\infty}^{\infty} dt \exp(-i\omega t)\exp\{-2W(\boldsymbol{\varkappa})\}\times$$
$$\times \sum_{l,l'} \exp\{-i\boldsymbol{\varkappa}.(\mathbf{l}-\mathbf{l}')\}\langle \boldsymbol{\varkappa}.\hat{\mathbf{u}}(l,0)\boldsymbol{\varkappa}.\hat{\mathbf{u}}(l',t)\rangle$$
$$= \frac{\sigma_c}{4\pi}\frac{k'}{k}\frac{(2\pi)^3}{v_0}\frac{1}{2M}\sum_{\boldsymbol{\tau}}\exp\{-2W(\boldsymbol{\varkappa})\}\sum_{j,\mathbf{q}}\frac{|\boldsymbol{\varkappa}.\boldsymbol{\sigma}^j(\mathbf{q})|^2}{\omega_j(\mathbf{q})}\times$$
$$\times[n_j(\mathbf{q})\,\delta\{\omega+\omega_j(\mathbf{q})\}\,\delta(\boldsymbol{\varkappa}+\mathbf{q}-\boldsymbol{\tau})+\{n_j(\mathbf{q})+1\}\,\delta\{\omega-\omega_j(\mathbf{q})\}\,\delta(\boldsymbol{\varkappa}-\mathbf{q}-\boldsymbol{\tau})]. \tag{4.71}$$

The cross-section (4.71) is the sum of two terms. The first, which contains $\delta\{\omega+\omega_j(\mathbf{q})\}\delta(\boldsymbol{\varkappa}+\mathbf{q}-\boldsymbol{\tau})$, represents a scattering process in which one phonon is annihilated and the second term, containing

$$\delta\{\omega-\omega_j(\mathbf{q})\}\,\delta(\boldsymbol{\varkappa}-\mathbf{q}-\boldsymbol{\tau}),$$

represents a process in which one phonon is created. In the limit $T \to 0$ only the second process occurs, since there are clearly no phonons to be annihilated at absolute zero. Let us consider in detail the structure of the one-phonon annihilation cross-section.

The two δ-functions associated with this scattering process represent conservation of energy and momentum and tell us that the scattering process obeys the conditions

$$E' = E+\hbar\omega_j(\mathbf{q}),$$
$$\mathbf{k}' = \mathbf{k}+\mathbf{q}-\boldsymbol{\tau}. \tag{4.72}$$

These conditions are so restrictive that for given scattering angle only phonons of a particular $\mathbf{q}$ and $\omega_j(\mathbf{q})$ can give scattering. We can make use of this to determine the phonon spectrum, $\omega_j(\mathbf{q})$, as a function of $\mathbf{q}$. Suppose we have a monochromatic beam incident on a crystal and we measure the energy of the neutrons scattered through a given angle by using a time of flight apparatus or by using another crystal as analyser. From (4.72) these scattered neutrons will in general have one of three energies corresponding to the three choices of j. Choose one of these energies; knowing the scattered energy we have $\mathbf{k}'$ and from the second of (4.72) we have $\mathbf{k}'-\mathbf{k} = \mathbf{q}-\boldsymbol{\tau}$. But $\mathbf{q}$ must lie in the first Brillouin zone, so we get both $\mathbf{q}$ and $\boldsymbol{\tau}$. Hence from this experiment we have both $\mathbf{q}$ and $\omega_j(\mathbf{q})$ for a phonon, i.e. we have a point on the phonon spectrum.

The conservation conditions (4.72) are illustrated in Fig. 4.4. Starting from the origin O we draw the vector $\mathbf{k}$ and the vector $-\boldsymbol{\tau}$ to give an end point at C. The dotted line through the origin is the direction of $\mathbf{k}'$ as determined by the position of the neutron counters. Then any vector such as OA', OA, or OA'' can represent $\mathbf{k}'$ while CA', CA, CA'' respectively represent $\mathbf{q}$. The momentum conservation law is obeyed automatically, but energy conservation is given only by a special choice of

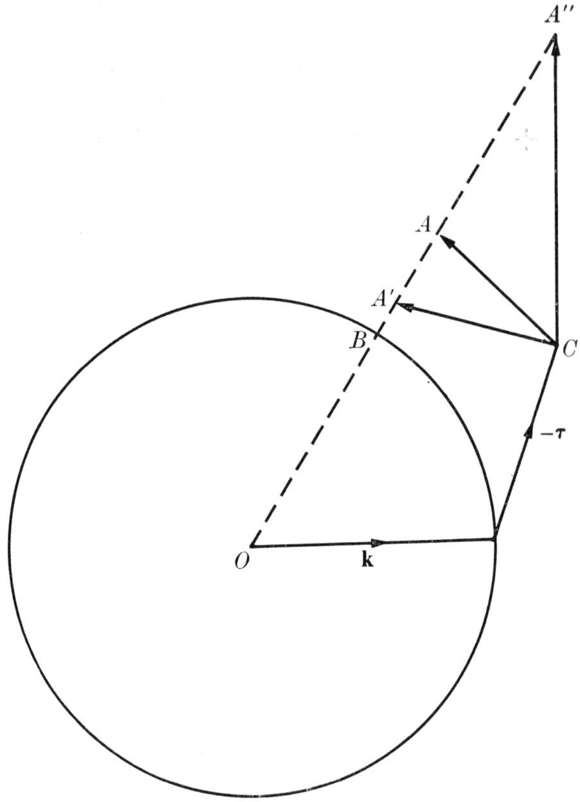

FIG. 4.4. Conservation of momentum and energy in a one-phonon scattering event.

the position A. In the figure a circle of radius $|\mathbf{k}|$ has been drawn about the origin; it intersects the direction of $\mathbf{k}'$ at B. The energy conservation condition then states that $(\hbar^2/2m)\{(OA)^2-(OB)^2\}$ must equal the energy associated with the vector $\mathbf{q}$, i.e. $\omega_j(\mathbf{q})$. For a position such as A' the neutron energy gain is too small compared with $\omega_j(\mathbf{q})$, and for a position such as A'' the neutron energy gain is too large. Hence it is only at some intermediate position, say A, that the conservation conditions can both be satisfied. The same construction has to be repeated for each choice of j. Therefore, in general, scattering at a fixed angle can correspond to annihilation of any one of the $3r$ phonon modes.

The numerical value of the cross-section can be obtained by first performing the sum over $\mathbf{q}$,

$$\left(\frac{d^2\sigma}{d\Omega dE'}\right)^{\text{inel}}_{\text{coh},-1} = \frac{N\sigma_c}{8\pi M}\frac{\hbar k'}{k}\sum_j \delta\{\hbar\omega+\hbar\omega_j(\mathbf{q})\}\exp\{-2W(\boldsymbol{\varkappa})\}\frac{n_j(\mathbf{q})}{\omega_j(\mathbf{q})}|\boldsymbol{\varkappa}.\boldsymbol{\sigma}^j(\mathbf{q})|^2, \quad (4.73)$$

where it is now understood that for each j, $\mathbf{q}$ and $\boldsymbol{\tau}$ are determined by (4.72).

We have chosen to denote the one-phonon annihilation cross-section by the suffix -1; the one-phonon creation cross-section will be denoted by the suffix $+1$.

When we integrate (4.73) over E' we must now remember that $\omega_j(\mathbf{q})$ is a function of E'. Since

$$\frac{\partial}{\partial E'}\{\hbar\omega+\hbar\omega_j(\mathbf{q})\} = -1+\frac{\hbar}{2E'}\mathbf{k}'.\nabla\omega_j(\mathbf{q}),$$

where $\nabla\omega_j(\mathbf{q})$ is the gradient of $\omega_j(\mathbf{q})$ with respect to $\mathbf{q}$, we have

$$\left(\frac{d\sigma}{d\Omega}\right)^{\text{inel}}_{\text{coh},-1} = \frac{N\sigma_c\hbar}{8\pi M}\sum_j \frac{k'}{k}|\boldsymbol{\varkappa}.\boldsymbol{\sigma}^j(\mathbf{q})|^2\frac{n_j(\mathbf{q})}{\omega_j(\mathbf{q})}\exp\{-2W(\boldsymbol{\varkappa})\}\left\{1-\frac{\hbar}{2E'}\mathbf{k}'.\nabla\omega_j(\mathbf{q})\right\}^{-1}. \quad (4.74)$$

Using $n_j(\mathbf{q}) \simeq k_B T/\hbar\omega_j(\mathbf{q})$ and putting $T = 300$ K we find this cross-section is typically of the order of 10 millibarns. Notice that for scattering near the forward direction, $\boldsymbol{\tau} = 0$ and the polarization factor $|\boldsymbol{\varkappa}.\boldsymbol{\sigma}^j(\mathbf{q})|^2$ becomes $|\mathbf{q}.\boldsymbol{\sigma}^j(\mathbf{q})|^2$. It follows that purely transverse modes give no scattering with $\boldsymbol{\tau} = 0$.

The coherent cross-section corresponding to the creation of a single phonon is evaluated similarly to give

$$\left(\frac{d\sigma}{d\Omega}\right)^{\text{inel}}_{\text{coh},+1} = \frac{N\sigma_c\hbar}{8\pi M}\sum_j \frac{k'}{k}|\boldsymbol{\varkappa}.\boldsymbol{\sigma}^j(\mathbf{q})|^2\frac{\{n_j(\mathbf{q})+1\}}{\omega_j(\mathbf{q})}\exp\{-2W(\boldsymbol{\varkappa})\}\left\{1+\frac{\hbar}{2E'}\mathbf{k}'.\nabla\omega_j(\mathbf{q})\right\}^{-1}, \quad (4.75)$$

where it is understood that $\mathbf{q}$ and E' are determined from

$$E' = E-\hbar\omega_j(\mathbf{q}), \qquad \mathbf{k}' = \mathbf{k}-\mathbf{q}-\boldsymbol{\tau}. \quad (4.76)$$

The $\boldsymbol{\tau} = 0$ part of the one-phonon creation process has a special property: eliminating $\mathbf{k}'$ from the eqns (4.76) gives

$$\frac{\hbar^2}{2m}(2\mathbf{k}.\mathbf{q}-q^2) = \hbar\omega_j(\mathbf{q}). \tag{4.77}$$

Now
$$\frac{\hbar^2}{2m}(2\mathbf{k}.\mathbf{q}-q^2) < \frac{\hbar^2}{m}\mathbf{k}.\mathbf{q} < \frac{\hbar^2 kq}{m},$$

and for small q
$$\hbar\omega_j(\mathbf{q}) \simeq \hbar cq,$$

where c is the velocity of sound. It follows that there is only a solution of (4.77) (other than the trivial solution $q = 0$) if

$$\frac{\hbar k}{m} > c,$$

i.e. if the velocity of the incoming neutron is greater than the velocity of sound in the material. Hence, if the incoming neutrons have such a low energy that their velocity is less than the velocity of sound in the crystal, then the one-phonon coherent creation process has zero cross-section for $\boldsymbol{\tau} = 0$, i.e. near the forward direction. It is possible to take advantage of this when looking for magnetic inelastic scattering; for long wavelength neutrons there is very little non-magnetic scattering in the forward direction (because the one-phonon creation process is forbidden and the one-phonon annihilation process can be made small by having a low temperature) and therefore this is a good place to look for magnetic scattering.

The one-phonon incoherent cross-section is given from (4.69) by

$$\left(\frac{\mathrm{d}^2\sigma}{\mathrm{d}\Omega\mathrm{d}E'}\right)^{\text{inel}}_{\text{incoh}} = \frac{\sigma_{\mathrm{i}}}{4\pi}\frac{k'}{k}\frac{1}{2\pi\hbar}\int_{-\infty}^{\infty} \mathrm{d}t\, \exp(-\mathrm{i}\omega t)\exp\{-2W(\boldsymbol{\varkappa})\}\times \\ \times \sum_l \langle\boldsymbol{\varkappa}.\hat{\mathbf{u}}(l,0)\boldsymbol{\varkappa}.\hat{\mathbf{u}}(l,t)\rangle,$$

which with the aid of the result (4.70) with $l = l'$ reduces to

$$\left(\frac{\mathrm{d}^2\sigma}{\mathrm{d}\Omega\mathrm{d}E'}\right)^{\text{inel}}_{\text{incoh}} = \frac{\sigma_{\mathrm{i}}}{8\pi M}\exp\{-2W(\boldsymbol{\varkappa})\}\frac{k'}{k}\sum_{j,\mathbf{q}}\frac{|\boldsymbol{\varkappa}.\boldsymbol{\sigma}^j(\mathbf{q})|^2}{\omega_j(\mathbf{q})}\times \\ \times[n_j(\mathbf{q})\,\delta\{\omega+\omega_j(\mathbf{q})\}+\{n_j(\mathbf{q})+1\}\,\delta\{\omega-\omega_j(\mathbf{q})\}]. \tag{4.78}$$

We notice that this formula contains δ-functions to ensure energy conservation, the first term representing one-phonon annihilation scattering and the second one-phonon creation scattering, but there are no momentum conservation conditions associated with them. Thus, one-phonon incoherent scattering cannot give us the detailed information that coherent scattering can give. Nevertheless it is valuable, because for Bravais lattices it can be used to determine the density of states $Z(\omega)$ directly. In terms of $Z(\omega)$ the creation part of (4.78) can be written

$$\left(\frac{d^2\sigma}{d\Omega dE'}\right)^{\text{inel}}_{\text{incoh},+1} = \frac{\sigma_i}{8\pi M} \exp\{-2W(\boldsymbol{\varkappa})\} \frac{k'}{k} 3N \int_0^{\omega_m} d\omega' \frac{Z(\omega')}{\omega'} \times$$
$$\times \{|\boldsymbol{\varkappa}.\boldsymbol{\sigma}^j(\mathbf{q})|^2\}_{\text{av}}\{n(\omega')+1\}\delta(\omega-\omega'), \qquad (4.79)$$

where the subscript av means that an average over modes with given ω' has been taken. For cubic crystals $|\boldsymbol{\varkappa}.\boldsymbol{\sigma}^j(\mathbf{q})|^2_{\text{av}}$ is rigorously $\frac{1}{3}\kappa^2$. Using this in (4.79),

$$\left(\frac{d^2\sigma}{d\Omega dE'}\right)^{\text{inel}}_{\text{incoh},+1} = \frac{N\sigma_i}{8\pi M}\frac{k'}{k}\kappa^2 \exp\{-2W(\boldsymbol{\varkappa})\}\frac{Z(\omega)}{\omega}\{n(\omega)+1\}. \qquad (4.80)$$

This formula tells us that incoherent one-phonon scattering takes place into all directions and that for any direction the incoherently scattered neutrons have a continuous energy distribution that is proportional to $Z(\omega)$ (where $\hbar\omega$ is the energy lost by the neutron) and other factors. If the final energy E' is measured, then k' and k are both known, ω is easily calculated, and the intensity leads directly to $Z(\omega)$ as a function of ω.

The one-phonon annihilation part of (4.78) can be written in a similar way:

$$\left(\frac{d^2\sigma}{d\Omega dE'}\right)^{\text{inel}}_{\text{incoh},-1} = \frac{N\sigma_i}{8\pi M}\frac{k'}{k}\kappa^2 \exp\{-2W(\boldsymbol{\varkappa})\}\frac{Z(-\omega)}{(-\omega)}n(-\omega)$$
$$= \frac{N\sigma_i}{8\pi M}\frac{k'}{k}\kappa^2 \exp\{-2W(\boldsymbol{\varkappa})\}\frac{Z(\omega)}{\omega}\{n(\omega)+1\}, \qquad (4.81)$$

if we define
$$Z(\omega) = Z(-\omega). \qquad (4.82)$$

Hence, from (4.80) and (4.81) it follows that the cross-section (4.78) for $\omega > 0$ and $\omega < 0$ is of exactly the same form, so that we can drop the suffixes ± 1 on (4.80) and (4.81).

The generalization of the preceding results of this section to non-Bravais lattices is relatively straightforward. First

$$\langle\exp\{-i\boldsymbol{\varkappa}.\hat{\mathbf{R}}_{ld}(0)\}\exp\{i\boldsymbol{\varkappa}.\hat{\mathbf{R}}_{l'd'}(t)\}\rangle = \exp\{-i\boldsymbol{\varkappa}(\mathbf{l}+\mathbf{d}-\mathbf{l}'-\mathbf{d}')\}\times$$
$$\times\left\langle\exp\left\{-i\boldsymbol{\varkappa}.\hat{\mathbf{u}}\binom{l}{d},0\right\}\exp\left\{i\boldsymbol{\varkappa}.\hat{\mathbf{u}}\binom{l'}{d'},t\right\}\right\rangle, \qquad (4.83)$$

and the correlation function occurring on the right-hand side is simply

$$\exp\{-W_d(\boldsymbol{\varkappa})\}\exp\{-W_{d'}(\boldsymbol{\varkappa})\}\exp\left\{\left\langle\boldsymbol{\varkappa}.\hat{\mathbf{u}}\binom{l}{d},0\right)\boldsymbol{\varkappa}.\hat{\mathbf{u}}\binom{l'}{d'},t\right)\right\rangle\right\}, \qquad (4.84)$$

as is clear from (4.66). $W_d(\varkappa)$ is given by eqn (4.51). Thus it follows from (4.83) and (4.84) that in place of eqn (4.68) for the inelastic coherent cross-section we obtain, for a general lattice structure,

$$\left(\frac{d^2\sigma}{d\Omega dE'}\right)^{\text{inel}}_{\text{coh}} = \frac{k'}{k}\frac{1}{2\pi\hbar}\int_{-\infty}^{\infty} dt \exp(-i\omega t)\sum_{d,d'} \bar{b}_d^* \bar{b}_{d'} \exp\{-W_d(\boldsymbol{\varkappa})\}\times$$
$$\times \exp\{-W_{d'}(\boldsymbol{\varkappa})\}\exp\{-i\boldsymbol{\varkappa}(\mathbf{d}-\mathbf{d}')\}\sum_{l,l'}\exp\{-i\boldsymbol{\varkappa}.(\mathbf{l}-\mathbf{l}')\}\times$$
$$\times\left[\exp\left\{\left\langle \boldsymbol{\varkappa}.\hat{\mathbf{u}}\binom{l}{d},0\right)\boldsymbol{\varkappa}.\hat{\mathbf{u}}\left(\binom{l'}{d'},t\right)\right\rangle\right\}-1\right]. \quad (4.85)$$

For the inelastic incoherent cross-section we find

$$\left(\frac{d^2\sigma}{d\Omega dE'}\right)^{\text{inel}}_{\text{incoh}} = \frac{k'}{k}\frac{1}{2\pi\hbar}\int_{-\infty}^{\infty} dt \exp(-i\omega t)\sum_{d}\{|\overline{b_d^2}|-|\bar{b}_d|^2\}\exp\{-2W_d(\boldsymbol{\varkappa})\}\times$$
$$\times\sum_l\left[\exp\left\{\left\langle \boldsymbol{\varkappa}.\hat{\mathbf{u}}\left(\binom{l}{d},0\right)\boldsymbol{\varkappa}.\hat{\mathbf{u}}\left(\binom{l}{d},t\right)\right\rangle\right\}-1\right]. \quad (4.86)$$

In order to obtain explicit expressions for the one-phonon cross-sections we have only to calculate the correlation function

$$\left\langle \boldsymbol{\varkappa}.\hat{\mathbf{u}}\left(\binom{l}{d},0\right)\boldsymbol{\varkappa}.\hat{\mathbf{u}}\left(\binom{l'}{d'},t\right)\right\rangle.$$

From eqn (4.30) it is easily shown that

$$\left\langle \boldsymbol{\varkappa}.\hat{\mathbf{u}}\left(\binom{l}{d},0\right)\boldsymbol{\varkappa}.\hat{\mathbf{u}}\left(\binom{l'}{d'},t\right)\right\rangle$$
$$= \frac{\hbar}{2N\sqrt{(M_d M_{d'})}}\sum_{j,\mathbf{q}}\omega_j^{-1}(\mathbf{q})[\boldsymbol{\varkappa}.\boldsymbol{\sigma}_d^j(\mathbf{q})\boldsymbol{\varkappa}.\boldsymbol{\sigma}_{d'}^{*j}(\mathbf{q})\exp\{i\mathbf{q}.(\mathbf{l}-\mathbf{l}')+i\omega_j(\mathbf{q})t\}\times$$
$$\times\{n_j(\mathbf{q})+1\}+\boldsymbol{\varkappa}.\boldsymbol{\sigma}_d^{*j}(\mathbf{q})\boldsymbol{\varkappa}.\boldsymbol{\sigma}_{d'}^j(\mathbf{q})\exp\{-i\mathbf{q}.(\mathbf{l}-\mathbf{l}')-i\omega_j(\mathbf{q})t\}n_j(\mathbf{q})]. \quad (4.87)$$

Thus the one-phonon inelastic coherent cross-section is

$$\left(\frac{d^2\sigma}{d\Omega dE'}\right)^{\text{inel}}_{\text{coh}} = \frac{k'}{k}\frac{(2\pi)^3}{2v_0}\sum_{\boldsymbol{\tau}}\sum_{j,\mathbf{q}}\omega_j^{-1}(\mathbf{q})\left|\sum_d \bar{b}_d\exp\{-W_d(\boldsymbol{\varkappa})\}\times\right.$$
$$\left.\times\exp(i\boldsymbol{\varkappa}.\mathbf{d})\frac{\boldsymbol{\varkappa}.\boldsymbol{\sigma}_d^j(\mathbf{q})}{\sqrt{M_d}}\right|^2[n_j(\mathbf{q})\,\delta\{\omega+\omega_j(\mathbf{q})\}\delta(\boldsymbol{\varkappa}+\mathbf{q}-\boldsymbol{\tau})+$$
$$+\{n_j(\mathbf{q})+1\}\,\delta\{\omega-\omega_j(\mathbf{q})\}\,\delta(\boldsymbol{\varkappa}-\mathbf{q}-\boldsymbol{\tau})] \quad (4.88)$$

and any further discussion of this would parallel that given of eqn (4.71).

The one-phonon inelastic incoherent cross-section is simply derived from (4.87) by merely setting $d = d'$ and $l = l'$. We find

$$\left(\frac{\mathrm{d}^2\sigma}{\mathrm{d}\Omega \mathrm{d}E'}\right)^{\text{inel}}_{\text{incoh}} = \frac{1}{2}\frac{k'}{k}\sum_{d}\{\overline{|b_d|^2} - |\bar{b}_d|^2\} \times$$
$$\times \exp\{-2W_d(\boldsymbol{\kappa})\}\frac{1}{M_d}\sum_{j,\mathbf{q}}\omega_j^{-1}(\mathbf{q})|\boldsymbol{\kappa}.\boldsymbol{\sigma}_d^j(\mathbf{q})|^2 \times$$
$$\times [n_j(\mathbf{q})\,\delta\{\omega+\omega_j(\mathbf{q})\} + \{n_j(\mathbf{q})+1\}\,\delta\{\omega-\omega_j(\mathbf{q})\}]. \quad (4.89)$$

This incoherent scattering is not as useful as that for a simple lattice because, in general, it is not possible to estimate $|\boldsymbol{\kappa}.\boldsymbol{\sigma}_d^j(\mathbf{q})|^2$ for the different d values and therefore the relative contributions made by each species of atom remain unknown.

4.4. Discussion

Thus far we have made no inquiry into the calculation of the second-order derivatives of the potential energy $A_{\alpha\beta}(l, l')$, the so-called force constants, and while a detailed answer to such a question is out of place here it is nonetheless worth while to see in general the line of reasoning employed in interpreting one-phonon cross-sections. Also, we have not said how the cross-section is modified when the harmonic approximation is no longer valid. Here again we cannot pause to give a detailed discussion but only indicate the relevant fundamental results. We begin our discussion with an outline of the properties required of the target sample to observe the various types of one-phonon scattering.

From the preceding section, inelastic one-phonon scattering can be used to determine directly two fundamental quantities: the coherent scattering gives the dispersion relation $\omega_j(\mathbf{q})$ and the incoherent scattering the density of states $Z(\omega)$. (Experiments have also been undertaken to determine the polarization vectors.) The latter is observed most conveniently in crystals with little or no coherent scattering (nor should the capture cross-section be large, of course), e.g. vanadium.

Experiments to determine the dispersion laws are most easily carried out with single crystals that have the following properties: (a) a relatively simple lattice structure, the number of dispersion curves we recall is equal to $3r$, (b) small incoherent and capture cross-sections, (c) almost perfect crystal structure and high purity, and, to obtain reasonable counting rates, a large volume. The identification of the various branches of the dispersion curves is simplest along symmetry directions because in this instance the polarization vectors $\sigma_\alpha^j(\mathbf{q})$ are either parallel or perpendicular to $\mathbf{q}$, and hence the presence of the polarization factor

$\boldsymbol{\varkappa}.\boldsymbol{\sigma}^j(\mathbf{q})$ in the cross-section can be utilized to make the contribution to the scattering from certain modes zero. Similarly, for a mirror plane at least one of the polarization vectors must be perpendicular to the plane, so that if $\boldsymbol{\varkappa}$ is arranged to lie in this plane there is no scattering from the modes corresponding to the perpendicular polarization vectors.

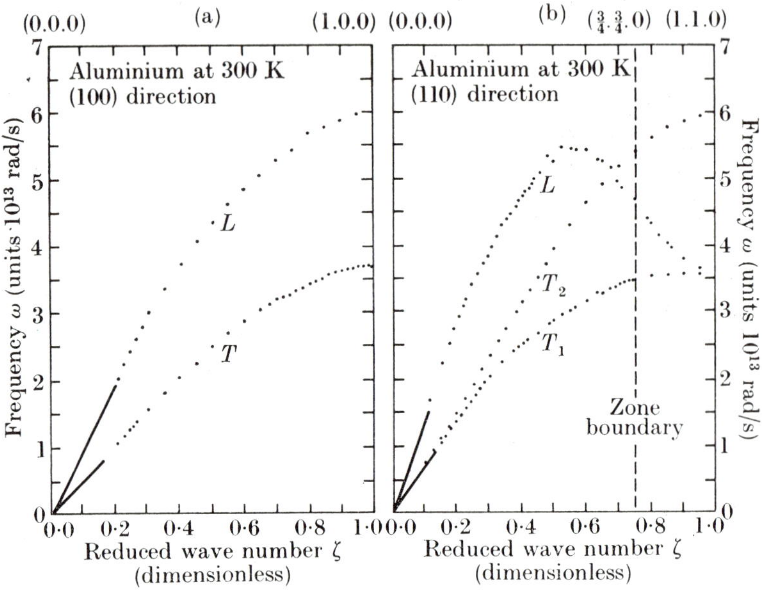

FIG. 4.5. The dispersion curves for Al (f.c.c.) at $T = 300$ K in two symmetry directions: (a) along the [100] direction and (b) along the [110] direction. Yarnell *et al. IAEA* 1963, vol. 1.

It is also true, of course, that the theoretical interpretation of the dispersion curves is much simpler for the special symmetry directions. An illustration of measured dispersion curves is given in Fig. 4.5.

The measured dispersion curves are usually analysed in terms of the Born–von Kármán model of (harmonic) lattice dynamics in which the interatomic force constants A are restricted only by the requirements of crystal symmetry. The simplest possible case is to assume that the only non-zero force constants are between nearest-neighbour atoms. These can then be determined from the elastic constants of the crystal, which are related to the slope of the dispersion curves as $\mathbf{q} \to 0$. If the force constants are allowed to extend to second neighbours, the additional parameters included are determined by fitting the dispersion curves to the observed values at the zone boundary. The number of force constants required to obtain an adequate fit to the data is often quite large. In metals, for instance, fifteen shells of neighbours may have to be taken into account. The density of states $Z(\omega)$ may, of course, be obtained from a theoretical model of the lattice vibrations so constructed.

Let us now consider how the one-phonon coherent cross-section derived in the preceding section is affected by taking into account the third- and higher-order interactions in the Hamiltonian of the translational motion of the nuclei, eqn (4.3).† (Macroscopically these higher interactions give rise to thermal expansion and a finite thermal conductivity.) The presence of these higher-order terms prevents the diagonalization of the Hamiltonian in terms of the phonon operators introduced in the discussion of the harmonic part of the Hamiltonian. Because the third- and higher-order derivatives of the potential energy are considerably smaller than the second it is still meaningful to use the concept of phonons, but they can no longer be regarded as forming a non-interacting Bose gas. The terms in the Hamiltonian involving three or more displacements represent interactions of the phonons and we must therefore associate with a given phonon a finite lifetime. Furthermore, in general we can expect the phonon interactions to lead to a shift in the dispersion relation from the non-interacting value $\omega_j(\mathbf{q})$. Thus, in the presence of phonon–phonon interactions the energy of a given phonon can be written

$$\omega_0 = \omega_j(\mathbf{q})+\Delta_j(\mathbf{q},\omega)+\mathrm{i}\Gamma_j(\mathbf{q},\omega), \tag{4.90}$$

where Δ represents the shift from the non-interacting phonon energy $\omega_j(\mathbf{q})$ and Γ the decay constant.

At temperatures $T \ll \hbar\omega_{\mathrm{m}}/k_{\mathrm{B}}$ both Δ and Γ are small because the magnitude of the displacements of the nuclei is small and hence the terms quadratic in the displacements give the dominant part of the Hamiltonian. Furthermore, we expect Δ and Γ to be small for small $\omega_j(\mathbf{q})$ because the density of states is small for $\omega_j(\mathbf{q}) \ll \omega_{\mathrm{m}}$ and thus the probability of the scattering of two phonons is itself also small.

The one-phonon coherent cross-section, eqn (4.71), can be rewritten

$$\left(\frac{\mathrm{d}^2\sigma}{\mathrm{d}\Omega\mathrm{d}E'}\right)^{\mathrm{inel}}_{\mathrm{coh}} = \frac{\sigma_{\mathrm{c}}}{8\pi M}\frac{k'}{k}\frac{(2\pi)^3}{v_0}\sum_{\boldsymbol{\tau}}\exp\{-2W(\boldsymbol{\varkappa})\}\{n(\omega)+1\}\times$$
$$\times\sum_{j,\mathbf{q}}\frac{|\boldsymbol{\varkappa}.\boldsymbol{\sigma}^j(\mathbf{q})|^2}{\omega_j(\mathbf{q})}[-\delta\{\omega+\omega_j(\mathbf{q})\}\delta(\boldsymbol{\varkappa}+\mathbf{q}-\boldsymbol{\tau})+$$
$$+\delta\{\omega-\omega_j(\mathbf{q})\}\,\delta(\boldsymbol{\varkappa}-\mathbf{q}-\boldsymbol{\tau})]. \tag{4.91}$$

† A detailed discussion has been given by A. A. Maradudin and A. E. Fein, *Phys. Rev.* **128**, 2589 (1962). A more direct calculation from Balcar, *Acta phys. austriaca* **31**, 300 (1970) is outlined in Appendix C.

In the presence of anharmonic forces we can still meaningfully seek the scattering from a single phonon of wave vector $\mathbf{q}$, so that the δ-functions expressing the conservation of momentum will remain unchanged. However, because the phonons now have a finite lifetime the δ-functions in energy must be modified, the process of creation or annihilation of a phonon now having an energy spread. We therefore write

$$\left(\frac{d^2\sigma}{d\Omega dE'}\right)^{\text{inel}}_{\text{coh}} = \frac{\sigma_c}{8\pi M}\frac{k'}{k}\frac{(2\pi)^3}{v_0}\sum_{\boldsymbol{\tau}}\exp\{-2W(\boldsymbol{\varkappa})\}\{n(\omega)+1\}\times$$

$$\times\sum_{j,\mathbf{q}}\frac{|\boldsymbol{\varkappa}.\boldsymbol{\sigma}^j(\mathbf{q})|^2}{\pi\omega_j(\mathbf{q})}\left[\frac{-\Gamma_j(\mathbf{q},\omega)}{\{\omega+\omega_j(\mathbf{q})+\Delta_j(\mathbf{q},\omega)\}^2+\Gamma_j^2(\mathbf{q},\omega)}\,\delta(\boldsymbol{\varkappa}+\mathbf{q}-\boldsymbol{\tau})+\right.$$

$$\left.+\frac{\Gamma_j(\mathbf{q},\omega)}{\{\omega-\omega_j(\mathbf{q})-\Delta_j(\mathbf{q},\omega)\}^2+\Gamma_j^2(\mathbf{q},\omega)}\,\delta(\boldsymbol{\varkappa}-\mathbf{q}-\boldsymbol{\tau})\right]. \quad (4.92)$$

In the limit of both Δ and Γ going to zero this expression is identical to (4.71). (The anharmonic forces will also modify the Debye–Waller factor from (4.38), but this is of secondary importance in discussing the modification of the cross-section.) From (4.92) the cross-section for $\omega = \pm\{\omega_j(\mathbf{q})+\Delta_j(\mathbf{q},\omega)\}$ now has a certain finite width associated with it, the magnitude of which is governed by $\Gamma_j(\mathbf{q},\omega)$. Near the zone boundary Γ is expected to be large so that here the width of the one-phonon cross-section peaks is also large. Also, the one-phonon peaks will broaden with temperature because of the increasing importance of the anharmonic terms in the Hamiltonian of the translation motion of the nuclei. For $T \sim \theta_D$ and $\mathbf{q}$ near the zone boundary the widths can be anticipated to be of the order 10^{-3} eV in most substances.

There are, of course, other mechanisms that lead to the scattering of phonons in a crystal and consequently increase widths in the one-phonon peaks in the cross-section. In particular we mention the electron–phonon† interaction, which is responsible for the electrical resistivity at high temperatures. In addition to contributing to the width of the phonon peaks the interaction also leads to the so-called Kohn effect.‡

This effect can be understood as follows. In a real metal the motions of the conduction electrons and ions are clearly not independent of each other; if the ions move about their equilibrium positions then the electrons are scattered by an array of vibrating potentials. Because typical

† S. K. Joshi and A. K. Rajagopal, *Solid St. Phys.* **22**, 159 (1968).
‡ W. Kohn, *Phys. Rev. Lett.* **2**, 393 (1959).

phonon frequencies are much less than electronic frequencies of importance the motion of the ions appears to the electrons as a static perturbation varying in space with a wave vector $\mathbf{q}$. As a consequence of this the Coulomb potential between the ions is screened by the electrons, the 'bare' potential being essentially multiplied by the susceptibility of the conduction electrons to the static perturbation. For a free electron gas this susceptibility possesses a singularity at $|\mathbf{q}|$ equal to twice the Fermi wave vector. It follows that the ion–ion potential, and therefore the phonon frequency, reflects this singularity; in fact as the screening abruptly decreases so the phonon frequencies increase. In real metals the effect on the phonon frequency depends both on the strength of the electron–phonon interaction and the sharpness and shape of the Fermi surface. However, in principle the Kohn effect affords a means of gaining information on the dimensions of the Fermi surface and has been observed in a number of metals. The most extensive study has been made for lead, which is a favourable example because of a high electron–phonon interaction.†

4.5. Multi-phonon cross-sections

There are two main reasons for wanting to know multi-phonon cross-sections. First, in many experiments it is useful to be able to estimate the 'background scattering' so that the more interesting parts of the scattering process can be isolated. Second, as the incident energy increases, more and more of the cross-section is dominated by multi-phonon terms; hence cross-sections at the high energies are dominated by multi-phonon effects.

However, any discussion of multi-phonon effects runs into an immediate difficulty. The conventional expansion of the cross-section into elastic terms, one-phonon terms, two-phonon terms, etc., is a rapidly convergent series for modest temperatures and low incident energies but is only poorly convergent for other conditions. In other words, whenever multi-phonon effects are important, the conventional theory gives only a poorly convergent series to describe them. This naturally gives rise to difficulties of calculation, which are considerably helped by two early suggestions of Placzek.

When a scattering process involves two or more phonons, the conservation of energy and crystal momentum are not restrictive, in sharp contrast to the conditions on the one-phonon cross-sections (see (4.71) and the following discussion). Hence the multi-phonon scattering tends to

† Cf. R. Stedman *et al.*, *Phys. Rev.* **163**, 567 (1967).

be a smoothly varying function of scattering angle and incident energy. In these circumstances the interference of the scattering from two different sites is not likely to be important. Hence we can treat all the scattering as though it were incoherent, i.e. in (4.68) ignore all the terms with $l' \neq l$. This is known as Placzek's *incoherent approximation.*

Placzek also noticed that if, in place of a phonon expansion, the cross-section was rearranged as an expansion in the ratio of neutron to nuclear mass, then a rapidly convergent series was obtained.†

Yet another idea, developed by Sjölander, is to rearrange the multi-phonon terms into a form that has the correct asymptotic behaviour at high incident energies.‡

We find it convenient to present in sequence:

(a) the conventional multi-phonon theory,

(b) Sjölander's suggestion of a Gaussian approximation for multi-phonon differential cross-sections,

(c) the Placzek mass expansion for the total scattering cross-section, and

(d) a brief discussion of corrections to the incoherent approximation.

We restrict ourselves throughout to cubic Bravais lattices.

(*a*) *Conventional multi-phonon expansion*

The starting-point for this is (4.69). Expanding the exponential gives the cross-section as a sum of terms involving one, two, three, etc., phonons. In general the pth term is

$$\left(\frac{\mathrm{d}^2\sigma_p}{\mathrm{d}\Omega\mathrm{d}E'}\right)^{\mathrm{inel}}_{\mathrm{incoh}} = \frac{\sigma_{\mathrm{i}}}{4\pi}\frac{k'}{k}\frac{N}{2\pi\hbar}\int_{-\infty}^{\infty}\mathrm{d}t\,\exp(-\mathrm{i}\omega t)\exp\{-2W(\boldsymbol{\kappa})\}\times$$
$$\times\frac{1}{p!}\langle\boldsymbol{\kappa}\,.\,\hat{\mathbf{u}}(l,0)\boldsymbol{\kappa}\,.\,\hat{\mathbf{u}}(l,t)\rangle^p, \quad (4.93)$$

which, by an argument analogous to that giving (4.79) and (4.80) and using a form similar to (4.41 a), gives

$$\left(\frac{\mathrm{d}^2\sigma_p}{\mathrm{d}\Omega\mathrm{d}E'}\right)^{\mathrm{inel}}_{\mathrm{incoh}} = \frac{\sigma_{\mathrm{i}}}{4\pi}\frac{k'}{k}\frac{N}{\hbar}\exp\{-2W(\boldsymbol{\kappa})\}\times$$
$$\times\frac{1}{p!}\left(\frac{\hbar\kappa^2}{2M}\right)^p\prod_{i=1}^{p}\left\{\int_{-\infty}^{\infty}\mathrm{d}\omega_i\,\frac{Z(\omega_i)}{\omega_i}\,n(\omega_i)\right\}\delta\Big\{\omega+\sum_i\omega_i\Big\}. \quad (4.94)$$

For p larger than unity this is not an easy expression to evaluate, nor is it particularly useful to do so because the sum over phonon partial

† G. Placzek, *Phys. Rev.* **86**, 377 (1952); ibid. **93**, 895 (1954); ibid. **105**, 1240 (1957).
‡ A. Sjölander, *Ark. Fys.* **14**, 315 (1958).

cross-sections converges only slowly with p. Placzek has given an instructive discussion of why this is so.

We know that for high incident energies it is a reasonable, though not exact, approximation to use the static approximation, which gives the total incoherent cross-section as σ_{i} in place of the exact result

$$\sigma_{\mathrm{i}}/(1+m/M)^2.$$

To obtain this static approximation we neglect the appearance of the terms $\sum_i \omega_i$ in the δ-function of (4.94) and hence obtain immediately

$$\left(\frac{\mathrm{d}^2\sigma_p}{\mathrm{d}\Omega\mathrm{d}E'}\right)^{\mathrm{inel}}_{\mathrm{incoh}} \simeq \frac{\sigma_{\mathrm{i}}}{4\pi} N\,\delta(\hbar\omega)\exp\{-2W(\boldsymbol{\varkappa})\}\frac{1}{p!}\{2W(\boldsymbol{\varkappa})\}^p. \tag{4.95}$$

Hence in this approximation we have

$$\frac{4\pi}{N\sigma_{\mathrm{i}}}\left(\frac{\mathrm{d}\sigma}{\mathrm{d}\Omega}\right)_{\mathrm{elas}} = \exp(-2W) = 1-2W+\tfrac{1}{2}(2W)^2-\frac{1}{3!}(2W)^3+\ldots,$$

$$\frac{4\pi}{N\sigma_{\mathrm{i}}}\left(\frac{\mathrm{d}\sigma}{\mathrm{d}\Omega}\right)_{p=1} = 2W\exp(-2W) = 2W-(2W)^2+\frac{1}{2!}(2W)^3-\ldots,$$

$$\frac{4\pi}{N\sigma_{\mathrm{i}}}\left(\frac{\mathrm{d}\sigma}{\mathrm{d}\Omega}\right)_{p=2} = \tfrac{1}{2}(2W)^2\exp(-2W) = \tfrac{1}{2}(2W)^2-\tfrac{1}{2}(2W)^3+\ldots,$$

and

$$\frac{4\pi}{N\sigma_{\mathrm{i}}}\sum_{p=0}^{\infty}\left(\frac{\mathrm{d}\sigma}{\mathrm{d}\Omega}\right)_p = 1.$$

This simple argument shows that, certainly for high energies, the phonon series is heavily compensating, and one rough method of summing the series from the nth term upwards is simply to replace the Debye–Waller factor by unity in the nth term. However, we use this static approximation only for illustrative purposes and turn immediately to more sophisticated discussions.

(b) *The Gaussian approximation*

We return to (4.70) and for ease of notation define for a cubic crystal

$$\langle\boldsymbol{\varkappa}\,.\,\hat{\mathbf{u}}(l,0)\boldsymbol{\varkappa}\,.\,\hat{\mathbf{u}}(l,t)\rangle = \frac{\hbar\kappa^2}{2M}\gamma(t), \tag{4.96}$$

where

$$\gamma(t) = \int_{-\infty}^{\infty}\mathrm{d}\omega\,\frac{Z(\omega)}{\omega}\,n(\omega)\exp(-\mathrm{i}\omega t). \tag{4.97}$$

Hence $\gamma(0)$ is directly related to the Debye–Waller factor by

$$2W(\kappa) = \frac{\hbar\kappa^2}{2M}\gamma(0). \tag{4.98}$$

For small times we can expand

$$\gamma(t) = \gamma(0) - \mathrm{i}t - \frac{2t^2}{3\hbar}\mathscr{E} + \ldots, \tag{4.99}$$

where

$$\mathscr{E} = \tfrac{3}{2}\int_0^\infty \mathrm{d}\omega\; Z(\omega)\hbar\omega\{n(\omega)+\tfrac{1}{2}\} \tag{4.100}$$

is the mean kinetic energy per atom and approaches $3k_\mathrm{B}T/2$ at high temperatures.

If we used (4.96) and (4.99) as an approximation to be used for all p in (4.93) we could, of course, sum the series over p once again and recover the formula

$$\left(\frac{\mathrm{d}^2\sigma}{\mathrm{d}\Omega\mathrm{d}E'}\right)_{\mathrm{incoh}} = \frac{N\sigma_\mathrm{i}}{4\pi}\frac{k'}{k}\frac{1}{2\pi\hbar}\int_{-\infty}^{\infty} \mathrm{d}t\,\exp(-\mathrm{i}\omega t)\exp\left[\frac{\hbar\kappa^2}{2M}\left\{-\mathrm{i}t - \frac{2t^2\mathscr{E}}{3\hbar}\right\}\right], \tag{4.101}$$

which was first given in Chapter 3. However, this would give a poor representation of the elastic and one-phonon terms particularly. We therefore treat these two exactly and use the approximation (4.99) for $p \geqslant 2$ only. We therefore need to consider

$$\sum_{p=2}^{\infty}\frac{1}{p!}\left(\frac{\hbar\kappa^2}{2M}\right)^p\left\{\gamma(0) - \mathrm{i}t - \frac{2t^2\mathscr{E}}{3\hbar} + \ldots\right\}^p = \sum_{p=2}^{\infty}\frac{1}{p!}\left(\frac{\hbar\kappa^2\gamma(0)}{2M}\right)^p\left\{1 - \frac{\mathrm{i}t}{\gamma(0)} - \frac{2t^2\mathscr{E}}{3\hbar\gamma(0)} + \ldots\right\}^p. \tag{4.102 a}$$

We now write

$$\mathrm{e}^x = 1 - \frac{\mathrm{i}t}{\gamma(0)} - \frac{2t^2\mathscr{E}}{3\hbar\gamma(0)} + \ldots$$

and solve for x to leading order in t. Then (4.102 a) becomes

$$\sum_{p=2}^{\infty}\frac{1}{p!}\left(\frac{\hbar\kappa^2\gamma(0)}{2M}\right)^p \mathrm{e}^{xp} = \sum_{p=2}^{\infty}\frac{1}{p!}\left(\frac{\hbar\kappa^2\gamma(0)}{2M}\right)^p \exp\left[p\left\{-\frac{\mathrm{i}t}{\gamma(0)} - \tfrac{1}{2}\Delta^2t^2 + \ldots\right\}\right], \tag{4.102 b}$$

where

$$\Delta^2 = \frac{4\mathscr{E}}{3\hbar\gamma(0)} - \frac{1}{\gamma^2(0)}. \tag{4.103}$$

Using (4.102 b) in (4.93) and performing the integral over t we get as the multi-phonon cross-section

$$\frac{\sigma_\mathrm{i}}{4\pi}\frac{k'}{k}\frac{N}{\hbar}\exp(-2W)\sum_{p=2}^{\infty}\frac{1}{\Delta}\frac{1}{p!}\frac{(2W)^p}{(2\pi p)^{\frac{1}{2}}}\exp\left(-\frac{\omega^2}{2\Delta^2 p} - \frac{p}{2\Delta^2\gamma^2(0)} - \frac{\omega}{\gamma(0)\Delta^2}\right). \tag{4.104}$$

Adding this contribution to the elastic ($p = 0$) and one-phonon ($p = 1$) terms of (4.93) gives the final result

$$\left(\frac{d^2\sigma}{d\Omega dE'}\right)_{\text{incoh}} = \frac{N\sigma_i}{4\pi}\frac{k'}{k}\exp\{-2W(\boldsymbol{\kappa})\}\times$$

$$\times\left[\delta(\hbar\omega)+\frac{\hbar^2\kappa^2}{2M}\frac{Z(\omega)}{\omega}\{1+n(\omega)\}+\frac{1}{\hbar\Delta}\exp\{-\omega/\Delta^2\gamma(0)\}F(x,y)\right], \quad (4.105)$$

where
$$x = \omega/\Delta, \quad (4.106)$$

$$y = 2W\exp\left\{\frac{-1}{2\Delta^2\gamma^2(0)}\right\}, \quad (4.107)$$

and
$$F(x,y) = \sum_{p=2}^{\infty}\frac{1}{p!(2\pi p)^{\frac{1}{2}}}y^p\exp\left(-\frac{x^2}{2p}\right). \quad (4.108)$$

This function is given in Table 4.1.

In this approximation each multi-phonon term in (4.104) is represented as a Gaussian function of ω with mean values and widths depending upon p. The final result (4.105) has the following merits.

(a) It has treated the elastic and one-phonon terms exactly and therefore gives good results for low incident energy.

(b) Although the multi-phonon terms have been approximated, the method has treated accurately the correlation terms linear and quadratic in t. Hence automatically the formula gives the correct asymptotic limit at high energies.

(c) Because the linear and quadratic terms in t are treated exactly, the mean energy transfer and mean square energy transfer are correctly given by the formula at all incident energies (see § 3.3).

The formula (4.105) has been quoted as for the incoherent cross-section. However, using the 'incoherent approximation' we get the equivalent coherent cross-section merely by replacing σ_i by σ_c in the multi-phonon part of the expression. The 'incoherent approximation' would be a poor representation of the elastic and one-phonon coherent terms, and therefore it is particularly important that they be treated exactly when the incident energy is not high.

It is worth noting the sensitivity of the three terms in (4.105) to the spectrum $Z(\omega)$. The second term, representing the one-phonon cross-section, is directly proportional to $Z(\omega)$ and therefore is dependent upon the precise shape of $Z(\omega)$. The first term, however, is just elastic scattering and, as we saw in § 4.2, depends only on the single parameter $\overline{\omega^{-2}}$. The third term, describing the multi-phonon cross-section, depends upon two parameters: $\gamma(0)$, which in turn involves only $\overline{\omega^{-2}}$, and $\mathscr{E}$, the mean kinetic energy of the target nuclei, which becomes independent of $Z(\omega)$ at high enough temperatures.

(c) *Mass expansion*

We now consider a procedure for obtaining an approximation to the total cross-section containing all multi-phonon terms. As in our discussion of the phonon expansion above, we shall concern ourselves only with crystals with a Bravais lattice structure and employ the incoherent approximation.

Consider the inelastic cross-section (4.69). Using the notation of (4.96) and (4.98) and denoting $\hbar^2\kappa^2/2M$ by E_R, this contains

$$\frac{1}{2\pi\hbar}\int_{-\infty}^{\infty} dt\, e^{-i\omega t}[\exp\{E_R(\gamma(t)-\gamma(0))/\hbar\}-\exp\{-E_R\gamma(0)/\hbar\}]$$

$$=\frac{1}{2\pi\hbar}\int_{-\infty}^{\infty} dt\, e^{-i\omega t}\exp\{-E_R\gamma(0)/\hbar\}\sum_{l=1}^{\infty}\frac{1}{l!}\left(\frac{E_R}{\hbar}\right)^l\gamma^l(t). \quad (4.109)$$

Now in this expression we write

$$\exp\{-E_R\gamma(0)/\hbar\} = \sum_{n=l}^{\infty}\frac{1}{(n-l)!}\left(\frac{E_R}{\hbar}\right)^{n-l}\{-\gamma(0)\}^{n-l}$$

and rearrange the resultant expression into an expansion in $E_R/\hbar$. It is simple to show that (4.109) can then be written

$$\sum_{n=1}^{\infty}(E_R/\hbar)^n\sum_{l=1}^{n}\frac{\{-\gamma(0)\}^{n-l}}{l!(n-l)!}\frac{1}{2\pi\hbar}\int_{-\infty}^{\infty} dt\, e^{-i\omega t}\gamma^l(t). \quad (4.110)$$

This is an expansion $1/M$ and, as pointed out by Placzek, it is a very rapidly convergent sum. In fact we shall find that only the third power of $1/M$ is required to obtain a good approximation to the total cross-section.

Now, from (4.97),

$$\frac{1}{2\pi}\int_{-\infty}^{\infty} dt\, e^{-i\omega t}\gamma^l(t) = \left\{\prod_{i=1}^{l}\int_{-\infty}^{\infty} d\omega_i\,\frac{Z(\omega_i)}{\omega_i}\,n(\omega_i)\right\}\delta\left(\omega+\sum_{i=1}^{l}\omega_i\right). \quad (4.111)$$

Hence (4.69) becomes

$$\left(\frac{\mathrm{d}^2\sigma}{\mathrm{d}\Omega\mathrm{d}E'}\right)^{\mathrm{inel}}_{\mathrm{incoh}} = \frac{N\sigma_{\mathrm{i}}}{4\pi}\frac{k'}{\hbar k}\sum_{n=1}^{\infty}\left(\frac{E_{\mathrm{R}}}{\hbar}\right)^n \sum_{l=1}^{n}\frac{\{-\gamma(0)\}^{n-l}}{l!(n-l)!}\times$$
$$\times\left\{\prod_{i=1}^{l}\int_{-\infty}^{\infty}\mathrm{d}\omega_i\,\frac{Z(\omega_i)}{\omega_i}\,n(\omega_i)\right\}\delta\left(\omega+\sum_{i=1}^{l}\omega_i\right). \qquad (4.112)$$

Integrating over E' gives

$$\left(\frac{\mathrm{d}\sigma}{\mathrm{d}\Omega}\right)^{\mathrm{inel}}_{\mathrm{incoh}} = \frac{N\sigma_{\mathrm{i}}}{4\pi}\frac{k'}{k}\sum_{n=1}^{\infty}\left(\frac{E_{\mathrm{R}}}{\hbar}\right)^n \sum_{l=1}^{n}\frac{\{-\gamma(0)\}^{n-l}}{l!(n-l)!}\left\{\prod_{i=1}^{l}\int_{-\infty}^{\infty}\mathrm{d}\omega_i\,\frac{Z(\omega_i)}{\omega_i}\,n(\omega_i)\right\}, \qquad (4.113)$$

where it is now understood that

$$\sum_{i=1}^{l}\omega_i > -\frac{\hbar^2k^2}{2m}, \qquad k' = \left\{k^2+\frac{2m}{\hbar}\sum_{i=1}^{l}\omega_i\right\}^{\frac{1}{2}}, \qquad (4.114)$$

and κ^2, which appears in E_{R}, is

$$\kappa^2 = k^2+k'^2-2kk'\cos\theta, \qquad (4.115)$$

where θ is the scattering angle.

Integrating again,

$$\sigma_{\mathrm{i}}^{\mathrm{inel}} = \frac{N\sigma_{\mathrm{i}}}{4k^2}\sum_{n=1}^{\infty}\left(\frac{\hbar}{2M}\right)^n\frac{1}{(n+1)}\sum_{l=1}^{n}\frac{\{-\gamma(0)\}^{n-l}}{l!(n-l)!}\times$$
$$\times\prod_{i=1}^{l}\int_{-\infty}^{\infty}\mathrm{d}\omega_i\,\frac{Z(\omega_i)}{\omega_i}\,n(\omega_i)\{(k+k')^{2n+2}-(k-k')^{2n+2}\}. \qquad (4.116)$$

The equation (4.112) is not very convenient for calculation, but (4.116) is easily computed, using, for example, the Debye model to give $Z(\omega)$.

Using $\quad \theta = \dfrac{T}{\theta_{\mathrm{D}}}, \qquad x = \sqrt{\left(\dfrac{\hbar^2k^2}{2mk_{\mathrm{B}}\theta_{\mathrm{D}}}\right)}, \qquad \xi_i = \dfrac{\hbar\omega_i}{k_{\mathrm{B}}\theta_{\mathrm{D}}}, \qquad (4.117)$

and, finally, $\quad \phi(\theta) = 6\left\{\dfrac{1}{4}+\displaystyle\int_0^1\dfrac{\mathrm{d}u\,u}{\mathrm{e}^{u/\theta}-1}\right\} = 3\displaystyle\int_{-1}^{1}\dfrac{\mathrm{d}u\,u}{\mathrm{e}^{u/\theta}-1}, \qquad (4.118)$

we obtain

$$\sigma_{\mathrm{i}}^{\mathrm{inel}} = \frac{N\sigma_{\mathrm{i}}}{4x^2}\sum_{n=1}^{\infty}\left(\frac{m}{M}\right)^n\frac{1}{(n+1)}\sum_{l=1}^{n}\frac{3^l\{-\phi(\theta)\}^{n-l}}{l!(n-l)!}\times$$
$$\times\prod_{i=1}^{l}\int_{-1}^{1}\frac{\mathrm{d}\xi_i\,\xi_i}{\mathrm{e}^{\xi_i/\theta}-1}\{(x+x')^{2n+2}-(x-x')^{2n+2}\}, \qquad (4.119)$$

where $$x' = \left\{x^2+\sum_{i=1}^{l}\xi_i\right\}^{\frac{1}{2}} \quad \text{for } \sum_{i=1}^{l}\xi_i > -x^2$$
$$= 0 \qquad\qquad \text{for } \sum_{i=1}^{l}\xi_i < -x^2. \tag{4.120}$$

In this notation the elastic incoherent cross-section is

$$\sigma_{\text{i}}^{\text{el}} = N\sigma_{\text{i}} \sum_{n=0}^{\infty} \left(\frac{m}{M}\right)^n \frac{1}{(n+1)!} \{-4x^2\phi(\theta)\}^n. \tag{4.121}$$

Thus if we add this to the total inelastic incoherent cross-section given by (4.119), we obtain an expression for the total incoherent cross-section.

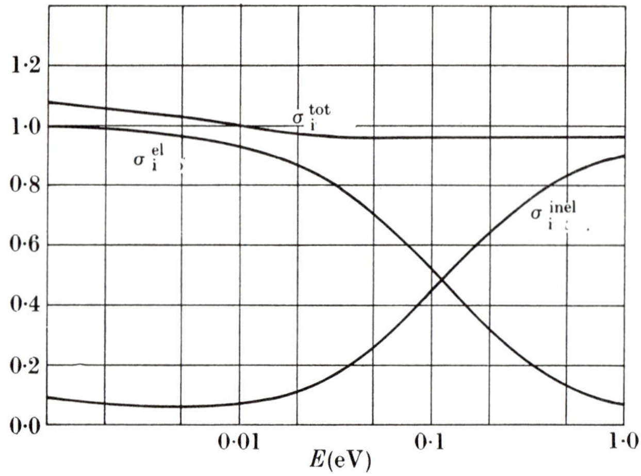

FIG. 4.6. Typical curves for cross-section as a function of energy.

It has been shown that it is adequate to truncate the expansion in n to include up to the term in $(m/M)^3$. The total incoherent cross-section is then

$$\sigma_i^{\text{tot}}(M, E, T, \theta_{\text{D}})$$
$$= N\sigma_{\text{i}}\{1+(m/M)A_1(x,\theta)+(m/M)^2A_2(x,\theta)+(m/M)^3A_3(x,\theta)\} \tag{4.122}$$

and the coefficients A_1, A_2, and A_3 are given for a wide range of x and θ in Table 4.2.

Typical curves of total incoherent cross-sections as functions of energy are shown in Fig. 4.6. It should be noted that the elastic scattering is appreciable for $E \lesssim 0{\cdot}5$ eV and is larger than the inelastic scattering for $E \lesssim 0{\cdot}1$ eV. However, the total incoherent cross-section varies only slowly with energy and this, of course, is one reason why a simple expression (4.122) can be obtained for it.

In summary, (4.119) gives the total inelastic incoherent cross-section for a Debye spectrum of lattice frequencies and, in the incoherent approximation, the same formula is used for the total inelastic coherent cross-section. Of course, in the first case we use σ_i in (4.119) and in the second σ_c. Eqn (4.122) is the total incoherent cross-section up to terms of order $(m/M)^3$.

We have shown that the total elastic coherent cross-section for a powder sample, eqn (4.47), has exactly the same form as the total elastic incoherent cross-section in the limit of high energies. Thus, in this limit we can obtain the total coherent cross-section from (4.122) merely by replacing σ_i by σ_c. However, this approximation is clearly a poor one at low energies where the total elastic coherent cross-section has much structure. In general it is given by

$$\sigma_c^{el} = \frac{N\sigma_c \pi^2\hbar^2}{2mEv_0} \sum_{\tau\neq 0}^{\tau<2k} \{z(\tau)/\tau\}\exp\left\{-\frac{\hbar\gamma(0)\tau^2}{2M}\right\}, \tag{4.123}$$

TABLE 4.3

$z(\tau)$, the reciprocal lattice weighting factor

Face-centred cubic crystals				*Body-centred cubic crystals*			
s	$z\left(\frac{2\pi}{a}\sqrt{s}\right)$	s	$z\left(\frac{2\pi}{a}\sqrt{s}\right)$	s	$z\left(\frac{2\pi}{a}\sqrt{2s}\right)$	s	$z\left(\frac{2\pi}{a}\sqrt{2s}\right)$
0	1	26	24	0	1	75	56
1	12	27	96	3	8	76	24
2	6	28	48	4	6	80	48
3	24	29	24	8	12	83	72
4	12	30	0	11	24	84	48
5	24	31	96	12	8	91	48
6	8	32	6	16	6	96	24
7	48	33	96	19	24	99	72
8	6	34	48	20	24	100	30
9	36	35	48	24	24	104	72
10	24	36	36	27	32	107	72
11	24	37	120	32	12	108	32
12	24	38	24	35	48	115	48
13	72	39	48	36	30	116	72
14	0	40	24	40	24	120	48
15	48	41	48	43	24	123	48
16	12	42	48	44	24	128	12
17	48	43	120	48	8	131	120
18	30	44	24	51	48	132	48
19	72	45	120	52	24	136	48
20	24	46	0	56	48		
21	48	47	96	59	72		
22	24	48	24	64	6		
23	48	49	108	67	24		
24	8	50	30	68	48		
25	84			72	36		

where $z(\tau)$ is the multiplicity of each τ and is listed in Table 4.3 for body-centred and face-centred lattice structures. In terms of a Debye spectrum (4.123) becomes

$$\sigma_c^{el} = \frac{N\sigma_c \pi^2\hbar^2}{2mEv_0} \sum_{\tau\neq 0}^{\tau<2k} \{z(\tau)/\tau\}\exp\left\{-\frac{3\hbar^2\beta\phi_1(\theta)\tau^2}{2M}\right\}, \tag{4.124}$$

where
$$\phi_1(\theta) = \int_{-1}^{1} \frac{u\,du}{(e^{u/\theta}-1)} = \tfrac{1}{3}\phi(\theta).$$

The function $\phi_1(\theta)$ is listed in Table 4.4.

TABLE 4.4

The function $\phi_1(\theta)$

θ	$\phi_1(\theta)$
0·05	0·508225
0·10	0·532889
0·15	0·573583
0·2	0·628350
0·3	0·767813
0·4	0·931032
0·5	1·106947
0·6	1·290136
0·7	1·477799
0·8	1·668387
0·9	1·860982
1·0	2·055009
1·1	2·250093
1·2	2·445978
1·3	2·642484
1·4	2·839481
1·5	3·036873
1·6	3·234587
1·8	3·630769
2·0	4·027708
2·2	4·425200
2·5	5·022186
3·0	6·018497
3·5	7·015859
4·0	8·013880
5·0	10·011106

(d) Corrections to incoherent approximation

At the beginning of this discussion of multi-phonon cross-sections we pointed out that it was not unreasonable to discuss all multi-phonon scattering as incoherent, i.e. for all phonon processes involving two or more phonons we calculate the incoherent cross-section and then simply replace $\sigma_{\rm i}$ by $\sigma_{\rm i}+\sigma_{\rm c}$. This is a poor approximation for the elastic coherent scattering, which we have therefore treated exactly. We now give a brief discussion of the validity of this incoherent approximation, although a detailed examination of the point would be outside the scope of this book.

The starting point is now (4.68), which, expanding in phonon partial cross-sections, becomes

$$\left(\frac{\mathrm{d}^2\sigma}{\mathrm{d}\Omega\mathrm{d}E'}\right)^{\rm inel}_{\rm coh}=\frac{\sigma_{\rm c}}{4\pi}\frac{k'}{k}\frac{N}{2\pi\hbar}\int\limits_{-\infty}^{\infty}\mathrm{d}t\,\exp(-\mathrm{i}\omega t)\exp\{-2W(\boldsymbol{\kappa})\}\times$$
$$\times\sum_{l}\exp(\mathrm{i}\boldsymbol{\kappa}.\mathbf{l})\sum_{p=1}^{\infty}\frac{1}{p!}\left[\frac{\hbar}{2NM}\sum_{j,\mathbf{q}}\frac{|\boldsymbol{\kappa}.\boldsymbol{\sigma}^{j}(\mathbf{q})|^2}{\omega_j(\mathbf{q})}\times\right.$$
$$\left.\times[\exp\{-\mathrm{i}\mathbf{q}.\mathbf{l}+\mathrm{i}t\omega_j(\mathbf{q})\}\{1+n_j(\mathbf{q})\}+\exp\{\mathrm{i}\mathbf{q}.\mathbf{l}-\mathrm{i}t\omega_j(\mathbf{q})\}n_j(\mathbf{q})]\right]^p. \tag{4.125}$$

Because we expect this term to be small we evaluate it using a simple Debye model. In this model, because $\omega_j(\mathbf{q})$ is assumed independent of j, the sum over j can be done immediately to give

$$\sum_{j}|\boldsymbol{\kappa}.\boldsymbol{\sigma}^{j}(\mathbf{q})|^2=\kappa^2.$$

Also $\omega_j(\mathbf{q})$ is assumed independent of the direction of $\mathbf{q}$, and hence the factors $\exp(-\mathrm{i}\mathbf{q}.\mathbf{l})$ can be replaced by their angular averages

$$\frac{\sin(ql)}{ql}.$$

Finally, if we are looking at a polycrystal we can average over the orientation of $\boldsymbol{\kappa}$ and so replace $\exp(\mathrm{i}\boldsymbol{\kappa}.\mathbf{l})$ by

$$\frac{\sin(\kappa l)}{\kappa l}.$$

Hence (4.126) becomes

$$\left(\frac{\mathrm{d}^2\sigma}{\mathrm{d}\Omega\mathrm{d}E'}\right)^{\rm inel}_{\rm coh}=\frac{\sigma_{\rm c}}{4\pi}\frac{k'}{k}\frac{N}{2\pi\hbar}\int\limits_{-\infty}^{\infty}\mathrm{d}t\,\exp(-\mathrm{i}\omega t)\exp\{-2W(\boldsymbol{\kappa})\}\times$$
$$\times\sum_{l}\frac{\sin(\kappa l)}{\kappa l}\sum_{p=1}^{\infty}\frac{1}{p!}\left(\frac{\hbar\kappa^2}{2M}\right)^p\times$$
$$\times\prod_{i=1}^{p}\frac{1}{N}\sum_{q_i}\frac{1}{\omega(q_i)}\frac{\sin(q_i l)}{q_i l}[\exp\{\mathrm{i}t\,\omega(q_i)\}\{1+n(q_i)\}+\exp\{-\mathrm{i}t\,\omega(q_i)\}n(q_i)]. \tag{4.126}$$

It is not hard to show that the term $\mathbf{l} = 0$ in (4.126) corresponds exactly to (4.94) with σ_c replacing σ_i. The incoherent approximation therefore involves neglecting all terms in (4.126) except the term with $\mathbf{l} = 0$. The size of these terms has been estimated by Marshall and Stuart (University of California Radiation Laboratory, UCRL 5568), who find that typically they amount to only a few per cent of the total cross-sections.

5

SCATTERING BY MAGNETIC INTERACTIONS

WHEREAS the preceding four chapters have dealt with purely nuclear scattering, this and the four chapters immediately following it are concerned with the scattering of neutrons through their magnetic interaction with unpaired electrons. The present chapter is specifically concerned, first, with deriving the cross-section in a general form, and then, representing it in forms appropriate for the discussion of the two cases (a) when the wave functions of the unpaired electrons are localized about the ions in the crystal and (b) when their itinerant behaviour has to be allowed for. The problems arising from the first of these two cases will only be partially resolved in this chapter, the full answers being given in Chapter 6.

5.1. Cross-section for scattering by unpaired electrons

The Hamiltonian of a single neutron (magnetic moment μ) in a magnetic field $\mathbf{H}$ and electric field $\mathbf{E}$ is, to order $1/m^2$,†

$$\mathscr{H} = \frac{1}{2M}\hat{\mathbf{p}}^2 - \hat{\boldsymbol{\mu}}.\mathbf{H} - \frac{1}{mc}\hat{\boldsymbol{\mu}}.(\mathbf{E}\times\hat{\mathbf{p}}) - \frac{\hbar\mu}{2mc}\boldsymbol{\nabla}.\mathbf{E}. \tag{5.1}$$

The first term in (5.1) is simply the kinetic energy of the neutron, the second is the energy of a dipole moment $\boldsymbol{\mu}$ in a magnetic field $\mathbf{H}$ and the remaining two terms, both of order $1/m^2$, are important when we come to consider scattering of polarized neutrons. For the present we consider just the magnetic part of the interaction, the magnetic field arising from unpaired electrons in a crystal.

It is more convenient to write this term in the form

$$-\gamma\mu_{\mathrm{N}}\hat{\boldsymbol{\sigma}}.\mathbf{H}, \tag{5.2}$$

where $\hat{\sigma}^{\alpha}$ are Pauli matrices and $\gamma = -1{\cdot}91$. The magnetic field due to a single electron moving with velocity $\mathbf{v}_{\mathrm{e}}$ is‡

$$\mathbf{H} = \mathrm{curl}\left\{\frac{\boldsymbol{\mu}_{\mathrm{e}}\times\mathbf{R}}{|\mathbf{R}|^3}\right\} + \frac{(-e)}{c}\frac{\mathbf{v}_{\mathrm{e}}\times\mathbf{R}}{|\mathbf{R}|^3}, \tag{5.3}$$

† L. L. Foldy, *Rev. mod. Phys.* **30**, 471 (1958). Notice that the electric fields appear half their strength for a charged particle.

‡ L. D. Landau and E. M. Lifshitz, *Electrodynamics of continuous media*, Pergamon Press.

where $\mathbf{R}$ is the distance from the electron to the point at which the field is measured. The magnetic moment operator of an electron $\boldsymbol{\mu}_e$ is

$$\hat{\boldsymbol{\mu}}_e = -2\mu_B\,\hat{\mathbf{s}}, \tag{5.4}$$

so that the interaction potential between the neutron and the electron becomes

$$-\gamma\mu_N\,\hat{\boldsymbol{\sigma}}.\mathbf{H} = \gamma\mu_N\left\{2\mu_B\,\hat{\boldsymbol{\sigma}}.\,\mathrm{curl}\left(\frac{\hat{\mathbf{s}}\times\mathbf{R}}{|\mathbf{R}|^3}\right)-\frac{e}{2m_e c}\left(\hat{\mathbf{p}}_e.\frac{\hat{\boldsymbol{\sigma}}\times\mathbf{R}}{|\mathbf{R}|^3}+\frac{\hat{\boldsymbol{\sigma}}\times\mathbf{R}}{|\mathbf{R}|^3}.\hat{\mathbf{p}}_e\right)\right\}. \tag{5.5}$$

Note that the second term has been written as one half of the sum of $\hat{\mathbf{p}}_e.(\hat{\boldsymbol{\sigma}}\times\mathbf{R}/|\mathbf{R}|^3)$ and $(\hat{\boldsymbol{\sigma}}\times\mathbf{R}/|\mathbf{R}|^3).\hat{\mathbf{p}}_e$ to take proper account of the operator nature of the momentum of the electron.

The first term on the right-hand side of (5.5) is classically the dipole–dipole interaction between the neutron and the electron while the second term is the interaction of the neutron with the electron that arises from the translational motion of the latter.

From our master formula for the neutron cross-section, eqn (1.10), the cross-section for the scattering of neutrons through the interaction potential given in eqn (5.5) is

$$\frac{d^2\sigma}{d\Omega dE'} = \left(\frac{m}{2\pi\hbar^2}\right)^2(2\gamma\mu_N\mu_B)^2\sum_{\lambda\lambda'\sigma}\frac{k'}{k}p_\lambda p_\sigma\left|\left\langle \mathbf{k}'\lambda'\sigma'\right|\sum_i\hat{\boldsymbol{\sigma}}.\,\mathrm{curl}\left(\frac{\hat{\mathbf{s}}_i\times\mathbf{R}}{|\mathbf{R}|^3}\right)-\right.$$
$$\left.-\frac{1}{2\hbar}\left(\hat{\mathbf{p}}_i.\frac{\hat{\boldsymbol{\sigma}}\times\mathbf{R}}{|\mathbf{R}|^3}+\frac{\hat{\boldsymbol{\sigma}}\times\mathbf{R}}{|\mathbf{R}|^3}.\hat{\mathbf{p}}_i\right)\left|\mathbf{k}\lambda\sigma\right\rangle\right|^2\delta(\hbar\omega+E_\lambda-E_{\lambda'}). \tag{5.6}$$

In (5.6) $|\lambda\rangle$ and $|\lambda'\rangle$ are the initial and final states of the target and E_λ and $E_{\lambda'}$ the corresponding energies.

The integration over the neutron coordinates in the matrix element of the interaction potential required in eqn (5.6) is quite straightforward if we make use of the well-known identities

$$\mathbf{R}/|\mathbf{R}|^3 = -\boldsymbol{\nabla}(1/|\mathbf{R}|) \tag{5.7}$$

and

$$\frac{1}{|\mathbf{R}|} = \frac{1}{2\pi^2}\int d\mathbf{q}\,\frac{1}{q^2}\exp(i\mathbf{q}.\mathbf{R}).$$

Thus, for example,

$$\mathrm{curl}(\hat{\mathbf{s}}\times\mathbf{R}/|\mathbf{R}|^3) = -\mathrm{curl}\left(\hat{\mathbf{s}}\times\boldsymbol{\nabla}\left(\frac{1}{|\mathbf{R}|}\right)\right)$$
$$= -\boldsymbol{\nabla}\times(\hat{\mathbf{s}}\times\boldsymbol{\nabla})\left(\frac{1}{|\mathbf{R}|}\right) = -\frac{1}{2\pi^2}\int d\mathbf{q}\,\frac{1}{q^2}\{\boldsymbol{\nabla}\times\hat{\mathbf{s}}\times\boldsymbol{\nabla})\}\exp(i\mathbf{q}.\mathbf{R})$$
$$= \frac{1}{2\pi^2}\int d\mathbf{q}\frac{1}{q^2}\{\mathbf{q}\times(\hat{\mathbf{s}}\times\mathbf{q})\}\exp(i\mathbf{q}.\mathbf{R}).$$

We find

$$\left\langle \mathbf{k}' \left| \hat{\boldsymbol{\sigma}}.\mathrm{curl}\left(\frac{\hat{\mathbf{s}}_i \times \mathbf{R}}{|\mathbf{R}|^3}\right) \right| \mathbf{k} \right\rangle = 4\pi \exp(\mathrm{i}\boldsymbol{\varkappa}.\mathbf{r}_i)\hat{\boldsymbol{\sigma}}.\{\tilde{\boldsymbol{\varkappa}} \times (\hat{\mathbf{s}}_i \times \tilde{\boldsymbol{\varkappa}})\} \quad (5.8\,\mathrm{a})$$

and

$$\left\langle \mathbf{k}' \left| \hat{\mathbf{p}}_i.\frac{\hat{\boldsymbol{\sigma}} \times \mathbf{R}}{|\mathbf{R}|^3} \right| \mathbf{k} \right\rangle = -\frac{4\pi\mathrm{i}}{|\boldsymbol{\varkappa}|} \exp(\mathrm{i}\boldsymbol{\varkappa}.\mathbf{r}_i)\hat{\boldsymbol{\sigma}}.(\tilde{\boldsymbol{\varkappa}} \times \hat{\mathbf{p}}_i), \quad (5.8\,\mathrm{b})$$

where $\mathbf{r}_i$ is the position vector of the ith electron and $\tilde{\boldsymbol{\varkappa}} = \boldsymbol{\varkappa}/|\boldsymbol{\varkappa}|$. The results (5.8 a), (5.8 b) together with the fact that $\tilde{\boldsymbol{\varkappa}} \times \hat{\mathbf{p}}_i$ commutes with $\exp(\mathrm{i}\boldsymbol{\varkappa}.\mathbf{r}_i)$ (this is trivially proved with the aid of the commutation relation (3.37)) enable us to write the neutron cross-section in the form

$$\frac{\mathrm{d}^2\sigma}{\mathrm{d}\Omega\mathrm{d}E'} = \left(\frac{m}{2\pi\hbar^2}\right)^2 (2\gamma\mu_\mathrm{N}\,\mu_\mathrm{B})^2(4\pi)^2 \sum_{\lambda\lambda'\sigma\sigma'} p_\lambda p_\sigma \frac{k'}{k} \times$$

$$\times \langle\lambda\sigma|\,(\hat{\boldsymbol{\sigma}}.\hat{\mathbf{Q}}_\perp)^+\,|\lambda'\sigma'\rangle\langle\lambda'\sigma'|\,\hat{\boldsymbol{\sigma}}.\hat{\mathbf{Q}}_\perp\,|\lambda\sigma\rangle\,\delta(\hbar\omega + E_\lambda - E_{\lambda'}), \quad (5.9)$$

where the operator $\hat{\mathbf{Q}}_\perp$ is defined by

$$\hat{\mathbf{Q}}_\perp = \sum_i \exp(\mathrm{i}\boldsymbol{\varkappa}.\mathbf{r}_i)\left\{\tilde{\boldsymbol{\varkappa}} \times (\hat{\mathbf{s}}_i \times \tilde{\boldsymbol{\varkappa}}) - \frac{\mathrm{i}}{\hbar|\boldsymbol{\varkappa}|}\,\tilde{\boldsymbol{\varkappa}} \times \hat{\mathbf{p}}_i\right\}. \quad (5.10)$$

For unpolarized neutrons

$$\sum_\sigma p_\sigma \langle\sigma|\,\hat{\sigma}_\alpha \hat{\sigma}_\beta\,|\sigma\rangle - \delta_{\alpha,\beta}.$$

Also, $\dfrac{m}{2\pi\hbar^2}.2\gamma\mu_\mathrm{N}\,\mu_\mathrm{B}.4\pi = \dfrac{m}{2\pi\hbar^2}.2\gamma\,\dfrac{e\hbar}{2m_\mathrm{p}c}.\dfrac{e\hbar}{2m_\mathrm{e}c}.4\pi = \gamma e^2/m_\mathrm{e}c^2.$

Thus we have, finally,

$$\frac{\mathrm{d}^2\sigma}{\mathrm{d}\Omega\mathrm{d}E'} = \left(\frac{\gamma e^2}{m_\mathrm{e}c^2}\right)^2 \frac{k'}{k} \sum_{\lambda\lambda'} p_\lambda \langle\lambda|\,\hat{\mathbf{Q}}_\perp^+\,|\lambda'\rangle.\langle\lambda'|\,\hat{\mathbf{Q}}_\perp\,|\lambda\rangle\,\delta(\hbar\omega + E_\lambda - E_{\lambda'}) \quad (5.11\,\mathrm{a})$$

$$\equiv \left(\frac{\gamma e^2}{m_\mathrm{e}c^2}\right)^2 \frac{k'}{k} \sum_{\alpha,\beta} (\delta_{\alpha,\beta} - \tilde{\kappa}_\alpha\tilde{\kappa}_\beta) \sum_{\lambda,\lambda'} p_\lambda \langle\lambda|\,\hat{Q}_\alpha^+\,|\lambda'\rangle\langle\lambda'|\,\hat{Q}_\beta\,|\lambda\rangle \times$$

$$\times\,\delta(\hbar\omega + E_\lambda - E_{\lambda'}). \quad (5.11\,\mathrm{b})$$

In the second form of the cross-section (5.11 b) the operator $\hat{\mathbf{Q}}$ is related to the operator $\hat{\mathbf{Q}}_\perp$ through

$$\hat{\mathbf{Q}}_\perp = \tilde{\boldsymbol{\varkappa}} \times (\hat{\mathbf{Q}} \times \tilde{\boldsymbol{\varkappa}}), \quad (5.12\,\mathrm{a})$$

and use has been made of the identity

$$\hat{\mathbf{Q}}_\perp^+.\hat{\mathbf{Q}}_\perp = \sum_{\alpha,\beta} (\delta_{\alpha,\beta} - \tilde{\kappa}_\alpha\tilde{\kappa}_\beta)\hat{Q}_\alpha^+\hat{Q}_\beta. \quad (5.12\,\mathrm{b})$$

5.2. Spin-only scattering

In this and the following two sections we focus our attention on those target systems in which the unpaired electrons possess wave functions localized about the sites of a crystal lattice, i.e. about sites defined by the vectors $\mathbf{R}_{ld} = \mathbf{l}+\mathbf{d}$. (The scattering cross-section when the ions are not stationary, as is the case here, but vibrate about the equilibrium positions $\mathbf{R}_{ld}$, is discussed in § 8.1.) In other words we assume the magnetic properties of the crystal to be well described by the Heitler–London model, each unpaired electron being associated with an atomic orbital of the form

$$f(r)Y^l_m(\tilde{\mathbf{r}})$$

(l = orbital quantum number, m = magnetic quantum number), where $f(r)$ is the radial part of the wave function and $Y^l_m(\tilde{\mathbf{r}})$ a spherical harmonic or some linear combination as dictated by the crystal field.

In many cases of interest the total orbital angular momentum of the magnetic ions is either zero or quenched by the crystalline field. Under these circumstances the second part of $\hat{\mathbf{Q}}_\perp$ is zero and so it follows from the definition of the operator $\hat{\mathbf{Q}}$ that in this particular instance it is

$$\hat{\mathbf{Q}} = \sum_i \exp(i\boldsymbol{\varkappa}.\mathbf{r}_i)\hat{\mathbf{s}}_i. \tag{5.13}$$

The sum in (5.13) runs over all sites in the crystal (i.e. all $\mathbf{R}_{ld}$) and all the unpaired electrons associated with the ions.

Each cell in the crystal is identical, by definition, so that we can rewrite (5.13) as

$$\sum_l \exp(i\boldsymbol{\varkappa}.\mathbf{l}) \sum_i{}_{\text{cell}} \exp(i\boldsymbol{\varkappa}.\mathbf{r}_i)\hat{\mathbf{s}}_i,$$

where now the second sum is over all the unpaired electrons in a single cell, as indicated by the suffix cell.

Within the cell the ions are situated at the sites defined by the vectors $\mathbf{d}$ so that if we denote the position of the νth electron of the ion at the site $\mathbf{d}$ relative to this site position by $\mathbf{r}_\nu$, we can write

$$\sum_i{}_{\text{cell}} \exp(i\boldsymbol{\varkappa}.\mathbf{r}_i)\hat{\mathbf{s}}_i = \sum_d \exp(i\boldsymbol{\varkappa}.\mathbf{d}) \sum_{\nu(d)} \exp(i\boldsymbol{\varkappa}.\mathbf{r}_\nu)\hat{\mathbf{s}}_\nu.$$

Thus (5.13) is

$$\hat{\mathbf{Q}} = \sum_{l,d} \exp(i\boldsymbol{\varkappa}.\mathbf{R}_{ld}) \sum_{\nu(d)} \exp(i\boldsymbol{\varkappa}.\mathbf{r}_\nu)\hat{\mathbf{s}}_\nu. \tag{5.14}$$

The unpaired electrons of each ion will couple together to give a ground state with a certain total spin $\hat{\mathbf{S}}_{ld}$, and the neutrons will certainly not have enough energy to break down this coupling. $\hat{\mathbf{S}}_{ld}$ being the only vector associated with $\hat{\mathbf{Q}}$ the matrix elements of the latter must be proportional to those of $\hat{\mathbf{S}}_{ld}$ (Wigner–Eckart theorem). We can therefore write

$$\langle\lambda'|\,\hat{\mathbf{Q}}\,|\lambda\rangle = \sum_{l,d} \exp(i\boldsymbol{\varkappa}.\mathbf{R}_{ld})F_d(\boldsymbol{\varkappa})\langle\lambda'\,|\hat{\mathbf{S}}_{ld}\,|\lambda\rangle, \tag{5.15}$$

where $F_d(\boldsymbol{\varkappa})$ is the form factor defined as the Fourier transform of the normalized spin density associated with the ion at the dth site in the unit cell, i.e.

$$F_d(\boldsymbol{\varkappa}) = \int \mathrm{d}\mathbf{r}\, \exp(\mathrm{i}\boldsymbol{\varkappa}.\mathbf{r})\mathscr{s}_d(\mathbf{r}), \tag{5.16}$$

where $\mathscr{s}_d(\mathbf{r})$ is the normalized spin density. Note that, by definition,

$$F_d(0) = 1. \tag{5.17}$$

If we insert (5.15) into (5.11 b), the partial differential cross-section for the magnetic scattering by ions with only spin angular momentum is

$$\frac{\mathrm{d}^2\sigma}{\mathrm{d}\Omega\mathrm{d}E'} = \left(\frac{\gamma e^2}{m_\mathrm{e} c^2}\right)^2 \frac{k'}{k} \sum_{\alpha,\beta} (\delta_{\alpha\beta} - \tilde{\kappa}_\alpha \tilde{\kappa}_\beta) \sum_{\lambda,\lambda'} p_\lambda \times$$
$$\times \sum_{l,d} \sum_{l',d'} F_d^*(\boldsymbol{\varkappa}) F_{d'}(\boldsymbol{\varkappa}) \exp\{\mathrm{i}\boldsymbol{\varkappa}.(\mathbf{R}_{l'd'} - \mathbf{R}_{ld})\} \times$$
$$\times \langle\lambda|\, \hat{S}^\alpha_{ld}\, |\lambda'\rangle\langle\lambda'|\, \hat{S}^\beta_{l'd'}\, |\lambda\rangle\, \delta(\hbar\omega + E_\lambda - E_{\lambda'}). \tag{5.18}$$

5.3. Approximate cross-section for scattering by ions with both spin and orbital angular momentum

In general a magnetic ion in a crystal possesses both spin and orbital angular momentum and so it is necessary to include both terms in eqn (5.10). The calculation of the matrix elements of $\hat{\mathbf{Q}}_\perp$ is then no longer as simple a task as for the spin-only case considered in the preceding section. A detailed discussion is given in Chapter 6. Here we merely quote certain results obtained there, for the sake of completeness.

If the mean radius of the wave function of the unpaired electrons is much less than $|\boldsymbol{\varkappa}|^{-1}$, as is often the case, then $\hat{\mathbf{Q}}$ has the form (dipole approximation),

$$\hat{\mathbf{Q}} \simeq \hat{\mathbf{Q}}^{(\mathrm{D})} = \sum_{l,d} \exp(\mathrm{i}\boldsymbol{\varkappa}.\mathbf{R}_{ld})\{\bar{\jmath}_0\, \hat{\mathbf{S}}_{ld} + \tfrac{1}{2}(\bar{\jmath}_0 + \bar{\jmath}_2)\hat{\mathbf{L}}_{ld}\}, \tag{5.19}$$

where $\hat{\mathbf{L}}_{ld}$ is the angular momentum operator for the ion at the site $\mathbf{R}_{ld}$. The coefficient $\bar{\jmath}_{\bar{K}}(\kappa)$, usually referred to as a radial integral, is given by

$$\bar{\jmath}_{\bar{K}}(\kappa) = \int_0^\infty \mathrm{d}r\, r^2 j_{\bar{K}}(\kappa r)|f(r)|^2, \tag{5.20}$$

where $j_{\bar{K}}$ is a spherical Bessel function of order $\bar{K}$. It has been assumed in (5.19) that all the ions in the crystal are identical, otherwise the radial integrals would depend upon d. In the limit $|\boldsymbol{\varkappa}| = 0$,

$$\bar{\jmath}_0\, \hat{\mathbf{S}} + \tfrac{1}{2}(\bar{\jmath}_0 + \bar{\jmath}_2)\hat{\mathbf{L}} \to \tfrac{1}{2}(\hat{\mathbf{L}} + 2\hat{\mathbf{S}}),\dagger \tag{5.21}$$

so that the *scattering amplitude in the forward direction is proportional to the total magnetic moment per ion.*

† (5.21) is correct for conventional experimental arrangements but needs to be modified for some special cases (Baryshevskii *et al.*, *J. exp. theor. Phys.* **29,** 685 (1969); ibid. **31,** 583 (1970)).

Some magnetic ions that would have no orbital angular momentum because of quenching by the crystalline field are found in some cases to have gyromagnetic ratios g that differ from the spin-only value $g = 2$, i.e. they have a finite orbital angular momentum. This arises because of the action of spin–orbit coupling, which, by mixing excited states into the ground state, aligns some orbital moment parallel or antiparallel to the spin. For these ions, by definition of the g value within the ground state manifold, $\hat{\mathbf{L}}$ may be replaced by $(g-2)\hat{\mathbf{S}}$. Hence,

$$\hat{\mathbf{Q}}^{(\mathrm{D})} = \sum_{l,d} \exp(\mathrm{i}\boldsymbol{\kappa}.\mathbf{R}_{ld})\{\bar{j}_0+\tfrac{1}{2}(g-2)(\bar{j}_0+\bar{j}_2)\}\hat{\mathbf{S}}_{ld}$$

$$= \tfrac{1}{2}gF(\boldsymbol{\kappa}) \sum_{l,d} \exp(\mathrm{i}\boldsymbol{\kappa}.\mathbf{R}_{ld})\hat{\mathbf{S}}_{ld}, \tag{5.22}$$

where $F(\boldsymbol{\kappa})$ is the form factor of the ion and is given by

$$F(\boldsymbol{\kappa}) = \bar{j}_0+\left(\frac{g-2}{g}\right)\bar{j}_2. \tag{5.23}$$

Examples where the orbital moment is unquenched are all the rare-earth ions. In these ions the spin–orbit coupling combines $\hat{\mathbf{S}}$ and $\hat{\mathbf{L}}$ to give various $\hat{\mathbf{J}}$ values. Within the states of given $\mathbf{J}$ we write

$$2\hat{\mathbf{S}} = g_S\hat{\mathbf{J}},$$

$$\hat{\mathbf{L}} = g_L\hat{\mathbf{J}},$$

and

$$g\hat{\mathbf{J}} = \hat{\mathbf{L}}+2\hat{\mathbf{S}}.$$

To calculate g_S, g_L, and g we take the scalar product of the defining equations with $\hat{\mathbf{J}}$ and use the relation

$$2\hat{\mathbf{L}}.\hat{\mathbf{S}} = \hat{\mathbf{J}}^2-\hat{\mathbf{L}}^2-\hat{\mathbf{S}}^2.$$

We then find

$$g_S = \frac{J(J+1)-L(L+1)+S(S+1)}{J(J+1)},$$

$$g_L = \frac{J(J+1)+L(L+1)-S(S+1)}{2J(J+1)}, \tag{5.24}$$

and finally g, the Landé splitting factor for the ion

$$g = 1+\frac{J(J+1)-L(L+1)+S(S+1)}{2J(J+1)}.$$

With these definitions,

$$\hat{\mathbf{Q}}^{(\mathrm{D})} = \tfrac{1}{2}gF(\boldsymbol{\kappa}) \sum_{l,d} \exp(\mathrm{i}\boldsymbol{\kappa}.\mathbf{R}_{ld})\hat{\mathbf{J}}_{ld}, \tag{5.25}$$

where

$$F(\boldsymbol{\kappa}) = \bar{j}_0\frac{g_S}{g}+(\bar{j}_0+\bar{j}_2)\frac{g_L}{g} \tag{5.26}$$

is the form factor of the ion.

We emphasize that the expression for $\hat{\mathbf{Q}}$ given by eqn (5.19) is only approximate, a detailed analysis of the effects of orbital angular momentum on the cross-section for magnetic scattering being given in Chapter 6.

The results (5.22) and (5.25) allow us to write, within the dipole approximation and for identical magnetic ions,

$$\frac{\mathrm{d}^2\sigma}{\mathrm{d}\Omega\mathrm{d}E'} = \left(\frac{\gamma e^2}{m_e c^2}\right)^2 \{\tfrac{1}{2}gF(\boldsymbol{\varkappa})\}^2 \frac{k'}{k} \sum_{\alpha\beta} (\delta_{\alpha\beta}-\tilde{\kappa}_\alpha\tilde{\kappa}_\beta) \sum_{\lambda,\lambda'} p_\lambda \times$$
$$\times \sum_{l,d}\sum_{l',d'} \exp\{\mathrm{i}\boldsymbol{\varkappa}.(\mathbf{R}_{l'd'}-\mathbf{R}_{ld})\}\langle\lambda|\,\hat{S}^\alpha_{ld}\,|\lambda'\rangle\langle\lambda'|\,\hat{S}^\beta_{l'd'}\,|\lambda\rangle\,\delta(\hbar\omega+E_\lambda-E_{\lambda'}), \tag{5.27}$$

where the spin operator $\hat{\mathbf{S}}$ may be the actual spin, or the total angular momentum $\hat{\mathbf{J}}$, or some effective spin operator in the case of scattering by ions with partially quenched orbital angular momentum. The form factor $F(\boldsymbol{\varkappa})$ in (5.27) is

$$F(\boldsymbol{\varkappa}) = \bar{j}_0\frac{g_S}{g}+(\bar{j}_0+\bar{j}_2)\left(\frac{g-g_S}{g}\right), \tag{5.28}$$

where $\mu_B g_S S$ is the spin contribution and $\mu_B(g-g_S)S$ is the orbital moment contribution to the total magnetic moment of an ion $\mu_B gS$.

5.4. Examples of magnetic cross-sections

We now illustrate the evaluation of the cross-section (5.27) by considering two simple examples (a) a perfect paramagnet and (b) a paramagnet in a magnetic field. In both examples we assume the ions to form a Bravais lattice.

(a) Paramagnets

In a perfect paramagnet the energy is independent of spin orientation and so the sum over λ' in (5.27) can be done immediately by closure and recognizing that the scattering must be elastic. We get

$$\frac{\mathrm{d}\sigma}{\mathrm{d}\Omega} = \left(\frac{\gamma e^2}{m_e c^2}\right)^2 \{\tfrac{1}{2}gF(\boldsymbol{\varkappa})\}^2 \sum_{\alpha\beta} (\delta_{\alpha\beta}-\tilde{\kappa}_\alpha\tilde{\kappa}_\beta) \sum_{l,l'} \exp\{\mathrm{i}\boldsymbol{\varkappa}(\mathbf{l}'-\mathbf{l})\times$$
$$\times \sum_\lambda p_\lambda\langle\lambda|\,\hat{S}^\alpha_l\hat{S}^\beta_{l'}\,|\lambda\rangle. \tag{5.29}$$

Now
$$\sum_\lambda p_\lambda\langle\lambda|\,\hat{S}^\alpha_l\hat{S}^\beta_{l'}\,|\lambda\rangle = \langle\hat{S}^\alpha_l\hat{S}^\beta_{l'}\rangle, \tag{5.30}$$

where, as usual, the angular brackets denote a thermal average at a temperature $T = 1/k_B\beta$. For a paramagnet there is no spatial correlation between spins, i.e.

$$\langle\hat{S}^\alpha_l\hat{S}^\beta_{l'}\rangle = \langle\hat{S}^\alpha_l\rangle\langle\hat{S}^\beta_{l'}\rangle, \quad \text{if } l \neq l'. \tag{5.31}$$

By definition $\langle\hat{\mathbf{S}}\rangle = 0$ so that there is no contribution to the cross-section from terms with $l \neq l'$.

For $l = l'$ we have

$$\langle \hat{S}_l^\alpha \hat{S}_l^\beta \rangle = \delta_{\alpha\beta} \langle (\hat{S}_l^\alpha)^2 \rangle$$
$$= \tfrac{1}{3}\delta_{\alpha\beta} \langle (\hat{\mathbf{S}})^2 \rangle = \tfrac{1}{3}\delta_{\alpha\beta}\, S(S+1). \tag{5.32}$$

Thus, with this result the cross-section (5.29) becomes

$$\frac{d\sigma}{d\Omega} = \left(\frac{\gamma e^2}{m_e c^2}\right)^2 \{\tfrac{1}{2} g F(\boldsymbol{\kappa})\}^2 \sum_{\alpha\beta} (\delta_{\alpha\beta} - \tilde{\kappa}_\alpha \tilde{\kappa}_\beta) N \tfrac{1}{3}\delta_{\alpha\beta}\, S(S+1)$$
$$= \left(\frac{\gamma e^2}{m_e c^2}\right)^2 \{\tfrac{1}{2} g F(\boldsymbol{\kappa})\}^2 N \tfrac{2}{3} S(S+1). \tag{5.33}$$

This formula shows that the cross-section is large for large spin values S, just as we would expect. There is no coherent scattering because the paramagnetic ions are randomly orientated, and the dependence on the scattering vector $\boldsymbol{\kappa}$ comes only through the form factor $F(\boldsymbol{\kappa})$.

(b) Paramagnets in a magnetic field

A perfect paramagnet at low enough temperatures will be appreciably polarized if a magnetic field **H** is applied. In this case, unless the neutron wavelength is exceptionally long, the energy changes on reversing a spin, $g\mu_B \sim 1$ K, are small compared to the energy of an incident neutron, ~ 300 K, and we may therefore ignore such energy changes and take the scattering processes to be purely elastic. Thus we can still use the cross-section (5.29), but the thermal average must be evaluated with care.

Clearly,
$$\langle \hat{S}_l^x \hat{S}_{l'}^y \rangle = \langle \hat{S}_l^x \hat{S}_{l'}^z \rangle = 0, \text{ etc.}, \tag{5.34 a}$$
and with **H** parallel to the z-axis,
$$\langle \hat{S}_l^x \hat{S}_{l'}^x \rangle = \langle \hat{S}_l^y \hat{S}_{l'}^y \rangle = \delta_{l,l'} \langle (\hat{S}^x)^2 \rangle. \tag{5.34 b}$$
Also,
$$\langle \hat{S}_l^z \hat{S}_{l'}^z \rangle = \langle \hat{S}^z \rangle^2 + \delta_{l,l'} \{\langle (\hat{S}^z)^2 \rangle - \langle \hat{S}^z \rangle^2\}. \tag{5.35}$$
Thus, with the aid of (5.34 a) and (5.34 b),
$$\sum_{\alpha,\beta} (\delta_{\alpha,\beta} - \tilde{\kappa}_\alpha \tilde{\kappa}_\beta) \langle \hat{S}_l^\alpha \hat{S}_{l'}^\beta \rangle = (1+\tilde{\kappa}_z^2) \langle \hat{S}_l^x \hat{S}_{l'}^x \rangle + (1-\tilde{\kappa}_z^2) \langle \hat{S}_l^z \hat{S}_{l'}^z \rangle, \tag{5.36}$$
and so
$$\sum_{l,l'} \exp\{i\boldsymbol{\kappa}.(\mathbf{l}'-\mathbf{l})\} \sum_{\alpha,\beta} (\delta_{\alpha\beta} - \tilde{\kappa}_\alpha \tilde{\kappa}_\beta) \langle \hat{S}_l^\alpha \hat{S}_{l'}^\beta \rangle$$
$$= (1-\tilde{\kappa}_z^2) \langle \hat{S}^z \rangle^2 \left| \sum_l \exp(i\boldsymbol{\kappa}.\mathbf{l}) \right|^2 +$$
$$+ N[(1+\tilde{\kappa}_z^2) \langle (\hat{S}^x)^2 \rangle + (1-\tilde{\kappa}_z^2) \{\langle (\hat{S}^z)^2 \rangle - \langle \hat{S}^z \rangle^2\}]$$
$$= (1-\tilde{\kappa}_z^2) \langle \hat{S}^z \rangle^2 N \frac{(2\pi)^3}{v_0} \sum_{\boldsymbol{\tau}} \delta(\boldsymbol{\kappa} - \boldsymbol{\tau}) +$$
$$+ N[\tfrac{1}{2} S(S+1) + \tfrac{1}{2} \langle (\hat{S}^z)^2 \rangle - \langle \hat{S}^z \rangle^2 + \tilde{\kappa}_z^2 \{\tfrac{1}{2} S(S+1) - \tfrac{3}{2} \langle (\hat{S}^z)^2 \rangle + \langle \hat{S}^z \rangle^2\}]. \tag{5.37}$$

In obtaining the second line of (5.37) we have used

$$2\langle(\hat{S}^x)^2\rangle = S(S+1)-\langle(\hat{S}^z)^2\rangle.$$

The structure of (5.37) is to be noted. The first term, which gives rise to coherent scattering, is proportional to the square of the average value of the z-component of spin; the second term, which gives incoherent scattering, is proportional to the square of the fluctuations of the transverse and the z-component of the spin.

The average values of $\hat{S}^z$ and $(\hat{S}^z)^2$ are readily calculated. The partition function Z is given by

$$Z = \sum_{m=-S}^{S} \exp(g\mu_B H\beta m).$$

Some simple algebra gives the result

$$Z(u) = \sinh u(S+\tfrac{1}{2})/\sinh(\tfrac{1}{2}u), \tag{5.38}$$

where we have defined $u = g\mu_B H\beta$, and

$$\langle\hat{S}^z\rangle = \frac{1}{Z(u)}\sum_{m=-S}^{S} m e^{mu} = \frac{1}{Z(u)}\frac{dZ(u)}{du},$$

which from (5.38) leads to the result

$$\langle\hat{S}^z\rangle = \tfrac{1}{2}[(2S+1)\coth\{\tfrac{1}{2}u(2S+1)\}-\coth(\tfrac{1}{2}u)]. \tag{5.39}$$

When $u = g\mu_B H\beta \ll 1$ we have

$$\langle\hat{S}^z\rangle \simeq \tfrac{1}{3}g\mu_B H\beta\, S(S+1). \tag{5.40}$$

Furthermore,

$$\begin{aligned}\langle(\hat{S}^z)^2\rangle &= \frac{1}{Z(u)}\sum_{m=-S}^{S} m^2 e^{mu} = \frac{1}{Z(u)}\frac{d^2Z(u)}{du^2}\\ &= \frac{1}{4}\left[(2S+1)^2+\frac{\cosh u+3}{\cosh u-1}-2(2S+1)\coth\{\tfrac{1}{2}u(2S+1)\}\coth(\tfrac{1}{2}u)\right].\end{aligned} \tag{5.41}$$

When $u \ll 1$,

$$\langle(\hat{S}^z)^2\rangle \simeq \tfrac{1}{12}[4S(S+1)+\tfrac{1}{30}(g\mu_B H\beta)^2\{4(2S+1)-5(2S+1)^2+1\}]. \tag{5.42}$$

In the limit of $H = 0$ (i.e. $u = 0$), the cross-section given by (5.37) becomes identical to that for the scattering from a perfect paramagnet, (5.33), as, of course, it must.

5.5. Cross-section in the formalism of band theory

In the preceding sections we have taken the electrons responsible for the magnetic field in (5.1) to possess wave functions that are strongly localized about the ions in the target crystal. While this is quite reasonable for ionic crystals and rare-earth metals it would seem a doubtful assumption for transition metals, for instance, whose magnetic properties are mainly determined by their $3d$ electrons, which are known to give a large contribution to the low-temperature specific heat. Thus we have also to consider the calculation of the cross-section for the scattering of neutrons from a solid that is more appropriately described in terms of band theory rather than the atomic, or Heitler–London theory considered hitherto.

The band theory of a solid is constructed by envisaging the solid as being built by taking free atoms and placing them at the sites of a perfect crystal lattice structure. The mutual interaction of the atoms causes the energy levels of the constituent atoms to broaden into bands of energy states, the width of these bands being determined by the degree of perturbation to which the levels are subject on forming the solid. Thus, for instance, the $3s$ electrons of sodium form a wide band because their corresponding wave function in the free atom has a large mean radius as compared, say, to the mean radius of the wave function of the $4f$ electrons in the rare earths, which are deeply buried in the atom and form a very narrow band in the solid, centred about the energy of the $4f$ level in the free atom. Each electron state of the free atom becomes a band in the solid but we can discard all but the bands coming from the outermost shells of the free atoms, since the core states play no part in the dynamical behaviour of the solid but go to form the field of ion potentials in which the remaining electrons move. Of course, it must also be remembered that the magnetic interaction of the neutrons with the solid involves only the unpaired electron spins, so that here also we do not need to consider the closed shells of electrons in the ion cores.

(It is true, of course, that there is an unpairing of electron spin in the ion cores due to configurational interaction within the ion, the unpaired s-electron spin density giving rise to the major part of the magnetic field at the nucleus of the ion. However, this unpaired spin gives a negligible contribution to the scattering amplitude because of its smallness.)

Let there be $\mathscr{N}$ electrons involved in determining the magnetic properties of the solid. We denote the different bands to which these belong by the index λ, e.g. for d electrons in a Bravais lattice there are $2l+1 = 2.2+1 = 5$ bands to be taken into account. We then need the eigenstates φ of the Hamiltonian

$$\mathscr{H} = \sum_{i=1}^{\mathscr{N}} \frac{1}{2m_{\mathrm{e}}} \hat{\mathbf{p}}_i^2 + \hat{V}(\mathbf{r}_i), \tag{5.43}$$

where $\hat{V}(\mathbf{r})$ is the periodic potential due to the ion cores. The periodicity of $\hat{V}(\mathbf{r})$ means that

$$\hat{V}(\mathbf{r}) = \hat{V}(\mathbf{r}+\mathbf{l}), \tag{5.44}$$

where $\mathbf{l}$ is any lattice vector (cf. § 2.1). The eigenfunctions of (5.43) with $\hat{V}$ satisfying (5.44) are Bloch functions,†

$$\varphi_{\mathbf{k}\lambda}(\mathbf{r}) = \exp(i\mathbf{k}.\mathbf{r})u_{\mathbf{k}\lambda}(\mathbf{r}), \tag{5.45 a}$$

where

$$u_{\mathbf{k}\lambda}(\mathbf{r}+\mathbf{l}) = u_{\mathbf{k}\lambda}(\mathbf{r}) \tag{5.45 b}$$

and

$$\left\{\frac{1}{2m_e}\hat{\mathbf{p}}^2+\hat{V}(\mathbf{r})\right\}\varphi_{\mathbf{k}\lambda}(\mathbf{r}) = \mathscr{E}_\lambda(\mathbf{k})\varphi_{\mathbf{k}\lambda}(\mathbf{r}). \tag{5.46}$$

The N values of $\mathbf{k}$ (N being the number of unit cells) are determined, as usual, by applying periodic boundary conditions to the solid, and have a density $V/(2\pi)^3$ in reciprocal space. The total number of one-particle eigenstates is therefore $2\lambda N$, the factor two coming from the two possible spin states of an electron. We shall use the index σ to denote spin states, σ taking the values $+1$ and -1, which we often represent by $\uparrow$ and $\downarrow$ respectively. The inclusion of the spin state of the electrons means that the Bloch function becomes a two-component spinor, $\varphi_{\mathbf{k}\lambda\sigma}$, but we often drop the spin index for brevity where this cannot cause any confusion. One last point concerning the Bloch states (5.45) is that since we shall always assume the Hamiltonian to be invariant under space inversion, we have

$$\varphi_{\mathbf{k}\lambda\sigma}(\mathbf{r}) = \varphi_{-\mathbf{k}\lambda\sigma}(-\mathbf{r}), \tag{5.47}$$

and, correspondingly, the band energies $\mathscr{E}_\lambda(\mathbf{k})$ obey

$$\mathscr{E}_\lambda(\mathbf{k}) = \mathscr{E}_\lambda(-\mathbf{k}). \tag{5.48}$$

We can express the operator $\hat{\mathbf{Q}}_\perp$, eqn (5.10), in terms of Bloch states, or for that matter any complete set of states, with the aid of second quantization formalism.‡ This method of describing many-particle systems is based upon the concept that if ψ_ν ($\nu = 1,\ldots, \mathscr{N}$) are any complete set of orthogonal normalized wave functions then operators $\hat{c}^+_\nu$ and $\hat{c}_\nu$ can be defined to create and annihilate particles in the state ψ_ν. For particles that obey Fermi–Dirac statistics $\hat{c}^+_\nu$ and $\hat{c}_\mu$ satisfy the following *anticommutation* rule:

$$\hat{c}_\nu\hat{c}^+_\mu+\hat{c}^+_\mu\hat{c}_\nu = [\hat{c}_\nu, \hat{c}^+_\mu]_+ = \delta_{\nu,\mu}, \tag{5.49}$$

† We denote the eigenvalues of (5.43) by the wave vector $\mathbf{k}$ in accord with current usage. This should cause no confusion with the wave vector associated with incident neutron. For an introduction to band theory we refer the reader to J. Callaway, *Energy band theory*, Academic Press (1964) and C. Kittel, *Quantum theory of solids*, Chapters 9 and 10, John Wiley.

‡ L. D. Landau and E. M. Lifshitz, *Quantum mechanics* (second edn), §§ 64 and 65, Pergamon Press.

all other anticommutators being zero. We also note that the particle-number operator for the state ψ_ν, namely $\hat{n}_\nu = \hat{c}_\nu^+ \hat{c}_\nu$, obeys the relation

$$\hat{n}_\nu \hat{n}_\nu = \hat{n}_\nu, \tag{5.50}$$

as a consequence of the condition

$$(\hat{c}_\nu^+)^2 = (\hat{c}_\nu)^2 = 0, \tag{5.51}$$

which is itself a direct consequence of the Pauli exclusion principle.

In order to express the Hamiltonian describing the particles, in our case electrons, in terms of the operators $\hat{c}$ and $\hat{c}^+$, we require prescriptions for operators that are the sum of one-particle operators, two-particle operators, etc. These are as follows:† If $\hat{f}_i^{(1)}$ acts only on the coordinates of the ith particle (by coordinates we mean both spin and space coordinates) and

$$\hat{F}^{(1)} = \sum_i \hat{f}_i^{(1)}, \tag{5.52 a}$$

then

$$\hat{F}^{(1)} = \sum_{\nu,\mu} f_{\nu\mu}^{(1)} \hat{c}_\nu^+ \hat{c}_\mu, \tag{5.52 b}$$

where

$$f_{\nu\mu}^{(1)} = \langle \psi_\nu | \hat{f}^{(1)} | \psi_\mu \rangle.$$

Similarly, if $\hat{f}_{ij}^{(2)}$ is an operator acting on particles i and j at once, and

$$\hat{F}^{(2)} = \sum_{i>j} \hat{f}_{ij}^{(2)}, \tag{5.53 a}$$

then

$$\hat{F}^{(2)} = \tfrac{1}{2} \sum_{\mu,\nu,\mu',\nu'} (f^{(2)})_{\mu'\nu'}^{\mu\nu} \hat{c}_\mu^+ \hat{c}_\nu^+ \hat{c}_{\nu'} \hat{c}_{\mu'}, \tag{5.53 b}$$

where

$$(f^{(2)})_{\mu'\nu'}^{\mu\nu} = \langle \psi_\mu, \psi_\nu | \hat{f}^{(2)} | \psi_{\mu'}, \psi_{\nu'} \rangle.$$

The procedure can be extended to the sum of three-particle operators, etc., but we shall not have occasion to use them.

As an example of an operator that is the sum of single-particle operators, we can take the Hamiltonian (5.43). From (5.52) it follows that this can be written as

$$\mathscr{H} = \sum_{\mu,\nu} \left\langle \psi_\mu \left| \frac{1}{2m_e} \hat{\mathbf{p}}^2 + \hat{V}(\mathbf{r}) \right| \psi_\nu \right\rangle \hat{c}_\mu^+ \hat{c}_\nu. \tag{5.54}$$

If we now choose the states ψ_ν to be Bloch states $\varphi_{\mathbf{k}\lambda\sigma}$, eqn (5.45 a), then it follows from (5.46) that (5.54) reduces the diagonal form,

$$\mathscr{H} = \sum_{\mathbf{k}\lambda\sigma} \mathscr{E}_\lambda(\mathbf{k}) \hat{c}_{\mathbf{k}\lambda\sigma}^+ \hat{c}_{\mathbf{k}\lambda\sigma}. \tag{5.55}$$

A useful concept in second quantization is that of a particle field operator. If ξ denotes the coordinates of a particle, then the particle

† Ibid.

field operator $\hat{\psi}(\xi)$ is defined to be

$$\hat{\psi}(\xi) = \sum_{\nu} \psi_{\nu}(\xi)\hat{c}_{\nu}$$

and

$$\hat{\psi}^{+}(\xi) = \sum_{\nu} \psi_{\nu}^{*}(\xi)\hat{c}_{\nu}^{+}. \tag{5.56}$$

Note the analogy between these expressions and the expansion of a wave function in terms of a complete set of eigenfunctions. Also, by what has been said concerning the creation and annihilation operators $\hat{c}_{\nu}^{+}$ and $\hat{c}_{\nu}$ it is evident that the operator $\hat{\psi}$ decreases the *total* number of particles in the system by one, while $\hat{\psi}^{+}$ increases it by one.

It follows from the anticommutation relations for $\hat{c}$ and $\hat{c}^{+}$ that $\hat{\psi}$ and $\hat{\psi}^{+}$ obey

$$\hat{\psi}(\xi)\hat{\psi}^{+}(\xi')+\hat{\psi}^{+}(\xi')\hat{\psi}(\xi) = [\hat{\psi}(\xi),\hat{\psi}^{+}(\xi')]_{+} = \delta(\xi-\xi'), \tag{5.57}$$

and all other anticommutators are zero. In (5.57), $\delta(\xi-\xi')$ is shorthand for $\delta(x-x')\delta(y-y')\delta(z-z')\delta_{\sigma,\sigma'}$.

The microscopic particle density $\hat{\rho}(\mathbf{r})$ introduced in Chapter 3 is given by

$$\hat{\rho}(\mathbf{r}) = \hat{\psi}^{+}(\mathbf{r})\hat{\psi}(\mathbf{r}). \tag{5.58}$$

Let us now proceed to express $\hat{\mathbf{Q}}_{\perp}$, eqn (5.10), in terms of operators referring to Bloch states, $\varphi_{\mathbf{k}\lambda\sigma}$.

Consider first the spin part of (5.10). Clearly we require an expression for

$$\sum_{i} \exp(\mathrm{i}\boldsymbol{\varkappa}.\mathbf{r}_{i})\hat{\mathbf{s}}_{i}.$$

The spin density $\hat{\mathfrak{s}}(\mathbf{r})$ is defined as

$$\hat{\mathfrak{s}}(\mathbf{r}) = \sum_{i} \delta(\mathbf{r}-\mathbf{r}_{i})\hat{\mathbf{s}}_{i}. \tag{5.59}$$

If we now write

$$\hat{\mathfrak{s}}(\mathbf{r}) = \frac{1}{V}\sum_{\mathbf{q}} \exp(\mathrm{i}\mathbf{q}.\mathbf{r})\hat{\mathfrak{s}}(\mathbf{q}), \tag{5.60}$$

with

$$\hat{\mathfrak{s}}(\mathbf{q}) = \int \mathrm{d}\mathbf{r}\, \exp(-\mathrm{i}\mathbf{q}.\mathbf{r})\hat{\mathfrak{s}}(\mathbf{r}), \tag{5.61}$$

then

$$\sum_{i} \exp(\mathrm{i}\boldsymbol{\varkappa}.\mathbf{r}_{i})\hat{\mathbf{s}}_{i} = \int \mathrm{d}\mathbf{r}\, \exp(\mathrm{i}\boldsymbol{\varkappa}.\mathbf{r}) \sum_{i} \delta(\mathbf{r}-\mathbf{r}_{i})\hat{\mathbf{s}}_{i}$$

$$= \int \mathrm{d}\mathbf{r}\, \exp(\mathrm{i}\boldsymbol{\varkappa}.\mathbf{r})\hat{\mathfrak{s}}(\mathbf{r}) = \hat{\mathfrak{s}}(-\boldsymbol{\varkappa}). \tag{5.62}$$

By analogy with (5.58), the spin density $\hat{\mathfrak{s}}(\mathbf{r})$ is given in the second quantized formalism by

$$\hat{\mathfrak{s}}(\mathbf{r}) = \hat{\psi}^{+}(\mathbf{r})\hat{\mathbf{s}}\hat{\psi}(\mathbf{r}), \tag{5.63}$$

which in terms of Bloch states reads

$$\hat{\mathfrak{s}}(\mathbf{r}) = \sum_{\substack{\mathbf{k},\lambda \\ \mathbf{k}',\lambda'}} \sum_{\sigma,\sigma'} \varphi_{\mathbf{k}\lambda\sigma}^{*}(\mathbf{r})\hat{c}_{\mathbf{k}\lambda\sigma}^{+}\, \hat{\mathbf{s}}\varphi_{\mathbf{k}'\lambda'\sigma'}(\mathbf{r})\hat{c}_{\mathbf{k}'\lambda'\sigma'}. \tag{5.64}$$

We now separate the spin spinor in the Bloch function $\varphi_{\mathbf{k}\lambda\sigma}$ from the spatial eigenfunction, i.e. write

$$\varphi_{\mathbf{k}\lambda\sigma} = \varphi_{\mathbf{k}\lambda}\,\chi_{\sigma}, \tag{5.65}$$

so that $\hat{\mathfrak{J}}(\mathbf{q})$ is

$$\hat{\mathfrak{J}}(\mathbf{q}) = \int d\mathbf{r}\,\exp(-i\mathbf{q}.\mathbf{r}) \sum_{\substack{\mathbf{k}\lambda\\ \mathbf{k}'\lambda'}} \varphi^*_{\mathbf{k}\lambda}(\mathbf{r})\varphi_{\mathbf{k}'\lambda'}(\mathbf{r}) \sum_{\sigma,\sigma'} \hat{c}^+_{\mathbf{k}\lambda\sigma}\,\chi^+_{\sigma}\,\hat{\mathbf{s}}\,\chi_{\sigma'}\,\hat{c}_{\mathbf{k}'\lambda'\sigma'}. \tag{5.66}$$

Let us evaluate the spin part of (5.66). We take

$$\chi_{\uparrow} = \begin{pmatrix}1\\0\end{pmatrix} \quad \text{and} \quad \chi_{\downarrow} = \begin{pmatrix}0\\1\end{pmatrix}$$

and write $\hat{\mathbf{s}} = \frac{1}{2}(\hat{\sigma}_x, \hat{\sigma}_y, \hat{\sigma}_z)$,

where $\hat{\sigma}_\alpha$ are Pauli matrices. Explicitly

$$\hat{\sigma}_x = \begin{pmatrix}0 & 1\\1 & 0\end{pmatrix}, \qquad \hat{\sigma}_y = \begin{pmatrix}0 & -i\\i & 0\end{pmatrix}, \qquad \hat{\sigma}_z = \begin{pmatrix}1 & 0\\0 & 1\end{pmatrix},$$

from which the following relations are readily derived:

$$\begin{aligned} \hat{\sigma}_z\chi_{\uparrow} &= \chi_{\uparrow} & \hat{\sigma}_z\chi_{\downarrow} &= -\chi_{\downarrow}\\ \hat{\sigma}_y\chi_{\uparrow} &= i\chi_{\downarrow} & \hat{\sigma}_y\chi_{\downarrow} &= -i\chi_{\uparrow}\\ \hat{\sigma}_x\chi_{\uparrow} &= \chi_{\downarrow} & \hat{\sigma}_x\chi_{\downarrow} &= \chi_{\uparrow}. \end{aligned}$$

It is then a trivial matter to show that

$$\sum_{\sigma,\sigma'} \hat{c}^+_{\mathbf{k}\sigma}\chi^+_{\sigma}\,\hat{\mathbf{s}}\,\chi_{\sigma'}\,\hat{c}_{\mathbf{k}'\sigma'}$$
$$= \left\{\tfrac{1}{2}(\hat{c}^+_{\mathbf{k}\uparrow}\hat{c}_{\mathbf{k}'\downarrow}+\hat{c}^+_{\mathbf{k}\downarrow}\hat{c}_{\mathbf{k}'\uparrow}),\ \frac{1}{2i}(\hat{c}^+_{\mathbf{k}\uparrow}\hat{c}_{\mathbf{k}'\downarrow}-\hat{c}^+_{\mathbf{k}\downarrow}\hat{c}_{\mathbf{k}'\uparrow}),\ \tfrac{1}{2}(\hat{c}^+_{\mathbf{k}\uparrow}\hat{c}_{\mathbf{k}'\uparrow}-\hat{c}^+_{\mathbf{k}\downarrow}\hat{c}_{\mathbf{k}'\downarrow})\right\}, \tag{5.67 a}$$

and hence (5.66) becomes

$$\hat{\mathfrak{J}}(\mathbf{q}) = \int d\mathbf{r}\,\exp(-i\mathbf{q}.\mathbf{r}) \sum_{\substack{\mathbf{k}\lambda\\ \mathbf{k}'\lambda'}} \varphi^*_{\mathbf{k}\lambda}(\mathbf{r})\varphi_{\mathbf{k}'\lambda'}(\mathbf{r})\times$$
$$\times\left\{\tfrac{1}{2}(\hat{c}^+_{\mathbf{k}\lambda\uparrow}\hat{c}_{\mathbf{k}'\lambda'\downarrow}+\hat{c}^+_{\mathbf{k}\lambda\downarrow}\hat{c}_{\mathbf{k}'\lambda'\uparrow}),\ \frac{1}{2i}(\hat{c}^+_{\mathbf{k}\lambda\uparrow}\hat{c}_{\mathbf{k}'\lambda'\downarrow}-\hat{c}^+_{\mathbf{k}\lambda\downarrow}\hat{c}_{\mathbf{k}'\lambda'\uparrow}),\right.$$
$$\left.\tfrac{1}{2}(\hat{c}^+_{\mathbf{k}\lambda\uparrow}\hat{c}_{\mathbf{k}'\lambda'\uparrow}-\hat{c}^+_{\mathbf{k}\lambda\downarrow}\hat{c}_{\mathbf{k}'\lambda'\downarrow})\right\}. \tag{5.67 b}$$

Since the spin angular momentum operators $\hat{S}^x$ and $\hat{S}^y$, when expressed in terms of raising and lowering operators $\hat{S}^+$ and $\hat{S}^-$, read

$$\hat{S}^x = \tfrac{1}{2}(\hat{S}^+ + \hat{S}^-) \quad \text{and} \quad \hat{S}^y = \frac{1}{2i}(\hat{S}^+ - \hat{S}^-), \tag{5.68}$$

(5.67) implies that the equivalent electron-spin raising and lowering operators (or spin-flip operators) are

$$\hat{c}^+_{\uparrow}\hat{c}_{\downarrow} \quad \text{and} \quad \hat{c}^+_{\downarrow}\hat{c}_{\uparrow}$$

respectively, while the operator for its z component of spin is

$$\tfrac{1}{2}(\hat{c}_{\uparrow}^{\dagger}\hat{c}_{\uparrow}-\hat{c}_{\downarrow}^{\dagger}\hat{c}_{\downarrow}).$$

To proceed further in our task of obtaining an expression for the neutron cross-section in terms of the formalism of band theory, we introduce the tight-binding approximation. This consists of expanding Bloch functions in terms of a set of wave functions that are strongly localized about the ion-core potentials, these functions being taken to resemble atomic wave functions closely. This approximation scheme is based upon the argument that if the radial part of the wave function of the unpaired electrons in the free atom is sufficiently small for them to suffer only a small distortion on forming the solid, then the wave function of these electrons in the solid must surely be well represented by some linear combination of the corresponding atomic wave functions, i.e. tight-binding theory is expected to yield a good description of electrons in narrow energy bands. The form of the linear combination is dictated by the requirement that the resultant wave function should satisfy Bloch's theorem, namely

$$\varphi_{\mathbf{k}}(\mathbf{r}+\mathbf{l}) = \exp(i\mathbf{k}.\mathbf{l})\varphi_{\mathbf{k}}(\mathbf{r}). \tag{5.69}$$

For a single band the combination

$$\sum_{\mathbf{l}} \exp(i\mathbf{k}.\mathbf{l})\phi(\mathbf{r}-\mathbf{l}) \tag{5.70}$$

satisfies this requirement, where ϕ are the localized or tight-binding wave functions. In order that we may introduce operators that create and annihilate electrons in the states ϕ, as we shall in Chapter 9, we must assume the wave functions ϕ to be orthogonal, i.e. to satisfy

$$\int d\mathbf{r}\, \phi^*(\mathbf{r})\phi(\mathbf{r}-\mathbf{l}) = 0 \quad (\mathbf{l} \neq 0). \tag{5.71}$$

If the tight-binding functions are taken to be atomic orbitals then, clearly, (5.71) is *not* true. However, for our purposes we shall merely assume that a set of wave functions ϕ that satisfy (5.71) exist, and that these can, if need be, be approximated to by replacing them in the formalism by atomic orbitals, this being the equivalent to the usual tight-binding theory. In the limit of zero bandwidth this replacement is, of course, an exact one. (In the theory of the transport properties of Bloch electrons in external fields the set of functions ϕ that we have defined are called Wannier functions.)

In the general case of several bands and also more than one atom per unit cell we have to replace the expansion coefficients $\exp(i\mathbf{k}.\mathbf{l})$ in

(5.70) by more complicated ones. We shall verify that the required (exact) expansion in this general case is

$$\varphi_{\mathbf{k}\lambda}(\mathbf{r}) = \frac{1}{\sqrt{N}} \sum_{l,d,\Lambda} \exp(i\mathbf{k}.\mathbf{l})\, a_\lambda^\Lambda(\mathbf{d};\mathbf{k})\phi_\Lambda(\mathbf{r}-\mathbf{R}_{ld}), \tag{5.72}$$

where Λ specifies the various wave functions that contribute to a given band. (We draw attention to the similarity between the following argument and that given in § 4.1 for the expansion of the displacement operator u for a general lattice structure.) For d electrons there would be five different types of wave function, so $\Lambda = 1, 2,..., 5$. The coefficients $a_\lambda^\Lambda(\mathbf{d};\mathbf{k})$ in (5.72) will be shown to be solutions of an eigenfunction equation (see (5.78)). The wave functions ϕ_Λ satisfy the orthonormality condition

$$\int d\mathbf{r}\; \phi_\Lambda^*(\mathbf{r}-\mathbf{R}_{ld})\phi_{\Lambda'}(\mathbf{r}-\mathbf{R}_{l'd'}) = \delta_{l,l'}\,\delta_{d,d'}\,\delta_{\Lambda,\Lambda'}. \tag{5.73}$$

The operator that creates an electron in the state ϕ_Λ localized about $\mathbf{R}_{ld}$ is given in terms of the operators for the Bloch states $\varphi_{\mathbf{k}\lambda}$ by

$$\hat{c}^+_{ld\Lambda\sigma} = \frac{1}{\sqrt{N}} \sum_{\mathbf{k}\lambda} \exp(-i\mathbf{k}.\mathbf{l})\{a_\lambda^\Lambda(\mathbf{d};\mathbf{k})\}^*\hat{c}^+_{\mathbf{k}\lambda\sigma}, \tag{5.74}$$

which, because (5.74) is a canonical transformation, satisfy

$$[\hat{c}_{ld\Lambda\sigma}, \hat{c}^+_{l'd'\Lambda'\sigma'}]_+ = \delta_{l,l'}\,\delta_{d,d'}\,\delta_{\Lambda,\Lambda'}\,\delta_{\sigma,\sigma'} \tag{5.75}$$

all other anticommutators being zero.

Finally, in terms of the operators $\hat{c}^+_{ld\Lambda\sigma}$ and $\hat{c}_{ld\Lambda\sigma}$ the Hamiltonian (5.55) is, from (5.52 b),

$$\mathscr{H} = \sum_{\substack{l,l'\\ d,d'}} \sum_{\Lambda,\Lambda'} \sum_{\sigma} T^{dd'}_{\Lambda\Lambda'}(\mathbf{l}-\mathbf{l}')\hat{c}^+_{ld\Lambda\sigma}\hat{c}_{l'd'\Lambda'\sigma}, \tag{5.76}$$

where the matrix elements $T^{dd'}_{\Lambda\Lambda'}(\mathbf{l}-\mathbf{l}')$ are†

$$T^{dd'}_{\Lambda\Lambda'}(\mathbf{l}-\mathbf{l}') = \int d\mathbf{r}\; \phi_\Lambda^*(\mathbf{r}-\mathbf{R}_{ld})\left\{\frac{1}{2m_e}\hat{\mathbf{p}}^2+\hat{V}(\mathbf{r})\right\}\phi_{\Lambda'}(\mathbf{r}-\mathbf{R}_{l'd'}). \tag{5.77}$$

The eigenvalue equation for the transformation coefficients is

$$\sum_{l'd'\Lambda'} T^{dd'}_{\Lambda\Lambda'}(\mathbf{l}-\mathbf{l}')a_\lambda^{\Lambda'}(\mathbf{d}';\mathbf{k})\exp(i\mathbf{k}.\mathbf{l}') = \mathscr{E}_\lambda(\mathbf{k})a_\lambda^\Lambda(\mathbf{d};\mathbf{k})\exp(i\mathbf{k}.\mathbf{l}). \tag{5.78}$$

We now review some of the algebra for the transformation from the Bloch states $\varphi_{\mathbf{k}\lambda}$ to the states $\phi_\Lambda(\mathbf{r}-\mathbf{R}_{ld})$.

First we observe that $T^{dd'}_{\Lambda\Lambda'}(\mathbf{l}-\mathbf{l}')$ has the property

$$T^{dd'}_{\Lambda\Lambda'}(\mathbf{l}-\mathbf{l}') = T^{d'd}_{\Lambda'\Lambda}(\mathbf{l}'-\mathbf{l}), \tag{5.79}$$

† $\hat{V}$ will be spin dependent in a magnetically ordered system and thus $T^{dd'}_{\Lambda\Lambda'}$ and a_λ^Λ, from eqn (5.78), will also depend on spin, i.e. $\lambda \to \lambda, \sigma$.

hence† $$a_\lambda^\Lambda(\mathbf{d};\mathbf{k}) = \{a_\lambda^\Lambda(\mathbf{d};-\mathbf{k})\}^*, \tag{5.80}$$

as is readily seen on using (5.79) in (5.78). To prove (5.79), note that because $T^{dd'}_{\Lambda\Lambda'}(\mathbf{l}-\mathbf{l}')$ is the matrix element of a Hermitian operator (the single-particle Hamiltonian) it must satisfy

$$T^{dd'}_{\Lambda\Lambda'}(\mathbf{l}-\mathbf{l}') = \{T^{d'd}_{\Lambda'\Lambda}(\mathbf{l}'-\mathbf{l})\}^*$$

and further, since the wave functions ϕ represent stationary states they can always be taken to be purely real; hence (5.79) follows.

Also, the coefficients a_λ^Λ must satisfy the orthogonality relations

$$\sum_{d,\Lambda}\{a_\lambda^\Lambda(\mathbf{d};\mathbf{k})\}^* a_{\lambda'}^\Lambda(\mathbf{d};\mathbf{k}) = \delta_{\lambda,\lambda'} \tag{5.81 a}$$

and $$\sum_{\lambda}\{a_\lambda^{\Lambda'}(\mathbf{d}';\mathbf{k})\}^* a_\lambda^\Lambda(\mathbf{d};\mathbf{k}) = \delta_{d,d'}\,\delta_{\Lambda,\Lambda'}. \tag{5.81 b}$$

The inverse transformation to (5.72) is obtained by multiplying both sides by $\{\exp(i\mathbf{k}.\mathbf{l}')a_\lambda^{\Lambda'}(\mathbf{d}';\mathbf{k})\}^*$ and summing over $\mathbf{k}$ and λ:

$$\sum_{\mathbf{k},\lambda}\exp(-i\mathbf{k}.\mathbf{l}')\{a_\lambda^{\Lambda'}(\mathbf{d}';\mathbf{k})\}^*\varphi_{\mathbf{k}\lambda}(\mathbf{r})$$

$$= \frac{1}{\sqrt{N}}\sum_{l,d,\Lambda}\sum_{\mathbf{k}}\exp\{i\mathbf{k}(\mathbf{l}-\mathbf{l}')\}\sum_{\lambda}a_\lambda^\Lambda(\mathbf{d};\mathbf{k})\{a_\lambda^{\Lambda'}(\mathbf{d}';\mathbf{k})\}^*\phi_\Lambda(\mathbf{r}-\mathbf{R}_{ld})$$

$$= \frac{1}{\sqrt{N}}\sum_{l,d,\Lambda}N\delta_{l,l'}\,\delta_{d,d'}\,\delta_{\Lambda,\Lambda'}\,\phi_\Lambda(\mathbf{r}-\mathbf{R}_{ld}) = \sqrt{N}\,\phi_{\Lambda'}(\mathbf{r}-\mathbf{R}_{l'd'}),$$

i.e. $$\phi_\Lambda(\mathbf{r}-\mathbf{R}_{ld}) = \frac{1}{\sqrt{N}}\sum_{\mathbf{k},\lambda}\exp(-i\mathbf{k}.\mathbf{l})\{a_\lambda^\Lambda(\mathbf{d};\mathbf{k})\}^*\varphi_{\mathbf{k}\lambda}(\mathbf{r}). \tag{5.82}$$

Similarly, $$\hat{c}^+_{\mathbf{k}\lambda\sigma} = \frac{1}{\sqrt{N}}\sum_{ld\Lambda}\exp(i\mathbf{k}.\mathbf{l})a_\lambda^\Lambda(\mathbf{d};\mathbf{k})\hat{c}^+_{ld\Lambda\sigma}. \tag{5.83}$$

It now remains to show that the Hamiltonian (5.76) corresponds to the Bloch Hamiltonian given by eqn (5.55). Substitute for the operators in (5.76) the expansion (5.74) and its conjugate. Thus,

$$\mathscr{H} = \frac{1}{N}\sum_{\mathbf{k},\lambda}\sum_{\mathbf{k}',\lambda'}\sum_{\substack{l,l'\\d,d'}}\sum_{\Lambda,\Lambda'}\sum_{\sigma}T^{dd'}_{\Lambda\Lambda'}(\mathbf{l}-\mathbf{l}')\times$$

$$\times\exp(-i\mathbf{k}.\mathbf{l})\{a_\lambda^\Lambda(\mathbf{d};\mathbf{k})\}^*\hat{c}^+_{\mathbf{k}\lambda\sigma}\exp(i\mathbf{k}'.\mathbf{l}')a_{\lambda'}^{\Lambda'}(\mathbf{d}';\mathbf{k}')\hat{c}_{\mathbf{k}'\lambda'\sigma},$$

† The atomic wave functions in (5.72) are often written as linear combinations that transform as the irreducible representation of the continuous group **l** in the crystal symmetry (cf. Slater and Koster, *Phys. Rev.* **94**, 1498 (1954)). In this instance the coefficients (5.80) are purely real.

and on using (5.78),

$$\mathscr{H} = \frac{1}{N}\sum_{\mathbf{k},\lambda}\sum_{\mathbf{k}',\lambda'}\sum_{l,d,\Lambda}\sum_{\sigma}\exp(-i\mathbf{k}.\mathbf{l})\{a_\lambda^\Lambda(\mathbf{d};\mathbf{k})\}^*\times$$

$$\times \hat{c}^+_{\mathbf{k}\lambda\sigma}\exp(i\mathbf{k}'.\mathbf{l})a_{\lambda'}^{\Lambda}(\mathbf{d};\mathbf{k}')\hat{c}_{\mathbf{k}'\lambda'\sigma}\mathscr{E}_{\lambda'}(\mathbf{k}')$$

$$= \sum_{\mathbf{k},\lambda}\sum_{\mathbf{k}',\lambda'}\delta_{\mathbf{k},\mathbf{k}'}\sum_{d,\Lambda}\sum_{\sigma}\{a_\lambda^\Lambda(\mathbf{d};\mathbf{k})\}^*a_{\lambda'}^{\Lambda}(\mathbf{d};\mathbf{k})\mathscr{E}_{\lambda'}(\mathbf{k})\hat{c}^+_{\mathbf{k}\lambda\sigma}\hat{c}_{\mathbf{k}\lambda'\sigma}$$

$$= \sum_{\mathbf{k},\lambda}\sum_{\mathbf{k}',\lambda'}\delta_{\mathbf{k},\mathbf{k}'}\delta_{\lambda,\lambda'}\sum_{\sigma}\mathscr{E}_\lambda(\mathbf{k})\hat{c}^+_{\mathbf{k}\lambda\sigma}\hat{c}_{\mathbf{k}\lambda\sigma} = \sum_{\mathbf{k}\lambda\sigma}\mathscr{E}_\lambda(\mathbf{k})\hat{c}^+_{\mathbf{k}\lambda\sigma}\hat{c}_{\mathbf{k}\lambda\sigma},$$

where to reach the penultimate formula we have used (5.81 a).

Consider the spatial part of $\hat{\mathfrak{s}}(\mathbf{q})$, eqn (5.66), namely

$$\int d\mathbf{r}\exp(-i\mathbf{q}.\mathbf{r})\sum_{\substack{\mathbf{k},\lambda\\ \mathbf{k}',\lambda'}}\varphi^*_{\mathbf{k}\lambda}(\mathbf{r})\varphi_{\mathbf{k}'\lambda'}(\mathbf{r}) \tag{5.84}$$

and transform to the localized functions ϕ. From (5.72) this is

$$\frac{1}{N}\int d\mathbf{r}\exp(-i\mathbf{q}.\mathbf{r})\sum_{\substack{\mathbf{k},\lambda\\ \mathbf{k}',\lambda'}}\sum_{l,d,\Lambda}\sum_{l',d',\Lambda'}\exp(-i\mathbf{k}.\mathbf{l})\{a_\lambda^\Lambda(\mathbf{d};\mathbf{k})\}^*\phi^*_\Lambda(\mathbf{r}-\mathbf{R}_{ld})\times$$

$$\times\exp(i\mathbf{k}'.\mathbf{l}')a_{\lambda'}^{\Lambda'}(\mathbf{d}';\mathbf{k}')\phi_{\Lambda'}(\mathbf{r}-\mathbf{R}_{l'd'}),$$

which, on changing the integration variable $\mathbf{r}\to\mathbf{r}+\mathbf{R}_{ld}$, becomes

$$\frac{1}{N}\sum_{l,d,\Lambda}\sum_{l',d',\Lambda'}\int d\mathbf{r}\sum_{\substack{\mathbf{k},\lambda\\ \mathbf{k}',\lambda'}}\exp(-i\mathbf{q}.\mathbf{r}-i\mathbf{q}.\mathbf{R}_{ld}-i\mathbf{k}.\mathbf{l}+i\mathbf{k}'.\mathbf{l}')\times$$

$$\times a_{\lambda'}^{\Lambda'}(\mathbf{d}';\mathbf{k}')\{a_\lambda^\Lambda(\mathbf{d};\mathbf{k})\}^*\phi^*_\Lambda(\mathbf{r})\phi_{\Lambda'}(\mathbf{r}+\mathbf{R}_{ld}-\mathbf{R}_{l'd'})$$

$$= \frac{1}{N}\sum_{l,l'}\sum_{d,\Lambda}\sum_{d',\Lambda'}\int d\mathbf{r}\sum_{\substack{\mathbf{k},\lambda\\ \mathbf{k}',\lambda'}}\exp(-i\mathbf{q}.\mathbf{r}-i\mathbf{q}.\mathbf{d})\exp\{-i\mathbf{l}'.(\mathbf{k}-\mathbf{k}'+\mathbf{q})\}\times$$

$$\times\exp\{-i\mathbf{l}(\mathbf{k}+\mathbf{q})\}a_{\lambda'}^{\Lambda'}(\mathbf{d}';\mathbf{k}')\{a_\lambda^\Lambda(\mathbf{d};\mathbf{k})\}^*\phi^*_\Lambda(\mathbf{r})\phi_{\Lambda'}(\mathbf{r}+\mathbf{l}+\mathbf{d}-\mathbf{d}'). \tag{5.85}$$

The sum over $\mathbf{l}'$ in (5.85) gives $N\delta_{\mathbf{k}',\mathbf{k}+\mathbf{q}}$ so that the whole expression reduces to

$$\sum_{l}\sum_{d,\Lambda}\sum_{d',\Lambda'}\sum_{\substack{\mathbf{k},\lambda\\ \mathbf{k}',\lambda'}}\delta_{\mathbf{k}',\mathbf{k}+\mathbf{q}}\int d\mathbf{r}\exp(-i\mathbf{q}.\mathbf{r})\exp(-i\mathbf{q}.\mathbf{d}-i\mathbf{k}'.\mathbf{l})\times$$

$$\times a_{\lambda'}^{\Lambda'}(\mathbf{d}',\mathbf{k}')\{a_\lambda^\Lambda(\mathbf{d};\mathbf{k})\}^*\phi^*_\Lambda(\mathbf{r})\phi_{\Lambda'}(\mathbf{r}+\mathbf{l}+\mathbf{d}-\mathbf{d}'). \tag{5.86}$$

If $\mathbf{q} = 0$ the orthonormality condition (5.73) gives for (5.86)

$$\sum_{l}\sum_{d,\Lambda}\sum_{d',\Lambda'}\sum_{\substack{\mathbf{k},\lambda\\ \mathbf{k}',\lambda'}}\delta_{\mathbf{k},\mathbf{k}'}\delta_{l,0}\delta_{d,d'}\delta_{\Lambda,\Lambda'}a_{\lambda'}^{\Lambda}(\mathbf{d};\mathbf{k})\{a_\lambda^\Lambda(\mathbf{d};\mathbf{k})\}^* = \sum_{\substack{\mathbf{k},\lambda\\ \mathbf{k}',\lambda'}}\delta_{\mathbf{k},\mathbf{k}'}\delta_{\lambda,\lambda'},$$

which is seen to agree with the result obtained from (5.84) for $\mathbf{q} = 0$, when it is recalled that the Bloch functions are orthogonal. This result means that for $\boldsymbol{\varkappa} = 0$ the scattering amplitude is proportional to the total spin density within a unit cell.

Eqn (5.86) is an *exact* expression for (5.84) in terms of the localized functions ϕ, its utility being that we can use the fact that the functions ϕ are well localized to obtain a meaningful approximation for $\mathfrak{Z}(\mathbf{q})$ that is simpler than that given by (5.66). We therefore introduce the assumption that the overlap of the ϕ's on different sites can be ignored in (5.86) and thus set $\mathbf{l} = 0$ and $\mathbf{d} = \mathbf{d}'$. In this approximation, the expression for $\mathfrak{Z}(\mathbf{q})$ becomes

$$\mathfrak{Z}(\mathbf{q}) = \sum_{\Lambda,\Lambda'} \int \mathrm{d}\mathbf{r}\, \exp(-\mathrm{i}\mathbf{q}.\mathbf{r})\phi^*_\Lambda(\mathbf{r})\phi_{\Lambda'}(\mathbf{r}) \sum_{\mathbf{k},\lambda,\lambda'} \sum_d \exp(-\mathrm{i}\mathbf{q}.\mathbf{d})\times$$
$$\times a^{\Lambda'}_{\lambda'}(\mathbf{d};\mathbf{k}+\mathbf{q})\{a^{\Lambda}_{\lambda}(\mathbf{d};\mathbf{k})\}^* \sum_{\sigma,\sigma'} \hat{c}^+_{\mathbf{k}\lambda\sigma}\chi^+_\sigma \hat{\mathbf{s}}\chi_{\sigma'}\hat{c}_{\mathbf{k}+\mathbf{q},\lambda'\sigma'}. \qquad (5.87)$$

Let us consider this expression for the specific case of a set of degenerate bands arising from electrons of a particular atomic shell, for instance the $3d$ electrons of a transition-metal ion. The labels Λ on the localized functions ϕ_Λ then refer to the magnetic quantum number and the radial parts of these functions will be very similar. Thus it might well be appropriate to ignore their dependence on Λ and replace the integral by the atomic form factor F. Under these circumstances we obtain for $\mathfrak{Z}(-\boldsymbol{\varkappa})$ the approximate expression

$$\mathfrak{Z}(-\boldsymbol{\varkappa}) = F(\boldsymbol{\varkappa}) \sum_{\mathbf{k}} \sum_{\substack{\Lambda,\Lambda'\\ \lambda,\lambda'}} \sum_d \exp(\mathrm{i}\boldsymbol{\varkappa}.\mathbf{d}) a^{\Lambda'}_{\lambda'}(\mathbf{d};\mathbf{k}-\boldsymbol{\varkappa})\times$$
$$\times\{a^{\Lambda}_{\lambda}(\mathbf{d};\mathbf{k})\}^* \sum_{\sigma,\sigma'} \hat{c}^+_{\mathbf{k},\lambda,\sigma}\chi^+_\sigma \hat{\mathbf{s}}\chi_{\sigma'}\hat{c}_{\mathbf{k}-\boldsymbol{\varkappa},\lambda'\sigma'}, \qquad (5.88)$$

so that the details of the band structure are embedded in the expansion coefficients a^{Λ}_{λ}. Note that only in the case of a *single* band and a *Bravais lattice* are these coefficients equal to unity, when the expression (5.87) for $\mathfrak{Z}$ reduces to the simple form

$$\mathfrak{Z}(\mathbf{q}) = \int \mathrm{d}\mathbf{r}\, \exp(-\mathrm{i}\mathbf{q}.\mathbf{r})|\phi(\mathbf{r})|^2 \sum_{\mathbf{k}} \sum_{\sigma,\sigma'} \hat{c}^+_{\mathbf{k}\sigma}\chi^+_\sigma \hat{\mathbf{s}}\chi^+_{\sigma'}\hat{c}_{\mathbf{k}+\mathbf{q},\sigma'}. \qquad (5.89)$$

For this particular case we observe that, in the light of the discussion following (5.67), the spin part of the cross-section for the scattering by itinerant electrons has a structure similar to that for spin-only scattering by localized electrons.

It is not appropriate for us to give a lengthy discussion here but we draw attention to one difference, which may be observable between form factors for the itinerant electron model and for the localized model. It is a characteristic approximation in the itinerant theory, as outlined here, to assume all integrals involving an overlap between different functions, say $\phi(\mathbf{r})$ and $\phi(\mathbf{r}-\mathbf{l})$, to be zero. In this approximation the effective form factor is

$$F(\boldsymbol{\varkappa}) = \int d\mathbf{r}\, \exp(i\boldsymbol{\varkappa}.\mathbf{r})|\phi(\mathbf{r})|^2.$$

However, in salts, the atomic form factor is defined as (see (5.16))

$$F_d(\boldsymbol{\varkappa}) = \int d\mathbf{r}\, \exp(i\boldsymbol{\varkappa}.\mathbf{r})\mathscr{s}_d(\mathbf{r}),$$

where $\mathscr{s}_d(\mathbf{r})$ is the normalized density on atom d. If we denote the atomic wave function as $A(\mathbf{r})$ then we have

$$F_d(\boldsymbol{\varkappa}) = \int d\mathbf{r}\, \exp(i\boldsymbol{\varkappa}.\mathbf{r})|A(\mathbf{r})|^2.$$

These formulae are not quite the same because $\phi(\mathbf{r})$ is a Wannier function (cf. p. 119) and differs a little from $A(\mathbf{r})$. In general

$$\phi(\mathbf{r}) = \sum_n W_n A(\mathbf{r}-\mathbf{n}),$$

where W_n are the Wannier expansion coefficients with W_0 almost, but not quite, equal to unity. Neglecting all overlap effects,

$$\sum_n W_n^2 = 1$$

and

$$F(\boldsymbol{\varkappa}) = F_d(\boldsymbol{\varkappa}) \sum_n |W_n|^2 \exp(i\boldsymbol{\varkappa}.\mathbf{n}).$$

For purposes of illustration let us suppose that the Wannier function spreads only to the r nearest neighbours labelled with vectors $\boldsymbol{\rho}$. Then

$$\phi(\mathbf{r}) \simeq (1-rW_1^2)^{\frac{1}{2}}A(\mathbf{r}) + \sum_\rho W_1 A(\mathbf{r}-\boldsymbol{\rho})$$

and, neglecting all overlap terms,

$$F(\boldsymbol{\varkappa}) \simeq F_d(\boldsymbol{\varkappa})\Big\{1-rW_1^2+\sum_\rho W_1^2 \exp(i\boldsymbol{\varkappa}.\boldsymbol{\rho})\Big\}.$$

In general, therefore, we might expect $F(\boldsymbol{\varkappa})$ to fall off more rapidly with $\boldsymbol{\varkappa}$ than $F_d(\boldsymbol{\varkappa})$. Note, however, that for any reciprocal lattice vector $\boldsymbol{\tau}$

$$F(\boldsymbol{\tau}) \equiv F_d(\boldsymbol{\tau}).$$

Therefore the difference between these functions cannot be observed in ferromagnets by Bragg scattering. In principle, however, a difference in form factor can exist and in some ways the problem is analogous to covalency effects in salts (to be discussed in Chapter 7).

The foregoing discussion is greatly oversimplified compared to the situation we expect to find in a real metal. We use it for illustrative purposes only, and to draw attention to the fact that the form factor for inelastic scattering in a metal need not be a smooth interpolation of the factor measured at reciprocal lattice points in Bragg scattering experiments.

Turning our attention to the second part of $\hat{\mathbf{Q}}_\perp$, that part of the neutron–electron interaction which arises from the translational motion of the latter, we can show that it is probable that in most cases of interest the contribution to the scattering amplitude from this term is much smaller than that arising from the spin part considered above.

Utilizing the fact that $\exp(i\boldsymbol{\kappa}.\mathbf{r}_i)$ commutes with $\tilde{\boldsymbol{\kappa}}\times\hat{\mathbf{p}}_i$,

$$\sum_i \exp(i\boldsymbol{\kappa}.\mathbf{r}_i)\tilde{\boldsymbol{\kappa}}\times\hat{\mathbf{p}}_i = \tfrac{1}{2}\sum_i \tilde{\boldsymbol{\kappa}}\times\{\hat{\mathbf{p}}_i\exp(i\boldsymbol{\kappa}.\mathbf{r}_i)+\exp(i\boldsymbol{\kappa}.\mathbf{r}_i)\hat{\mathbf{p}}_i\}$$

$$= \tfrac{1}{2}\sum_i \tilde{\boldsymbol{\kappa}}\times\int d\mathbf{r}\,\exp(i\boldsymbol{\kappa}.\mathbf{r})\{\hat{\mathbf{p}}_i\delta(\mathbf{r}-\mathbf{r}_i)+\delta(\mathbf{r}-\mathbf{r}_i)\hat{\mathbf{p}}_i\}; \tag{5.90}$$

so let us consider

$$\hat{\mathscr{p}}(\mathbf{r}) = \tfrac{1}{2}\sum_i \hat{\mathbf{p}}_i\delta(\mathbf{r}-\mathbf{r}_i)+\delta(\mathbf{r}-\mathbf{r}_i)\hat{\mathbf{p}}_i, \tag{5.91}$$

which is the momentum density operator.

In second quantized form

$$\hat{\mathscr{p}}(\mathbf{r}) = \tfrac{1}{2}\{\hat{\psi}^+(\mathbf{r})\hat{\mathbf{p}}\hat{\psi}(\mathbf{r})+\hat{\psi}(\mathbf{r})\hat{\mathbf{p}}\hat{\psi}^+(\mathbf{r})\}, \tag{5.92}$$

which for Bloch states

$$= \tfrac{1}{2}\sum_{\mathbf{k},\lambda}\sum_{\mathbf{k}',\lambda'}\{\hat{c}^+_{\mathbf{k}\lambda}\hat{c}_{\mathbf{k}'\lambda'}\varphi^*_{\mathbf{k}\lambda}(\mathbf{r})\hat{\mathbf{p}}\varphi_{\mathbf{k}'\lambda'}(\mathbf{r})+\hat{c}_{\mathbf{k}'\lambda'}\hat{c}^+_{\mathbf{k}\lambda}\varphi_{\mathbf{k}'\lambda'}(\mathbf{r})\hat{\mathbf{p}}\varphi^*_{\mathbf{k}\lambda}(\mathbf{r})\} \tag{5.93}$$

(we obviously have $\sigma = \sigma'$ because $\hat{\mathscr{p}}$ is independent of electron spin).

Since the diagonal matrix elements of $\hat{\mathbf{p}}$ are zero, and using the anti-commutation relation for the Bloch-state operators, (5.93) can be rewritten

$$\hat{\mathscr{p}}(\mathbf{r}) = \tfrac{1}{2}\sum_{\mathbf{k},\lambda}\sum_{\mathbf{k}',\lambda'}\hat{c}^+_{\mathbf{k}\lambda}\hat{c}_{\mathbf{k}'\lambda'}\{\varphi^*_{\mathbf{k}\lambda}(\mathbf{r})\hat{\mathbf{p}}\varphi_{\mathbf{k}'\lambda'}(\mathbf{r})-\varphi_{\mathbf{k}'\lambda'}(\mathbf{r})\hat{\mathbf{p}}\varphi^*_{\mathbf{k}\lambda}(\mathbf{r})\}, \tag{5.94}$$

and thus

$$\hat{\mathscr{p}}(\mathbf{q}) = \int d\mathbf{r}\,\exp(-i\mathbf{q}.\mathbf{r})\hat{\mathscr{p}}(\mathbf{r}) = \tfrac{1}{2}\sum_{\mathbf{k},\lambda}\sum_{\mathbf{k}',\lambda'}\hat{c}^+_{\mathbf{k}\lambda}\hat{c}_{\mathbf{k}'\lambda'}\int d\mathbf{r}\,\exp(-i\mathbf{q}.\mathbf{r})\times$$

$$\times\{\varphi^*_{\mathbf{k}\lambda}(\mathbf{r})\hat{\mathbf{p}}\varphi_{\mathbf{k}'\lambda'}(\mathbf{r})-\varphi_{\mathbf{k}'\lambda'}(\mathbf{r})\hat{\mathbf{p}}\varphi^*_{\mathbf{k}\lambda}(\mathbf{r})\}. \tag{5.95}$$

If $\mathbf{q} = 0$ the integrals involved in (5.95) for $\mathbf{k} = \mathbf{k}'$ and $\lambda = \lambda'$ are equal in magnitude but opposite in sign, and further,

$$\int d\mathbf{r}\,\varphi^*_{\mathbf{k}\lambda}(\mathbf{r})\hat{\mathbf{p}}\varphi_{\mathbf{k}\lambda}(\mathbf{r})$$

is the average value of the momentum of an electron in the band λ and with wave vector $\mathbf{k}$, and this is equal to†

$$\frac{m_e}{\hbar}\operatorname{grad}_{\mathbf{k}}\mathscr{E}_\lambda(\mathbf{k}).$$

† See for example, C. Kittel, *Quantum theory of solids*, Chapter 9, John Wiley (1963).

For a spherical band with an effective mass m_λ^*

$$\mathscr{E}_\lambda(\mathbf{k}) = \frac{\hbar^2 k^2}{2m_\lambda^*},$$

so that, if we are willing to neglect the terms for which $\mathbf{k}, \lambda \neq \mathbf{k}', \lambda'$ in (5.95),

$$\hat{p}(\mathbf{q}=0) \simeq \sum_{\mathbf{k},\lambda} \frac{m_\mathrm{e}}{m_\lambda^*} \hat{c}_{\mathbf{k}\lambda}^+ \hat{c}_{\mathbf{k}\lambda} \hbar\mathbf{k}. \tag{5.96}$$

For narrow bands the ratio m_e/m_λ^* can be of order of 10^{-1} (tight-binding theory leads to the result $m^*/m_\mathrm{e} \propto 1/w$, where w is a measure of the bandwidth: for Fe, $m^*/m_\mathrm{e} \sim 9$) so that the contribution to the scattering amplitude from the interaction between the neutron and the translational motion of the electrons is likely to be smaller than that from the spin–spin part for all $\boldsymbol{\kappa}$.

This comment is supported by a study of the orbital contribution to the inelastic paramagnetic cross-section calculated for a tight-binding band model of nickel.† The spin contribution dominates the cross-section for $\hbar\omega \lesssim 0{\cdot}05$ eV for all $\boldsymbol{\kappa}$ examined along the true high symmetry directions. But at higher energies the situation is markedly different. For example, in the [100] direction the scattering for $\boldsymbol{\kappa} \sim 0$ and $\boldsymbol{\kappa} \sim$ zone boundary comes solely from the orbital interactions for energies exceeding 0·2 eV and 2 eV respectively. The relative importance of the orbital contribution depends on the direction of $\boldsymbol{\kappa}$, and it is found to increase on going from the [100] direction to [110] and [111]. However, in the energy and wave-vector range used in conventional inelastic neutron scattering studies of metals we may conclude that the orbital contribution is indeed small compared with the spin contribution.

† The authors are pleased to acknowledge collaboration with Dr. C. G. Windsor in the study described here.

6

MAGNETIC SCATTERING BY SALTS: THEORY

THIS chapter is devoted to a detailed discussion of the cross-section for the scattering by magnetic ions in crystals, the ions being described by the atomic or Heitler–London model. In particular an exact analysis is given of the contribution to the cross-section arising from the orbital motion of the unpaired electrons, a topic briefly discussed in Chapter 5.

6.1. Introduction

It is clear from the outset that the neutron cross-section must depend on all the factors that determine the magnetic state of an ion in a crystal, i.e. it must be sensitive to the crystal-field environment of the ion, the degree of covalency, and exchange interactions. The two latter effects may well be small. In nearly all cases of interest the unpaired electrons of an ion in a crystal couple according to Hund's rule (the cyanides form an example of salts where Hund's rule breaks down) and the neutrons will certainly not have enough energy to break this coupling.

For transition-metal ions, which are characterized by having an incomplete $3d$ shell, the crystal field is a small perturbation compared with the intra-atomic electrostatic repulsion but large compared with the spin–orbit coupling (intermediate crystal-field case). (Experiments show that the main crystal-field splittings in the transition-metal salts are of the order of 1 eV, whereas the spin–orbit coupling parameter is at most 0·1 eV.) The ion is therefore described first in terms of its total orbital angular momentum L in the crystal field. If this field removes completely the orbital degeneracy of the ground-state wave function then the expectation value of the orbital angular momentum operator is zero; the crystal field is said to quench the orbital moment. When the spin–orbit coupling is introduced, however, it mixes higher energy states into the ground state, with the result that the g value of the ion differs from its spin-only value $g = 2$. Physically, the spin, acting through the spin–orbit coupling, induces an orbital moment. For example, the orbital angular momentum of Ni^{2+} is completely quenched by a cubic crystal field but the measured value of g is 2·2, so that spin–orbit coupling gives rise to a 10 per cent orbital moment. The behaviour of the rare earths in a crystal is somewhat simpler. The mean radius of the $4f$ shell is smaller than the overall size of the ion and the perturbation of the crystal field is smaller than both the intra-atomic electrostatic repulsion between the electrons and the spin–orbit coupling (weak-field case). This means that the 'finished' ion, with L and S coupled to give a total angular momentum J, is examined in the environment of the crystal field, which has the effect of partially or totally removing the $(2J+1)$-fold degeneracy. The total spread in energy of these levels is found to be of the order 5–20 meV. If the incident neutron energy is much greater than this, the cross-section will be independent of the detailed structure of the crystal-field splittings. If, on the other hand, less energetic neutrons are used and an energy analysis is performed on the scattered neutrons, then it is possible to determine the energy-level splittings.

The amplitude of the magnetic scattering by an ion with both spin and orbital angular momentum is proportional to the Fourier transform of the total magnetization density. To prove this we need to express

$$\hat{\mathbf{Q}}_{\perp} = \sum_i \exp(i\boldsymbol{\varkappa}.\mathbf{r}_i)\left\{\tilde{\boldsymbol{\varkappa}}\times(\hat{\mathbf{s}}_i\times\tilde{\boldsymbol{\varkappa}})-\frac{i}{\hbar|\boldsymbol{\varkappa}|}\tilde{\boldsymbol{\varkappa}}\times\hat{\mathbf{p}}_i\right\}$$

(cf. (5.10)) in terms of the spin and orbital angular momentum density operators.

The spin magnetization density operator $\hat{\mathbf{M}}_{\mathrm{S}}(\mathbf{r})$ is defined through the relationship

$$\sum_i \exp(i\boldsymbol{\varkappa}.\mathbf{r}_i)\hat{\mathbf{s}}_i = \int \mathrm{d}\mathbf{r}\, \exp(i\boldsymbol{\varkappa}.\mathbf{r}) \sum_i \hat{\mathbf{s}}_i\,\delta(\mathbf{r}-\mathbf{r}_i)$$

$$= -\frac{1}{2\mu_{\mathrm{B}}}\int \mathrm{d}\mathbf{r}\, \exp(i\boldsymbol{\varkappa}.\mathbf{r})\,\hat{\mathbf{M}}_{\mathrm{S}}(\mathbf{r}), \tag{6.1}$$

i.e.

$$\hat{\mathbf{M}}_{\mathrm{S}}(\mathbf{r}) = -2\mu_{\mathrm{B}} \sum_i \hat{\mathbf{s}}_i\,\delta(\mathbf{r}-\mathbf{r}_i). \tag{6.2}$$

The orbital part can be expressed in terms of the orbital current density operator $\hat{\mathbf{j}}(\mathbf{r})$, where (cf. p. 125)

$$\hat{\mathbf{j}}(\mathbf{r}) = -\frac{e}{2m_{\mathrm{e}}}\sum_i \hat{\mathbf{p}}_i\,\delta(\mathbf{r}-\mathbf{r}_i)+\delta(\mathbf{r}-\mathbf{r}_i)\hat{\mathbf{p}}_i. \tag{6.3}$$

For

$$-\frac{m_{\mathrm{e}}}{e}\int \mathrm{d}\mathbf{r}\, \exp(i\boldsymbol{\varkappa}.\mathbf{r})\hat{\mathbf{j}}(\mathbf{r}) = \tfrac{1}{2}\int \mathrm{d}\mathbf{r}\, \exp(i\boldsymbol{\varkappa}.\mathbf{r}) \sum_i \hat{\mathbf{p}}_i\,\delta(\mathbf{r}-\mathbf{r}_i)+\delta(\mathbf{r}-\mathbf{r}_i)\hat{\mathbf{p}}_i$$

$$= \tfrac{1}{2}\sum_i \hat{\mathbf{p}}_i \exp(i\boldsymbol{\varkappa}.\mathbf{r}_i)+\exp(i\boldsymbol{\varkappa}.\mathbf{r}_i)\hat{\mathbf{p}}_i,$$

so that

$$-\frac{\mathrm{i}}{\hbar|\boldsymbol{\varkappa}|}\tilde{\boldsymbol{\varkappa}}\times\tfrac{1}{2}\sum_i \exp(\mathrm{i}\boldsymbol{\varkappa}.\mathbf{r}_i)\hat{\mathbf{p}}_i+\hat{\mathbf{p}}_i\exp(\mathrm{i}\boldsymbol{\varkappa}.\mathbf{r}_i)$$
$$=\frac{\mathrm{i}m_\mathrm{e}}{\hbar e|\boldsymbol{\varkappa}|}\tilde{\boldsymbol{\varkappa}}\times\int \mathrm{d}\mathbf{r}\,\exp(\mathrm{i}\boldsymbol{\varkappa}.\mathbf{r})\hat{\mathbf{j}}(\mathbf{r}). \qquad (6.4)$$

The classical mean current density $\mathbf{j}$ in a conductor carrying a constant (conduction) current density $\mathbf{j}'$ is given by†

$$\mathbf{j} = c\,\mathrm{curl}\,\mathbf{M}+\mathbf{j}', \qquad (6.5)$$

where $\mathbf{M}$ is the magnetic moment density. The conduction current density is proportional to the electric field and can, therefore, be expressed as the gradient of a scalar function $\phi(\mathbf{r})$, say. Clearly

$$\tilde{\boldsymbol{\varkappa}}\times\int \mathrm{d}\mathbf{r}\,\exp(\mathrm{i}\boldsymbol{\varkappa}.\mathbf{r})\nabla\phi(\mathbf{r}) = 0,$$

so that the second term in (6.5) cannot contribute to the right-hand side of (6.4). We introduce an orbital magnetic moment density operator $\hat{\mathbf{M}}_\mathrm{L}$ by analogy with $\mathbf{M}$ in eqn (6.5), in terms of which

$$\int \mathrm{d}\mathbf{r}\,\exp(\mathrm{i}\boldsymbol{\varkappa}.\mathbf{r})\hat{\mathbf{j}}(\mathbf{r}) = c\int \mathrm{d}\mathbf{r}\,\exp(\mathrm{i}\boldsymbol{\varkappa}.\mathbf{r})\mathrm{curl}\,\hat{\mathbf{M}}_\mathrm{L}(\mathbf{r})$$
$$= c\int \mathrm{d}\mathbf{r}\,\mathrm{curl}\{\exp(\mathrm{i}\boldsymbol{\varkappa}.\mathbf{r})\hat{\mathbf{M}}_\mathrm{L}(\mathbf{r})\}-$$
$$-c\int \mathrm{d}\mathbf{r}\,\mathrm{grad}\{\exp(\mathrm{i}\boldsymbol{\varkappa}.\mathbf{r})\}\times\hat{\mathbf{M}}_\mathrm{L}(\mathbf{r}).$$

The first term integrates to zero, hence

$$\int \mathrm{d}\mathbf{r}\,\exp(\mathrm{i}\boldsymbol{\varkappa}.\mathbf{r})\hat{\mathbf{j}}(\mathbf{r}) = -\mathrm{i}c\boldsymbol{\varkappa}\times\int \mathrm{d}\mathbf{r}\,\exp(\mathrm{i}\boldsymbol{\varkappa}.\mathbf{r})\hat{\mathbf{M}}_\mathrm{L}(\mathbf{r}). \qquad (6.6)$$

Whence, using (6.1), (6.4), and (6.6), we have

$$\hat{\mathbf{Q}}_\perp = \sum_i \exp(\mathrm{i}\boldsymbol{\varkappa}.\mathbf{r}_i)\left\{\tilde{\boldsymbol{\varkappa}}\times(\hat{\mathbf{s}}_i\times\tilde{\boldsymbol{\varkappa}})-\frac{\mathrm{i}}{\hbar|\boldsymbol{\varkappa}|}\tilde{\boldsymbol{\varkappa}}\times\hat{\mathbf{p}}_i\right\}$$
$$=\int \mathrm{d}\mathbf{r}\,\exp(\mathrm{i}\boldsymbol{\varkappa}.\mathbf{r})\left\{-\frac{1}{2\mu_\mathrm{B}}\tilde{\boldsymbol{\varkappa}}\times\{\hat{\mathbf{M}}_\mathrm{S}(\mathbf{r})\times\tilde{\boldsymbol{\varkappa}}\}+\frac{\mathrm{i}m_\mathrm{e}}{\hbar e|\boldsymbol{\varkappa}|}\tilde{\boldsymbol{\varkappa}}\times(-\mathrm{i}c)\{\boldsymbol{\varkappa}\times\hat{\mathbf{M}}_\mathrm{L}(\mathbf{r})\}\right\}$$
$$=-\frac{1}{2\mu_\mathrm{B}}\int \mathrm{d}\mathbf{r}\,\exp(\mathrm{i}\boldsymbol{\varkappa}.\mathbf{r})\tilde{\boldsymbol{\varkappa}}\times\{\hat{\mathbf{M}}(\mathbf{r})\times\tilde{\boldsymbol{\varkappa}}\}, \qquad (6.7)$$

where $\hat{\mathbf{M}}(\mathbf{r}) = \hat{\mathbf{M}}_\mathrm{L}(\mathbf{r})+\hat{\mathbf{M}}_\mathrm{S}(\mathbf{r})$. This is the required result.‡

This result could have been anticipated because the neutrons are scattered through the interaction of their magnetic moment with the magnetic field arising from the unpaired electrons and this, in turn, is

† Landau and E. Lifshitz, *Electrodynamics of continuous media*, Pergamon Press.

‡ G. T. Trammell, *Phys. Rev.* **92**, 1387 (1953), and O. Steinsvoll *et al.*, *Phys. Rev.* **161**, 499 (1967).

determined by their total magnetization. The fact that the scattering amplitude contains the Fourier transform of the magnetization density is a consequence of the Born approximation, for in this approximation the scattering amplitude is given by the Fourier transform of the interaction potential.

Let us now examine the precise form of the matrix elements of $\hat{\mathbf{Q}}_\perp$ taken between the target states $|\lambda\rangle$ and $|\lambda'\rangle$ that occur in the cross-section. (The discussion parallels that given in § 2.3 but is repeated here for completeness.)

The allowed states of the unpaired electrons associated with the jth ion ($j = 1, 2,..., rN$) are defined by wave functions ϕ_j, ϕ'_j,... . The functions ϕ_i and ϕ_j are assumed orthogonal.

Because we are using the Heitler–London model the configurational part of the wave function is constructed from a simple product of the individual wave functions ϕ_j. In the simple version of this model it is assumed that these configurational wave functions are unaffected by the immediate environment of spin ordering. It follows immediately that the quantity we need for Bragg scattering

$$\sum_\lambda p_\lambda \langle\lambda|\, \hat{\mathbf{Q}}_\perp \,|\lambda\rangle$$

involves simply the sum of single-atom contributions. Thus

$$\sum_\lambda p_\lambda\langle\lambda|\, \hat{\mathbf{Q}}_\perp \,|\lambda\rangle = \sum_l \exp(\mathrm{i}\boldsymbol{\kappa}.\mathbf{l}) \sum_d \exp(\mathrm{i}\boldsymbol{\kappa}.\mathbf{d})\langle\phi_d|\, \hat{\mathbf{Q}}_\perp(d) \,|\phi_d\rangle_{\mathrm{av}}, \quad (6.8)$$

where the subscript av means that the average is taken over the occupied wave functions ϕ_d at the appropriate temperature. Since the configurational part of ϕ_d is always the same in this model, the averaging process is, in most cases, trivial. Thus for a ferromagnet the matrix element can be computed for the ground state and the averaging process simply introduces a factor $\langle\hat{\mathbf{S}}\rangle/S$.

For an antiferromagnet of simple type we need to compute

$$\langle\phi_d|\, \hat{\mathbf{Q}}_\perp(d) \,|\phi_d\rangle,$$

first assuming the spin on the atom is fully aligned and then multiplying by $\langle\hat{\mathbf{S}}\rangle/S$—which for an antiferromagnet is not unity even at absolute zero. For more complex spin-ordering patterns, helical arrangements for example, the precise nature of the averaging process must be thought out carefully; particularly when orbital moments are contributing to $\hat{\mathbf{Q}}_\perp$. Nevertheless in all cases the calculation follows two steps: first calculate the individual atom matrix elements; secondly average these in some way dictated by the thermodynamics of the total spin system of the target.

In the following sections we shall be concerned with the evaluation of the matrix elements $\langle\phi_d|\,\hat{\mathbf{Q}}_\perp(d)\,|\phi_d\rangle$. For the sake of brevity we will drop the explicit reference to the dth ion within a unit cell except where the omission is likely to cause confusion.

6.2. Small $|\boldsymbol{\varkappa}|$ approximation

The evaluation of the matrix elements of $\hat{\mathbf{Q}}_\perp$ in the general case is very complicated. However, we can obtain a simple result for small $|\boldsymbol{\varkappa}|$, which is a meaningful approximation to examine since $|\boldsymbol{\varkappa}|^{-1}$ is often much greater than the mean radius of the orbital wave function of the unpaired electrons.

For small $|\boldsymbol{\varkappa}|$ and elastic scattering we can take

$$\exp(\mathrm{i}\boldsymbol{\varkappa}.\mathbf{r})\hat{\mathbf{p}}+\hat{\mathbf{p}}\exp(\mathrm{i}\boldsymbol{\varkappa}.\mathbf{r}) \simeq \mathrm{i}\hbar(\hat{\mathbf{l}}\times\boldsymbol{\varkappa}), \tag{6.9}$$

where $\hat{\mathbf{l}}$ is the orbital angular momentum operator of the electron with respect to the position of the parent nucleus. The proof of (6.9) is as follows.

On expanding the exponentials in the left-hand side of (6.9) we require to evaluate

$$(\boldsymbol{\varkappa}.\mathbf{r})\hat{\mathbf{p}}+\hat{\mathbf{p}}(\boldsymbol{\varkappa}.\mathbf{r}), \tag{6.10}$$

which we divide into two parts by both adding and subtracting

$$\tfrac{1}{2}\{(\boldsymbol{\varkappa}.\hat{\mathbf{p}})\mathbf{r}+\mathbf{r}(\boldsymbol{\varkappa}.\hat{\mathbf{p}})\}$$

to give

$$\begin{aligned}&\tfrac{1}{2}\{(\boldsymbol{\varkappa}.\mathbf{r})\hat{\mathbf{p}}+\hat{\mathbf{p}}(\boldsymbol{\varkappa}.\mathbf{r})\}+\tfrac{1}{2}\{(\boldsymbol{\varkappa}.\mathbf{r})\hat{\mathbf{p}}+\hat{\mathbf{p}}(\boldsymbol{\varkappa}.\mathbf{r})\}+\\ &\quad+\tfrac{1}{2}\{(\boldsymbol{\varkappa}.\hat{\mathbf{p}})\mathbf{r}+\mathbf{r}(\boldsymbol{\varkappa}.\hat{\mathbf{p}})\}-\tfrac{1}{2}\{(\boldsymbol{\varkappa}.\hat{\mathbf{p}})\mathbf{r}+\mathbf{r}(\boldsymbol{\varkappa}.\hat{\mathbf{p}})\}.\end{aligned} \tag{6.11 a}$$

We now choose to group the first and third terms together and the second and fourth terms together to give

$$\begin{aligned}&\tfrac{1}{2}\{(\boldsymbol{\varkappa}.\mathbf{r})\hat{\mathbf{p}}+(\boldsymbol{\varkappa}.\hat{\mathbf{p}})\mathbf{r}+\hat{\mathbf{p}}(\boldsymbol{\varkappa}.\mathbf{r})+\mathbf{r}(\boldsymbol{\varkappa}.\hat{\mathbf{p}})\}+\\ &\quad+\tfrac{1}{2}\{(\boldsymbol{\varkappa}.\mathbf{r})\hat{\mathbf{p}}+\hat{\mathbf{p}}(\boldsymbol{\varkappa}.\mathbf{r})-(\boldsymbol{\varkappa}.\hat{\mathbf{p}})\mathbf{r}-\mathbf{r}(\boldsymbol{\varkappa}.\hat{\mathbf{p}})\}.\end{aligned} \tag{6.11 b}$$

In the first line we write $\hat{\mathbf{p}} = m_\mathrm{e}\dfrac{\partial}{\partial t}\mathbf{r}$ to get

$$2m_\mathrm{e}\frac{\partial}{\partial t}\{(\boldsymbol{\varkappa}.\mathbf{r})\mathbf{r}\} \tag{6.11 c}$$

and the second line becomes

$$\begin{aligned}(\boldsymbol{\varkappa}.\mathbf{r})\hat{\mathbf{p}}-\mathbf{r}(\boldsymbol{\varkappa}.\hat{\mathbf{p}}) &= -\boldsymbol{\varkappa}\times(\mathbf{r}\times\hat{\mathbf{p}})\\ &= -\hbar\boldsymbol{\varkappa}\times\hat{\mathbf{l}}.\end{aligned} \tag{6.11 d}$$

Hence, inserting these results into the expansion of the left-hand side of (6.9) we get

$$\exp(\mathrm{i}\boldsymbol{\varkappa}.\mathbf{r})\hat{\mathbf{p}}+\hat{\mathbf{p}}\exp(\mathrm{i}\boldsymbol{\varkappa}.\mathbf{r}) \simeq 2m_{\mathrm{e}}\frac{\partial}{\partial t}\{\mathbf{r}+\mathrm{i}(\boldsymbol{\varkappa}.\mathbf{r})\mathbf{r}\}+\mathrm{i}\hbar\hat{\mathbf{l}}\times\boldsymbol{\varkappa}$$

$$= \mathrm{i}\hbar\hat{\mathbf{l}}\times\boldsymbol{\varkappa}+\frac{2m_{\mathrm{e}}}{\mathrm{i}\hbar}[(1+\mathrm{i}\boldsymbol{\varkappa}.\mathbf{r})\mathbf{r},\mathscr{H}]. \quad (6.12)$$

For elastic scattering the second term is zero and it is feasible to neglect it also for inelastic scattering.† We thus arrive at eqn (6.9).

With the use of this result

$$\hat{\mathbf{Q}}_{\perp} = \tilde{\boldsymbol{\varkappa}}\times(\hat{\mathbf{s}}\times\tilde{\boldsymbol{\varkappa}})-\frac{\mathrm{i}}{2\hbar|\boldsymbol{\varkappa}|}\tilde{\boldsymbol{\varkappa}}\times(\mathrm{i}\hbar\hat{\mathbf{l}}\times\boldsymbol{\varkappa}) = \tfrac{1}{2}\tilde{\boldsymbol{\varkappa}}\times\{(\hat{\mathbf{l}}+2\hat{\mathbf{s}})\times\tilde{\boldsymbol{\varkappa}}\}. \quad (6.13)$$

Thus, if the unpaired electrons of the ion couple to form a total spin $\hat{\mathbf{S}}$ and orbital angular momentum $\hat{\mathbf{L}}$ the elastic magnetic cross-section (per ion) is approximately given, for small $\boldsymbol{\varkappa}$, by

$$\frac{\mathrm{d}\sigma}{\mathrm{d}\Omega} = \left(\frac{\gamma e^2}{2m_{\mathrm{e}}c^2}\right)^2 \sum_{\alpha,\beta}(\delta_{\alpha\beta}-\tilde{\kappa}_{\alpha}\tilde{\kappa}_{\beta})\sum_{\lambda}p_{\lambda}\langle\lambda|(\hat{\mathbf{L}}+2\hat{\mathbf{S}})_{\alpha}|\lambda\rangle\langle\lambda|(\hat{\mathbf{L}}+2\hat{\mathbf{S}})_{\beta}|\lambda\rangle. \quad (6.14)$$

The evaluation of the matrix elements of $(\hat{\mathbf{L}}+2\hat{\mathbf{S}})_{\alpha}$ that occur in (6.14) for the case of scattering by transition-metal ions requires, in general, a detailed evaluation of the wave functions of the ion in the presence of the crystal field. For those ions in which $\hat{\mathbf{S}}$ is approximately a constant of motion (e.g. Ni^{2+} and Cr^{3+} normally belong to this class), $\hat{\mathbf{L}}$ can be replaced by $(g-2)\hat{\mathbf{S}}$, by definition of the g value of the ion.

To discuss the rare earths we recall that

$$\hat{\mathbf{L}}+2\hat{\mathbf{S}} = g\hat{\mathbf{J}}, \quad (6.15)$$

where g is the Landé splitting factor (cf. § 5.3)

$$g = 1+\frac{J(J+1)-L(L+1)+S(S+1)}{2J(J+1)}. \quad (6.16)$$

Under the condition that the incident neutron energy is much greater than the spread in energy of the $2J+1$ levels, the scattering is elastic. For a single, free ion the cross-section around the forward direction ($\boldsymbol{\varkappa}$ small) is

$$\frac{\mathrm{d}\sigma}{\mathrm{d}\Omega} = \left(\frac{\gamma e^2}{2m_{\mathrm{e}}c^2}\right)^2 g^2\overline{\hat{\mathbf{J}}_{\perp}.\hat{\mathbf{J}}_{\perp}} \quad (6.17)$$

and we can usually take

$$\overline{\hat{\mathbf{J}}_{\perp}.\hat{\mathbf{J}}_{\perp}} = \overline{\hat{\mathbf{J}}^2-(\tilde{\boldsymbol{\varkappa}}.\hat{\mathbf{J}})^2} = \mathbf{J}^2-\tfrac{1}{3}\mathbf{J}^2 = \tfrac{2}{3}J(J+1) \quad (6.18)$$

† G. T. Trammel, *Phys. Rev.* **92**, 1387 (1953).

The cross-section (6.17) is in this instance entirely independent of crystal-field effects and also the temperature.

6.3. General case of ions with both spin and orbital angular momentum

Our discussion of the evaluation of the matrix elements of $\hat{\mathbf{Q}}_{\perp}$ in the general case follows closely that given by Johnston† and makes extensive use of Racah algebra. The conventions and notations we adopt adhere closely to those used in the books by Edmonds and Judd.‡ It is convenient for the purpose of presentation to break the algebra down into a discussion of (a) the orbital term and (b) the spin term, though our ultimate aim is to dovetail these two parts together. In the following we consider a single ion, the cross-section for an array being obtained as in § 6.1.

The algebra involved in this section is particularly tedious, and must inevitably be so. Because it is hard to retain a perspective on the calculation amidst complicated algebraic manipulations, the main results are collected together in § 6.4, which can be read independently of this derivation.

(a) Orbital term

We assume that all n electrons in the unfilled shell of orbital angular momentum l have the same radial wave function $f(r)$. Thus, the wave functions of the individual electrons are

$$|lm\rangle = Y^{l}_{m}(\tilde{\mathbf{r}})f(r), \tag{6.19}$$

where $Y^{l}_{m}(\tilde{\mathbf{r}})$ is a spherical harmonic of rank l and order m.

It is convenient to work in terms of the spherical components of a vector $\mathbf{r}$ (instead of the cartesian components) and, furthermore, it is convenient to define these components so that they look exactly like the spherical harmonics $Y^{1}_{q}(\tilde{\mathbf{r}})$. Hence we define

$$r_{\pm 1} = \mp\frac{1}{\sqrt{2}}(x \pm iy), \qquad r_{0} = z. \tag{6.20 a}$$

Then we have

$$\tilde{r}_{q} = r_{q}/|\mathbf{r}| = \left(\frac{4\pi}{3}\right)^{\frac{1}{2}} Y^{1}_{q}(\tilde{\mathbf{r}}), \quad \text{for } q = -1, 0, +1. \tag{6.20 b}$$

It is well known that product eigenstates, say of orbital and spin

† D. F. Johnston, *Proc. phys. Soc.* 88, 37 (1966).

‡ A. R. Edmonds, *Angular momentum in quantum mechanics*, Princeton University Press; B. R. Judd, *Operator techniques in atomic spectroscopy*, Advanced Physics Monograph Series, McGraw-Hill (1963). See also A. De-Shalit and I. Tahni, *Nuclear shell theory*, Academic Press (1963).

angular momentum, labelled by L and S respectively, can be constructed using Clebsch–Gordan coefficients. Thus L and S can be combined to give a total state of angular momentum J, where J ranges from $|L-S|$ to $L+S$. By exact analogy to this we can combine together tensors of rank k_1 and k_2, say, to give one of rank K, using these same Clebsch–Gordan coefficients. The formula for doing this is simply

$$T_Q^K = \sum_{q_1,q_2} z_{q_1}^{k_1} y_{q_2}^{k_2} (k_1 q_1 k_2 q_2 \mid KQ), \tag{6.21}$$

where the spherical tensors are z^{k_1} and y^{k_2}, which combine to give the spherical tensor T_Q^K.

In actual calculations it is convenient to express the Clebsch–Gordan coefficient $(k_1 q_1 k_2 q_2 \mid KQ)$ in terms of the $3j$ symbol $\begin{pmatrix} k_1 & k_2 & K \\ q_1 & q_2 & -Q \end{pmatrix}$, which possesses a number of extremely useful symmetry properties, which are reviewed by Edmonds. In terms of the $3j$ symbol the Clebsch–Gordan coefficient is

$$(k_1 q_1 k_2 q_2 \mid K, -Q) = (-1)^{k_2-k_1+Q}(2K+1)^{\frac{1}{2}} \begin{pmatrix} k_1 & k_2 & K \\ q_1 & q_2 & Q \end{pmatrix}. \tag{6.22}$$

The $3j$ symbol automatically equals zero unless both $q_1+q_2+Q = 0$ and k_1, k_2, K satisfy the triangle conditions

$$k_1+k_2-K \geqslant 0, \qquad k_1-k_2+K \geqslant 0, \qquad -k_1+k_2+K \geqslant 0.$$

The sum k_1+k_2+K must be an integer.

The $3j$ symbol is invariant under an even permutation of the columns and is multiplied by $(-1)^{k_1+k_2+K}$ for an odd permutation. Further,

$$\begin{pmatrix} k_1 & k_2 & K \\ q_1 & q_2 & Q \end{pmatrix} = (-1)^{k_1+k_2+K} \begin{pmatrix} k_1 & k_2 & K \\ -q_1 & -q_2 & -Q \end{pmatrix}, \tag{6.23}$$

and from this it follows as a special case that

$$\begin{pmatrix} k_1 & k_2 & K \\ 0 & 0 & 0 \end{pmatrix}$$

is zero unless k_1+k_2+K is an even integer.

For a single electron our immediate task is to find the matrix element

$$\langle lm |\exp(i\boldsymbol{\kappa}.\mathbf{r})(\tilde{\boldsymbol{\kappa}}\times\boldsymbol{\nabla})_q| lm'\rangle. \tag{6.24}$$

As an example of the above notation we obtain an alternative form for $\tilde{\boldsymbol{\kappa}}\times\boldsymbol{\nabla}$. We know this is a vector, i.e. a tensor of rank one, and hence by analogy with (6.21),

$$(\tilde{\boldsymbol{\kappa}}\times\boldsymbol{\nabla})_q = c \sum_{q_1,q_2} \tilde{\kappa}_{q_1} \nabla_{q_2} (1q_1\, 1q_2 \mid 1q).$$

To determine the constant of proportionality c we put $q = 0$ and rewrite the left-hand side as

$$(\tilde{\boldsymbol{\kappa}}\times\boldsymbol{\nabla})_0 = (\tilde{\boldsymbol{\kappa}}\times\boldsymbol{\nabla})_z = \tilde{\kappa}_x\nabla_y-\tilde{\kappa}_y\nabla_x = -\mathrm{i}(\tilde{\kappa}_{-1}\nabla_{+1}-\tilde{\kappa}_{+1}\nabla_{-1}).$$

But the right-hand side becomes

$$c\{\tilde{\kappa}_{-1}\nabla_{+1}(1,-111\mid 10)+\tilde{\kappa}_0\nabla_0(1010\mid 10)+\tilde{\kappa}_{+1}\nabla_{-1}(111,-1\mid 10)\}$$
$$= \frac{c}{\sqrt{2}}(\tilde{\kappa}_{-1}\nabla_{+1}-\tilde{\kappa}_{+1}\nabla_{-1}).$$

Hence $c = -\mathrm{i}\sqrt{2}$ and we get

$$(\tilde{\boldsymbol{\kappa}}\times\boldsymbol{\nabla})_q = \mathrm{i}(-1)^{1+q}\sqrt{6}\sum_{q_1,q_2}\tilde{\kappa}_{q_1}\nabla_{q_2}\begin{pmatrix}1 & 1 & 1\\ q_1 & q_2 & -q\end{pmatrix}. \tag{6.25}$$

Before using this result in (6.24) we rearrange the latter as follows.

If we make use of the fact that

$$\int \mathrm{d}\mathbf{r}\ \boldsymbol{\nabla}\{fY^{*l}_m\exp(\mathrm{i}\boldsymbol{\kappa}.\mathbf{r})fY^l_{m'}\} = 0,$$

and also

$$(\boldsymbol{\kappa}\times\boldsymbol{\nabla})\exp(\mathrm{i}\boldsymbol{\kappa}.\mathbf{r}) = 0,$$

the matrix element (6.24) can be written

$$\langle lm|\exp(\mathrm{i}\boldsymbol{\kappa}.\mathbf{r})(\tilde{\boldsymbol{\kappa}}\times\boldsymbol{\nabla})\,|lm'\rangle$$
$$= \tfrac{1}{2}\int \mathrm{d}\mathbf{r}\ \exp(\mathrm{i}\boldsymbol{\kappa}.\mathbf{r})[fY^{*l}_m\{(\tilde{\boldsymbol{\kappa}}\times\boldsymbol{\nabla})fY^l_{m'}\}-fY^l_{m'}\{(\tilde{\boldsymbol{\kappa}}\times\boldsymbol{\nabla})fY^{*l}_m\}]$$
$$= \tfrac{1}{2}\int \mathrm{d}\mathbf{r}\ \exp(\mathrm{i}\boldsymbol{\kappa}.\mathbf{r})f^2[Y^{*l}_m\{(\tilde{\boldsymbol{\kappa}}\times\boldsymbol{\nabla})Y^l_{m'}\}-Y^l_{m'}\{(\tilde{\boldsymbol{\kappa}}\times\boldsymbol{\nabla})Y^{*l}_m\}]. \tag{6.26}$$

To proceed further we note that

$$\boldsymbol{\nabla} = \tfrac{1}{2}[\nabla^2,\mathbf{r}] \tag{6.27}$$

and

$$\nabla^2\{Y^K_Q(\tilde{\mathbf{r}})f(r)\} = Y^K_Q(\tilde{\mathbf{r}})\left\{\frac{1}{r}\frac{\mathrm{d}^2}{\mathrm{d}r^2}r-\frac{K(K+1)}{r^2}\right\}f(r), \tag{6.28}$$

which enables us to rewrite (6.26) in the form

$$\langle lm|\exp(\mathrm{i}\boldsymbol{\kappa}.\mathbf{r})(\tilde{\boldsymbol{\kappa}}\times\boldsymbol{\nabla})\,|lm'\rangle$$
$$= \tfrac{1}{4}\tilde{\boldsymbol{\kappa}}\times\int \mathrm{d}\mathbf{r}\, f^2\exp(\mathrm{i}\boldsymbol{\kappa}.\mathbf{r})\{Y^{*l}_m\nabla^2(\mathbf{r}\,Y^l_{m'})-Y^l_{m'}\nabla^2(\mathbf{r}Y^{*l}_m)\}. \tag{6.29}$$

Now, from (6.20 b),

$$r_q = r\tilde{r}_q = r\left(\frac{4\pi}{3}\right)^{\frac{1}{2}}Y^1_q(\tilde{\mathbf{r}}).$$

The product of two spherical harmonics with the same argument can be expressed as an expansion in single spherical harmonics, namely

$$Y^K_Q(\tilde{\mathbf{r}})Y^{K'}_{Q'}(\tilde{\mathbf{r}}) = \sum_{\bar{K}\bar{Q}}\left\{\frac{(2K+1)(2K'+1)(2\bar{K}+1)}{4\pi}\right\}^{\frac{1}{2}}\times$$
$$\times\begin{pmatrix}K & K' & \bar{K}\\ Q & Q' & \bar{Q}\end{pmatrix}Y^{*\bar{K}}_{\bar{Q}}(\tilde{\mathbf{r}})\begin{pmatrix}K & K' & \bar{K}\\ 0 & 0 & 0\end{pmatrix}. \tag{6.30}$$

From this we obtain

$$\tilde{r}_q Y^l_{m'}(\tilde{\mathbf{r}}) = \sum_Q (-1)^{l+Q}\left\{l^{\frac{1}{2}} Y^{l-1}_Q(\tilde{\mathbf{r}})\begin{pmatrix}1 & l & l-1\\ q & m' & -Q\end{pmatrix} - \right.$$
$$\left. -(l+1)^{\frac{1}{2}} Y^{l+1}_Q(\tilde{\mathbf{r}})\begin{pmatrix}1 & l & l+1\\ q & m' & -Q\end{pmatrix}\right\}. \tag{6.31}$$

Note that with our phase convention

$$Y^{*K}_Q(\tilde{\mathbf{r}}) = (-1)^Q Y^K_{-Q}(\tilde{\mathbf{r}}). \tag{6.32}$$

If we use (6.31) in conjunction with (6.28)

$$\nabla^2(r_q Y^l_{m'}) = \frac{1}{r}\sum_Q (-1)^{l+Q+1}\left\{l^{\frac{1}{2}}(l+1)(l-2) Y^{l-1}_Q(\tilde{\mathbf{r}})\begin{pmatrix}1 & l & l-1\\ q & m' & -Q\end{pmatrix} - \right.$$
$$\left. -(l+1)^{\frac{1}{2}} l(l+3) Y^{l+1}_Q(\tilde{\mathbf{r}})\begin{pmatrix}1 & l & l+1\\ q & m' & -Q\end{pmatrix}\right\}. \tag{6.33}$$

Hence $\qquad Y^{*l}_m \nabla^2(r_q Y^l_{m'}) - Y^l_{m'} \nabla^2(r_q Y^{*l}_m),$

as is required in (6.29),

$$= \frac{1}{r}(-1)^{l+1+m}\sum_Q (-1)^Q\left[Y^l_{-m}\left\{l^{\frac{1}{2}}(l+1)(l-2)Y^{l-1}_Q\begin{pmatrix}1 & l & l-1\\ q & m' & -Q\end{pmatrix} - \right.\right.$$
$$\left. -(l+1)^{\frac{1}{2}} l(l+3) Y^{l+1}_Q\begin{pmatrix}1 & l & l+1\\ q & m' & -Q\end{pmatrix}\right\} -$$
$$-Y^l_{m'}\left\{l^{\frac{1}{2}}(l+1)(l-2)Y^{l-1}_Q\begin{pmatrix}1 & l & l-1\\ q & -m & -Q\end{pmatrix} - \right.$$
$$\left.\left. -(l+1)^{\frac{1}{2}} l(l+3) Y^{l+1}_Q\begin{pmatrix}1 & l & l+1\\ q & -m & -Q\end{pmatrix}\right\}\right]. \tag{6.34}$$

The next step is to multiply (6.34) by $f^2 \exp(\mathrm{i}\boldsymbol{\varkappa}.\mathbf{r})$ and integrate over $\mathbf{r}$. To separate the radial and angular parts of the integral, $\exp(\mathrm{i}\boldsymbol{\varkappa}.\mathbf{r})$ is expressed as

$$\exp(\mathrm{i}\boldsymbol{\varkappa}.\mathbf{r}) = (4\pi)\sum_{K'=0}^{\infty}\sum_{Q'=-K'}^{K'} \mathrm{i}^{K'} j_{K'}(\kappa r) Y^{K'}_{Q'}(\tilde{\mathbf{r}}) Y^{*K'}_{Q'}(\tilde{\boldsymbol{\varkappa}}). \tag{6.35}$$

Here $j_{K'}(x)$ is a spherical Bessel function of order K', and we note the relation

$$j_{K'}(x) = \frac{x}{(2K'+1)}\{j_{K'-1}(x) + j_{K'+1}(x)\}. \tag{6.36}$$

From (6.29), (6.34), (6.35), and (6.36) the radial part of the integration is seen to be

$$\int_0^{\infty} r^2\,\mathrm{d}r\,\frac{1}{r} f^2 j_{K'}(\kappa r) = \frac{|\boldsymbol{\varkappa}|}{(2K'+1)}\{\bar{j}_{K'-1}(\kappa) + \bar{j}_{K'+1}(\kappa)\}, \tag{6.37}$$

where the radial integrals $\bar{j}_K(\kappa)$ are defined as

$$\bar{j}_K(\kappa) = \int_0^\infty r^2 f^2 j_K(\kappa r)\, dr. \tag{6.38}$$

Turning now to the angular part of the integration over $\mathbf{r}$ it is evident from (6.34) and (6.35) that this involves the integral of the product of three spherical harmonics. For this we have the result

$$\int d\tilde{\mathbf{r}}\, Y^K_Q(\tilde{\mathbf{r}}) Y^{K'}_{Q'}(\tilde{\mathbf{r}}) Y^{K''}_{Q''}(\tilde{\mathbf{r}}) = \left\{\frac{(2K+1)(2K'+1)(2K''+1)}{4\pi}\right\}^{\frac{1}{2}} \begin{pmatrix} K & K' & K'' \\ 0 & 0 & 0 \end{pmatrix} \begin{pmatrix} K & K' & K'' \\ Q & Q' & Q'' \end{pmatrix}. \tag{6.39}$$

Thus, on multiplying (6.34) by $Y^{K'}_{Q'}(\tilde{\mathbf{r}})$, which occurs in (6.35), and using (6.39),

$$\begin{aligned} &\int d\tilde{\mathbf{r}}\, Y^{K'}_{Q'}\{Y^{*l}_m \nabla^2(r_q Y^l_{m'}) - Y^l_{m'} \nabla^2(r_q Y^{*l}_m)\} \\ &= \frac{1}{r}(-1)^{l+1+m}\left\{\frac{(2K'+1)(2l+1)}{4\pi}\right\}^{\frac{1}{2}} \sum_Q (-1)^Q \Bigg[(l+1)(l-2)\{l(2l-1)\}^{\frac{1}{2}} \\ &\times \begin{pmatrix} K' & l & l-1 \\ 0 & 0 & 0 \end{pmatrix} \left\{\begin{pmatrix} 1 & l & l-1 \\ q & m' & Q \end{pmatrix}\begin{pmatrix} K' & l & l-1 \\ Q' & -m & -Q \end{pmatrix} - \right. \\ &\left. - \begin{pmatrix} 1 & l & l-1 \\ q & -m & Q \end{pmatrix}\begin{pmatrix} K' & l & l-1 \\ Q' & m' & -Q \end{pmatrix}\right\} - l(l+3)\{(l+1)(2l+3)\}^{\frac{1}{2}} \times \\ &\times \begin{pmatrix} K' & l & l+1 \\ 0 & 0 & 0 \end{pmatrix} \left\{\begin{pmatrix} 1 & l & l+1 \\ q & m' & Q \end{pmatrix}\begin{pmatrix} K' & l & l+1 \\ Q' & -m & -Q \end{pmatrix} - \right. \\ &\left. - \begin{pmatrix} 1 & l & l+1 \\ q & -m & Q \end{pmatrix}\begin{pmatrix} K' & l & l+1 \\ Q' & m' & -Q \end{pmatrix}\right\}\Bigg]. \end{aligned} \tag{6.40}$$

We now note that (6.40) can be greatly simplified by using the relation

$$\begin{aligned} &\sum_{m_3} \begin{pmatrix} j_1 & j_2 & j_3 \\ m_1 & m_2 & m_3 \end{pmatrix} \begin{pmatrix} l_1 & l_2 & j_3 \\ n_1 & n_2 & -m_3 \end{pmatrix} \\ &= \sum_{l_3 n_3} (-1)^{j_3+l_3+m_1+n_1}(2l_3+1) \times \\ &\times \begin{Bmatrix} j_1 & j_2 & j_3 \\ l_1 & l_2 & l_3 \end{Bmatrix} \begin{pmatrix} l_1 & j_2 & l_3 \\ n_1 & m_2 & n_3 \end{pmatrix} \begin{pmatrix} j_1 & l_2 & l_3 \\ m_1 & n_2 & -n_3 \end{pmatrix}, \end{aligned} \tag{6.41}$$

where $\begin{Bmatrix} j_1 & j_2 & j_3 \\ l_1 & l_2 & l_3 \end{Bmatrix}$ is a $6j$ symbol. The latter is automatically zero unless j_1, j_2, j_3, l_1, l_2, and l_3 satisfy the four triangle conditions represented by

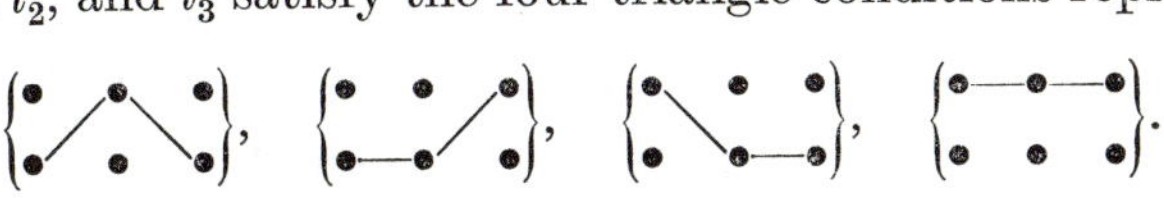.

Further, the $6j$ symbol is invariant under either an even or odd permutation of its columns. It also possesses the symmetry

$$\begin{Bmatrix} j_1 & j_2 & j_3 \\ l_1 & l_2 & l_3 \end{Bmatrix} = \begin{Bmatrix} j_1 & l_2 & l_3 \\ l_1 & j_2 & j_3 \end{Bmatrix}.$$

On making use of (6.41) to perform the sum over Q in (6.40) we obtain for the latter

$$\int \mathrm{d}\tilde{\mathbf{r}}\, Y_{Q'}^{K'}\{Y_m^{*l}\nabla^2(r_q Y_{m'}^l) - Y_{m'}^l \nabla^2(r_q Y_m^{*l})\}$$
$$= \frac{1}{r}\left\{\frac{(2K'+1)}{4\pi}\right\}^{\frac{1}{2}}(2l+1)\sum_{KQ}(2K+1)(-1)^{K+m'}\times$$
$$\times\begin{pmatrix} K' & 1 & K \\ Q' & q & Q \end{pmatrix}\left\{\begin{pmatrix} l & l & K \\ m' & -m & -Q \end{pmatrix} - \begin{pmatrix} l & l & K \\ -m & m' & -Q \end{pmatrix}\right\}\times$$
$$\times\{(l+1)(l-2)(-1)^l B - l(l+3)(-1)^{l-1}A\}, \quad (6.42)$$

where

$$A(K,K',l) = (-1)^{l-1}\left\{\frac{(l+1)(2l+3)}{(2l+1)}\right\}^{\frac{1}{2}}\begin{pmatrix} l & K' & l+1 \\ 0 & 0 & 0 \end{pmatrix}\begin{Bmatrix} l & 1 & l+1 \\ K' & l & K \end{Bmatrix} \quad (6.43)$$

and

$$B(K,K',l) = (-1)^l\left\{\frac{l(2l-1)}{(2l+1)}\right\}^{\frac{1}{2}}\begin{pmatrix} l & K' & l-1 \\ 0 & 0 & 0 \end{pmatrix}\begin{Bmatrix} l & 1 & l-1 \\ K' & l & K \end{Bmatrix}.$$

Because A and B contain the $3j$ symbols

$$\begin{pmatrix} l & K' & l\pm 1 \\ 0 & 0 & 0 \end{pmatrix},$$

they are both non-zero only if K' is an odd integer. The triangle condition on K and K' requires

$$K = K' \quad \text{or} \quad K'\pm 1.$$

However,

$$\begin{pmatrix} l & l & K \\ m' & -m & -Q \end{pmatrix} - \begin{pmatrix} l & l & K \\ -m & m' & -Q \end{pmatrix} = \begin{pmatrix} l & l & K \\ m' & -m & -Q \end{pmatrix}\{1-(-1)^{2l+K}\},$$

and this is zero unless K is odd. Hence $K = K'$.

A direct calculation shows that

$$A(K',K',l) = -B(K',K',l)$$

with

$$A(K',K',l) = (-1)^{\frac{1}{2}(K'-1)}\frac{1}{2(2l+1)}\times$$
$$\times\left\{\frac{2l+1-K'}{2K'+1}\right\}^{\frac{1}{2}}\frac{\{\frac{1}{2}(2l+1+K')\}!}{\{\frac{1}{2}(K'-1)\}!\,\{\frac{1}{2}(K'+1)\}!\,\{\frac{1}{2}(2l+1-K')\}!}\times$$
$$\times\left\{\frac{(K'-1)!\,(K'+1)!\,(2l+1-K')!}{(2l+1+K')!}\right\}^{\frac{1}{2}}. \quad (6.44)$$

Thus we obtain from (6.42) the result

$$\int d\tilde{\mathbf{r}}\, Y_{Q'}^{K'}\{Y_m^{*l}\nabla^2(r_q Y_{m'}^l)-Y_{m'}^l\nabla^2(r_q Y_m^{*l})\}$$
$$=\frac{4}{r}\left(\frac{1}{4\pi}\right)^{\frac{1}{2}}\{(2K'+1)(2l+1)\}^{\frac{3}{2}}A(K',K',l)\sum_Q(-1)^Q(K'Qlm'\mid lm)\times$$
$$\times\begin{pmatrix}K' & 1 & K'\\ Q' & q & -Q\end{pmatrix}. \qquad (6.45)$$

From (6.25), (6.29), (6.35), and (6.37), together with (6.45), we obtain

$$\langle lm|\exp(\mathrm{i}\boldsymbol{\varkappa}.\mathbf{r})(\tilde{\boldsymbol{\varkappa}}\times\boldsymbol{\nabla})_q\,|lm'\rangle$$
$$=\tfrac{1}{4}|\boldsymbol{\varkappa}|(4\pi)\sqrt{6}(-1)^{1+q}\sum_{q_1q_2}\begin{pmatrix}1 & 1 & 1\\ q_1 & q_2 & -q\end{pmatrix}\sum_{K'Q'}\mathrm{i}^{K'+1}\frac{1}{(2K'+1)}\times$$
$$\times(j_{K'-1}+j_{K'+1})\tilde{\kappa}_{q_1}Y_{Q'}^{*K'}(\tilde{\boldsymbol{\varkappa}})\int d\tilde{\mathbf{r}}\,Y_{Q'}^{K'}\{Y_m^{*l}\nabla^2(r_{q_2}Y_{m'}^l)-Y_{m'}^l\nabla^2(r_{q_2}Y_m^{*l})\}$$
$$=|\boldsymbol{\varkappa}|(4\pi)^{\frac{1}{2}}\sqrt{6}(-1)^{1+q}(2l+1)^{\frac{3}{2}}\sum_{K'Q}\sum_{\bar{K}\bar{Q}}\mathrm{i}^{K'+1}(j_{K'-1}+j_{K'+1})(2K'+1)\times$$
$$\times(-1)^{Q+\bar{Q}}A(K',K',l)(K'Qlm'\mid lm)(2\bar{K}+1)^{\frac{1}{2}}Y_{\bar{Q}}^{\bar{K}}(\tilde{\boldsymbol{\varkappa}})\begin{pmatrix}1 & K' & \bar{K}\\ 0 & 0 & 0\end{pmatrix}\times$$
$$\times\sum_{q_1q_2Q'}(-1)^{Q'}\begin{pmatrix}1 & 1 & 1\\ q_1 & q_2 & -q\end{pmatrix}\begin{pmatrix}K' & 1 & K'\\ Q' & q_2 & -Q\end{pmatrix}\begin{pmatrix}1 & K' & \bar{K}\\ q_1 & -Q' & \bar{Q}\end{pmatrix}, \qquad (6.46)$$

where in obtaining the last line we have made use of (6.20 b) and (6.30).

The sums over q_1, q_2, and Q' that involve the product of the last three $3j$ symbols on the right-hand side of (6.46) can be replaced by the product of a $3j$ and a $6j$ symbol through the relation

$$\begin{pmatrix}j_1 & j_2 & j_3\\ m_1 & m_2 & m_3\end{pmatrix}\begin{Bmatrix}j_1 & j_2 & j_3\\ l_1 & l_2 & l_3\end{Bmatrix}=\sum_{\mathrm{all}\,n}(-1)^{l_1+l_2+l_3+n_1+n_2+n_3}\times$$
$$\times\begin{pmatrix}j_1 & l_2 & l_3\\ m_1 & n_2 & -n_3\end{pmatrix}\begin{pmatrix}l_1 & j_2 & l_3\\ -n_1 & m_2 & n_3\end{pmatrix}\begin{pmatrix}l_1 & l_2 & j_3\\ n_1 & -n_2 & m_3\end{pmatrix}. \qquad (6.47)$$

If in addition to using (6.47) we make the change in notation $Q\to Q'$ and $\bar{K}$, $\bar{Q}\to K$, Q, respectively, then we have finally from (6.46) the result

$$\langle lm|\exp(\mathrm{i}\boldsymbol{\varkappa}.\mathbf{r})(\tilde{\boldsymbol{\varkappa}}\times\boldsymbol{\nabla})_q\,|lm'\rangle$$
$$=-|\boldsymbol{\varkappa}|(8\pi)^{\frac{1}{2}}(2l+1)^{\frac{3}{2}}\sum_{KQ}\sum_{K'Q'}\mathrm{i}^{K'+1}Y_Q^K(\tilde{\boldsymbol{\varkappa}})(j_{K'-1}+j_{K'+1})\times$$
$$\times(2K'+1)(2K+1)^{\frac{1}{2}}A(K',K',l)\begin{pmatrix}1 & K & K'\\ 0 & 0 & 0\end{pmatrix}\begin{Bmatrix}1 & 1 & 1\\ K' & K & K'\end{Bmatrix}\times$$
$$\times(K'Q'lm'\mid lm)(KQK'Q'\mid 1q). \qquad (6.48)$$

We recall that the integer $K' = 1, 3,..., (2l-1)$. The $3j$ symbol

$$\begin{pmatrix} 1 & K & K' \\ 0 & 0 & 0 \end{pmatrix}$$

is non-zero only if K is even and therefore $K = K'\pm 1$, i.e. $K = 0, 2,..., 2l$.

The result (6.48) applies to the case of a single electron, but it is not difficult to generalize it to the configuration of l^n, the n electrons coupling to give a total spin S and orbital angular momentum L. Since Hund's rule for the coupling is assumed to apply, $2S$ is either the number of electrons or the number of holes, $2(2l+1)-n$, whichever is the smaller.

In general, if the operator $\hat{F}$ is the sum of single-particle operators $\hat{f}$, then

$$\begin{aligned}
&\langle l^n vSM_S\, LM_L \,|\, \hat{F} \,|\, l^n v'S'M'_S\, L'M'_L\rangle \\
&\quad = n \sum_{\bar{\theta}} (\theta\{\,|\,\bar{\theta})(\theta'\{\,|\,\bar{\theta}) \sum_{m,\bar{M}_L,m'} (\bar{L}\bar{M}_L\, lm \,|\, LM_L)(\bar{L}\bar{M}_L\, lm' \,|\, L'M'_L)\times \\
&\qquad \times \sum_{m_s,\bar{M}_S,m'_s} (\bar{S}\bar{M}_S\, sm_s \,|\, SM_S)(\bar{S}\bar{M}_S\, sm'_s \,|\, S'M'_S)\times \\
&\qquad \times \langle sm_s\, lm \,|\, \hat{f} \,|\, sm'_s\, lm'\rangle. \qquad (6.49)
\end{aligned}$$

The notation used in eqn (6.49) is as follows: the symbol v stands for any other quantum numbers that are needed when the set $SM_S\, LM_L$ fails to define the state uniquely; θ, θ', and $\bar{\theta}$ are shorthand for

$$\theta \equiv vSL, \qquad \theta' \equiv v'S'L', \qquad \bar{\theta} \equiv \bar{v}\bar{S}\bar{L},$$

and it is to be understood that θ and θ' define terms of l^n, while $\bar{\theta}$ defines a term of l^{n-1}. $(\theta\{\,|\,\bar{\theta})$ are Racah's coefficients of fractional parentage and are purely real.†

In (6.48) the dependence of $\langle lm|\exp(i\boldsymbol{\kappa}.\mathbf{r})(\tilde{\boldsymbol{\kappa}}.\boldsymbol{\nabla})_q\,|lm'\rangle$ on the quantum numbers m and m' is contained solely in the Clebsch–Gordan coefficient $(K'Q'lm' \,|\, lm)$. If we make use of (6.49) then

$$\Big\langle vSM_S\, LM_L \Big| \sum_{\text{electrons}} \exp(i\boldsymbol{\kappa}.\mathbf{r})(\tilde{\boldsymbol{\kappa}}\times\boldsymbol{\nabla})_q \Big| v'S'M'_S\, L'M'_L\Big\rangle \qquad (6.50)$$

is obtained from (6.48) merely by replacing $(K'Q'lm' \,|\, lm)$ by

$$\begin{aligned}
&\sqrt{\{(2l+1)(2L+1)(2L'+1)\}}\,\delta_{S,S'}\,\delta_{M_S,M'_S}(-)^{l-K'+Q'-L+L'-M'_L}\times \\
&\quad\times \begin{pmatrix} L' & K' & L \\ M'_L & Q' & -M_L \end{pmatrix} n \sum_{\bar{\theta}} (-)^{\bar{L}} (\theta\{\,|\,\bar{\theta})(\theta'\{\,|\,\bar{\theta}) \begin{Bmatrix} L' & K' & L \\ l & \bar{L} & l \end{Bmatrix}. \qquad (6.51)
\end{aligned}$$

† Values of Racah's coefficients for $l = 1$, 2, and 3 are given by C. W. Nielson and G. F. Koster, *Spectroscopic coefficients for p^n, d^n and f^n*, M.I.T. Press (1963).

If $J = L \pm S$ is a good quantum number then we need to calculate

$$\left\langle \theta JM \middle| \sum_{\text{electrons}} \exp(\mathrm{i}\boldsymbol{\varkappa}.\mathbf{r})(\tilde{\boldsymbol{\varkappa}} \times \boldsymbol{\nabla})_q \middle| \theta' J' M' \right\rangle. \tag{6.52}$$

To do this it is only necessary to multiply (6.50) by $(SM_{\mathrm{s}}\, LM_{\mathrm{L}} \mid JM)$ and $(S'M'_{\mathrm{s}}\, L'M'_{\mathrm{L}} \mid J'M')$ and sum over M_{s}, M'_{s}, M_{L}, and M'_{L}. The result is that (6.52) is obtained from (6.48) by replacing $(K'Q'lm' \mid lm)$ by

$$\sqrt{\{(2l+1)(2L+1)(2L'+1)(2J'+1)\}}\delta_{S,S'}(-)^{S+l+J'+L+L'} \times$$
$$\times \begin{Bmatrix} J' & K' & J \\ L & S & L' \end{Bmatrix} n \sum_{\bar{\theta}} (-)^{\bar{L}} (\theta\{ \mid \bar{\theta})(\theta'\{ \mid \bar{\theta}) \begin{Bmatrix} L' & K' & L \\ l & \bar{L} & l \end{Bmatrix} \times$$
$$\times (K'Q'J'M' \mid JM). \tag{6.53}$$

Now combine (6.48) and (6.53) and write (6.52) in the form

$$\left\langle \theta JM \middle| \sum_{\text{electrons}} \exp(\mathrm{i}\boldsymbol{\varkappa}.\mathbf{r})(\tilde{\boldsymbol{\varkappa}} \times \boldsymbol{\nabla})_q \middle| \theta' J' M' \right\rangle$$
$$= -(4\pi)^{\frac{1}{2}} |\boldsymbol{\varkappa}| \sum_{K,K'} A(K,K') \sum_{Q,Q'} Y_Q^K(\tilde{\boldsymbol{\varkappa}})(K'Q'J'M' \mid JM)(KQK'Q' \mid 1q). \tag{6.54}$$

In eqn (6.54),

$$A(K,K') = (2l+1)^2\sqrt{\{2(2L+1)(2L'+1)(2J'+1)\}}\delta_{S,S'}(-)^{S+l+J'+L+L'} \times$$
$$\times \mathrm{i}^{K'+1}(2K'+1)(2K+1)^{\frac{1}{2}} \begin{pmatrix} 1 & K & K' \\ 0 & 0 & 0 \end{pmatrix} \begin{Bmatrix} 1 & 1 & 1 \\ K' & K & K' \end{Bmatrix} \times$$
$$\times A(K',K',l)\{\bar{\jmath}_{K'-1} + \bar{\jmath}_{K'+1}\} \begin{Bmatrix} J' & K' & J \\ L & S & L' \end{Bmatrix} \times$$
$$\times n \sum_{\bar{\theta}} (-)^{\bar{L}} (\theta\{ \mid \bar{\theta})(\theta'\{ \mid \bar{\theta}) \begin{Bmatrix} L' & K' & L \\ l & \bar{L} & l \end{Bmatrix}, \tag{6.55}$$

and it is a simple matter to show that

$$\frac{A(K'-1,K')}{A(K'+1,K')} = \left(\frac{K'+1}{K'}\right)^{\frac{1}{2}}. \tag{6.56}$$

(*b*) *Spin term*

If we write

$$\hat{\mathbf{Q}} = \sum_i \exp(\mathrm{i}\boldsymbol{\varkappa}.\mathbf{r}_i)\left(\hat{\mathbf{s}}_i - \frac{1}{|\boldsymbol{\varkappa}|}\tilde{\boldsymbol{\varkappa}} \times \boldsymbol{\nabla}_i\right) +$$
$$+ (\text{any function proportional to } \tilde{\boldsymbol{\varkappa}}), \tag{6.57}$$

then the interaction operator

$$\hat{\mathbf{Q}}_\perp = \tilde{\boldsymbol{\varkappa}} \times (\hat{\mathbf{Q}} \times \tilde{\boldsymbol{\varkappa}})$$

and
$$\hat{\mathbf{Q}}_\perp . \hat{\mathbf{Q}}_\perp = \sum_{\alpha,\beta} (\delta_{\alpha,\beta} - \tilde{\kappa}_\alpha \tilde{\kappa}_\beta) \hat{Q}^\alpha \hat{Q}^\beta, \tag{6.58}$$

with α, $\beta = x$, y, or z.

In the previous section we evaluated the matrix elements of the orbital contribution to $\hat{\mathbf{Q}}$ and we turn now to the evaluation of the spin part

$$\sum_i \exp(\mathrm{i}\boldsymbol{\varkappa}\,.\,\mathbf{r}_i)\hat{\mathbf{s}}_i.$$

We emphasize that $\hat{\mathbf{Q}}$ is an intermediate operator in as much as the cross-section involves $\hat{\mathbf{Q}}_\perp$. However, $\hat{\mathbf{Q}}$ has the conceptual value that it enables us to retain the same form for the cross-section in the general case as in the spin-only case and, in addition, it is easier to use in calculations. Note that it is defined only to within any function that is proportional to the scattering vector, a property that we shall later use to simplify the algebra.

To evaluate the matrix elements of $\sum \exp(\mathrm{i}\boldsymbol{\varkappa}\,.\,\mathbf{r})\hat{\mathbf{s}}$ we again make use of (6.35) to expand $\exp(\mathrm{i}\boldsymbol{\varkappa}\,.\,\mathbf{r})$ and obtain

$$\begin{aligned}\exp(\mathrm{i}\boldsymbol{\varkappa}\,.\,\mathbf{r})\hat{s}_q &= (4\pi)\sum_{K,Q} \mathrm{i}^K j_K(\kappa r) Y_Q^{*K}(\tilde{\boldsymbol{\varkappa}}) Y_Q^K(\tilde{\mathbf{r}})\hat{s}_q \\ &= (4\pi)^{\frac{1}{2}} \sum_{K,Q}\sum_{K',Q'} \mathrm{i}^K Y_Q^{*K}(\tilde{\boldsymbol{\varkappa}}) f_{Q'}^{K'}(1qKQ\mid K'Q'),\end{aligned} \tag{6.59}$$

where the tensor $f_{Q'}^{K'}$ is defined by

$$f_{Q'}^{K'} = (4\pi)^{\frac{1}{2}} j_K(\kappa r) \sum_{\bar{q},\bar{Q}} \hat{s}_{\bar{q}} Y_{\bar{Q}}^K(\tilde{\mathbf{r}})(1\bar{q}K\bar{Q}\mid K'Q'). \tag{6.60}$$

In rewriting the spin term in this way we have utilized the relation

$$\sum_{K',Q'} (2K'+1)\begin{pmatrix}1 & K & K'\\ \bar{q} & \bar{Q} & -Q'\end{pmatrix}\begin{pmatrix}1 & K & K'\\ q & Q & -Q'\end{pmatrix} = \delta_{q,\bar{q}}\delta_{Q,\bar{Q}}. \tag{6.61}$$

Since $f_{Q'}^{K'}$ is a spherical tensor, of rank K', the Wigner–Eckart theorem applies to its matrix elements, i.e.

$$\begin{aligned}&\langle \theta JM \mid \textstyle\sum f_{Q'}^{K'} \mid \theta' J' M'\rangle \\ &\quad = (-1)^{J-M}\begin{pmatrix} J & K' & J'\\ -M & Q' & M'\end{pmatrix}[\theta J \| \textstyle\sum f^{K'} \| \theta' J'].\end{aligned} \tag{6.62}$$

Here $[\theta J\|\sum f^{K'}\|\theta' J']$ is called a reduced matrix element and it is independent of the quantum numbers M and M'.

In evaluating this reduced matrix element it is convenient to obtain some general results for the matrix elements of a mixed tensor, $z_{q_1}^{k_1} y_{q_2}^{k_2}$, where z acts only on the spin part of the wave function and y on the orbital part.

The Wigner–Eckart theorem for double tensors is

$$\begin{aligned}&\langle vSM_{\mathrm{S}}\,LM_{\mathrm{L}} \mid z_{q_1}^{k_1} y_{q_2}^{k_2} \mid v'S'M_{\mathrm{S}}'\,L'M_{\mathrm{L}}'\rangle \\ &\quad = (-)^{S-M_{\mathrm{S}}+L-M_{\mathrm{L}}}\begin{pmatrix} S & k_1 & S'\\ -M_{\mathrm{S}} & q_1 & M_{\mathrm{S}}'\end{pmatrix}\begin{pmatrix} L & k_2 & L'\\ -M_{\mathrm{L}} & q_2 & M_{\mathrm{L}}'\end{pmatrix}[\theta\|z^{k_1}y^{k_2}\|\theta'].\end{aligned} \tag{6.63}$$

From (6.47) and (6.49)

$$\sum_{\text{electrons}} [\theta\|z^{k_1}y^{k_2}\|\theta'] = (-)^{l+s+L+S+k_1+k_2}\times$$
$$\times\sqrt{\{(2S+1)(2S'+1)(2L+1)(2L'+1)\}}[s\|z^{k_1}\|s][l\|y^{k_2}\|l]\times$$
$$\times n\sum_{\bar\theta}(\theta\{|\bar\theta)(\theta'\{|\bar\theta)(-)^{\bar S+\bar L}\begin{Bmatrix} S' & k_1 & S \\ s & \bar S & s\end{Bmatrix}\begin{Bmatrix} L' & k_2 & L \\ l & \bar L & l\end{Bmatrix}. \quad (6.64)$$

If
$$F^{K'}_{Q'} = \sum_{\text{electrons}}\sum_{q_1,q_2} z^{k_1}_{q_1}y^{k_2}_{q_2}(k_1q_1k_2q_2\,|\,K'Q'), \quad (6.65)$$

then (cf. eqn (6.62))

$$\langle\theta JM|\,F^{K'}_{Q'}\,|\theta'J'M'\rangle = (-)^{J-M}\begin{pmatrix} J & K' & J' \\ -M & Q' & M'\end{pmatrix}[\theta J\|F^{K'}\|\theta'J'] \quad (6.66)$$

by the Wigner–Eckart theorem. Hence, on using (6.64),

$$[\theta J\|F^{K'}\|\theta'J']$$
$$= \sum_{M,M',Q'}(-)^{M-J}\begin{pmatrix} J & K' & J' \\ -M & Q' & M'\end{pmatrix}\times$$
$$\times\sum_{M_S,M_L}\sum_{M'_S,M'_L}(SM_S\,LM_L\,|\,JM)(S'M'_S\,L'M'_L\,|\,J'M')\times$$
$$\times\sum_{q_1,q_2}(k_1q_1k_2q_2\,|\,K'Q')\langle vSM_S\,LM_L|\,z^{k_1}_{q_1}y^{k_2}_{q_2}\,|v'S'M'_S\,L'M'_L\rangle$$
$$= \sqrt{\{(2K'+1)(2S+1)(2S'+1)(2L+1)(2L'+1)(2J+1)(2J'+1)\}}\times$$
$$\times(-)^{L+k_1+k_2+l+s+S}[s\|z^{k_1}\|s][l\|y^{k_2}\|l]\times$$
$$\times n\sum_{\bar\theta}(\theta\{|\bar\theta)(\theta'\{|\bar\theta)(-)^{\bar S+\bar L}\begin{Bmatrix} S' & k_1 & S \\ s & \bar S & s\end{Bmatrix}\begin{Bmatrix} L' & k_2 & L \\ l & \bar L & l\end{Bmatrix}\times$$
$$\times\begin{Bmatrix} S & S' & k_1 \\ L & L' & k_2 \\ J & J' & K'\end{Bmatrix}. \quad (6.67)$$

In arriving at the last line of (6.67) we have used the relation

$$\begin{Bmatrix} j_{11} & j_{12} & j_{13} \\ j_{21} & j_{22} & j_{23} \\ j_{31} & j_{32} & j_{33}\end{Bmatrix} = \sum_{\text{all } m}\begin{pmatrix} j_{11} & j_{12} & j_{13} \\ m_{11} & m_{12} & m_{13}\end{pmatrix}\begin{pmatrix} j_{21} & j_{22} & j_{23} \\ m_{21} & m_{22} & m_{23}\end{pmatrix}\times$$
$$\times\begin{pmatrix} j_{31} & j_{32} & j_{33} \\ m_{31} & m_{32} & m_{33}\end{pmatrix}\begin{pmatrix} j_{11} & j_{21} & j_{31} \\ m_{11} & m_{21} & m_{31}\end{pmatrix}\times$$
$$\times\begin{pmatrix} j_{12} & j_{22} & j_{32} \\ m_{12} & m_{22} & m_{32}\end{pmatrix}\begin{pmatrix} j_{13} & j_{23} & j_{33} \\ m_{13} & m_{23} & m_{33}\end{pmatrix} \quad (6.68)$$

to reduce the sum over the six $3j$ symbols to a $9j$ symbol

$$\begin{Bmatrix} S & S' & k_1 \\ L & L' & k_2 \\ J & J' & K' \end{Bmatrix}.$$

From (6.68) it is clear that the arguments of each row and column of a $9j$ symbol must satisfy a triangle condition. The $9j$ symbol is invariant under an even permutation of either rows or columns and is multiplied by a factor $(-1)^N$, where N is the sum of the nine parameters, for an odd permutation. It is also invariant under reflection about either diagonal.

We now set $$z_{q_1}^{k_1} = \hat{s}_{q_1} \quad (k_1 = 1)$$

and $$y_{q_2}^{k_2} = Y_Q^K(\tilde{\mathbf{r}}) \quad (k_2 = K,\ q_2 = Q),$$

so that the reduced matrix element required in (6.62) is

$$[\theta J\|\textstyle\sum f^{K'}\|\theta' J'] = (4\pi)^{\frac{1}{2}}\bar{j}_K(\kappa)[\theta J\|F^{K'}\|\theta' J']. \tag{6.69}$$

Since $$[s\|\hat{s}\|s] = \sqrt{(\tfrac{3}{2})}$$

and $$[l\|Y^K\|l] = (-1)^l(2l+1)\left(\frac{2K+1}{4\pi}\right)^{\frac{1}{2}}\begin{pmatrix} l & K & l \\ 0 & 0 & 0 \end{pmatrix}$$

we obtain from (6.67) and (6.69) the result

$$\begin{aligned}
&[\theta J\|\textstyle\sum f^{K'}\|\theta' J'] \\
&= (-1)^{K'+J'-J+S'+L'}(2l+1)\bar{j}_K(\kappa)\{\tfrac{3}{2}(2S+1)(2S'+1)(2L+1)(2L'+1)\times \\
&\qquad \times(2J+1)(2J'+1)(2K+1)(2K'+1)\}^{\frac{1}{2}}\begin{pmatrix} l & K & l \\ 0 & 0 & 0 \end{pmatrix}\begin{Bmatrix} 1 & K & K' \\ S' & L' & J' \\ S & L & J \end{Bmatrix}\times \\
&\qquad \times n\sum_{\bar{\theta}}(\theta\{|\bar{\theta})(\theta'\{|\bar{\theta})(-1)^{\bar{S}+\bar{L}+\frac{1}{2}}\begin{Bmatrix} S & 1 & S' \\ \frac{1}{2} & \bar{S} & \frac{1}{2} \end{Bmatrix}\begin{Bmatrix} L & K & L' \\ l & \bar{L} & l \end{Bmatrix} \\
&= (-1)^{K'+J'-J}(2J+1)^{\frac{1}{2}}C(K,K').
\end{aligned} \tag{6.70}$$

The $3j$ symbol $$\begin{pmatrix} l & K & l \\ 0 & 0 & 0 \end{pmatrix}$$

in $C(K,K')$ is non-zero only if K is an even integer, i.e. $K = 0, 2,..., 2l$. In general $K' = K$ or $K\pm 1$. If, however, $S = S'$, $L = L'$, $J = J'$ interchange of the second and third row of the $9j$ symbol in $C(K,K')$ multiplies it by

$$(-1)^{1+K+K'},$$

while leaving the $9j$ symbol unchanged. Hence we conclude that the $9j$ is non-zero only if K' is odd, i.e. we have $K' = K\pm 1$.

If we combine (6.59), (6.62), and (6.70)

$$\langle \theta JM|\sum \exp(\mathrm{i}\boldsymbol{\varkappa}.\mathbf{r})\hat{s}_q\,|\theta' J' M'\rangle$$

$$= (4\pi)^{\frac{1}{2}} \sum_{K,Q}\sum_{K',Q'} \mathrm{i}^K Y_Q^{*K}(\tilde{\boldsymbol{\varkappa}})(1qKQ\mid K'Q')\times$$

$$\times(-1)^{J-M}\begin{pmatrix} J & K' & J' \\ -M & Q' & M' \end{pmatrix}[\theta J\|\sum f^{K'}\|\theta' J']$$

$$= (4\pi)^{\frac{1}{2}} \sum_{K,Q}\sum_{K',Q'} \mathrm{i}^K Y_Q^{K}(\tilde{\boldsymbol{\varkappa}})\left(\frac{2K'+1}{3}\right)^{\frac{1}{2}} C(K,K')\times$$

$$\times(K'Q'J'M'\mid JM)(KQK'Q'\mid 1q). \quad (6.71)$$

Finally, if we combine (6.54) and (6.71) the matrix elements of $\hat{\mathbf{Q}}$ in the SLJ basis are

$$\langle \theta JM|\,\hat{Q}_q\,|\theta' J' M'\rangle$$

$$= \left\langle \theta JM \middle| \sum_{\text{electrons}} \exp(\mathrm{i}\boldsymbol{\varkappa}.\mathbf{r})\left\{\hat{s}_q - \frac{1}{|\boldsymbol{\varkappa}|}(\tilde{\boldsymbol{\varkappa}}\times\boldsymbol{\nabla})_q\right\}\middle| \theta' J' M'\right\rangle$$

$$= (4\pi)^{\frac{1}{2}} \sum_{K,Q}\sum_{K',Q'} Y_Q^{K}(\tilde{\boldsymbol{\varkappa}})\left\{\mathrm{i}^K\left(\frac{2K'+1}{3}\right)^{\frac{1}{2}} C(K,K') + A(K,K')\right\}\times$$

$$\times(K'Q'J'M'\mid JM)(KQK'Q'\mid 1q). \quad (6.72)$$

Let us now evaluate the matrix elements of $\hat{\mathbf{Q}}_\perp$, eqn (6.58). This merely entails the derivation of

$$\langle \theta JM|\sum \exp(\mathrm{i}\boldsymbol{\varkappa}.\mathbf{r})\{\tilde{\boldsymbol{\varkappa}}\times(\hat{\mathbf{s}}\times\tilde{\boldsymbol{\varkappa}})\}_q\,|\theta' J' M'\rangle \quad (6.73)$$

from (6.71).

Since $\tilde{\boldsymbol{\varkappa}}\times(\hat{\mathbf{s}}\times\tilde{\boldsymbol{\varkappa}}) = \hat{\mathbf{s}} - \tilde{\boldsymbol{\varkappa}}(\hat{\mathbf{s}}.\tilde{\boldsymbol{\varkappa}})$ we look first at $\tilde{\kappa}_q(\hat{\mathbf{s}}.\tilde{\boldsymbol{\varkappa}})$

$$= \tilde{\kappa}_q \sum_{\bar{q}} (-1)^{\bar{q}} \tilde{\kappa}_{\bar{q}} \hat{s}_{-\bar{q}} = \left(\frac{4\pi}{3}\right) \sum_{\bar{q}} (-1)^{\bar{q}} Y_q^1(\tilde{\boldsymbol{\varkappa}}) Y_{\bar{q}}^1(\tilde{\boldsymbol{\varkappa}}) \hat{s}_{-\bar{q}}. \quad (6.74)$$

Since the dependence on q of $\langle \theta JM|\sum \hat{s}_q\,|\theta' J' M'\rangle$ is in the Clebsch–Gordan coefficient $(KQK'Q'\mid 1q)$ we need to evaluate

$$\left(\frac{4\pi}{3}\right) \sum_{\bar{q}} (-1)^{\bar{q}} Y_q^1(\tilde{\boldsymbol{\varkappa}}) Y_{\bar{q}}^1(\tilde{\boldsymbol{\varkappa}})(KQK'Q'\mid 1q). \quad (6.75)$$

We wish to combine this term with the matrix elements of

$$\sum \exp(\mathrm{i}\boldsymbol{\varkappa}.\mathbf{r})\hat{s}_q.$$

Therefore we use (6.30) to put the $\boldsymbol{\varkappa}$ dependence of (6.75) into a single spherical harmonic, and obtain for the latter the result

$$(12\pi)^{\frac{1}{2}}(-1)^{K+K'}\sum_{\bar{K},\bar{Q}}(2\bar{K}+1)^{\frac{1}{2}}Y_{\bar{Q}}^{*\bar{K}}(\tilde{\boldsymbol{\varkappa}})\begin{pmatrix}1 & 1 & \bar{K}\\ 0 & 0 & 0\end{pmatrix}\times$$
$$\times\sum_{\bar{q}}\begin{pmatrix}1 & 1 & \bar{K}\\ q & \bar{q} & \bar{Q}\end{pmatrix}\begin{pmatrix}K & K' & 1\\ Q & Q' & \bar{q}\end{pmatrix}.$$

If this result is used in (6.73) then we must use (6.30) again to combine $Y_Q^K(\tilde{\boldsymbol{\varkappa}})$ and $Y_{\bar{Q}}^{*\bar{K}}(\tilde{\boldsymbol{\varkappa}})$. On rearranging the sum to produce an expansion in $Y_Q^K(\tilde{\boldsymbol{\varkappa}})$ we find, on using (6.71),

$$\langle\theta JM|\sum\exp(\mathrm{i}\boldsymbol{\varkappa}.\mathbf{r})\{\tilde{\boldsymbol{\varkappa}}\times(\hat{\mathbf{s}}\times\tilde{\boldsymbol{\varkappa}})\}_q|\theta'J'M'\rangle$$
$$=(4\pi)^{\frac{1}{2}}\sum_{K,Q}\sum_{K',Q'}Y_Q^K(\tilde{\boldsymbol{\varkappa}})\left(\frac{2K'+1}{3}\right)^{\frac{1}{2}}(K'Q'J'M'\mid JM)\times$$
$$\times\Big[\mathrm{i}^K C(K,K')(KQK'Q'\mid 1q)-\sqrt{3}\sum_{\bar{K},\bar{Q}}\sum_{K'',Q''}(-1)^{\bar{Q}+K'+Q}\times$$
$$\times\mathrm{i}^{K''}(2\bar{K}+1)C(K'',K')\{(2K+1)(2K''+1)\}^{\frac{1}{2}}\times$$
$$\times\begin{pmatrix}1 & 1 & \bar{K}\\ 0 & 0 & 0\end{pmatrix}\begin{pmatrix}K'' & \bar{K} & K\\ Q'' & -\bar{Q} & -Q\end{pmatrix}\begin{pmatrix}K'' & \bar{K} & K\\ 0 & 0 & 0\end{pmatrix}\times$$
$$\times\sum_{\bar{q}}\begin{pmatrix}1 & 1 & \bar{K}\\ q & \bar{q} & \bar{Q}\end{pmatrix}\begin{pmatrix}K'' & K' & 1\\ Q'' & Q' & \bar{q}\end{pmatrix}\Big].\tag{6.76}$$

The sums over $\bar{q}$, $\bar{Q}$, and Q'' in the second part of (6.76) can be performed with (6.47). This part then becomes

$$(-1)^{K'+K+1}\sum_{K''}\mathrm{i}^{K''}C(K'',K')\{(2K+1)(2K''+1)\}^{\frac{1}{2}}(KQK'Q'\mid 1q)\times$$
$$\times\sum_{\bar{K}}(2\bar{K}+1)\begin{pmatrix}1 & 1 & \bar{K}\\ 0 & 0 & 0\end{pmatrix}\begin{pmatrix}K'' & \bar{K} & K\\ 0 & 0 & 0\end{pmatrix}\begin{Bmatrix}1 & K & K'\\ K'' & 1 & \bar{K}\end{Bmatrix}.\tag{6.77}$$

The triangle condition on $\bar{K}$ allows the values $\bar{K}=0$, 1, 2, but since $\bar{K}$ must be even for a non-zero contribution $\bar{K}=0$ and 2.

For $\bar{K}=0$, $K''=K$ and

$$\begin{pmatrix}1 & 1 & 0\\ 0 & 0 & 0\end{pmatrix}\begin{pmatrix}K & K & 0\\ 0 & 0 & 0\end{pmatrix}\begin{Bmatrix}1 & K & K'\\ K & 1 & 0\end{Bmatrix}=\frac{(-1)^{K'}}{3(2K+1)},$$

so the $\bar{K}=0$ term in (6.77) is simply

$$-\tfrac{1}{3}\mathrm{i}^K C(K,K')(KQK'Q'\mid 1q).$$

For the contribution from $\bar{K}=2$ we use

$$(2\bar{K}+1)\begin{pmatrix}1 & 1 & \bar{K}\\ 0 & 0 & 0\end{pmatrix}=\left(\frac{10}{3}\right)^{\frac{1}{2}},$$

and thus finally obtain from (6.76) the result

$$\langle \theta JM|\sum \exp(\mathrm{i}\boldsymbol{\varkappa}.\mathbf{r})\{\tilde{\boldsymbol{\varkappa}}\times(\hat{\mathbf{s}}\times\tilde{\boldsymbol{\varkappa}})\}_q|\theta' J' M'\rangle$$
$$= (4\pi)^{\frac{1}{2}} \sum_{K,Q} \sum_{K',Q'} Y_Q^K(\tilde{\boldsymbol{\varkappa}}) B(K,K')(K'Q'J'M' \mid JM)(KQK'Q' \mid 1q), \tag{6.78}$$

where

$$B(K,K') = \left(\frac{2K'+1}{3}\right)^{\frac{1}{2}}\Big[\tfrac{2}{3}\mathrm{i}^K C(K,K') - (-1)^{K'}\{\tfrac{10}{3}(2K+1)\}^{\frac{1}{2}}\times$$
$$\times \sum_{\bar{K}} \mathrm{i}^{\bar{K}}(2\bar{K}+1)^{\frac{1}{2}} C(\bar{K},K') \begin{pmatrix} 2 & \bar{K} & K \\ 0 & 0 & 0 \end{pmatrix} \begin{Bmatrix} K' & \bar{K} & 1 \\ 2 & 1 & K \end{Bmatrix}\Big]. \tag{6.79}$$

It is a simple matter to show that

$$B(K',K') = \mathrm{i}^{K'}\{\tfrac{1}{3}(2K'+1)\}^{\frac{1}{2}} C(K',K'), \tag{6.80}$$

$$B(K'-1,K')$$
$$= \mathrm{i}^{K'-1}\frac{(K'+1)}{\{3(2K'+1)\}^{\frac{1}{2}}}\left\{C(K'-1,K') - \left(\frac{K'}{K'+1}\right)^{\frac{1}{2}} C(K'+1,K')\right\}, \tag{6.81}$$

and

$$\frac{B(K'-1,K')}{B(K'+1,K')} = \left(\frac{K'+1}{K'}\right)^{\frac{1}{2}}.$$

If we combine the result (6.54) for the orbital part with (6.78) for the spin part of the neutron–ion interaction we obtain the result

$$\langle \theta JM|\hat{Q}_{\perp,q}|\theta' J' M'\rangle$$
$$= \left\langle l^n vSLJM\middle|\sum_{\text{electrons}} \exp(\mathrm{i}\boldsymbol{\varkappa}.\mathbf{r})\left\{\tilde{\boldsymbol{\varkappa}}\times(\hat{\mathbf{s}}\times\tilde{\boldsymbol{\varkappa}}) - \frac{\mathrm{i}}{\hbar|\boldsymbol{\varkappa}|}\tilde{\boldsymbol{\varkappa}}\times\hat{\mathbf{p}}\right\}_q\middle| l^n v' S' L' J' M'\right\rangle$$
$$= (4\pi)^{\frac{1}{2}} \sum_{K,Q}\sum_{K',Q'} Y_Q^K(\tilde{\boldsymbol{\varkappa}})\{A(K,K') + B(K,K')\}(K'Q'J'M' \mid JM)\times$$
$$\times (KQK'Q' \mid 1q). \tag{6.82}$$

Let us now return to and simplify eqn (6.72) for the matrix elements of $\hat{\mathbf{Q}}$. For $C(K,K')$, $K = K'$ and $K'\pm 1$, and for $A(K,K')$, $K = K'\pm 1$. We can add to (6.72) a function proportional to $\tilde{\boldsymbol{\varkappa}}$ (which leaves $\hat{\mathbf{Q}}_{\perp}$ unchanged) and choose this function so that it cancels the contributions in (6.72) from $K = K'+1$.

Consider first the calculation for the orbital term. Let us add the function

$$\tilde{\kappa}_q \sum_{K',Q'} Y_{Q'}^{*K'}(\tilde{\boldsymbol{\varkappa}}) F_{Q'}^{K'}$$
$$= \sum_{\substack{K',Q'\\Q}} (-1)^{K'+Q'+Q} F_{Q'}^{K'}\left\{(K')^{\frac{1}{2}} Y_Q^{K'-1}\begin{pmatrix} 1 & K' & K'-1 \\ q & -Q' & -Q \end{pmatrix} - \right.$$
$$\left. - (K'+1)^{\frac{1}{2}} Y_Q^{K'+1}\begin{pmatrix} 1 & K' & K'+1 \\ q & -Q' & -Q \end{pmatrix}\right\}. \tag{6.83}$$

In arriving at the second line we have used (6.20 b) and (6.31).

We find that if we set

$$F^{K'}_{Q'} = (-1)^{1+q+K'+Q'+Q}(4\pi)^{\frac{1}{2}}\frac{(3K')^{\frac{1}{2}}}{(K'+1)}A(K'+1,K')(K'Q'J'M'\mid JM),$$

then

$$(4\pi)^{\frac{1}{2}}\sum_{K,Q}\sum_{K',Q'}Y^{K}_{Q}(\tilde{\varkappa})A(K,K')(K'Q'J'M'\mid JM)\times \\ \times(KQK'Q'\mid 1q)+\tilde{\kappa}_q\sum_{K',Q'}Y^{*K'}_{Q'}(\tilde{\varkappa})F^{K'}_{Q'} \\ = (4\pi)^{\frac{1}{2}}\sum_{\substack{K',Q'\\Q}}Y^{K'-1}_{Q}(\tilde{\varkappa})\left(\frac{2K'+1}{K'+1}\right)A(K'-1,K')(K'Q'J'M'\mid JM)\times \\ \times(K'-1QK'Q'\mid 1q). \qquad (6.84)$$

Similarly, if for the spin contribution to (6.72) we set

$$F^{K'}_{Q'} = (-1)^{1+q+K'+Q'+Q}(4\pi)^{\frac{1}{2}}\left(\frac{3}{K'+1}\right)^{\frac{1}{2}}\mathrm{i}^{K'+1}C(K'+1,K')\times \\ \times(K'Q'J'M'\mid JM),$$

then

$$(4\pi)^{\frac{1}{2}}\sum_{K,Q}\sum_{K',Q'}Y^{K}_{Q}(\tilde{\varkappa})\mathrm{i}^{K}\left(\frac{2K'+1}{3}\right)^{\frac{1}{2}}C(K,K')(K'Q'J'M'\mid JM)\times \\ \times(KQK'Q'\mid 1q)+\tilde{\kappa}_q\sum_{K',Q'}Y^{*K'}_{Q'}(\tilde{\varkappa})F^{K'}_{Q'} \\ = (4\pi)^{\frac{1}{2}}\sum_{\substack{K',Q'\\Q}}\left\{Y^{K'-1}_{Q}(\tilde{\varkappa})\left(\frac{2K'+1}{K'+1}\right)B(K'-1,K')(K'-1QK'Q'\mid 1q)+\right. \\ \left.+Y^{K'}_{Q}(\tilde{\varkappa})B(K',K')(K'QK'Q'\mid 1q)\right\}(K'Q'J'M'\mid JM). \qquad (6.85)$$

On combining (6.84) and (6.85) we have the desired expression for the matrix elements of $\hat{Q}_q$ in the $SLJM$ basis, namely

$$\langle\theta JM\mid\hat{Q}_q\mid\theta'J'M'\rangle \\ = (4\pi)^{\frac{1}{2}}\sum_{\substack{K',Q'\\Q}}\left[Y^{K'-1}_{Q}(\tilde{\varkappa})\left(\frac{2K'+1}{K'+1}\right)\{A(K'-1,K')+B(K'-1,K')\}\times\right. \\ \left.\times(K'-1QK'Q'\mid 1q)+Y^{K'}_{Q}(\tilde{\varkappa})B(K',K')(K'QK'Q'\mid 1q)\right]\times \\ \times(K'Q'J'M'\mid JM). \qquad (6.86)$$

We recall the fact that if $S = S'$, $L = L'$, $J = J'$, then $B(K',K') = 0$.

6.4. Summary of formulae

Since the algebra encountered in § 6.3 has been heavy we gather together the principal results derived therein.

For a particular ion in a unit cell the matrix elements of the qth spherical component of the interaction operator $\hat{\mathbf{Q}}_{\perp}$ in the SLJ basis are

$$\langle l^n vSLJM|\,\hat{Q}_{\perp,q}\,|l^n v'S'L'J'M'\rangle = (4\pi)^{\frac{1}{2}} \sum_{KK'} \{A(K,K')+B(K,K')\}\times$$

$$\times \sum_{QQ'} Y_Q^K(\tilde{\varkappa})(K'Q'J'M'\,|\,JM)(KQK'Q'\,|\,1q). \quad (6.87\text{ a})$$

The coefficients $A(K,K')$ and $B(K,K')$ are as follows:

$$A(K,K') = (-)^{l+L+L'+S+J'}i^{K'+1}\delta_{S,S'}(2l+1)^2\times$$

$$\times\{(2L+1)(2L'+1)(2J'+1)/3\}^{\frac{1}{2}}\times$$

$$\times\begin{Bmatrix} K' & L' & L \\ S & J & J' \end{Bmatrix}\Gamma(K,K')A(K',K',l)(\bar{j}_{K'+1}+\bar{j}_{K'-1})\times$$

$$\times n\sum_{\bar{\theta}}(-)^{\bar{L}}(\theta\{|\bar{\theta})(\theta'\{|\bar{\theta})\begin{Bmatrix} K' & l & l \\ \bar{L} & L & L' \end{Bmatrix}. \quad (6.88)$$

The summation is over all $n-1$ electron states $\bar{\theta}$ that are common to $|\theta\rangle$ and $|\theta'\rangle$, and $(\{\,|\,)$ denotes the corresponding fractional parentage coefficients.† $\Gamma(K,K')$ has the values

$$\Gamma(K,K') = (K')^{\frac{1}{2}}, \qquad \text{if } K = K'+1,$$

$$= (K'+1)^{\frac{1}{2}}, \quad \text{if } K = K'-1,$$

$$= 0, \qquad \text{otherwise.} \quad (6.89)$$

$A(K',K',l)$ is zero unless K' is one of the integers 1, 3, 5,..., $(2l-1)$. If this condition is satisfied and we set

$$K' = 2m+1,$$

then

$$A(K',K',l) = \frac{(-)^m}{(2l+1)}\frac{(m+l+1)!}{m!\,(m+1)!\,(l-m)!}\times$$

$$\times\left\{\frac{(l-m)(2m)!\,(2m+2)!\,(2l-2m)!}{2(4m+3)(2l+2m+2)!}\right\}^{\frac{1}{2}}. \quad (6.90)$$

The requirement that K' be odd, together with eqn (6.89), implies that the integer K must be even and not greater than $2l$. Hence, for a given K', the only non-zero coefficients are $A(K'\pm1,K')$, and these are related by

$$\frac{A(K'+1,K')}{A(K'-1,K')} = \left(\frac{K'}{K'+1}\right)^{\frac{1}{2}}. \quad (6.91)$$

† op. cit., p. 140.

The spin factor $B(K,K')$ is slightly more cumbersome, namely

$$B(K,K') = \tfrac{2}{3}\mathrm{i}^K\left(\frac{2K'+1}{3}\right)^{\frac{1}{2}}C(K,K')+$$
$$+\tfrac{1}{3}(-)^{K'+1}\{10(2K+1)(2K'+1)\}^{\frac{1}{2}}\sum_{\bar{K}}\mathrm{i}^{\bar{K}}(2\bar{K}+1)^{\frac{1}{2}}\times$$
$$\times\begin{pmatrix}2 & \bar{K} & K\\ 0 & 0 & 0\end{pmatrix}\begin{Bmatrix}K' & \bar{K} & 1\\ 2 & 1 & K\end{Bmatrix}C(\bar{K},K'), \quad (6.92)$$

where

$$C(\dot{K},K') = (-1)^{L'+S'}\times$$
$$\times\{\tfrac{3}{2}(2\dot{K}+1)(2K'+1)(2S+1)(2S'+1)(2L+1)(2L'+1)(2J'+1)\}^{\frac{1}{2}}\times$$
$$\times(2l+1)\begin{pmatrix}l & \dot{K} & l\\ 0 & 0 & 0\end{pmatrix}\begin{Bmatrix}1 & \dot{K} & K'\\ S' & L' & J'\\ S & L & J\end{Bmatrix}\bar{\jmath}_{\dot{K}}(\kappa)\times$$
$$\times n\sum_{\bar{\theta}}(-1)^{\frac{1}{2}+\bar{S}+\bar{L}}(\theta\{|\bar{\theta})(\theta'\{|\bar{\theta})\begin{Bmatrix}S & 1 & S'\\ \frac{1}{2} & \bar{S} & \frac{1}{2}\end{Bmatrix}\begin{Bmatrix}L & \dot{K} & L'\\ l & \bar{L} & l\end{Bmatrix}. \quad (6.93)$$

The $3j$ symbol in (6.93) makes $\dot{K}$, and hence $\bar{K}$ and K in (6.92), even for a non-zero $B(K,K')$. The maximum value of $\dot{K}$ is $2l$. Hence triangular conditions ensure that the upper limit of K' is $2l+1$ and of K, $2l+2$. If K' is even the only non-zero coefficient occurs when $K = K'$, in which case

$$B(K',K') = \mathrm{i}^{K'}\left(\frac{2K'+1}{3}\right)^{\frac{1}{2}}C(K',K'). \quad (6.94)$$

If K' is odd then $K = K'\pm1$. For $K = K'-1$,

$$B(K'-1,K')$$
$$= \mathrm{i}^{K'-1}[(K'+1)C(K'-1,K')-\{K'(K'+1)\}^{\frac{1}{2}}C(K'+1,K')]\times$$
$$\times\{3(2K'+1)\}^{-\frac{1}{2}} \quad (6.95)$$

and the coefficient $B(K'+1,K')$ is got from this through

$$B(K'+1,K') = \{K'/(K'+1)\}^{\frac{1}{2}}B(K'-1,K'). \quad (6.96)$$

For the particular case $S = S'$, $L = L'$, and $J = J'$ the $9j$ coefficient in $C(\dot{K},K')$ requires that $B(K,K')$ is non-zero for $K = K'\pm1$ with $K' = 1, 3,\ldots, (2l+1)$. Thus in this instance the conditions on K and K' for non-zero $B(K,K')$ coincide with those for $A(K,K')$, except for the fact that for $B(K,K')$ the maximum value of K' is $(2l+1)$ whereas for $A(K,K')$ it is $(2l-1)$.

The above formulae for the matrix elements $\hat{\mathbf{Q}}_\perp$ refer to the $|SLJM\rangle$ basis. However, we remarked in the introduction that often the magnetic state of an ion is appropriately described in terms of the $|SM_\mathrm{s}\,LM_\mathrm{L}\rangle$

basis. The cross-section can easily be evaluated in this basis through the transformation

$$|SM_s LM_L\rangle = \sum_{JM} (SM_s LM_L \mid JM)|SLJM\rangle, \tag{6.97}$$

noting that $A(K, K')$ and $C(\dot{K}, K')$ are functions of J.

In some instances it proves more convenient to calculate the cross-section in terms of the matrix elements of the operator $\hat{\mathbf{Q}}$, which is related to $\hat{\mathbf{Q}}_\perp$ through $\hat{\mathbf{Q}}_\perp = \tilde{\boldsymbol{\varkappa}} \times (\hat{\mathbf{Q}} \times \tilde{\boldsymbol{\varkappa}})$ and has matrix elements

$$\langle l^n vSLJM|\, \hat{Q}_q\, |l^n v'S'L'J'M'\rangle = (4\pi)^{\frac{1}{2}} \sum_{\substack{K',Q'\\Q}} \left[Y_Q^{K'-1}(\tilde{\boldsymbol{\varkappa}}) \left(\frac{2K'+1}{K'+1}\right) \times \right.$$
$$\times \{A(K'-1, K') + B(K'-1, K')\}(K'-1\, QK'Q'|1q) +$$
$$\left. + Y_Q^{K'}(\tilde{\boldsymbol{\varkappa}}) B(K', K')(K'QK'Q' \mid 1q) \right] (K'Q'J'M' \mid JM). \tag{6.87 b}$$

The operator $\hat{\mathbf{Q}}$ has the conceptual value of retaining the same form for the cross-section in the general case of both spin and orbital scattering as in the spin-only case.

Though the detailed structure of the cross-section appears very complicated, involving summations over $3j$, $6j$, and $9j$ symbols,† it is very amenable to computing. There is just one set of parameters, the radial integrals $\bar{j}_K$, which are calculated from a given radial wave function. By comparing the cross-section computed from (6.87 a) or (6.87 b) with that measured one can determine the detailed nature of the unpaired electrons in the ion, provided, of course, one is able to make proper allowance for covalency.

The relative significance of the terms in eqns (6.88) and (6.92) for $A(K, K')$ and $B(K, K')$ depends upon the relative size of the radial integrals $\bar{j}_K(\kappa)$

$$= \int_0^\infty \mathrm{d}r\; r^2 f^2 j_K(\kappa r).$$

We note that in general more than one order K of $\bar{j}_K(\kappa)$ appears in a given term in (6.87). Since the small argument expansion of the spherical Bessel function is

$$j_K(x) = \frac{x^K}{1.3.5.\ldots(2K+1)}, \tag{6.98}$$

it is often sufficient to consider only the first few terms in the sum over K. The *dipole approximation* $K' = 1$, $K = 0, 2$, is considered in detail in the next section.

† Tables of $3j$ and $6j$ symbols have been compiled by Rotenberg *et al.* M.I.T. Press. A table of $9j$ symbols has been made by Howell, University of Southampton, Maths Dept., Research Report No. 59–2 (1959).

6.5. Dipole approximation

In this approximation ($K' = 1$, $K = 0$, 2) we assume that $|\boldsymbol{\varkappa}|^{-1}$ is much larger than the mean radius of the wave function of the unpaired electrons.

First, consider the cross-section for the scattering of unpolarized neutrons by a system of free ions. In this instance

$$|\lambda\rangle = |JM\rangle, \qquad |\lambda'\rangle = |JM'\rangle, \tag{6.99}$$

and we must average over all the initial orientations of the ion and sum over all possible final orientations, i.e.

$$\sum_{\lambda,\lambda'} p_\lambda \equiv \sum_{M,M'} (2J+1)^{-1}, \tag{6.100}$$

and so, from eqn (6.87 a),

$$\begin{aligned}
&\sum_q \sum_{\lambda\lambda'} p_\lambda |\langle\lambda'|\,\hat{Q}_{\perp,q}\,|\lambda\rangle|^2 \\
&= \sum_{MM'} (2J+1)^{-1} \sum_{\substack{qK,K'\\ K_1,K_1'}} \{A(K,K')+B(K,K')\}\{A(K_1,K_1')+B(K_1,K_1')\}\times \\
&\quad\times \sum_{QQ'} \sum_{Q_1Q_1'} 4\pi Y^K_Q(\tilde{\boldsymbol{\varkappa}}) Y^{*K_1}_{Q_1}(\tilde{\boldsymbol{\varkappa}})(K'Q'JM' \mid JM)(K_1'Q_1'JM' \mid JM)\times \\
&\qquad\qquad\times(KQK'Q' \mid 1q)(K_1\,Q_1\,K_1'Q_1' \mid 1q) \\
&= \sum_q \sum_{KK'} \sum_{QQ'} \frac{4\pi}{2K'+1} |Y^K_Q(\tilde{\boldsymbol{\varkappa}})|^2 \{A(K,K')+B(K,K')\}^2\times \\
&\qquad\qquad\times(KQK'Q' \mid 1q)^2.
\end{aligned} \tag{6.101}$$

Since

$$\sum_{qQ'} (KQK'Q' \mid 1q)^2 = \frac{3}{2K+1}$$

and

$$\sum_Q |Y^K_Q(\tilde{\boldsymbol{\varkappa}})|^2 = \frac{2K+1}{4\pi},$$

the required cross-section is

$$\frac{\mathrm{d}\sigma}{\mathrm{d}\Omega} = \left(\frac{\gamma e^2}{m_e c^2}\right)^2 \sum_{K,K'} \frac{3}{2K'+1} \{A(K,K')+B(K,K')\}^2. \tag{6.102}$$

Taking $K' = 1$, $K = 0$, 2,

$$\frac{\mathrm{d}\sigma}{\mathrm{d}\Omega} = \left(\frac{\gamma e^2}{m_e c^2}\right)^2 \tfrac{3}{2}\{A(0,1)+B(0,1)\}^2 \tag{6.103}$$

on using (6.91) and (6.96).

To evaluate $A(0,1)$ from (6.88) note that

$$n \sum_{\bar{\theta}} (\theta\{|\bar{\theta})^2 (-)^{\bar{L}} \begin{Bmatrix} L & 1 & L \\ l & \bar{L} & l \end{Bmatrix} = (-)^{-l+1-L} \left\{\frac{L(L+1)}{(2L+1)l(l+1)(2l+1)}\right\}^{\frac{1}{2}} \tag{6.104 a}$$

and

$$\begin{Bmatrix} L & L & 1 \\ J & J & S \end{Bmatrix} = \tfrac{1}{2}(-)^{J+L+S+1} \frac{J(J+1)+L(L+1)-S(S+1)}{\{J(J+1)(2J+1)L(L+1)(2L+1)\}^{\frac{1}{2}}}. \tag{6.104 b}$$

Also from (6.90)

$$A(1,1,l) = \frac{1}{2l+1}\left\{\frac{l(l+1)}{6(2l+1)}\right\}^{\frac{1}{2}}. \tag{6.104 c}$$

With the results (6.104) the value of $A(0,1)$ is found to be

$$A(0,1) = -\tfrac{1}{6}(\bar{j}_0+\bar{j}_2)\frac{1}{\sqrt{\{J(J+1)\}}}\{J(J+1)+L(L+1)-S(S+1)\}. \tag{6.105}$$

In evaluating $B(0,1)$ from (6.92) and (6.93) we assume, for the present, that the coefficient of $\bar{j}_2$ in $C(2,1)$ is small compared to the coefficient of $\bar{j}_2$ in $A(0,1)$. From (6.93),

$$C(0,1) = \bar{j}_0\,\tfrac{3}{2}\sqrt{(2)(2S+1)(2L+1)}\sqrt{\{(2J+1)(2l+1)\}}\times$$
$$\times(-)^{L+l+\frac{1}{2}+S}\begin{Bmatrix} S & S & 1 \\ L & L & 0 \\ J & J & 1 \end{Bmatrix}\times$$
$$\times n\sum_{\bar{\theta}}(\theta\{|\bar{\theta})^2(-)^{\bar{S}+\bar{L}}\begin{Bmatrix} S & 1 & S \\ \frac{1}{2} & \bar{S} & \frac{1}{2} \end{Bmatrix}\begin{Bmatrix} L & 0 & L \\ l & \bar{L} & l \end{Bmatrix}. \tag{6.106}$$

The $9j$ symbol in (6.106) reduces to

$$\begin{Bmatrix} S & S & 1 \\ L & L & 0 \\ J & J & 1 \end{Bmatrix} = \frac{(-)^{S+L+J+1}}{\sqrt{\{3(2L+1)\}}}\begin{Bmatrix} S & S & 1 \\ J & J & L \end{Bmatrix} \tag{6.107 a}$$

and

$$\begin{Bmatrix} L & 0 & L \\ l & \bar{L} & l \end{Bmatrix} = \frac{(-)^{\bar{L}+l+L}}{\sqrt{\{(2l+1)(2L+1)\}}}. \tag{6.107 b}$$

The sum over the coefficients of fractional parentage in (6.106) can be written down by analogy with (6.104 a) and the $6j$ symbol in (6.107 a) is obtained from (6.104 b). The result for $C(0,1)$ is then

$$C(0,1) = -\tfrac{1}{2}\bar{j}_0\frac{1}{\sqrt{\{J(J+1)\}}}\{J(J+1)+S(S+1)-L(L+1)\} \tag{6.108 a}$$

and we shall take

$$B(0,1) \simeq \tfrac{2}{3}C(0,1) \tag{6.108 b}$$

from (6.92).

On combining (6.105) and (6.108 b) in (6.103) the neutron cross-section in the dipole approximation for the scattering from an isolated ion is

$$\frac{d\sigma}{d\Omega} = \left(\frac{\gamma e^2}{m_e c^2}\right)^2 \frac{J(J+1)}{6} g^2F^2(\kappa), \tag{6.109}$$

where g is the Landé splitting factor (6.16), and

$$F(\kappa) = \bar{j}_0+\bar{j}_2\left\{\frac{J(J+1)+L(L+1)-S(S+1)}{3J(J+1)+S(S+1)-L(L+1)}\right\} \tag{6.110}$$

is the free-ion form factor in the dipole approximation. In the limit $|\boldsymbol{\kappa}| \to 0$, $F \to 1$, and (6.109) becomes identical with (6.17).

In deriving (6.109) we neglected $C(2,1)$, which contains $\bar{j}_2$. We now examine the validity of this step in more detail. From (6.93),

$$C(2,1) = -\bar{j}_2\left[(2S+1)(2L+1)(2l+1)\begin{pmatrix} l & 2 & l \\ 0 & 0 & 0\end{pmatrix}\times\right.$$
$$\left.\times 3\{\tfrac{5}{2}(2J+1)\}^{\frac{1}{2}}(-)^{L+S-\frac{1}{2}}\begin{Bmatrix} S & S & 1 \\ L & L & 2 \\ J & J & 1\end{Bmatrix}\right]\times$$
$$\times n\sum_{\bar\theta}(\theta\{|\bar\theta)^2(-)^{\bar S+\bar L}\begin{Bmatrix} S & 1 & S \\ \frac{1}{2} & \bar S & \frac{1}{2}\end{Bmatrix}\begin{Bmatrix} L & 2 & L \\ l & \bar L & l\end{Bmatrix}. \tag{6.111}$$

The $9j$ symbol occurring in this expression can easily be evaluated from

$$\begin{Bmatrix} S & S & 1 \\ L & L & 2 \\ J & J & 1\end{Bmatrix} = \frac{1}{60}\left\{\frac{1}{4L(L+1)-3}\right\}\times$$
$$\times\left\{\frac{30(2L-1)(2L+3)}{S(S+1)(2S+1)L(L+1)(2L+1)J(J+1)(2J+1)}\right\}^{\frac{1}{2}}\times$$
$$\times[L(L+1)\{L(L+1)+2J(J+1)+2S(S+1)\}-$$
$$-3\{J(J+1)-S(S+1)\}^2]. \tag{6.112}$$

As an example we consider d^3. The ground state is ${}^4F$, for which the term in square brackets in (6.111) has the value ($L = 3$, $S = J = 3/2$)

$$-\frac{4.3^2}{5}\sqrt{2}. \tag{6.113 a}$$

The parent states $\bar\theta$ of d^3 are $\bar S = 1$, $\bar L = 1$ and 3, for which the coefficients $(\theta\{|\bar\theta)^2$ have the values 1/5 and 4/5, respectively. Thus

$$n\sum_{\bar\theta}(\theta\{|\bar\theta)^2(-)^{\bar S+\bar L}\begin{Bmatrix} S & 1 & S \\ \frac{1}{2} & \bar S & \frac{1}{2}\end{Bmatrix}\begin{Bmatrix} L & 2 & L \\ 2 & \bar L & 2\end{Bmatrix} = -\frac{1}{14\sqrt{15}}. \tag{6.113 b}$$

With the results (6.113 a) and (6.113 b) we find, for d^3,

$$C(2,1) = -\frac{2.3^2}{5.7}\sqrt{\left(\frac{2}{15}\right)}\bar{j}_2.$$

The coefficient of $C(2,1)$ in $B(0,1)$ is $-\frac{1}{3}\sqrt{2}$, so that in neglecting $C(2,1)$ in this instance we have dropped a term

$$\frac{4.3}{35}\frac{1}{\sqrt{15}}\bar{j}_2.$$

The coefficient of $\bar{j}_2$ in $A(0,1)$ for d^3 is

$$-4\frac{1}{\sqrt{15}}$$

so that we conclude that, at least for this example, the error involved in neglecting $C(2,1)$ in $B(0,1)$ is indeed small, the ratio of the $\bar{j}_2$ term in $A(0,1)$ to that in $B(0,1)$ being 3/35; for d^7 the ratio is 4/105.

We now consider the consequences of the dipole approximation in more detail.

In writing $\hat{Q}_{\perp,q}$ in the form (6.87 a) we have implicitly assumed that the wave function of the unpaired electrons in a crystal environment takes the form

$$|\lambda\rangle = \sum_M |JM\rangle\langle JM\,|\lambda\rangle, \tag{6.114 a}$$

where the matrix elements $\langle JM\,|\,\lambda\rangle$ are determined by the crystal field.† Thus

$$(K'Q'JM'\,|\,JM) \Rightarrow \sum_{M,M'} \langle\lambda|JM\rangle(K'Q'JM'\,|\,JM)\langle JM'\,|\,\lambda\rangle. \tag{6.114 b}$$

In the dipole approximation we are concerned only with the $K' = 1$ term. With the formalism of § 6.3 it can be shown that

$$\begin{aligned}[\theta J\|\hat{\mathbf{L}}\|\theta J] &= (-)^{S+L+J+1}(2J+1)\begin{Bmatrix} J & J & 1 \\ L & L & S\end{Bmatrix}[L\|\hat{\mathbf{L}}\|L] \\ &= (-)^{S+L+J+1}(2J+1)\begin{Bmatrix} J & J & 1 \\ L & L & S\end{Bmatrix}\sqrt{\{L(L+1)(2L+1)\}} \\ &= \frac{1}{2}\left\{\frac{(2J+1)}{J(J+1)}\right\}^{\frac{1}{2}}\{J(J+1)+L(L+1)-S(S+1)\}\end{aligned}$$

and, similarly,

$$[\theta J\|\hat{\mathbf{S}}\|\theta J] = \frac{1}{2}\left\{\frac{(2J+1)}{J(J+1)}\right\}^{\frac{1}{2}}\{J(J+1)+S(S+1)-L(L+1)\}.$$

Hence
$$A(0,1) = -\frac{1}{3\sqrt{(2J+1)}}(\bar{j}_0+\bar{j}_2)[\theta J\|\hat{\mathbf{L}}\|\theta J]$$

and, neglecting $C(2,1)$,

$$B(0,1) \simeq -\frac{2}{3\sqrt{(2J+1)}}\bar{j}_0[\theta J\|\hat{\mathbf{S}}\|\theta J].$$

† Alternatively, we could, of course, base our discussion on $\hat{\mathbf{Q}}$, eqn (6.86 b).

The latter two results imply the relations

$$(1Q'JM' \mid JM)A(0,1) \Rightarrow \tfrac{1}{3}(\bar{j}_0+\bar{j}_2)\langle\lambda|\,\hat{L}_{Q'}\,|\lambda\rangle$$

and

$$(1Q'JM' \mid JM)B(0,1) \Rightarrow \tfrac{2}{3}\bar{j}_0\,\langle\lambda|\,\hat{S}_{Q'}\,|\lambda\rangle,$$

i.e.

$$3(1Q'JM' \mid JM)\{A(0,1)+B(0,1)\} \Rightarrow \langle\lambda|\,\{(\bar{j}_0+\bar{j}_2)\hat{\mathbf{L}}+2\bar{j}_0\,\hat{\mathbf{S}}\}_{Q'}\,|\lambda\rangle. \tag{6.115}$$

In the limit $|\boldsymbol{\kappa}| \to 0$ the right-hand side of (6.115) becomes

$$\langle\lambda|\,(\hat{\mathbf{L}}+2\hat{\mathbf{S}})_{Q'}\,|\lambda\rangle.$$

Returning now to the evaluation of the cross-section for ions with wave functions of the form (6.114 a), we make the assumption that on the right-hand side we can take $Q' = 0$ (this means that only those states for which $M = M'$ are retained). Hence

$$\langle\lambda|\,\hat{Q}_{\perp,q}\,|\lambda\rangle \simeq (4\pi)^{\frac{1}{2}}\tfrac{1}{3}\langle\lambda|\,\{(\bar{j}_0+\bar{j}_2)\hat{\mathbf{L}}+2\bar{j}_0\,\hat{\mathbf{S}}\}_z\,|\lambda\rangle\times$$
$$\times\left\{Y_0^0(\tilde{\boldsymbol{\kappa}})(0010 \mid 10)\delta_{q,0}+\frac{1}{\sqrt{2}}\,Y_q^2(\tilde{\boldsymbol{\kappa}})(2q10 \mid 1q)\right\}.$$

On making use of the results

$$(0010 \mid 10) = 1,$$

$$(2q10 \mid 1q) = -\sqrt{\left(\frac{4-q^2}{10}\right)},$$

$$Y_0^0(\tilde{\boldsymbol{\kappa}}) = 1/(4\pi)^{\frac{1}{2}},$$

$$Y_0^2(\tilde{\boldsymbol{\kappa}}) = \left(\frac{5}{4\pi}\right)^{\frac{1}{2}}(1-\tfrac{3}{2}\sin^2\theta),$$

and

$$Y_{\pm 1}^2(\tilde{\boldsymbol{\kappa}}) = \mp\left(\frac{15}{2\,.\,4\pi}\right)^{\frac{1}{2}}\sin\theta\cos\theta\exp(\pm\mathrm{i}\phi),$$

we obtain

$$\sum_q |\langle\lambda|\,\hat{Q}_{\perp,q}\,|\lambda\rangle|^2 \simeq |\tfrac{1}{2}\sin\theta\,\langle\lambda|\{(\bar{j}_0+\bar{j}_2)\hat{\mathbf{L}}+2\bar{j}_0\,\hat{\mathbf{S}}\}_z\,|\lambda\rangle|^2, \tag{6.116}$$

where θ is the angle between $\tilde{\boldsymbol{\kappa}}$ and the direction of quantization $\tilde{\mathbf{z}}$.

From (6.116) we see that in the dipole approximation the elastic cross-section for unpolarized neutrons is

$$\frac{\mathrm{d}\sigma}{\mathrm{d}\Omega} \simeq \left(\frac{\gamma e^2}{2m_{\mathrm{e}}c^2}\right)^2 \sum_\lambda p_\lambda\,\langle\lambda|\,\{(\bar{j}_0+\bar{j}_2)\hat{\mathbf{L}}+2\bar{j}_0\,\hat{\mathbf{S}}\}_\perp\,|\lambda\rangle\,.$$
$$.\,\langle\lambda|\,\{(\bar{j}_0+\bar{j}_2)\hat{\mathbf{L}}+2\bar{j}_0\,\hat{\mathbf{S}}\}_\perp\,|\lambda\rangle. \tag{6.117}$$

More generally we have shown that in the dipole approximation (which can be said to be valid when $\bar{j}_0$ is larger than $\bar{j}_2$) the operator $\hat{\mathbf{Q}}$ (cf. (6.58)) can be replaced by

$$\hat{\mathbf{Q}}^{(\mathrm{D})} = \tfrac{1}{2}\{(\bar{j}_0(\kappa)+\bar{j}_2(\kappa))\hat{\mathbf{L}}+2\bar{j}_0(\kappa)\hat{\mathbf{S}}\}. \tag{6.118}$$

Two cases are of particular importance.

(i) When $\hat{\mathbf{L}}$ can be replaced by $(g-2)\hat{\mathbf{S}}$,

$$\hat{\mathbf{Q}}^{(\mathrm{D})} = \tfrac{1}{2}gF(\kappa)\hat{\mathbf{S}}, \tag{6.119}$$

where $F(\kappa)$ is the form factor of the ion and is given by

$$F(\kappa) = \bar{j}_0(\kappa)+(1-2/g)\bar{j}_2(\kappa). \tag{6.120}$$

(ii) When the ion is characterized by a total angular momentum J,

$$\hat{\mathbf{Q}}^{(\mathrm{D})} = \tfrac{1}{2}g\,F(\kappa)\hat{\mathbf{J}}, \tag{6.121}$$

where g is the Landé splitting factor (6.16) and the form factor $F(\kappa)$ is given by eqn (6.110).

6.6. Elastic scattering from rare earths†

In the rare-earth ions the contribution to the cross-section from the orbital angular momentum cannot in general be disregarded. Further, as we go to larger values of $|\boldsymbol{\kappa}|$ so the validity of the dipole approximation breaks down. We now give a detailed discussion of the application of the formalism derived in § 6.3 and summarized in § 6.4 to rare-earth ions, first when they are fully saturated ($J = M$) and secondly when the crystal field partially and wholly removes the $2J+1$ degeneracy. (Our discussion in principle covers all ions with unpaired f electrons, both actinide and rare-earth elements.)

(a) *Saturated rare earths*

We expect the state $|SLJJ\rangle$ to be a good description of the ions in a ferromagnetically or antiferromagnetically ordered rare-earth compound or for ions fully aligned by an external magnetic field. The ions are assumed aligned parallel to the z-axis.

Our starting point is eqn (6.87 b) for $\langle SLJM|\,\hat{Q}_q\,|S'L'J'M'\rangle$. We set $S = S'$, $L = L'$, $J = J'$ and consider first $q = \pm 1$. Thus

$$\begin{aligned}\langle SLJM|\,\hat{Q}_{\pm 1}\,|SLJM\rangle &\\ = (-1)^{J-M+1}\{12\pi(2J+1)\}^{\frac{1}{2}} &\sum_{K'} Y^{K'-1}_{\pm 1}(\tilde{\boldsymbol{\kappa}})\left(\frac{2K'+1}{K'+1}\right)\times\\ \times\{A(K'-1,K')+B(K'-1,K')\}&\begin{pmatrix}K' & J & J\\ 0 & M & -M\end{pmatrix}\begin{pmatrix}K'-1 & K' & 1\\ \pm 1 & 0 & \mp 1\end{pmatrix}.\end{aligned} \tag{6.122}$$

Note that $K' \leqslant 2J-1$ if J is an integer or $K' \leqslant 2J$ if J is half integer.

† Balcar *et al.*, *J. Phys.* C, **3**, 1292 (1970).

By definition of the spherical components of $\hat{\mathbf{Q}}$ (cf. (6.20))

$$\hat{Q}_x = -\frac{1}{\sqrt{2}}(\hat{Q}_{+1}-\hat{Q}_{-1}), \qquad \hat{Q}_y = \frac{\mathrm{i}}{\sqrt{2}}(\hat{Q}_{+1}+\hat{Q}_{-1}). \tag{6.123}$$

If we write $\tilde{\boldsymbol{\varkappa}} = (\sin\theta\cos\phi, \sin\theta\sin\phi, \cos\theta)$ and use the results

$$Y_1^{K'-1}-Y_{-1}^{K'-1} = 2\cos\phi\,\mathrm{e}^{-\mathrm{i}\phi}Y_1^{K'-1}$$

and

$$Y_1^{K'-1}+Y_{-1}^{K'-1} = 2\mathrm{i}\sin\phi\,\mathrm{e}^{-\mathrm{i}\phi}\,Y_1^{K'-1}, \tag{6.124}$$

then, from (6.122),

$$\langle SLJJ|\,\hat{Q}_x\,|SLJJ\rangle = M(\theta)\cos\phi \tag{6.125 a}$$

and

$$\langle SLJJ|\,\hat{Q}_y\,|SLJJ\rangle = M(\theta)\sin\phi. \tag{6.125 b}$$

In eqns (6.125 a) and (6.125 b) the function $M(\theta)$ is defined by

$$M(\theta) = (4\pi)^{\frac{1}{2}}\sum_{K'}\mathrm{e}^{-\mathrm{i}\phi}Y_1^{K'-1}(\theta,\phi)\left(\frac{K'-1}{K'}\right)^{\frac{1}{2}}Z(K') \tag{6.126}$$

and

$$Z(K') = \{3(2J+1)\}^{\frac{1}{2}}\left(\frac{2K'+1}{K'+1}\right)\{A(K'-1,K')+B(K'-1,K')\}\times$$

$$\times\begin{pmatrix}K' & J & J\\ 0 & J & -J\end{pmatrix}\begin{pmatrix}K'-1 & K' & 1\\ 0 & 0 & 0\end{pmatrix}. \tag{6.127}$$

There are several properties of $M(\theta)$ that are worth noting at this point. The smallest value of K' in eqn (6.126) is 3, and it follows from the structure of $A(K'-1,K')$ and $B(K'-1,K')$, eqns (6.88) and (6.95), that the lowest possible order radial integral in $M(\theta)$ is $\bar{j}_2(\kappa)$. Since only $\bar{j}_0(\kappa)$ is finite for $|\boldsymbol{\varkappa}| \to 0$, $M(\theta)$ must be zero for zero wave vector. It is also zero for $|\boldsymbol{\varkappa}| \to \infty$. Consider now its dependence on θ, the angle between the axis of quantization and the direction of $\tilde{\boldsymbol{\varkappa}}$. First, $M(0) = 0$ and a simple calculation shows that $M(\pi-\theta) = -M(\theta)$, so that $M(\frac{1}{2}\pi) = 0$. $M(\theta)$ for Pr^{3+} is shown in Fig. 6.1 as a function of $|\boldsymbol{\varkappa}|$ for various values of θ.

With $L = 0$, $A(K'-1,K')$ is zero, of course, and the only contribution from $B(K'-1,K')$ is for $K' = 1$, and this gives $Z(K') = S\bar{j}_0(\kappa)$. On the other hand, if $S = 0$, then $B(K'-1,K') = 0$ but $A(K'-1,K')$ is in general non-zero for all K'. Hence $M(\theta)$ is a manifestation of the orbital moment of the rare-earth ions.

We can also express $\langle SLJJ|\,\hat{Q}_z\,|SLJJ\rangle$ in terms of the functions $Z(K')$. From (6.87 b) and (6.127)

$$\langle SLJJ|\,\hat{Q}_z\,|SLJJ\rangle = (4\pi)^{\frac{1}{2}}\sum_{K'}Y_0^{K'-1}(\theta,\phi)Z(K'). \tag{6.128}$$

Clearly (6.128) is independent of ϕ. In contrast to $M(\theta)$, the smallest value of $K' = 1$ and hence $\langle SLJJ|\,\hat{Q}_z\,|SLJJ\rangle$ is non-zero in the limit $|\boldsymbol{\varkappa}| \to 0$.

It is instructive to compare the above results for the diagonal matrix elements of $\hat{Q}_x$, $\hat{Q}_y$, and $\hat{Q}_z$ with those we would derive in the dipole approximation, eqn (6.121). The latter case only involves the matrix elements of $\hat{J}_\alpha$, for which there are no matrix elements diagonal in M for $\alpha = x$ or y, i.e. both $\langle SLJJ|\,\hat{Q}_x\,|SLJJ\rangle$ and $\langle SLJJ|\,\hat{Q}_y\,|SLJJ\rangle$ are zero in the dipole approximation. Further, from (6.121),

$$\langle SLJJ|\,\hat{Q}_z^{(\mathrm{D})}\,|SLJJ\rangle = \tfrac{1}{2}gF(\kappa)J, \tag{6.129}$$

with $F(\kappa)$ given by eqn (6.110). On comparing (6.128) and (6.129) it is evident that the coefficient of $\bar{j}_0(\kappa)$ in $Z(1)$ must equal $\frac{1}{2}gJ$. (Note that $Y_0^0(\theta,\phi) = (4\pi)^{-\frac{1}{2}}$.)

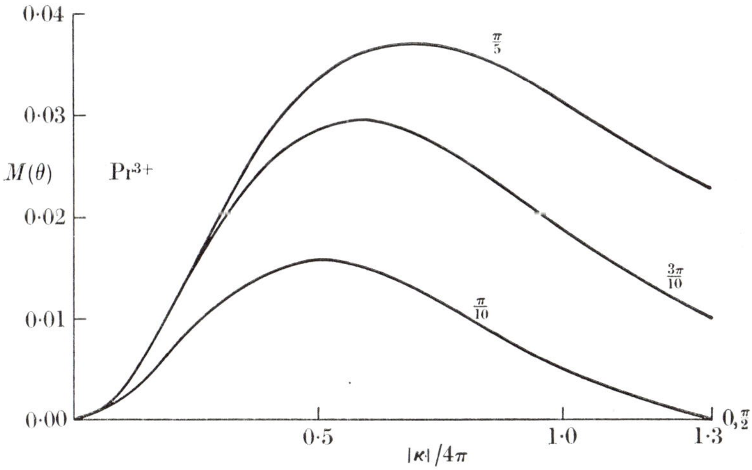

FIG. 6.1. The function $M(\theta)$ defined by eqns (6.126) and (6.127) is shown for the tripositive rare earth Pr^{3+} as a function of $|\kappa|$ for various values of θ.

The functions $Z(K')$, which can be regarded as the 'characteristic' functions for the scattering from rare-earth ions, are tabulated in Table 6.1 for the tripositive rare-earth ions. It is of value to know the orbital contribution to these functions, and in Table 6.2 we give just this part of $Z(K')$, i.e. eqn (6.127) with $B(K'-1, K')$ set equal to zero. Inspection of eqn (6.88) for $A(K, K')$ shows that the orbital contribution to $Z(K')$ is of the form $c_{K'}^{(L)}(\bar{j}_{K'-1}+\bar{j}_{K'+1})$ and it is the coefficients $c_{K'}^{(L)}$ that are listed in Table 6.2. Further, the conditions on K and K' in $A(K, K')$ and $B(K, K')$ shows that $Z(7)$ comes entirely from $B(K, K')$, the spin part of the matrix elements. Note that for $L = 5$ there is no orbital contribution to $Z(3)$, so that the leading order

TABLE 6.1

The coefficients $Z(K')$ defined by eqn (6.127) *are given for the tripositive rare-earth ions. From* (6.127) *it follows that*

$$Z(K') = c_{K'-1}\bar{j}_{K'-1}(\kappa)+c_{K'+1}\bar{j}_{K'+1}(\kappa)$$

and it is the coefficients $c_{K'\pm1}$ that are tabulated

	$\bar{j}_0$	$\bar{j}_2$	$\bar{j}_4$	$\bar{j}_6$
		Ce^{3+} f^1 $^2F_{5/2}$		
$Z(1)$	1·07142857	1·71428571		
$Z(3)$		0·09583148	0·31943828	
$Z(5)$			0·00360750	0·04329004
		Pr^{3+} f^2 $^3H_4$		
$Z(1)$	1·60000000	2·63111111		
$Z(3)$		−0·09865803	−0·13453368	
$Z(5)$			−0·01836547	−0·11678557
$Z(7)$				0·00233133
		Nd^{3+} f^3 $^4I_{9/2}$		
$Z(1)$	1·63636364	2·95041322		
$Z(3)$		−0·20896503	−0·25329095	
$Z(5)$			0·03820789	0·14258681
$Z(7)$				−0·00614959
		Pm^{3+} f^4 $^5I_4$		
$Z(1)$	1·20000000	2·71515152		
$Z(3)$		−0·10866182	−0·07024602	
$Z(5)$			0·00509439	−0·06449156
$Z(7)$				0·00232318
		Sm^{3+} f^5 $^6H_{5/2}$		
$Z(1)$	0·35714286	1·93650794		
$Z(3)$		0·04614109	0·06291966	
$Z(5)$			−0·00426341	0·00335243
		Tb^{3+} f^8 $^7F_6$		
$Z(1)$	4·50000000	1·66666666		
$Z(3)$		0·00000000	0·28459047	
$Z(5)$			−0·05050505	0·03496503
$Z(7)$				−0·01131382
		Dy^{3+} f^9 $^6H_{15/2}$		
$Z(1)$	5·00000000	2·66666667		
$Z(3)$		−0·22360680	−0·08131156	
$Z(5)$			0·02525253	−0·12432012
$Z(7)$				0·05656908

TABLE 6.1 (*cont.*)

	$\bar{j}_0$	$\bar{j}_2$	$\bar{j}_4$	$\bar{j}_6$
		Ho^{3+} f^{10} $^5I_8$		
$Z(1)$	5·00000000	3·06666667		
$Z(3)$		−0·31304952	−0·28459047	
$Z(5)$			0·12626263	0·14763015
$Z(7)$				−0·11313815
		Er^{3+} f^{11} $^4I_{15/2}$		
$Z(1)$	4·50000000	2·93333333		
$Z(3)$		−0·13416408	−0·16262313	
$Z(5)$			−0·02525253	−0·04662005
$Z(7)$				0·11313815
		Tm^{3+} f^{12} $^3H_6$		
$Z(1)$	3·50000000	2·33333333		
$Z(3)$		0·22360680	0·08131156	
$Z(5)$			−0·17676768	−0·02719503
$Z(7)$				−0·05656908
		Yb^{3+} f^{13} $^2F_{7/2}$		
$Z(1)$	2·00000000	1·33333333		
$Z(3)$		0·44721360	0·16262313	
$Z(5)$			0·10101010	0·01554002
$Z(7)$				0·01131382

TABLE 6.2

The orbital contribution $Z_L(K') = c_{K'}^{(L)}(\bar{j}_{K'-1}+\bar{j}_{K'+1})$ *of the coefficients* $Z(K')$ *given in Table* 6.1

	$\bar{j}_0, \bar{j}_2$	$\bar{j}_2, \bar{j}_4$	$\bar{j}_4, \bar{j}_6$
		Ce^{3+} f^1 $^2F_{5/2}$	
$Z_L(1)$	1·42857143		
$Z_L(3)$		0·15971914	
$Z_L(5)$			0·00721501
		Pr^{3+} f^2 $^3H_4$	
$Z_L(1)$	2·40000000		
$Z_L(3)$		0·0	
$Z_L(5)$			−0·03673095

TABLE 6.2 (*cont.*)

	$\bar{j}_0, \bar{j}_2$	$\bar{j}_2, \bar{j}_4$	$\bar{j}_4, \bar{j}_6$
		Nd^{3+} f^3 $^4I_{9/2}$	
$Z_L(1)$	2·86363636		
$Z_L(3)$		−0·16717202	
$Z_L(5)$			0·02292473
		Pm^{3+} f^4 $^5I_4$	
$Z_L(1)$	2·80000000		
$Z_L(3)$		−0·14488242	
$Z_L(5)$			0·01528316
		Sm^{3+} f^5 $^6H_{5/2}$	
$Z_L(1)$	2·14285714		
$Z_L(3)$		0·0	
$Z_L(5)$			−0·00284228
		Tb^{3+} f^8 $^7F_6$	
$Z_L(1)$	1·50000000		
$Z_L(3)$		0·22360680	
$Z_L(5)$			0·02525253
		Dy^{3+} f^9 $^6H_{15/2}$	
$Z_L(1)$	2·50000000		
$Z_L(3)$		0·0	
$Z_L(5)$			−0·07575758
		Ho^{3+} f^{10} $^5I_8$	
$Z_L(1)$	3·00000000		
$Z_L(3)$		−0·22360680	
$Z_L(5)$			0·05050505
		Er^{3+} f^{11} $^4I_{15/2}$	
$Z_L(1)$	3·00000000		
$Z_L(3)$		−0·22360680	
$Z_L(5)$			0·05050505
		Tm^{3+} f^{12} $^3H_6$	
$Z_L(1)$	2·50000000		
$Z_L(3)$		0·0	
$Z_L(5)$			−0·07575758
		Yb^{3+} f^{13} $^2F_{7/2}$	
$Z_L(1)$	1·50000000		
$Z_L(3)$		0·22360680	
$Z_L(5)$			0·02525253

part of $M(\theta)$ in these cases arises solely from the spin contribution to the scattering.

The cross-section for scattering of unpolarized neutrons is proportional to

$$\hat{\mathbf{Q}}_{\perp}.\hat{\mathbf{Q}}_{\perp} = \sum_{\alpha,\beta} (\delta_{\alpha,\beta} - \tilde{\kappa}_{\alpha}\tilde{\kappa}_{\beta})\hat{Q}_{\alpha}\hat{Q}_{\beta}.$$

From (6.125) it follows that

$$\sum_{\alpha,\beta} (\delta_{\alpha,\beta} - \tilde{\kappa}_{\alpha}\tilde{\kappa}_{\beta})\langle SLJJ|\,\hat{Q}_{\alpha}\,|SLJJ\rangle\langle SLJJ|\,\hat{Q}_{\beta}\,|SLJJ\rangle$$
$$= \sin^2\theta[\langle SLJJ|\,\hat{Q}_z\,|SLJJ\rangle - \cot\theta\, M(\theta)]^2, \qquad (6.130)$$

which is independent of ϕ, i.e. the cross-section has cylindrical symmetry.

In the limit of zero wave vector the argument of the square bracket on the right-hand side of (6.130) has the value $\frac{1}{2}gJ$. Thus we identify the function

$$F(\kappa, \theta) = \frac{2}{gJ}[\langle SLJJ|\,\hat{Q}_z\,|SLJJ\rangle - \cot\theta\, M(\theta)] \qquad (6.131)$$

as the atomic form factor.

Fig. 6.2 shows $F(\kappa, \theta)$ for various values of θ as a function of $|\boldsymbol{\kappa}|/4\pi$ for the tripositive rare-earth ions Pr^{3+}, Sm^{3+}, and Er^{3+}. For Pr^{3+} and Er^{3+} the atomic form factor contracts as θ decreases, the effect being more pronounced for Pr^{3+} (first half of series) than for Er^{3+} (second half). It is also evident that the atomic form factor for Er^{3+} at $\theta = \frac{1}{2}\pi$ is contracted with respect to that for Pr^{3+} for the same value of θ. This can be understood in terms of the dipole approximation.

There is probably little difference in the values of $\bar{j}_2(\kappa)$ between the rare earths, except perhaps at large values of $|\boldsymbol{\kappa}|$. Hence let us examine the coefficient of $\bar{j}_2(\kappa)$ as given by the dipole approximation, eqn (6.110), namely

$$\frac{J(J+1)+L(L+1)-S(S+1)}{3J(J+1)+S(S+1)-L(L+1)}.$$

In the first half of the series spin–orbit interaction couples **L** and **S** so as to give $J = L-S$, for which the coefficient is

$$\{1-2S/(L+1)\}^{-1},$$

whereas in the second half $J' = L'+S'$, say, and the coefficient is

$$\{1+2S'/L'\}^{-1}.$$

Hence we see that, in the dipole approximation, the coefficient of $\bar{j}_2(\kappa)$ in $\langle SLJJ|\,\hat{Q}_z\,|SLJJ\rangle$ is always smaller in the second half of the rare earth series than in the first half, as shown in Fig. 6.2, for Pr^{3+} and Er^{3+}.

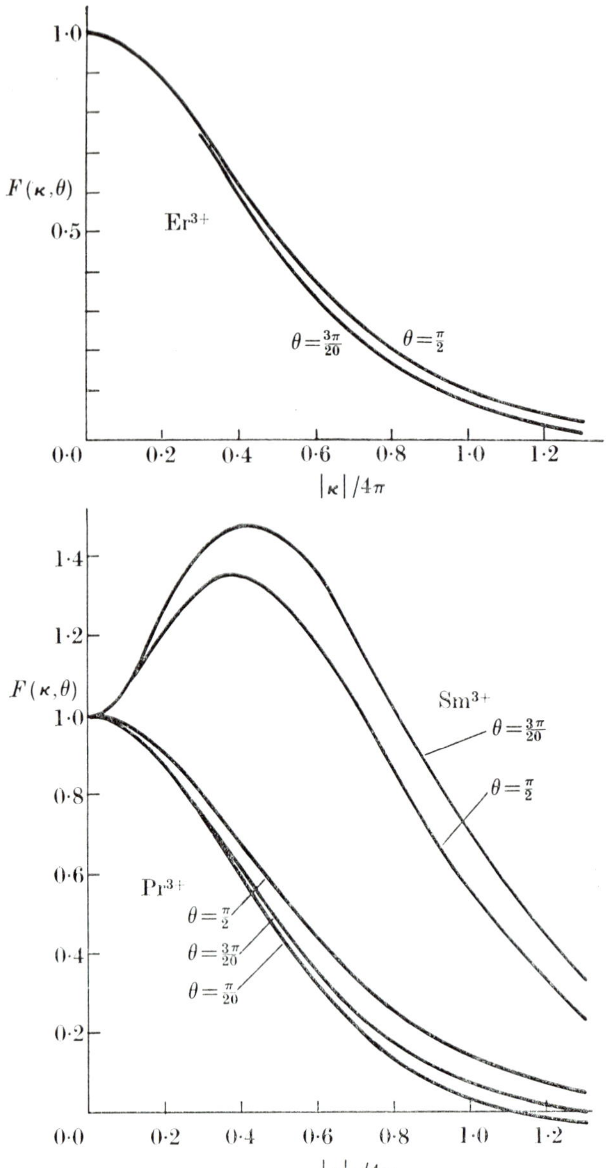

FIG. 6.2. The atomic form factor defined by eqn (6.131) is shown for the three rare-earth ions Pr^{3+}, Sm^{3+}, and Er^{3+} as a function of $|\boldsymbol{\kappa}|/4\pi$ for various values of $\boldsymbol{\theta}$.

The calculation of $F(\kappa,\theta)$, as defined by (6.131), from the functions $Z(K')$, eqn (6.127) is greatly facilitated by making use of the following expression for

$$\sin\theta\langle SLJJ|\,\hat{Q}_z\,|SLJJ\rangle - \cos\theta\, M(\theta). \tag{6.132}$$

From (6.126) and (6.128) we see that (6.132) is equal to

$$(4\pi)^{\frac{1}{2}}\sum_{K'} Z(K')\mathrm{e}^{-\mathrm{i}\phi}\left\{\mathrm{e}^{\mathrm{i}\phi}\sin\theta\, Y_0^{K'-1}(\theta,\phi) - \cos\theta\left(\frac{K'-1}{K'}\right)^{\frac{1}{2}} Y_1^{K'-1}(\theta,\phi)\right\}.$$

Since
$$\mathrm{e}^{\mathrm{i}\phi}\sin\theta = -\left(\frac{8\pi}{3}\right)^{\frac{1}{2}} Y_1^1(\theta,\phi)$$

and
$$\cos\theta = \left(\frac{4\pi}{3}\right)^{\frac{1}{2}} Y_0^1(\theta,\phi)$$

we can use (6.30) to show that

$$\mathrm{e}^{\mathrm{i}\phi}\sin\theta\, Y_0^{K'-1}(\theta,\phi) = -(2K'-1)^{-\frac{1}{2}}\times$$
$$\times\left[\left\{\frac{K'(K'+1)}{(2K'+1)}\right\}^{\frac{1}{2}} Y_1^{K'}(\theta,\phi) - \left\{\frac{(K'-1)(K'-2)}{(2K'-3)}\right\}^{\frac{1}{2}} Y_1^{K'-2}(\theta,\phi)\right]$$

and
$$\cos\theta\, Y_1^{K'-1}(\theta,\phi) = (2K'-1)^{-\frac{1}{2}}\times$$
$$\times\left[\left\{\frac{(K'-1)(K'+1)}{(2K'+1)}\right\}^{\frac{1}{2}} Y_1^{K'}(\theta,\phi) + \left\{\frac{K'(K'-2)}{(2K'-3)}\right\}^{\frac{1}{2}} Y_1^{K'-2}(\theta,\phi)\right].$$

Hence

$$\mathrm{e}^{\mathrm{i}\phi}\sin\theta\, Y_0^{K'-1}(\theta,\phi) - \cos\theta\left(\frac{K'-1}{K'}\right)^{\frac{1}{2}} Y_1^{K'-1}(\theta,\phi)$$
$$= -\left\{\frac{(K'+1)(2K'-1)}{K'(2K'+1)}\right\}^{\frac{1}{2}} Y_1^{K'}(\theta,\phi),$$

and therefore

$$\sin\theta\,\langle SLJJ|\,\hat{Q}_z\,|SLJJ\rangle - \cos\theta\, M(\theta)$$
$$= -(4\pi)^{\frac{1}{2}}\mathrm{e}^{-\mathrm{i}\phi}\sum_{K'} Z(K')\left\{\frac{(K'+1)(2K'-1)}{K'(2K'+1)}\right\}^{\frac{1}{2}} Y_1^{K'}(\theta,\phi), \tag{6.133}$$

which is the desired relationship.

In those cases where it is appropriate to average the cross-section over all directions of the scattering vector, the relation

$$\int_0^{2\pi}\mathrm{d}\phi\int_0^{\pi}\sin\theta\,\mathrm{d}\theta\, Y_{Q'}^{*K'}(\theta,\phi) Y_{Q''}^{K''}(\theta,\phi) = \delta_{K',K''}\delta_{Q',Q''} \tag{6.134}$$

in conjunction with (6.133) gives the result

$$\frac{1}{4\pi}\int d\tilde{\boldsymbol{\kappa}} \sum_{\alpha,\beta} (\delta_{\alpha,\beta}-\tilde{\kappa}_\alpha\tilde{\kappa}_\beta)\langle SLJJ|\,\hat{Q}_\alpha\,|SLJJ\rangle\langle SLJJ|\,\hat{Q}_\beta\,|SLJJ\rangle$$
$$= \sum_{K'} Z^2(K')\,\frac{(K'+1)(2K'-1)}{K'(2K'+1)}. \tag{6.135}$$

(*b*) *Crystal field effects*

When an ion is placed in a crystal it does, of course, experience many interactions that are absent for a free ion. For instance, there are electric and magnetic interactions with other ions in the crystal and to some degree a coupling to the crystal lattice. Even so, many of the properties of rare-earth ions have been well explained on the simple model of only taking into account the interaction arising from the electrostatic field created by neighbouring ions.

The crystal field, as it is called, possesses the point-group symmetry of the crystal lattice. Hence, to determine the allowed levels in the presence of a crystal field it is only necessary to determine what irreducible representations of the point group are contained in the reducible representation of the full rotation group of order† J. Consider for example a normal lattice site in a sodium chloride structure. Here the appropriate point group is the cubic group O whose irreducible representations we denote by Γ_n (Bethe's notation) with $n = 1$–8. Of these only Γ_1 and Γ_2 are non-degenerate. Γ_3, Γ_6, and Γ_7 are doubly degenerate, Γ_4 and Γ_5 are triply, and Γ_8 is fourfold degenerate. The decompositions for values of J appropriate to tripositive rare-earth ions are given in Table 6.3.‡ From this table we see that for none of the tripositive rare-earth ions does a cubic crystal field completely remove the $2J+1$ degeneracy of the ground-state wave function.

Let us now turn to the evaluation of the elastic cross-section for rare earths in a crystal field. The ground-state wave functions $|\Gamma_n v\rangle$ are given by

$$|\Gamma_n v\rangle = \sum_M |SLJM\rangle\langle SLJM|\Gamma_n v\rangle, \tag{6.136}$$

where the label v distinguishes between the functions that belong to the representation Γ_n. In most cases the magnetic field experienced by the ions (here assumed small compared to the crystal field) will remove the degeneracy and make one of the $|\Gamma_n v\rangle$ the appropriate ground state.

† For a more complete study of the group theoretic aspect, see, for example, R. S. Knox and A. Gold, *Symmetry in the solid state*, W. A. Benjamin (1964).

‡ For other values of J see K. R. Lea, M. J. M. Leask, and W. P. Wolf, *Physics Chem. Solids*, **23**, 1381 (1962).

TABLE 6.3

Decomposition of reducible representation of full rotation group of order J into irreducible representations of the cubic group O as appropriate for tripositive rare-earth ions in cubic crystals.

J	5/2	7/2	4	9/2
Decomposition into irred. rep.	$\Gamma_7+\Gamma_8$	$\Gamma_6+\Gamma_7+\Gamma_8$	$\Gamma_1+\Gamma_3+\Gamma_4+\Gamma_5$	$\Gamma_6+2\Gamma_8$

J	6	15/2	8
Decomposition into irred. rep.	$\Gamma_1+\Gamma_2+\Gamma_3+\Gamma_4+2\Gamma_5$	$\Gamma_6+\Gamma_7+3\Gamma_8$	$\Gamma_1+2\Gamma_3+2\Gamma_4+2\Gamma_5$

The contribution to the cross-section is determined by

$$\begin{aligned}\langle\lambda|\,\hat{\mathbf{Q}}_{\perp}^{+}\,|\lambda\rangle.\langle\lambda|\,\hat{\mathbf{Q}}_{\perp}\,|\lambda\rangle \\ &\equiv \langle\Gamma_n v|\,\hat{\mathbf{Q}}_{\perp}\,|\Gamma_n v\rangle.\langle\Gamma_n v|\,\hat{\mathbf{Q}}_{\perp}\,|\Gamma_n v\rangle \\ &= \sum_{\alpha,\beta}(\delta_{\alpha,\beta}-\tilde{\kappa}_\alpha\tilde{\kappa}_\beta)\langle\Gamma_n v|\,\hat{Q}_\alpha^{+}\,|\Gamma_n v\rangle\langle\Gamma_n v|\,\hat{Q}_\beta\,|\Gamma_n v\rangle.\end{aligned} \tag{6.137}$$

Hence, in view of (6.136), we require the matrix elements

$$\langle SLJM|\,\hat{Q}_\alpha\,|SLJM'\rangle.$$

These can also be expressed in terms of the characteristic functions Z introduced in § 6.6 (a).

From eqn (6.87 b)

$$\begin{aligned}\langle SLJM|\,\hat{Q}_q\,|SLJM'\rangle \\ &= (-1)^{J-M-q}\{12\pi(2J+1)\}^{\frac{1}{2}}\sum_{K',Q'}\binom{2K'+1}{K'+1}Y_{Q'+q}^{K'-1}(\tilde{\boldsymbol{\kappa}})\times \\ &\times\{A(K'-1,K')+B(K'-1,K')\}\times \\ &\times\begin{pmatrix}K' & J & J\\ -Q' & M' & -M\end{pmatrix}\begin{pmatrix}K'-1 & K' & 1\\ Q'+q & -Q' & -q\end{pmatrix},\end{aligned} \tag{6.138}$$

where $Q' = M'-M$. The matrix elements of $\hat{Q}_x$ and $\hat{Q}_y$ contain

$$Y_{Q'+1}^{K'-1}\begin{pmatrix}K'-1 & K' & 1\\ Q'+1 & -Q' & -1\end{pmatrix}\pm Y_{Q'-1}^{K'-1}\begin{pmatrix}K'-1 & K' & 1\\ Q'-1 & -Q' & 1\end{pmatrix}. \tag{6.139}$$

Only one of the $3j$ symbols need be evaluated since the other can be obtained merely by reversing the sign of Q'. We find

$$\begin{pmatrix}K'-1 & K' & 1\\ Q'+1 & -Q' & -1\end{pmatrix} = (-1)^{K'-Q'}\left\{\frac{(K'-Q')(K'-Q'-1)}{(2K'-1)2K'(2K'+1)}\right\}^{\frac{1}{2}},$$

so that (6.139) becomes

$$(-1)^{K'-Q'}\{(2K'-1)2K'(2K'+1)\}^{-\frac{1}{2}}\times$$
$$\times[Y^{K'-1}_{Q'+1}\{(K'-Q')(K'-Q'-1)\}^{\frac{1}{2}}\pm Y^{K'-1}_{Q'-1}\{(K'+Q')(K'+Q'-1)\}^{\frac{1}{2}}].$$

Hence, for example,

$$\langle SLJM|\,\hat{Q}_x\,|SLJM'\rangle = \left(-\frac{1}{\sqrt{2}}\right)(4\pi)^{\frac{1}{2}}\sum_{K'Q'}Z(K')P(K',Q')\times$$
$$\times[Y^{K'-1}_{Q'+1}(\tilde{\boldsymbol{\varkappa}})\{(K'-Q')(K'-Q'-1)\}^{\frac{1}{2}}-$$
$$-Y^{K'-1}_{Q'-1}(\tilde{\boldsymbol{\varkappa}})\{(K'+Q')(K'+Q'-1)\}^{\frac{1}{2}}]\times$$
$$\times\{(2K'-1)2K'(2K'+1)\}^{-\frac{1}{2}}\begin{pmatrix}K'-1 & K' & 1\\ 0 & 0 & 0\end{pmatrix}^{-1},$$

where we have defined

$$P(K',Q') = (-1)^{J-M'}\begin{pmatrix}K' & J & J\\ -Q' & M' & -M\end{pmatrix}\begin{pmatrix}K' & J & J\\ 0 & J & -J\end{pmatrix}^{-1}. \quad (6.140)$$

Since $\{(2K'-1)2K'(2K'+1)\}^{-\frac{1}{2}}\begin{pmatrix}K'-1 & K' & 1\\ 0 & 0 & 0\end{pmatrix}^{-1} = -\dfrac{1}{K'\sqrt{2}},$

we obtain for the matrix element of $\hat{Q}_x$ the result

$$\langle SLJM|\,\hat{Q}_x\,|SLJM'\rangle = \frac{\mathrm{i}}{2}(4\pi)^{\frac{1}{2}}\sum_{K',Q'}\frac{Z(K')}{K'}P(K',Q')\times$$
$$\times[Y^{K'-1}_{Q'+1}(\tilde{\boldsymbol{\varkappa}})\{(K'-Q')(K'-Q'-1)\}^{\frac{1}{2}}-Y^{K'-1}_{Q'-1}(\tilde{\boldsymbol{\varkappa}})\{(K'+Q')(K'+Q'-1)\}^{\frac{1}{2}}] \quad (6.141)$$

and for $\hat{Q}_y$

$$\langle SLJM|\,\hat{Q}_y\,|SLJM'\rangle = -\tfrac{1}{2}(4\pi)^{\frac{1}{2}}\sum_{K',Q'}\frac{Z(K')}{K'}P(K',Q')\times$$
$$\times[Y^{K'-1}_{Q'+1}(\tilde{\boldsymbol{\varkappa}})\{(K'-Q')(K'-Q'-1)\}^{\frac{1}{2}}+$$
$$+Y^{K'-1}_{Q'-1}(\tilde{\boldsymbol{\varkappa}})\{(K'+Q')(K'+Q'-1)\}^{\frac{1}{2}}]. \quad (6.142)$$

Finally, for $\hat{Q}_0$, we note that

$$\begin{pmatrix}K'-1 & K' & 1\\ Q' & -Q' & 0\end{pmatrix}\begin{pmatrix}K'-1 & K' & 1\\ 0 & 0 & 0\end{pmatrix}^{-1} = \frac{(-1)^{Q'}}{K'}\{(K'-Q')(K'+Q')\}^{\frac{1}{2}}$$

and hence

$$\langle SLJM|\,\hat{Q}_z\,|SLJM'\rangle = (4\pi)^{\frac{1}{2}}\sum_{K',Q'}\frac{Z(K')}{K'}P(K',Q')Y^{K'-1}_{Q'}(\tilde{\boldsymbol{\varkappa}})\times$$
$$\times\{(K'-Q')(K'+Q')\}^{\frac{1}{2}}. \quad (6.143)$$

The following property of the matrix elements (6.141), (6.142), and (6.143) are of help in actual calculations. If J is an integer,

$$\langle SLJM|\,\hat{Q}_\alpha\,|SLJM'\rangle = (-1)^{M'-M+1}\langle SLJ-M'|\,\hat{Q}_\alpha\,|SLJ-M\rangle.$$

Another property that it is worth while to recall is that if $|\lambda\rangle$ is a singlet state $\langle\lambda|\,\hat{Q}_\alpha\,|\lambda\rangle = 0$, i.e. the crystal field quenches out the magnetization.

6.7. Cross-section for transitions between crystal-field levels of a rare-earth ion

As we mentioned in the preceding section, the effect of the crystal field on a rare-earth ion is to partially or totally remove the $2J+1$ degeneracy of the ground state.

Transitions between the crystal-field levels can in principle be observed in neutron scattering. If the exchange interaction between the ions is weak it is sufficient to consider the cross-section for scattering by a single isolated ion, in which case scattering will be present at all values of $|\boldsymbol{\kappa}|$, as in paramagnetic scattering. One condition that must be satisfied if the crystal-field level scheme is to be observed by neutron scattering is that the total spread in energy of the levels must be comparable to the energy of the incident neutrons. If this is not so only an average over the level structure will be observed.

In discussing the cross-section for transitions between crystal-field levels within the isolated ion model we shall pay particular attention to the case where the crystal field has cubic symmetry. The irreducible representations of the cubic group are here denoted by Γ_n (Bethe's notation) and the corresponding wave functions by $|\Gamma_n v\rangle$, where the label v distinguishes the degenerate wave functions. It follows from eqn (5.11 b) that the partial differential cross-section per ion for the transition $\Gamma_n \to \Gamma_{n'}$ is

$$\frac{d^2\sigma}{d\Omega dE'} = \left(\frac{\gamma e^2}{m_e c^2}\right)^2 \frac{k'}{k} \sum_{\alpha,\beta} (\delta_{\alpha,\beta} - \tilde{\kappa}_\alpha \tilde{\kappa}_\beta) \times$$

$$\times \sum_{v,v'} p_n \langle\Gamma_n v|\,\hat{Q}_\alpha^+\,|\Gamma_{n'} v'\rangle\langle\Gamma_{n'} v'|\,\hat{Q}_\beta\,|\Gamma_n v\rangle\,\delta(\hbar\omega + E_n - E_{n'}). \quad (6.144)$$

Here p_n is the product of the probability that the Γ_n state is occupied and the appropriate degeneracy weighting factor, and E_n and $E_{n'}$ are, respectively, the energies of the Γ_n and $\Gamma_{n'}$ levels.

In general all the transitions will have a non-zero amplitude at some value of the scattering vector. However, just as in optical spectroscopy, there are strong and weak transitions. The strong transitions are those that occur for zero wave vector, that is to say, those allowed in the dipole approximation (dipole transitions). In this instance (cf. eqn (6.121)

$$\hat{\mathbf{Q}} \simeq \tfrac{1}{2} g F(\kappa) \hat{\mathbf{J}} \quad (6.145)$$

and we have therefore to determine for what levels Γ_n and $\Gamma_{n'}$ there is a non-zero matrix element $\langle\Gamma_n|\hat{J}^\alpha|\Gamma_{n'}\rangle$. This is a well-known question in optical spectroscopy and it is easily answered with a little group theory.†

In a cubic crystal field $\hat{\mathbf{J}}$ transforms as Γ_4. The matrix element $\langle\Gamma_n|\hat{J}^\alpha|\Gamma_{n'}\rangle$ is non-zero if the representation $\Gamma_n\times\Gamma_4\times\Gamma_{n'}$ contains the unit representation Γ_1. To be specific, consider the simple case of

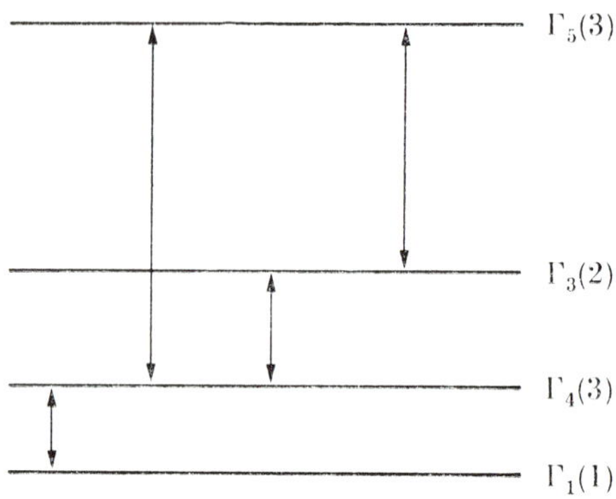

FIG. 6.3. Schematic diagram of crystal-field levels of Pr^{3+} in a cubic crystal field. The allowed dipole transitions are marked, together with the degeneracies of the various crystal-field levels.

$Pr^{3+}(f^2\,{}^2H_4)$. Reference to Table 6.3 shows that the state $J = 4$ splits into four levels in a cubic crystal field, as illustrated in Fig. 6.3. To determine whether or not the $\Gamma_3\leftrightarrow\Gamma_5$ transition is allowed or not in the dipole approximation we proceed as follows. The direct products of the representations $\Gamma_n\times\Gamma_{n'}$ are listed in Table 6.4, from which we see that $\Gamma_4\times\Gamma_5 = \Gamma_2+\Gamma_3+\Gamma_4+\Gamma_5$. Since $\Gamma_n\times\Gamma_n$ contains the unit representation the presence of Γ_3 in $\Gamma_4\times\Gamma_5$ means that $\Gamma_3\leftrightarrow\Gamma_5$ is indeed an allowed dipole transition. Similar calculations for the other possible transitions show that the dipole transitions for Pr^{3+} in a cubic crystal field are $\Gamma_1\leftrightarrow\Gamma_4$, $\Gamma_4\leftrightarrow\Gamma_5$, $\Gamma_3\leftrightarrow\Gamma_4$, and $\Gamma_3\leftrightarrow\Gamma_5$.

As far as the two remaining transitions, we know that they have zero amplitude at $|\boldsymbol{\varkappa}| = 0$ and also $|\boldsymbol{\varkappa}|\to\infty$. In order to determine whether or not they are significant it is necessary to evaluate the cross-section in full. The results for Pr^{3+} are shown in Table 6.5.‡ The actual quantity

† See, for example, L. D. Landau and E. M. Lifshitz, *Quantum mechanics* (Pergamon Press).

‡ The authors are pleased to acknowledge collaboration with Dr. E. Balcar in making the calculations.

TABLE 6.4

Direct products $\Gamma_n \times \Gamma_{n'}$ of one- and two-valued representations of the cubic group O

	Γ_1	Γ_2	Γ_3	Γ_4	Γ_5	Γ_6	Γ_7	Γ_8
Γ_1	Γ_1	Γ_2	Γ_3	Γ_4	Γ_5	Γ_6	Γ_7	Γ_8
Γ_2	Γ_2	Γ_1	Γ_3	Γ_5	Γ_4	Γ_7	Γ_6	Γ_8
Γ_3	Γ_3	Γ_3	$\Gamma_1+\Gamma_2+\Gamma_3$	$\Gamma_4+\Gamma_5$	$\Gamma_4+\Gamma_5$	Γ_8	Γ_8	$\Gamma_6+\Gamma_7+\Gamma_8$
Γ_4	Γ_4	Γ_5	$\Gamma_4+\Gamma_5$	$\Gamma_1+\Gamma_3+\Gamma_4+\Gamma_5$	$\Gamma_2+\Gamma_3+\Gamma_4+\Gamma_5$	$\Gamma_6+\Gamma_8$	$\Gamma_7+\Gamma_8$	$\Gamma_6+\Gamma_7+2\Gamma_8$
Γ_5	Γ_5	Γ_4	$\Gamma_4+\Gamma_5$	$\Gamma_2+\Gamma_3+\Gamma_4+\Gamma_5$	$\Gamma_1+\Gamma_3+\Gamma_4+\Gamma_5$	$\Gamma_7+\Gamma_8$	$\Gamma_6+\Gamma_8$	$\Gamma_6+\Gamma_7+2\Gamma_8$
Γ_6	Γ_6	Γ_7	Γ_8	$\Gamma_6+\Gamma_8$	$\Gamma_7+\Gamma_8$	$\Gamma_1+\Gamma_4$	$\Gamma_2+\Gamma_5$	$\Gamma_3+\Gamma_4+\Gamma_5$
Γ_7	Γ_7	Γ_6	Γ_8	$\Gamma_7+\Gamma_8$	$\Gamma_6+\Gamma_8$	$\Gamma_2+\Gamma_5$	$\Gamma_1+\Gamma_4$	$\Gamma_3+\Gamma_4+\Gamma_5$
Γ_8	Γ_8	Γ_8	$\Gamma_6+\Gamma_7+\Gamma_8$	$\Gamma_6+\Gamma_7+2\Gamma_8$	$\Gamma_6+\Gamma_7+2\Gamma_3$	$\Gamma_3+\Gamma_4+\Gamma_5$	$\Gamma_3+\Gamma_4+\Gamma_5$	$\Gamma_1+\Gamma_2+\Gamma_3+2\Gamma_4+2\Gamma_5$

TABLE 6.5

The quantity defined by eqn (6.146), which determines the cross-section for transitions between crystal-field levels, is tabulated for all six transitions appropriate to Pr^{3+} *in a cubic crystal field*

$n \ n'$	$\bar{j}_0^2$	$\bar{j}_0\bar{j}_2$	$\bar{j}_2^2$	$\bar{j}_2\bar{j}_4$	$\bar{j}_4^2$	$\bar{j}_4\bar{j}_6$	$\bar{j}_6^2$
1—4	2·1333	7·0163	5·7805	0·0316	0·0272	0·0716	0·2292
3—5	1·2800	4·2098	3·4861	0·0673	0·0556	0·1232	0·3976
4—5	1·1200	3·6836	3·0813	0·1436	0·1067	0·1117	0·3659
4—3	2·9867	9·8228	8·1077	0·0849	0·0659	0·1021	0·3295
1—5	0·0	0·0	0·0248	0·0677	0·0464	0·0032	0·0144
1—3	0·0	0·0	0·0	0·0	0·0060	0·0758	0·2428

listed for each of the six transitions is

$$\frac{1}{4\pi}\int \mathrm{d}\tilde{\boldsymbol{\varkappa}} \sum_{\alpha\beta} (\delta_{\alpha,\beta}-\tilde{\kappa}_\alpha\tilde{\kappa}_\beta) \sum_{v,v'} \langle\Gamma_n v|\, \hat{Q}_\alpha^+ \,|\Gamma_{n'} v'\rangle\langle\Gamma_{n'} v'|\, \hat{Q}_\beta \,|\Gamma_n v\rangle, \tag{6.146}$$

which determines the cross-section averaged over all directions of the scattering vector. For each pair of levels (6.146) is in general the sum of products of two radial integrals $\bar{j}_K$, defined by eqn (6.38), and Table 6.5 contains the numerical coefficients associated with the various products. We note that for the two transitions not allowed in the dipole approximation, $\Gamma_1 \leftrightarrow \Gamma_5$ and $\Gamma_1 \leftrightarrow \Gamma_3$, there are no terms involving $\bar{j}_0$. Since this is the only radial integral that is non-zero for $|\boldsymbol{\varkappa}| = 0$, this result is in accord with our previous findings. A comparison of the coefficients of the remaining terms clearly shows that, in this particular case, the only transitions of importance are the dipole transitions shown in Fig. 6.3.

7

ELASTIC MAGNETIC SCATTERING

THE elastic magnetic scattering of thermal neutrons has been used in two fields. First, the determination of magnetic structures, i.e. the location and orientation of unpaired spins in crystal structures, and second, to gain information on the properties of the unpaired spin by measuring magnetic form factors. A discussion of either of these two fields of research almost warrants a book to itself, so we can only cover the fundamental aspects of each in one chapter. We do not discuss structure and spin density determinations separately, but choose instead to subdivide the chapter into sections devoted to different magnetic states, the whole being divided into essentially two parts, namely metal and ionic crystals. First, however, we examine the cross-section for elastic magnetic scattering.

7.1. Cross-section for elastic magnetic scattering

Our starting-point is eqn (5.27) for the partial differential cross-section for purely magnetic scattering. For elastic scattering the final target state $|\lambda'\rangle$ must have the same energy as the initial state $|\lambda\rangle$, so we obtain from (5.27) under this condition

$$\frac{\mathrm{d}\sigma}{\mathrm{d}\Omega}=\left(\frac{\gamma e^2}{m_\mathrm{e}c^2}\right)^2\{\tfrac{1}{2}gF(\boldsymbol{\varkappa})\}^2\sum_{\alpha,\beta}(\delta_{\alpha,\beta}-\tilde{\varkappa}_\alpha\tilde{\varkappa}_\beta)\sum_\lambda p_\lambda\sum_{l,l'}\exp\{\mathrm{i}\boldsymbol{\varkappa}.(\mathbf{l}'-\mathbf{l})\}\times$$
$$\times\langle\lambda|\,\hat{S}^\alpha_l\,|\lambda\rangle\langle\lambda|\,\hat{S}^\beta_{l'}\,|\lambda\rangle. \qquad (7.1)$$

For the present we just consider target systems with a Bravais lattice structure.

Clearly, the first task is to evaluate

$$\sum_\lambda p_\lambda\langle\lambda|\,\hat{S}^\alpha_l\,|\lambda\rangle\langle\lambda|\,\hat{S}^\beta_{l'}\,|\lambda\rangle. \qquad (7.2)$$

We emphasize that this is not equal to $\langle\hat{S}^\alpha\hat{S}^\beta\rangle$.

Before proceeding with the evaluation of (7.2) we draw attention to the fact that we choose in all our subsequent discussions to have our axis of quantization parallel to the mean direction of the spins. Of course, in principle we are free to choose any quantization axis we like but only if we make this particular choice is the density matrix diagonal, as we shall assume.

For a ferromagnet with a Bravais crystal structure the target states $|\lambda\rangle$ can be chosen so that for each state $\langle\lambda|\,\hat{S}^{\alpha}_{l}\,|\lambda\rangle$ is independent of the site position **l**. Hence, for this choice of states,

$$\langle\lambda|\,\hat{S}^{\alpha}_{l}\,|\lambda\rangle = \left\langle\lambda\left|\frac{1}{N}\,\hat{S}^{\alpha}_{\text{tot}}\right|\lambda\right\rangle = \frac{1}{N}\,\delta_{\alpha,z}\langle\lambda|\,\hat{S}^{z}_{\text{tot}}\,|\lambda\rangle, \tag{7.3}$$

where we have assumed the (ferromagnetic) spins to be aligned parallel to the z direction and S^{z}_{tot} is the total z component of spin, viz.

$$\hat{S}^{z}_{\text{tot}} = \sum_{l}\hat{S}^{z}_{l}. \tag{7.4}$$

Substituting (7.3) in (7.2) gives for the latter

$$\frac{1}{N^2}\,\delta_{\alpha,z}\,\delta_{\beta,z}\sum_{\lambda}p_{\lambda}\,\langle\lambda|\,\hat{S}^{z}_{\text{tot}}\,|\lambda\rangle^2. \tag{7.5}$$

We now write $$\hat{S}^{z}_{\text{tot}} = \langle\hat{S}^{z}_{\text{tot}}\rangle+(\hat{S}^{z}_{\text{tot}}-\langle\hat{S}^{z}_{\text{tot}}\rangle), \tag{7.6}$$
where, as usual, the angular brackets denote the expectation value of the enclosed operator at temperature T:

$$\langle\hat{S}^{z}_{\text{tot}}\rangle = \sum_{\lambda}p_{\lambda}\langle\lambda|\,\hat{S}^{z}_{\text{tot}}\,|\lambda\rangle. \tag{7.7}$$

Using (7.6) in (7.5), we get

$$\frac{1}{N^2}\,\delta_{\alpha,z}\,\delta_{\beta,z}\{\langle\hat{S}^{z}_{\text{tot}}\rangle^2+\langle(\hat{S}^{z}_{\text{tot}}-\langle\hat{S}^{z}_{\text{tot}}\rangle)^2\rangle\}. \tag{7.8}$$

But, by general statistical mechanics,

$$\langle(\hat{S}^{z}_{\text{tot}}-\langle\hat{S}^{z}_{\text{tot}}\rangle)^2\rangle = \langle(\hat{S}^{z}_{\text{tot}})^2\rangle-\langle\hat{S}^{z}_{\text{tot}}\rangle^2 = \frac{Nk_{\text{B}}\,T}{g^2\mu_{\text{B}}^2}\,\chi, \tag{7.9}$$

where χ is the differential susceptibility per ion. This term is only of order N, negligible compared to the leading term in (7.8). Hence we get for (7.2),

$$\delta_{\alpha,z}\,\delta_{\beta,z}\langle\hat{S}\rangle^2. \tag{7.10}$$

(This foregoing discussion is a special version of the general discussion given in Chapter 2.) Using this result in the cross-section (7.1),

$$\frac{\mathrm{d}\sigma}{\mathrm{d}\Omega} = \left(\frac{\gamma e^2}{m_{\text{e}}\,c^2}\right)^2\{\tfrac{1}{2}gF(\boldsymbol{\varkappa})\}^2(1-\tilde{\kappa}_z^2)\exp\{-2W(\boldsymbol{\varkappa})\}\langle\hat{S}\rangle^2\sum_{l,l'}\exp\{\mathrm{i}\boldsymbol{\varkappa}\,.\,(\mathbf{l}'-\mathbf{l})\}$$
$$= \left(\frac{\gamma e^2}{m_{\text{e}}\,c^2}\right)^2\{\tfrac{1}{2}g\,F(\boldsymbol{\varkappa})\}^2\exp\{-2W(\boldsymbol{\varkappa})\}\langle\hat{S}\rangle^2N\,\frac{(2\pi)^3}{v_0}\,\times$$
$$\times\sum_{\boldsymbol{\tau}}\delta(\boldsymbol{\varkappa}-\boldsymbol{\tau})\{1-(\tilde{\boldsymbol{\tau}}\,.\,\tilde{\boldsymbol{\eta}})^2\}_{\text{av}}, \tag{7.11 a}$$

where to be more general we have denoted the direction of the spin orientation by the unit vector $\tilde{\boldsymbol{\eta}}$. Hence

$$\langle\hat{\mathbf{S}}\rangle = \langle\hat{S}\rangle\tilde{\boldsymbol{\eta}}, \tag{7.11 b}$$

and we have inserted the Debye–Waller factor to take account of lattice vibrations (cf. Chapter 4).

The subscript 'av' on the orientation factor in (7.11 a) signifies that an average over domain orientations $\tilde{\boldsymbol{\eta}}$ is to be taken.

We see immediately from the expression (7.11 a) for the cross-section for elastic magnetic scattering from a ferromagnet that the magnetic Bragg scattering occurs at exactly the same reciprocal lattice points as nuclear scattering, and, because the magnetic diffraction peaks exactly coincide with the nuclear diffraction peaks, difficulty arises in distinguishing one effect from the other. For a Bravais lattice structure the ratio of the magnetic to nuclear contributions for any given Bragg peak is

$$\frac{1}{|\bar{b}|^2}\left(\frac{\gamma e^2}{m_e c^2}\right)^2 \langle \hat{S}\rangle^2 |F(\boldsymbol{\tau})|^2 \{1-(\tilde{\boldsymbol{\tau}}.\tilde{\boldsymbol{\eta}})^2\}_{\text{av}}.$$

For iron $\langle \hat{S}\rangle$ at $T = 0$ is 1·1 and this ratio is reasonable, but for nickel $\langle \hat{S}\rangle_{T=0} = 0{\cdot}3$ and the ratio is very small, certainly too small for the purely magnetic scattering to be easily seen in nickel. Magnetic scattering can be observed even for nickel by using polarized neutrons. Polarization phenomena in general are fully discussed in Chapter 10, though we shall consider the application of the polarized-beam technique in magnetic scattering on several occasions in the remaining part of this chapter.

Because the atomic form factor $F(\boldsymbol{\tau})$ in (7.11 a) decreases rapidly with increasing $\boldsymbol{\tau}$ the magnetic scattering is negligible for all except the first few Bragg peaks. It is possible to distinguish it from nuclear scattering in three ways.

(a) The magnetic scattering falls off rapidly when the temperature of the ferromagnetic target is raised near its Curie temperature T_c because the magnetic cross-section is proportional to $\langle \hat{\mathbf{S}}\rangle^2$. When comparing cross-sections at different temperatures it is important to make allowance for the small temperature dependence of the nuclear scattering. The latter results from the vibrations of the crystal lattice, as we saw in Chapter 4. The cross-section (7.11 a) contains the Debye–Waller factor $\exp\{-2W(\boldsymbol{\varkappa})\}$.

(b) The magnetic scattering has a characteristic dependence on $\boldsymbol{\tau}$ through $F(\boldsymbol{\tau})$, whereas the nuclear scattering varies only slightly with $\boldsymbol{\tau}$ as a consequence of lattice vibrations.

(c) The magnetic scattering depends on the orientation of the spin $\langle \hat{\mathbf{S}}\rangle$ relative to the vector $\boldsymbol{\tau}$ through the last factor in (7.11 a).

In the absence of an external magnetic field a ferromagnet splits into domains and to obtain the cross-section we must average over the orientations of these domains, i.e. over the 'easy axes' for $\tilde{\boldsymbol{\eta}}$. If all directions in space were equally likely, then clearly $(\tilde{\boldsymbol{\tau}}.\tilde{\boldsymbol{\eta}})^2$ would average to $\frac{1}{3}$ so that

$$\{1-(\tilde{\boldsymbol{\tau}}.\tilde{\boldsymbol{\eta}})^2\}_{\text{av}} = \tfrac{2}{3}. \tag{7.12}$$

This result is also correct for cubic symmetry. Thus when (7.12) is valid the cross-section (7.11 a) becomes

$$\frac{d\sigma}{d\Omega} = \left(\frac{\gamma e^2}{m_e c^2}\right)^2 \{\tfrac{1}{2}gF(\boldsymbol{\varkappa})\}^2\langle\bar{S}\rangle^2 N\frac{(2\pi)^3}{v_0}\frac{2}{3}\sum_{\boldsymbol{\tau}}\delta(\boldsymbol{\varkappa}-\boldsymbol{\tau})\exp\{-2W(\boldsymbol{\tau})\}. \tag{7.13}$$

Now if an external magnetic field is applied in the direction of $\boldsymbol{\varkappa}$, i.e. in the direction of $\boldsymbol{\tau}$ for a given Bragg peak, the ferromagnetic spins can be made parallel to $\boldsymbol{\tau}$ so that $\tilde{\boldsymbol{\tau}}.\tilde{\boldsymbol{\eta}}$ is unity and thus the elastic magnetic cross-section vanishes. Hence the difference in cross-section for the external field off and the field on gives directly the magnetic scattering alone. This is a very simple and convenient technique for measuring the elastic magnetic scattering from ferromagnets.

It must be emphasized that the simple result (7.13) is strictly valid only for cubic crystals. For cobalt, for instance, the easy directions are along the hexagonal z-axis so there are just two 'easy' directions,

$$\tilde{\boldsymbol{\eta}} = (0, 0, \pm 1).$$

Thus in zero field the cross-section (7.11 a) becomes

$$\frac{d\sigma}{d\Omega} = \left(\frac{\gamma e^2}{m_e c^2}\right)^2 \{\tfrac{1}{2}gF(\boldsymbol{\varkappa})\}^2\langle\bar{S}\rangle^2 N\frac{(2\pi)^3}{v_0}\sum_{\boldsymbol{\tau}}\delta(\boldsymbol{\varkappa}-\boldsymbol{\tau})(1-\tilde{\tau}_z^2)\exp\{-2W(\boldsymbol{\tau})\}. \tag{7.14}$$

Of course, it is also possible to make this cross-section zero by applying a magnetic field but, because of the larger anisotropy energy in a non-cubic crystal, it would have to be a larger field.

The discussion given above was based on eqn (7.1) for the cross-section for elastic magnetic scattering and this was itself derived from (5.27). Eqn (5.27), however, was obtained from arguments based on a model in which the wave functions of the unpaired electron spins were well localized about their respective nuclei. We have previously noted that such a picture is not correct for the transition metals, for example. Hence we must reconsider the cross-section derived from (7.1), eqn (7.11 a), and ask for a more general expression, valid for both metals and ionic systems.

Let the spin density at position $\mathbf{r}$ in the target system have a magnitude $\mathcal{s}(\mathbf{r})$. In terms of this function the magnetic cross-section is

$$\frac{d\sigma}{d\Omega} = \left(\frac{\gamma e^2}{m_e c^2}\right)^2 \left|\int d\mathbf{r}\, \exp(i\boldsymbol{\varkappa}.\mathbf{r})\{\tilde{\boldsymbol{\varkappa}}\times\mathcal{s}(\mathbf{r})\times\tilde{\boldsymbol{\varkappa}}\}\right|^2. \tag{7.15}$$

Because $\mathscr{S}(\mathbf{r})$ is periodic we can write (cf. § 2.3)

$$\int_V \mathrm{d}\mathbf{r}\, \exp(\mathrm{i}\boldsymbol{\varkappa}.\mathbf{r})\mathscr{S}(\mathbf{r}) = \frac{(2\pi)^3}{v_0}\sum_{\boldsymbol{\tau}} \delta(\boldsymbol{\varkappa}-\boldsymbol{\tau})\mathscr{F}(\boldsymbol{\tau}), \tag{7.16}$$

where $\mathscr{F}(\boldsymbol{\tau})$ is the *magnetic unit-cell vector structure factor*. Inverting (7.16) we get

$$\mathscr{S}(\mathbf{r}) = \frac{1}{v_0}\sum_{\boldsymbol{\tau}} \exp(-\mathrm{i}\boldsymbol{\tau}.\mathbf{r})\mathscr{F}(\boldsymbol{\tau}). \tag{7.17}$$

Multiplying (7.17) by $\exp(\mathrm{i}\boldsymbol{\tau}'.\mathbf{r})$ and integrating over a cell of volume v_0 gives

$$\int_{v_0} \mathrm{d}\mathbf{r}\, \exp(\mathrm{i}\boldsymbol{\tau}'.\mathbf{r})\mathscr{S}(\mathbf{r}) = \frac{1}{v_0}\sum_{\boldsymbol{\tau}} \mathscr{F}(\boldsymbol{\tau})\int_{v_0} \mathrm{d}\mathbf{r}\, \exp\{\mathrm{i}(\boldsymbol{\tau}'-\boldsymbol{\tau}).\mathbf{r}\}.$$

The integral on the right-hand side vanishes unless $\boldsymbol{\tau}'$ and $\boldsymbol{\tau}$ coincide and, in that case, has value v_0. Hence

$$\mathscr{F}(\boldsymbol{\tau}) = \int_{v_0} \mathrm{d}\mathbf{r}\, \exp(\mathrm{i}\boldsymbol{\tau}.\mathbf{r})\mathscr{S}(\mathbf{r}). \tag{7.18}$$

Finally we note that the cross-section (7.15) involves

$$\int \mathrm{d}\mathbf{r}\, \exp(\mathrm{i}\boldsymbol{\varkappa}.\mathbf{r})\mathscr{S}^{\alpha}(\mathbf{r}) \int \mathrm{d}\mathbf{r}'\, \exp(-\mathrm{i}\boldsymbol{\varkappa}.\mathbf{r}')\mathscr{S}^{\beta}(\mathbf{r}')$$

$$= \sum_{l,l'} \exp\{\mathrm{i}\boldsymbol{\varkappa}(\mathbf{l}-\mathbf{l}')\} \int_{\text{cell}} \mathrm{d}\mathbf{r}\, \exp(\mathrm{i}\boldsymbol{\varkappa}.\mathbf{r})\mathscr{S}^{\alpha}(\mathbf{r}) \int_{\text{cell}} \mathrm{d}\mathbf{r}'\, \exp(-\mathrm{i}\boldsymbol{\varkappa}.\mathbf{r}')\mathscr{S}^{\beta}(\mathbf{r}')$$

$$= N\frac{(2\pi)^3}{v_0}\sum_{\boldsymbol{\tau}} \delta(\boldsymbol{\varkappa}-\boldsymbol{\tau})\mathscr{F}^{\alpha}(\boldsymbol{\tau})\mathscr{F}^{\beta}(\boldsymbol{\tau}).$$

Hence (7.15) becomes

$$\frac{\mathrm{d}\sigma}{\mathrm{d}\Omega} = \left(\frac{\gamma e^2}{m_e c^2}\right)^2 N\frac{(2\pi)^3}{v_0}\sum_{\boldsymbol{\tau}} \delta(\boldsymbol{\varkappa}-\boldsymbol{\tau})|\tilde{\boldsymbol{\varkappa}}\times\{\mathscr{F}(\boldsymbol{\tau})\times\tilde{\boldsymbol{\varkappa}}\}|^2. \tag{7.19}$$

These results are to be contrasted with those obtained when we imagine $\mathscr{S}(\mathbf{r})$ to be made up of the superposition of individual atomic distributions. Thus, if

$$\mathscr{S}(\mathbf{r}) = \sum_{l,d} \mathbf{S}_d(\mathbf{r}-\mathbf{l}-\mathbf{d}), \tag{7.20}$$

then for each atom there exists a *vector form factor*

$$\mathscr{F}_d(\boldsymbol{\varkappa}) = \int \mathrm{d}\mathbf{r}\, \exp(\mathrm{i}\boldsymbol{\varkappa}.\mathbf{r})\mathbf{S}_d(\mathbf{r}) \tag{7.21 a}$$

with the inverse

$$\mathbf{S}_d(\mathbf{r}) = \frac{1}{(2\pi)^3}\int \mathrm{d}\boldsymbol{\varkappa}\, \exp(-\mathrm{i}\boldsymbol{\varkappa}.\mathbf{r})\mathscr{F}_d(\boldsymbol{\varkappa}). \tag{7.21 b}$$

(Note this is not the atomic form factor, which is introduced in (7.23).)

In terms of these quantities

$$\mathscr{F}(\boldsymbol{\tau}) = \sum_{d} \exp(i\boldsymbol{\tau}\,.\,\mathbf{d})\mathscr{F}_d(\boldsymbol{\tau}). \tag{7.22}$$

From (7.21 b) we notice that to deduce $\mathbf{S}_d(\mathbf{r})$ we need to know $\mathscr{F}_d(\boldsymbol{\varkappa})$ for all $\boldsymbol{\varkappa}$. In principle this could be obtained from paramagnetic scattering, or indeed inelastic scattering, but it cannot be obtained from Bragg scattering, which gives only the form factors evaluated at the reciprocal lattice vectors $\boldsymbol{\tau}$.

On the other hand we notice from (7.17) that $\mathscr{s}(\mathbf{r})$ can be deduced rigorously from a knowledge of $\mathscr{F}(\boldsymbol{\tau})$ at the reciprocal lattice points only. In other words $\mathscr{s}(\mathbf{r})$ is a rigorously defined quantity, which can be determined from Bragg scattering; it tells us exactly what the spin density is at every point $\mathbf{r}$ without specifying which atom each part of the density 'belongs' to. In contrast the quantity $\mathbf{S}_d(\mathbf{r}-\mathbf{l}-\mathbf{d})$ in (7.20) gives us more information about the system, it describes exactly the spin density 'belonging' to the atom $\mathbf{l}$, $\mathbf{d}$ but it cannot be deduced from Bragg scattering because, by (7.21 b), this involves a full Fourier integral. But we must also recall that $\mathbf{S}_d(\mathbf{r})$ is defined only by a model of the systems and has a well-defined meaning only to the extent that the model faithfully represents the real system.

We conclude that Bragg scattering gives $\mathscr{F}(\boldsymbol{\tau})$ and that we can determine $\mathscr{s}(\mathbf{r})$ through (7.17). Now, as we have seen, $\mathscr{s}(\mathbf{r})$ is a rigorously defined quantity of great importance; it is not dependent on any model of the target system and the experiments to measure it can be performed with a high precision using polarized neutrons. It is very rare indeed in the study of materials, and especially in a study of transition metals, that such precise experimental results have such a direct and meaningful interpretation. There is only one small uncertainty in the interpretation and that is the contribution to the scattering arising from the orbital moment. The orbital moment is not quite quenched in these metals and to the extent that the scattering due to orbital moment has a different form factor from that due to the spin the interpretation is in doubt. However, we demonstrated in § 5.5 that for spins in a narrow energy band the orbital scattering is probably small, so that the error involved for the transition metals is not serious. We discuss the experimental determinations of $\mathscr{s}(\mathbf{r})$ in more detail in § 7.3.

In those materials where it is meaningful to speak of a spin density belonging to a particular ion in a unit cell we can rewrite (7.21 a) as

$$\mathscr{F}_d(\boldsymbol{\varkappa}) = \tfrac{1}{2}g_d\langle\hat{S}_d\rangle F_d(\boldsymbol{\varkappa})\tilde{\boldsymbol{\eta}}_d(\boldsymbol{\varkappa}). \tag{7.23}$$

Here $\tilde{\boldsymbol{\eta}}_d(\boldsymbol{\varkappa})$ is a unit vector in the direction of the integral and $F_d(\boldsymbol{\varkappa})$ is the *atomic form factor* for the ion and is defined such that $F_d(\boldsymbol{\varkappa} = 0) = 1$. Further, $\langle \hat{S}_d \rangle$ is the average value of the spin associated with the ion and g_d the corresponding gyromagnetic ratio.

We note that for $\boldsymbol{\varkappa} = 0$, (7.23) becomes

$$\mathscr{F}_d(0) = \int \mathrm{d}\mathbf{r}\, \boldsymbol{s}(\mathbf{r}) = \tfrac{1}{2} g_d \langle \hat{S}_d \rangle \tilde{\boldsymbol{\eta}}_d(0)$$

and from this it follows that $\tilde{\boldsymbol{\eta}}_d(0)$ gives the direction of the spin moment on the dth site. For most spin structures $\tilde{\boldsymbol{\eta}}_d(\boldsymbol{\varkappa})$ is independent of $\boldsymbol{\varkappa}$. This is the case for all the examples discussed below.

Using (7.23) in (7.22) and taking account of the vibration of the ions about their equilibrium positions we obtain for $\mathscr{F}(\boldsymbol{\tau})$

$$\mathscr{F}(\boldsymbol{\tau}) = \tfrac{1}{2} \sum_d \exp(\mathrm{i}\boldsymbol{\tau}\,.\,\mathbf{d}) \exp\{-W_d(\boldsymbol{\tau})\} g_d \langle \hat{S}_d \rangle F_d(\boldsymbol{\tau}) \tilde{\boldsymbol{\eta}}_d(\boldsymbol{\tau}). \tag{7.24}$$

Hence the corresponding expression for the cross-section for elastic magnetic scattering is

$$\frac{\mathrm{d}\sigma}{\mathrm{d}\Omega} = \left(\frac{\gamma e^2}{m_e c^2}\right)^2 N \frac{(2\pi)^3}{v_0} \sum_{\boldsymbol{\tau}} \delta(\boldsymbol{\varkappa}-\boldsymbol{\tau}) \times$$
$$\times \left| \sum_d \exp(\mathrm{i}\boldsymbol{\tau}\,.\,\mathbf{d}) \exp\{-W_d(\boldsymbol{\tau})\} \tfrac{1}{2} g_d \langle \hat{S}_d \rangle F_d(\boldsymbol{\tau}) [\tilde{\boldsymbol{\tau}} \times \{\tilde{\boldsymbol{\eta}}_d(\boldsymbol{\tau}) \times \tilde{\boldsymbol{\tau}}\}] \right|^2. \tag{7.25}$$

As a particular example of (7.25) consider a simple antiferromagnet, in which (i) all the spins in a unit cell are either parallel or antiparallel to a single direction $\tilde{\boldsymbol{\eta}}$, and (ii) all the sublattices are identical. In this instance (7.25) reduces to

$$\frac{\mathrm{d}\sigma}{\mathrm{d}\Omega} = \left(\frac{\gamma e^2}{m_e c^2}\right)^2 N \frac{(2\pi)^3}{v_0} \sum_{\boldsymbol{\tau}} \delta(\boldsymbol{\varkappa}-\boldsymbol{\tau}) \exp\{-2W(\boldsymbol{\tau})\} |F_M(\boldsymbol{\tau})|^2 \{1-(\tilde{\boldsymbol{\tau}}\,.\,\tilde{\boldsymbol{\eta}})^2\}_{\mathrm{av}}. \tag{7.26}$$

Here $F_M(\boldsymbol{\tau})$ is the *magnetic unit-cell structure factor* defined by

$$F_M(\boldsymbol{\tau}) = \sum_d \tfrac{1}{2} g_d \langle \hat{S}_d \rangle F_d(\boldsymbol{\tau}) \sigma_d \exp(\mathrm{i}\boldsymbol{\tau}\,.\,\mathbf{d}), \tag{7.27 a}$$

which in the particular case under discussion simplifies to

$$\tfrac{1}{2} g \langle \hat{S} \rangle F(\boldsymbol{\tau}) \sum_d \sigma_d \exp(\mathrm{i}\boldsymbol{\tau}\,.\,\mathbf{d}). \tag{7.27 b}$$

(If we adhered strictly to the definition of structure factors given in § 2.3, (7.27) would include a factor of $(\gamma e^2/m_e c^2)$. However, it proves convenient in the notation to pull this factor out of $F_M(\boldsymbol{\tau})$ as is done in (7.26).)

In (7.27) σ_d is either ± 1 according to whether the spin at site $\mathbf{d}$ is on the 'plus' or 'minus' sublattice and the suffix av on the last term of (7.26) indicates an average over the domain orientations of $\tilde{\boldsymbol{\eta}}$.

The cross-section (7.26) will be zero when the temperature of the antiferromagnet is raised above its Néel temperature, for then $\langle \hat{S} \rangle$ will be zero. The high Bragg peaks have a weak intensity because of the atomic form factor term $|F(\boldsymbol{\tau})|^2$. Sometimes the orientation factor $\{1-(\tilde{\boldsymbol{\tau}}.\tilde{\boldsymbol{\eta}})^2\}_{\mathrm{av}}$ enables us to deduce the absolute orientation of the spin relative to the crystal axes. It is important to remember the distinction between determining the 'configuration' and 'orientation' of the spins. The 'configuration' is the array of plus and minus spins signifying the spins are parallel or antiparallel to $\tilde{\boldsymbol{\eta}}$ and gives no information on the orientation of $\tilde{\boldsymbol{\eta}}$ relative to the crystal axes. The 'configuration' is usually determined easily by finding out just what reciprocal vectors $\boldsymbol{\tau}$ give a contribution to (7.26). In many cases the magnetic unit cell is larger than the chemical unit cell because two atoms can be chemically equivalent but have opposite spins and sometimes this means that the unit cell has to be enlarged to get a repeating unit.

We have argued in this section that the cross-section for elastic magnetic scattering by either the localized or itinerant model is of the form (7.15) with $\int \mathrm{d}\mathbf{r}\, \mathfrak{s}(\mathbf{r})\exp(\mathrm{i}\boldsymbol{\varkappa}.\mathbf{r})$ given by (7.16). It is worth while to pause before embarking on a detailed examination of the formulae derived above, to examine explicitly elastic magnetic scattering from an itinerant electron system.

7.2. Elastic scattering from itinerant electron model

In § 5.5 we demonstrated that the cross-section for the scattering from an itinerant model is expressible in the form

$$\frac{\mathrm{d}^2\sigma}{\mathrm{d}\Omega \mathrm{d}E'} = \left(\frac{\gamma e^2}{m_\mathrm{e} c^2}\right)^2 \frac{k'}{k} \sum_{\alpha,\beta} (\delta_{\alpha\beta}-\tilde{\kappa}_\alpha \tilde{\kappa}_\beta) \sum_{\lambda,\lambda'} p_\lambda \langle\lambda|\,\hat{\mathfrak{s}}^\alpha(\boldsymbol{\varkappa})\,|\lambda'\rangle\langle\lambda'|\,\hat{\mathfrak{s}}^\beta(-\boldsymbol{\varkappa})\,|\lambda\rangle \times$$
$$\times\, \delta(\hbar\omega+E_\lambda-E_{\lambda'}). \qquad (7.28)$$

Here we have taken only the scattering due to the spin part of the neutron–electron interaction.

For elastic scattering from a ferromagnet, an argument parallel to that leading to (7.11 a) gives for the cross-section

$$\frac{\mathrm{d}\sigma}{\mathrm{d}\Omega} = \left(\frac{\gamma e^2}{m_\mathrm{e} c^2}\right)^2 (1-\tilde{\kappa}_z^2)\langle\hat{\mathfrak{s}}^z(\boldsymbol{\varkappa})\rangle\langle\hat{\mathfrak{s}}^z(-\boldsymbol{\varkappa})\rangle. \qquad (7.29)$$

Now in § 5.5 it was shown that for a single narrow energy band the operator $\hat{\mathfrak{s}}^z(\boldsymbol{\varkappa})$ is approximately given by

$$\hat{\mathfrak{s}}^z(\boldsymbol{\varkappa}) = \int \mathrm{d}\mathbf{r}\, \exp(-\mathrm{i}\boldsymbol{\varkappa}.\mathbf{r})|\phi(\mathbf{r})|^2 \sum_{\mathbf{k}} \tfrac{1}{2}(\hat{c}^+_{\boldsymbol{\varkappa}+\mathbf{k}\uparrow}\hat{c}_{\mathbf{k}\uparrow}-\hat{c}^+_{\boldsymbol{\varkappa}+\mathbf{k}\downarrow}\hat{c}_{\mathbf{k}\downarrow}), \qquad (7.30)$$

where $c^+_{\mathbf{k}\sigma}$ and $c_{\mathbf{k}\sigma}$ are the creation and annihilation operators for an electron in the Bloch state with wave vector $\mathbf{k}$ and spin σ. Transforming these operators to the localized representation, we find

$$\hat{s}^z(\boldsymbol{\varkappa}) = \int \mathrm{d}\mathbf{r}\, \exp(-\mathrm{i}\boldsymbol{\varkappa}.\mathbf{r})|\phi(\mathbf{r})|^2 \tfrac{1}{2} \sum_{l} \exp(-\mathrm{i}\boldsymbol{\varkappa}.\mathbf{l})(\hat{c}^{+}_{l\uparrow}\hat{c}_{l\uparrow} - \hat{c}^{+}_{l\downarrow}\hat{c}_{l\downarrow}), \quad (7.31)$$

Hence

$$\langle\hat{s}^z(\boldsymbol{\varkappa})\rangle\langle\hat{s}^z(-\boldsymbol{\varkappa})\rangle$$

$$= \left|\int \mathrm{d}\mathbf{r}\, \exp(-\mathrm{i}\boldsymbol{\varkappa}.\mathbf{r})|\phi(\mathbf{r})|^2\right|^2 \{\tfrac{1}{2}\langle\hat{n}_{\uparrow}-\hat{n}_{\downarrow}\rangle\}^2 \sum_{l,l'} \exp\{\mathrm{i}\boldsymbol{\varkappa}.(\mathbf{l}-\mathbf{l}')\}, \quad (7.32)$$

because, as a consequence of translational symmetry, $\langle\hat{c}^{+}_{l\uparrow}\hat{c}_{l\uparrow}\rangle = \langle\hat{n}_{l\uparrow}\rangle$ is independent of the site label $\mathbf{l}$.

On inserting (7.32) in (7.29) we get immediately for the elastic magnetic cross-section the result

$$\frac{\mathrm{d}\sigma}{\mathrm{d}\Omega} = \left(\frac{\gamma e^2}{m_e c^2}\right)^2 (1-\tilde{\kappa}_z^2)\left|\int \mathrm{d}\mathbf{r}\, \exp(-\mathrm{i}\boldsymbol{\varkappa}.\mathbf{r})|\phi(\mathbf{r})|^2\right|^2 \times$$

$$\times \{\tfrac{1}{2}\langle\hat{n}_{\uparrow}-\hat{n}_{\downarrow}\rangle\}^2 N \frac{(2\pi)^3}{v_0} \sum_{\boldsymbol{\tau}} \delta(\boldsymbol{\varkappa}-\boldsymbol{\tau}). \quad (7.33)$$

Each term in this expression corresponds to a term in the expression (7.11 a) derived on the basis of the localized model. For $\frac{1}{2}\langle\hat{n}_{\uparrow}-\hat{n}_{\downarrow}\rangle$ is the average spin at any site and

$$\int \mathrm{d}\mathbf{r}\, \exp(\mathrm{i}\boldsymbol{\varkappa}.\mathbf{r})|\phi(\mathbf{r})|^2$$

is the Fourier transform of the spin density at each site, i.e. the atomic form factor.

7.3. Determination of spin densities and patterns in metals

(a) *Elastic magnetic cross-section for polarized neutrons*

The elastic magnetic scattering from ferromagnetic metals is generally small in comparison with the nuclear scattering, e.g. for iron and nickel it is about 8 per cent and 0·6 per cent respectively, so that accurate measurements would seem out of the question. However, the cross-section for the elastic scattering of polarized neutrons contains a magnetic–nuclear interference term, which, because it is linear in the magnetic scattering amplitude, makes it possible to measure very small magnetic cross-sections with great accuracy. This fact is the reason why most spin density determinations have been made with polarized neutrons.

There are, in addition to the high sensitivity of the polarized-beam technique, two other advantages that are of great value. First, the fact that the interference term is linear in the magnetic scattering amplitude means that its absolute sign can be determined for a particular reflection. Second, the change in the polarization of the neutrons can be used to give additional information on the magnetic structure. This second point is more fully discussed in the chapter on polarization phenomena, Chapter 10.

Consider the cross-section for the elastic scattering of polarized neutrons by a ferromagnet. In eqn (10.95) it is shown that this is

$$\frac{d\sigma}{d\Omega} = \Big|\sum_l \exp(i\boldsymbol{\varkappa}.\mathbf{l})\Big|^2 \Big|\sum_d \exp(i\boldsymbol{\varkappa}.\mathbf{d})\bar{b}_d\Big|^2 + N\sum_d (\overline{b_d^2} - \bar{b}_d^2) +$$
$$+\Big(\frac{\gamma e^2}{m_e c^2}\Big)\Big|\sum_l \exp(i\boldsymbol{\varkappa}.\mathbf{l})\Big|^2 \Big[\tfrac{1}{2}gF(\boldsymbol{\varkappa})\langle\hat{S}\rangle\mathbf{P}.\tilde{\boldsymbol{\varkappa}}\times(\tilde{\boldsymbol{\eta}}\times\tilde{\boldsymbol{\varkappa}})\times$$
$$\times\sum_{d,d'} 2\bar{b}_d \cos\{\boldsymbol{\varkappa}.(\mathbf{d}-\mathbf{d}')\}+$$
$$+\Big(\frac{\gamma e^2}{m_e c^2}\Big)|\tfrac{1}{2}gF(\boldsymbol{\varkappa})|^2\langle\hat{S}\rangle^2\{1-(\tilde{\boldsymbol{\varkappa}}.\tilde{\boldsymbol{\eta}})^2\}\Big|\sum_d \exp(i\boldsymbol{\varkappa}.\mathbf{d})\Big|^2\Big], \quad (7.34)$$

where $\mathbf{P}$ is the polarization of the incident beam.

Of the four terms in this expression only the third has not already been explicitly encountered; this is the magnetic–nuclear interference term referred to above. Because we are concerned here with the scattering at the Bragg peaks, the second term in the cross-section, the incoherent nuclear contribution, will be neglected. If the experimental arrangement is such that the direction of the ferromagnetic spins $\tilde{\boldsymbol{\eta}}$ is perpendicular to the scattering vector $\boldsymbol{\varkappa}$, then for a Bravais crystal (7.34) reduces to

$$\frac{d\sigma}{d\Omega} = \Big|\sum_l \exp(i\boldsymbol{\varkappa}.\mathbf{l})\Big|^2\Big\{\bar{b}^2 + 2\bar{b}\Big(\frac{\gamma e^2}{m_e c^2}\Big)\tfrac{1}{2}gF(\boldsymbol{\varkappa})\langle\hat{S}\rangle\mathbf{P}.\tilde{\boldsymbol{\eta}} +$$
$$+\Big(\frac{\gamma e^2}{m_e c^2}\Big)^2|\tfrac{1}{2}gF(\boldsymbol{\varkappa})|^2\langle\hat{S}\rangle^2\Big\}, \quad (7.35)$$

i.e.

$$\frac{d\sigma}{d\Omega} = \Big|\sum_l \exp(i\boldsymbol{\varkappa}.\mathbf{l})\Big|^2\Big\{\bar{b} + \Big(\frac{\gamma e^2}{m_e c^2}\Big)\tfrac{1}{2}gF(\boldsymbol{\varkappa})\langle\hat{S}\rangle\Big\}^2, \quad \text{for } \mathbf{P}.\tilde{\boldsymbol{\eta}} = 1,$$
$$= \Big|\sum_l \exp(i\boldsymbol{\varkappa}.\mathbf{l})\Big|^2\Big\{\bar{b} - \Big(\frac{\gamma e^2}{m_e c^2}\Big)\tfrac{1}{2}gF(\boldsymbol{\varkappa})\langle\hat{S}\rangle\Big\}^2, \quad \text{for } \mathbf{P}.\tilde{\boldsymbol{\eta}} = -1.$$
$$(7.36)$$

Hence, if the scattering length $\bar{b}$ for the material is known, a measurement of the ratio of the two intensities at a Bragg peak for $\mathbf{P}$ parallel to the spin direction and for **P** antiparallel to the spin direction gives the magnetic scattering amplitude

$$\left(\frac{\gamma e^2}{m_e c^2}\right)\tfrac{1}{2}gF(\boldsymbol{\varkappa})\langle\hat{S}\rangle.$$

The ratio of these two intensities is usually called the flipping ratio and denoted by the letter R, i.e.

$$R = \left(1+\frac{\gamma e^2}{\bar{b}m_e c^2}\tfrac{1}{2}gF(\boldsymbol{\varkappa})\langle\hat{S}\rangle\right)^2\Big/\left(1-\frac{\gamma e^2}{\bar{b}m_e c^2}\tfrac{1}{2}gF(\boldsymbol{\varkappa})\langle\hat{S}\rangle\right)^2. \tag{7.37}$$

The superiority of the polarized-beam method in sensitivity over the direct measurement of the magnetic scattering with unpolarized neutrons is simply demonstrated. Suppose the magnetic scattering amplitude for a given reflection is equal to $\bar{b}/100$. For this reflection

$$R = \left(\frac{1+0{\cdot}01}{1-0{\cdot}01}\right)^2 \simeq 1{\cdot}04, \tag{7.38}$$

whereas the magnetic contribution to the total intensity for the scattering of unpolarized neutrons would be 1 part in 10^4.

Before passing on to discuss examples of the use of the polarized-beam method in determining spin densities we pause to make a few comments on the production of polarized neutrons. It is clear from the cross-section (7.36) that if, for a given reflection, $\bar{b}$ was close to the magnetic scattering amplitude, then the scattered beam would consist primarily of neutrons in one particular state of polarization. This condition on the coherent nuclear and magnetic scattering amplitudes is closely fulfilled in the (111) reflection from a cobalt–iron crystal consisting of 92 atomic per cent cobalt, and is the system most frequently used at present for both polarizing and analysing crystals. There are, besides the fulfilment of this condition, other requirements demanded of a crystal for it to be an efficient polarizer or analyser.

For instance, extinction and also depolarization of the beam as it passes through the crystal can obviously seriously affect the efficiency of a crystal as a polarizer. A detailed discussion of the properties required for an efficient polarizing or analysing crystal is given by Nathans† *et al.*

(b) *Spin density in nickel*‡

The spin density and atomic magnetic form factor for nickel have been determined with great accuracy by Mook using the polarized-beam

† R. Nathans, C. G. Shull, G. Shirane, and A. Andresen, *Physics Chem. Solids*, **10**, 138 (1959).

‡ H. A. Mook, *Phys. Rev.* **148**, 495 (1966).

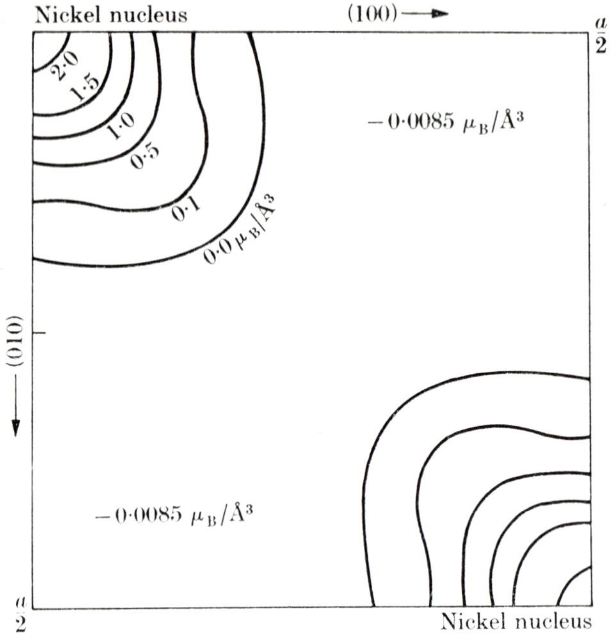

FIG. 7.1 (a). The magnetic moment distribution in the [100] plane for Ni.

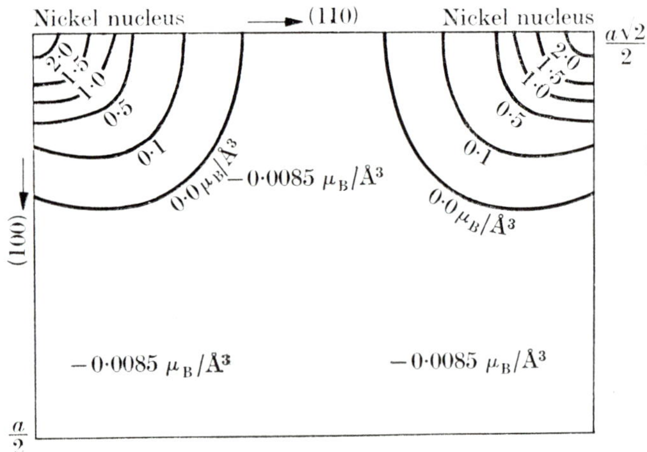

FIG. 7.1 (b). The magnetic moment distribution in the [110] plane for Ni.

technique described above. The spin density maps for the [100] and [110] planes are shown in Fig. 7.1 (a) and (b). The most striking feature of these maps is that they clearly demonstrate that the Ni spin density is quite asymmetric about the lattice sites. To understand the full implication of this result we ask how $3d$ wave functions behave in a cubic crystal field.†

The cubic field partially removes the five-fold orbital degeneracy by splitting the wave functions according to the representations E_g and T_{2g} of the cubic group‡ O. The two orbitals of the E_g representation behave like (Fig. 7.2)

$$(3z^2-r^2)f(r) \quad \text{and} \quad (x^2-y^2)f(r),$$

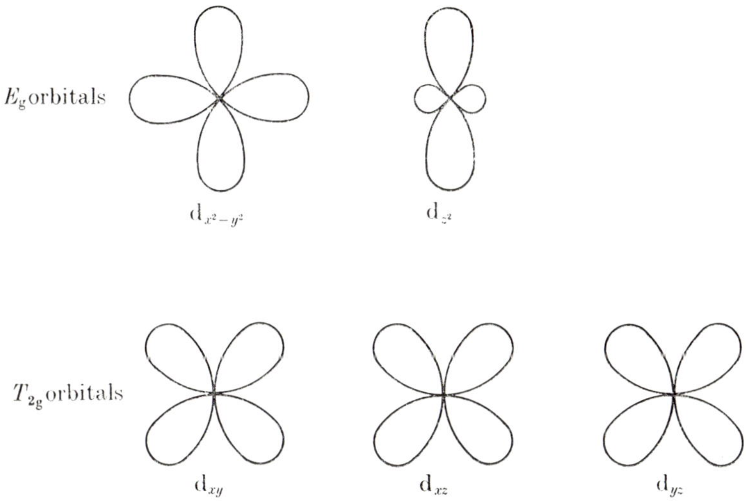

FIG. 7.2. d orbitals in a cubic crystal field.

where $f(r)$ is the radial part of the wave function. The three orbitals of the T_{2g} representation are proportional to

$$xyf(r), \quad yzf(r), \quad \text{and } zxf(r).$$

Electrons occupying wave functions of E_g symmetry result in a spin density along the [100] cube edges and electrons occupying wave functions of T_{2g} symmetry in a spin density along the [111] cube diagonals. However, if all these orbitals are equally occupied, the resulting form factor $F(\boldsymbol{\varkappa})$ is spherically symmetric. In this instance the occupation of the T_{2g} and E_g orbitals would be in the ratio 60 per cent to 40 per cent.

† See, for instance, L. D. Landau and E. M. Lifshitz, *Quantum mechanics* (second edition), Chapter 12, and R. S. Knox and A. Gold, *Symmetry in the solid state*, W. A. Benjamin (1964).

‡ The notation E_g and T_{2g} for the representations of the cubic group O is that adopted by chemists. In the frequently used notations due to Bouckaert, Smoluchowski, and Wigner (*Phys. Rev.* **50**, 58 (1936)) and Bethe (*Annln Phys.* **3**, 133 (1929)) E_g, T_{2g} correspond to Γ_{12}, Γ'_{25} and Γ_3, Γ_5, respectively.

With these facts in mind we infer from Fig. 7.1 (a) and (b) that the occupancy of unpaired electrons in the T_{2g} orbitals in Ni is greater than that required for spherical symmetry; Mook in fact finds that 81 ± 1 per cent of the $3d$ magnetic electrons occupy T_{2g} orbitals. The difference in the radial distribution of the spin density along the symmetry axes is clearly illustrated in Fig. 7.3.

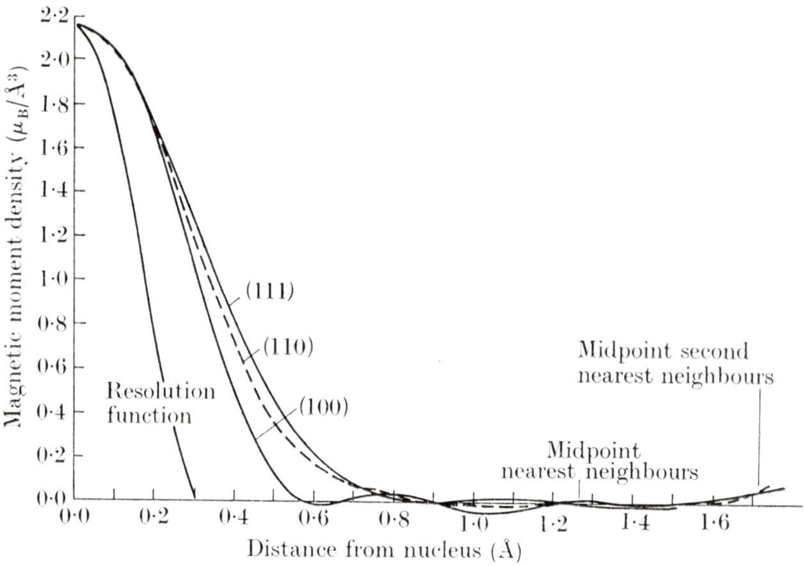

FIG. 7.3. Radial magnetic moment distributions along principal symmetry axes for Ni.

Because the flipping ratio R cannot be measured for all Bragg reflections there must be errors in the evaluation of the spin density from eqn (7.17). Thus the radial spin density distributions given in Fig. 7.3 are, more correctly, the true spin density distributions as seen with the finite resolution function arising from measuring, in this case, the first 27 Bragg reflections. Any detail finer than the resolution function given in Fig. 7.3 cannot be resolved.

A further consequence of having R only for a limited number of Bragg reflections is that $s(\mathbf{r})$ obtained from the series (7.17) will oscillate for large values of $\mathbf{r}$. This would seem to prevent an accurate determination of the spin density between atoms because the oscillations in the calculated $s(\mathbf{r})$ obscure its true small value in this region. To overcome this difficulty we can ask for the average value of $s(\mathbf{r})$ over a small volume. The results given in Fig. 7.1 for the spin density far away from the lattice points were derived in this way. If we take the volume to be a cube with edges parallel to the cell axes and of length 2δ then the spin density averaged over this volume, $\overline{\mathscr{s}(\mathbf{r})}$, is from (7.17)

$$\overline{\mathscr{s}(\mathbf{r})} = \frac{1}{v_0}\sum_{t_1,t_2,t_3}\mathscr{F}(\boldsymbol{\tau})\left(\frac{1}{2\delta}\right)^3\int\limits_{x-\delta}^{x+\delta}\mathrm{d}x\int\limits_{y-\delta}^{y+\delta}\mathrm{d}y\int\limits_{z-\delta}^{z+\delta}\mathrm{d}z\exp\left\{-\frac{2\pi\mathrm{i}}{a}(t_1x+t_2y+t_3z)\right\}$$

$$= \frac{1}{(2\pi\delta/a)^3v_0}\sum_{t_1,t_2,t_3}\{\mathscr{F}(\boldsymbol{\tau})/t_1t_2t_3\}\sin\left(\frac{2\pi t_1\delta}{a}\right)\sin\left(\frac{2\pi t_2\delta}{a}\right)\sin\left(\frac{2\pi t_3\delta}{a}\right). \quad (7.39)$$

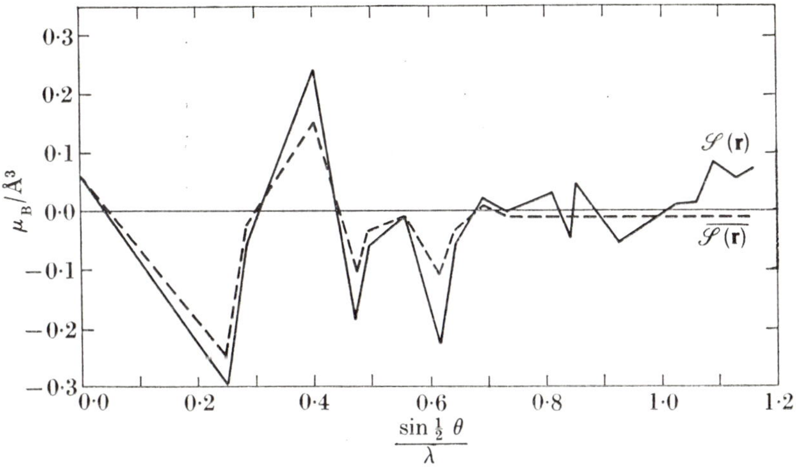

FIG. 7.4. The functions $\mathscr{s}(\mathbf{r})$ and $\overline{\mathscr{s}(\mathbf{r})}$ defined by eqns (7.17) and (7.39) for nickel at the position $(\frac{1}{2}a, 0, 0)$.

The presence of the factor $(t_1t_2t_3)^{-1}$ in this series for $\overline{\mathscr{s}(\mathbf{r})}$ makes it converge much more rapidly than the series for $\mathscr{s}(\mathbf{r})$. In calculating $\overline{\mathscr{s}(\mathbf{r})}$ we are asking a less detailed question than when we calculate $\mathscr{s}(\mathbf{r})$ and thus, with the limited data, we obtain a more accurate result. The criterion for $\overline{\mathscr{s}(\mathbf{r})}$ to be a good approximation to the true spin density is that it should be almost independent of the choice of δ for a given value of $\mathbf{r}$. The effect of the averaging procedure can be seen in detail by examining Fig. 7.4, which shows $\mathscr{s}(\mathbf{r})$ and $\overline{\mathscr{s}(\mathbf{r})}$ for Ni at the position $(\frac{1}{2}a, 0, 0)$. The size of the oscillations is a measure of convergence and $\mathscr{s}(\mathbf{r})$ is seen to be oscillating widely at the maximum number of terms included in its corresponding series, whereas $\overline{\mathscr{s}(\mathbf{r})}$ has long settled down to a constant value. Fig. 7.5 shows $\overline{\mathscr{s}(\mathbf{r})}$ along the [100] direction.

As a result of his calculations of $\overline{\mathscr{s}(\mathbf{r})}$ Mook was able to conclude that regions of negative (i.e. oppositely directed to the total spin) spin density exist in Ni between neighbouring ions, as shown in Fig. 7.1.

In summary, the careful measurements by Mook of the spin density in Ni have furnished three important facts: (i) the major part of the total spin density is very well localized about the Ni nuclei, (ii) there is a distinct asymmetry in the spin density, the occupancy of the T_{2g} orbitals exceeding that required for spherical symmetry by some 20 per

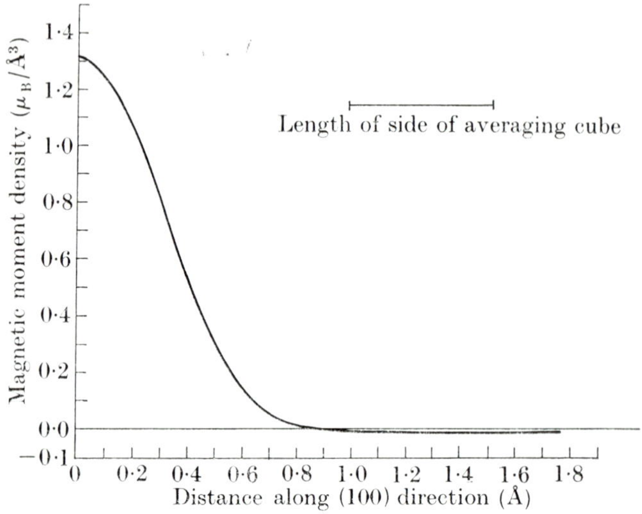

FIG. 7.5. The average magnetic moment density $\overline{\mathscr{s}(\mathbf{r})}$ for Ni for a cubic block 0·15 lattice constants on a side plotted along the [100] direction.

cent, and (iii) there is a region of almost constant negative spin density between neighbouring Ni atoms.†

Before we consider the atomic form factor for Ni we point out an alternative approach to obtaining an average spin density $\overline{\mathscr{s}(\mathbf{r})}$ than that given by (7.39). Instead of averaging $\mathscr{s}(\mathbf{r})$ over a small cube, let us make a Gaussian average about $\mathbf{r}$, i.e.

$$\overline{\mathscr{s}(\mathbf{r})} = \int \mathrm{d}\mathbf{r}' \exp\{-\delta(\mathbf{r}-\mathbf{r}')^2\}\mathscr{s}(\mathbf{r}')\Big/\int \mathrm{d}\mathbf{r}' \exp(-\delta r'^2). \tag{7.40}$$

With the defining equation for $\mathscr{s}(\mathbf{r})$ in terms of $\mathscr{F}(\boldsymbol{\tau})$, eqn (7.17), we obtain from (7.40) a very rapidly converging series:

$$\begin{aligned}\overline{\mathscr{s}(\mathbf{r})} &= \left(\frac{\delta}{\pi}\right)^{\frac{3}{2}} \frac{1}{v_0} \sum_{\boldsymbol{\tau}} \mathscr{F}(\boldsymbol{\tau}) \int \mathrm{d}\mathbf{r}' \exp\{-\delta(\mathbf{r}-\mathbf{r}')^2 - \mathrm{i}\boldsymbol{\tau}.\mathbf{r}'\} \\ &= \left(\frac{\delta}{\pi}\right)^{\frac{3}{2}} \frac{1}{v_0} \sum_{\boldsymbol{\tau}} \mathscr{F}(\boldsymbol{\tau}) \exp(-\mathrm{i}\boldsymbol{\tau}.\mathbf{r}) \int \mathrm{d}\mathbf{r}' \exp(\mathrm{i}\boldsymbol{\tau}.\mathbf{r}' - \delta r'^2) \\ &= \frac{1}{v_0} \sum_{\boldsymbol{\tau}} \mathscr{F}(\boldsymbol{\tau}) \exp(-\mathrm{i}\boldsymbol{\tau}.\mathbf{r}) \exp(-\tau^2/4\delta). \end{aligned} \tag{7.40 a}$$

† A first-principles calculation for nickel has been attempted by Hodges *et al.*, *Phys. Rev.* **152**, 505 (1966).

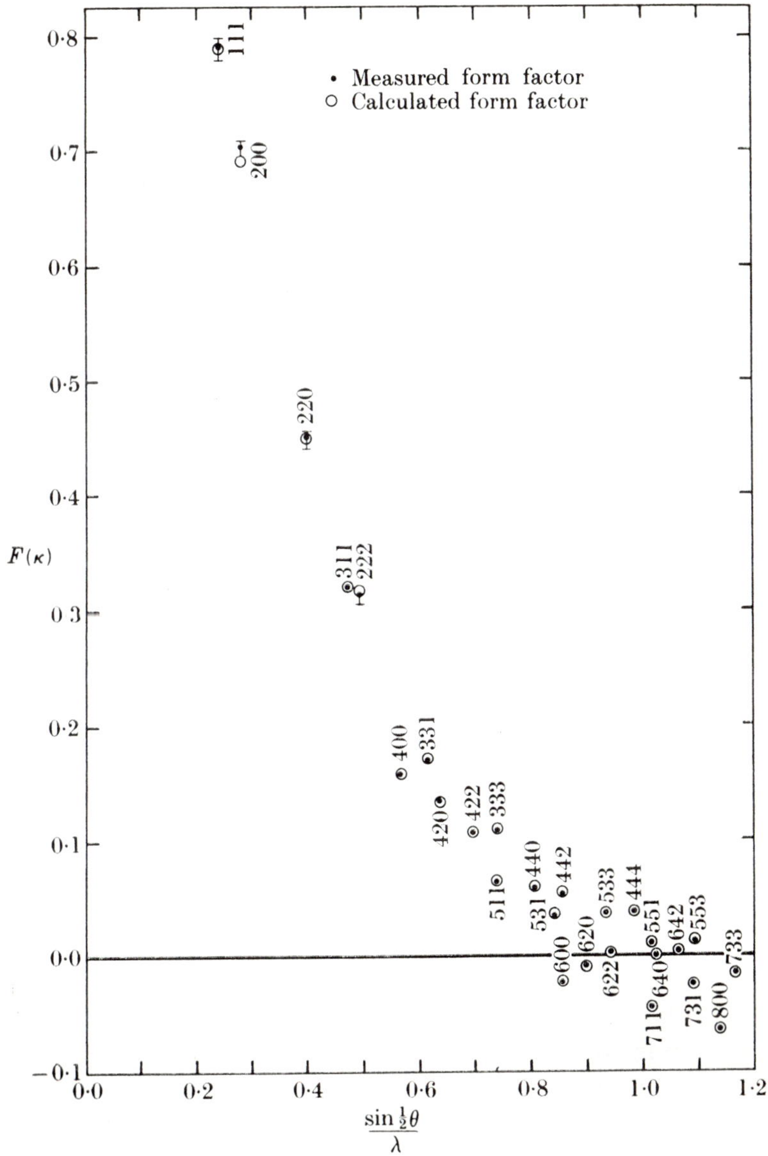

FIG. 7.6. Measured and calculated (eqn (7.41)) atomic form factors for Ni.

Mook's atomic form factor $F(\varkappa)$ for Ni is shown in Fig. 7.6. The experimental points plotted in this figure were obtained by normalizing the magnetic scattering amplitude derived from the measured value of R with the calculated [000] magnetic amplitude, which Mook took to be (at room temperature) $0{\cdot}155\times10^{-14}$ m.

The asymmetry in the spin distribution is vividly demonstrated by the difference in the reported measurements for pairs of reflections with the same value of $(\sin\frac{1}{2}\theta)/\lambda$. There are four such pairs included in Fig. 7.6, namely

$$(511) \text{ and } (333)\colon (\sin\tfrac{1}{2}\theta)/\lambda = 0{\cdot}7387$$
$$(600) \text{ and } (442)\colon (\sin\tfrac{1}{2}\theta)/\lambda = 0{\cdot}8530$$
$$(551) \text{ and } (771)\colon (\sin\tfrac{1}{2}\theta)/\lambda = 1{\cdot}0153$$
$$(731) \text{ and } (553)\colon (\sin\tfrac{1}{2}\theta)/\lambda = 1{\cdot}0920.$$

From Chapters 5 and 6 we anticipate that the atomic form factor of Ni can be expressed as

$$F(\varkappa) = \frac{2}{g}F_{\text{spin}}(\varkappa)+\left(\frac{g-2}{g}\right)F_{\text{orbit}}(\varkappa)+F_{\text{core}}(\varkappa). \qquad (7.41)$$

The value of the gyromagnetic ratio g for nickel is 2·2. The third term in (7.41) represents the almost negligibly small contribution to $F(\varkappa)$ from the polarization of the argon core due to configuration interaction. By far the largest contribution to $F(\varkappa)$ comes from the spin, the orbital contribution being only about 10 per cent. Taking note of the constant negative spin density observed in Fig. 7.1, F_{spin} can be written as

$$F_{\text{spin}}(\varkappa) = (1+\alpha)F_{3d}(\varkappa)-\alpha\delta(\varkappa), \qquad (7.42)$$

where α is a constant and $F_{3d}(\varkappa)$ is the spin part of the form factor for $3d$ electrons as calculated by Weiss and Freeman, and Watson and Freeman.† The numerical calculations were made with the unrestricted Hartree–Fock approximation for the free Ni^{2+} ion. For the orbital part of the calculated atomic form factor (open circles in Fig. 7.6) Mook used unpublished work by Watson and Freeman. Taking $\alpha = 0{\cdot}19$ and assuming 19 per cent occupancy of the E_g orbitals, as opposed to 40 per cent for spherical symmetry, Mook achieved the extraordinarily fine agreement between his measured and calculated form factors seen in Fig. 7.6. From this agreement we may conclude that the moment distribution in Ni is almost identical in form to that obtained from the model of free-atom distributions, treated within the unrestricted Hartree–Fock approximation, imposed on a constant, negative background.

The value 0·19 of the parameter α, introduced in (7.42), corresponds to including in the moment distribution a uniform background of $-0{\cdot}0091\ \mu_B/\text{Å}^3$ so that this value of α is consistent with the negative contribution given in Fig. 7.1.

† R. J. Weiss and A. J. Freeman, *Physics Chem. Solids*, **10**, 147 (1959). R. E. Watson and A. J. Freeman, *Phys. Rev.* **120**, 1125 (1960) and *Acta crystallogr.* **14**, 27 (1961).

Taking account of all the facts derived from his experiment Mook arrived at the following model for the spin density distribution in Ni:

$$\begin{aligned} &3d \text{ spin } \; 0{\cdot}656\,\mu_{\mathrm{B}}/\text{atom} \\ &3d \text{ orbit } \; 0{\cdot}055\,\mu_{\mathrm{B}}/\text{atom} \\ &\text{negative contribution } \; -0{\cdot}105\,\mu_{\mathrm{B}}. \end{aligned} \tag{7.43}$$

These total to the observed saturation magnetization of 0·606 μ_{B}/atom.

Having looked in some detail at Mook's determination of the spin density in Ni it is as well to appraise the inherent sources of error in the polarized-beam technique. $\mathscr{s}(\mathbf{r})$ is obtained via eqn (7.17) from $\mathscr{F}(\boldsymbol{\tau})$, and the polarized-beam technique yields values of $\mathscr{F}(\boldsymbol{\tau})$ for $\boldsymbol{\tau} \neq 0$. $\mathscr{F}(0)$ is determined by conventional magnetization measurements. To extract $\mathscr{F}(\boldsymbol{\tau})$ from the flipping ratio R, the quantity actually measured in the neutron experiment, the coherent scattering length $\bar{b}$ must be known. Thus, *three* experiments in all are involved in the determination of $\mathscr{s}(\mathbf{r})$. An error in the measurement of the magnetization will result in constant shift in $\mathscr{s}(\mathbf{r})$ from its true value, while an error in the determination of $\bar{b}$ will result in an error in the scale of the spatially varying part of the spin density.

(c) *Spin densities in iron and cobalt*

In addition to Ni, the spin densities of two other transition metals have been investigated in detail by the polarized-beam technique. Shull and Yamada† investigated iron and Moon‡ hexagonal cobalt. The results for both these elements have features that are in marked contrast to the results obtained for Ni.

Figs. 7.7, 7.8, and 7.9 are the spin density maps and distributions along the three principal crystal axes for Fe. While it is at once evident that Fe, like Ni, has an asymmetric spin distribution, a closer examination shows that there is a pronounced concentration of density along the [100] cube edges, which is characteristic of E_{g} orbitals. By matching calculated and observed form factors Shull and Yamada arrived at the value 53 ± 1 per cent for the occupancy of E_{g} orbitals. This is to be compared to the value of 19 ± 1 per cent for Ni. While this difference is not yet fully understood, it is probable that it is concerned with the different crystal lattice structures of Ni (f.c.c.) and Fe (b.c.c.). Fig. 7.10 shows a model of the magnetization distribution in the unit cell of iron.§

† C. G. Shull and Y. Yamada, *J. phys. Soc. Japan*, **17**, BIII, 1 (1962).
‡ R. M. Moon, *Phys. Rev.* **136**, A195 (1964).
§ C. G. Shull and H. A. Mook, *Phys. Rev. Lett.* **16**, 184 (1966).

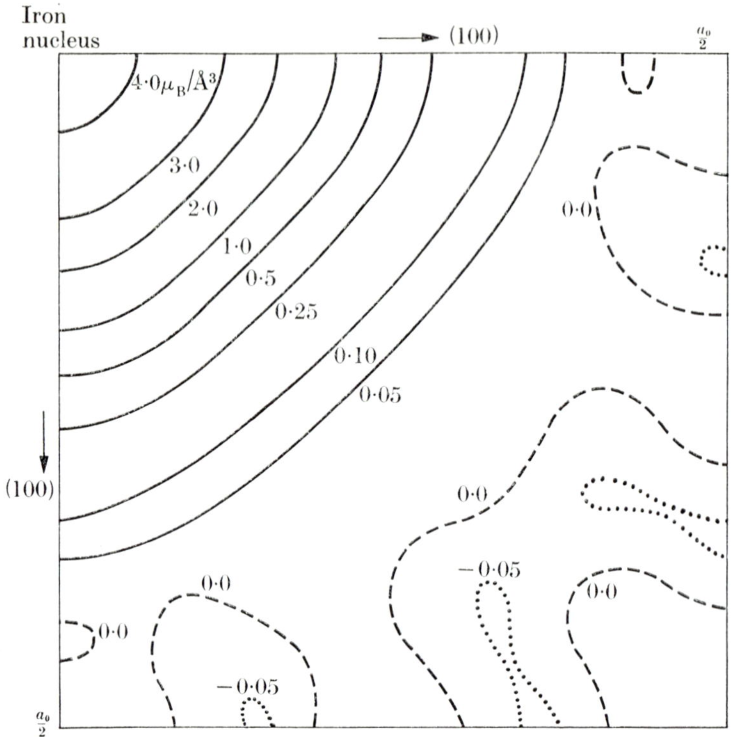

Fig. 7.7

Figs. 7.7, 7.8, and 7.9. Spin density maps and distributions along the three principal crystal axes for Fe.

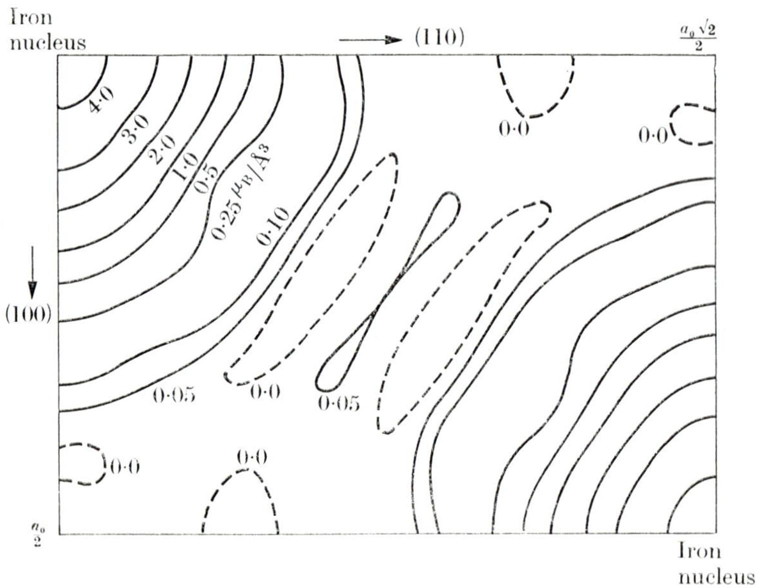

Fig. 7.8

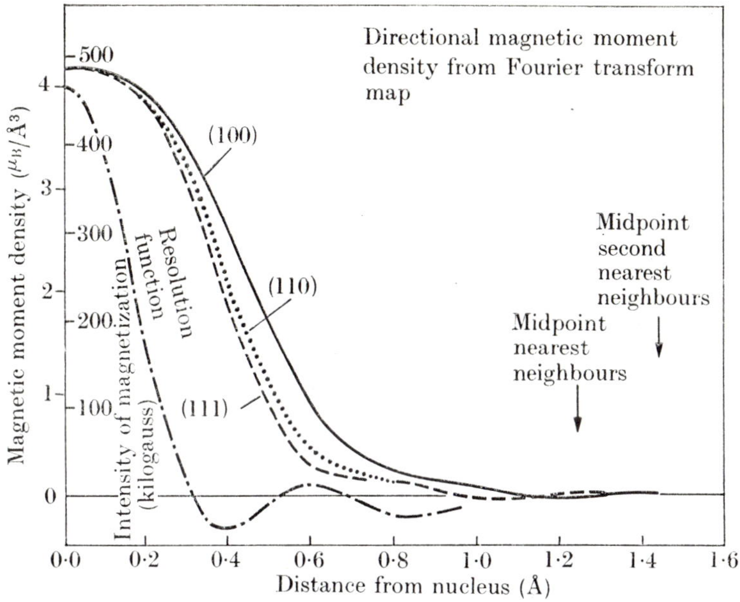

FIG. 7.9

The concentration along the cube edges is over-emphasized to show its relationship to the rings of negative magnetization.

In contrast to the results for both Ni and Fe the experiment by Moon on hexagonal cobalt indicates complete spherical symmetry in the spin density distribution. The form factor of hexagonal cobalt is given in Fig. 7.11, together with measurements by Nathans and Paoletti on f.c.c. cobalt.† The measurements on hexagonal cobalt are seen to lie on a smooth curve, while those for f.c.c. cobalt lie off the curve, which indicates that the spin density distribution is asymmetric.

Again an almost constant negative spin density was observed in regions farthest from the atoms. At first sight the spin density map, Fig. 7.12, might be interpreted as showing an overlap between the two atoms. However, the map is a projection onto the basal plane and the two atoms are, in fact, on different levels, resulting in the apparent overlap of the distributions. The true separation of nearest-neighbour atoms is greater than the projected distance by a factor of $\sqrt{3}$. From all his measurements Moon gave the following picture of the spin density distribution in hexagonal cobalt. The bulk of the density is localized and spherical and there is an almost constant negative moment density

† R. Nathans and A. Paoletti, *Phys. Rev. Lett.* **2**, 254 (1959).

in the regions farthest from the atoms. The figures derived are:

$3d$ spin	$1{\cdot}86 \pm 0{\cdot}07\ \mu_B$
$3d$ orbit	$0{\cdot}13 \pm 0{\cdot}01\ \mu_B$
constant density	$-0{\cdot}28 \pm 0{\cdot}07\ \mu_B$.

Note that the orbital moment is about the same fraction of the spin moment as in the case of Ni; g for metallic cobalt is 2·17.

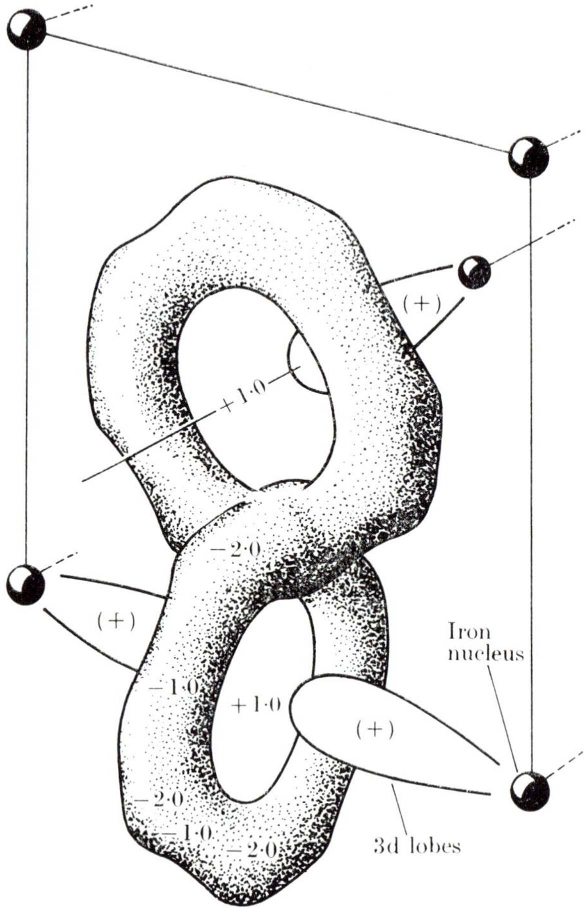

FIG. 7.10. A model of the magnetic moment density in Fe.

(*d*) *Spin configuration of chromium*

Having considered several examples of the determination of spin densities in metals let us look at an example of determining a spin configuration. We choose to examine an element that has proved particularly thorny, namely chromium. This forms a body-centred cubic structure and was first investigated by Shull and Wilkinson in 1953.†

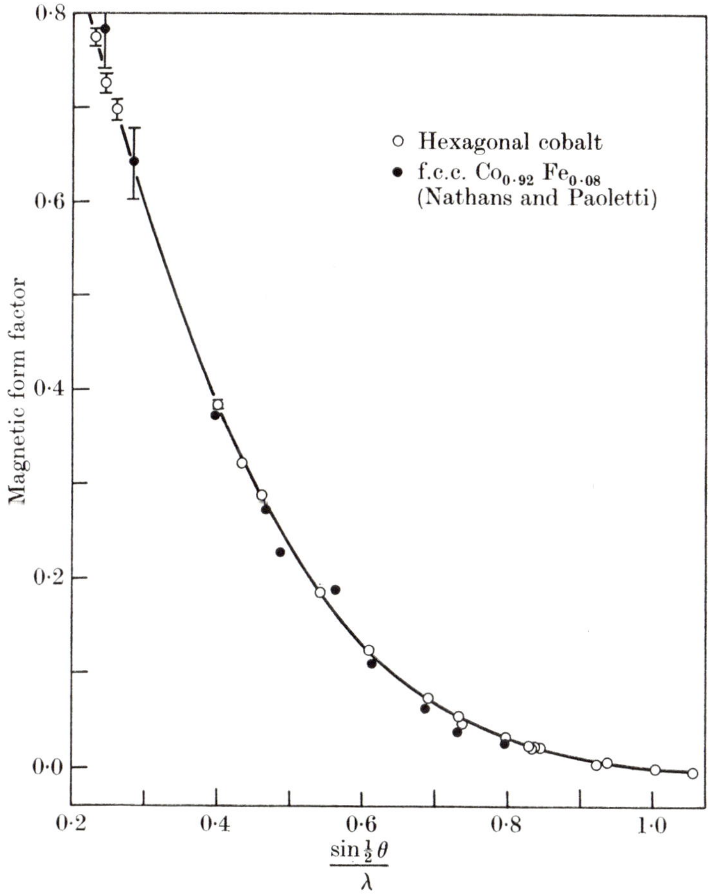

FIG. 7.11. Magnetic form factors for hexagonal and face-centred cobalt.

They derived the spin configuration shown in Fig. 7.13 from a powder pattern. Experiments on single crystals by Shirane and Takei‡ and Brown *et al.*§ have shown this configuration to be erroneous, the correct one being much more complicated. It proves a very worthwhile exercise to trace the understanding of chromium from the experiment by Shull and Wilkinson to the present day.

† C. G. Shull and M. K. Wilkinson, *Rev. mod. Phys.* **25**, 100 (1953).
‡ G. Shirane and W. J. Takei, *J. phys. Soc. Japan*, **17**, BIII, 35 (1962).
§ P. J. Brown *et al. Proc. phys. Soc.* **85**, 1185 (1965).

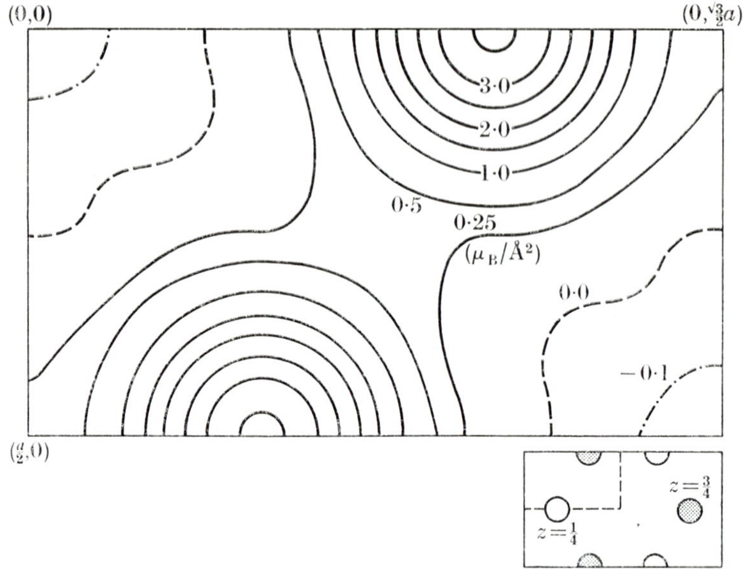

FIG. 7.12. Magnet moment density of hexagonal cobalt projected onto the basal plane.

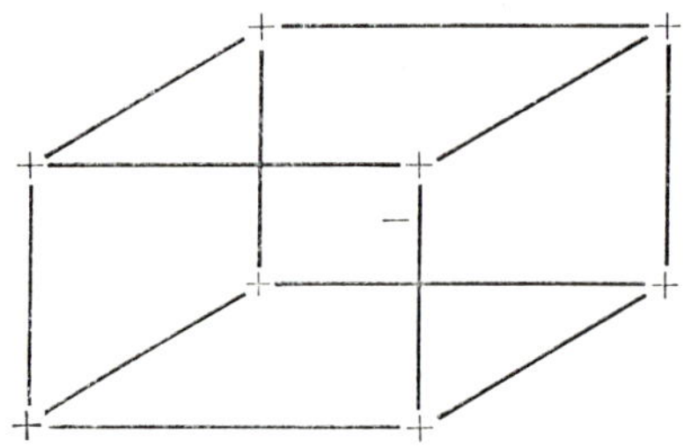

FIG. 7.13. Spin configuration for Cr proposed by Shull and Wilkinson from a powder pattern.

The first few non-zero reciprocal lattice vectors $\boldsymbol{\tau}$ for a body-centred cubic crystal with a unit-cell side a are

$$(1,1,0) \quad |\boldsymbol{\tau}| = \frac{2\pi}{a}\sqrt{2} \qquad (2,2,0) \quad |\boldsymbol{\tau}| = \frac{2\pi}{a}\sqrt{8}$$

$$(2,0,0) \quad |\boldsymbol{\tau}| = \frac{2\pi}{a}\sqrt{4} \qquad (3,1,0) \quad |\boldsymbol{\tau}| = \frac{2\pi}{a}\sqrt{10}$$

$$(2,1,1) \quad |\boldsymbol{\tau}| = \frac{2\pi}{a}\sqrt{6} \qquad (2,2,2) \quad |\boldsymbol{\tau}| = \frac{2\pi}{a}\sqrt{12}. \tag{7.44}$$

(For a b.c.c. crystal $t_1+t_2+t_3$ is an even integer, cf. § 2.1.)

Hence in a powder pattern we get *nuclear* Debye–Scherrer cones at scattering angles θ given by substituting these values of $|\boldsymbol{\tau}|$ into the Bragg scattering condition

$$\sin(\tfrac{1}{2}\theta) = \tfrac{1}{2}|\boldsymbol{\tau}|/|\mathbf{k}|. \tag{7.45}$$

Thus, the first nuclear Bragg peak occurs at an angle

$$\theta = 2\sin^{-1}\left(\frac{1}{2|\mathbf{k}|}\frac{2\pi}{a}\sqrt{2}\right). \tag{7.46}$$

Experimentally this peak, and all other nuclear peaks, are seen, and, in addition, a small peak at a smaller angle than (7.46) was reported. From the value of θ for this peak it is possible to deduce from (7.45) that it corresponds to a $|\boldsymbol{\tau}|$ value of $2\pi/a$, i.e. one member of the magnetic reciprocal lattice is

$$\boldsymbol{\tau} = \frac{2\pi}{a}(1,0,0). \tag{7.47}$$

This reciprocal lattice vector is not amongst the allowed set for nuclear scattering and it follows that the magnetic unit cell is larger than the chemical unit cell. We may assume this magnetic unit cell coincides with that of a simple cubic lattice. The magnetic reciprocal lattice vectors are therefore given by

$$\boldsymbol{\tau} = \frac{2\pi}{a}(t_1, t_2, t_3), \tag{7.48}$$

where t_i are any integers.

For this model (Fig. 7.13) the magnetic unit-cell structure factor, defined by eqn (7.27), has the behaviour

$$\begin{aligned} F_{\mathrm{M}}(\boldsymbol{\tau}) \propto 1-\exp(\mathrm{i}\boldsymbol{\tau}.\mathbf{d}) &= 1-\exp \mathrm{i}\pi(t_1+t_2+t_3) \\ &= 0 \quad \text{if } t_1+t_2+t_3 \text{ is an even integer} \\ \text{and} \qquad &= 2 \quad \text{if } t_1+t_2+t_3 \text{ is an odd integer.} \end{aligned} \tag{7.49}$$

Because Cr has cubic symmetry, the orientation factor in the cross-section is 2/3 whatever the allowed orientations of $\langle\hat{\mathbf{S}}\rangle$ may be. Hence, using the result (7.49), the elastic magnetic cross-section (7.26) becomes (apart from the Debye–Waller factor)

$$\frac{\mathrm{d}\sigma}{\mathrm{d}\Omega} = \left(\frac{\gamma e^2}{m_e c^2}\right)^2 \langle \hat{S}\rangle^2 N\frac{(2\pi)^3}{v_0}4.\tfrac{2}{3}\sum_{\boldsymbol{\tau}}\delta(\boldsymbol{\varkappa}-\boldsymbol{\tau})|\tfrac{1}{2}gF(\boldsymbol{\tau})|^2, \tag{7.50}$$

with the condition $\quad t_1+t_2+t_3 = \text{odd integer},$

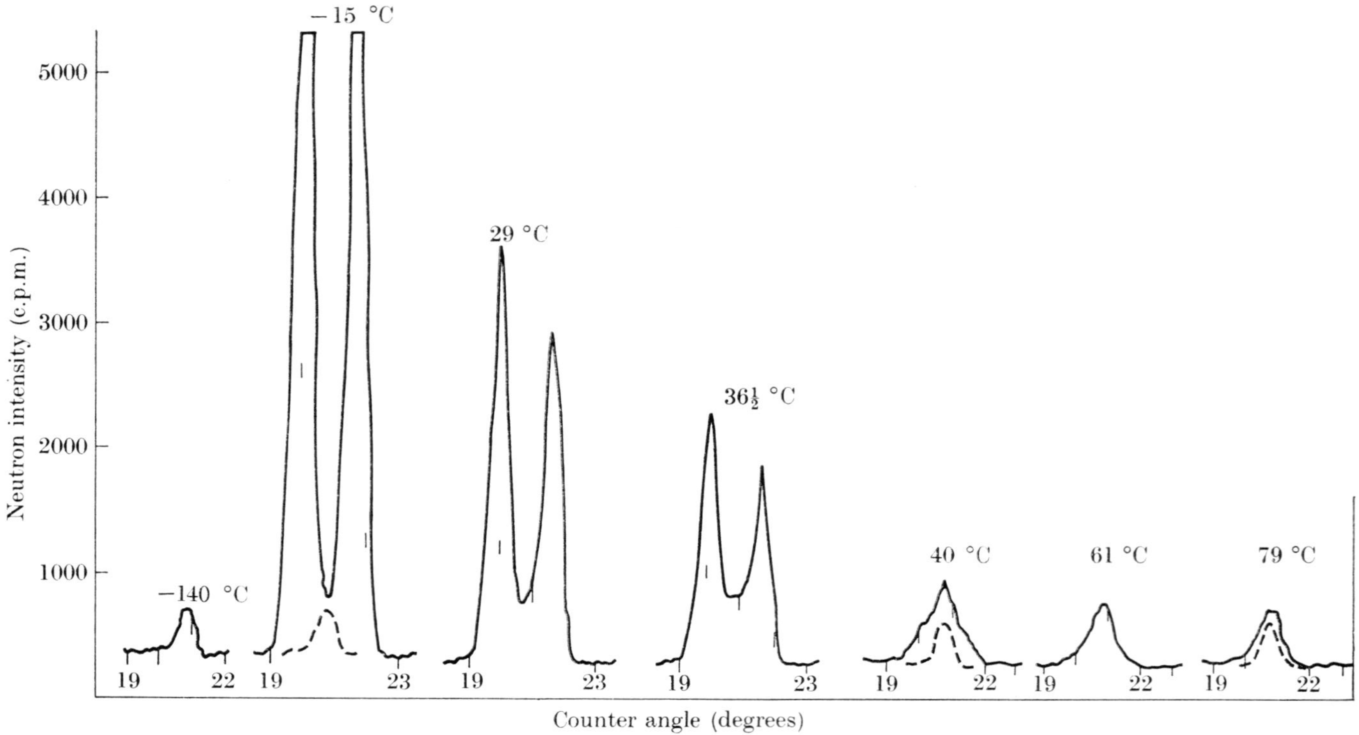

FIG. 7.14. The shape of the (100) reflection at various temperatures for a large single crystal of Cr, with counter and crystal moving in synchronism after alignment on the (200) nuclear reflection. Dashed curves are the $\frac{1}{2}\lambda$ component from the (200) reflection.

whereas the nuclear scattering cross-section is given by

$$\frac{d\sigma}{d\Omega} = N\frac{(2\pi)^3}{v_0}|\bar{b}|^2 4 \sum_{\boldsymbol{\tau}} \delta(\boldsymbol{\varkappa}-\boldsymbol{\tau}), \tag{7.51}$$

with $t_1+t_2+t_3 = \text{even integer}.$

For this model, therefore, the nuclear and magnetic peaks appear quite separately from one another. Note that the orientations of $\langle\hat{\mathbf{S}}\rangle$ are undetermined by this powder experiment.

Shull and Wilkinson deduced from the observed intensity of the (100) peak that the Cr atoms possess a magnetic moment of $0{\cdot}4\ \mu_B$, by assuming the atomic form factor to be identical to that for Fe. The peak disappeared at a temperature of 460 K; this they interpreted to be the Néel temperature. The next magnetic peak is expected to occur at $\boldsymbol{\tau} = (2\pi/a)(1, 1, 1)$, i.e. $|\boldsymbol{\tau}| = (2\pi/a)\sqrt{3}$. The form factor dependence makes this peak much weaker than the (100) peak, which itself was only just observable, so that it is not surprising that the (111) peak was not detected in the experiment.

Measurements of the electrical resistivity and other physical properties of Cr indicate anomalies at 40 °C, a temperature far removed from the Néel temperature proposed by Shull and Wilkinson. A study of Cr single crystals by Corliss *et al.* in (1959) led to a Néel temperature of 38 °C, a value much more akin to the other experimental evidence.† Corliss *et al.* also reported that the (100) reflection consisted of two satellite lines. The measurements by Bacon of the (100) reflection at various temperatures are shown in Fig. 7.14 and the integrated intensities are given in Fig. 7.15 as a function of temperature.‡

To explain their results, Corliss *et al.* proposed that chromium is made up of 'antiphase domains' of about eleven atoms on a side. These domains are not of the type we usually mean because they are regularly repeating. Effectively the unit cell is no longer of size $a\times a\times a$ but $11a\times 11a\times 11a$. In detail the body-centred spins are taken antiparallel to the corner spins and the splitting of the (100) peak is taken to be a consequence of antiphase domains, namely a spin rotation of π about every eleven unit cells normal to the boundaries of the domains. The domain boundaries are parallel to the crystallographic cube faces, but the crystal as a whole consists of a *random* arrangement of regions in any one of which the domain boundary is parallel to a particular cube face. The antiphase domain unit is shown in Fig. 7.16 together with the

† L. M. Corliss, J. M. Hastings, and R. J. Weiss, *Phys. Rev. Lett.* **3**, 211 (1959).
‡ G. E. Bacon, *Acta crystallogr.* **14**, 823 (1961).

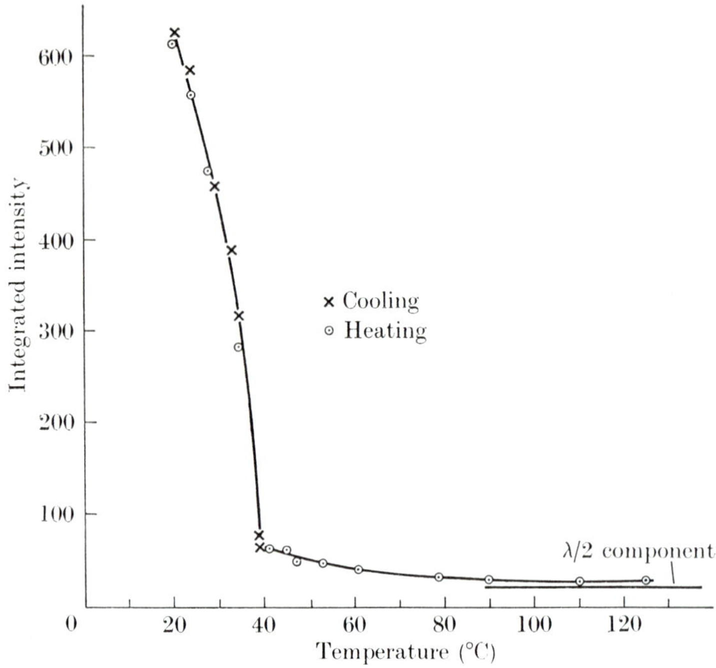

FIG. 7.15. The variation with temperature of the integrated intensity of the (100) reflection of a single crystal of pure chromium, near 40 °C. Points taken during cooling and heating, respectively, show no thermal hysteresis. The saturation value of the intensity, at low temperatures, is 1600 on the arbitrary scale used.

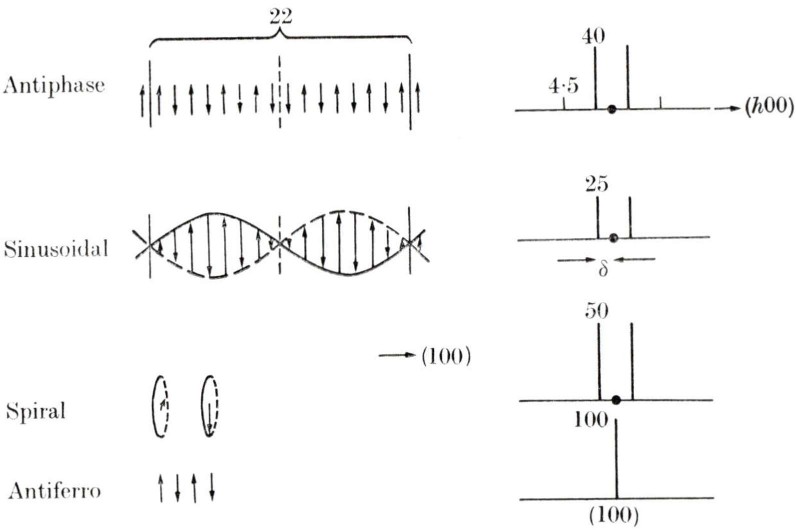

FIG. 7.16. Models for Cr spin arrangement with their magnetic intensities, calculated with the periodicity of 22 unit cells. This spin direction corresponds to the high temperature phase.

corresponding form of the (100) reflection. A consequence of the antiphase domain is the existence of additional peaks corresponding to odd harmonics, the third satellites being estimated to possess an intensity of about 11 per cent of the first pair.

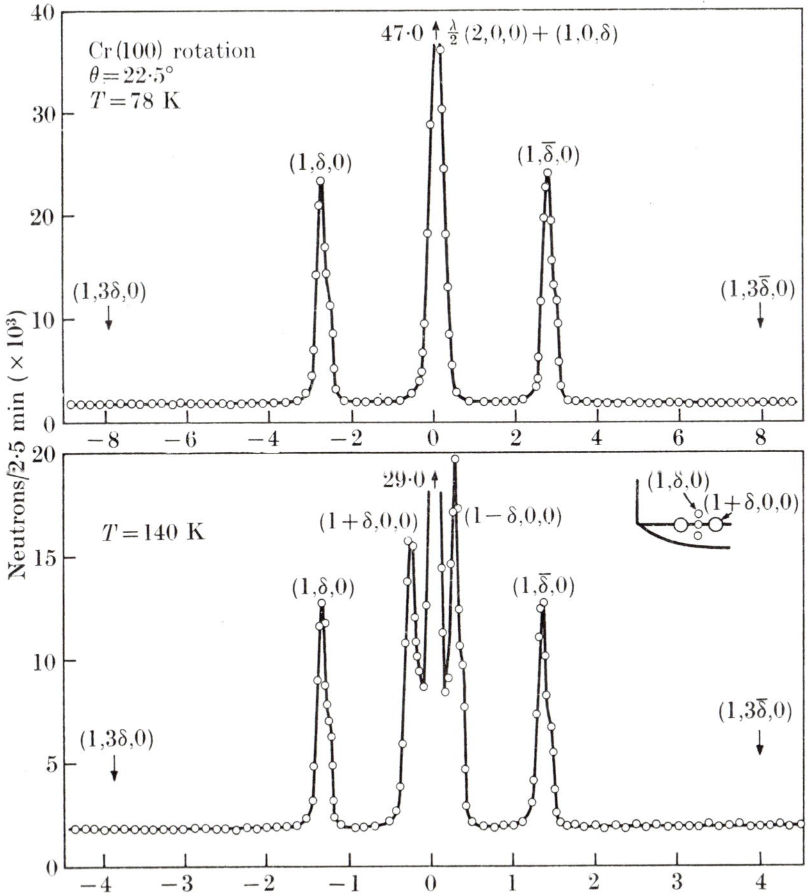

FIG. 7.17. Neutron diffraction patterns from Cr single crystal at 78 K and 140 K, with $\lambda = 1{\cdot}14$ Å. The crystal is rotated around the (001) axis while the counter is set at $2\theta = 22{\cdot}5°$. The slight asymmetry of the peak shape is due to a lineage structure of the monochromator. The beam is not filtered, and thus contains $\sim 0{\cdot}5\%$ of the $\frac{1}{2}\lambda$ component.

Also included in Fig. 7.16 are two other spin arrangements that result in a pair of satellite lines as reported by Corliss *et al.*, together with their corresponding magnetic intensities. One is a transverse sinusoidal spin arrangement and the other a simple spiral model. (Spiral spin structures are discussed in § 7.6.) In both models the direction of propagation of the periodic spin structure is (1, 0, 0).

Shirane and Takei showed that if there was a third satellite line, as required by the antiphase domain model, then its intensity was less than 1 per cent of the first ones. Moreover, the absence of the detailed structure in the (100) peak in their diffraction pattern taken at 78 K, Fig. 7.17, shows that at this temperature the spins are parallel to the direction of propagation of the periodic spin structure. These two pieces of evidence led Shirane and Takei to postulate what is now believed to be the correct low-temperature spin configuration of Cr, namely a modulation in the magnitude of the spins whose direction is parallel to the propagation vector.

At 121 K the spins become perpendicular to the propagation of the periodic spin arrangement, and hence the observed modified (100) reflection is observed by Shirane and Takei at 140 K. That is to say we have the transverse sinusoidal spin arrangement of Fig. 7.16. This spin configuration above the low-temperature transition of Cr was verified by an experiment by Brown *et al.*, who, from an analysis of the polarization of the scattered neutrons, were able to rule out a spiral structure for the spins. When the incident polarization was made parallel to the (010) cube axis and the Bragg reflection $(0, 0, 1-\delta)$ observed, the flipping ratio R was found to be $1{\cdot}20 \pm 0{\cdot}03$, which they interpreted to imply domains of transverse sinusoidal spins parallel to (1, 0, 0) and (0, 1, 0), in the ratio 1 : 1·2.

We do not consider in this book the origin of the spin configuration observed in chromium; the various theories are discussed by Arrott.†

7.4. Ionic crystals

All the examples we consider in this section are antiferromagnets. The primary reason for this is, simply, that many magnetic ionic crystals are antiferromagnets; and some of the ionic ferromagnets that are known cannot at present be studied by neutron scattering, e.g. EuO.

(*a*) *Manganese ditelluride*

As a first example let us discuss manganese ditelluride. Fig. 7.18 shows the neutron diffraction pattern found by Hastings *et al.* for manganese ditelluride‡ ($MnTe_2$) at 4·2 K. This crystal is known to possess a pyrite

† A. Arrott, *Magnetism* (ed. Rado and Suhl) **IIB**, 295 (1966).
‡ J. M. Hastings, N. Elliott, and L. M. Corliss, *Phys. Rev.* **115**, 13 (1959).

structure. Hastings *et al.* proved that the magnetic unit cell shown in Fig. 7.19 is consistent with the neutron diffraction pattern. Let us demonstrate that this is so.

The magnetic unit cell in Fig. 7.19 is based on a f.c.c. lattice in which eight of the twelve nearest neighbours of a manganese ion have a spin

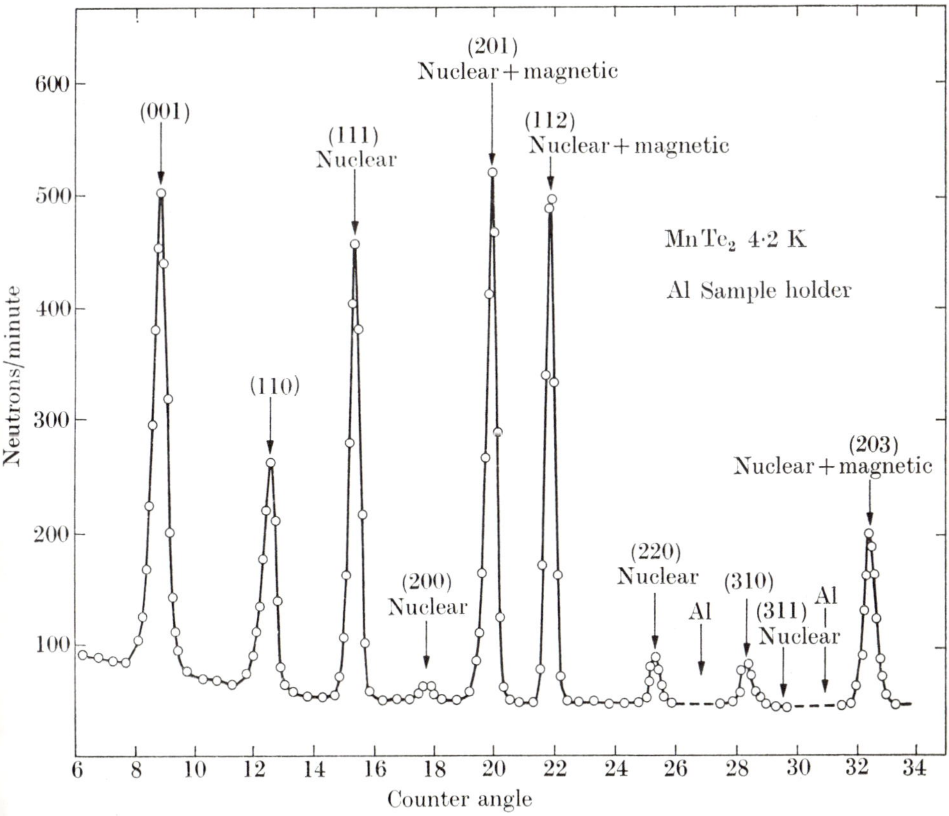

FIG. 7.18. Neutron diffraction pattern of MnTe$_2$.

orientated antiparallel to it and the remaining four possess spin parallel to it. There are four atoms within the cell, two of 'up' spin and two of 'down' spin, but it is not possible to determine the direction of the spin moments within the plane. The magnetic unit cell structure factor $F_M(\boldsymbol{\tau})$ is therefore simply

$$F_M(\boldsymbol{\tau}) \propto \sum_d \sigma_{\mathbf{d}} \exp(\mathrm{i}\boldsymbol{\tau}.\mathbf{d})$$
$$= 1-\exp \mathrm{i}\pi(t_2+t_3)-\exp \mathrm{i}\pi(t_1+t_3)+\exp \mathrm{i}\pi(t_1+t_2), \quad (7.52)$$

where the reciprocal lattice vectors $\boldsymbol{\tau}$ are those for simple cubic lattice of cell side a, i.e.

$$\boldsymbol{\tau} = \frac{2\pi}{a}(t_1, t_2, t_3) \quad (t_i \text{ any integers}). \tag{7.53}$$

Those vectors $\boldsymbol{\tau}$ that coincide with the reciprocal lattice vectors for a f.c.c. lattice will give $F_{\mathrm{M}}(\boldsymbol{\tau}) = 0$ because for each of these reciprocal lattice vectors $\exp i\boldsymbol{\tau}.\mathbf{d} = 1$. Hence there is no magnetic scattering at

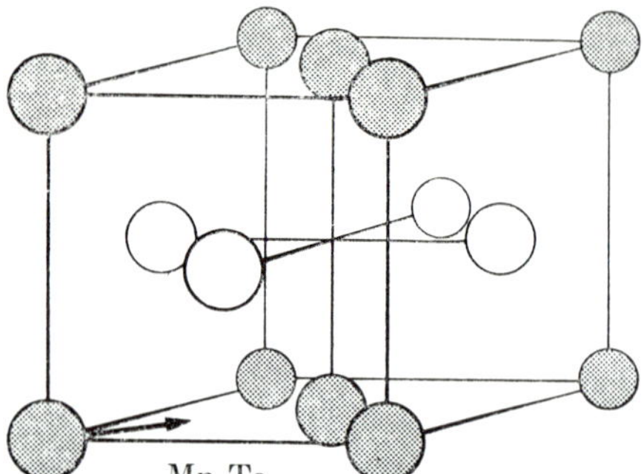

FIG. 7.19. Magnetic structure of $MnTe_2$. Positive and negative orientations of the dipoles are shown by black and white spheres. The direction of the magnetic axis relative to the crystallographic axes is shown by an arrow. For $MnTe_2$ the spin direction lies in the ferromagnetic sheets, but orientation within the plane is unspecified.†

the nuclear reflections arising from the nuclei of the manganese ions. It is easily verified that $F_{\mathrm{M}}(\boldsymbol{\tau})$ is non-zero only for those $\boldsymbol{\tau}$ given in (7.53) for which either

(a) t_1 and t_2 are both even and t_3 odd

or (b) t_1 and t_2 are both odd and t_3 even,

and that this indexing corresponds to that given in Fig. 7.18.

There are several magnetic reflections shown in Fig. 7.18 that coincide with nuclear reflections. As we have just shown, the nuclear scattering cannot arise from the nuclei of the manganese ions but comes, of course, from the scattering from the nuclei of the anions. The magnetic contributions to the mixed peaks were obtained by subtracting the room temperature pattern with the appropriate Debye–Waller correction.

(b) *Manganese fluoride*

As a second example let us consider the neutron diffraction pattern for MnF_2 obtained by Erickson.‡ MnF_2 has a lattice such that the Mn

† Mössbauer measurements support a non-collinear spin arrangement which is consistent with the present neutron measurements. Hastings *et al.*, *Phys. Rev.* **1**, 3209 (1970).

‡ R. A. Erickson, *Phys. Rev.* **90**, 779 (1953).

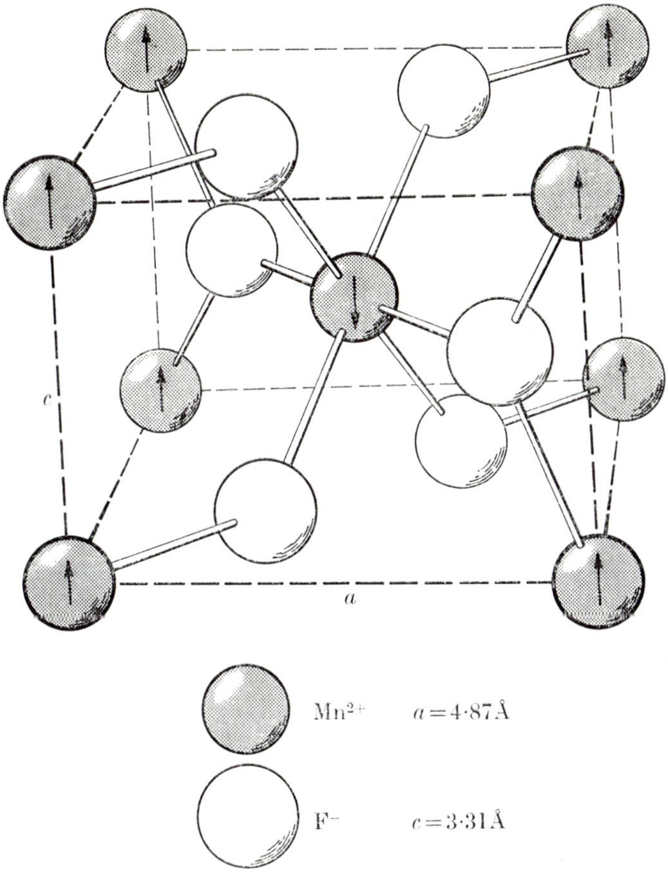

FIG. 7.20. Unit cell of MnF_2.

atoms form a body-centred tetragonal lattice as shown in Fig. 7.20. The unit cell has a volume a^2c with two Mn and four F in it. The reciprocal lattice is given by

$$\boldsymbol{\tau} = \frac{2\pi}{a}\left(t_1, t_2, \frac{a}{c}t_3\right) \tag{7.54}$$

and the Mn and F occupy the positions:

$$\text{Mn} \quad (0,0,0) \quad \text{and} \quad \tfrac{1}{2}(a,a,c) \tag{7.55}$$

$$\text{F} \quad a(u,u,0),\ a(1-u,1-u,0),\ a(\tfrac{1}{2}+u,\tfrac{1}{2}-u,c/2a),$$

and $$a(\tfrac{1}{2}-u,\tfrac{1}{2}+u,c/2a),$$

where $u \simeq 0{\cdot}31$.

The unit-cell structure factor for nuclear scattering is

$$
\begin{aligned}
F_{\mathrm{N}}(\boldsymbol{\tau}) &= \bar{b}_{\mathrm{Mn}}\{1+\exp i\pi(t_1+t_2+t_3)\}+\\
&\quad +\bar{b}_{\mathrm{F}}[\exp 2\pi i u(t_1+t_2)+\exp 2\pi i(1-u)(t_1+t_2)+\\
&\quad +\exp i\pi(t_1+t_2+t_3)\{\exp 2\pi i u(t_1-t_2)+\exp -2\pi i u(t_1-t_2)\}]\\
&= \bar{b}_{\mathrm{Mn}}\{1+\exp i\pi(t_1+t_2+t_3)\}+2\bar{b}_{\mathrm{F}}[\cos\{2\pi u(t_1+t_2)\}+\\
&\quad +\exp i\pi(t_1+t_2+t_3)\cos\{2\pi u(t_1-t_2)\}].
\end{aligned}
$$

Hence

$$
\begin{aligned}
F_{\mathrm{N}}(\boldsymbol{\tau}) &= 2\bar{b}_{\mathrm{Mn}}+4\bar{b}_{\mathrm{F}}\cos(2\pi u t_1)\cos(2\pi u t_2), && \text{if } t_1+t_2+t_3 \text{ is an even integer,}\\
&= -4\bar{b}_{\mathrm{F}}\sin(2\pi u t_1)\sin(2\pi u t_2), && \text{if } t_1+t_2+t_3 \text{ is an odd integer.}
\end{aligned} \tag{7.56}
$$

If we recall that the reciprocal lattice vectors of a body-centred tetragonal lattice coincide with (7.54) if $t_1+t_2+t_3$ is even, then we understand why in (7.56) the scattering from the nuclei of the manganese ions occurs only when this condition is satisfied. The nuclei of the fluorine ions contribute to all reflections except those of the form $(0, t_2, t_3)$ with t_2+t_3 odd.

The first few reciprocal lattice vectors with their multiplicity and structure factors are

$\boldsymbol{\tau}$	$Z(\boldsymbol{\tau})$	$F_{\mathrm{N}}(\boldsymbol{\tau})$
(1, 0, 0)	4	0
(0, 0, 1)	2	0
(1, 1, 0)	4	$2\bar{b}_{\mathrm{Mn}}+4\bar{b}_{\mathrm{F}}\cos^2(2\pi u)$
(1, 0, 1)	8	$2\bar{b}_{\mathrm{Mn}}+4\bar{b}_{\mathrm{F}}\cos(2\pi u)$

Therefore we expect the first nuclear peak to occur at an angle corresponding to the (1, 1, 0) vector because, for this particular symmetry, the first two non-zero reciprocal lattice vectors happen to have zero unit-cell structure factors.

In addition to the expected nuclear peaks at angles corresponding to the (1, 1, 0), (1, 0, 1), and other vectors, Erickson observed peaks at angles corresponding to the (1, 0, 0), (1, 1, 1), (2, 1, 0), and (2, 0, 1) vectors (Fig. 7.21). The first point to note is that the magnetic peaks can be indexed on the same unit cell as the nuclear peaks, i.e. peaks do not appear at points other than those given by (7.54). The magnetic unit cell is therefore of the same size as the chemical unit cell. There are only two Mn atoms in the unit cell; these must have opposite spins. Hence the spin pattern is that with 'up' spins on the corner atoms and 'down' spins on the body centres. The magnetic structure factor is proportional to

$$F_{\mathrm{M}} \propto 1-\exp i\pi(t_1+t_2+t_3) = 0, \quad \text{if } t_1+t_2+t_3 \text{ is an even integer}$$
$$= 2, \quad \text{if } t_1+t_2+t_3 \text{ is an odd integer.} \qquad (7.57)$$

We now notice that we expect magnetic peaks to appear at both (100) and (001). But actually no (001) peak is observed. It follows that the

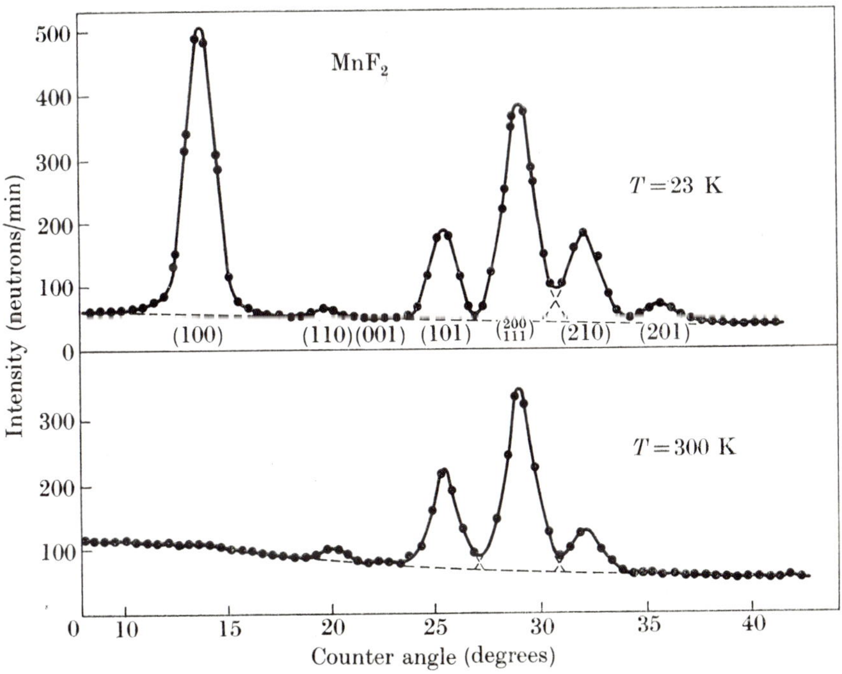

FIG. 7.21. Neutron diffraction patterns for MnF$_2$ in the paramagnetic state (300 K) and in the antiferromagnetic state (23 K). The unit cells for antiferromagnetic and nuclear scattering are of the same size.

spins must be oriented in this direction, i.e. $\tilde{\eta} = (0,0,1)$, so then the orientation factor $\{1-(\tilde{\tau}.\tilde{\eta})^2\}$ is zero for this peak. Hence, for MnF$_2$ we can easily deduce both the spin configuration and orientation. The magnetic peaks disappear at the Néel temperature of 75 K. The intensity of the peaks fits very closely to (7.26) with $\langle \hat{S} \rangle$ equal to 5/2 for very low T and g equal to 2 (no orbital contribution). We conclude that for this model of MnF$_2$ there are three types of Bragg reflection:

(a) $(0, t_2, t_3)$ reflections with t_2+t_3 odd will be purely magnetic,

(b) all other reflections with $t_1+t_2+t_3$ odd will be mixed nuclear and magnetic, the nuclear scattering coming solely from the nuclei of the F ions, and

(c) reflections with $t_1+t_2+t_3$ even will be purely nuclear.

Let us now seek the conditions required of an antiferromagnet for it to be sensitive to a study with polarized neutrons. If the antiferromagnet has a collinear spin configuration and it is arranged so that $\tilde{\boldsymbol{\varkappa}}.\tilde{\boldsymbol{\eta}} = 0$ and that **P** coincides with the spin axis, then the flipping ratio R for a particular Bragg reflection will differ from unity provided (i) neither the magnetic nor the nuclear structure factors for the reflection are zero, and (ii) the two structure factors are not out of phase by $\frac{1}{2}\pi$. The MnF_2 structure satisfies both these requirements. The fundamental aspect is a lack of translational symmetry within the unit cell of the environment of the magnetic ions. For MnF_2 the environment of the Mn^{2+} ions differs by a rotation of $\frac{1}{2}\pi$ about the (001) crystal axis and it is the coherence between the nuclear scattering from the non-magnetic F ions and the magnetic scattering from the magnetic Mn^{2+} ions (b type reflections) that produces the polarization dependence of the cross-section.

Even if the above requirements are satisfied it is still necessary to be able to prepare a sample with an overall predominance of one antiferromagnetic domain over the other. This requirement for an almost single antiferromagnetic domain single crystal obviously restricts the usefulness of the polarized-beam technique with antiferromagnets. Nathans *et al.*† showed that a single crystal of MnF_2 cooled to helium temperatures turned out to be predominantly of one domain type. This is determined by a measurement of the polarization ratio of the (210) reflection for which the nuclear and magnetic structure factors are known, eqns (7.56) and (7.57).

Nathans *et al.* also found that the flipping ratio R differed from unity for some of the supposedly pure nuclear reflections in MnF_2 (c type reflections). The measurement of so-called 'forbidden' magnetic scattering demonstrates the presence of unpaired electron spin on the ligand F ions. The measured structure factors for the 'forbidden' reflections in MnF_2 are shown in Fig. 7.22. The presence of unpaired electron spin on the ligand ions in MnF_2 has also been demonstrated by measuring the hyperfine field at the F nuclei by nuclear magnetic resonance (NMR). At $T \sim 0$ K the NMR measurements give a hyperfine field of ~ 40 kG. Now if the magnetic field at a F nucleus was due solely to the dipole

† R. Nathans, H. A. Alperin, S. J. Pickart, and P. J. Brown, *J. appl. Phys.* **34**, 1182 (1963).

field of the Mn^{2+} ions the observed field would be $\sim$ 12·5 kG, the major part of which is from the nearest-neighbour magnetic ions. The observed additional magnetic field, called the transferred hyperfine field, is believed to be due to the interaction of the unpaired electron spin on the F^- ion with the F nucleus. In essence, what is being observed in both the neutron and NMR experiments is the change in the ions from their state

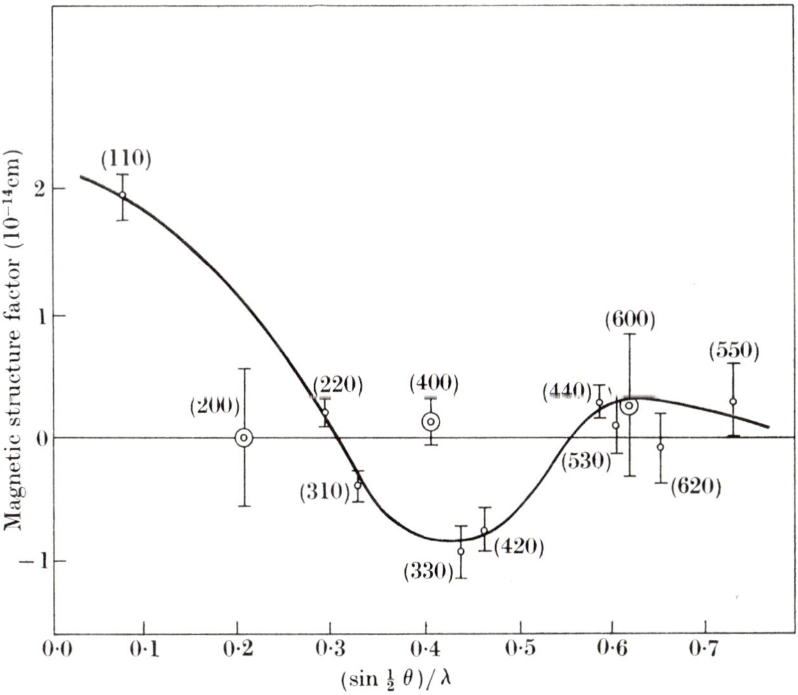

Fig. 7.22. The structure factor for the 'forbidden' magnetic reflections in MnF_2. The points enclosed by a circle are expected to have zero amplitude. R. Nathans *et al.*, *J. app. Phys.* **34**, 1182 (1963).

in the truly ionic state, i.e. covalency effects. Clearly an understanding of these experiments is fundamental to our understanding of magnetic ionic crystals and we therefore consider in more detail in the following section the role of covalency in elastic magnetic scattering.

7.5. Covalency effects in neutron diffraction from ferromagnetic and antiferromagnetic salts†

We confine our attention in a study of covalency to transition-metal ions, which possess unpaired electrons in d orbits, and we shall for the

† J. Owen and J. H. M. Thornley, *Rep. Prog. Phys.* **29**, 675 (1966), and J. Hubbard and W. Marshall, *Proc. phys. Soc.* **86**, 561 (1965).

most part consider the ligand ions to be either F^- or O^{2-}, each of which has the configuration $1s^2 2s^2 2p^6$. In particular we are interested in the two antiferromagnetic salts MnF_2, which we briefly discussed in the previous section, and NiO. Manganese in a purely ionic substance (no covalency) would be in the configuration $Mn^{2+}:3d^5$ and nickel in the configuration $Ni^{2+}:3d^8$. Hence, in a purely ionic crystal the atomic form factor $F(\varkappa)$ would closely resemble the Fourier transform of a $3d$ orbital. In a real crystal the atomic form factor will deviate from this because of covalency, and it is our task to investigate just how the form factor is modified. We can readily see what the major modifications will be without resorting to any particular model calculation.

First, we can neglect in discussing the magnetic state of a transition-metal ion in a crystal both its $3s$ and $3p$ orbitals, because of their low energy. We also neglect the $4s$ and $4p$ orbitals. As for the ligand ions, we can surely neglect their $1s$ orbitals.

In a complex consisting of a single magnetic ion and its neighbouring ligand ions there is a transfer of electrons from the ligands to the magnetic ion, mainly p-electron transfer to the $3d$ orbitals, creating an unpaired spin density on the ligand ions. Because of Pauli's principle, the spin of the transferred electron must be opposite to that of the single electron which occupies the $3d$ state. Hence, for a single $3d$ orbital, the (back) transfer results in unpaired spin in a $2p$ state of the ligand that is parallel to the unpaired spin of the magnetic ion. In this instance the moment of the magnetic ion is decreased as a consequence of covalency. Neighbouring magnetic ions in a simple antiferromagnet have opposite spin moments and thus the net unpaired spin density created on the intervening ligand ions can be cancelled, depending on the symmetry of the crystal structure. Since the cross-section for magnetic Bragg scattering is proportional to the square of the magnetic moment associated with the magnetic ions, covalency reduces the intensity of the peaks relative to those from a purely ionic crystal. Now the spatial distribution of unpaired spin is contained in the form factor. If the structure possesses the symmetry necessary to effect cancellation of the unpaired spin on the ligand ions, then it follows that this unpaired spin is 'lost' to the neutrons and the form factor is uniformly diminished with respect to its free ion value. However, if a net spin density resides on the ligands, as it does for MnF_2, then because the distance between the magnetic and ligand ions is large compared to the mean radius of a $3d$ orbit and because the $2p$ orbitals themselves are diffuse, the unpaired spin will manifest itself primarily in a peak about $\varkappa = 0$ in the atomic form factor.

Whether or not the bulk of this peak lies within the first Bragg reflection for an antiferromagnet depends on the size of the unit cell relative to the distance between the magnetic and ligand ions. For ferromagnets, of course, there is no question of cancellation of unpaired spin created on the ligand ions.

To justify these physical arguments and obtain a more detailed understanding of covalency effects in neutron diffraction, and also to enable us to tie up the results obtained from neutron experiments with those from resonance experiments, we must consider a model in some detail. We choose to base our model calculations on the molecular orbital method. To see how this works we start by looking at a linear chain consisting of alternate single magnetic and ligand orbitals as shown in Fig. 7.23. Each d orbit contains just one electron, and these are taken

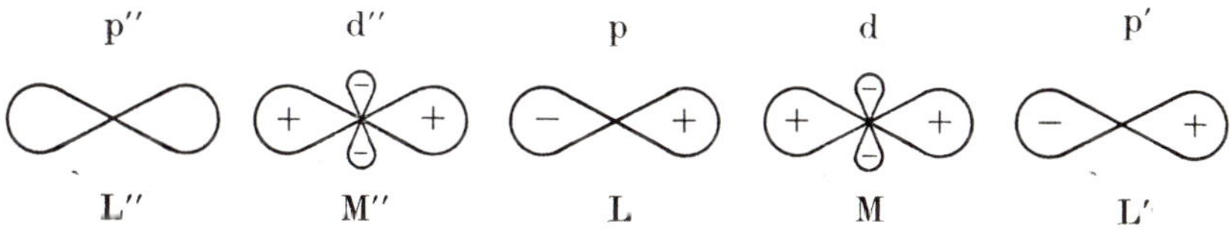

FIG. 7.23. Orbitals of a linear chain antiferromagnet.

to be antiferromagnetically ordered in the chain, while each p orbit contains two electrons. In the approximation of taking a crystal to be ionic, electrons leave the magnetic ion to fill the p orbital of the ligands, so the latter are lower in energy than the d orbital, typically by about 10 eV.

Fig. 7.24 shows the overlap between neighbouring p and d wave functions for Mn^{2+}–F^- bond. From this diagram it is clear that a wave function of the form

$$\psi_B(\mathbf{r}) = N_B\{p(\mathbf{r})+Bd(\mathbf{r})\}: \text{bonding} \tag{7.58}$$

piles up electronic charge midway between the nuclei, whereas a wave function of the form

$$\psi_A(\mathbf{r}) = N_A\{d(\mathbf{r})-Ap(\mathbf{r})\}: \text{antibonding} \tag{7.59}$$

reduces the charge between the nuclei. In fact the wave function (7.59), the anti-bonding wave function, possesses a node between the two nuclei at which it has zero amplitude. The normalization constants N_A and N_B are given by

$$N_A^{-2} = 1+A^2-2AS,$$
$$N_B^{-2} = 1+B^2+2BS,$$

where S is the overlap integral

$$S = \langle d|p\rangle.$$

Because the antibonding and bonding orbitals must be orthogonal, we have the relationship

$$A = B+S-ABS \simeq B+S.$$

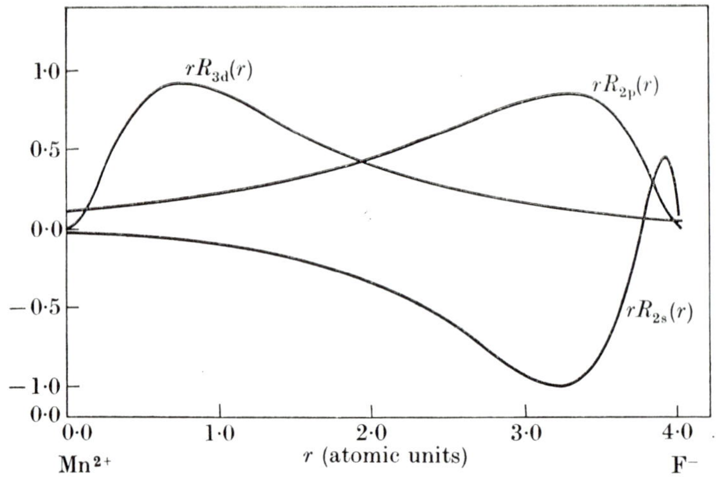

FIG. 7.24. The radial part of the Hartree–Fock free-ion wave functions of Mn^{2+} and F^-. The ions are placed at the interatomic spacing in MnF_2.

Note that if $B = 0$, which means that there is no *covalent transfer* of spin onto the ligand ions,

$$\psi_A = \frac{1}{\sqrt{(1-S^2)}}\{d(\mathbf{r})-S\,p(\mathbf{r})\}$$

and $$\psi_B = p(\mathbf{r}).$$

Thus there remains an *overlap transfer* of spin, which is here seen to be a consequence of orthogonalization of the two states ψ_A and ψ_B. The magnitude of the overlap transfer is given by the square of the overlap integral S. If, for any reason, it was considered necessary in a calculation to include the core states of the magnetic ion then, because they have a very low energy, it would be sufficient to describe their presence simply in terms of overlap transfer.

The bonding orbital (7.58) must lie lower in energy than the antibonding orbital (the bonding orbital is predominantly p-like and the p orbit lies lower in energy than the d) and consequently it can be assumed that the spin up and spin down bonding orbitals are completely full. Thus the bonding orbitals have no unpaired spin associated with them and we therefore need consider only the antibonding orbitals in connection with neutron scattering and resonance experiments. In this way we describe covalency as the spreading of the magnetic orbital on to the ligands, the actual process being a transfer of down spin holes to the ligands.

For the linear chain given in Fig. 7.23 the antibonding orbital to be associated with the ion M is

$$\psi_M(\mathbf{r}) = N_\sigma\{d(\mathbf{r})-A_\sigma p(\mathbf{r})+A_\sigma p'(\mathbf{r})\}. \tag{7.60}$$

(The letters σ and π denote zero and unit angular momentum about the bond axis, respectively.) In (7.60), N_σ is determined by the normalization condition

$$N_\sigma^{-2} = 1-4A_\sigma S_\sigma+2A_\sigma^2, \tag{7.61}$$

where S_σ is the overlap integral between the p and d orbits. The spin density associated with M is

$$\begin{aligned} D(\mathbf{r}) &= \psi_M^2(\mathbf{r}) \\ &\simeq N_\sigma^2 d^2(\mathbf{r})-2A_\sigma N_\sigma^2 d(\mathbf{r})\{p(\mathbf{r})-p'(\mathbf{r})\}+A_\sigma^2 N_\sigma^2\{p^2(\mathbf{r})+p'^2(\mathbf{r})\}, \end{aligned} \tag{7.62}$$

where we have neglected the overlap between p and p'.

For convenience of notation in this section we use unnormalized form factors defined by analogy with

$$F(\varkappa) = \int d\mathbf{r}\, D(\mathbf{r})\exp(i\varkappa.\mathbf{r}). \tag{7.63}$$

Correct to second order we have from (7.61) and (7.62)

$$\begin{aligned} D(\mathbf{r}) = d^2(\mathbf{r})(1+4A_\sigma S_\sigma-2A_\sigma^2)-2A_\sigma d(\mathbf{r})\{p(\mathbf{r})-p'(\mathbf{r})\}+ \\ +A_\sigma^2\{p^2(\mathbf{r})+p'^2(\mathbf{r})\}. \end{aligned} \tag{7.64}$$

The first term in (7.64) involves $d^2(\mathbf{r})$ and is confined to the parent magnetic ion M; the second term is an overlap density and is also confined to the immediate vicinity of M; but the third term gives a density entirely on the ligands and is equally distant from other ions as from the parent ion. The spin density given by M'' is similar to that by M, but in an antiferromagnet is of opposite sign. In particular, we notice that the moment density $A_\sigma^2 p^2(\mathbf{r})$ due to M is exactly cancelled by an equal and opposite contribution from M'. It follows that Bragg scattering can detect only the first two terms of (7.64) and, in effect, in the neighbourhood of M, defined as extending from the nucleus L, through M, to the nucleus L', we may take

$$\begin{aligned} \mathscr{D}(\mathbf{r}) &= \sum_{\mathbf{d}} \sigma_d D(\mathbf{r}-\mathbf{d}) \\ &\simeq d^2(\mathbf{r})(1+4A_\sigma S_\sigma-2A_\sigma^2)-2A_\sigma d(\mathbf{r})\{p(\mathbf{r})-p'(\mathbf{r})\} \end{aligned} \tag{7.65}$$

and use a similar expression with opposite sign in the neighbourhood of M''. The expressions (7.64) and (7.65) should be contrasted with one another and with the conventional result that ignores covalency and takes $\mathscr{D}(\mathbf{r})$ as simply $d^2(\mathbf{r})$.

We notice that the total density in (7.64), and also the conventional result, is unity, whereas the total density in (7.65) is $1-2A_\sigma^2$. The 'lost' density of $2A_\sigma^2$ has been exactly cancelled off with similar contributions of opposite sign coming from other magnetic ions. Thus, the Bragg scattering from such an antiferromagnet is that appropriate to a smaller spin density per ion than we would naïvely have supposed.

Furthermore, the form factor is not that appropriate to d functions alone but involves the overlap density, as indicated in (7.65). This overlap density makes the form factor appear flatter than would have been expected. To see this it is useful to rearrange (7.65) in the form

$$\mathscr{D}(\mathbf{r}) = d^2(\mathbf{r})(1-2A_\sigma^2)+2A_\sigma\{2S_\sigma d^2(\mathbf{r})-d(\mathbf{r})p(\mathbf{r})+d(\mathbf{r})p'(\mathbf{r})\}.$$

The first term corresponds to a moment of $(1-2A_\sigma^2)$ distributed in the d orbital and the second term represents a correction to the moment that is positive near the centre of M and negative near the boundaries of M, and overall has zero moment. This second term has a form factor that is zero at $\varkappa = 0$ and passes through a maximum as $\varkappa$ increases. The total form factor is, therefore, slightly larger than $(1-2A_\sigma^2)d^2(\mathbf{r})$, except at $\varkappa = 0$, and this makes the form factor appear flatter.

A third consequence follows from (7.65). Because $d^2(\mathbf{r})$ falls off rapidly with $\mathbf{r}$, the spin density near the ligand nuclei must be dominated by the second term of (7.65) and this is negative. Hence, the spin density, which is positive in the immediate neighbourhood of M, must pass through zero and become negative at some point intermediate between M and the neighbouring ligand L. By symmetry the spin density is zero at the ligand nucleus itself. Hence, each magnetic ion of positive spin is surrounded by a region of negative spin and vice versa. This last conclusion is most unexpected because naïvely we would have argued that as we moved along the chain from the M'' nucleus to the M nucleus we would have only one node in the moment density and that this would be at nucleus L. Instead we conclude there are three nodes, one at nucleus L and one each side of it.

This last conclusion, and also the conclusion that the form factor should appear flatter than expected, depends on the sign of A_σ. We now turn to a discussion of three-dimensional systems and verify the choice of sign for A_σ.

The three-dimensional system we consider consists of a transition-metal ion surrounded by a regular octahedron of ligand ions. This is the magnetic complex most commonly found. Earlier in this chapter we noted that the fivefold degenerate $3d$ wave functions split in a cubic environment into two groups, the orbitals of one transforming as the E_g representation of the cubic group O and the orbitals of the other like

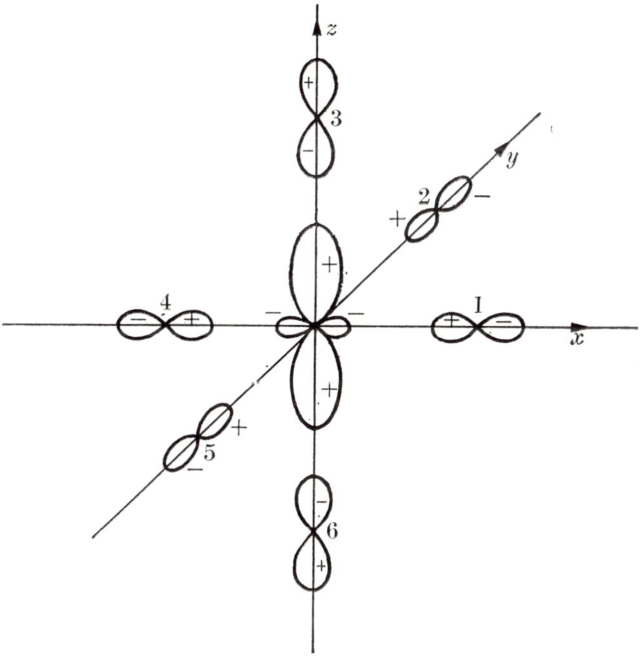

FIG. 7.25. Schematic diagram of the antibonding wave function ψ_{z^2} (eqn (7.66)).

T_{2g}. In the present case it is necessary to find also the combinations of the ligand orbitals that transform according to these representations. The required combinations are shown schematically in Figs. 7.25 and 7.26, and the corresponding orbitals are

$$\psi_{z^2} = N_\sigma[d_{z^2} - A_\sigma\{p_3^\sigma + p_6^\sigma - \tfrac{1}{2}p_1^\sigma - \tfrac{1}{2}p_2^\sigma - \tfrac{1}{2}p_4^\sigma - \tfrac{1}{2}p_5^\sigma\} -$$
$$- A_s\{s_3 + s_6 - \tfrac{1}{2}s_1 - \tfrac{1}{2}s_2 - \tfrac{1}{2}s_4 - \tfrac{1}{2}s_5\}] \qquad (7.66)$$

and

$$\psi_{zx} = N_\pi[d_{zx} - A_\pi\{p_1^z + p_3^x - p_4^z - p_6^x\}], \qquad (7.67)$$

where, for example, p_3^σ stands for that p orbital of ligand number 3 and has a positive lobe pointing towards the magnetic ion. These equations define the parameters A_σ, A_π, and A_s. The normalization factors N_σ and N_π are given by

$$N_\sigma^{-2} = 1+3A_\sigma^2-6A_\sigma S_\sigma+3A_s^2-6A_s S_s,$$
$$N_\pi^{-2} = 1+4A_\pi^2-8A_\pi S_\pi, \qquad (7.68)$$

where S_σ, S_π, and S_s are the overlap integrals, of positive sign, defined by

$$S_\sigma = \langle d_{z^2} \,|\, p_3^\sigma \rangle, \qquad S_\pi = \langle d_{zx} \,|\, p_3^x \rangle, \qquad S_s = \langle d_{z^2} \,|\, s_3 \rangle. \qquad (7.69)$$

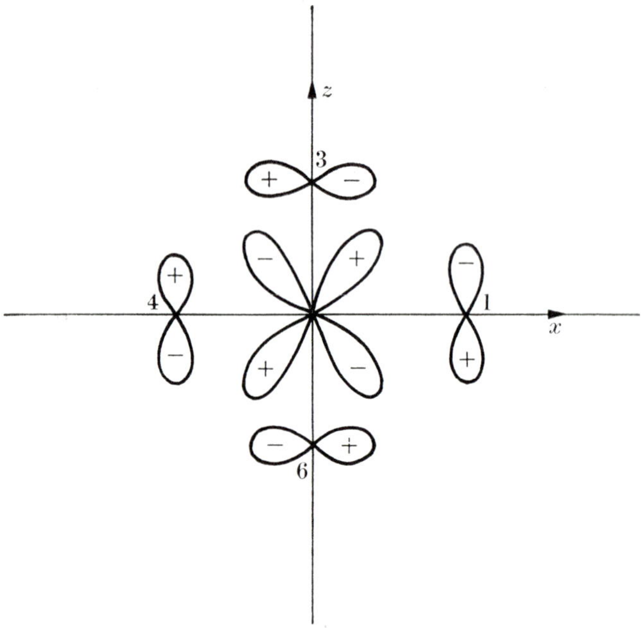

FIG. 7.26. Schematic diagram of the antibonding wave function ψ_{zx} (eqn (7.67)).

In those complexes with ligands, such as F^-, O^{2-}, Cl^-, the parameters A_σ, A_s, A_π must be positive by the following argument. Because the p orbitals are lower in energy than the d orbitals, the bonding orbitals must be predominantly ligand p orbitals but will have some admixture of d orbitals; in order to get bonding the electron charge must be increased in the region between the nuclei, hence the admixture must be of a sign to give addition of d and p orbitals in the overlap region. Because the antibonding orbital must be orthogonal to the bonding orbital it follows that, for such cases, the antibonding orbital must give subtraction of d and p orbitals in the overlap region and, furthermore, that the antibonding orbital must be predominantly a d orbital; hence A_σ, A_s, A_π are positive and small with the sign choices indicated by (7.66) and (7.67).

In those complexes with unsaturated ligands such as CN^- the sign of the parameters A_σ, A_s, A_π is not so clear. According to Orgel† the main admixture of the ion d_{zx} type orbitals is to ligand p_π orbitals, which are empty in the absence of covalency. In this case the d orbital has lower energy than the important p orbital and so it is the bonding orbital that is predominantly d orbital; in this case, therefore, the magnetically active orbitals are as shown by (7.67) but with A_π negative.

The wave functions (7.66) and (7.67) are two of the five antibonding orbitals, the other three can be written down in a similar way. The distribution $D(\mathbf{r})$ associated with orbital ψ_{z^2} is simply $|\psi_{z^2}|^2$ if it is singly occupied, and zero if it is empty or doubly occupied. The total distribution $D(\mathbf{r})$ can be obtained by placing electrons one by one in successive energy levels and summing up the densities due to the unpaired electrons.

Antiferromagnets and ferromagnets are made up of many such complexes and to a good approximation we can take the total spin density as the sum of that due to each individual complex.

For three-dimensional antiferromagnets of the rock-salt type (MnO, NiO), the modification of the spin density by covalency follows exactly the argument for the linear chain. The formulae required follow from (7.66) and (7.67). We shall not write them all down because they are exactly analogous to the linear chain formulae given above. Three conclusions, analogous to those we derived for the linear chain, are as follows.

(1) The absolute intensities are weaker than a simple model neglecting covalency would predict. The effective moment that can be seen is written down by analogy with the linear chain discussion. The moment associated with Bragg scattering from an E_g orbital is

$$1-3A_\sigma^2-3A_s^2, \tag{7.70}$$

where A_σ and A_s are defined by (7.66). The moment associated with Bragg scattering from a T_{2g} orbital is

$$1-4A_\pi^2, \tag{7.71}$$

where A_π is defined by (7.67).

(2) Although the absolute intensities of the Bragg peaks are low, as $\boldsymbol{\tau}$ increases the form factor falls off more slowly than would be expected. The 'overlap density' that causes this is exactly analogous to that discussed for the linear-chain model. This conclusion depends on the sign of A_σ, A_s, and A_π. Therefore, for cyanide complexes where we expect

† L. E. Orgel, *J. chem. Phys.* **23**, 1819 (1955).

the opposite sign for A_π, the form factor should fall off more rapidly than expected. However, in a cyanide complex the bonding is so strong that the atomic orbitals themselves may be seriously perturbed and so obscure this effect.

(3) Moving along any of the three cube directions, the spin density does not change sign simply according to the sign of the moment on the nearest ion: there are nodes at points other than those dictated by symmetry and the variation of the spin density along and near to these cube directions is as described in the linear-chain model. This result concerning nodes of the spin density may be sensitive to other effects, and is also dependent on the sign of A_σ, A_s, A_π.

So far we have discussed in detail only very simple antiferromagnets where the effects of covalency could be seen quite easily. In the more general case the exact cancellation of that spin density entirely on the ligands does not take place: for example, in the ferromagnet no cancellation at all can take place and, in an antiferromagnet such as MnF_2, the environment of each F^- ligand is asymmetrical, so exact cancellation would not be expected. To discuss the more general case we must, therefore, use the full form for $D(\mathbf{r})$, i.e. the three-dimensional analogue of (7.64).

In the full form for $D(\mathbf{r})$, the terms additional to those we considered previously are those entirely on the ligands. We notice that this additional spin density is at such a large distance from the central ion that the form factor for it must have a strong and sharp peak in the forward direction. There are other peaks in the form factor of this ligand spin density but because the ligand wave functions are themselves widespread these peaks are much weaker than the forward peak. In effect, therefore, except near the forward direction, the intensities and form factors should behave like those of the very simple antiferromagnets. In particular the absolute intensities measured for all Bragg peaks except those very close to the forward direction should appear low compared to that expected on a model that neglected covalency. The form factors measured will appear to extrapolate to the values (7.70) or (7.71), or a suitably weighted mean, as appropriate.

However, Bragg scattering can, in principle, show up two other effects in the general case, which were absent for the very simple antiferromagnets. First, the spin density on the ligands gives a sharp forward peak to the form factor and for some crystals the unit cell may be large enough to give Bragg peaks even at angles close to the forward direction: this would make the forward peak directly observable by Bragg scattering.

Secondly, the form factor of the ligand density is not entirely concentrated near the forward direction, it has a small value elsewhere and therefore, in principle, can be detected at these larger angles. In practice this detection would be done most easily by observing the asymmetry in the form factor produced by the departure from spherical symmetry of the ligand spin density or by observing Bragg scattering from peaks which, in the absence of covalency, would be forbidden (Nathans *et al.*).†

In the discussion given above, we ignored another factor that makes antiferromagnetic diffraction peaks low, namely, the zero-point motion. The deviation of the mean spin from the fully aligned value is small and spin-wave theory does not give a thoroughly good estimate of it (see Chapter 9). When more accurate experiments are available, this factor will have to be taken into account.

We have argued above that the main effect of covalency is to reduce the absolute value of the form factor by factors (7.70), (7.71), or some suitable mean, as appropriate. The order of magnitude of these factors can be estimated from the values of A_σ, A_s, and A_π, which are known for a few cases.

In $KNiF_3$ the spin density of the Ni^{2+} ions is associated with the E_g orbitals and the factor (7.70) is appropriate. The parameters A_σ^2 and A_s^2 have been determined by Shulman and Knox‡ by observing the NMR of the F^{19} nucleus. They find

$$\left.\begin{aligned} A_\sigma^2 &= 3{\cdot}7\times 10^{-2} \\ A_s^2 &= 0{\cdot}5\times 10^{-2} \end{aligned}\right\} \quad (\text{for } Ni^{2+} \text{ in } KNiF_3). \tag{7.72}$$

The reduction is, therefore, 0·874; i.e. the form factor should be low by 13 per cent and the scattering intensity by 26 per cent. Absolute intensity measurements for Bragg scattering from a Ni^{2+} salt have been made by Alperin,§ who examined NiO. The Shulman and Knox results are those appropriate for F^- ligands: we would expect the parameters to be similar but possibly a little larger for the O^{2-} ligands in NiO. Alperin observed that extrapolating the form factor to the forward direction indicated a moment of $1{\cdot}81\pm0{\cdot}20$; typically we could expect Ni^{2+} to have a moment of 2·2 so the observed reduction factor is $0{\cdot}82\pm0{\cdot}10$. This is in satisfactory agreement with the reduction factor to be expected.

In $KMnF_3$ and in MnF_2 the spin density of the Mn^{2+} ions is associated

† loc. cit, p. 208.

‡ R. Shulman, K. Knox, *Phys. Rev. Lett.* **4**, 603 (1960).

§ H. A. Alperin, *J. phys. Soc. Japan*, **17** BIII, 12 (1962).

with both the E_g and T_{2g} orbitals. The appropriate reduction factor is, from (7.70) and (7.71),

$$\tfrac{1}{5}\{2(1-3A_\sigma^2-3A_s^2)+3(1-4A_\pi^2)\}$$
$$= 1-1{\cdot}2(A_\sigma^2+A_s^2+2A_\pi^2) \quad \text{for } Mn^{2+} \text{ in } MnF_2. \tag{7.73}$$

For these salts A_σ and A_π cannot be measured directly by NMR. Only the difference $A_\sigma^2-A_\pi^2$ can be measured, and Shulman and Knox found this to be small: $A_\sigma^2-A_\pi^2 = (0{\cdot}18\pm0{\cdot}10)\times10^{-2}$. In view of the value they found for A_σ^2 in a Ni salt and the similar value for A_π^2 in a Cr^{3+} salt, Shulman and Knox made the reasonable interpretation that in $A_\sigma^2-A_\pi^2$ there was almost an exact cancellation within experimental error. We can test this interpretation of Shulman and Knox by evaluating (7.73). Using $A_\sigma^2 = A_\pi^2 \simeq 3{\cdot}7\times10^{-2}$ and $A_s^2 \simeq 0{\cdot}5\times10^{-2}$, we find a reduction of 14 per cent in the form factor and 28 per cent in the neutron diffraction intensity. There have been many neutron diffraction experiments on Mn^{2+} salts and in none of them to date has any reduction of scattering intensity been noticed. Certainly, therefore, we may assume that the intensity reduction is considerably smaller than the 28 per cent expected by the argument of Shulman and Knox. A small value for both A_σ^2 and A_π^2 would be consistent with both the NMR and the neutron diffraction experiments on Mn^{2+} salts, but such an explanation is not in line with the results, both by NMR and the neutron diffraction experiments, that in Ni^{2+} and Cr^{3+} the parameters A_σ^2 and A_π^2 are of appreciable magnitude. However, that covalency is indeed small in manganous salts is substantiated by measurements by Fender *et al.*† on MnO; they found

$$A_\sigma^2+A_s^2+2A_\pi^2 = 3{\cdot}5 \text{ per cent.} \tag{7.74}$$

Turning now to the question of the detailed shape of the form-factor curves, Alperin's measurements on NiO led to a form-factor curve that differs considerably from the theoretical curve, calculated from self-consistent wave functions of Freeman and Watson.‡ Indeed, Alperin's experimental form-factor curve corresponds to a $3d$-like charge distribution that is contracted some 17 per cent as compared with the charge distribution predicted from the calculated wave functions.

Now we may note that the overlap charge density given by the second term of (7.65) is of such a shape and sign as to simulate a radial contraction of the d-electron charge distribution given by the first term of (7.65). That the overlap charge simulates a contraction is a consequence

† B. E. F. Fender, A. J. Jacobson, and F. A. Wedgwood, *J. chem. Phys.* **48,** 990 (1968).

‡ A. J. Freeman and R. Watson, *Phys. Rev.* **120,** 1125 (1960) and *Acta crystallogr.* **14,** 27 (1961).

of the fact that we are dealing with an antibonding orbital, as may be seen by sketching the shapes of the bonding and antibonding wave functions.

We may now inquire whether we can explain the deviation of Alperin's measured form-factor curve from the theoretical prediction based upon the free atom wave functions as a consequence of covalency effects. From the charge distribution given by (7.65) generalized to the three-dimensional NiO case, we obtain for the spherically averaged form factor the expression

$$F(\varkappa) = \{1+6A_\sigma S_\sigma+6A_s S_s-3A_\sigma^2-3A_s^2\}F_d(\varkappa)- \\ -6A_\sigma F_{0\sigma}(\varkappa)-6A_s F_{0s}(\varkappa), \qquad (7.75)$$

where $F_d(\varkappa)$ is the spherically averaged form factor for a d electron alone, and $F_{0\sigma}(\varkappa)$ is the spherically averaged form factor corresponding to the overlap between the d_σ and $2p_\sigma$ orbitals, given by

$$F_{0\sigma}(\varkappa) = \int j_0(\kappa r)d_\sigma(\mathbf{r})p_\sigma(\mathbf{r})\,d\mathbf{r}, \qquad (7.76)$$

where $d_\sigma(\mathbf{r})$ is a σ orbital centred on the origin, $p_\sigma(\mathbf{r})$ a $2p_\sigma$ orbital on one of the ligands, and $j_0(x)$ the spherical Bessel function of zero order; F_{0s} is the corresponding overlap form factor between d_σ and $2s$ orbitals. Note that $F_{0\sigma}(0) = S_\sigma$, $F_{0s}(0) = S_s$. If in (7.75) we now neglect the small term in A_s, and take $A_\sigma^2 = 0{\cdot}055$, which we deduce from the observed reduction in absolute intensity by a factor 0·82 as discussed above, then we find that (7.75) gives a good fit to Alperin's experimental curve for the lower-order reflections, but not so good a fit for higher-order reflections.

For the higher-order reflections there are two further small and unrelated corrections that are of importance. One of these is a correction to allow for the fact that part of the moment of the Ni^{2+} ion is orbital and this part has a slightly different form factor; this effect has been discussed by Blume.† The other correction allows for the effect of spin polarization of the fully occupied shells, especially the fully occupied T_{2g} orbitals of the Ni^{2+} ion; this correction has been discussed by Watson and Freeman. These two effects may be allowed for by correcting the d-electron form factors $F_d(\varkappa)$ entering into (7.75). When these corrections have been made we obtain from (7.75) the theoretical form-factor curve of Fig. 7.27. In calculating this curve an adaptation of the Barnet–Coulson method was used to evaluate the two-centre integrals of the type shown in (7.76); the wave functions used are those calculated by Watson for Ni^{2+} and O^{2-}.

† M. Blume, *Phys. Rev.* **124**, 96 (1961).

It will be seen from Fig. 7.27 that the inclusion of covalency effects has brought the theoretical curve into good agreement with Alperin's experimental curve for small $\varkappa$, although there is still some discrepancy for large $\varkappa$, where covalency effects are unimportant. It should be noted

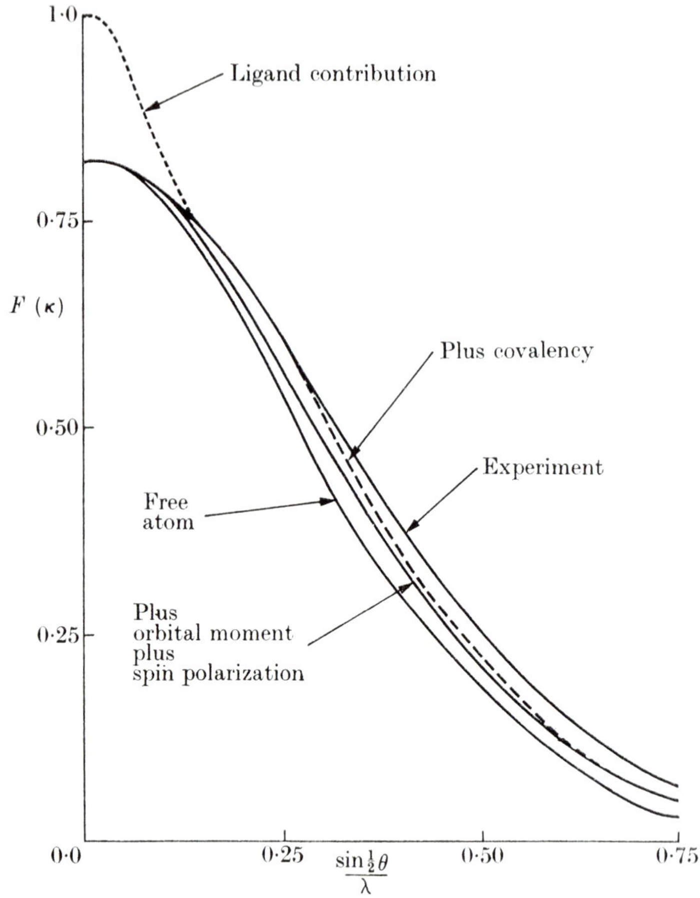

FIG. 7.27. Experimental and theoretical form factors for NiO.

that the latter curve has been scaled by a factor 0·82 to allow for the observed reduction in absolute intensity. We may, therefore, conclude that taking proper account of covalency effects together with the other small corrections mentioned above goes a long way towards explaining Alperin's experimental form-factor curve for NiO.

If we were now to apply the above theory to the case of MnO, we should be in some difficulty. For, contrary to the NiO case, the measured form factors of MnO (and other Mn^{2+} salts) correspond to a charge distribution that is expanded relative to the theoretical charge distribution calculated from Watson's wave functions. Even if we assumed A_σ^2 and A_π^2 to be small for Mn salts, as indicated by the absolute intensity measurements (see discussion above), we could not explain the expansion of the charge distribution.†

Consider a single Ni^{2+} magnetic complex in a diamagnetic host crystal. Since for an isolated Ni^{2+} complex there is no cancellation of spin density on the ligand atoms of the type discussed for antiferromagnets, the spherically averaged diffuse elastic form factor will be given by

$$\begin{aligned} F'(\varkappa) &= \int j_0(\kappa r) D(\mathbf{r})\,d\mathbf{r} \\ &= (1+6A_\sigma S_\sigma+6A_\mathrm{s} S_\mathrm{s}-3A_\sigma^2-3A_\mathrm{s}^2)F_\mathrm{d}(\varkappa)-6A_\sigma F_{0\sigma}(\varkappa)- \\ &\qquad -6A_\mathrm{s} F_{0\mathrm{s}}(\varkappa)+3A_\sigma^2 F_{\mathrm{L}\sigma}(\varkappa)+3A_\mathrm{s}^2 F_{\mathrm{Ls}}(\varkappa), \end{aligned} \tag{7.77}$$

where $D(\mathbf{r})$ is the charge density distribution given by (7.64) and $F_{\mathrm{L}\sigma}(\varkappa)$ and $F_{\mathrm{Ls}}(\varkappa)$ are the spherical averages of the terms arising from the spin-density on the ligands of the complex; $F_{\mathrm{L}\sigma}(\varkappa)$ is given by

$$F_{\mathrm{L}\sigma}(\varkappa) = \Big\langle \tfrac{1}{6} \sum_{\rho} \int L_{\rho\sigma}(\mathbf{r}) \exp(i\boldsymbol{\varkappa}.\mathbf{r})\,d\mathbf{r} \Big\rangle_{\substack{\text{spherical}\\ \text{average}}}, \tag{7.78}$$

with a similar formula for $F_{\mathrm{Ls}}(\varkappa)$, where $L_{\rho\sigma}(\mathbf{r})$ is the charge density of a $2p_\sigma$ orbital on the ligand centre $\boldsymbol{\rho}$, and the $\boldsymbol{\rho}$ sum runs over all six neighbouring ligands and the spherical average is over all directions of $\boldsymbol{\varkappa}$. To a good approximation,

$$F_{\mathrm{L}\sigma}(\varkappa) \simeq \frac{\sin(\kappa R)}{\kappa R} \int j_0(\kappa r) p^2(\mathbf{r})\,d\mathbf{r}, \tag{7.79}$$

where R is the ion–ligand bond length and $p^2(\mathbf{r})$ is the charge density of a $2p$-orbital centred on the origin; there is a similar approximate formula for $F_{\ \mathrm{s}}(\varkappa)$.

The first factor in (7.79) falls off very rapidly with increasing $\varkappa$, so the terms involving $F_{\mathrm{L}\sigma}$, F_{Ls} in (7.77) only contribute for small $\varkappa$, giving $F'(\varkappa)$ a narrow forward peak of width $\sim \pi/R$. It is only in the region of this peak that $F(\varkappa)$ and $F'(\varkappa)$ differ. The curve of Fig. 7.27 labelled 'ligand contribution' gives $F'(\varkappa)$ for Ni^{2+} in an O^{2-} complex. It will be seen that $F'(\varkappa) \to 1$ as $\varkappa \to 0$; this is because for an isolated Ni^{2+} complex the cancellation effect discussed for antiferromagnets does not occur.

It would be valuable to have experimental observations of this forward peak. To obtain evidence for it from Bragg scattering, it would be necessary to use a crystal with a large enough unit cell to have magnetic reflections inside $(\sin\frac{1}{2}\theta)/\lambda \sim 0{\cdot}1$; i.e. there should be several complexes per unit cell. To observe it by diffuse elastic scattering (cf. Chapter 14)

† But see J. Hubbard, J. Hopgood, and D. Rimmer, *Proc. phys. Soc.* **88**, 13 (1966).

it would be necessary to use a crystal dilute enough to avoid short-range order: as an alternative, the elastic diffuse scattering from random distribution of ions on the octahedral sides of a ferrite could be observed.

In summary it must be emphasized that the modifications introduced in elastic magnetic scattering by covalency are, by their very nature, very small and in systems more complex than those discussed in this section it is difficult to unravel contributions from various effects additional to covalency.† The other possible effects include the following.

(a) If the orbital moment of the magnetic ions is not zero its presence will tend to expand the form factor as compared to the form factor for the spin density alone.

(b) Magnetic ordering in the crystal can establish different radial wave functions for the spin up and spin down states. This is a very complicated effect but it will probably lead to an expansion in the form factor over the free-ion value.

(c) The radial wave functions in a crystal are likely to differ from those in the free ion, mainly because the electrons of the ligand ions help shield the $3d$ electrons of the magnetic ion from the charge of the nucleus of the magnetic ion. This will clearly result in an expansion of the radial $3d$ functions from their free-ion state and thereby cause a contraction in the atomic form factor.

(d) For antiferromagnets there is a reduction in the magnitude of the magnetic moment because of zero-point motion.

7.6. Helical spin ordering

In 1959 Yoshimori‡ proposed that the spins in antiferromagnetic MnO_2, which possesses the same crystal structure as MnF_2, i.e. body-centred orthorhombic, form a helical configuration with a pitch that is not commensurate with the lattice. His arguments were based on considerations of the classical energy of the spin system.

The helical configuration for MnO_2 is shown in Fig. 7.28 and Yoshimori demonstrated that such a configuration was consistent with the diffraction pattern obtained for MnO_2 by Erickson.§ Since Yoshimori's paper, many examples of helical and more complex spin configurations have been found, particularly in the rare earths,|| and neutron diffraction has

† P. J. Brown and J. B. Forsyth, *Proc. phys. Soc.* **92**, 125 (1967).

‡ Yoshimori, *J. phys. Soc. Japan*, **14**, 807 (1959).

§ R. A. Erickson, unpublished work quoted by Yoshimori.

|| Several review articles have been written on helical spin configurations and the spin ordering in the rare earths, e.g. R. J. Elliott, *Magnetism* (ed. Rado and Suhl) IIA, 385 and T. Nagamiya, *Solid St. Phys.* **20**, 305 (1967).

been the principal tool in the investigations. We begin our discussion of helical configurations with an examination of the underlying features of Yoshimori's paper on MnO_2, which remains one of the best introductions to the subject of helical spin configurations, and then consider the form of the cross-section in a more general case.

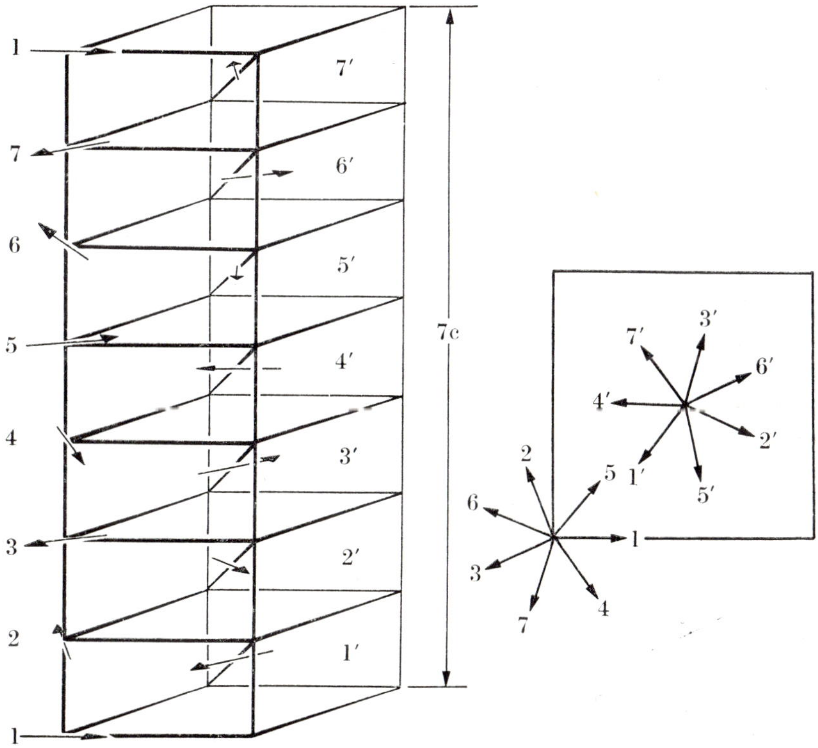

FIG. 7.28. Helical arrangement of magnetic moments in MnO_2.

We consider a simple lattice with just one magnetic atom per unit cell and inquire, by considering just the classical exchange forces (i.e. no anisotropy terms), what spin configurations are stable. The total exchange energy is

$$-\sum_{m,n} J(\mathbf{m}-\mathbf{n})\mathbf{S}_m.\mathbf{S}_n, \tag{7.80}$$

where $\mathbf{S}_m$ is the classical spin vector for the ion at the lattice site $\mathbf{m}$ and $J(\mathbf{m}-\mathbf{n})$ is the exchange parameter between spins at sites $\mathbf{m}$ and $\mathbf{n}$. Writing

$$J(\mathbf{m}) = \frac{1}{N}\sum_{\mathbf{q}} \mathscr{T}(\mathbf{q})\exp(\mathrm{i}\mathbf{q}.\mathbf{m}), \qquad \mathscr{T}(\mathbf{q}) = \mathscr{T}(-\mathbf{q}) \tag{7.81}$$

and
$$\mathbf{S}_m = \frac{1}{\sqrt{N}} \sum_{\mathbf{q}} \mathbf{S}_{\mathbf{q}} \exp(i\mathbf{q}.\mathbf{m}), \qquad \mathbf{S}_{-\mathbf{q}} = \mathbf{S}^*_{\mathbf{q}}, \tag{7.82}$$

where N is the number of unit cells in the crystal (= the number of magnetic ions), we obtain for (7.80)

$$-\sum_{\mathbf{q}} \mathscr{T}(\mathbf{q})\mathbf{S}_{\mathbf{q}}.\mathbf{S}_{-\mathbf{q}}. \tag{7.83}$$

We now seek the lowest minimum of (7.83) subject to the condition that
$$\sum_m \mathbf{S}_m^2 = \text{constant}.$$

From (7.82) this condition is equivalent to

$$\sum_{\mathbf{q}} \mathbf{S}_{\mathbf{q}}.\mathbf{S}_{-\mathbf{q}} = \text{constant}, \tag{7.84}$$

whence the lowest minimum of (7.83) is achieved for that $\mathbf{q}$ for which $\mathscr{T}(\mathbf{q})$ is the highest maximum. We denote this $\mathbf{q}$ by $\mathbf{Q}$ so that, since $-\mathbf{Q}$ is also allowed, we have for the minimum of (7.83) the value

$$-\mathscr{T}(\mathbf{Q})(\mathbf{S}_{\mathbf{Q}}.\mathbf{S}_{-\mathbf{Q}}+\mathbf{S}_{-\mathbf{Q}}.\mathbf{S}_{\mathbf{Q}}). \tag{7.85}$$

The corresponding $\mathbf{S}_m$ is, from (7.82),

$$\mathbf{S}_m = \frac{1}{N^{\frac{1}{2}}}\{\mathbf{S}_{\mathbf{Q}} \exp(i\mathbf{Q}.\mathbf{m})+\mathbf{S}_{-\mathbf{Q}} \exp(-i\mathbf{Q}.\mathbf{m})\}. \tag{7.86}$$

In component form eqn (7.86) reads

$$S_m^\alpha = A_\alpha \cos(\mathbf{Q}.\mathbf{m}+a_\alpha), \tag{7.87}$$

where A_α and a_α are constants. These equations show that as $\mathbf{m}$ advances in the direction of $\mathbf{Q}$ the spin vector $\mathbf{S}_m$ describes an ellipse. Since, however, the condition $\mathbf{S}_m^2$ = constant must be satisfied for all $\mathbf{m}$ it follows that the ellipse must be a circle. Taking the z axis perpendicular to the plane of the circle, (7.87) reduces to

$$\begin{aligned} S_m^x &= S\cos(\mathbf{Q}.\mathbf{m}+a) \\ S_m^y &= S\sin(\mathbf{Q}.\mathbf{m}+a) \\ S_m^z &= 0. \end{aligned} \tag{7.88}$$

In general the direction of $\mathbf{Q}$ is not perpendicular to the plane of rotation of the spin vectors, the relative orientation being fixed by anisotropy energy. The magnitude of $\mathbf{Q}$ is determined solely by the exchange parameters and thus, in general, it is incommensurate with the lattice period.

The structure of MnO_2 is shown in Fig. 7.29. Three exchange parameters J_1, J_2, and J_3 are believed to be important. Yoshimori showed that the helical spin configuration illustrated in Fig. 7.28 was favoured in MnO_2 if these parameters satisfy

$$J_2 > J_1 \quad \text{with } J_1 > 0, \tag{7.89}$$

the boundary conditions for stability being

$$J_2 = J_1 \quad \text{and} \quad 2J_2 J_3 = J_1^2.$$

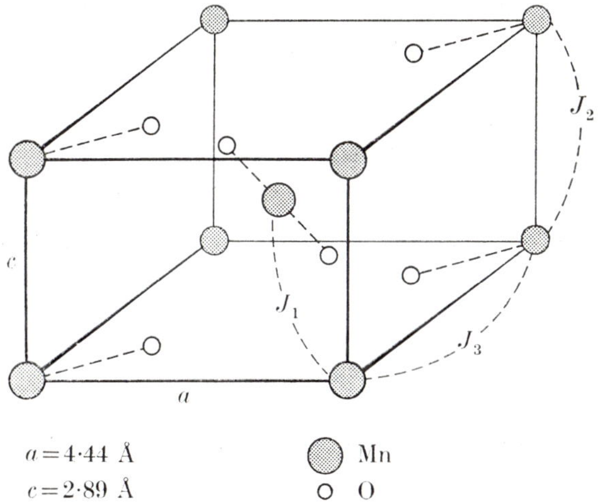

FIG. 7.29. Structure of MnO_2.

The spin configuration of Fig. 7.28 is described by

$$\mathbf{S}_l = S\exp(\mathrm{i}\mathbf{w}\,.\,\mathbf{l})\{\tilde{\mathbf{x}}\cos(\mathbf{Q}\,.\,\mathbf{l})+\tilde{\mathbf{y}}\sin(\mathbf{Q}\,.\,\mathbf{l})\}. \tag{7.90}$$

In (7.90) the reciprocal vector $\mathbf{w}$ is defined by

$$\mathbf{w} = \frac{2\pi}{a}(1, 1, a/c), \tag{7.91}$$

so that

$$\begin{aligned}\exp(\mathrm{i}\mathbf{w}\,.\,\mathbf{l}) &= +1 \quad \text{for Mn ions at the corner sites,}\\ &= -1 \quad \text{for Mn ions at the body centre sites.}\end{aligned}$$

The spins screw along the direction of $\mathbf{Q}$ (which coincides with the z axis) with a pitch $2\pi/|\mathbf{Q}|$.

To calculate the cross-section for the spin configuration described by (7.90) we need to evaluate

$$\begin{aligned}\sum_{\alpha,\beta}(\delta_{\alpha\beta}-\tilde{\kappa}_\alpha\tilde{\kappa}_\beta)(S_l^\alpha)^* S_{l'}^\beta\\ &= S^2\exp\{\mathrm{i}\mathbf{w}\,.\,(\mathbf{l}'-\mathbf{l})\}[(1-\tilde{\kappa}_x^2)\cos(\mathbf{Q}\,.\,\mathbf{l})\cos(\mathbf{Q}\,.\,\mathbf{l}')+\\ &\qquad +(1-\tilde{\kappa}_y^2)\sin(\mathbf{Q}\,.\,\mathbf{l})\sin(\mathbf{Q}\,.\,\mathbf{l}')-\tilde{\kappa}_x\tilde{\kappa}_y\sin\{\mathbf{Q}\,.\,(\mathbf{l}+\mathbf{l}')\}]\\ &= S^2\exp\{\mathrm{i}\mathbf{w}\,.\,(\mathbf{l}'-\mathbf{l})\}[\tfrac{1}{2}(1+\tilde{\kappa}_z^2)\cos\{\mathbf{Q}\,.\,(\mathbf{l}-\mathbf{l}')\}+\\ &\qquad +\tfrac{1}{4}\{(\tilde{\kappa}_y+\mathrm{i}\tilde{\kappa}_x)^2\exp\{\mathrm{i}\mathbf{Q}\,.\,(\mathbf{l}+\mathbf{l}')\}+\text{c.c.}\}].\end{aligned} \tag{7.92}$$

The cross-section involves (7.92) multiplied by $\exp\{i\boldsymbol{\kappa}.(\mathbf{l}'-\mathbf{l})\}$ and summed over all lattice sites $\mathbf{l}$ and $\mathbf{l}'$. If $2\mathbf{Q}$ does not coincide with a reciprocal lattice vector

$$\boldsymbol{\tau} = \frac{2\pi}{a}\left(t_1, t_2, \frac{a}{c}t_3\right) \quad (t_1+t_2+t_3 = \text{even integer}),$$

there is no contribution to the cross-section from the second term within the square bracket in (7.92); whence

$$\frac{d\sigma}{d\Omega} = \left(\frac{\gamma e^2}{m_e c^2}\right)^2 (\tfrac{1}{2}S)^2 |\tfrac{1}{2}gF(\boldsymbol{\kappa})|^2 N \frac{(2\pi)^3}{v_0}(1+\tilde{\kappa}_z^2) \times$$
$$\times \sum_{\boldsymbol{\tau}} \{\delta(\boldsymbol{\kappa}+\mathbf{w}-\mathbf{Q}-\boldsymbol{\tau})+\delta(\boldsymbol{\kappa}+\mathbf{w}+\mathbf{Q}-\boldsymbol{\tau})\}. \quad (7.93)$$

Hence we see that there are Bragg reflections when either of the two conditions

$$\boldsymbol{\kappa} = \boldsymbol{\tau}-\mathbf{w}\pm\mathbf{Q} \quad (7.94)$$

are satisfied.

If we take

$$\mathbf{Q} = \frac{2\pi}{c}(0, 0, 2/7) \quad (7.95)$$

so that the pitch of the helical structure $2\pi/|\mathbf{Q}| = 7c/2$, and make our unit cell seven times the length of the normal unit cell in the (001) direction, Bragg reflections occur at

$$(t_1-1, t_2-1, 7(t_3-1)\pm 2), \quad \text{with } t_1+t_2+t_3 = \text{even integer}. \quad (7.96)$$

These coincide exactly with the reflections reported by Erickson in his study of MnO_2.

Some of the spin configurations that have been found since Yoshimori's analysis for MnO_2 are very complicated. Besides yielding the spin configuration, neutron diffraction has also enabled the turn angle between spins in adjacent planes to be measured as a function of temperature.

In general, for one atom per unit cell the spin vectors obey

$$\begin{aligned}\mathbf{S}_n &= \sum_{\lambda=-\infty}^{\infty} \mathbf{A}_{\lambda\mathbf{Q}} \exp(i\lambda\mathbf{Q}.\mathbf{n}) \\ &= \mathbf{A}_0+\mathbf{A}_{\mathbf{Q}}\exp(i\mathbf{Q}.\mathbf{n})+\mathbf{A}_{-\mathbf{Q}}\exp(-i\mathbf{Q}.\mathbf{n})+ \\ &\quad +\mathbf{A}_{2\mathbf{Q}}\exp(i2\mathbf{Q}.\mathbf{n})+\mathbf{A}_{-2\mathbf{Q}}\exp(-i2\mathbf{Q}.\mathbf{n})+ \\ &\quad +\dots. \end{aligned} \quad (7.97)$$

Because $\mathbf{S}_n$ is real,

$$\mathbf{A}_{-\lambda\mathbf{Q}} = \mathbf{A}^*_{\lambda\mathbf{Q}}, \quad (7.98)$$

(7.97) is a reasonable description of the spin pattern provided none of the $\lambda\mathbf{Q}$ is equal to $\boldsymbol{\tau}$—because if, say, $\lambda_1\mathbf{Q}$ were equal to $\boldsymbol{\tau}_1$, then

$\mathbf{A}_{\lambda_1 Q}\exp(i\lambda_1\mathbf{Q}.\mathbf{n})$ would be just $\mathbf{A}_{\lambda_1 Q}$ for all $\mathbf{n}$ and could equally well be absorbed in $\mathbf{A}_0$. With this proviso it is easy to show that the elastic magnetic neutron cross-section becomes

$$\frac{d\sigma}{d\Omega} = \left(\frac{\gamma e^2}{m_e c^2}\right)^2 \frac{g^2 N(2\pi)^3}{4v_0} \sum_{\boldsymbol{\tau}} \sum_{\lambda} \delta(\lambda\mathbf{Q}+\boldsymbol{\varkappa}-\boldsymbol{\tau})|F(\boldsymbol{\varkappa})|^2 \times \\ \times\{\mathbf{A}^*_{\lambda Q}.\mathbf{A}_{\lambda Q}-|\tilde{\boldsymbol{\varkappa}}.\mathbf{A}_{\lambda Q}|^2\}. \quad (7.99)$$

If the sample has more than one crystallographically distinct atom per unit cell and (7.97) is taken to describe the spin variation from one cell to another, then each term of (7.99) is to be multiplied by the square of the modulus of the unit cell form factor and N is to be taken as the number of unit cells each of volume v_0.

Notice from (7.99) that the uniform component, $\lambda = 0$, produces scattering at the reciprocal lattice vectors only. The other components produce satellites at scattering vectors $\boldsymbol{\tau}-\lambda\mathbf{Q}$. In practice it usually happens that $\mathbf{Q}$ itself is small or the deviation of $\mathbf{Q}$ from a vector that would describe a simple antiferromagnet is small. In these cases the satellites appear clustered around the Bragg peaks or the superlattice peaks respectively. These satellites can only be observed clearly in single-crystal experiments; they have not been properly resolved in powder-diffraction experiments. Thus, for example, the spin patterns in metallic chromium were not correctly deduced until good pure single crystals of chromium were available.

The orientation factor in (7.99) is a great help in determining the details of the spin pattern. We shall not at this time give a discussion of particular cases, but it is worth noting that the $\lambda\mathbf{Q}$ satellite of $\boldsymbol{\tau} = 0$ appears with an orientation factor

$$|\mathbf{A}_{\lambda Q}|^2-|\tilde{\mathbf{Q}}.\mathbf{A}_{\lambda Q}|^2,$$

which is zero for a longitudinal wave, i.e. $\mathbf{A}_{\lambda Q}$ parallel to $\mathbf{Q}$. Hence such longitudinal waves can only be observed as satellites to $\boldsymbol{\tau}$ not zero.

Before leaving this subject it is worth noting that (7.97) describes the spin pattern on a 'rigid ion' model. We could also have spin density waves in the conduction-electron gas of a type described by Overhauser.† For such spin density waves one would, as a first approximation, describe the spin density at $\mathbf{r}$ by

$$\mathscr{s}(\mathbf{r}) = \sum_{\lambda} \mathbf{A}_{\lambda Q}\exp(i\lambda\mathbf{Q}.\mathbf{r}) \sum_{n} |\phi(\mathbf{r}-\mathbf{n})|^2, \quad (7.100)$$

† A. W. Overhauser and A. Arrott, *Phys. Rev. Lett.* **4**, 226 (1960).

which leads to the result

$$\frac{\mathrm{d}\sigma}{\mathrm{d}\Omega} = \left(\frac{\gamma e^2}{m_e c^2}\right)^2 \frac{g^2 N(2\pi)^3}{4v_0} \sum_{\boldsymbol{\tau}} \sum_{\lambda} \delta(\lambda \mathbf{Q} + \boldsymbol{\varkappa} - \boldsymbol{\tau}) |F(\boldsymbol{\tau})|^2 \{|\mathbf{A}_{\lambda \mathbf{Q}}|^2 - |\tilde{\boldsymbol{\varkappa}} . \mathbf{A}_{\lambda \mathbf{Q}}|^2\}. \tag{7.101}$$

The form factor appearing in (7.101) is $F(\boldsymbol{\tau})$, not $F(\boldsymbol{\varkappa})$ as in (7.99). The difference comes about because of the different physical descriptions given by (7.97) and (7.100). In (7.97) the spin of an atom is supposed to rotate as a rigid unit, but in (7.100) the spin density is modulated within the atom itself. The true situation must be somewhere intermediate between these two extremes; for chromium and for the rare-earth metals the rigid-ion approximation is probably close to the truth, but if spin density waves exist in conduction electrons, for example in Na or Cu, then (7.100) is probably a more appropriate description. In the latter case $|F(\boldsymbol{\tau})|^2$ would be expected to fall off with $\boldsymbol{\tau}$ so rapidly that, in effect, only $\boldsymbol{\tau} = 0$ made a contribution to (7.101); this would make longitudinal spin density waves unobservable by these methods.

8

CORRELATION FUNCTIONS IN MAGNETIC SCATTERING

IN Chapter 3 we showed that the neutron cross-section for purely nuclear scattering can be reduced to the time Fourier transform of a correlation function constructed from operators belonging to the target system. We mentioned there that a similar reduction could also be made for magnetic scattering, though the algebraic details were more involved because of the more complex form of the interaction between a neutron and the magnetic field of unpaired electrons.

In this chapter we first carry through this programme for magnetic scattering from both localized magnetic moments (see § 8.1) and also itinerant electrons (see § 8.2).

In order to handle the mathematical difficulties presented by the dynamical properties of magnetic systems the correlation functions are themselves expressed in terms of various functions that are more amenable to calculation. In presenting this part of the work we draw upon the material in Appendix B on linear response theory. Though our discussion in § 8.3 is virtually self-contained the reader will probably wish, in parallel, to consult Appendix B to gain a full understanding of the approach. § 8.4 generalizes the theory to consider more complex lattices and § 8.5 gives the equivalent discussion needed for itinerant electrons.

8.1. Correlation functions: localized model

In Chapter 5 we found that the neutron cross-section for the scattering by identical magnetic ions situated at the sites $\mathbf{R}_{ld}$ of a crystal is

$$\frac{\mathrm{d}^2\sigma}{\mathrm{d}\Omega\mathrm{d}E'}=\left(\frac{\gamma e^2}{m_e c^2}\right)^2\{\tfrac{1}{2}gF(\boldsymbol{\varkappa})\}^2\frac{k'}{k}\sum_{\alpha,\beta}(\delta_{\alpha\beta}-\tilde{\kappa}_\alpha\tilde{\kappa}_\beta)\times$$

$$\times\sum_{\lambda,\lambda'}p_\lambda\sum_{l,d}\sum_{l',d'}\langle\lambda|\exp(-\mathrm{i}\boldsymbol{\varkappa}.\mathbf{R}_{ld})\hat{S}^{\alpha}_{ld}\,|\lambda'\rangle\times$$

$$\times\langle\lambda'|\exp(\mathrm{i}\boldsymbol{\varkappa}.\mathbf{R}_{l'd'})\hat{S}^{\beta}_{l'd'}\,|\lambda\rangle\delta(\hbar\omega+E_\lambda-E_{\lambda'}).\quad(8.1)$$

Here the orbital motion of the electrons is partially taken into account via the dipole approximation to the form factor (compare §§ 5.3, 6.5).

To express (8.1) in terms of a correlation function we proceed as in Chapter 3, writing the delta function in terms of its integral representation. Thus

$$\sum_{\lambda,\lambda'} p_\lambda \langle\lambda|\exp(-\mathrm{i}\boldsymbol{\varkappa}.\mathbf{R}_{ld})\hat{S}^{\alpha}_{ld}\,|\lambda'\rangle\langle\lambda'|\exp(\mathrm{i}\boldsymbol{\varkappa}.\mathbf{R}_{l'd'})\hat{S}^{\beta}_{l'd'}\,|\lambda\rangle\delta(\hbar\omega+E_\lambda-E_{\lambda'})$$

$$=\sum_{\lambda,\lambda'} p_\lambda \frac{1}{2\pi\hbar}\int_{-\infty}^{\infty} \mathrm{d}t\,\exp(-\mathrm{i}\omega t)\langle\lambda|\exp(-\mathrm{i}\boldsymbol{\varkappa}.\mathbf{R}_{ld})\hat{S}^{\alpha}_{ld}|\lambda'\rangle\langle\lambda'|\exp(\mathrm{i}\mathscr{H}t/\hbar)\times$$
$$\times\exp(\mathrm{i}\boldsymbol{\varkappa}.\mathbf{R}_{l'd'})\exp(-\mathrm{i}\mathscr{H}t/\hbar+\mathrm{i}\mathscr{H}t/\hbar)\hat{S}^{\beta}_{l'd'}\exp(-\mathrm{i}\mathscr{H}t/\hbar)|\lambda\rangle$$

$$=\frac{1}{2\pi\hbar}\int_{-\infty}^{\infty} \mathrm{d}t\,\exp(-\mathrm{i}\omega t)\langle\exp\{-\mathrm{i}\boldsymbol{\varkappa}.\hat{\mathbf{R}}_{ld}(0)\}\hat{S}^{\alpha}_{ld}(0)\exp\{\mathrm{i}\boldsymbol{\varkappa}.\hat{\mathbf{R}}_{l'd'}(t)\}\hat{S}^{\beta}_{l'd'}(t)\rangle. \tag{8.2}$$

It is convenient to rewrite (8.2) in the form

$$\frac{1}{2\pi\hbar}\int_{-\infty}^{\infty} \mathrm{d}t\,\exp(-\mathrm{i}\omega t)\Big\langle\exp\{-\mathrm{i}\boldsymbol{\varkappa}.\hat{\mathbf{R}}_{ld}(0)\}\hat{S}^{\alpha}_{ld}(0)\int \mathrm{d}\mathbf{r}'\,\exp(\mathrm{i}\boldsymbol{\varkappa}.\mathbf{r}')\times$$
$$\times\delta\{\mathbf{r}'-\hat{\mathbf{R}}_{l'd'}(t)\}\hat{S}^{\beta}_{l'd'}(t)\Big\rangle$$

$$=\frac{1}{2\pi\hbar}\int_{-\infty}^{\infty} \mathrm{d}t\int \mathrm{d}\mathbf{r}\,\exp(\mathrm{i}\boldsymbol{\varkappa}.\mathbf{r}-\mathrm{i}\omega t)\int \mathrm{d}\mathbf{r}'\,\langle\delta\{\mathbf{r}+\hat{\mathbf{R}}_{ld}(0)-\mathbf{r}'\}\hat{S}^{\alpha}_{ld}(0)\times$$
$$\times\delta\{\mathbf{r}'-\hat{\mathbf{R}}_{l'd'}(t)\}\hat{S}^{\beta}_{l'd'}(t)\rangle, \tag{8.3}$$

and hence the partial differential cross-section is

$$\frac{\mathrm{d}^2\sigma}{\mathrm{d}\Omega\mathrm{d}E'}=\left(\frac{\gamma e^2}{m_\mathrm{e}c^2}\right)^2\{\tfrac{1}{2}gF(\boldsymbol{\varkappa})\}^2\frac{k'}{k}\sum_{\alpha,\beta}(\delta_{\alpha\beta}-\tilde{\varkappa}_\alpha\tilde{\varkappa}_\beta)\times$$
$$\times\frac{N}{2\pi\hbar}\int_{-\infty}^{\infty}\mathrm{d}t\int \mathrm{d}\mathbf{r}\,\exp(\mathrm{i}\boldsymbol{\varkappa}.\mathbf{r}-\mathrm{i}\omega t)\Gamma_{\alpha\beta}(\mathbf{r},t), \tag{8.4}$$

where

$$\Gamma_{\alpha\beta}(\mathbf{r},t)=\frac{1}{N}\sum_{l,d}\sum_{l',d'}\int \mathrm{d}\mathbf{r}'\,\langle\delta\{\mathbf{r}+\hat{\mathbf{R}}_{ld}(0)-\mathbf{r}'\}\hat{S}^{\alpha}_{ld}(0)\delta\{\mathbf{r}'-\hat{\mathbf{R}}_{l'd'}(t)\}\hat{S}^{\beta}_{l'd'}(t)\rangle. \tag{8.5}$$

In (8.4), N is the number of unit cells in the crystal.

Now to a good approximation the motion of an ion is independent of its spin orientation because the spin-dependent forces are small. Hence (8.5) becomes

$$\Gamma_{\alpha\beta}(\mathbf{r},t)=\frac{1}{N}\sum_{l,d}\sum_{l',d'}\langle\hat{S}^{\alpha}_{ld}(0)\hat{S}^{\beta}_{l'd'}(t)\rangle\int \mathrm{d}\mathbf{r}'\,\langle\delta\{\mathbf{r}+\hat{\mathbf{R}}_{ld}(0)-\mathbf{r}'\}\times$$
$$\times\delta\{\mathbf{r}'-\hat{\mathbf{R}}_{l'd'}(t)\}\rangle. \tag{8.6}$$

Let us now take the particular case of a Bravais lattice and write for (8.6)

$$\Gamma_{\alpha\beta}(\mathbf{r},t) = \sum_{l} \gamma_{\alpha\beta}(l,t)G_l(\mathbf{r},t), \tag{8.7}$$

where

$$\gamma_{\alpha\beta}(l,t) = \frac{1}{N}\sum_{m} \langle \hat{S}^{\alpha}_{m}(0)\hat{S}^{\beta}_{m+l}(t)\rangle = \gamma_{\alpha\beta}(l,\infty)+\gamma'_{\alpha\beta}(l,t) \tag{8.8}$$

and

$$\begin{aligned} G_l(\mathbf{r},t) &= \int \mathrm{d}\mathbf{r}' \,\langle \delta\{\mathbf{r}+\hat{\mathbf{R}}_0(0)-\mathbf{r}'\}\delta\{\mathbf{r}'-\hat{\mathbf{R}}_l(t)\}\rangle \\ &= G_l(\mathbf{r},\infty)+G'_l(\mathbf{r},t). \end{aligned} \tag{8.9}$$

Therefore

$$\begin{aligned} \Gamma_{\alpha\beta}(\mathbf{r},t) = &\sum_{l} \gamma_{\alpha\beta}(l,\infty)G_l(\mathbf{r},\infty)+\sum_{l} \gamma_{\alpha\beta}(l,\infty)G'_l(\mathbf{r},t)+ \\ &+\sum_{l} \gamma'_{\alpha\beta}(l,t)G_l(\mathbf{r},\infty)+\sum_{l} \gamma'_{\alpha\beta}(l,t)G'_l(\mathbf{r},t). \end{aligned} \tag{8.10}$$

There is a contribution to the cross-section (8.4) from each of the four terms in (8.10). The first of these gives elastic magnetic scattering; the second scattering that is elastic in the spin system but inelastic in the phonon system (magnetovibrational scattering); the third term gives inelastic magnetic scattering; the fourth term gives scattering inelastic in both the spin and phonon systems.

From (8.9) we readily evaluate

$$G_l(\mathbf{r},\infty) = \left(\frac{1}{2\pi}\right)^3 \int \mathrm{d}\mathbf{q}\, \exp\{\mathrm{i}\mathbf{q}\,.\,(\mathbf{r}-\mathbf{l})\}\exp\{-2W(\mathbf{q})\}. \tag{8.11}$$

Using this result in the first and third terms in (8.10), the corresponding cross-sections are given by

$$\begin{aligned} \left(\frac{\mathrm{d}\sigma}{\mathrm{d}\Omega}\right)_{\mathrm{el}} = &\left(\frac{\gamma e^2}{m_\mathrm{e}c^2}\right)^2 \{\tfrac{1}{2}gF(\boldsymbol{\varkappa})\}^2 \sum_{\alpha,\beta} (\delta_{\alpha\beta}-\tilde{\kappa}_\alpha\tilde{\kappa}_\beta)N\times \\ &\times \sum_{l} \exp(\mathrm{i}\boldsymbol{\varkappa}\,.\,\mathbf{l})\exp\{-2W(\boldsymbol{\varkappa})\}\gamma_{\alpha\beta}(l,\infty) \end{aligned} \tag{8.12}$$

and the inelastic magnetic scattering

$$\begin{aligned} \left(\frac{\mathrm{d}^2\sigma}{\mathrm{d}\Omega\mathrm{d}E'}\right)_{\mathrm{inel}} = &\left(\frac{\gamma e^2}{m_\mathrm{e}c^2}\right)^2 \{\tfrac{1}{2}gF(\boldsymbol{\varkappa})\}^2 N\frac{k'}{k}\sum_{\alpha,\beta} (\delta_{\alpha\beta}-\tilde{\kappa}_\alpha\tilde{\kappa}_\beta)\times \\ &\times \sum_{l} \exp(\mathrm{i}\boldsymbol{\varkappa}\,.\,\mathbf{l})\exp\{-2W(\boldsymbol{\varkappa})\}\frac{1}{2\pi\hbar}\int_{-\infty}^{\infty} \mathrm{d}t\,\exp(-\mathrm{i}\omega t)\gamma'_{\alpha\beta}(l,t). \end{aligned} \tag{8.13}$$

The magnetovibrational cross-section is easily rearranged into the form, using (4.64) and (4.67),

$$\left(\frac{d^2\sigma}{d\Omega dE'}\right)_{\text{m.v.}} = \left(\frac{\gamma e^2}{m_e c^2}\right)^2 \{\tfrac{1}{2}gF(\boldsymbol{\varkappa})\}^2 \frac{k'}{k} \sum_{\alpha,\beta} (\delta_{\alpha\beta} - \tilde{\kappa}_\alpha \tilde{\kappa}_\beta) \times$$
$$\times \frac{1}{2\pi\hbar} \int_{-\infty}^{\infty} dt \exp(-i\omega t) \sum_{l,l'} \gamma_{\alpha\beta}(l'-l, \infty) \exp\{i\boldsymbol{\varkappa}.(\mathbf{l}'-\mathbf{l})\} \times$$
$$\times \exp\{-2W(\boldsymbol{\varkappa})\}\{\exp\langle\boldsymbol{\varkappa}.\hat{\mathbf{u}}(l,0)\boldsymbol{\varkappa}.\hat{\mathbf{u}}(l',t)\rangle - 1\}, \qquad (8.14)$$

and similarly the cross-section inelastic in both the magnetic and vibrational system is

$$\left(\frac{d^2\sigma}{d\Omega dE'}\right) = \left(\frac{\gamma e^2}{m_e c^2}\right)^2 \{\tfrac{1}{2}gF(\boldsymbol{\varkappa})\}^2 \frac{k'}{k} \sum_{\alpha,\beta} (\delta_{\alpha\beta} - \tilde{\kappa}_\alpha \tilde{\kappa}_\beta) \times$$
$$\times \frac{1}{2\pi\hbar} \int_{-\infty}^{\infty} dt \exp(-i\omega t) \sum_{l,l'} \gamma'_{\alpha\beta}(l'-l, t) \exp\{i\boldsymbol{\varkappa}.(\mathbf{l}'-\mathbf{l})\} \times$$
$$\times \exp\{-2W(\boldsymbol{\varkappa})\}\{\exp\langle\boldsymbol{\varkappa}.\hat{\mathbf{u}}(l,0)\boldsymbol{\varkappa}.\hat{\mathbf{u}}(l',t)\rangle - 1\}. \qquad (8.15)$$

For a ferromagnet

$$\gamma_{\alpha\beta}(l'-l, \infty) = \delta_{\alpha\beta}\delta_{\alpha z}\langle\hat{S}^z\rangle^2, \qquad (8.16\,\text{a})$$

and then a comparison between (4.68) and (8.14) shows that the latter is simply (4.68) with the substitution

$$\bar{b} \to \left(\frac{\gamma e^2}{m_e c^2}\right)\tfrac{1}{2}gF(\boldsymbol{\varkappa})(1-\tilde{\kappa}_z^2)\langle\hat{S}^z\rangle. \qquad (8.16\,\text{b})$$

In other words the magnetovibrational scattering is exactly like the coherent inelastic nuclear scattering, but with the difference that the neutron is interacting with the magnetic moment of the atom through magnetic forces rather than with the nucleus of the atom through nuclear forces.

The cross-section (8.15), which is inelastic in both the magnetic and phonon systems, is not usually very interesting or important, because it is generally of a diffuse character with no distinct features in momentum or energy conservation (compare with the comments at the beginning of § 4.5 concerned with multi-phonon cross-sections). One special case is worth noting, however: for a paramagnet we can neglect the time dependence of $\gamma'_{\alpha\beta}(t)$, hence

$$\gamma'_{\alpha\beta}(l'-l, t) = \delta_{\alpha\beta}\delta_{l',l}\tfrac{1}{3}S(S+1). \qquad (8.17\,\text{a})$$

(More precisely, we can use (8.17 a) as a reasonable assumption when the exchange energies in the target are negligible compared to the neutron energy. Notice that (8.17 a) is *not* applicable to a magnetic system with substantial exchange effects but which is at high enough temperature to be in the paramagnetic phase.) So (8.15) becomes

$$\frac{\mathrm{d}^2\sigma}{\mathrm{d}\Omega\mathrm{d}E'} = \left(\frac{\gamma e^2}{m_\mathrm{e}c^2}\right)^2 \{\tfrac{1}{2}gF(\boldsymbol{\varkappa})\}^2 \frac{k'}{k}\,\tfrac{2}{3}S(S+1)\times$$
$$\times\frac{N}{2\pi\hbar}\int\limits_{-\infty}^{\infty}\mathrm{d}t\,\exp(-\mathrm{i}\omega t)\exp\{-2W(\boldsymbol{\varkappa})\}\{\exp\langle\boldsymbol{\varkappa}.\hat{\mathbf{u}}(0)\boldsymbol{\varkappa}.\hat{\mathbf{u}}(t)\rangle-1\}, \tag{8.17 b}$$

which, comparing with (4.69), is simply the nuclear incoherent inelastic scattering with the substitution

$$\frac{\sigma_\mathrm{i}}{4\pi}\to\left(\frac{\gamma e^2}{m_\mathrm{e}c^2}\right)^2\{\tfrac{1}{2}gF(\boldsymbol{\varkappa})\}^2\,\tfrac{2}{3}S(S+1). \tag{8.18}$$

8.2. Correlation functions: itinerant model

The preceding discussion has been specifically concerned with the scattering by magnetic ions whose unpaired electrons are well localized about the lattice sites of the crystal. It is also, of course, possible to express the cross-section derived in § 5.5 for the scattering by itinerant electrons in terms of correlation functions. By proceeding in an analogous manner to that leading to (8.2) the cross-section for the scattering by itinerant electrons can be written

$$\frac{\mathrm{d}^2\sigma}{\mathrm{d}\Omega\mathrm{d}E'} = \left(\frac{\gamma e^2}{m_\mathrm{e}c^2}\right)^2\frac{k'}{k}\sum_{\alpha,\beta}(\delta_{\alpha\beta}-\tilde{\kappa}_\alpha\tilde{\kappa}_\beta)\frac{1}{2\pi\hbar}\int\limits_{-\infty}^{\infty}\mathrm{d}t\,\exp(-\mathrm{i}\omega t)\times$$
$$\times\langle\hat{\mathscr{s}}^\alpha(\boldsymbol{\varkappa},0)\hat{\mathscr{s}}^\beta(-\boldsymbol{\varkappa},t)\rangle. \tag{8.19}$$

In (8.19) only that part of the scattering arising from the neutron–electron spin interaction is included, since, as we saw in the latter part of § 5.5, the magnitude of the scattering amplitude due to the interaction of a neutron and the translational motion of the electrons is, for electrons in narrow energy bands, much smaller.

We recall that

$$\hat{\mathscr{s}}(\mathbf{q}) = \int\mathrm{d}\mathbf{r}\,\exp(-\mathrm{i}\mathbf{q}.\mathbf{r})\sum_{\mathbf{k},\lambda}\sum_{\mathbf{k}',\lambda'}\phi^*_{\mathbf{k}\lambda}(\mathbf{r})\phi_{\mathbf{k}'\lambda'}(\mathbf{r})\sum_{\sigma,\sigma'}\hat{c}^+_{\mathbf{k}\lambda\sigma}\chi^+_\sigma\,\hat{\mathbf{s}}\chi_{\sigma'}\hat{c}_{\mathbf{k}'\lambda'\sigma'}$$
$$= \int\mathrm{d}\mathbf{r}\,\exp(-\mathrm{i}\mathbf{q}.\mathbf{r})\sum_{\mathbf{k},\lambda}\sum_{\mathbf{k}',\lambda'}\phi^*_{\mathbf{k}\lambda}(\mathbf{r})\phi_{\mathbf{k}'\lambda'}(\mathbf{r})\Big(\tfrac{1}{2}[\hat{c}^+_{\mathbf{k}\lambda\uparrow}\hat{c}_{\mathbf{k}'\lambda'\downarrow}+\mathrm{c.c.}],$$
$$\frac{1}{2\mathrm{i}}[\hat{c}^+_{\mathbf{k}\lambda\downarrow}\hat{c}_{\mathbf{k}'\lambda'\downarrow}-\mathrm{c.c.}],\tfrac{1}{2}[\hat{c}^+_{\mathbf{k}\lambda\uparrow}\hat{c}_{\mathbf{k}'\lambda'\uparrow}-\hat{c}^+_{\mathbf{k}\lambda\downarrow}\hat{c}_{\mathbf{k}'\lambda'\downarrow}]\Big). \tag{8.20}$$

In § 5.5 it was shown that this expression can be simplified by making a canonical transformation to a localized representation, and utilizing the fact that in narrow energy bands it is a good approximation to neglect the overlap of wave functions centred about different lattice sites. When we come to consider the dynamical behaviour of electrons in narrow energy bands we shall simplify the analysis by choosing to examine a model based upon a single band. It is, therefore, pertinent to consider the cross-section for the scattering by electrons in a single, narrow, energy band. In this instance we find that

$$\int d\mathbf{r}\, \exp(-i\mathbf{q}.\mathbf{r}) \sum_{\mathbf{k},\lambda} \sum_{\mathbf{k}',\lambda'} \phi^*_{\mathbf{k}\lambda}(\mathbf{r})\phi_{\mathbf{k}'\lambda'}(\mathbf{r})$$

can be replaced by

$$\int d\mathbf{r}\, \exp(-i\mathbf{q}.\mathbf{r})\, |\phi(\mathbf{r})|^2 \sum_{\mathbf{k},\mathbf{k}'} \delta_{\mathbf{k}',\mathbf{k}+\mathbf{q}}$$

in the approximation in which the overlap of the wave functions $\phi(\mathbf{r})$ centred about the lattice points of a Bravais crystal is neglected. Thus in this scheme

$$\mathfrak{z}(\mathbf{q}) = \int d\mathbf{r}\, \exp(-i\mathbf{q}.\mathbf{r})\, |\phi(\mathbf{r})|^2 \sum_{\mathbf{k}} \Big(\tfrac{1}{2}[\hat{c}^+_{\mathbf{k}\uparrow}\hat{c}_{\mathbf{k}+\mathbf{q}\downarrow}+\text{c.c.}],$$

$$\frac{1}{2i}[\hat{c}^+_{\mathbf{k}\uparrow}\hat{c}_{\mathbf{k}+\mathbf{q}\downarrow}-\text{c.c.}],\ \tfrac{1}{2}[\hat{c}^+_{\mathbf{k}\uparrow}\hat{c}_{\mathbf{k}+\mathbf{q}\uparrow}-\hat{c}^+_{\mathbf{k}\downarrow}\hat{c}_{\mathbf{k}+\mathbf{q}\downarrow}]\Big), \tag{8.21}$$

and we shall in future denote

$$F(\varkappa) = \int d\mathbf{r}\, \exp(i\varkappa.\mathbf{r})\, |\phi(\mathbf{r})|^2, \tag{8.22}$$

for this is the form factor appropriate to the model.

8.3. Relaxation functions and generalized susceptibilities: localized model and Bravais lattices

In this section we turn our attention to representing the inelastic magnetic cross-section (8.13) in terms of functions that are of more direct significance in statistical mechanics. We shall be careful to use a notation which in § 8.4 is easily generalized to non-Bravais lattices.

First let us represent $\gamma_{\alpha\beta}$ in terms of the operators $\hat{S}^\alpha_{\mathbf{q}}$ defined by

$$\hat{S}^\alpha_l = \frac{1}{N}\sum_{\mathbf{q}} \hat{S}^\alpha_{\mathbf{q}} \exp(i\mathbf{q}.\mathbf{l}), \tag{8.23}$$

i.e.

$$\hat{S}^\alpha_{\mathbf{q}} = \sum_l \exp(-i\mathbf{q}.\mathbf{l})\hat{S}^\alpha_l. \tag{8.24}$$

Then

$$\sum_l \exp(i\varkappa.\mathbf{l})\gamma_{\alpha\beta}(l,t) = \frac{1}{N}\sum_{lm} \exp(i\varkappa.\mathbf{l})\langle \hat{S}^\alpha_m(0)\hat{S}^\beta_{m+l}(t)\rangle$$

$$= \frac{1}{N}\langle \hat{S}^\alpha_{\varkappa}(0)\hat{S}^\beta_{-\varkappa}(t)\rangle. \tag{8.25}$$

Taking the *sum* of (8.12) and (8.13) the result (8.25) leads to

$$\frac{\mathrm{d}^2\sigma}{\mathrm{d}\Omega\mathrm{d}E'} = \left(\frac{\gamma e^2}{m_e c^2}\right)^2 \{\tfrac{1}{2}gF(\boldsymbol{\varkappa})\}^2 \frac{k'}{k} \exp\{-2W(\boldsymbol{\varkappa})\} \times$$

$$\times \sum_{\alpha\beta} (\delta_{\alpha\beta} - \tilde{\kappa}_\alpha \tilde{\kappa}_\beta) \frac{1}{2\pi\hbar} \int_{-\infty}^{\infty} \mathrm{d}t \exp(-\mathrm{i}\omega t) \langle \hat{S}^\alpha_{\boldsymbol{\varkappa}}(0) \hat{S}^\beta_{-\boldsymbol{\varkappa}}(t) \rangle. \tag{8.26 a}$$

Note that it can be proved that (8.26 a) satisfies the condition of detailed balance,

$$\left(\frac{\mathrm{d}^2\sigma}{\mathrm{d}\Omega\mathrm{d}E'}\right)_{\boldsymbol{\varkappa},\omega} = \exp(\hbar\omega\beta) \left(\frac{\mathrm{d}^2\sigma}{\mathrm{d}\Omega\mathrm{d}E'}\right)_{-\boldsymbol{\varkappa},-\omega}. \tag{8.26 b}$$

The correlation function in (8.26 a) can in many simple cases be calculated directly from a theory of the magnetic system. But it is related to quantities of more direct physical interpretation and it is therefore important to bring out these relations clearly—especially for discussions of problems where no strictly rigorous theory can be constructed.

In Appendix B we have calculated the linear response of a magnetic system to a magnetic field **H** that varies in both space and time. We found that the magnetization is given by (cf. (B.79) and (B.80))

$$M^\alpha_l(t) = -g\mu_B \langle \hat{S}^\alpha_l \rangle + \sum_{l'} \int_{-\infty}^{t} \mathrm{d}t' \sum_\beta \phi^{\alpha\beta}(l-l', t-t') H^\beta_{l'}(t'), \tag{8.27 a}$$

where
$$\phi^{\alpha\beta}(l,t) = (g\mu_B)^2 \frac{\mathrm{i}}{\hbar} \langle [\hat{S}^\alpha_l(t), \hat{S}^\beta_0(0)] \rangle. \tag{8.27 b}$$

If the time dependence of **H** is of the form $\exp(-\mathrm{i}\omega t)$ then

$$M^\alpha_l(t) = -g\mu_B \langle \hat{S}^\alpha_l \rangle + \exp(-\mathrm{i}\omega t) \sum_{l'} \sum_\beta \chi^{\alpha\beta}[l-l', \omega] H^\beta_{l'}, \tag{8.28 a}$$

where
$$\chi^{\alpha\beta}[l,\omega] = \int_0^\infty \mathrm{d}t \exp(\mathrm{i}\omega t) \phi^{\alpha\beta}(l,t) \tag{8.28 b}$$

is a generalized susceptibility tensor. Because the operators $\hat{S}^\alpha_l(t)$ are Hermitian, $\phi^{\alpha\beta}(l,t)$ is real. Hence from (8.28 b) the real and imaginary parts of $\chi^{\alpha\beta}[l,\omega]$ are given by

$$\chi^{\alpha\beta}[l,\omega]' = \int_0^\infty \mathrm{d}t \cos(\omega t) \phi^{\alpha\beta}(l,t)$$

and
$$\chi^{\alpha\beta}[l,\omega]'' = \int_0^\infty \mathrm{d}t \sin(\omega t) \phi^{\alpha\beta}(l,t),$$

and hence these functions are related by the Kramers–Krönig relations (see (B.14)).

However, it is much more convenient to work in terms of the Fourier transforms of these operators, which are *not* Hermitian. Thus

$$\{\hat{S}^{\alpha}_{\mathbf{q}}(t)\}^{+} = \hat{S}^{\alpha}_{-\mathbf{q}}(t) \quad \text{for } \alpha = x, y, z.$$

For a magnetic field varying in space like

$$H^{\beta}_{\mathbf{q}}(t')\exp(\mathrm{i}\mathbf{q}\,.\,\mathbf{l})$$

we get
$$M^{\alpha}_{\mathbf{q}}(t) = -g\mu_{\mathrm{B}}\langle\hat{S}^{\alpha}_{\mathbf{q}}\rangle + \int\limits_{-\infty}^{t} \mathrm{d}t' \sum_{\beta} \phi^{\alpha\beta}_{\mathbf{q}}(t-t')H^{\beta}_{\mathbf{q}}(t'), \tag{8.29 a}$$

where
$$\phi^{\alpha\beta}_{\mathbf{q}}(t) = \frac{(g\mu_{\mathrm{B}})^2}{N}\frac{\mathrm{i}}{\hbar}\langle[\hat{S}^{\alpha}_{\mathbf{q}}(t), \hat{S}^{\beta}_{-\mathbf{q}}(0)]\rangle. \tag{8.29 b}$$

If the time dependence of $H^{\beta}_{\mathbf{q}}(t')$ is of the form $\exp(-\mathrm{i}\omega t')$ then

$$M^{\alpha}_{\mathbf{q}}(t) = -g\mu_{\mathrm{B}}\langle\hat{S}^{\alpha}_{\mathbf{q}}\rangle + \exp(-\mathrm{i}\omega t)\sum_{\beta}\chi^{\alpha\beta}_{\mathbf{q}}[\omega]H^{\beta}_{\mathbf{q}}, \tag{8.30 a}$$

where
$$\chi^{\alpha\beta}_{\mathbf{q}}[\omega] = \int\limits_{0}^{\infty} \mathrm{d}t\, \exp(\mathrm{i}\omega t)\,\phi^{\alpha\beta}_{\mathbf{q}}(t) \tag{8.30 b}$$

is a generalized susceptibility tensor.

We choose to define the relaxation function through

$$\frac{\partial}{\partial t}\mathscr{R}^{\alpha\beta}_{\mathbf{q}}(t) = \frac{-N}{(g\mu_{\mathrm{B}})^2}\phi^{\alpha\beta}_{\mathbf{q}}(t). \tag{8.31}$$

We note that this differs by a simple factor from the function $R^{\alpha\beta}_{\mathbf{q}}(t)$ defined in Appendix B, i.e.

$$\mathscr{R}^{\alpha\beta}_{\mathbf{q}}(t) = \frac{N}{(g\mu_{\mathrm{B}})^2}R^{\alpha\beta}_{\mathbf{q}}(t). \tag{8.31 a}$$

Then
$$\mathscr{R}^{\alpha\beta}_{\mathbf{q}}(t) = \int\limits_{0}^{\beta}\mathrm{d}\lambda\,\langle\hat{S}^{\beta}_{-\mathbf{q}}(-\mathrm{i}\hbar\lambda)\hat{S}^{\alpha}_{\mathbf{q}}(t)\rangle - \beta\langle\hat{S}^{\beta}_{-\mathbf{q}}(0)\hat{S}^{\alpha}_{\mathbf{q}}(\infty)\rangle. \tag{8.32}$$

Eqn (8.32) is best verified by simply taking the derivative with respect to time. The Fourier transform is simply

$$\mathscr{R}^{\alpha\beta}_{\mathbf{q}}(\omega) = \frac{1}{2\pi}\int\limits_{-\infty}^{\infty}\mathrm{d}t\,\exp(\mathrm{i}\omega t)\mathscr{R}^{\alpha\beta}_{\mathbf{q}}(t), \tag{8.33}$$

and then it is straightforward to prove that

$$\frac{1}{2\pi}\int\limits_{-\infty}^{\infty}\mathrm{d}t\,\exp(\mathrm{i}\omega t)\langle\hat{S}^{\beta}_{-\mathbf{q}}(0)\hat{S}^{\alpha}_{\mathbf{q}}(t)\rangle$$
$$= \frac{\hbar\omega}{\exp(\beta\hbar\omega)-1}\mathscr{R}^{\alpha\beta}_{\mathbf{q}}(\omega) + \delta(\omega)\langle\hat{S}^{\beta}_{-\mathbf{q}}(0)\hat{S}^{\alpha}_{\mathbf{q}}(\infty)\rangle. \tag{8.34}$$

The symmetry properties of these functions are as follows. From (8.29 b)

$$\phi_{\mathbf{q}}^{\alpha\beta}(t)^* = \phi_{-\mathbf{q}}^{\alpha\beta}(t) = -\phi_{\mathbf{q}}^{\beta\alpha}(-t), \tag{8.35 a}$$

and hence

$$\chi_{\mathbf{q}}^{\alpha\beta}[\omega]^* = \chi_{-\mathbf{q}}^{\alpha\beta}[-\omega], \tag{8.35 b}$$

$$\mathscr{R}_{\mathbf{q}}^{\alpha\beta}(t)^* = \mathscr{R}_{-\mathbf{q}}^{\alpha\beta}(t) = \mathscr{R}_{\mathbf{q}}^{\beta\alpha}(-t), \tag{8.35 c}$$

$$\mathscr{R}_{\mathbf{q}}^{\alpha\beta}(\omega)^* = \mathscr{R}_{-\mathbf{q}}^{\alpha\beta}(-\omega) = \mathscr{R}_{\mathbf{q}}^{\beta\alpha}(\omega). \tag{8.35 d}$$

Using these relations we obtain from (8.26 a)

$$\frac{\mathrm{d}^2\sigma}{\mathrm{d}\Omega\mathrm{d}E'} = \left(\frac{\gamma e^2}{m_\mathrm{e}c^2}\right)^2 \{\tfrac{1}{2}gF(\boldsymbol{\varkappa})\}^2 \frac{k'}{k} \exp\{-2W(\boldsymbol{\varkappa})\} \times \\ \times \sum_{\alpha\beta} (\delta_{\alpha\beta}-\tilde{\kappa}_\alpha\tilde{\kappa}_\beta) \frac{\omega\exp(\hbar\omega\beta)}{\exp(\hbar\omega\beta)-1} \mathscr{R}_{\boldsymbol{\varkappa}}^{\alpha\beta}(\omega) \tag{8.36 a}$$

$$= \left(\frac{\gamma e^2}{m_\mathrm{e}c^2}\right)^2 \{\tfrac{1}{2}gF(\boldsymbol{\varkappa})\}^2 \frac{k'}{k} \exp\{-2W(\boldsymbol{\varkappa})\} \times \\ \times \frac{\omega\exp(\hbar\omega\beta)}{\exp(\hbar\omega\beta)-1} \sum_{\alpha\beta} (\delta_{\alpha\beta}-\tilde{\kappa}_\alpha\tilde{\kappa}_\beta)\tfrac{1}{2}\{\mathscr{R}_{\boldsymbol{\varkappa}}^{\alpha\beta}(\omega)+\mathscr{R}_{\boldsymbol{\varkappa}}^{\beta\alpha}(\omega)\}. \tag{8.36 b}$$

From (8.35 d) we notice that

$$\mathscr{R}_{-\boldsymbol{\varkappa}}^{\alpha\beta}(-\omega)+\mathscr{R}_{-\boldsymbol{\varkappa}}^{\beta\alpha}(-\omega) = \mathscr{R}_{\boldsymbol{\varkappa}}^{\alpha\beta}(\omega)+\mathscr{R}_{\boldsymbol{\varkappa}}^{\beta\alpha}(\omega); \tag{8.37a}$$

hence from (8.36 b) we recover the detailed balance condition (8.26 b).

From (8.35 c) and (8.35 d) we see that, in general, $\mathscr{R}_{\mathbf{q}}^{\alpha\beta}(t)$ and $\mathscr{R}_{\mathbf{q}}^{\alpha\beta}(\omega)$ are complex, but the symmetrized form

$$\overline{\mathscr{R}_{\mathbf{q}}^{\alpha\beta}}(\omega) = \tfrac{1}{2}\{\mathscr{R}_{\mathbf{q}}^{\alpha\beta}(\omega)+\mathscr{R}_{\mathbf{q}}^{\beta\alpha}(\omega)\}, \tag{8.37 b}$$

which is all that is required in the cross-section formula (8.36 b), is always real. Furthermore we notice that the symmetrical form

$$\overline{\mathscr{R}_{\mathbf{q}}^{\alpha\beta}}(t) = \tfrac{1}{2}\{\mathscr{R}_{\mathbf{q}}^{\alpha\beta}(t)+\mathscr{R}_{\mathbf{q}}^{\beta\alpha}(t)\} \tag{8.37 c}$$

is, in general, complex with a real part that is even in t and an imaginary part that is odd in t. (For many simple lattices, and all Bravais lattices, each magnetic atom is a centre of symmetry, and then $\mathbf{q}$ and $-\mathbf{q}$ must be equivalent: in this case $\mathscr{R}_{\mathbf{q}}^{\alpha\beta}(t)$ is itself real, $\overline{\mathscr{R}_{\mathbf{q}}^{\alpha\beta}}(t)$ is real and even in t, and $\mathscr{R}_{\mathbf{q}}^{\alpha\beta}(\omega)$ is real and even in ω.)

Now, starting from (8.33) and using (8.31) and (8.30 b), we can prove that

$$\mathscr{R}_{\boldsymbol{\varkappa}}^{\alpha\beta}(\omega) = \frac{N}{2\pi\mathrm{i}\omega(g\mu_\mathrm{B})^2}\{\chi_{\boldsymbol{\varkappa}}^{\alpha\beta}[\omega]-\chi_{\boldsymbol{\varkappa}}^{\beta\alpha}[\omega]^*\}; \tag{8.38}$$

hence

$$\mathscr{R}_{\boldsymbol{\varkappa}}^{\alpha\beta}(\omega)+\mathscr{R}_{\boldsymbol{\varkappa}}^{\beta\alpha}(\omega) = \frac{N}{\pi\omega(g\mu_\mathrm{B})^2}\,\mathrm{Im}\{\chi_{\boldsymbol{\varkappa}}^{\alpha\beta}[\omega]+\chi_{\boldsymbol{\varkappa}}^{\beta\alpha}[\omega]\}, \tag{8.39}$$

and an alternative form for the cross-section (8.36) is therefore

$$\frac{d^2\sigma}{d\Omega dE'} = \left(\frac{\gamma e^2}{m_e c^2}\right)^2 \{\tfrac{1}{2}gF(\varkappa)\}^2 \frac{k'}{k} \exp\{-2W(\varkappa)\} \times$$
$$\times \frac{N}{\pi(g\mu_B)^2} \frac{\exp(\hbar\omega\beta)}{\exp(\hbar\omega\beta)-1} \sum_{\alpha\beta} (\delta_{\alpha\beta}-\tilde{\kappa}_\alpha \tilde{\kappa}_\beta) \operatorname{Im} \chi_{\varkappa}^{\alpha\beta}[\omega]. \quad (8.40)$$

If we insert (8.31) into (8.30 b) we obtain

$$\chi_{\mathbf{q}}^{\alpha\beta}[\omega] = \frac{(g\mu_B)^2}{N} \mathscr{R}_{\mathbf{q}}^{\alpha\beta}(t=0) + \frac{(g\mu_B)^2}{N} i\omega \int_0^\infty dt \exp(i\omega t) \mathscr{R}_{\mathbf{q}}^{\alpha\beta}(t). \quad (8.41)$$

If in (8.41) we take the limit $\omega = 0$ we see that $\chi_{\mathbf{q}}^{\alpha\beta}[0]$ is proportional to $\mathscr{R}_{\mathbf{q}}^{\alpha\beta}(t=0)$. It is to be noted that in the general case $\chi_{\mathbf{q}}^{\alpha\beta}[0]$ is *not* the same as the *isothermal susceptibility* $\chi_{\mathbf{q}}^{\alpha\beta}$ evaluated in Appendix B. The difference between the two functions is removed if we replace $\langle \hat{S}_{-\mathbf{q}}^{\beta}(0)\hat{S}_{\mathbf{q}}^{\alpha}(\infty)\rangle$ by $\langle \hat{S}_{-\mathbf{q}}^{\beta}\rangle\langle \hat{S}_{\mathbf{q}}^{\alpha}\rangle$. This one assumes to be a valid procedure, as indeed it is in most cases of practical importance, and thus we write

$$\chi_{\mathbf{q}}^{\alpha\beta} = \chi_{\mathbf{q}}^{\alpha\beta}[\omega=0] = \frac{g^2\mu_B^2}{N} \mathscr{R}_{\mathbf{q}}^{\alpha\beta}(t=0). \quad (8.42)$$

We now define the *spectral weight function*

$$F_{\mathbf{q}}^{\alpha\beta}(\omega) = \frac{1}{2\pi} \int_{-\infty}^{\infty} dt \exp(i\omega t) \left\{\frac{\mathscr{R}_{\mathbf{q}}^{\alpha\beta}(t)}{\mathscr{R}_{\mathbf{q}}^{\alpha\beta}(t=0)}\right\}$$
$$= \mathscr{R}_{\mathbf{q}}^{\alpha\beta}(\omega)/\mathscr{R}_{\mathbf{q}}^{\alpha\beta}(t=0). \quad (8.43)$$

Then

$$\mathscr{R}_{\mathbf{q}}^{\alpha\beta}(\omega) = \frac{N}{(g\mu_B)^2} \chi_{\mathbf{q}}^{\alpha\beta} F_{\mathbf{q}}^{\alpha\beta}(\omega), \quad (8.44)$$

and (8.36) can be rewritten as

$$\frac{d^2\sigma}{d\Omega dE'} = \left(\frac{\gamma e^2}{m_e c^2}\right)^2 \{\tfrac{1}{2}gF(\varkappa)\}^2 \frac{k'}{k} \exp\{-2W(\varkappa)\} \times$$
$$\times \frac{\omega \exp(\hbar\omega\beta)}{\exp(\hbar\omega\beta)-1} \frac{N}{(g\mu_B)^2} \sum_{\alpha\beta} (\delta_{\alpha\beta}-\tilde{\kappa}_\alpha \tilde{\kappa}_\beta) \chi_{\varkappa}^{\alpha\beta} F_{\varkappa}^{\alpha\beta}(\omega). \quad (8.45)$$

From (8.43) we notice that

$$\int_{-\infty}^{\infty} d\omega\, F_{\varkappa}^{\alpha\beta}(\omega) = 1. \quad (8.46)$$

Our purpose in quoting so many alternative forms for the cross-section is that the quantities introduced, particularly $\chi_{\mathbf{q}}^{\alpha\beta}[\omega]$, $\mathscr{R}_{\mathbf{q}}^{\alpha\beta}(t)$, and $\mathscr{R}_{\mathbf{q}}^{\alpha\beta}(\omega)$, have a well-defined physical interpretation (see Appendix B) and are more amenable to calculation than are the correlation functions themselves.

The isothermal susceptibility $\chi_{\mathbf{q}}^{\alpha\beta}$ is of considerable significance in statistical mechanics and (8.42) shows how it is related to the functions we have defined. From (8.45) and (8.46) together we note that

$$\int_{-\infty}^{\infty} \mathrm{d}\omega \frac{k}{k'} \frac{1-\exp(-\hbar\omega\beta)}{\omega} \left(\frac{\mathrm{d}^2\sigma}{\mathrm{d}\Omega\mathrm{d}E'}\right)_{\text{inelas}}$$

$$= \left(\frac{\gamma e^2}{m_e c^2}\right)^2 \{\tfrac{1}{2}gF(\boldsymbol{\varkappa})\}^2 \exp\{-2W(\boldsymbol{\varkappa})\} \frac{N}{g^2\mu_B^2} \sum_{\alpha\beta} (\delta_{\alpha\beta}-\tilde{\kappa}_\alpha\tilde{\kappa}_\beta)\chi_{\boldsymbol{\varkappa}}^{\alpha\beta}, \quad (8.47)$$

which means that the isothermal susceptibility can be extracted from the neutron cross-section directly and without a knowledge of the spectral weight $F_{\boldsymbol{\varkappa}}^{\alpha\beta}(\omega)$.

In many theoretical studies it is convenient to work in terms of retarded and advanced Green functions. We therefore now examine how the cross-sections are related to such functions. The *retarded* Green function is defined as

$$G_{\boldsymbol{\varkappa}}^{\alpha\beta}(t) = -\mathrm{i}\theta(t)\langle[\hat{S}_{\boldsymbol{\varkappa}}^{\alpha}(t), \hat{S}_{-\boldsymbol{\varkappa}}^{\beta}(0)]\rangle \quad (8.48)$$

and satisfies the equation of motion

$$\mathrm{i}\hbar\frac{\partial}{\partial t} G_{\boldsymbol{\varkappa}}^{\alpha\beta}(t) = \hbar\,\delta(t)\langle[\hat{S}_{\boldsymbol{\varkappa}}^{\alpha}, \hat{S}_{-\boldsymbol{\varkappa}}^{\beta}]\rangle - \mathrm{i}\,\theta(t)\langle[[\hat{S}_{\boldsymbol{\varkappa}}^{\alpha}(t), \mathscr{H}], \hat{S}_{-\boldsymbol{\varkappa}}^{\beta}(0)]\rangle. \quad (8.49)$$

Then

$$G_{\boldsymbol{\varkappa}}^{\alpha\beta}(\omega) = \frac{1}{2\pi}\int_{-\infty}^{\infty} \mathrm{d}t \exp(\mathrm{i}\omega t)\, G_{\boldsymbol{\varkappa}}^{\alpha\beta}(t)$$

$$= \frac{-\mathrm{i}}{2\pi}\int_{0}^{\infty} \mathrm{d}t \exp(\mathrm{i}\omega t)\, \langle[\hat{S}_{\boldsymbol{\varkappa}}^{\alpha}(t), \hat{S}_{-\boldsymbol{\varkappa}}^{\beta}(0)]\rangle$$

$$= \frac{-N\hbar}{2\pi(g\mu_B)^2}\chi_{\boldsymbol{\varkappa}}^{\alpha\beta}[\omega]. \quad (8.50)$$

Hence from (8.40)

$$\left(\frac{\mathrm{d}^2\sigma}{\mathrm{d}\Omega\mathrm{d}E'}\right)_{\text{inelas}} = \left(\frac{\gamma e^2}{m_e c^2}\right)^2 \{\tfrac{1}{2}gF(\boldsymbol{\varkappa})\}^2 \frac{k'}{k} \exp\{-2W(\boldsymbol{\varkappa})\}\times$$

$$\times \frac{-\hbar}{2} \frac{\exp(\hbar\omega\beta)}{\exp(\hbar\omega\beta)-1} \sum_{\alpha\beta} (\delta_{\alpha\beta}-\tilde{\kappa}_\alpha\tilde{\kappa}_\beta)\,\mathrm{Im}\{G_{\boldsymbol{\varkappa}}^{\alpha\beta}(\omega)\}. \quad (8.51)$$

So far the results we have given in this section have been quite general. However, when the total z component of spin is a constant of the motion, as it is for a simple Heisenberg model, there is a considerable simplification in the formulae. We now examine this point.

If $\hat{S}^z_{\text{tot}}$ is a constant of motion, i.e.

$$\sum_l [\hat{S}^z_l, \mathscr{H}] = 0, \tag{8.52}$$

then the raising and lowering operators

$$\hat{S}^+_l = \hat{S}^x_l + \mathrm{i}\,\hat{S}^y_l, \qquad \hat{S}^-_l = \hat{S}^x_l - \mathrm{i}\,\hat{S}^y_l \tag{8.53}$$

change the total z component by plus or minus one unit. Hence

$$\langle \hat{S}^+_l(0)\hat{S}^+_{l'}(t)\rangle = \langle \hat{S}^-_l(0)\hat{S}^-_{l'}(t)\rangle = 0.$$

Because of this, terms in the cross-section like

$$-\tilde{\kappa}_x\tilde{\kappa}_y\langle \hat{S}^x_{\boldsymbol{\kappa}}(0)\hat{S}^y_{-\boldsymbol{\kappa}}(t) + \hat{S}^y_{\boldsymbol{\kappa}}(0)\hat{S}^x_{\boldsymbol{\kappa}}(t)\rangle$$

vanish, as also do cross terms like

$$-\tilde{\kappa}_x\tilde{\kappa}_z\langle \hat{S}^x_{\boldsymbol{\kappa}}(0)\hat{S}^z_{-\boldsymbol{\kappa}}(t)\rangle.$$

Also the x and y axes are equivalent, so

$$\langle \hat{S}^x_{\boldsymbol{\kappa}}(0)\hat{S}^x_{-\boldsymbol{\kappa}}(t)\rangle = \langle \hat{S}^y_{\boldsymbol{\kappa}}(0)\hat{S}^y_{-\boldsymbol{\kappa}}(t)\rangle.$$

Hence (8.26 a) simplifies to

$$\frac{\mathrm{d}^2\sigma}{\mathrm{d}\Omega\mathrm{d}E'} = \left(\frac{\gamma e^2}{m_{\mathrm{e}}c^2}\right)^2 \{\tfrac{1}{2}gF(\boldsymbol{\kappa})\}^2\frac{k'}{k}\exp\{-2W(\boldsymbol{\kappa})\}\times$$

$$\times\left\{(1-\tilde{\kappa}_z^2)\frac{1}{2\pi\hbar}\int_{-\infty}^{\infty}\mathrm{d}t\,\exp(-\mathrm{i}\omega t)\,\langle \hat{S}^z_{\boldsymbol{\kappa}}(0)\hat{S}^z_{-\boldsymbol{\kappa}}(t)\rangle + \right.$$

$$\left. +(1+\tilde{\kappa}_z^2)\frac{1}{2\pi\hbar}\int_{-\infty}^{\infty}\mathrm{d}t\,\exp(-\mathrm{i}\omega t)\,\langle \hat{S}^x_{\boldsymbol{\kappa}}(0)\hat{S}^x_{-\boldsymbol{\kappa}}(t)\rangle\right\}. \tag{8.54}$$

From (8.29 b) we find that the cross terms

$$\phi^{xz}_{\mathbf{q}}(t),\ \phi^{zx}_{\mathbf{q}}(t),\ \phi^{zy}_{\mathbf{q}}(t),\ \text{and}\ \phi^{yz}_{\mathbf{q}}(t)\ \text{are zero,}$$

and that

$$\phi^{xy}_{\mathbf{q}}(t) + \phi^{yx}_{\mathbf{q}}(t) = 0.$$

The non-zero terms which concern us are then just

$$\phi^{xx}_{\mathbf{q}}(t) = \phi^{yy}_{\mathbf{q}}(t) \quad \text{and} \quad \phi^{zz}_{\mathbf{q}}(t).$$

From (8.30 b) and (8.31) it follows that similar remarks apply also to $\chi^{\alpha\beta}_{\mathbf{q}}[\omega]$, $\mathscr{R}^{\alpha\beta}_{\mathbf{q}}(t)$, and $\mathscr{R}^{\alpha\beta}_{\mathbf{q}}(\omega)$. The cross-section (8.36 a) now becomes the sum of two terms, which we denote as longitudinal and transverse respectively, where

$$\left(\frac{\mathrm{d}^2\sigma}{\mathrm{d}\Omega\mathrm{d}E'}\right)^{\text{long}}_{\text{inelas}} = \left(\frac{\gamma e^2}{m_{\mathrm{e}}c^2}\right)^2 \{\tfrac{1}{2}gF(\boldsymbol{\kappa})\}^2\frac{k'}{k}\exp\{-2W(\boldsymbol{\kappa})\}\times$$

$$\times(1-\tilde{\kappa}_z^2)\,\frac{\omega\exp(\hbar\omega\beta)}{\exp(\hbar\omega\beta)-1}\,\mathscr{R}^{zz}_{\boldsymbol{\kappa}}(\omega) \tag{8.55}$$

and

$$\left(\frac{d^2\sigma}{d\Omega dE'}\right)^{\text{trans}}_{\text{inelas}} = \left(\frac{\gamma e^2}{m_e c^2}\right)^2 \{\tfrac{1}{2}gF(\boldsymbol{\varkappa})\}^2 \frac{k'}{k} \exp\{-2W(\boldsymbol{\varkappa})\} \times$$

$$\times (1+\kappa_z^2) \frac{\omega \exp(\hbar\omega\beta)}{\exp(\hbar\omega\beta)-1} \mathscr{R}^{xx}_{\varkappa}(\omega). \tag{8.56}$$

From (8.39), (8.44), and (8.50) we can write down the cross-section in terms of the generalized susceptibility, spectral weight, or retarded Green function by the simple substitution into (8.55) and (8.56) of the relations

$$\mathscr{R}^{zz}_{\varkappa}(\omega) = \frac{N}{\pi\omega(g\mu_B)^2} \operatorname{Im}\{\chi^{zz}_{\varkappa}[\omega]\} \tag{8.57}$$

$$= \frac{N}{(g\mu_B)^2} \chi^{zz}_{\varkappa} F^{zz}_{\varkappa}(\omega) \tag{8.58}$$

$$= -\frac{2}{\hbar\omega} \operatorname{Im}\{G^{zz}_{\varkappa}(\omega)\} \tag{8.59}$$

and exactly the same equations with superscripts xx replacing zz.

To take full advantage of all these symmetries it is customary in theoretical calculations to work in terms of raising and lowering operators which are Hermitian conjugates of one another. Thus

$$\hat{S}^+_{\mathbf{q}} = \sum_n \exp(-i\mathbf{q}.\mathbf{n})\hat{S}^+_n = \hat{S}^x_{\mathbf{q}} + i\,\hat{S}^y_{\mathbf{q}},$$

$$\hat{S}^-_{\mathbf{q}} = \sum_n \exp(i\mathbf{q}.\mathbf{n})\,\hat{S}^-_n = \hat{S}^x_{-\mathbf{q}} - i\,\hat{S}^y_{-\mathbf{q}}. \tag{8.60}$$

Hence

$$\hat{S}^x_{\mathbf{q}} = \tfrac{1}{2}(\hat{S}^+_{\mathbf{q}} + \hat{S}^-_{-\mathbf{q}}),$$

$$\hat{S}^y_{\mathbf{q}} = \frac{1}{2i}(\hat{S}^+_{\mathbf{q}} - \hat{S}^-_{-\mathbf{q}}). \tag{8.61}$$

In terms of these variables (8.54) can be rewritten as

$$\frac{d^2\sigma}{d\Omega dE'} = \left(\frac{\gamma e^2}{m_e c^2}\right)^2 \{\tfrac{1}{2}gF(\boldsymbol{\varkappa})\}^2 \frac{k'}{k} \exp\{-2W(\boldsymbol{\varkappa})\} \times$$

$$\times \Bigg\{(1-\kappa_z^2) \frac{1}{2\pi\hbar} \int_{-\infty}^{\infty} dt \exp(-i\omega t)\, \langle \hat{S}^z_{\varkappa}(0)\hat{S}^z_{-\varkappa}(t)\rangle +$$

$$+ (1+\kappa_z^2) \frac{1}{2\pi\hbar} \int_{-\infty}^{\infty} dt \exp(-i\omega t)\, \tfrac{1}{4}\langle \hat{S}^+_{\varkappa}(0)\hat{S}^-_{\varkappa}(t) + \hat{S}^-_{-\varkappa}(0)\hat{S}^+_{-\varkappa}(t)\rangle\Bigg\}. \tag{8.62}$$

Now defining

$$\phi_{\mathbf{q}}^{(0)}(t) = \phi_{\mathbf{q}}^{zz}(t) = \frac{(g\mu_{\mathrm{B}})^2}{N}\frac{\mathrm{i}}{\hbar}\langle[\hat{S}_{\mathbf{q}}^{z}(t), \hat{S}_{-\mathbf{q}}^{z}(0)]\rangle,$$

$$\phi_{\mathbf{q}}^{(+)}(t) = \frac{(g\mu_{\mathrm{B}})^2}{N}\frac{\mathrm{i}}{\hbar}\langle[\hat{S}_{\mathbf{q}}^{+}(t), \hat{S}_{\mathbf{q}}^{-}(0)]\rangle,$$

$$\phi_{\mathbf{q}}^{(-)}(t) = \phi_{\mathbf{q}}^{(+)}(t)^* = \frac{(g\mu_{\mathrm{B}})^2}{N}\frac{\mathrm{i}}{\hbar}\langle[\hat{S}_{\mathbf{q}}^{-}(t), \hat{S}_{\mathbf{q}}^{+}(0)]\rangle, \tag{8.63}$$

and $\chi_{\mathbf{q}}^{(\nu)}[\omega]$ and $\mathscr{R}_{\mathbf{q}}^{(\nu)}(t)$ by the analogues of (8.30 b) and (8.31), i.e.

$$\chi_{\mathbf{q}}^{(\nu)}[\omega] = \int_0^\infty \mathrm{d}t\, \exp(\mathrm{i}\omega t)\, \phi_{\mathbf{q}}^{(\nu)}(t), \tag{8.64}$$

$$\frac{\partial}{\partial t}\mathscr{R}_{\mathbf{q}}^{(\nu)}(t) = \frac{-N}{(g\mu_{\mathrm{B}})^2}\phi_{\mathbf{q}}^{(\nu)}(t), \tag{8.65}$$

we find

$$\mathscr{R}_{\mathbf{q}}^{(0)}(t) = \int_0^\beta \mathrm{d}\lambda\, \langle \hat{S}_{-\mathbf{q}}^{z}(-\mathrm{i}\hbar\lambda)\hat{S}_{\mathbf{q}}^{z}(t)\rangle - \beta\langle \hat{S}_{-\mathbf{q}}^{z}(0)\hat{S}_{\mathbf{q}}^{z}(\infty)\rangle,$$

$$\mathscr{R}_{\mathbf{q}}^{(+)}(t) = \int_0^\beta \mathrm{d}\lambda\, \langle \hat{S}_{\mathbf{q}}^{-}(-\mathrm{i}\hbar\lambda)\hat{S}_{\mathbf{q}}^{+}(t)\rangle - \beta\langle \hat{S}_{\mathbf{q}}^{-}(0)\hat{S}_{\mathbf{q}}^{+}(\infty)\rangle,$$

$$\mathscr{R}_{\mathbf{q}}^{(-)}(t) = \int_0^\beta \mathrm{d}\lambda\, \langle \hat{S}_{\mathbf{q}}^{+}(-\mathrm{i}\hbar\lambda)\hat{S}_{\mathbf{q}}^{-}(t)\rangle - \beta\langle \hat{S}_{\mathbf{q}}^{+}(0)\hat{S}_{\mathbf{q}}^{-}(\infty)\rangle, \tag{8.66}$$

and

$$\mathscr{R}_{\mathbf{q}}^{xx}(t) = \tfrac{1}{4}\{\mathscr{R}_{\mathbf{q}}^{(+)}(t) + \mathscr{R}_{-\mathbf{q}}^{(-)}(t)\}. \tag{8.67}$$

Hence

$$\mathscr{R}_{\mathbf{q}}^{xx}(\omega) = \tfrac{1}{4}\{\mathscr{R}_{\mathbf{q}}^{(+)}(\omega) + \mathscr{R}_{-\mathbf{q}}^{(-)}(\omega)\}. \tag{8.68}$$

Now from the definitions (8.63) we have

$$\phi_{\mathbf{q}}^{(0)}(t) = \phi_{-\mathbf{q}}^{(0)}(t)^* = -\phi_{-\mathbf{q}}^{(0)}(-t), \tag{8.69}$$

$$\phi_{\mathbf{q}}^{(\nu)}(t) = \phi_{\mathbf{q}}^{(-\nu)}(t)^* = -\phi_{\mathbf{q}}^{(-\nu)}(-t); \quad \nu = + \text{ or } - \tag{8.70}$$

and hence

$$\chi_{\mathbf{q}}^{(0)}[\omega]^* = \chi_{-\mathbf{q}}^{(0)}[-\omega], \tag{8.71 a}$$

$$\chi_{\mathbf{q}}^{(\nu)}[\omega]^* = \chi_{\mathbf{q}}^{(-\nu)}[-\omega]; \quad \nu = + \text{ or } - \tag{8.71 b}$$

$$\mathscr{R}_{\mathbf{q}}^{(0)}(t) = \mathscr{R}_{-\mathbf{q}}^{(0)}(t)^* = \mathscr{R}_{-\mathbf{q}}^{(0)}(-t), \tag{8.72 a}$$

$$\mathscr{R}_{\mathbf{q}}^{(\nu)}(t) = \mathscr{R}_{\mathbf{q}}^{(-\nu)}(t)^* = \mathscr{R}_{\mathbf{q}}^{(-\nu)}(-t); \quad \nu = + \text{ or } - \tag{8.72 b}$$

$$\mathscr{R}_{\mathbf{q}}^{(0)}(\omega) = \mathscr{R}_{\mathbf{q}}^{(0)}(\omega)^* = \mathscr{R}_{-\mathbf{q}}^{(0)}(-\omega), \tag{8.73 a}$$

$$\mathscr{R}_{\mathbf{q}}^{(\nu)}(\omega) = \mathscr{R}_{\mathbf{q}}^{(\nu)}(\omega)^* = \mathscr{R}_{\mathbf{q}}^{(-\nu)}(-\omega); \quad \nu = + \text{ or } -, \tag{8.73 b}$$

and

$$\mathrm{Im}\{\chi_{\mathbf{q}}^{(\nu)}[\omega]\} = \frac{\pi(g\mu_{\mathrm{B}})^2}{N}\omega\mathscr{R}_{\mathbf{q}}^{(\nu)}(\omega); \quad \nu = 0, +, \text{ or } -. \tag{8.74}$$

Finally, defining retarded Green functions as

$$G_{\mathbf{q}}^{(0)}(t) = -\mathrm{i}\theta(t)\langle[\hat{S}_{\mathbf{q}}^{z}(t), \hat{S}_{-\mathbf{q}}^{z}(0)]\rangle$$

and

$$G_{\mathbf{q}}^{(\nu)}(t) = -\mathrm{i}\theta(t)\langle[\hat{S}_{\mathbf{q}}^{\nu}(t), \hat{S}_{\mathbf{q}}^{-\nu}(0)]\rangle; \quad \nu = + \text{ or } -, \tag{8.75}$$

$$\mathrm{Im}\{G_{\mathbf{q}}^{(\nu)}(\omega)\} = -\tfrac{1}{2}\hbar\omega\mathscr{R}_{\mathbf{q}}^{(\nu)}(\omega); \quad \nu = 0, +, \text{ or } -. \tag{8.76}$$

These new variables do not affect the form for the longitudinal cross-section, which remains as (8.55); for ease of reference we repeat the nomenclature

$$\mathscr{R}_{\boldsymbol{\varkappa}}^{zz}(\omega) = \mathscr{R}_{\boldsymbol{\varkappa}}^{(0)}(\omega)$$
$$= \frac{N}{\pi\omega(g\mu_{\mathrm{B}})^2}\,\mathrm{Im}\,\chi_{\boldsymbol{\varkappa}}^{(0)}(\omega)$$
$$= \frac{-2}{\hbar\omega}\,\mathrm{Im}\,G_{\boldsymbol{\varkappa}}^{(0)}(\omega), \tag{8.77}$$

whereas the transverse cross-section (8.56) can now be re-expressed using

$$\mathscr{R}_{\boldsymbol{\varkappa}}^{xx}(\omega) = \tfrac{1}{4}\{\mathscr{R}_{\boldsymbol{\varkappa}}^{(+)}(\omega) + \mathscr{R}_{-\boldsymbol{\varkappa}}^{(-)}(\omega)\}$$
$$= \frac{N}{4\pi(g\mu_{\mathrm{B}})^2\omega}\,\mathrm{Im}\{\chi_{\boldsymbol{\varkappa}}^{(+)}[\omega] + \chi_{-\boldsymbol{\varkappa}}^{(-)}[\omega]\}$$
$$= \frac{-1}{2\hbar\omega}\,\mathrm{Im}\{G_{\boldsymbol{\varkappa}}^{(+)}(\omega) + G_{-\boldsymbol{\varkappa}}^{(-)}(\omega)\}. \tag{8.78 a}$$

An alternative form to (8.78 a) is given by using (8.73 b) and (8.74), namely

$$\mathscr{R}_{\boldsymbol{\varkappa}}^{xx}(\omega) = \tfrac{1}{4}\{\mathscr{R}_{\boldsymbol{\varkappa}}^{(+)}(\omega) + \mathscr{R}_{-\boldsymbol{\varkappa}}^{(+)}(-\omega)\}$$
$$= \frac{N}{4\pi(g\mu_{\mathrm{B}})^2\omega}\,\mathrm{Im}\{\chi_{\boldsymbol{\varkappa}}^{(+)}[\omega] - \chi_{-\boldsymbol{\varkappa}}^{(+)}[-\omega]\}$$
$$= \frac{-1}{2\hbar\omega}\,\mathrm{Im}\{G_{\boldsymbol{\varkappa}}^{(+)}(\omega) - G_{-\boldsymbol{\varkappa}}^{(+)}(-\omega)\}, \tag{8.78 b}$$

which, when substituted into (8.56) and defining

$$\mathscr{A}(\mathbf{k}, \mathbf{k}') = \left(\frac{\gamma e^2}{m_{\mathrm{e}} c^2}\right)^2 \{\tfrac{1}{2} g F(\boldsymbol{\varkappa})\}^2 \frac{k'}{k} \exp\{-2W(\boldsymbol{\varkappa})\}(1 + \tilde{\kappa}_z^2), \tag{8.79}$$

gives

$$\left(\frac{\mathrm{d}^2\sigma}{\mathrm{d}\Omega\mathrm{d}E'}\right)_{\text{inelas}}^{\text{trans}}$$
$$= \mathscr{A}(\mathbf{k}, \mathbf{k}')\tfrac{1}{4}[\{n(\omega)+1\}\omega\mathscr{R}_{\boldsymbol{\varkappa}}^{(+)}(\omega) - n(-\omega)\omega\mathscr{R}_{-\boldsymbol{\varkappa}}^{(+)}(-\omega)]$$
$$= \frac{\mathscr{A}(\mathbf{k}, \mathbf{k}')N}{4\pi(g\mu_{\mathrm{B}})^2}[\{n(\omega)+1\}\mathrm{Im}\{\chi_{\boldsymbol{\varkappa}}^{(+)}[\omega]\} + n(-\omega)\mathrm{Im}\{\chi_{-\boldsymbol{\varkappa}}^{(+)}[-\omega]\}]$$
$$= \mathscr{A}(\mathbf{k}, \mathbf{k}')\left(\frac{-1}{2\hbar}\right)[\{n(\omega)+1\}\mathrm{Im}\{G_{\boldsymbol{\varkappa}}^{(+)}(\omega)\} + n(-\omega)\mathrm{Im}\{G_{-\boldsymbol{\varkappa}}^{(+)}(-\omega)\}], \tag{8.80}$$

where
$$n(\omega) = \frac{1}{\exp(\beta\hbar\omega)-1},$$
and (8.80) brings out the similarity, for simple processes, of creating and annihilating magnons just as phonons could be created or annihilated. (To see this similarity unambiguously we can anticipate a result from Chapter 9 which shows for non-interacting spin waves (magnons) that $\mathrm{Im}\{G_{\boldsymbol{\kappa}}^{(+)}(\omega)\}$ is simply proportional to $\delta(\omega-\omega_{\boldsymbol{\kappa}})$, where $\omega_{\boldsymbol{\kappa}}$ is the frequency of a spin wave of wave vector $\boldsymbol{\kappa}$.)

Earlier, in connection with (8.37), we remarked that the general formulae simplified when every magnetic atom was a centre of symmetry, so that $\mathbf{q}$ and $-\mathbf{q}$ were equivalent. As we would expect when we have both this symmetry condition and also the condition that $\hat{S}^z_{\mathrm{tot}}$ be a constant of motion, the formulae become exceptionally simple. We shall now review these formulae when both conditions are satisfied.

(8.35 a) simplifies to
$$\phi^{zz}_{\mathbf{q}}(t)^* = \phi^{zz}_{\mathbf{q}}(t) = -\phi^{zz}_{\mathbf{q}}(-t) = \phi^{zz}_{-\mathbf{q}}(t) \tag{8.81 a}$$
and
$$\phi^{xx}_{\mathbf{q}}(t)^* = \phi^{xx}_{\mathbf{q}}(t) = -\phi^{xx}_{\mathbf{q}}(-t) = \phi^{xx}_{-\mathbf{q}}(t). \tag{8.81 b}$$
Hence $\phi^{zz}_{\mathbf{q}}(t)$ and $\phi^{xx}_{\mathbf{q}}(t)$ are real and odd functions of t.

Because $\phi^{zz}_{\mathbf{q}}(t)$ and $\phi^{xx}_{\mathbf{q}}(t)$ are real it follows that the real and imaginary parts of $\chi^{zz}_{\mathbf{q}}[\omega]$ and $\chi^{xx}_{\mathbf{q}}[\omega]$ are related by Kramers–Krönig relations (see (B.14)). Furthermore, from (8.35 b) we have
$$\chi^{zz}_{\mathbf{q}}[\omega]^* = \chi^{zz}_{\mathbf{q}}[-\omega], \tag{8.82 a}$$
$$\chi^{xx}_{\mathbf{q}}[\omega]^* = \chi^{xx}_{\mathbf{q}}[-\omega]. \tag{8.82 b}$$
Hence the real parts of $\chi^{zz}_{\mathbf{q}}[\omega]$ and $\chi^{xx}_{\mathbf{q}}[\omega]$ are both even in ω, whereas the imaginary parts are both odd in ω.

(8.35 c) now simplifies to
$$\mathscr{R}^{zz}_{\mathbf{q}}(t)^* = \mathscr{R}^{zz}_{\mathbf{q}}(t) = \mathscr{R}^{zz}_{\mathbf{q}}(-t) = \mathscr{R}^{zz}_{-\mathbf{q}}(t), \tag{8.82 c}$$
and
$$\mathscr{R}^{xx}_{\mathbf{q}}(t)^* = \mathscr{R}^{xx}_{\mathbf{q}}(t) = \mathscr{R}^{xx}_{\mathbf{q}}(-t) = \mathscr{R}^{xx}_{-\mathbf{q}}(t). \tag{8.82 d}$$
Hence both functions are real and even in t.

(8.35 d) simplifies to
$$\mathscr{R}^{zz}_{\mathbf{q}}(\omega)^* = \mathscr{R}^{zz}_{\mathbf{q}}(\omega) = \mathscr{R}^{zz}_{\mathbf{q}}(-\omega) = \mathscr{R}^{zz}_{-\mathbf{q}}(\omega), \tag{8.82 e}$$
$$\mathscr{R}^{xx}_{\mathbf{q}}(\omega)^* = \mathscr{R}^{xx}_{\mathbf{q}}(\omega) = \mathscr{R}^{xx}_{\mathbf{q}}(-\omega) = \mathscr{R}^{xx}_{-\mathbf{q}}(\omega). \tag{8.82 f}$$
Hence both functions, which are always real, now become even in ω.

It follows immediately from (8.43) that the spectral weights $F^{zz}_{\mathbf{q}}(\omega)$ and $F^{xx}_{\mathbf{q}}(\omega)$, which are always real, are also even in ω.

The functions $\phi^{(0)}_{\mathbf{q}}(t)$, $\chi^{(0)}_{\mathbf{q}}[\omega]$, $\mathscr{R}^{(0)}_{\mathbf{q}}(t)$, $\mathscr{R}^{(0)}_{\mathbf{q}}(\omega)$, and $G^{0}_{\mathbf{q}}(\omega)$ are identical to $\phi^{zz}_{\mathbf{q}}(t)$,..., and therefore simplify as outlined in eqns (8.81) and (8.82). The functions $\phi_{\mathbf{q}}^{(+)}(t)$, $\phi_{\mathbf{q}}^{(-)}(t)$,... do not simplify any further than as described by (8.70), (8.71), (8.72), and (8.73).

8.4. Relaxation functions and generalized susceptibilities: localized model and non-Bravais lattice

In this section we generalize the results of § 8.3 to a non-Bravais lattice. This is not too difficult because in § 8.3 we were careful not to use the symmetry properties of the Bravais lattice, except in a few comments which we noted explicitly.

We begin by defining the true spin density in this model. It is

$$\hat{\mathfrak{s}}(\mathbf{r},t) = \sum_{l,d} \tfrac{1}{2} g_d\, \hat{\mathfrak{s}}_{l+d}(t) \rho_d(\mathbf{r}-\mathbf{l}-\mathbf{d}), \tag{8.83}$$

where $\rho_d(\mathbf{R})$ is a normalized density centred on atom $\mathbf{l}+\mathbf{d}$ and with Fourier transform $F_d(\mathbf{q})\exp\{-W_d(\mathbf{q})\}$. Then the spin magnetization density is

$$\hat{\mathbf{M}}(\mathbf{r},t) = -2\mu_{\mathrm{B}}\, \hat{\mathfrak{s}}(\mathbf{r},t) \tag{8.84}$$

and the interaction energy with a magnetic field is

$$-\int d\mathbf{r}\; \hat{\mathbf{M}}(\mathbf{r},t).\mathbf{H}(\mathbf{r},t).$$

Hence if $\mathbf{H}(\mathbf{r},t)$ is of the form $\tilde{\boldsymbol{\beta}} H_{\mathbf{q}}^{\beta}(t)\exp(i\mathbf{q}.\mathbf{r})$ this interaction energy becomes

$$\sum_{l,d} \mu_{\mathrm{B}} g_d F_d(\mathbf{q}) \exp\{-W_d(\mathbf{q})\}\, \hat{S}_{l+d}^{\beta}(t) \exp\{-i\mathbf{q}.(\mathbf{l}+\mathbf{d})\}\, H_{\mathbf{q}}^{\beta}(t).$$

From these expressions we see immediately that when we ask for the response of the Fourier transform of $\hat{\mathbf{M}}(\mathbf{r},t)$, i.e.

$$\begin{aligned}\hat{\mathbf{M}}_{\mathbf{q}}(t) &= \int d\mathbf{r}\, \exp(-\mathbf{q}.\mathbf{r})\hat{\mathbf{M}}(\mathbf{r},t) \\ &= -\sum_{l,d} g_d \mu_{\mathrm{B}} \hat{S}_{l+d}(t) F_d(\mathbf{q}) \exp\{-W_d(\mathbf{q})\} \exp\{-i\mathbf{q}.(\mathbf{l}+\mathbf{d})\}, \end{aligned} \tag{8.85}$$

to an applied field like $\tilde{\boldsymbol{\beta}} H_{\mathbf{q}}^{\beta} \exp(i\mathbf{q}.\mathbf{r})$, then additional factors like $F_d(\mathbf{q})\exp\{-W_d(\mathbf{q})\}$ are going to occur. This suggests that we define operators

$$\hat{T}_{\mathbf{q}}^{\alpha}(t) = \sum_{l,d} \tfrac{1}{2} g_d F_d(\mathbf{q}) \exp\{-W_d(\mathbf{q})\} \exp\{-i\mathbf{q}.(\mathbf{l}+\mathbf{d})\} \hat{S}_{l+d}^{\alpha} \tag{8.86}$$

and a response function

$$\tilde{\phi}_{\mathbf{q}}^{\alpha\beta}(t) = \frac{(2\mu_{\mathrm{B}})^2}{N}\frac{i}{\hbar} \langle[\hat{T}_{\mathbf{q}}^{\alpha}(t), \hat{T}_{-\mathbf{q}}^{\beta}(0)]\rangle, \tag{8.87}$$

where N is the number of unit cells in the crystal and the normalization of (8.87) is chosen so as to keep a close analogy with (8.29 b). We notice that the response function $\tilde{\phi}_{\mathbf{q}}^{\alpha\beta}(t)$ defined by (8.87) is not the same as $\phi_{\mathbf{q}}^{\alpha\beta}(t)$ defined by (8.29 b), but we shall find they have many symmetry properties in common.

We then define a generalized susceptibility as

$$\bar{\chi}^{\alpha\beta}_{\mathbf{q}}[\omega] = \int_0^\infty \mathrm{d}t \exp(\mathrm{i}\omega t)\, \bar{\phi}^{\alpha\beta}_{\mathbf{q}}(t) \tag{8.88}$$

and a relaxation function through

$$\frac{\partial}{\partial t} \bar{\mathscr{R}}^{\alpha\beta}_{\mathbf{q}}(t) = \frac{-N}{(2\mu_{\mathrm{B}})^2} \bar{\phi}^{\alpha\beta}_{\mathbf{q}}(t) \tag{8.89}$$

with Fourier transform

$$\bar{\mathscr{R}}^{\alpha\beta}_{\mathbf{q}}(\omega) = \frac{1}{2\pi} \int_{-\infty}^{\infty} \mathrm{d}t \exp(\mathrm{i}\omega t)\, \bar{\mathscr{R}}^{\alpha\beta}_{\mathbf{q}}(t). \tag{8.90}$$

Then in place of (8.32) we have

$$\bar{\mathscr{R}}^{\alpha\beta}_{\mathbf{q}}(t) = \int_0^\beta \mathrm{d}\lambda\, \langle \hat{T}^{\beta}_{-\mathbf{q}}(-\mathrm{i}\hbar\lambda)\hat{T}^{\alpha}_{\mathbf{q}}(t)\rangle - \beta\langle \hat{T}^{\beta}_{-\mathbf{q}}(0)\hat{T}^{\alpha}_{\mathbf{q}}(\infty)\rangle. \tag{8.91}$$

In terms of these new operators, the magnetic cross-section becomes, in place of (8.26 a),

$$\frac{\mathrm{d}^2\sigma}{\mathrm{d}\Omega \mathrm{d}E'} = \left(\frac{\gamma e^2}{m_{\mathrm{e}} c^2}\right)^2 \frac{k'}{k} \sum_{\alpha\beta} (\delta_{\alpha\beta} - \tilde{\kappa}_\alpha \tilde{\kappa}_\beta) \frac{1}{2\pi\hbar} \int_{-\infty}^{\infty} \mathrm{d}t \exp(-\mathrm{i}\omega t)\, \langle \hat{T}^{\alpha}_{\boldsymbol{\kappa}}(0)\hat{T}^{\beta}_{-\boldsymbol{\kappa}}(t)\rangle, \tag{8.92}$$

and in place of (8.34) we get

$$\frac{1}{2\pi} \int_{-\infty}^{\infty} \mathrm{d}t \exp(\mathrm{i}\omega t)\, \langle \hat{T}^{\beta}_{-\mathbf{q}}(0)\hat{T}^{\alpha}_{\mathbf{q}}(t)\rangle = \frac{\hbar\omega}{\exp(\beta\hbar\omega)-1} \bar{\mathscr{R}}^{\alpha\beta}_{\mathbf{q}}(\omega) + \delta(\omega)\langle \hat{T}^{\beta}_{-\mathbf{q}}(0)\hat{T}^{\alpha}_{\mathbf{q}}(\infty)\rangle. \tag{8.93}$$

Now in (8.86)

$$F_d(\mathbf{q})\exp\{-W_d(\mathbf{q})\} = \int \mathrm{d}\mathbf{r} \exp\{-\mathrm{i}\mathbf{q}.\mathbf{r}\}\rho(\mathbf{r})$$
$$= \langle \exp(-\mathrm{i}\mathbf{q}.\hat{\mathbf{u}}_d)\rangle \int \mathrm{d}\mathbf{r} \exp(-\mathrm{i}\mathbf{q}.\mathbf{r})\, |\phi(\mathbf{r})|^2. \tag{8.94 a}$$

Hence $\quad F_d(\mathbf{q})^* \exp\{-W_d(\mathbf{q})^*\} = F_d(-\mathbf{q})\exp\{-W_d(-\mathbf{q})\}. \qquad (8.94\,\mathrm{b})$

It follows that the Hermitian conjugate of $\hat{T}^{\alpha}_{\mathbf{q}}$ is $\hat{T}^{\alpha}_{-\mathbf{q}}$ and that all the symmetry relations for ϕ, χ, and $\mathscr{R}$ listed in (8.35) apply also to the functions $\bar{\phi}$, $\bar{\chi}$, $\bar{\mathscr{R}}$ defined in this section. In particular, using (8.35 d), we find that (8.92) and (8.93) can be combined to give

$$\left(\frac{\mathrm{d}^2\sigma}{\mathrm{d}\Omega \mathrm{d}E'}\right)_{\mathrm{inelas}} = \left(\frac{\gamma e^2}{m_{\mathrm{e}} c^2}\right)^2 \frac{k'}{k} \sum_{\alpha\beta} (\delta_{\alpha\beta} - \tilde{\kappa}_\alpha \tilde{\kappa}_\beta) \frac{\omega \exp(\hbar\omega\beta)}{\exp(\hbar\omega\beta)-1} \bar{\mathscr{R}}^{\alpha\beta}_{\boldsymbol{\kappa}}(\omega). \tag{8.95}$$

Now in place of (8.38) we get

$$\bar{\mathscr{R}}_{\kappa}^{\alpha\beta}(\omega) = \frac{N}{2\pi i\omega(2\mu_B)^2}\{\bar{\chi}_{\mathbf{q}}^{\alpha\beta}[\omega] - \bar{\chi}_{\mathbf{q}}^{\beta\alpha}[\omega]^*\}, \tag{8.96}$$

and hence, in place of (8.40),

$$\left(\frac{d^2\sigma}{d\Omega dE'}\right)_{\text{inelas}} = \left(\frac{\gamma e^2}{m_e c^2}\right)^2 \frac{k'}{k} \sum_{\alpha\beta} (\delta_{\alpha\beta} - \tilde{\kappa}_\alpha \tilde{\kappa}_\beta) \times$$

$$\times \frac{N}{\pi(2\mu_B)^2} \frac{\exp(\hbar\omega\beta)}{\exp(\hbar\omega\beta)-1} \text{Im}\{\bar{\chi}_{\kappa}^{\alpha\beta}[\omega]\}. \tag{8.97}$$

In place of (8.41) we now get

$$\bar{\chi}_{\mathbf{q}}^{\alpha\beta}[\omega] = \frac{(2\mu_B)^2}{N} \bar{\mathscr{R}}_{\mathbf{q}}^{\alpha\beta}(t=0) + \frac{(2\mu_B)^2}{N} i\omega \int_0^\infty dt \exp(i\omega t)\, \bar{\mathscr{R}}_{\mathbf{q}}^{\alpha\beta}(t). \tag{8.98}$$

Now $\bar{\mathscr{R}}_{\mathbf{q}}^{\alpha\beta}(t=0)$ is proportional to the total isothermal (see p. 240) susceptibility. Hence (8.98) tells us that the susceptibility, $\bar{\chi}_{\mathbf{q}}^{\alpha\beta}[\omega]$, is defined as the susceptibility per unit cell. The isothermal susceptibility per unit cell is

$$\bar{\chi}_{\mathbf{q}}^{\alpha\beta} = \bar{\chi}_{\mathbf{q}}^{\alpha\beta}[\omega=0] = \frac{(2\mu_B)^2}{N} \bar{\mathscr{R}}_{\mathbf{q}}^{\alpha\beta}(t=0). \tag{8.99}$$

In our new variables we define the retarded Green function, in place of (8.48), as

$$\bar{G}_{\kappa}^{\alpha\beta}(t) = -i\theta(t)\langle[\hat{T}_{\kappa}^{\alpha}(t), \hat{T}_{-\kappa}^{\beta}(0)]\rangle \tag{8.100}$$

and then find

$$\bar{G}_{\kappa}^{\alpha\beta}(\omega) = \frac{-N\hbar}{2\pi(2\mu_B)^2} \bar{\chi}_{\kappa}^{\alpha\beta}[\omega], \tag{8.101}$$

which, using (8.97), we can use to give

$$\left(\frac{d^2\sigma}{d\Omega dE'}\right)_{\text{inelas}} = \left(\frac{\gamma e^2}{m_e c^2}\right)^2 \frac{k'}{k} \sum_{\alpha\beta} (\delta_{\alpha\beta} - \tilde{\kappa}_\alpha \tilde{\kappa}_\beta) \frac{-\hbar}{2} \frac{\exp(\hbar\omega\beta)}{\exp(\hbar\omega\beta)-1} \text{Im}\, \bar{G}_{\kappa}^{\alpha\beta}(\omega). \tag{8.102}$$

When the total z component of spin is a constant of motion the cross-section again splits into a longitudinal and transverse part, where (in place of (8.55))

$$\left(\frac{d^2\sigma}{d\Omega dE'}\right)_{\text{inelas}}^{\text{long}} = \left(\frac{\gamma e^2}{m_e c^2}\right)^2 \frac{k'}{k} (1-\tilde{\kappa}_z^2) \frac{\omega \exp(\hbar\omega\beta)}{\exp(\hbar\omega\beta)-1} \bar{\mathscr{R}}_{\kappa}^{zz}(\omega) \tag{8.103}$$

and (in place of (8.56))

$$\left(\frac{\mathrm{d}^2\sigma}{\mathrm{d}\Omega\,\mathrm{d}E'}\right)^{\text{trans}}_{\text{inelas}} = \left(\frac{\gamma e^2}{m_e c^2}\right)^2 \frac{k'}{k}(1+\tilde{\kappa}_z^2)\,\frac{\omega\exp(\hbar\omega\beta)}{\exp(\hbar\omega\beta)-1}\,\bar{\mathscr{R}}^{xx}_{\kappa}(\omega). \quad (8.104)$$

To make analogy to the formulation in terms of raising and lowering operators we take care to define

$$\hat{T}^{+}_{\mathbf{q}}(t) = \sum_{l,d} g_d F_d(\mathbf{q})\exp\{-W_d(\mathbf{q})\}\exp\{-\mathrm{i}\mathbf{q}\,.\,(\mathbf{l}+\mathbf{d})\}\hat{S}^{+}_{l+d}$$

$$= \hat{T}^{x}_{\mathbf{q}}(t)+\mathrm{i}\hat{T}^{y}_{\mathbf{q}}(t) \quad (8.105)$$

and

$$\hat{T}^{-}_{\mathbf{q}}(t) = \sum_{l,d} g_d F_d(\mathbf{q})^{*}\exp\{-W_d(\mathbf{q})^{*}\}\exp\{\mathrm{i}\mathbf{q}\,.\,(\mathbf{l}+\mathbf{d})\}\,\hat{S}^{+}_{l+d}$$

$$= \{T^{+}_{\mathbf{q}}(t)\}^{+} = \hat{T}^{x}_{-\mathbf{q}}-\mathrm{i}\hat{T}^{y}_{-\mathbf{q}}. \quad (8.106)$$

Then all the discussion in § 8.3 that starts from (8.60) follows through as before: in particular, the symmetry relations in eqns (8.69)–(8.73) remain valid provided we define $\bar{\phi}^{\nu}_{\mathbf{q}}(t)$ and $\bar{\mathscr{R}}^{(\nu)}_{\mathbf{q}}(t)$ in terms of the $\hat{T}^{\nu}_{\mathbf{q}}(t)$ operators instead of using $\phi^{(\nu)}_{\mathbf{q}}(t)$, $\mathscr{R}^{(\nu)}_{\mathbf{q}}(t)$, and the $\hat{S}^{\nu}_{\mathbf{q}}(t)$ operators as in (8.63) and (8.66). To be quite explicit,

$$\bar{\phi}^{(0)}_{\mathbf{q}}(t) = \bar{\phi}^{zz}_{\mathbf{q}}(t), \quad (8.107\,\text{a})$$

$$\bar{\phi}^{(\nu)}_{\mathbf{q}}(t) = \frac{(2\mu_{\mathrm{B}})^2}{N}\frac{\mathrm{i}}{\hbar}\langle[\hat{T}^{\nu}_{\mathbf{q}}(t), \hat{T}^{-\nu}_{\mathbf{q}}(0)]\rangle \quad \text{for } \nu = + \text{ or } -; \quad (8.107\,\text{b})$$

$$\bar{\mathscr{R}}^{(0)}_{\mathbf{q}}(t) = \bar{\mathscr{R}}^{zz}_{\mathbf{q}}(t), \quad (8.107\,\text{c})$$

$$\bar{\mathscr{R}}^{(\nu)}_{\mathbf{q}}(t) = \int_0^{\beta}\mathrm{d}\lambda\,\langle\hat{T}^{-\nu}_{\mathbf{q}}(-\mathrm{i}\hbar\lambda)\hat{T}^{\nu}_{\mathbf{q}}(t)\rangle-\beta\langle T^{-\nu}_{\mathbf{q}}(0)T^{\nu}_{\mathbf{q}}(\infty)\rangle \quad \text{for } \nu = + \text{ or } -. \quad (8.107\,\text{d})$$

In the longitudinal cross-section (8.103) we can now work with $\bar{\mathscr{R}}^{zz}_{\kappa}(\omega)$ directly as defined by (8.90) and (8.91) or alternatively we can express it as

$$\bar{\mathscr{R}}^{zz}_{\kappa}(\omega) = \bar{\mathscr{R}}^{(0)}_{\kappa}(\omega)$$

$$= \frac{N}{(2\mu_{\mathrm{B}})^2}\bar{\chi}^{zz}_{\kappa}\,\bar{F}^{zz}_{\kappa}(\omega)$$

$$= \frac{N}{\pi(2\mu_{\mathrm{B}})^2\omega}\,\mathrm{Im}\,\bar{\chi}^{(0)}_{\kappa}[\omega]$$

$$= \frac{-2}{\hbar\omega}\,\mathrm{Im}\,\bar{G}^{0}_{\kappa}(\omega). \quad (8.108)$$

In the transverse cross-section we have the following alternatives for $\bar{\mathscr{R}}^{xx}_{\varkappa}(\omega)$:

$$\bar{\mathscr{R}}^{xx}_{\varkappa}(\omega) = \frac{N}{(2\mu_{\mathrm{B}})^2}\bar{\chi}^{xx}_{\varkappa}\,\bar{F}^{xx}_{\varkappa}(\omega)$$

$$= \frac{N}{\pi(2\mu_{\mathrm{B}})^2\omega}\,\mathrm{Im}\,\bar{\chi}^{xx}_{\varkappa}[\omega]$$

$$= \frac{-2}{\hbar\omega}\,\mathrm{Im}\,\bar{G}^{xx}_{\varkappa}(\omega)$$

$$= \tfrac{1}{4}\{\bar{\mathscr{R}}^{(+)}_{\varkappa}(\omega)+\bar{\mathscr{R}}^{(-)}_{-\varkappa}(\omega)\}$$

$$= \frac{N}{16\pi\mu_{\mathrm{B}}^2\,\omega}\,\mathrm{Im}\{\bar{\chi}^{(+)}_{\varkappa}[\omega]+\bar{\chi}^{(-)}_{-\varkappa}[\omega]\}$$

$$= \frac{-1}{2\hbar\omega}\,\mathrm{Im}\{\bar{G}^{(+)}_{\varkappa}(\omega)+\bar{G}^{(-)}_{-\varkappa}(\omega)\}. \qquad (8.109)$$

This is as much discussion as we need give for the general problem. It is evident that the functions used here, $\bar{\mathscr{R}}(t)$, $\bar{\mathscr{R}}(\omega)$, $\bar{\chi}[\omega]$, etc., reduce to the simpler forms $\mathscr{R}(t)$, $\mathscr{R}(\omega)$, $\chi[\omega]$, etc., in special cases when g is replaced by 2, the Debye–Waller factor is ignored, and we assume one atom per unit cell.

8.5. Generalized susceptibility: itinerant model

The analysis given in the previous two sections can be repeated for the itinerant model using eqns (8.19) and (8.20) as starting point. We shall only sketch the results here assuming an electron gas with straightforward exchange effects so that $\hat{S}^z_{\mathrm{tot}}$ is a constant of the motion.

From (8.21) we note that

$$\mathfrak{J}^z(\mathbf{q},t) = \tfrac{1}{2}F(\mathbf{q})\sum_{\mathbf{k}}\{\hat{c}^{+}_{\mathbf{k}\uparrow}(t)\hat{c}_{\mathbf{k}+\mathbf{q}\uparrow}(t)-\hat{c}^{+}_{\mathbf{k}\downarrow}(t)\hat{c}_{\mathbf{k}+\mathbf{q}\downarrow}(t)\},$$

$$\mathfrak{J}^{(+)}(\mathbf{q},t) = F(\mathbf{q})\sum_{\mathbf{k}}\hat{c}^{+}_{\mathbf{k}\uparrow}(t)\hat{c}_{\mathbf{k}+\mathbf{q}\downarrow}(t),$$

$$\mathfrak{J}^{(-)}(\mathbf{q},t) = F(\mathbf{q})\sum_{\mathbf{k}}\hat{c}^{+}_{\mathbf{k}+\mathbf{q}\downarrow}(t)\hat{c}_{\mathbf{k}\uparrow}(t). \qquad (8.110)$$

Then by analogy to (8.87) and (8.107) we define response functions

$$\bar{\phi}^{(0)}_{\mathbf{q}}(t) = \frac{(2\mu_{\mathrm{B}})^2}{N}\frac{\mathrm{i}}{\hbar}\langle[\mathfrak{J}^z(\mathbf{q},t),\mathfrak{J}^z(-\mathbf{q},0)]\rangle,$$

$$\bar{\phi}^{(\nu)}_{\mathbf{q}}(t) = \frac{(2\mu_{\mathrm{B}})^2}{N}\frac{\mathrm{i}}{\hbar}\langle[\mathfrak{J}^{\nu}(\mathbf{q},t),\mathfrak{J}^{-\nu}(\mathbf{q},0)]\rangle. \qquad (8.111)$$

From these we can define relaxation functions and generalized susceptibility functions in exact analogy to the previous sections. The cross-section is then again given by (8.103) and (8.104) in conjunction with (8.108) and (8.109). One example of the use of this notation is given in Chapter 9, where we discuss spin waves in an electron gas.

9

SCATTERING BY SPIN WAVES

9.1. Introduction

If we draw analogies between the scattering by the purely nuclear and purely magnetic interactions of neutrons, then the present chapter on scattering by spin waves corresponds to scattering by phonons. There is, however, an important difference between the two in that the dynamical behaviour of magnetic systems is more complex, even in the linear region where the spin waves in the magnetic system are non-interacting, this region corresponding in the nuclear problem to the harmonic approximation to lattice vibrations.

While spin waves have long been understood for magnetic systems based on the localized model, or the Heisenberg model, the theory of spin waves in a system of itinerant electrons has been formulated only in recent years. The techniques for dealing with the latter are perhaps not so familiar as those for describing spin waves in the Heisenberg model. In §§ 9.2 and 9.3 the cross-section for inelastic scattering by ferromagnets at low temperatures is discussed, first for the Heisenberg model and then, in § 9.3, for an itinerant electron model of ferromagnetism. The discussion of the Heisenberg model in § 9.2 is for crystals with a Bravais lattice structure and, in addition, dipolar forces are not included. § 9.4 therefore contains a discussion of dipolar forces and in § 9.5 the spin-wave spectrum and corresponding neutron cross-sections are studied for a Heisenberg Hamiltonian appropriate to some rare-earth metals. In the remaining sections, §§ 9.6 and 9.7, the cross-section for the scattering by antiferromagnets and ferrimagnet systems are derived, again for the Heisenberg model.

That we discuss spin waves in an itinerant electron model only for the ferromagnetic case is essentially a reflection of the present understanding of the whole problem, for, although some work has been done on other itinerant systems, the full development of the subject is yet to come.

There are many review articles on spin waves in the Heisenberg model, those by Van Kranendonk and Van Vleck† and Keffer‡ being perhaps closest in outlook to the approach employed here. For further details

† J. Van Kranendonk and J. H. Van Vleck, *Rev. mod. Phys.* **1**, **30** (1958).
‡ F. Keffer, *Handb. Phys.* **18**, 2 (1967).

and references on spin waves in itinerant electron systems we refer the reader to the book by Herring.†

As with the discussion of scattering by phonons, space prevents a detailed discussion of the experimental techniques used and the problems encountered in measuring spin-wave dispersion curves. Lowde has presented detailed discussions of the cross-sections for scattering by spin waves‡ and we would also cite those given by Collins§ and by Shirane *et al.*|| for further details on some points.

9.2. Spin waves in a Heisenberg ferromagnet

(*a*) *Linear spin-wave theory*

The Heisenberg Hamiltonian for a ferromagnetic system with an external magnetic field H in the z direction is

$$\mathscr{H} = -\sum_{\mathbf{l},\mathbf{l}'} J(\mathbf{l}-\mathbf{l}')\hat{\mathbf{S}}_{\mathbf{l}}.\hat{\mathbf{S}}_{\mathbf{l}'} - g\mu_{\mathrm{B}} H \sum_{\mathbf{l}} \hat{S}^z_{\mathbf{l}}. \tag{9.1}$$

In (9.1), $J(\mathbf{l}-\mathbf{l}')$ is the exchange parameter between spins situated at the Bravais lattice sites $\mathbf{l}$ and $\mathbf{l}'$ and is defined such that $J(0) = 0$.

Because the total z component of spin commutes with $\mathscr{H}$ it follows from our discussion in § 8.3 that the cross-section for the scattering by this system contains only the longitudinal term $\langle \hat{S}^z_{\mathbf{l}}(0)\hat{S}^z_{\mathbf{l}'}(t)\rangle$ and the transverse terms $\langle \hat{S}^{\pm}_{\mathbf{l}}(0)\hat{S}^{\mp}_{\mathbf{l}'}(t)\rangle$. At low temperatures the longitudinal term gives no inelastic scattering, as we shall presently confirm, and the transverse terms give rise to one-magnon creation and annihilation scattering processes. (The energy of a spin wave is quantized, and the unit of energy of a spin wave is called a magnon.)

On introducing the spin angular momentum raising and lowering operators $\hat{S}^{\pm} = \hat{S}^x \pm \mathrm{i}\,\hat{S}^y$ in (9.1), the Hamiltonian can be written

$$\mathscr{H} = -\sum_{\mathbf{l},\mathbf{l}'} J(\mathbf{l}-\mathbf{l}')\{\hat{S}^z_{\mathbf{l}}\hat{S}^z_{\mathbf{l}'} + \hat{S}^+_{\mathbf{l}}\hat{S}^-_{\mathbf{l}'}\} - g\mu_{\mathrm{B}} H \sum_{\mathbf{l}} \hat{S}^z_{\mathbf{l}}. \tag{9.2}$$

We recall that the operators $\hat{S}^{\pm}$ satisfy the commutation relations

$$[\hat{S}^+_{\mathbf{l}}, \hat{S}^-_{\mathbf{l}'}] = \delta_{\mathbf{l},\mathbf{l}'} 2\hat{S}^z_{\mathbf{l}} \tag{9.3 a}$$

and

$$[\hat{S}^z_{\mathbf{l}}, \hat{S}^+_{\mathbf{l}'}] = \delta_{\mathbf{l},\mathbf{l}'}\hat{S}^+_{\mathbf{l}}, \qquad [\hat{S}^z_{\mathbf{l}}, \hat{S}^-_{\mathbf{l}'}] = -\delta_{\mathbf{l},\mathbf{l}'}\hat{S}^-_{\mathbf{l}}. \tag{9.3 b}$$

† C. Herring, *Magnetism* **4**, ed. by G. T. Rado and H. Suhl (Academic Press, 1963).

‡ R. D. Lowde, lectures in *An advanced course on magnetic exchange interactions*, Rept. KR93 of Institut for Atomenergi, Kjeller, Norway and *J. appl. Phys.* **36**, 884 (1965).

§ M. F. Collins, lectures in *Elementary excitations and their interactions in solids*, NATO Advanced Study Institute, Cortina D'Ampezzo, Italy (1966), ed. A. A. Maradudin.

|| G. Shirane, V. J. Minkiewicz, and R. Nathans, *J. appl. Phys.* **39**, 383 (1968).

With the aid of these commutation relations the equation of motion for the operator $\hat{S}^+$ is readily obtained. We find

$$i\hbar\,\dot{\hat{S}}^+_{\mathbf{l}_1} = [\hat{S}^+_{\mathbf{l}_1},\mathscr{H}]$$
$$= g\mu_B H\hat{S}^+_{\mathbf{l}_1} - \sum_{\mathbf{l},\mathbf{l}'} J(\mathbf{l}-\mathbf{l}')\{2\delta_{\mathbf{l}_1,\mathbf{l}'}\hat{S}^+_{\mathbf{l}}\hat{S}^z_{\mathbf{l}_1} - \delta_{\mathbf{l}_1,\mathbf{l}}\hat{S}^+_{\mathbf{l}_1}\hat{S}^z_{\mathbf{l}'} - \delta_{\mathbf{l}_1,\mathbf{l}'}\hat{S}^z_{\mathbf{l}}\hat{S}^+_{\mathbf{l}_1}\}. \quad (9.4)$$

If $\mathbf{l}$ and $\mathbf{l}'$ are interchanged in the third term of the second part of (9.4) then, on account of the fact that

$$J(\mathbf{l}-\mathbf{l}') = J(\mathbf{l}'-\mathbf{l}) \quad (9.5)$$

because the crystal is Bravais, we get

$$i\hbar\dot{\hat{S}}^+_{\mathbf{l}} = g\mu_B H\hat{S}^+_{\mathbf{l}} + 2\sum_{\mathbf{l}'} J(\mathbf{l}-\mathbf{l}')\{\hat{S}^z_{\mathbf{l}'}\hat{S}^+_{\mathbf{l}} - \hat{S}^+_{\mathbf{l}'}\hat{S}^z_{\mathbf{l}}\}. \quad (9.6)$$

This equation of motion is made linear in $\hat{S}^+$ by making the replacement $\hat{S}^z \rightarrow S$. Thus, in this linear approximation,

$$i\hbar\dot{\hat{S}}^+_{\mathbf{l}} = g\mu_B H\hat{S}^+_{\mathbf{l}} + 2S\sum_{\mathbf{l}'} J(\mathbf{l}-\mathbf{l}')(\hat{S}^+_{\mathbf{l}} - \hat{S}^+_{\mathbf{l}'}). \quad (9.7)$$

If we introduce the operators $\hat{S}^\pm_{\mathbf{q}}$ through the canonical transformation

$$\hat{S}^\pm_{\mathbf{l}} = \frac{1}{N}\sum_{\mathbf{q}} \exp(\pm i\mathbf{q}.\mathbf{l})\hat{S}^\pm_{\mathbf{q}} \quad (9.8)$$

as in § 8.3, N being, as usual, the number of unit cells, the equation of motion for $\hat{S}^+_{\mathbf{q}}$ is, from (9.7),

$$i\hbar\dot{\hat{S}}^+_{\mathbf{q}} = g\mu_B H\hat{S}^+_{\mathbf{q}} + 2S\{\mathscr{J}(0) - \mathscr{J}(\mathbf{q})\}\hat{S}^+_{\mathbf{q}}. \quad (9.9)$$

Here
$$\mathscr{J}(\mathbf{q}) = \sum_{\mathbf{l}} J(\mathbf{l})\exp(-i\mathbf{q}.\mathbf{l}). \quad (9.10\,\text{a})$$

Note that
$$\mathscr{J}(\mathbf{q}) = \mathscr{J}(-\mathbf{q}) \quad (9.10\,\text{b})$$

because the lattice is Bravais, and, also,

$$\mathscr{J}(\mathbf{q}+\boldsymbol{\tau}) = \mathscr{J}(\mathbf{q}) \quad (9.10\,\text{c})$$

by definition of the reciprocal lattice vectors $\boldsymbol{\tau}$ (cf. § 2.1).

To solve (9.9), set

$$\hat{S}^+_{\mathbf{q}}(t) = \exp(it\mathscr{H}/\hbar)\,\hat{S}^+_{\mathbf{q}}\exp(-it\mathscr{H}/\hbar) = \exp(-i\omega_{\mathbf{q}}t)\hat{S}^+_{\mathbf{q}}, \quad (9.11)$$

then, clearly,
$$\hbar\omega_{\mathbf{q}} = g\mu_B H + 2S\{\mathscr{J}(0) - \mathscr{J}(\mathbf{q})\}, \quad (9.12)$$

and this is the dispersion relation for linear spin waves in a Heisenberg ferromagnet.

Before discussing (9.12) and its consequences, we note that the linearized equation of motion (9.7) could have been obtained directly by modifying the commutation relation (9.3 a) to read

$$[\hat{S}^+_{\mathbf{l}}, \hat{S}^-_{\mathbf{l}'}] \simeq \delta_{\mathbf{l},\mathbf{l}'}2S, \quad (9.13)$$

i.e. by replacing $\hat{S}^z$ by the c-number S, as is appropriate near saturation. We also need an expression for $\hat{S}^z$ itself, and an appropriate form is

$$\hat{S}^z_{\mathbf{l}} \simeq S - \frac{1}{2S}\hat{S}^-_{\mathbf{l}}\hat{S}^+_{\mathbf{l}}, \tag{9.14}$$

because (9.14) is valid both when $S^z_{\mathbf{l}} = S$ and when there is a single deviation on the atom at $\mathbf{l}$, i.e. when $S^z_{\mathbf{l}} = S-1$. The approximations (9.13) and (9.14) preserve the commutation relations (9.3). (Clearly (9.14) is exact for the special case of $S = \frac{1}{2}$.) This approximation scheme is very convenient to employ in deriving linearized equations of motion for more complex magnetic systems than that described by (9.1) and we shall adopt it throughout for this purpose.

Let us examine the dispersion relation (9.12). First, if the exchange parameter is between nearest-neighbour spins only, and of magnitude J,

$$\mathscr{J}(\mathbf{q}) = J\sum_{\boldsymbol{\rho}} \exp(i\mathbf{q}.\boldsymbol{\rho}) = rJ\gamma_{\mathbf{q}}, \tag{9.15}$$

where r is the number of nearest neighbours. Thus

$$\hbar\omega_{\mathbf{q}} = g\mu_{\mathrm{B}} H + 2rJS(1-\gamma_{\mathbf{q}}). \tag{9.16}$$

For small $\mathbf{q}$ and cubic (Bravais) lattices

$$\gamma_{\mathbf{q}} = \frac{1}{r}\sum_{\boldsymbol{\rho}} \exp(i\mathbf{q}.\boldsymbol{\rho}) = \frac{1}{r}\sum_{\boldsymbol{\rho}} \{1+i\mathbf{q}.\boldsymbol{\rho}+\tfrac{1}{2}(i\mathbf{q}.\boldsymbol{\rho})^2+...\}$$

$$\simeq \frac{1}{r}\sum_{\boldsymbol{\rho}} \{1-\tfrac{1}{2}\tfrac{1}{3}q^2\rho^2\} = 1-\tfrac{1}{6}q^2\rho^2. \tag{9.17}$$

Now

$$\rho^2 = \begin{cases} a^2 & \text{s.c.} \quad (r=6) \\ 3a^2/4 & \text{b.c.c.} \quad (r=8) \\ a^2/2 & \text{f.c.c.} \quad (r=12), \end{cases}$$

where a is the lattice parameter (cf. § 2.1), so $r\frac{1}{6}\rho^2 = a^2$ for all three cubic lattices. Hence

$$\hbar\omega_{\mathbf{q}} \simeq g\mu_{\mathrm{B}} H + Dq^2 \quad (|\mathbf{q}|\,|\boldsymbol{\rho}| \ll 1), \tag{9.18}$$

where

$$D = 2JSa^2. \tag{9.19}$$

In Fig. 9.1 the dispersion curve (9.16) is shown together with the approximation ($H = 0$) $\hbar\omega_{\mathbf{q}} \simeq Dq^2$.

More generally, when the exchange parameter is of longer range than just to nearest neighbours,

$$D = \tfrac{1}{3}S\sum_{\mathbf{l}} l^2 J(\mathbf{l}). \tag{9.20}$$

For terms up to $(\mathbf{q}.\mathbf{l})^6$

$$\hbar\omega_{\mathbf{q}} = g\mu_{\mathrm{B}}H + 2S\{\tfrac{1}{6}q^2 \sum_{\mathbf{l}} l^2 J(\mathbf{l}) - \tfrac{1}{24}\sum_{\mathbf{l}} (\mathbf{q}.\mathbf{l})^4 J(\mathbf{l}) + \tfrac{1}{720}\sum_{\mathbf{l}} (\mathbf{q}.\mathbf{l})^6 J(\mathbf{l})\}. \tag{9.21}$$

We shall presently find it convenient to express the results obtained from this expansion in terms of the moments

$$\overline{l^n} = S\Big\{\sum l^{n+2} J(\mathbf{l})\Big\}\Big/3D. \tag{9.22}$$

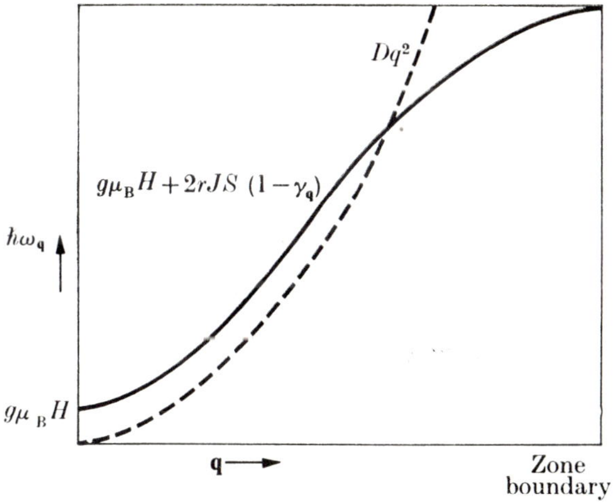

FIG. 9.1. The dispersion curve (9.16) is shown together with the small $\mathbf{q}$ approximation (for $H = 0$) $\hbar\omega_{\mathbf{q}} \cong Dq^2$.

To calculate the average value of the z component of spin at the site $\mathbf{l}$ we proceed as follows. From (9.8) and (9.14)

$$\langle \hat{S}^z_{\mathbf{l}} \rangle = S - \frac{1}{2S}\langle \hat{S}^-_{\mathbf{l}} \hat{S}^+_{\mathbf{l}} \rangle$$

$$= S - \frac{1}{2SN^2}\sum_{\mathbf{q}} \langle \hat{S}^-_{\mathbf{q}} \hat{S}^+_{\mathbf{q}} \rangle. \tag{9.23}$$

By definition

$$\langle \hat{S}^-_{\mathbf{q}} \hat{S}^+_{\mathbf{q}} \rangle = \mathrm{Tr}\exp(-\beta\mathscr{H})\,\hat{S}^-_{\mathbf{q}}\hat{S}^+_{\mathbf{q}}/\mathrm{Tr}\exp(-\beta\mathscr{H})$$

$$= \mathrm{Tr}\exp(-\beta\mathscr{H})\exp(\beta\mathscr{H})\,\hat{S}^+_{\mathbf{q}}\exp(-\beta\mathscr{H})\,\hat{S}^-_{\mathbf{q}}/\mathrm{Tr}\exp(-\beta\mathscr{H}), \tag{9.24}$$

the second line following by virtue of the invariance of the trace under a cyclic permutation of operators. Since

$$\exp(\beta\mathscr{H})\,\hat{S}^+_{\mathbf{q}}\exp(-\beta\mathscr{H}) = \hat{S}^+_{\mathbf{q}}(-\mathrm{i}\hbar\beta) \tag{9.25}$$

it follows from (9.11) that, in the linear approximation,

$$\exp(\beta\mathscr{H})\,\hat{S}_{\mathbf{q}}^{+}\exp(-\beta\mathscr{H}) = \hat{S}_{\mathbf{q}}^{+}\exp\{-\mathrm{i}\omega_{\mathbf{q}}(-\mathrm{i}\hbar\beta)\}$$
$$= \hat{S}_{\mathbf{q}}^{+}\exp(-\hbar\omega_{\mathbf{q}}\beta). \tag{9.26}$$

Substituting (9.26) in (9.24), we get

$$\langle\hat{S}_{\mathbf{q}}^{-}\hat{S}_{\mathbf{q}}^{+}\rangle = \exp(-\hbar\omega_{\mathbf{q}}\beta)\,\langle\hat{S}_{\mathbf{q}}^{+}\hat{S}_{\mathbf{q}}^{-}\rangle. \tag{9.27}$$

Now it is readily verified from (9.13) that $\hat{S}_{\mathbf{q}}^{\pm}$ satisfy the commutation relation

$$[\hat{S}_{\mathbf{q}}^{+},\hat{S}_{\mathbf{q}'}^{-}] = \delta_{\mathbf{q},\mathbf{q}'}\,2SN, \tag{9.28}$$

which when used in (9.27) leads to the result

$$\langle\hat{S}_{\mathbf{q}}^{-}\hat{S}_{\mathbf{q}}^{+}\rangle = \exp(-\hbar\omega_{\mathbf{q}}\beta)(2SN+\langle\hat{S}_{\mathbf{q}}^{-}\hat{S}_{\mathbf{q}}^{+}\rangle),$$

i.e.

$$\langle\hat{S}_{\mathbf{q}}^{-}\hat{S}_{\mathbf{q}}^{+}\rangle = \frac{2SN}{\exp(\hbar\omega_{\mathbf{q}}\beta)-1} = 2SNn_{\mathbf{q}}, \tag{9.29}$$

where

$$n_{\mathbf{q}} = \{\exp(\hbar\omega_{\mathbf{q}}\beta)-1\}^{-1}.$$

Obviously $n_{\mathbf{q}}$ is the occupation number for spin waves of vector $\mathbf{q}$ and, using the theory set out above, some straightforward algebra gives the results

$$\langle\hat{S}_{\mathbf{l}}^{x}(0)\hat{S}_{\mathbf{l}+\mathbf{R}}^{x}(t)\rangle = \langle\hat{S}_{\mathbf{l}}^{y}(0)\hat{S}_{\mathbf{l}+\mathbf{R}}^{y}(t)\rangle$$
$$= \frac{S}{2N}\sum_{\mathbf{q}}\{(n_{\mathbf{q}}+1)\exp(-\mathrm{i}\mathbf{q}.\mathbf{R}+\mathrm{i}\omega_{\mathbf{q}}t)+n_{\mathbf{q}}\exp(\mathrm{i}\mathbf{q}.\mathbf{R}-\mathrm{i}\omega_{\mathbf{q}}t)\}$$

and

$$\tfrac{1}{2}\langle\hat{S}_{\mathbf{l}}^{x}(0)\hat{S}_{\mathbf{l}+\mathbf{R}}^{y}(t)+\hat{S}_{\mathbf{l}+\mathbf{R}}^{y}(t)\hat{S}_{\mathbf{l}}^{x}(0)\rangle = \frac{S}{2N}\sum_{\mathbf{q}}(2n_{\mathbf{q}}+1)\sin(\mathbf{q}.\mathbf{R}-\omega_{\mathbf{q}}t).$$

From these formulae it follows that the non-interacting spin wave is an excitation described by the correlation functions

$$\langle\hat{S}_{\mathbf{l}}^{x}(0)\hat{S}_{\mathbf{l}+\mathbf{R}}^{x}(t)\rangle = \langle\hat{S}_{\mathbf{l}}^{y}(0)\hat{S}_{\mathbf{l}+\mathbf{R}}^{y}(t)\rangle$$
$$\sim \frac{S}{N}\cos(\mathbf{q}.\mathbf{R}-\omega_{\mathbf{q}}t)$$

and

$$\tfrac{1}{2}\langle\hat{S}_{\mathbf{l}}^{x}(0)\hat{S}_{\mathbf{l}+\mathbf{R}}^{y}(t)+\hat{S}_{\mathbf{l}+\mathbf{R}}^{y}(t)\hat{S}_{\mathbf{l}}^{x}(0)\rangle \sim \frac{S}{N}\sin(\mathbf{q}.\mathbf{R}-\omega_{\mathbf{q}}t).$$

These equations describe a precession of the spins around the z axis, the precession propagating with wave vector $\mathbf{q}$ and frequency $\omega_{\mathbf{q}}$.

With the result (9.29) the thermal average of the z component of spin (9.23) is

$$\langle\hat{S}_{\mathbf{l}}^{z}\rangle = S-\frac{1}{N}\sum_{\mathbf{q}}n_{\mathbf{q}}$$
$$= S-\frac{1}{N}\sum_{\mathbf{q}}\{\exp(\hbar\omega_{\mathbf{q}}\beta)-1\}^{-1}. \tag{9.30}$$

For the quadratic dispersion law (9.18), with D in the general case given by (9.20), we may write

$$\frac{1}{N}\sum_{\mathbf{q}}\{\exp(\hbar\omega_{\mathbf{q}}\beta)-1\}^{-1} = \frac{1}{N}\frac{V}{(2\pi)^3}\int \mathrm{d}\mathbf{q}\,\{\exp(\hbar\omega_{\mathbf{q}}\beta)-1\}^{-1}$$
$$= \frac{v_0}{(2\pi)^3}\int \mathrm{d}\mathbf{q}\,[\exp\{(g\mu_\mathrm{B}H+Dq^2)\beta\}-1]^{-1}.$$

In this form it gives negligible error if, for low temperatures, the integral is extended over all space, hence

$$S-\langle\hat{S}_\mathrm{l}^z\rangle = \frac{v_0}{4\pi^2}\left(\frac{k_\mathrm{B}T}{D}\right)^{\frac{3}{2}}\int\limits_0^\infty \frac{\mathrm{d}x\,x^{\frac{1}{2}}}{\exp(x+g\mu_\mathrm{B}H\beta)-1}, \tag{9.31}$$

so that it is evident that the average value of $\hat{S}^z$ decreases from its maximum value S, obtained at absolute zero, as $T^{\frac{3}{2}}$.

The integral
$$\zeta(n) = \frac{1}{\Gamma(n)}\int\limits_0^\infty \frac{\mathrm{d}x\,x^{n-1}}{\mathrm{e}^x-1} \tag{9.32}$$

(where $\Gamma(n)$ is the gamma function of order n) is Riemann's zeta function.† In particular,

$$\zeta(\tfrac{3}{2}) = 2{\cdot}612, \qquad \zeta(\tfrac{5}{2}) = 1{\cdot}341, \quad \text{and} \quad \zeta(\tfrac{7}{2}) = 1{\cdot}127. \tag{9.33}$$

Using (9.32) and $\Gamma(\frac{3}{2}) = \frac{1}{2}\sqrt{\pi}$ we obtain from (9.31) the result ($H = 0$)

$$\langle\hat{S}_\mathrm{l}^z\rangle = S-v_0\left(\frac{k_\mathrm{B}T}{4\pi D}\right)^{\frac{3}{2}}\zeta(\tfrac{3}{2}). \tag{9.34}$$

When H is not zero, but such that $g\mu_\mathrm{B}H \leqslant k_\mathrm{B}T$,

$$\langle\hat{S}_\mathrm{l}^z\rangle = S-v_0\left(\frac{k_\mathrm{B}T}{4\pi D}\right)^{\frac{3}{2}}F(h), \tag{9.35}$$

where $h = g\mu_\mathrm{B}H\beta$ and

$$F(h) = \zeta(\tfrac{3}{2})-3{\cdot}54h^{\frac{1}{2}}+1{\cdot}64h-0{\cdot}104h^2+0{\cdot}00425h^3 \tag{9.36}$$

to within an accuracy of at least 1 per cent.

If in place of the simple quadratic dispersion law (9.18) we use (9.21) in (9.30), then in place of (9.34) it can be shown that

$$\langle\hat{S}_\mathrm{l}^z\rangle = S-v_0\left(\frac{k_\mathrm{B}T}{4\pi D}\right)^{\frac{3}{2}}\zeta(\tfrac{3}{2})-v_0\left(\frac{k_\mathrm{B}T}{4\pi D}\right)^{\frac{5}{2}}\zeta(\tfrac{5}{2})\frac{3\pi}{4}\overline{l^2}-$$
$$-v_0\left(\frac{k_\mathrm{B}T}{4\pi D}\right)^{\frac{7}{2}}\zeta(\tfrac{7}{2})\frac{\pi^2}{32}\{33(\overline{l^2})^2-8\overline{l^4}+24(\overline{l^2})^2\overline{\cos^4\theta}\}, \tag{9.37}$$

† E. T. Whittaker and G. N. Watson, *Modern analysis*, C.U.P., Chapter 13 (1947).

where the moments $\overline{l^n}$ are defined by (9.22) and

$$\overline{\cos^4\theta} = \frac{\sum\limits_{\mathbf{l},\mathbf{l}'} J(\mathbf{l})J(\mathbf{l}')(\mathbf{l}.\mathbf{l}')^4}{\left\{\sum\limits_{\mathbf{l}} J(\mathbf{l})\mathbf{l}^4\right\}^2}. \tag{9.38}$$

For the particular case when $J(\mathbf{l})$ extends only to nearest-neighbour sites,

$$\overline{l^2} = a^2 \tag{9.39}$$

and

$$\tfrac{1}{32}\{33(\overline{l^2})^2 - 8\overline{l^4} + 24(\overline{l^2})^2\,\overline{\cos^4\theta}\} = a^4 \times \begin{cases} 33/32 & \text{s.c.} \\ 281/288 & \text{b.c.c.} \\ 15/16 & \text{f.c.c.} \end{cases}$$

The result (9.37) shows that the $T^{\frac{5}{2}}$ and $T^{\frac{7}{2}}$ terms in the expansion of $\langle \hat{S}^z \rangle$ do not have coefficients simply related to that of the $T^{\frac{3}{2}}$ term, but instead have an importance that depends upon the range of the exchange parameter $J(\mathbf{l})$, their relative importance increasing with the range of J.

(*b*) *One-magnon cross-sections*

Now that we have pursued some of the physical predictions of linear spin-wave theory let us evaluate the neutron cross-section. As we have already remarked, for the system described by the Hamiltonian (9.1) we need only evaluate $\langle \hat{S}^z_{\mathbf{l}}(0)\hat{S}^z_{\mathbf{l}'}(t)\rangle$ and $\langle \hat{S}^{\pm}_{\mathbf{l}}(0)\hat{S}^{\mp}_{\mathbf{l}'}(t)\rangle$. However, in the linear spin-wave approximation, $\hat{S}^z(t) = \hat{S}^z(0)$, as is easily verified from (9.11) and (9.14), and, therefore, the longitudinal cross-section leads only to elastic scattering, which does not concern us here. The evaluation of the correlation functions $\langle \hat{S}^{\pm}_{\mathbf{l}}(0)\hat{S}^{\mp}_{\mathbf{l}'}(t)\rangle$ is straightforward.

We have

$$\langle \hat{S}^-_{\mathbf{l}}(0)\hat{S}^+_{\mathbf{l}'}(t)\rangle = \frac{1}{N^2}\sum_{\mathbf{q}} \exp\{i\mathbf{q}.(\mathbf{l}'-\mathbf{l})\}\langle \hat{S}^-_{\mathbf{q}}(0)\hat{S}^+_{\mathbf{q}}(t)\rangle,$$

which on using (9.11) and (9.29) gives

$$\langle \hat{S}^-_{\mathbf{l}}(0)\hat{S}^+_{\mathbf{l}'}(t)\rangle = \frac{2S}{N}\sum_{\mathbf{q}} \exp\{i\mathbf{q}.(\mathbf{l}'-\mathbf{l}) - i\omega_{\mathbf{q}}t\}n_{\mathbf{q}}. \tag{9.40}$$

Also

$$\begin{aligned}\langle \hat{S}^+_{\mathbf{l}}(0)\hat{S}^-_{\mathbf{l}'}(t)\rangle &= \frac{1}{N^2}\sum_{\mathbf{q}} \exp\{-i\mathbf{q}.(\mathbf{l}'-\mathbf{l})\}\langle \hat{S}^+_{\mathbf{q}}(0)\hat{S}^-_{\mathbf{q}}(t)\rangle \\ &= \frac{2S}{N}\sum_{\mathbf{q}} \exp\{-i\mathbf{q}.(\mathbf{l}'-\mathbf{l}) + i\omega_{\mathbf{q}}t\}(1+n_{\mathbf{q}}),\end{aligned} \tag{9.41}$$

by virtue of (9.27).

From eqn (8.62) the transverse cross-section is given by

$$\frac{\mathrm{d}^2\sigma}{\mathrm{d}\Omega\mathrm{d}E'} = \left(\frac{\gamma e^2}{m_\mathrm{e}c^2}\right)^2 \{\tfrac{1}{2}gF(\boldsymbol{\varkappa})\}^2 \frac{k'}{k}\tfrac{1}{4}(1+\tilde{\kappa}_z^2)\exp\{-2W(\boldsymbol{\varkappa})\}\times$$

$$\times\frac{1}{2\pi\hbar}\int\limits_{-\infty}^{\infty}\mathrm{d}t\exp(-\mathrm{i}\omega t)\sum_{\mathbf{l},\mathbf{l}'}\exp\{\mathrm{i}\boldsymbol{\varkappa}.(\mathbf{l}'-\mathbf{l})\}\langle\hat{S}_\mathbf{l}^+(0)\hat{S}_{\mathbf{l}'}^-(t)+\hat{S}_\mathbf{l}^-(0)\hat{S}_{\mathbf{l}'}^+(t)\rangle \tag{9.42}$$

$$= \left(\frac{\mathrm{d}^2\sigma}{\mathrm{d}\Omega\mathrm{d}E'}\right)^{(+)} + \left(\frac{\mathrm{d}^2\sigma}{\mathrm{d}\Omega\mathrm{d}E'}\right)^{(-)}. \tag{9.43}$$

In (9.43), $(\mathrm{d}^2\sigma/\mathrm{d}\Omega\mathrm{d}E')^{(+)}$ is the cross-section in which one spin wave or magnon is created and $(\mathrm{d}^2\sigma/\mathrm{d}\Omega\mathrm{d}E')^{(-)}$ the corresponding cross-section for the annihilation process. From the results (9.40) and (9.41), (9.42) yields

$$\left(\frac{\mathrm{d}^2\sigma}{\mathrm{d}\Omega\mathrm{d}E'}\right)^{(\pm)} = \left(\frac{\gamma e^2}{m_\mathrm{e}c^2}\right)^2 \{\tfrac{1}{2}gF(\boldsymbol{\varkappa})\}^2 \frac{k'}{k}(1+\tilde{\kappa}_z^2)\exp\{-2W(\boldsymbol{\varkappa})\}\tfrac{1}{2}S\times$$

$$\times\frac{(2\pi)^3}{v_0}\sum_{\mathbf{q},\boldsymbol{\tau}}(n_\mathbf{q}+\tfrac{1}{2}\pm\tfrac{1}{2})\delta(\hbar\omega_\mathbf{q}\mp\hbar\omega)\delta(\boldsymbol{\varkappa}\mp\mathbf{q}-\boldsymbol{\tau}). \tag{9.44}$$

Because these formulae for the cross-sections in which a single magnon is either created or annihilated contain both a momentum and an energy conservation condition, inelastic experiments can determine the whole spin-wave spectrum, just as the one-phonon cross-sections permit determination of phonon spectra. These one-magnon cross-sections are of the order of millibarns per magnon.

There are two differences between the one-magnon cross-sections and the one-phonon cross-sections to which we draw attention. First, the nature of the interaction between the spin of the neutron and the spin of the electrons gives rise to the orientation factor $(1+\tilde{\kappa}_z^2)$ in (9.44). We remind the reader that we have defined the axes of quantization in our calculation to coincide with the z axes. Now if a magnetic field is applied to the target sample so as to make its magnetization parallel to the scattering vector $\boldsymbol{\varkappa}$, the orientation factor takes the value 2; if the field is applied so as to make the magnetization perpendicular to $\boldsymbol{\varkappa}$ it has the value 1. Since the nuclear scattering is independent of an external magnetic field, the orientation factor can clearly be employed to distinguish the purely magnetic part of the total scattering.

If the target crystal is made up of several different ferromagnetic domains then we must average the orientation factor over all possible directions of the magnetization. For a cubic crystal this averaging gives $(1+\tilde{\kappa}_z^2)_\mathrm{av} = 1+\frac{1}{3} = \frac{4}{3}$.

The second point we wish to make is concerned with the range of wave vector $\mathbf{q}$ over which the spin-wave dispersion curve can be measured in the well-known ferromagnets Fe, Co, and Ni, which have high Curie temperatures. The total width of the linear spin-wave spectrum in the case of nearest-neighbour coupling is $2rJS$ and we can estimate the magnitude of this by a somewhat crude argument. The molecular field approximation gives the critical temperature T_c of a ferromagnet:

$$T_c = \frac{2rJS(S+1)}{3k_B},$$

so that $2rJS = 3k_B T_c/(S+1) \simeq 2k_B T_c$ for $S = \frac{1}{2}$. If $T_c \sim 10^3$ K, as it is for Fe, the width of the spin-wave band is of order 0·2 eV, a value much in excess of the neutron energy changes that can at present be measured. For this reason only a limited part of the dispersion curve can be measured, in practice the maximum value of $\mathbf{q}$ reached is typically about one-third of the magnitude of the wave vectors defining the Brillouin zone. This point is illustrated in Fig. 9.2, which shows part of the dispersion curve for Fe, measured at room temperature. The (110) direction is the one with the shortest distance to the zone boundary ($q_{max} = 1{\cdot}55$ Å^{-1}) and it can be seen that the (110) measurements extend out to $0{\cdot}4q_{max}$. Naturally this point is linked closely to the high Curie temperature of Fe, Co, and Ni, the ferromagnets most frequently examined.†

(c) *Two-magnon interaction: Hartree–Fock approximation*

The spin-wave theory presented above does not take account of the interaction between two or more magnons, i.e. it is a linear theory. The inclusion of spin-wave interactions will clearly modify the dispersion law. Since neutron scattering affords a means of determining a substantial part of the dispersion curve it is important to our understanding of the dynamical behaviour of magnetic systems that we should derive the form of the modification as it appears in the linear spectrum of the Heisenberg ferromagnet, eqn (9.12). We only consider the effects due to two-magnon scattering processes.

The interaction between spin waves arises in two parts, called the kinematic and the dynamic interactions. The kinematic interaction is a consequence of spin statistics, namely that the maximum number of spin deviations that can occur at any site with a spin S is $2S$. Take as

† Mook *et. al.*, *J. appl. Phys.* **40**, 1450 (1969), report an abrupt decrease in the spin-wave intensity in Ni above 110 MeV, which they interpret as due to broadening through interaction with Stoner excitations. A similar effect occurs for Fe (Dr. H. Mook, private communication).

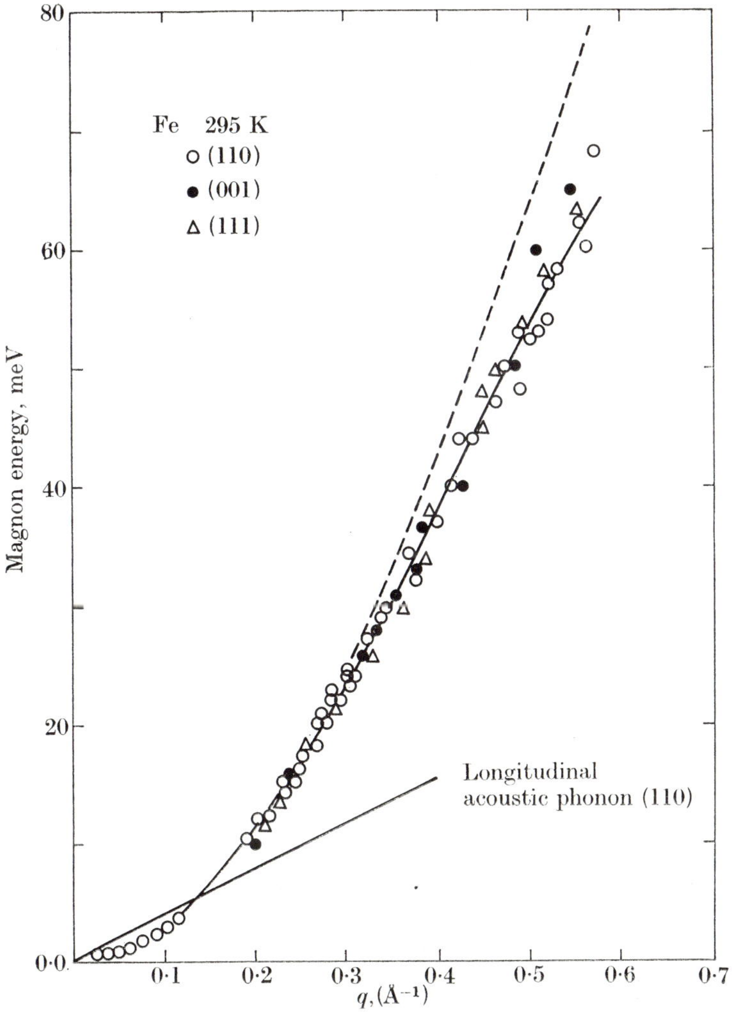

FIG. 9.2. Spin-wave dispersion curve for iron measured at room temperature. Dotted line is the dispersion curve Dq^2 with $D = 281$ meV Å^2. G. Shirane, V. J. Minkiewicz, and R. Nathans, *J. appl. Phys.* **39**, 383 (1968).

an example spins of magnitude $\frac{1}{2}$; then, clearly, two spin deviations cannot reside at the same site and the interaction that prevents this from occurring, the kinematic interaction as we have called it, is a repulsive one. The second type of interaction, the dynamic interaction, arises because it costs less energy for a spin to suffer a deviation if the spins with which it directly interacts via $J(\mathbf{l}-\mathbf{l}')$ have also undergone deviations from their fully aligned state. The dynamic interaction is, clearly, attractive.

The terminology of kinematic and dynamic interactions was introduced by Dyson† in his analysis of two spin-wave interactions in the Heisenberg ferromagnet. He showed that at low temperatures the kinematic interaction is small. In the discussion of two spin-wave interactions given below we shall neglect it from the outset.

To discuss the dynamical interaction it is convenient to introduce a representation of the spin operators in terms of Bose operators a and a^+. We take‡

$$\hat{S}^z_{\mathbf{l}} = S - \hat{a}^+_{\mathbf{l}}\hat{a}_{\mathbf{l}},$$
$$\hat{S}^+_{\mathbf{l}} = (2S)^{\frac{1}{2}}\hat{a}_{\mathbf{l}} \quad \text{and} \quad \hat{S}^-_{\mathbf{l}} = (2S)^{\frac{1}{2}}\hat{a}^+_{\mathbf{l}}\left(1-\frac{\hat{a}^+_{\mathbf{l}}\hat{a}_{\mathbf{l}}}{2S}\right), \tag{9.45}$$

where
$$[\hat{a}_{\mathbf{l}}, \hat{a}^+_{\mathbf{l}'}] = \delta_{\mathbf{l},\mathbf{l}'}. \tag{9.46}$$

The transformation (9.45) satisfies the spin angular momentum commutation rules (9.3) but does not preserve spin statistics because the Bose operators have an infinite spectrum. Thus by making the transformation (9.45) the kinematic interaction is ignored.

If (9.45) is used in the Heisenberg Hamiltonian (9.1) and the magnetic field set equal to zero we obtain, apart from a constant term,

$$\mathscr{H} = \sum_{\mathbf{q}} 2S\{\mathscr{J}(0)-\mathscr{J}(\mathbf{q})\}\hat{a}^+_{\mathbf{q}}\hat{a}_{\mathbf{q}} -$$
$$-\frac{1}{N}\sum_{\mathbf{q}_1,\mathbf{q}_2,\mathbf{q}_3,\mathbf{q}_4} \hat{a}^+_{\mathbf{q}_1}\hat{a}^+_{\mathbf{q}_2}\hat{a}_{\mathbf{q}_3}\hat{a}_{\mathbf{q}_4}\delta_{\mathbf{q}_1+\mathbf{q}_2,\mathbf{q}_3+\mathbf{q}_4}\{\mathscr{J}(\mathbf{q}_1-\mathbf{q}_3)-\mathscr{J}(\mathbf{q}_3)\}, \tag{9.47}$$

where
$$\hat{a}_{\mathbf{l}} = \frac{1}{N^{\frac{1}{2}}}\sum_{\mathbf{q}} \exp(i\mathbf{q}.\mathbf{l})\hat{a}_{\mathbf{q}}, \text{ etc.}, \tag{9.48}$$

and
$$[\hat{a}_{\mathbf{q}}, \hat{a}^+_{\mathbf{q}'}] = \delta_{\mathbf{q},\mathbf{q}'}.$$

The first term in (9.47) represents the non-interacting spin waves, the second the interaction between them. The latter is seen to contain a product of four Bose operators, which will result in products of three operators appearing in the equation of motion for $\hat{a}_{\mathbf{q}}$, say. We can, however, obtain linear equations of motion if we approximate the four operators in the interaction term by products of two operators times a thermal average.

Explicitly, we write the product

$$\hat{a}^+_{\mathbf{q}_1}\hat{a}^+_{\mathbf{q}_2}\hat{a}_{\mathbf{q}_3}\hat{a}_{\mathbf{q}_4}$$

as the sum of all non-zero pairings of two operators, i.e.

$$\hat{a}^+_{\mathbf{q}_1}\hat{a}^+_{\mathbf{q}_2}\hat{a}_{\mathbf{q}_3}\hat{a}_{\mathbf{q}_4} \to \langle\hat{a}^+_{\mathbf{q}_1}\hat{a}_{\mathbf{q}_3}\rangle\hat{a}^+_{\mathbf{q}_2}\hat{a}_{\mathbf{q}_4} + \langle\hat{a}^+_{\mathbf{q}_1}\hat{a}_{\mathbf{q}_4}\rangle\hat{a}^+_{\mathbf{q}_2}\hat{a}_{\mathbf{q}_3} + \langle\hat{a}^+_{\mathbf{q}_2}\hat{a}_{\mathbf{q}_3}\rangle\hat{a}^+_{\mathbf{q}_1}\hat{a}_{\mathbf{q}_4} +$$
$$+\langle\hat{a}^+_{\mathbf{q}_2}\hat{a}_{\mathbf{q}_4}\rangle\hat{a}^+_{\mathbf{q}_1}\hat{a}_{\mathbf{q}_3} - \langle\hat{a}^+_{\mathbf{q}_1}\hat{a}_{\mathbf{q}_3}\rangle\langle\hat{a}^+_{\mathbf{q}_2}\hat{a}_{\mathbf{q}_4}\rangle - \langle\hat{a}^+_{\mathbf{q}_1}\hat{a}_{\mathbf{q}_4}\rangle\langle\hat{a}^+_{\mathbf{q}_2}\hat{a}_{\mathbf{q}_3}\rangle. \tag{9.49}$$

† F. J. Dyson, *Phys. Rev.* **102**, 1217 and 1230.

‡ This transformation of spin operators to Boson operators is usually referred to as the Dyson–Maleev transformation. S. V. Maleev, *Soviet Phys. JETP* **7**, 1048 (1958).

(The approximation of the product of four operators as represented by (9.49) is usually referred to as the Hartree–Fock approximation. Note that the relative signs of the operator terms on the right-hand side of this expression can only be positive because the operators obey Bose statistics.)

The last two terms in (9.49) are required if we are not to count the remaining terms twice, as is made evident if the thermal average of the whole expression is taken. Since, however, these terms result only in an additional constant energy in the Hamiltonian they will not enter the equation of motion for $\hat{a}_{\mathbf{q}}$.

We also have the result

$$\langle \hat{a}_{\mathbf{q}}^{+}\hat{a}_{\mathbf{q}'}\rangle = \delta_{\mathbf{q},\mathbf{q}'}\langle \hat{a}_{\mathbf{q}}^{+}\hat{a}_{\mathbf{q}}\rangle. \tag{9.50}$$

With (9.50), the approximation (9.49) reduces to

$$n_{\mathbf{q}_1}(\delta_{\mathbf{q}_1,\mathbf{q}_3}\hat{a}_{\mathbf{q}_2}^{+}\hat{a}_{\mathbf{q}_4}+\delta_{\mathbf{q}_1,\mathbf{q}_4}\hat{a}_{\mathbf{q}_2}^{+}\hat{a}_{\mathbf{q}_3})+n_{\mathbf{q}_2}(\delta_{\mathbf{q}_2,\mathbf{q}_3}\hat{a}_{\mathbf{q}_1}^{+}\hat{a}_{\mathbf{q}_4}+\delta_{\mathbf{q}_2,\mathbf{q}_4}\hat{a}_{\mathbf{q}_1}^{+}\hat{a}_{\mathbf{q}_3})-$$
$$-n_{\mathbf{q}_1}n_{\mathbf{q}_2}(\delta_{\mathbf{q}_1,\mathbf{q}_3}\delta_{\mathbf{q}_2,\mathbf{q}_4}+\delta_{\mathbf{q}_1,\mathbf{q}_4}\delta_{\mathbf{q}_2,\mathbf{q}_3}), \tag{9.51}$$

where

$$n_{\mathbf{q}} = \langle \hat{a}_{\mathbf{q}}^{+}\hat{a}_{\mathbf{q}}\rangle = \langle \hat{n}_{\mathbf{q}}\rangle. \tag{9.52}$$

Inserting (9.51) for the product of operators in the interaction term in (9.47) the latter reduces to

$$\mathscr{H} = \sum_{\mathbf{q}} 2S\{\mathscr{J}(0)-\mathscr{J}(\mathbf{q})\}\hat{n}_{\mathbf{q}}-\frac{2}{N}\sum_{\mathbf{q},\mathbf{q}'}\{\mathscr{J}(0)+\mathscr{J}(\mathbf{q}-\mathbf{q}')-$$
$$-\mathscr{J}(\mathbf{q})-\mathscr{J}(\mathbf{q}')\}n_{\mathbf{q}}\hat{n}_{\mathbf{q}'}+\frac{1}{N}\sum_{\mathbf{q},\mathbf{q}'}\{\mathscr{J}(0)+\mathscr{J}(\mathbf{q}-\mathbf{q}')-\mathscr{J}(\mathbf{q})-\mathscr{J}(\mathbf{q}')\}n_{\mathbf{q}}n_{\mathbf{q}'}. \tag{9.53}$$

If we use this approximate Hamiltonian to obtain the equation of motion for $\hat{a}_{\mathbf{q}}$ it will, clearly, result in an equation containing only operators of the same kind. In fact,

$$i\hbar\dot{\hat{a}}_{\mathbf{q}} = [\hat{a}_{\mathbf{q}},\mathscr{H}] = 2S\{\mathscr{J}(0)-\mathscr{J}(\mathbf{q})\}\hat{a}_{\mathbf{q}}-$$
$$-\frac{2}{N}\sum_{\mathbf{q}}\{\mathscr{J}(0)+\mathscr{J}(\mathbf{q}-\mathbf{q}')-\mathscr{J}(\mathbf{q})-\mathscr{J}(\mathbf{q}')\}n_{\mathbf{q}'}\hat{a}_{\mathbf{q}}. \tag{9.54}$$

The spin-wave dispersion law in our approximate handling of the dynamical interaction is, therefore,

$$\hbar\omega_{\mathbf{q}} = 2S\{\mathscr{J}(0)-\mathscr{J}(\mathbf{q})\}-\frac{2}{N}\sum_{\mathbf{q}'}\{\mathscr{J}(0)+\mathscr{J}(\mathbf{q}-\mathbf{q}')-\mathscr{J}(\mathbf{q})-\mathscr{J}(\mathbf{q}')\}n_{\mathbf{q}'}. \tag{9.55}$$

The second term on the right-hand side of (9.55) represents a temperature- and $\mathbf{q}$-dependent shift from the linear spin-wave spectrum

$2S\{\mathscr{J}(0)-\mathscr{J}(\mathbf{q})\}$. This shift in energy is purely real, so that our approximation scheme (9.51) does not predict a lifetime for the spin waves.

To evaluate the temperature dependence of the spin-wave spectrum (9.55) we must calculate the thermal average $n_{\mathbf{q}'}$. Since the operators $\hat{a}_{\mathbf{q}}$ and $\hat{a}^{+}_{\mathbf{q}}$ are bosons it follows that

$$n_{\mathbf{q}} = \{\exp(\hbar\omega_{\mathbf{q}}\beta)-1\}^{-1}, \tag{9.56}$$

and thus (9.55) and (9.56) together give a transcendental equation for the energy spectrum $\hbar\omega_{\mathbf{q}}$, and then, from (9.56), we get $n_{\mathbf{q}}$.

We also require the magnetization. From (9.45),

$$\langle\hat{S}^{z}_{\mathbf{l}}\rangle = S-\langle\hat{a}^{+}_{\mathbf{l}}\hat{a}_{\mathbf{l}}\rangle = S-\frac{1}{N}\sum_{\mathbf{q}} n_{\mathbf{q}}. \tag{9.57}$$

Before proceeding any further we introduce a simplification in the algebra for isotropic exchange. If we denote the exchange parameter between nearest neighbours by J, that between next-nearest neighbours by J' and so on, and the corresponding numbers of neighbours by r, r', etc., then

$$\mathscr{J}(\mathbf{q}) = \sum_{\mathbf{l}} \exp(-i\mathbf{q}.\mathbf{l})J(\mathbf{l}) = rJ\gamma_{\mathbf{q}}+r'J'\gamma'_{\mathbf{q}}+\dots. \tag{9.58}$$

In (9.58) the functions $\gamma_{\mathbf{q}}$, $\gamma'_{\mathbf{q}}$, etc. are defined by

$$\gamma_{\mathbf{q}} = \frac{1}{r}\sum_{\boldsymbol{\rho}} \exp(i\mathbf{q}.\boldsymbol{\rho}) = \gamma_{-\mathbf{q}}, \qquad \gamma'_{\mathbf{q}} = \frac{1}{r'}\sum_{\boldsymbol{\rho}'} \exp(i\mathbf{q}.\boldsymbol{\rho}') = \gamma'_{-\mathbf{q}}, \tag{9.59}$$

and so on, the vectors $\boldsymbol{\rho}$, $\boldsymbol{\rho}'$,... joining nearest neighbours, next-nearest neighbours, and so forth. For example, in the case of a body-centred cubic lattice the two functions $\gamma_{\mathbf{q}}$ and $\gamma'_{\mathbf{q}}$ are

$$\gamma_{\mathbf{q}} = \cos\tfrac{1}{2}aq_x\cos\tfrac{1}{2}aq_y\cos\tfrac{1}{2}aq_z: \qquad r = 8$$

and

$$\gamma'_{\mathbf{q}} = \tfrac{1}{3}[\cos aq_x+\cos aq_y+\cos aq_z]: \quad r' = 6. \tag{9.60}$$

The functions $\gamma_{\mathbf{q}}$, $\gamma'_{\mathbf{q}}$,... defined by (9.59) all have the easily verified property that

$$\sum_{\mathbf{q}} \gamma_{\mathbf{q}-\mathbf{q}'} n_{\mathbf{q}} = \gamma_{\mathbf{q}'}\sum_{\mathbf{q}} \gamma_{\mathbf{q}} n_{\mathbf{q}}, \tag{9.61}$$

etc.

In view of this fact we can simplify the algebra involved in obtaining $\hbar\omega_{\mathbf{q}}$ and $\langle\hat{S}^{z}\rangle$ when only nearest-neighbour interactions are involved, though it is possible, of course, to carry through the analysis for an arbitrary range of $J(\mathbf{l})$.

For nearest-neighbour interactions only, (9.55) can be rewritten in the form

$$\hbar\omega_{\mathbf{q}}(T) = 2rJS(1-\gamma_{\mathbf{q}})\{1-C(T)/S\}, \tag{9.62}$$

where

$$C(T) = \frac{1}{N}\sum_{\mathbf{q}} (1-\gamma_{\mathbf{q}})n_{\mathbf{q}}. \tag{9.63}$$

It is plausible to regard (9.62) as the energy spectrum for linear spin waves with a temperature-dependent exchange integral $J(T)$, given by

$$J(T) = J\{1-C(T)/S\}. \tag{9.64}$$

The factor $\{1-C(T)/S\}$ is usually referred to as a renormalization factor.

For small $\mathbf{q}$ we can write for (9.62)

$$\hbar\omega_{\mathbf{q}} = D(T)q^2, \tag{9.65}$$

where $D(T)$ is the D of eqn (9.19) multiplied by the renormalization factor. If we use (9.65) and also expand the factor $(1-\gamma_{\mathbf{q}})$ in (9.63) to the first non-vanishing order, the equation for $C(T)$ reads, to leading order,

$$C(T) \simeq \frac{a^2}{r}\frac{1}{N}\sum_{\mathbf{q}}\frac{q^2}{\exp\{\beta D(0)q^2\}-1} = \frac{a^2}{r}\frac{v_0}{2\pi^2}\int_0^\infty \mathrm{d}q\,\frac{q^4}{\exp\{\beta D(0)q^2\}-1}, \tag{9.66}$$

i.e.

$$C(T) \simeq \frac{v_0 a^2}{4\pi^2 r}\left(\frac{k_{\mathrm{B}}T}{D(0)}\right)^{\frac{5}{2}}\frac{3\sqrt{\pi}}{4}\zeta(\tfrac{5}{2}), \tag{9.67}$$

on using the Riemann zeta integral defined by eqn (9.32). Eqn (9.67) for $C(T)$ shows that $D(T)$ is approximately given by

$$D(T) = D\left\{1-\frac{v_0\,6a^2\pi}{rS}\left(\frac{k_{\mathrm{B}}T}{4\pi D}\right)^{\frac{5}{2}}\zeta(\tfrac{5}{2})\right\}. \tag{9.68}$$

We have therefore found that the dynamical interaction between the spin waves gives, to leading order, a temperature dependence to D such that it decreases with T as $T^{\frac{5}{2}}$. More generally, when the exchange interaction extends beyond nearest neighbours,

$$D(T) = D\left\{1-\frac{v_0\,\overline{l^2}\pi}{S}\left(\frac{k_{\mathrm{B}}T}{4\pi D}\right)^{\frac{5}{2}}\zeta(\tfrac{5}{2})\right\}, \tag{9.69}$$

where D is given by (9.20) and the moment $\overline{l^2}$ is defined by (9.22).

In Fig. 9.3 the function $D(T)$, measured by inelastic neutron scattering, is plotted as a function of $(T/T_{\mathrm{c}})^{\frac{5}{2}}$ for Fe and Ni. It is clear from this figure that for both cases the $T^{\frac{5}{2}}$ dependence appears correct over a quite large temperature range, though the behaviour for iron at small T shows a pronounced departure from the $T^{\frac{5}{2}}$ law. We shall return to this point in § 9.3.

Finally, we calculate the temperature dependence of $\langle\hat{S}^z\rangle$ from (9.57).

We need

$$\frac{1}{N}\sum_{\mathbf{q}} n_{\mathbf{q}} = \frac{1}{N}\sum_{\mathbf{q}} \{\exp(\hbar\omega_{\mathbf{q}}\beta)-1\}^{-1} \simeq v_0\left(\frac{k_{\mathrm{B}} T}{4\pi D}\right)^{\frac{3}{2}}\zeta(\tfrac{3}{2})\{1-C(T)/S\}^{-\frac{3}{2}}$$

$$\simeq v_0\left(\frac{k_{\mathrm{B}} T}{4\pi D}\right)^{\frac{3}{2}}\zeta(\tfrac{3}{2})+\frac{v_0^2\, 9a^2\pi}{rS}\left(\frac{k_{\mathrm{B}} T}{4\pi D}\right)^{4}\zeta(\tfrac{3}{2})\zeta(\tfrac{5}{2}). \quad (9.70)$$

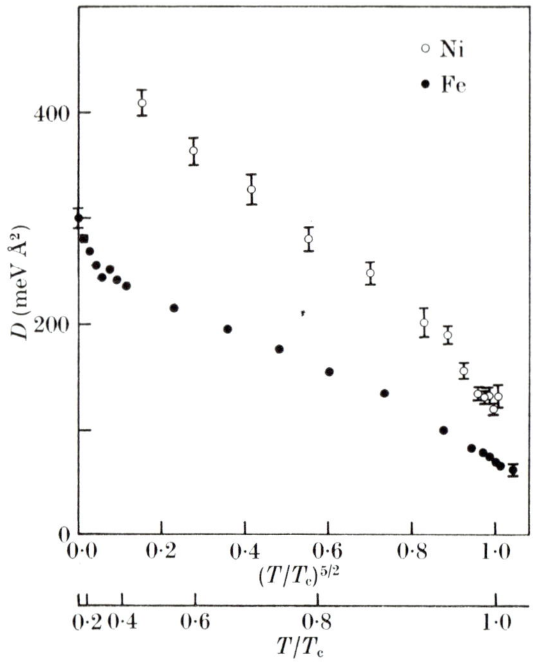

FIG. 9.3. The function $D(T)$ for iron and nickel plotted against $(T/T_{\mathrm{c}})^{\frac{5}{2}}$. M. W. Stringfellow, *Proc. phys. Soc.* (C) **1**, 950 (1968).

Eqn (9.70) tells us that the dynamical interaction introduces a T^4 term in the magnetization.† This result is in accord with the conclusion reached by Dyson. A comparison of the coefficient of T^4 in (9.70) with that obtained by Dyson shows that we have only obtained part of the correct answer. In terms of scattering formalism, our result corresponds to the s-wave part of the correct coefficient; the remaining terms come from higher order partial waves and as such depend on the point-group symmetry of the crystal. Naturally, it is not surprising that our rather simple analysis has not given the full answer.

† A detailed comparison between a renormalized spin-wave theory and the experimental data available for the ferromagnets EuO and EuS has been made by G. G. E. Low, *Proc. phys. Soc.* **82**, 922 (1963).

The T^4 term in (9.70) when the exchange parameter extends beyond nearest neighbours is

$$v_0^2 \frac{3\bar{l^2}\pi}{2S}\left(\frac{k_B T}{4\pi D}\right)^4 \zeta(\tfrac{3}{2})\zeta(\tfrac{5}{2}). \tag{9.71}$$

We will not discuss multi-spin-wave scattering processes because their contribution to the cross-section is known to be small. It is worth remarking, however, that there is one major difference between multi-phonon and multi-magnon scattering. In the former all processes are permitted, i.e. a two-phonon process can involve the creation of two phonons, annihilation of two phonons, or creation of one and the annihilation of another. But in a multi-spin process the total z component of spin can change at most by one unit; thus, for example, the two-spin-wave process in which two spin waves are annihilated or created is forbidden and only the process where one is created and one destroyed is allowed.

(*d*) *First-order approximation for magnon lifetime*

To go beyond the approximation scheme used in § 9.2 (*c*) in the discussion of two-spin-wave collisions and evaluate the lifetime of a spin wave it is very convenient to introduce the formalism of two-time thermal Green functions. The properties of these functions are discussed in Appendix B.

In terms of the Boson operators introduced in the transformations (9.45) and (9.48) we define the Green function $G_{\mathbf{q},\mathbf{q}'}(t)$ as

$$G_{\mathbf{q},\mathbf{q}'}(t) = -\mathrm{i}\theta(t)\langle[\hat{a}_{\mathbf{q}}(t), \hat{a}^+_{\mathbf{q}'}(0)]\rangle = \langle\langle \hat{a}_{\mathbf{q}}(t); \hat{a}^+_{\mathbf{q}'}(0)\rangle\rangle. \tag{9.72}$$

Here $\theta(t)$ is the unit step function. The Fourier transform of $G_{\mathbf{q},\mathbf{q}'}(t)$ satisfies the equation of motion

$$\hbar\omega G_{\mathbf{q},\mathbf{q}'}(\omega) = \frac{1}{2\pi}\langle[\hat{a}_{\mathbf{q}}, \hat{a}^+_{\mathbf{q}'}]\rangle + \langle\langle[\hat{a}_{\mathbf{q}}, \mathscr{H}]; \hat{a}^+_{\mathbf{q}'}\rangle\rangle = \frac{1}{2\pi}\delta_{\mathbf{q},\mathbf{q}'} + \langle\langle[\hat{a}_{\mathbf{q}}, \mathscr{H}]; \hat{a}^+_{\mathbf{q}'}\rangle\rangle, \tag{9.73}$$

where $\mathscr{H}$ is given by (9.47).

The value of studying $G_{\mathbf{q},\mathbf{q}'}(\omega)$ lies in the fact that its poles give the energies of the elementary excitations of the system described by $\mathscr{H}$ (in our case spin waves).

Clearly the Green function on the right-hand side of eqn (9.73) involves a Green function containing four operators. In fact for only nearest-neighbour coupling

$$\langle\langle[\hat{a}_{\mathbf{q}}, \mathscr{H}]; \hat{a}^+_{\mathbf{q}'}\rangle\rangle = 2rJS(1-\gamma_{\mathbf{q}})G_{\mathbf{q},\mathbf{q}'}(\omega) - \\ -\frac{rJ}{N}\sum_{\mathbf{q}_1,\mathbf{q}_2,\mathbf{q}_3} \delta_{\mathbf{q}+\mathbf{q}_1,\mathbf{q}_2+\mathbf{q}_3}\{\gamma_{\mathbf{q}-\mathbf{q}_2} - 2\gamma_{\mathbf{q}_2} + \gamma_{\mathbf{q}_1-\mathbf{q}_2}\}\langle\langle \hat{a}^+_{\mathbf{q}_1}\hat{a}_{\mathbf{q}_2}\hat{a}_{\mathbf{q}_3}; \hat{a}^+_{\mathbf{q}'}\rangle\rangle. \tag{9.74}$$

If the second term on the right-hand side of (9.74) is neglected altogether then

$$G_{\mathbf{q},\mathbf{q}'}(\omega) = \frac{\delta_{\mathbf{q},\mathbf{q}'}}{2\pi}\{\hbar\omega - 2rJS(1-\gamma_{\mathbf{q}})\}^{-1} \tag{9.75}$$

and the poles of $G_{\mathbf{q},\mathbf{q}'}(\omega)$ coincide with the energy of a spin wave in the linear approximation, i.e. eqn (9.16) with $H = 0$.

To go beyond this we write in (9.74)

$$\langle\langle \hat{a}^{+}_{\mathbf{q}_1}\hat{a}_{\mathbf{q}_2}\hat{a}_{\mathbf{q}_3}; \hat{a}^{+}_{\mathbf{q}'}\rangle\rangle = n_{\mathbf{q}_1}\{\delta_{\mathbf{q}_1,\mathbf{q}_2}G_{\mathbf{q}_3,\mathbf{q}'} + \delta_{\mathbf{q}_1,\mathbf{q}_3}G_{\mathbf{q}_2,\mathbf{q}'}\} + A(\mathbf{q}_1,\mathbf{q}_2,\mathbf{q}_3), \tag{9.76}$$

and obtain

$$\{\hbar\omega - \hbar\omega_{\mathbf{q}}(T)\}G_{\mathbf{q},\mathbf{q}'}$$
$$= \frac{1}{2\pi}\delta_{\mathbf{q},\mathbf{q}'} - \frac{rJ}{N}\sum_{\mathbf{q}_1,\mathbf{q}_2,\mathbf{q}_3}\delta_{\mathbf{q}+\mathbf{q}_1,\mathbf{q}_2+\mathbf{q}_3}[\gamma_{\mathbf{q}-\mathbf{q}_2} - 2\gamma_{\mathbf{q}_2} + \gamma_{\mathbf{q}_1-\mathbf{q}_2}]A(\mathbf{q}_1,\mathbf{q}_2,\mathbf{q}_3). \tag{9.77}$$

In (9.77), $\hbar\omega_{\mathbf{q}}(T)$ is the Hartree–Fock energy given by eqn (9.62). In general we expect $G_{\mathbf{q},\mathbf{q}'}(\omega)$ to have the form

$$G_{\mathbf{q},\mathbf{q}'}(\omega) = \frac{(1/2\pi)\delta_{\mathbf{q},\mathbf{q}'}}{\hbar\omega - \hbar\omega_{\mathbf{q}}(T) - \Sigma_{\mathbf{q}}(\omega)}, \tag{9.78}$$

where $\Sigma_{\mathbf{q}}(\omega)$ represents the higher order corrections to the Hartree–Fock approximation. Note that we anticipate that $\Sigma_{\mathbf{q}}(\omega)$, which is usually called a self energy, will be a function of ω.

We choose to express the denominator of (9.78) in the form

$$\hbar\omega - \hbar\omega_{\mathbf{q}} - \Delta(\mathbf{q},\omega) + \tfrac{1}{2}\mathrm{i}\,\Gamma(\mathbf{q},\omega). \tag{9.79}$$

Here $\Delta(\mathbf{q},\omega)$ is the real part of the self energy (including the Hartree–Fock term) for $\omega \to \omega + \mathrm{i}0^{+}$ and

$$\Gamma(\mathbf{q},\omega) = -2\,\mathrm{Im}\,\Sigma_{\mathbf{q}}(\omega + \mathrm{i}0^{+}). \tag{9.80}$$

$\Delta(\mathbf{q},\omega)$ is the shift in the energy from $\hbar\omega_{\mathbf{q}}$ due to two-spin-wave interactions and $\Gamma(\mathbf{q},\omega)$ is the width in energy, or the reciprocal of the lifetime. The latter statement is easily understood on noting that the one-spin-wave cross-section is proportional to the imaginary part of $G_{\mathbf{q},\mathbf{q}'}$ (cf. (8.80)). (The one-spin-wave cross-sections are determined by the correlation function $\langle \hat{S}^{+}_{\mathbf{q}}(t)\hat{S}^{-}_{\mathbf{q}'}(0)\rangle$ and this can in turn be derived from the Green function $\langle\langle \hat{S}^{+}_{\mathbf{q}}(t); \hat{S}^{-}_{\mathbf{q}'}(0)\rangle\rangle$. The latter cannot be derived rigorously from the Bose Green function $\langle\langle \hat{a}_{\mathbf{q}}(t); \hat{a}^{+}_{\mathbf{q}'}(0)\rangle\rangle$ alone, for it is evident from (9.45) that this would involve a Bose Green function of four operators. However, in the approximation to which we are working $\langle\langle \hat{S}^{+}_{\mathbf{q}}(t); \hat{S}^{-}_{\mathbf{q}'}(0)\rangle\rangle$ is proportional to $[\hbar\omega - \hbar\omega_{\mathbf{q}}(T) - \Sigma_{\mathbf{q}}(\omega)]^{-1}$. This point is more fully discussed by J. F. Cooke and H. Hahn, *Phys. Rev.* **1**, 1243 (1970).

To obtain the first-order approximation for $\Gamma(\mathbf{q}, \omega)$ we have to derive an equation for the function $A(\mathbf{q}_1, \mathbf{q}_2, \mathbf{q}_3)$ introduced in (9.76). For ease of notation in the calculation we choose to write the arguments of A as $A(\mathbf{p}, \mathbf{q}, \mathbf{q}')$.

The equation of motion requires the evaluation of the commutator $[\hat{a}_{\mathbf{p}}^{+}\hat{a}_{\mathbf{q}}\hat{a}_{\mathbf{q}'}, \mathscr{H}]$ and we find that

$$\{\hbar\omega - E(\mathbf{p}, \mathbf{q}, \mathbf{q}')\}[n_{\mathbf{p}}\{\delta_{\mathbf{p},\mathbf{q}}G_{\mathbf{q}',\mathbf{k}} + \delta_{\mathbf{p},\mathbf{q}'}G_{\mathbf{q},\mathbf{k}}\} + A(\mathbf{p}, \mathbf{q}, \mathbf{q}')]$$

$$= \frac{1}{2\pi}\langle[\hat{a}_{\mathbf{p}}^{+}\hat{a}_{\mathbf{q}}\hat{a}_{\mathbf{q}'}, \hat{a}_{\mathbf{k}}^{+}]\rangle -$$

$$-\frac{rJ}{N}\sum_{\substack{\mathbf{q}_1,\mathbf{q}_2\\ \mathbf{q}_3,\mathbf{q}_4}} \delta_{\mathbf{q}_1+\mathbf{q}_2,\mathbf{q}_3+\mathbf{q}_4}[\{\gamma_{\mathbf{q}_1-\mathbf{q}_3} - 2\gamma_{\mathbf{q}_3} + \gamma_{\mathbf{q}_2-\mathbf{q}_3}\}\times$$

$$\times\{\delta_{\mathbf{q}_1,\mathbf{q}}\langle\langle\hat{a}_{\mathbf{p}}^{+}\hat{a}_{\mathbf{q}_2}^{+}\hat{a}_{\mathbf{q}'}\hat{a}_{\mathbf{q}_3}\hat{a}_{\mathbf{q}_4}; \hat{a}_{\mathbf{k}}^{+}\rangle\rangle + \delta_{\mathbf{q}_1,\mathbf{q}'}\langle\langle\hat{a}_{\mathbf{p}}^{+}\hat{a}_{\mathbf{q}}\hat{a}_{\mathbf{q}_2}^{+}\hat{a}_{\mathbf{q}_3}\hat{a}_{\mathbf{q}_4}; \hat{a}_{\mathbf{k}}^{+}\rangle\rangle\} -$$

$$-\delta_{\mathbf{q}_3,\mathbf{p}}\{\gamma_{\mathbf{q}_1-\mathbf{q}_3} - \gamma_{\mathbf{q}_3} + \gamma_{\mathbf{q}_1-\mathbf{q}_4} - \gamma_{\mathbf{q}_4}\}\langle\langle\hat{a}_{\mathbf{q}_1}^{+}\hat{a}_{\mathbf{q}_2}^{+}\hat{a}_{\mathbf{q}_4}\hat{a}_{\mathbf{q}}\hat{a}_{\mathbf{q}'}; \hat{a}_{\mathbf{k}}^{+}\rangle\rangle]. \tag{9.81}$$

In (9.81)

$$E(\mathbf{p}, \mathbf{q}, \mathbf{q}') = \hbar\{\omega_{\mathbf{q}} + \omega_{\mathbf{q}'} - \omega_{\mathbf{p}}\}. \tag{9.82}$$

The first-order Green functions $G_{\mathbf{q},\mathbf{k}}$ and $G_{\mathbf{q}',\mathbf{k}}$ that appear on the left-hand side of (9.81), through the use of (9.76), are now transferred to the right-hand side of the equation and expressed through their equations of motion (9.74) in terms of the Green functions $\langle\langle\hat{a}_{\mathbf{q}_2}^{+}\hat{a}_{\mathbf{q}_3}\hat{a}_{\mathbf{q}_4}; \hat{a}_{\mathbf{k}}^{+}\rangle\rangle$. Having done this we break up the first two sixth-order Green functions and observe that the Green functions coming from $G_{\mathbf{q},\mathbf{k}}$ and $G_{\mathbf{q}',\mathbf{k}}$ are exactly cancelled. As a result of this the second term on the right-hand side of (9.81) simplifies to

$$-\frac{rJ}{N}\sum_{\mathbf{q}_1,\mathbf{q}_2,\mathbf{q}_3} \delta_{\mathbf{q}_1+\mathbf{q}_2,\mathbf{p}+\mathbf{q}_3}\{\gamma_{\mathbf{q}_1-\mathbf{p}} + \gamma_{\mathbf{q}_1-\mathbf{q}_3} + \gamma_{\mathbf{q}_2-\mathbf{p}} + \gamma_{\mathbf{q}_2-\mathbf{q}_3} - 2\gamma_{\mathbf{p}} - 2\gamma_{\mathbf{q}_3}\}\times$$

$$\times[n_{\mathbf{p}}\{\delta_{\mathbf{q}_1,\mathbf{q}'}\langle\langle\hat{a}_{\mathbf{q}}\hat{a}_{\mathbf{q}_2}^{+}\hat{a}_{\mathbf{q}_3}; \hat{a}_{\mathbf{k}}^{+}\rangle\rangle + \delta_{\mathbf{q}_1,\mathbf{q}}\langle\langle\hat{a}_{\mathbf{q}_2}^{+}\hat{a}_{\mathbf{q}'}\hat{a}_{\mathbf{q}_3}; \hat{a}_{\mathbf{k}}^{+}\rangle\rangle\} -$$

$$-\tfrac{1}{2}\langle\langle\hat{a}_{\mathbf{q}_1}^{+}\hat{a}_{\mathbf{q}_2}^{+}\hat{a}_{\mathbf{q}_3}\hat{a}_{\mathbf{q}}\hat{a}_{\mathbf{q}'}; \hat{a}_{\mathbf{k}}^{+}\rangle\rangle]. \tag{9.83}$$

The Green functions that remain in (9.83) we approximate in terms of first-order Green functions, i.e. we use the Hartree–Fock approximation. There is extensive cancellation of the terms: in fact all terms containing either $\delta_{\mathbf{q}_1,\mathbf{q}_3}$ or $\delta_{\mathbf{q}_2,\mathbf{q}_3}$ vanish. The remaining terms condense to

$$-\frac{2rJ}{N}G_{\mathbf{q}+\mathbf{q}'-\mathbf{p},\mathbf{k}}\{n_{\mathbf{p}}(1+n_{\mathbf{q}}+n_{\mathbf{q}'}) - n_{\mathbf{q}}n_{\mathbf{q}'}\}\{\gamma_{\mathbf{p}-\mathbf{q}} + \gamma_{\mathbf{p}-\mathbf{q}'} - \gamma_{\mathbf{p}} - \gamma_{\mathbf{q}+\mathbf{q}'-\mathbf{p}}\}. \tag{9.84}$$

Hence, from (9.81) and (9.84) we obtain the desired result for $A(\mathbf{p}, \mathbf{q}, \mathbf{q}')$, namely

$$\{\hbar\omega - E(\mathbf{p}, \mathbf{q}, \mathbf{q}')\}A(\mathbf{p}, \mathbf{q}, \mathbf{q}') = -\frac{2rJ}{N} G_{\mathbf{q}+\mathbf{q}'-\mathbf{p},\mathbf{k}} \times$$

$$\times \{n_{\mathbf{p}}(1+n_{\mathbf{q}}+n_{\mathbf{q}'}) - n_{\mathbf{q}}\, n_{\mathbf{q}'}\}\{\gamma_{\mathbf{p}-\mathbf{q}} + \gamma_{\mathbf{p}-\mathbf{q}'} - \gamma_{\mathbf{p}} - \gamma_{\mathbf{q}+\mathbf{q}'-\mathbf{p}}\}. \quad (9.85)$$

The inhomogeneous term in (9.81) is exactly cancelled by the corresponding contributions from the first-order Green functions $G_{\mathbf{q},\mathbf{k}}$ and $G_{\mathbf{q}',\mathbf{k}}$.

All there remains now is to substitute (9.85) into eqn (9.77) for $G_{\mathbf{q},\mathbf{q}'}$. It is readily observed that the resulting expression for $G_{\mathbf{q},\mathbf{q}'}$ is of the form (9.78), with

$$\Sigma_{\mathbf{q}}(\omega) = \frac{2r^2J^2}{N^2} \sum_{\mathbf{q}_1,\mathbf{q}_2,\mathbf{q}_3} \delta_{\mathbf{q}+\mathbf{q}_1,\mathbf{q}_2+\mathbf{q}_3}\{n_{\mathbf{q}_1}(1+n_{\mathbf{q}_2}+n_{\mathbf{q}_3}) - n_{\mathbf{q}_2}\, n_{\mathbf{q}_3}\} \times$$

$$\times \{\gamma_{\mathbf{q}-\mathbf{q}_2} + \gamma_{\mathbf{q}-\mathbf{q}_3} - \gamma_{\mathbf{q}_2} - \gamma_{\mathbf{q}_3}\}\{\gamma_{\mathbf{q}-\mathbf{q}_2} + \gamma_{\mathbf{q}-\mathbf{q}_3} - \gamma_{\mathbf{q}_1} - \gamma_{\mathbf{q}}\}\{\hbar\omega - E(\mathbf{q}_1, \mathbf{q}_2, \mathbf{q}_3)\}^{-1}. \quad (9.86)$$

Let us now evaluate from (9.86) the width in energy $\Gamma(\mathbf{q}, \omega)$, as defined by eqn (9.80), in the limit of low temperatures and small $\mathbf{q}$. In this limit we can surely replace $\hbar\omega$ by $\hbar\omega_{\mathbf{q}}$ and if we assume $\hbar\omega_{\mathbf{q}} \gg k_{\mathrm{B}} T$ the products of Bose factors in (9.86) can be neglected. Since

$$\mathrm{Im}\{\hbar\omega + \mathrm{i}0^+ - E(\mathbf{q}_1, \mathbf{q}_2, \mathbf{q}_3)\}^{-1} = -\pi\delta\{\hbar\omega - E(\mathbf{q}_1, \mathbf{q}_2, \mathbf{q}_3)\}$$

and

$$\delta(ax) = \frac{1}{a}\delta(x) \quad (a > 0),$$

we obtain the result (we drop $\omega = \omega_{\mathbf{q}}$ in the argument of Γ)

$$\Gamma(\mathbf{q}) = \frac{2\pi rJ}{SN^2} \sum_{\mathbf{q}_1,\mathbf{q}_2} n_{\mathbf{q}_1} \delta(\gamma_{\mathbf{q}_2} + \gamma_{\mathbf{q}+\mathbf{q}_1-\mathbf{q}_2} - \gamma_{\mathbf{q}} - \gamma_{\mathbf{q}_1})\{\gamma_{\mathbf{q}-\mathbf{q}_2} + \gamma_{\mathbf{q}_1-\mathbf{q}_2} - \gamma_{\mathbf{q}} - \gamma_{\mathbf{q}_1}\}^2. \quad (9.87)$$

Note that the presence of the δ-function allows us to write

$$\{\gamma_{\mathbf{q}-\mathbf{q}_2} + \gamma_{\mathbf{q}_1-\mathbf{q}_2} - \gamma_{\mathbf{q}_2} - \gamma_{\mathbf{q}+\mathbf{q}_1-\mathbf{q}_2}\} \to \{\gamma_{\mathbf{q}-\mathbf{q}_2} + \gamma_{\mathbf{q}_1-\mathbf{q}_2} - \gamma_{\mathbf{q}} - \gamma_{\mathbf{q}_1}\}.$$

At low temperatures the Bose factor $n_{\mathbf{q}_1}$ makes the dominant contribution from the sum over $\mathbf{q}_1$ come from small $\mathbf{q}_1$. Since we have specified that $\mathbf{q}$ is small the δ-function requires that $\mathbf{q}_2$ should also be small. Thus we require to expand $\gamma_{\mathbf{q}_1-\mathbf{q}_2}$ say, for small $\mathbf{q}_1$ and $\mathbf{q}_2$. For a simple cubic crystal with a unit cell of side a,

$$\gamma_{\mathbf{q}_1-\mathbf{q}_2} \simeq 1 - \frac{a^2}{6}(q_1^2 + q_2^2 - 2\mathbf{q}_1 \cdot \mathbf{q}_2).$$

On making use of this expansion we have from (9.87) for simple cubic crystals

$$\Gamma(\mathbf{q}) \simeq \frac{4\pi J a^2}{SN^2} \sum_{\mathbf{q}_1,\mathbf{q}_2} n_{\mathbf{q}_1}(\mathbf{q}\cdot\mathbf{q}_1)^2 \delta\{q_2^2-\mathbf{q}_2\cdot(\mathbf{q}+\mathbf{q}_1)+\mathbf{q}\cdot\mathbf{q}_1\}. \tag{9.88}$$

We now make use of the result

$$\frac{1}{N}\sum_{\mathbf{q}_2} \delta(q_2^2-\mathbf{q}_2\cdot(\mathbf{q}+\mathbf{q}_1)+\mathbf{q}\cdot\mathbf{q}_1) = \frac{v_0}{8\pi^2}|\mathbf{q}-\mathbf{q}_1|$$

to obtain

$$\Gamma(\mathbf{q}) \simeq \left(\frac{Ja^2 v_0}{2\pi S}\right)\frac{1}{N}\sum_{\mathbf{q}_1} n_{\mathbf{q}_1}(\mathbf{q}\cdot\mathbf{q}_1)^2|\mathbf{q}-\mathbf{q}_1|. \tag{9.89}$$

Although this result has been derived for the specific case of a simple cubic lattice it is in fact correct (within the initial assumptions) for all the cubic Bravais lattices.†

Proceeding as in § 9.2 (*a*) we evaluate (9.89) to leading order in temperature and find

$$\Gamma(\mathbf{q}) \simeq \left(\frac{6}{\nu r}\right)^3\left(\frac{J}{S}\right)(aq)^3\left\{\frac{k_B T}{8\pi J S}\right\}^{\frac{5}{2}}\zeta(\tfrac{5}{2}) \quad (\hbar\omega_{\mathbf{q}} \gg k_B T), \tag{9.90}$$

where

$$\nu = \begin{cases} 1 & \text{s.c.} \\ 2^{\frac{1}{3}} & \text{f.c.c.} \\ 3\times 2^{-\frac{4}{3}} & \text{b.c.c.} \end{cases}$$

Thus at low temperatures and long wavelengths the width in energy of a spin wave is proportional $q^3T^{\frac{5}{2}}$.

In neutron scattering experiments the condition $\hbar\omega_{\mathbf{q}} \gg k_B T$ is nearly always satisfied. However, it is worth while to note the form of the width in energy Γ in the opposite limit $\hbar\omega_{\mathbf{q}} \ll k_B T$. In this instance we can no longer neglect the products of Bose functions in (9.86) as we did in obtaining (9.87). Harris‡ has shown that the width in energy calculated from (9.86) is in this instance given by (and with, of course, $\hbar\omega = \hbar\omega_{\mathbf{q}}$).

$$\Gamma(\mathbf{q}) \propto q^4T^2\left\{\tfrac{1}{6}\ln^2\left(\frac{k_B T}{2JSa^2q^2}\right)+\tfrac{5}{9}\ln\left(\frac{k_B T}{2JSa^2q^2}\right)-0{\cdot}05\right\}. \tag{9.91}$$

The above calculation of the energy width was based upon the Hamiltonian (9.47), which does not take account of the kinematical interaction. It is therefore prudent to ask what modifications will be made if the latter is included. First, it is quite clear how one can proceed in this case using the formalism of two-time thermal Green functions. Instead

† The result (9.89) was first given (for spin $\frac{1}{2}$) by J. F. Cooke and H. A. Gersch *Phys. Rev.* **153,** 641 (1967).

‡ A. B. Harris, *Phys. Rev.* **175,** 674 (1968) and ibid. **184,** 606 (1969). See also E. Balcar, *J. Phys.* C, **2,** 1174 (1969).

of using the Hamiltonian (9.47) we must go back to the exact Hamiltonian, eqn (9.1), and seek the Green function

$$\langle\langle \hat{S}_{\mathbf{l}}^{+}; \hat{S}_{\mathbf{l}'}^{-}\rangle\rangle.$$

For this it is clear that the inhomogeneous term in the equation of motion must contain

$$\langle[\hat{S}_{\mathbf{l}}^{+}, \hat{S}_{\mathbf{l}'}^{-}]\rangle = 2\langle \hat{S}^{z}\rangle\delta_{\mathbf{l},\mathbf{l}'}.$$

Hence the Green function calculated to any approximation must itself be proportional to $\langle \hat{S}^{z}\rangle$.

Writing

$$\langle\langle \hat{S}_{\mathbf{l}}^{+}; \hat{S}_{\mathbf{l}'}^{-}\rangle\rangle = \frac{1}{N}\sum_{\mathbf{q}} \exp\{i\mathbf{q}.(\mathbf{l}-\mathbf{l}')\}\langle\langle \hat{S}_{\mathbf{q}}^{+}; \hat{S}_{\mathbf{q}}^{-}\rangle\rangle,$$

it can be easily proved from the equation of motion for $\langle\langle \hat{S}_{\mathbf{l}}^{+}; \hat{S}_{\mathbf{l}'}^{-}\rangle\rangle$ that

$$\lim_{\mathbf{q}\to 0}\langle\langle \hat{S}_{\mathbf{q}}^{+}; \hat{S}_{\mathbf{q}}^{-}\rangle\rangle = \frac{\langle \hat{S}^{z}\rangle}{\hbar\pi\omega}.$$

We therefore conclude that irrespective of the actual approximate form obtained for $\langle\langle \hat{S}_{\mathbf{q}}^{+}; \hat{S}_{\mathbf{q}}\rangle\rangle$ the inelastic spin-wave cross-section must vary with temperature as the magnetization for small spin-wave momenta.†

9.3. Itinerant electron model of ferromagnetism‡

The Heisenberg model of ferromagnetism discussed in the preceding section represents the localized theory of magnetism in solids. It is now our task to discuss the itinerant or band theory of ferromagnetism. Such a theory would certainly seem more appropriate than the Heisenberg model for the transition-metal elements and their alloys. Our discussion is designed to bring out the fundamental aspects of the itinerant model and we shall for this purpose adopt many approximations that are open to argument. In particular, we consider only a single band of electrons. This might well seem odd in view of the fact that we always have in the back of our minds the transition metals with their five d bands. However, the arguments we present below for a single band differ little from those that would have to be employed in the case of degenerate bands, and the latter are algebraically more complicated.§ Chapter 5 contains all the formalism required to extend the arguments to degenerate

† W. Marshall and G. Murray, *J. Phys.* C, **2**, 539 (1969), and S. W. Lovesey, *Phys. Lett.* **31A**, 53 (1970). The argument given here is due to Dr. J. Cooke.

‡ The authors are indebted to Dr. J. Hubbard for allowing them access to unpublished lectures on ferromagnetism in narrow energy bands given at Advanced Nato Summer School on Magnetism, McGill University, Canada (1967). Director, A. J. Freeman.

§ J. B. Sokoloff, *Phys. Rev.* **180**, 613 (1969).

bands. It also contains all the mathematical formalism (that we require below) for describing the many-body aspects of band electrons.†

We must first derive a Hamiltonian that we believe incorporates the salient features of electrons in a narrow energy band. Having done this we can investigate its properties, paying particular regard to the possibility of a collective mode of electron spin propagating in the band, i.e. to the existence of a spin wave. Before embarking on this scheme we ask what physically we must expect an itinerant electron model of ferromagnetism to contain. Experiments show that the d electrons of transition metals have a behaviour that in some instances makes them appear well localized and in other cases itinerant. For example, spin waves in ferromagnetic metals obey the fundamental predictions of the Heisenberg ferromagnet, while, on the other hand, the large d-electron contribution to the low-temperature specific heat and the occurrence of magnetic moments per atom that are far from integral numbers of Bohr magnetons μ_B are properties that are explainable by band theory. The key to understanding these seemingly opposing characteristics of localized and itinerant electron states is correlation between the electrons. For if, in spite of their band motion, the d electrons on an atom are strongly correlated with each other but only weakly with electrons on other atoms, then such intra-atomic correlation leads inevitably to a behaviour that is to some degree characteristic of an atomic electron model. We shall in fact find that our approximate Hamiltonian for electrons in a narrow band contains a repulsive interaction that operates between electrons of *opposite* spin on the same atom. The correlation caused by this 'on site' repulsion is that which leads to the possibility of a ferromagnetic state and the existence of spin-wave modes.

(*a*) *Hubbard Hamiltonian*‡

The Hamiltonian of electrons in a solid is of the form

$$\mathscr{H} = \sum_i \frac{1}{2m_e}\hat{\mathbf{p}}_i^2 + \hat{V}(\mathbf{r}_i) + \sum_{i \neq j} \frac{e^2}{|\mathbf{r}_i - \mathbf{r}_j|}. \tag{9.92}$$

Here $\hat{\mathbf{p}}_i$ and $\mathbf{r}_i$ are, respectively, the momentum operator and position vector of the ith electron. The first part of (9.92) consists of the kinetic energy of the ith electron and the periodic crystal potential $\hat{V}(\mathbf{r})$ due to the ion cores. As we noted in Chapter 5, the eigenfunctions of this part of the Hamiltonian, the band theory Hamiltonian, are Bloch functions

† Some features are discussed by R. Brout, *Phase transitions*, chap. 6, Benjamin (1965).

‡ J. Hubbard, *Proc. Roy. Soc.* **A276**, 238 (1963), **A277** (1963), and **A281**, 401 (1964).

$\varphi_{\mathbf{k}\sigma}(\mathbf{r})$. The second part of (9.92) is the Coulomb interaction of the electrons. It is this part that we seek to approximate and to do this we exploit the fact that the band is very narrow.

As in Chapter 5, we employ the second quantization formalism to write $\mathscr{H}$ in the form

$$\mathscr{H} = \sum_{\sigma}\sum_{\mathbf{l},\mathbf{l}'} T(\mathbf{l}-\mathbf{l}')\hat{c}^{+}_{\mathbf{l}\sigma}\hat{c}_{\mathbf{l}'\sigma}+\tfrac{1}{2}\sum_{\mathbf{l},\mathbf{m}}\sum_{\mathbf{l}',\mathbf{m}'}\sum_{\sigma,\sigma'}(\mathbf{lm}|1/r|\mathbf{l}'\mathbf{m}')\hat{c}^{+}_{\mathbf{l}\sigma}\hat{c}^{+}_{\mathbf{m}\sigma'}\hat{c}_{\mathbf{m}'\sigma'}\hat{c}_{\mathbf{l}'\sigma}. \tag{9.93}$$

We recall that the matrix $T(\mathbf{l}-\mathbf{l}')$ is

$$T(\mathbf{l}-\mathbf{l}') = \int \mathrm{d}\mathbf{r}\, \phi^*(\mathbf{r}-\mathbf{l})\left\{\frac{1}{2m_{\mathrm{e}}}\hat{\mathbf{p}}^2+\hat{V}(\mathbf{r})\right\}\phi(\mathbf{r}-\mathbf{l}')$$

and

$$(\mathbf{lm}|1/r|\mathbf{l}'\mathbf{m}') = e^2\int \mathrm{d}\mathbf{r}\int \mathrm{d}\mathbf{r}'\,\frac{\phi^*(\mathbf{r}-\mathbf{l})\phi(\mathbf{r}-\mathbf{l}')\phi^*(\mathbf{r}'-\mathbf{m})\phi(\mathbf{r}'-\mathbf{m}')}{|\mathbf{r}-\mathbf{r}'|}. \tag{9.94}$$

In (9.94) the $\phi(\mathbf{r}-\mathbf{l})$ are a set of orthonormal wave functions centred about the lattice sites $\mathbf{l}$. The operators $\hat{c}^{+}_{\mathbf{l}\sigma}$ and $\hat{c}_{\mathbf{l}\sigma}$, respectively create and destroy an electron in the state $\phi(\mathbf{r}-\mathbf{l})$ with spin σ.

In a narrow energy band the overlap of the functions ϕ centred about different sites is small. Therefore, the largest of the matrix elements $(\mathbf{lm}|1/r|\mathbf{l}'\mathbf{m}')$ is that with $\mathbf{l} = \mathbf{m} = \mathbf{l}' = \mathbf{m}'$; we denote this matrix element by I and neglect all others in (9.93). Hence our approximate Hamiltonian is

$$\mathscr{H} = \sum_{\mathbf{l},\mathbf{l}',\sigma} T(\mathbf{l}-\mathbf{l}')\hat{c}^{+}_{\mathbf{l}\sigma}\hat{c}_{\mathbf{l}'\sigma}+\tfrac{1}{2}I\sum_{\mathbf{l},\sigma,\sigma'}\hat{c}^{+}_{\mathbf{l}\sigma}\hat{c}^{+}_{\mathbf{l}\sigma'}\hat{c}_{\mathbf{l}\sigma'}\hat{c}_{\mathbf{l}\sigma}. \tag{9.95}$$

The sum over the four operators in the second term of (9.95) can be simplified by the following manipulations. Remembering that the $\hat{c}$'s are Fermion operators,

$$\hat{c}^{+}_{\sigma}\hat{c}^{+}_{\sigma'}\hat{c}_{\sigma'}\hat{c}_{\sigma} = -\hat{c}^{+}_{\sigma}\hat{c}^{+}_{\sigma'}\hat{c}_{\sigma}\hat{c}_{\sigma'} = -\hat{c}^{+}_{\sigma}(\delta_{\sigma,\sigma'}-\hat{c}_{\sigma}\hat{c}^{+}_{\sigma'})\hat{c}_{\sigma'}.$$

Thus,

$$\begin{aligned}\sum_{\sigma,\sigma'}\hat{c}^{+}_{\sigma}\hat{c}^{+}_{\sigma'}\hat{c}_{\sigma'}\hat{c}_{\sigma} &= -\hat{n}_{\uparrow}-\hat{n}_{\downarrow}+\hat{n}_{\uparrow}\hat{n}_{\downarrow}+\hat{n}_{\downarrow}\hat{n}_{\downarrow}+\hat{n}_{\uparrow}\hat{n}_{\uparrow}+\hat{n}_{\downarrow}\hat{n}_{\uparrow}\\ &= \hat{n}_{\uparrow}\hat{n}_{\downarrow}+\hat{n}_{\downarrow}\hat{n}_{\uparrow}\\ &\equiv \sum_{\sigma}\hat{n}_{\sigma}\hat{n}_{-\sigma}.\end{aligned} \tag{9.96}$$

In reaching the second line of (9.96) we have made use of the property

$$\hat{n}_{\sigma}\hat{n}_{\sigma} = \hat{c}^{+}_{\sigma}\hat{c}_{\sigma}\hat{c}^{+}_{\sigma}\hat{c}_{\sigma} = \hat{c}^{+}_{\sigma}\hat{c}_{\sigma} = \hat{n}_{\sigma}.$$

Using (9.96) in (9.95) we finally obtain the desired result,

$$\mathscr{H} = \sum_{\mathbf{l},\mathbf{l}',\sigma} T(\mathbf{l}-\mathbf{l}')\hat{c}^{+}_{\mathbf{l}\sigma}\hat{c}_{\mathbf{l}'\sigma}+\tfrac{1}{2}I\sum_{\mathbf{l},\sigma}\hat{n}_{\mathbf{l}\sigma}\hat{n}_{\mathbf{l}-\sigma}. \tag{9.97}$$

This Hamiltonian and a discussion of its properties was first given by Hubbard† (1963).

The first term in (9.97) is

$$\sum_{\mathbf{k},\sigma} \mathscr{E}_{\mathbf{k}} \hat{c}^{+}_{\mathbf{k}\sigma} \hat{c}_{\mathbf{k}\sigma} \tag{9.98}$$

and represents the band energy. The second term represents the Coulomb interaction of electrons of opposite spin at the same site only.

In deriving (9.97) we have neglected all the matrix elements

$$(\mathbf{lm}|1/r|\mathbf{l'm'})$$

except the one for which $\mathbf{l} = \mathbf{m} = \mathbf{l'} = \mathbf{m'}$ and it is only reasonable to question how good such an approximation is. First, I has the order of magnitude 20 eV for $3d$ electrons in transition metals, and these have a bandwidth of order 4–5 eV. The largest of the neglected terms is $(\mathbf{lm}|1/r|\mathbf{lm})$, where $\mathbf{l}$ and $\mathbf{m}$ are nearest-neighbour lattice sites. These integrals are of order 6 eV, but this figure is reduced by the screening of the interactions of electrons on different atoms by the conduction-band electrons. It is probable that a more realistic figure for these matrix elements is 2–3 eV, and therefore neglecting them in comparison with I would appear quite reasonable. There are two effects, however, that will certainly go some way to reducing I in an actual metal from the figure of 20 eV. The first of these effects arises from the production of a correlation hole in the conduction-electron gas about an atom with an extra d electron. The presence of the correlation hole reduces the electrostatic potential at the atom by about 5 eV, which is the same as reducing I by the same figure. Also, correlation between the d electrons at any site will go towards reducing I. The total reduction of I by these effects is a matter of some debate, but it is probable that if we choose an 'effective' I of order 5 eV, this would not be too far from the truth. We therefore regard the Hamiltonian (9.97) as an effective Hamiltonian for the description of electrons in a narrow energy band and investigate, in a simple manner, some of its consequences, treating I as an adjustable parameter of the order of several eV.

The interaction term in (9.97) is the product of four Fermion operators and thus to derive a linear equation of motion for $\hat{c}_{\mathbf{k}\sigma}$, say, we invoke the same approximation scheme (the Hartree–Fock approximation) as we employed in the previous section when discussing the dynamical interaction of ferromagnetic spin waves. That is to say, we take

$$\hat{n}_{\sigma}\hat{n}_{-\sigma} \Rightarrow \langle\hat{n}_{\sigma}\rangle\hat{n}_{-\sigma}+\hat{n}_{\sigma}\langle\hat{n}_{-\sigma}\rangle-\langle\hat{n}_{\sigma}\rangle\langle\hat{n}_{-\sigma}\rangle. \tag{9.99}$$

† J. Hubbard, ibid.

The terms on the right-hand side of (9.99) represent all the non-zero pairings of two operators that can be found from the product $\hat{n}_\sigma \hat{n}_{-\sigma}$. It is a relatively simple matter to see that we have chosen the correct relative sign of the two operators on the right-hand side. Furthermore, we shall in the main part of this section be concerned only with the zero-temperature properties of the system described by (9.97).

With the approximation (9.99) (we neglect the constant term) the Hamiltonian (9.97) becomes

$$\mathscr{H} = \sum_{\mathbf{l},\mathbf{l}',\sigma} T(\mathbf{l}-\mathbf{l}')\hat{c}^+_{\mathbf{l}\sigma}\hat{c}_{\mathbf{l}'\sigma} + I\sum_{\mathbf{l},\sigma}\langle\hat{n}_{\mathbf{l}-\sigma}\rangle\hat{n}_{\mathbf{l}\sigma}. \tag{9.100}$$

We now assume $\langle\hat{n}_{\mathbf{l}-\sigma}\rangle$ to be independent of the site label $\mathbf{l}$, in which case (9.100) in terms of Bloch state operators $\hat{c}^+_{\mathbf{k}\sigma}$ and $\hat{c}_{\mathbf{k}\sigma}$ is

$$\mathscr{H} = \sum_{\mathbf{k},\sigma}\{\mathscr{E}_\mathbf{k} + I\langle\hat{n}_{-\sigma}\rangle\}\hat{c}^+_{\mathbf{k}\sigma}\hat{c}_{\mathbf{k}\sigma}. \tag{9.101}$$

This Hamiltonian simply represents a band of non-interacting electrons with spin-dependent energies

$$\mathscr{E}_\mathbf{k} + I\langle\hat{n}_{-\sigma}\rangle. \tag{9.102}$$

Let $g(\mathscr{E})$ be the density of states per atom for the Bloch states $\psi_{\mathbf{k}\sigma}$. (9.102) tells us that the effective density of states for σ spin electrons in the presence of the interaction, $\rho_\sigma(\mathscr{E})$, is given by

$$\rho_\sigma(\mathscr{E}) = g(\mathscr{E} - I\langle\hat{n}_{-\sigma}\rangle). \tag{9.103}$$

If μ is the chemical potential then at absolute zero:

$$\langle\hat{n}_\sigma\rangle = \int_{-\infty}^{\mu}\rho_\sigma(\mathscr{E})\,\mathrm{d}\mathscr{E} = \int_{-\infty}^{\mu - I\langle\hat{n}_{-\sigma}\rangle} g(\mathscr{E})\,\mathrm{d}\mathscr{E} = \int_{-\infty}^{\mu - In + I\langle\hat{n}_\sigma\rangle} g(\mathscr{E})\,\mathrm{d}\mathscr{E}, \tag{9.104}$$

since $\langle\hat{n}_\sigma\rangle + \langle\hat{n}_{-\sigma}\rangle = n$, where n is the number of electrons per atom.

In the paramagnetic state there are as many spins up as there are down at any site, i.e. $\langle\hat{n}_\sigma\rangle = \frac{1}{2}n$. However, this state may be unstable to a ferromagnetic ordering. For if $\langle\hat{n}_\sigma\rangle = \frac{1}{2}n + \delta n$, say, the right-hand side of (9.104) changes by

$$I\,\delta n\,g(\mathscr{E}_\mathrm{f}),$$

where $\mathscr{E}_\mathrm{f} = \mu - \frac{1}{2}In$, and thus the paramagnetic state will be *magnetically* stable if

$$Ig(\mathscr{E}_\mathrm{f}) = 1.$$

On the other hand it is unstable to ferromagnetic order if

$$Ig(\mathscr{E}_\mathrm{f}) > 1. \tag{9.105}$$

(9.105) is the Stoner condition for ferromagnetism. It shows that conditions favourable to ferromagnetism are achieved with a large density of states at the Fermi energy $\mathscr{E}_\mathrm{f}$ and a large I. When (9.105) holds, we seek $\langle\hat{n}_\uparrow\rangle$ and $\langle\hat{n}_\downarrow\rangle$ that satisfy $\langle\hat{n}_\uparrow\rangle + \langle\hat{n}_\downarrow\rangle = n$, the band now splitting into two bands, one of spin up electrons and one of spin down electrons separated by an energy $\Delta = I(\langle\hat{n}_{\downarrow}\rangle - \langle\hat{n}_{\uparrow}\rangle)$.

(*b*) *Transverse neutron scattering cross-section*

In calculating the cross-section for the scattering of neutrons by the band electrons described by the Hamiltonian (9.97), we need only consider the scattering that arises from the spin part of the neutron–electron interaction because in Chapter 5 we found the amplitude for the scattering via that part of the interaction arising from the motion of the electrons to be smaller than the spin part for electrons in narrow energy bands. Furthermore, we saw in Chapter 8 that the spin-only scattering cross-section for band electrons can be expressed in terms of transverse $\chi_{\mathbf{k}}^{(\pm)}[\omega]$ and longitudinal $\chi_{\mathbf{k}}^{(0)}[\omega]$ generalized susceptibilities. Thus the calculation of the cross-section for the scattering by the electrons described by (9.97) reduces to calculating the linear response of the latter to an external magnetic field. We shall calculate first the transverse susceptibility.

To this end we add to the Hamiltonian describing the electrons the interaction term

$$g\mu_{\mathrm{B}} \sum_{\mathbf{l},\nu} (\hat{S}_{\mathbf{l}}^{\nu})^{+} H_{\mathbf{l}}^{\nu}(t) = g\mu_{\mathrm{B}} \sum_{\mathbf{l},\nu} \hat{S}_{\mathbf{l}}^{-\nu} H_{\mathbf{l}}^{\nu}(t), \tag{9.106}$$

where $H_{\mathbf{l}}^{\nu}(t)$ is the νth component ($\nu = +,-$) of a magnetic field that varies in both space and time. Remembering that $\hat{S}_{\mathbf{l}}^{+} = \hat{c}_{\mathbf{l}\uparrow}^{+}\hat{c}_{\mathbf{l}\downarrow}$ and $\hat{S}_{\mathbf{l}}^{-} = \hat{c}_{\mathbf{l}\downarrow}^{+}\hat{c}_{\mathbf{l}\uparrow}$ and with the approximate form (9.100) of the Hamiltonian (9.97) we have now the total Hamiltonian

$$\mathscr{H} = \sum_{\mathbf{l},\mathbf{l}',\sigma} T(\mathbf{l}-\mathbf{l}')\hat{c}_{\mathbf{l}\sigma}^{+}\hat{c}_{\mathbf{l}'\sigma} + I\sum_{\mathbf{l},\sigma}\langle\hat{n}_{-\sigma}\rangle\hat{n}_{\mathbf{l}\sigma} + \\ + g\mu_{\mathrm{B}}\sum_{\mathbf{l}} \hat{c}_{\mathbf{l}\uparrow}^{+}\hat{c}_{\mathbf{l}\downarrow}H_{\mathbf{l}}^{-}(t) + \hat{c}_{\mathbf{l}\downarrow}^{+}\hat{c}_{\mathbf{l}\uparrow}H_{\mathbf{l}}^{+}(t). \tag{9.107}$$

If

$$H_{\mathbf{l}}^{+}(t) = \frac{1}{N}\sum_{\mathbf{q}} \exp(i\mathbf{q}.\mathbf{l})\, H_{\mathbf{q}}^{+}(t), \tag{9.108}$$

then (9.107) in terms of Bloch state operators

$$\hat{c}_{\mathbf{k}\sigma} = \frac{1}{N^{\frac{1}{2}}}\sum_{\mathbf{l}} \exp(i\mathbf{k}.\mathbf{l})\hat{c}_{\mathbf{l}\sigma}, \text{ etc.,}$$

is

$$\mathscr{H} = \sum_{\mathbf{k},\sigma}\{\mathscr{E}_{\mathbf{k}} + I\langle\hat{n}_{-\sigma}\rangle\}\hat{c}_{\mathbf{k}\sigma}^{+}\hat{c}_{\mathbf{k}\sigma} + g\mu_{\mathrm{B}}\frac{1}{N}\sum_{\mathbf{k},\mathbf{q}} \hat{c}_{\mathbf{k}\downarrow}^{+}\hat{c}_{\mathbf{k}+\mathbf{q}\uparrow}H_{\mathbf{q}}^{+} + \text{c.c.} \tag{9.109}$$

and the equation of motion for the operator

$$\hat{c}_{\mathbf{k}+\mathbf{q}\uparrow}^{+}\hat{c}_{\mathbf{q}\downarrow},$$

in terms of which we can obtain $\hat{S}_{\mathbf{k}}^{+}(t)$, and therefore $\chi_{\mathbf{k}}^{(+)}[\omega]$, is

$$i\hbar\frac{\partial}{\partial t}\hat{c}_{\mathbf{k}+\mathbf{q}\uparrow}^{+}\hat{c}_{\mathbf{q}\downarrow} = [\hat{c}_{\mathbf{k}+\mathbf{q}\uparrow}^{+}\hat{c}_{\mathbf{q}\downarrow}, \mathscr{H}]$$

$$= \{\mathscr{E}_{\mathbf{q}}-\mathscr{E}_{\mathbf{k}+\mathbf{q}}-\Delta\}\hat{c}_{\mathbf{k}+\mathbf{q}\uparrow}^{+}\hat{c}_{\mathbf{q}\downarrow}+$$

$$+\frac{g\mu_{\mathrm{B}}}{N}\sum_{\mathbf{q}'} H_{\mathbf{q}'}^{+}(t)[\hat{c}_{\mathbf{k}+\mathbf{q}\uparrow}^{+}\hat{c}_{\mathbf{q}-\mathbf{q}'\uparrow}-\hat{c}_{\mathbf{k}+\mathbf{q}-\mathbf{q}'\downarrow}^{+}\hat{c}_{\mathbf{q}\downarrow}]. \quad (9.110\,\mathrm{a})$$

Here $$\Delta = I(\langle\hat{n}_{\downarrow}\rangle-\langle\hat{n}_{\uparrow}\rangle) \quad (9.110\,\mathrm{b})$$

is the splitting of the two bands.

For the particular choice of $H_{\mathbf{l}}^{+}(t)$

$$H_{\mathbf{l}}^{+}(t) = \exp(i\mathbf{k}.\mathbf{l})H^{+}(t), \quad (9.111)$$

we have $$H_{\mathbf{q}}^{+}(t) = N\delta_{\mathbf{k},\mathbf{q}}H^{+}(t).$$

Hence

$$i\hbar\frac{\partial}{\partial t}\hat{c}_{\mathbf{k}+\mathbf{q}\uparrow}^{+}\hat{c}_{\mathbf{q}\downarrow} = \{\mathscr{E}_{\mathbf{q}}-\mathscr{E}_{\mathbf{k}+\mathbf{q}}-\Delta\}\hat{c}_{\mathbf{k}+\mathbf{q}\uparrow}^{+}\hat{c}_{\mathbf{q}\downarrow}+g\mu_{\mathrm{B}}H^{+}(t)(\hat{n}_{\mathbf{k}+\mathbf{q}\uparrow}-\hat{n}_{\mathbf{q}\downarrow}), \quad (9.112)$$

where $$\hat{n}_{\mathbf{k}\sigma} = \hat{c}_{\mathbf{k}\sigma}^{+}\hat{c}_{\mathbf{k}\sigma}.$$

We now choose the time dependence of the external field $H^{+}(t)$ to be

$$H^{+}(t) = H\exp(-i\omega t+\epsilon t). \quad (9.113)$$

In (9.113) ϵ is a small positive constant which is included to make $H^{+}(-\infty) = 0$.

To first order in the applied field H^{ν}, $\hat{n}_{\mathbf{k}\sigma}$ is a constant independent of time. Since we only require the solution of (9.112) to terms linear in H^{+} this means that in the second term the time-dependent operators can be replaced by their values in the absence of the field. The solution of the resultant equation that is zero for $t \to -\infty$ is then readily shown to be

$$\hat{c}_{\mathbf{k}+\mathbf{q}\uparrow}^{+}\hat{c}_{\mathbf{q}\downarrow} = \frac{-g\mu_{\mathrm{B}}\exp(-i\omega t+\epsilon t)H^{+}(\hat{n}_{\mathbf{k}+\mathbf{q}\uparrow}-\hat{n}_{\mathbf{q}\downarrow})}{\mathscr{E}_{\mathbf{q}}-\mathscr{E}_{\mathbf{k}+\mathbf{q}}-\Delta-\hbar\omega-i} \quad (9.114)$$

and from this one can get

$$M_{\mathbf{k}}^{+}(t) = -g\mu_{\mathrm{B}}\sum_{\mathbf{l}}\exp(-i\mathbf{k}.\mathbf{l})\mathrm{Tr}\,\hat{\rho}(t)\hat{S}_{\mathbf{l}}^{+}$$

$$= -g\mu_{\mathrm{B}}\sum_{\mathbf{l}}\exp(-i\mathbf{k}.\mathbf{l})\mathrm{Tr}\,\hat{\rho}_{0}\hat{S}_{\mathbf{l}}^{+}(t)$$

$$= -g\mu_{\mathrm{B}}\,\mathrm{Tr}\,\hat{\rho}_{0}\sum_{\mathbf{q}}\hat{c}_{\mathbf{k}+\mathbf{q}\uparrow}^{+}(t)\hat{c}_{\mathbf{q}\downarrow}(t),$$

where $\hat{\rho}_0$ is the density matrix for the unperturbed system.

Now by definition (cf. Appendix B, § B.9) the transverse component of the magnetization is given in terms of $\chi_{\mathbf{k}}^{(+)}[\omega]$ by

$$M_{\mathbf{k}}^{+}(t) = \exp(-\mathrm{i}\omega t)\chi_{\mathbf{k}}^{(+)}[\omega]H_{\mathbf{k}}^{+}, \tag{9.115}$$

so that
$$\chi_{\mathbf{k},0}^{(+)}[\omega] = -(g\mu_{\mathrm{B}})^2\frac{1}{N}\sum_{\mathbf{q}}\frac{\langle\hat{n}_{\mathbf{k}+\mathbf{q}\uparrow}\rangle-\langle\hat{n}_{\mathbf{q}\downarrow}\rangle}{\mathscr{E}_{\mathbf{k}+\mathbf{q}}-\mathscr{E}_{\mathbf{q}}+\Delta+\hbar\omega+\mathrm{i}\epsilon}. \tag{9.116}$$

In (9.116) we may use
$$\langle\hat{n}_{\mathbf{k}\uparrow}\rangle = f_{\mathbf{k}\uparrow}, \tag{9.117}$$

where $f_{\mathbf{k}\uparrow}$ is the Fermi–Dirac distribution function for the Bloch state $\mathbf{k}_{\uparrow}$.

We have attached a suffix 0 to $\chi^{(+)}$ given in eqn (9.116) because it represents what we shall call the *non-interacting* susceptibility. The reason for this is as follows. In the Hamiltonian (9.107) we have employed the approximation (9.99) for $\hat{n}_{\sigma}\hat{n}_{-\sigma}$. However, with the addition of the applied field $H_{\mathbf{l}}^{\nu}(t)$ the total z component of spin is no longer a constant of motion and thus terms such as $\langle\hat{c}_{\mathbf{l}\uparrow}^{+}\hat{c}_{\mathbf{l}\downarrow}\rangle$ are no longer zero; this is clearly so because from (9.114) we can get the first order, or non-interacting, value of $\langle\hat{c}_{\mathbf{l}\uparrow}^{+}\hat{c}_{\mathbf{l}\downarrow}\rangle$. We have therefore to return to (9.99) and take into account the terms $\langle\hat{c}_{\mathbf{l}\uparrow}^{+}\hat{c}_{\mathbf{l}\downarrow}\rangle$ and $\langle\hat{c}_{\mathbf{l}\downarrow}^{+}\hat{c}_{\mathbf{l}\uparrow}\rangle$ that are non-zero in the presence of $H_{\mathbf{l}}^{\nu}(t)$ and include them in the approximation scheme for $\hat{n}_{\mathbf{l}\sigma}\hat{n}_{\mathbf{l}-\sigma}$ to be used in the Hamiltonian consisting of the sum of (9.97) and (9.106).

In place of (9.99) we now write for $\hat{n}_{\uparrow}\hat{n}_{\downarrow}$, say,

$$\begin{aligned}\hat{c}_{\uparrow}^{+}\hat{c}_{\uparrow}\hat{c}_{\downarrow}^{+}\hat{c}_{\downarrow} &= \hat{n}_{\uparrow}\langle\hat{n}_{\downarrow}\rangle+\langle\hat{n}_{\uparrow}\rangle\hat{n}_{\downarrow}-\langle\hat{n}_{\uparrow}\rangle\langle\hat{n}_{\downarrow}\rangle+\\&\quad+\hat{c}_{\uparrow}^{+}\hat{c}_{\downarrow}\langle\hat{c}_{\uparrow}\hat{c}_{\downarrow}^{+}\rangle+\langle\hat{c}_{\uparrow}^{+}\hat{c}_{\downarrow}\rangle\hat{c}_{\uparrow}\hat{c}_{\downarrow}^{+}-\langle\hat{c}_{\uparrow}^{+}\hat{c}_{\downarrow}\rangle\langle\hat{c}_{\uparrow}\hat{c}_{\downarrow}^{+}\rangle,\end{aligned} \tag{9.118}$$

so that

$$\begin{aligned}\sum_{\sigma}\hat{n}_{\mathbf{l}\sigma}\hat{n}_{\mathbf{l}-\sigma} &\Rightarrow -2\langle\hat{n}_{\mathbf{l}\uparrow}\rangle\langle\hat{n}_{\mathbf{l}\downarrow}\rangle+2\sum_{\sigma}\hat{n}_{\mathbf{l}\sigma}\langle\hat{n}_{\mathbf{l}-\sigma}\rangle+\\&\quad+2\{\hat{c}_{\mathbf{l}\uparrow}^{+}\hat{c}_{\mathbf{l}\downarrow}\langle\hat{c}_{\mathbf{l}\uparrow}\hat{c}_{\mathbf{l}\downarrow}^{+}\rangle+\langle\hat{c}_{\mathbf{l}\uparrow}^{+}\hat{c}_{\mathbf{l}\downarrow}\rangle\hat{c}_{\mathbf{l}\uparrow}\hat{c}_{\mathbf{l}\downarrow}^{+}-\langle\hat{c}_{\mathbf{l}\uparrow}^{+}\hat{c}_{\mathbf{l}\downarrow}\rangle\langle\hat{c}_{\mathbf{l}\uparrow}\hat{c}_{\mathbf{l}\downarrow}^{+}\rangle\}.\end{aligned} \tag{9.119}$$

We can again ignore the dependence of $\langle\hat{n}_{\mathbf{l}-\sigma}\rangle$ on $\mathbf{l}$ because the first corrections to $\langle\hat{n}_{\mathbf{l}-\sigma}\rangle$ due to the external field H^{ν} are of second order in H^{ν}.

Taking the approximation (9.119) in (9.97) we have in the interacting case the Hamiltonian (apart from constant terms)

$$\mathscr{H} = \sum_{\mathbf{l},\mathbf{l}',\sigma}T(\mathbf{l}-\mathbf{l}')\hat{c}_{\mathbf{l}\sigma}^{+}\hat{c}_{\mathbf{l}'\sigma}+I\sum_{\mathbf{l},\sigma}\langle\hat{n}_{-\sigma}\rangle\hat{n}_{\mathbf{l}\sigma}-I\sum_{\mathbf{l}}\hat{c}_{\mathbf{l}\uparrow}^{+}\hat{c}_{\mathbf{l}\downarrow}\langle\hat{c}_{\mathbf{l}\downarrow}^{+}\hat{c}_{\mathbf{l}\uparrow}\rangle+\hat{c}_{\mathbf{l}\downarrow}^{+}\hat{c}_{\mathbf{l}\uparrow}\langle\hat{c}_{\mathbf{l}\uparrow}^{+}\hat{c}_{\mathbf{l}\downarrow}\rangle.$$

It follows that, compared to (9.112), we get an additional contribution on the right-hand side of the equation of motion for $\hat{c}_{\mathbf{k}+\mathbf{q}\uparrow}^{+}\hat{c}_{\mathbf{q}\downarrow}$ which is

$$-I[\hat{c}_{\mathbf{k}+\mathbf{q}\uparrow}^{+}\hat{c}_{\mathbf{q}\downarrow},\sum_{\mathbf{l}}\hat{c}_{\mathbf{l}\downarrow}^{+}\hat{c}_{\mathbf{l}\uparrow}\langle\hat{c}_{\mathbf{l}\uparrow}^{+}\hat{c}_{\mathbf{l}\downarrow}\rangle].$$

Now $$M_{\mathbf{l}}^{+} = -g\mu_{\mathrm{B}}\langle \hat{S}_{\mathbf{l}}^{+}\rangle = -g\mu_{\mathrm{B}}\langle \hat{c}_{\mathbf{l}\uparrow}^{+}\hat{c}_{\mathbf{l}\downarrow}\rangle,$$

and also, by definition of the interacting susceptibility, $\chi_{\mathbf{q}}^{(+)}[\omega]$,

$$\begin{aligned} M_{\mathbf{l}}^{+} &= \frac{1}{N}\sum_{\mathbf{q}} \exp(\mathrm{i}\mathbf{q}\,.\,\mathbf{l})M_{\mathbf{q}}^{+}(t) \\ &= \frac{1}{N}\sum_{\mathbf{q}} \exp(\mathrm{i}\mathbf{q}\,.\,\mathbf{l})\chi_{\mathbf{q}}^{(+)}[\omega]H_{\mathbf{q}}^{+}\exp(-\mathrm{i}\omega t+\epsilon t). \end{aligned} \tag{9.120}$$

Therefore

$$\begin{aligned} \langle \hat{c}_{\mathbf{l}\uparrow}\hat{c}_{\mathbf{l}\downarrow}\rangle &= \frac{-1}{Ng\mu_{\mathrm{B}}}\sum_{\mathbf{q}} \exp(\mathrm{i}\mathbf{q}\,.\,\mathbf{l})\chi_{\mathbf{q}}^{(+)}[\omega]H_{\mathbf{q}}^{+}\exp(-\mathrm{i}\omega t+\epsilon t) \\ &= \frac{-1}{g\mu_{\mathrm{B}}} \exp(\mathrm{i}\mathbf{k}\,.\,\mathbf{l})\chi_{\mathbf{k}}^{(+)}[\omega]H^{+}\exp(-\mathrm{i}\omega t+\epsilon t). \end{aligned} \tag{9.121}$$

Hence the additional term in the equation of motion is

$$\begin{aligned} &\frac{I}{Ng\mu_{\mathrm{B}}}\,\chi_{\mathbf{k}}^{(+)}[\omega]H^{+}\exp(-\mathrm{i}\omega t+\epsilon t)\sum_{\mathbf{l},\mathbf{k}',\mathbf{q}'}\exp\{\mathrm{i}\mathbf{l}\,.\,(\mathbf{k}+\mathbf{k}'-\mathbf{q}')\}\times \\ &\qquad\qquad\times[\hat{c}_{\mathbf{k}+\mathbf{q}\uparrow}^{+}\hat{c}_{\mathbf{q}\downarrow},\,\hat{c}_{\mathbf{k}'\downarrow}^{+}\hat{c}_{\mathbf{q}'\uparrow}] \\ &= \frac{I}{g\mu_{\mathrm{B}}}\,\chi_{\mathbf{k}}^{(+)}[\omega]H^{+}\exp(-\mathrm{i}\omega t+\epsilon t)\sum_{\mathbf{k}'}[\hat{c}_{\mathbf{k}+\mathbf{q}\uparrow}^{+}\hat{c}_{\mathbf{q}\downarrow},\,\hat{c}_{\mathbf{k}'\downarrow}^{+}\hat{c}_{\mathbf{k}+\mathbf{k}'\uparrow}] \\ &= \frac{I}{g\mu_{\mathrm{B}}}\,\chi_{\mathbf{k}}^{(+)}[\omega]H^{+}\exp(-\mathrm{i}\omega t+\epsilon t)(\hat{n}_{\mathbf{k}+\mathbf{q}\uparrow}-\hat{n}_{\mathbf{q}\downarrow}). \end{aligned} \tag{9.122}$$

Adding this term to (9.112) we notice that the analysis follows through as before provided $H^{+}(t)$ is replaced by

$$\left\{1+\frac{I}{(g\mu_{\mathrm{B}})^{2}}\,\chi_{\mathbf{k}}^{(+)}[\omega]\right\}H^{+}(t).$$

Hence we get $$\chi_{\mathbf{k}}^{+}[\omega] = \left\{1+\frac{I}{(g\mu_{\mathrm{B}})^{2}}\,\chi_{\mathbf{k}}^{(+)}[\omega]\right\}\chi_{\mathbf{k},0}^{(+)}[\omega]. \tag{9.123}$$

Solving (9.123) for $\chi_{\mathbf{k}}^{(+)}[\omega]$, we obtain†

$$\chi_{\mathbf{k}}^{(+)}[\omega] = \frac{\chi_{\mathbf{k},0}^{(+)}[\omega]}{1-\{I/(g\mu_{\mathrm{B}})^{2}\}\chi_{\mathbf{k},0}^{(+)}[\omega]}. \tag{9.124}$$

In terms of $\chi_{\mathbf{k}}^{(\pm)}[\omega]$ the transverse part of the neutron cross-section for the scattering from the band electrons is given in eqn (8.80) to be

$$\begin{aligned} \frac{\mathrm{d}^{2}\sigma}{\mathrm{d}\Omega\mathrm{d}E'} &= \left(\frac{\gamma e^{2}}{m_{\mathrm{e}}c^{2}}\right)^{2}|F(\varkappa)|^{2}\,\frac{1}{4}\,\frac{k'}{k}\,(1+\tilde{\kappa}_{z}^{2})N/(g\mu_{\mathrm{B}})^{2}\times \\ &\qquad\times\frac{1}{\pi}\,\frac{\exp(\hbar\omega\beta)}{\exp(\hbar\omega\beta)-1}\,\{\mathrm{Im}\,\chi_{\varkappa}^{(+)}[\omega]+\mathrm{Im}\,\chi_{\varkappa}^{(-)}[\omega]\}. \end{aligned} \tag{9.125}$$

Because $$\mathrm{Im}\,\chi_{\mathbf{k}}^{(-)}[\omega] = -\mathrm{Im}\,\chi_{\mathbf{k}}^{(+)}[-\omega], \tag{9.126}$$

† A higher-order approximation has been derived by Hubbard and Jain, *J. Phys.* C, **1**, 1650 (1968).

eqn (9.124) for $\chi_{\mathbf{k}}^{(+)}[\omega]$ completely determines the transverse neutron cross-section (9.125).

Since $\chi_{\mathbf{k}}^{(+)}[\omega]$ is expressed in terms of the non-interacting susceptibility $\chi_{\mathbf{k},0}^{(+)}[\omega]$ we begin our discussion of the transverse part of the neutron cross-section by first discussing the properties of $\chi_{\mathbf{k},0}^{(+)}[\omega]$.

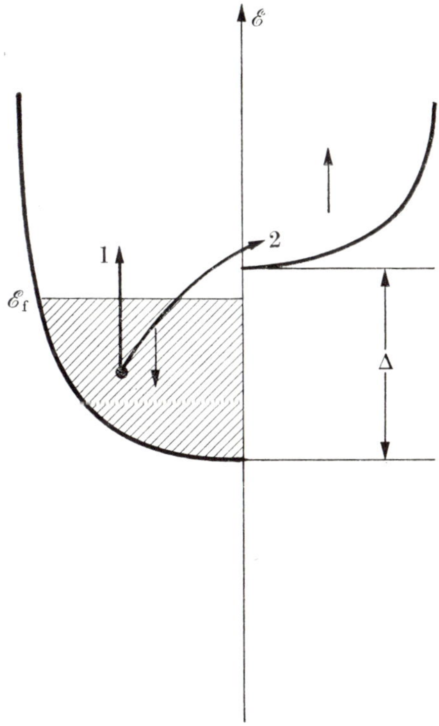

FIG. 9.4. Schematic diagram of two bands of ↓ and ↑ spin in the ferromagnetic state, for the particular case when the two bands do not overlap ($\Delta > \mathscr{E}_f$). The excitation of type 1 is non-spin flip, i.e. $\downarrow \to \downarrow$, whereas type 2 is spin flip, namely $\downarrow \to \uparrow$.

(c) *Non-interacting susceptibility*

First we recall that the poles of the generalized susceptibility functions $\chi^{(\pm)}$ determine the energy spectra of the excitations in the system. Let us therefore answer the question, what types of excitations can we obtain in our model of a single band split into two sub-bands of spin ↓ and spin ↑, as shown in Fig. 9.4? The simplest type of excitation is that in which a ↓ spin electron is merely promoted in energy. In Fig. 9.4 this is denoted as process (1). In constructing the energy spectrum for these excitations, Fig. 9.5, it must be borne in mind that because of the Pauli exclusion principle the electrons can only be promoted to empty states, i.e. the final energy of the spin must be greater than $\mathscr{E}_{f\sigma}$. This type of excitation does not involve a change of spin state and does not therefore interest us.

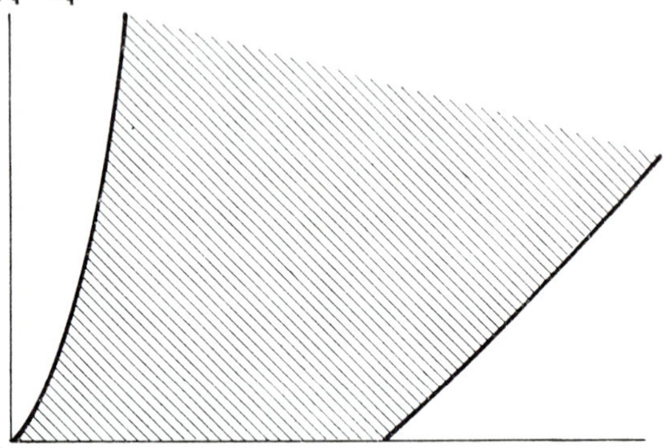

FIG. 9.5. Excitation of type 1 shown in Fig. 9.4 (non-spin flip). The energy spectrum is $E_{\mathbf{k}} = \mathscr{E}_{\mathbf{k+q}}-\mathscr{E}_{\mathbf{q}}$.

However, another type of process is that in which a $\downarrow$ spin electron is excited into the $\uparrow$ spin band. The minimum energy of this excitation is equal to the band splitting Δ and the energy spectrum of the excitations is $E_{\mathbf{k}} = \mathscr{E}_{\mathbf{k+q}}-\mathscr{E}_{\mathbf{q}}+\Delta$. In Fig. 9.6 (a) and (b) we show the excitation spectra, the former figure being for the case when the $\uparrow$ and $\downarrow$ spin bands do not overlap, as in Fig. 9.4, and Fig. 9.6 (b) for the case when the two bands overlap ($\Delta < \mathscr{E}_f$). If we now refer back to the expression we derived for $\chi^{(+)}_{\mathbf{k},0}[\omega]$, eqn (9.116), we can immediately understand its structure in the light of this discussion.

The denominator of the integrand is seen to have zeros when $-\hbar\omega$ is equal to the energy of the type 2 excitation of Fig. 9.4, and the numerator gives the appropriate weight. Also note that

$$\operatorname{Im}\chi^{(\pm)}_{\mathbf{k},0}[\omega] = (\mp)\pi(g\mu_B)^2\frac{1}{N}\sum_{\mathbf{q}}(f_{\mathbf{q}\downarrow}-f_{\mathbf{k+q}\uparrow})\delta(\mathscr{E}_{\mathbf{k+q}}-\mathscr{E}_{\mathbf{q}}+\Delta\pm\hbar\omega), \tag{9.127}$$

as follows from (9.116) and the identity

$$\operatorname{Im}\lim_{\epsilon\to 0^+}\frac{1}{x+i\epsilon} = -\pi\delta(x). \tag{9.128}$$

The above discussion and its relationship to the non-interacting

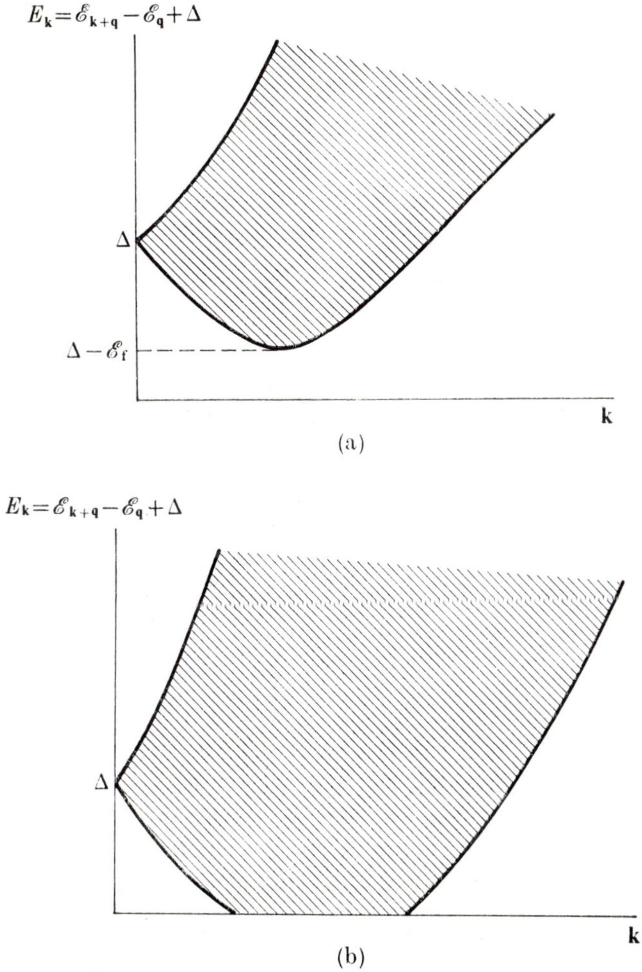

FIG. 9.6. These two figures show the excitation spectra for processes labelled type 2 in Fig. 9.4 (spin-flip excitations). (a) Corresponds to the configuration shown in Fig. 9.4 in which the two bands do no overlap ($\Delta > \mathscr{E}_{\mathrm{f}}$). (b) Shows the excitation spectrum for overlapping bands ($\Delta < \mathscr{E}_{\mathrm{f}}$).

transverse susceptibility is perhaps best illustrated for the particular case of $\mathbf{k} = 0$. From (9.127) we obtain for $\chi^{(-)}_{\mathbf{k}=0,0}$ the result

$$\operatorname{Im} \chi^{(-)}_{\mathbf{k}=0,0}[\omega] = \pi(g\mu_{\mathrm{B}})^2 \delta(\hbar\omega - \Delta) \frac{1}{N} \sum_{\mathbf{q}} (f_{\mathbf{q}\downarrow} - f_{\mathbf{q}\uparrow})$$
$$= \pi(g\mu_{\mathrm{B}})^2 \delta(\hbar\omega - \Delta)(\langle \hat{n}_{\downarrow} \rangle - \langle \hat{n}_{\uparrow} \rangle),$$

or

$$\operatorname{Im} \chi^{(-)}_{\mathbf{k}=0,0}[\omega] = \pi(g\mu_{\mathrm{B}})^2 \delta(\hbar\omega - \Delta) \frac{\Delta}{I}, \tag{9.129}$$

on using the definition of $\Delta = I(\langle\hat{n}_\downarrow\rangle - \langle\hat{n}_\uparrow\rangle)$. Thus the particular case $\mathbf{k} = 0$ merely corresponds to promoting an electron of spin $\downarrow$ to the bottom of the spin $\uparrow$ band, the energy cost to the external field being $\hbar\omega = \Delta$ (cf. Fig. 9.4).

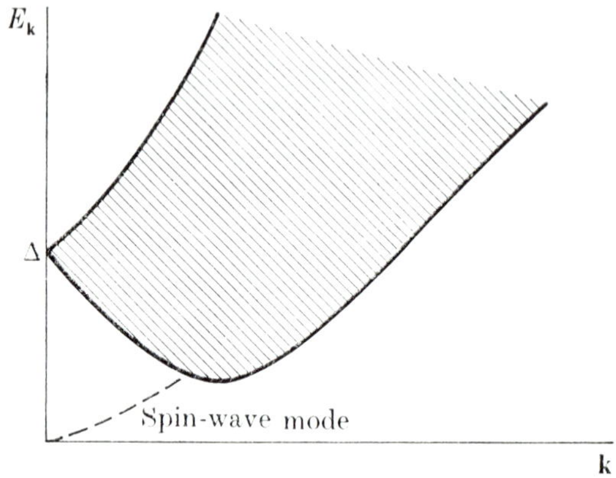

FIG. 9.7. Schematic diagram of spin wave dispersion curve around $\mathbf{k} = 0$ for a narrow s band of electron described by a Hubbard Hamiltonian, eqn (9.97).

(*d*) *Interacting susceptibility and spin waves*†

Now let us discuss the properties of $\chi_\mathbf{k}^{(+)}[\omega]$, eqn (9.124). Its poles are determined by the equation

$$1 = I\frac{1}{(g\mu_\mathrm{B})^2}\,\mathrm{Re}\,\chi_{\mathbf{k},0}^{(+)}[\omega],$$

and this reads in detail

$$1 = \frac{I}{N}\,\mathrm{Re}\sum_\mathbf{q}\frac{f_{\mathbf{q}\downarrow} - f_{\mathbf{k}+\mathbf{q}\uparrow}}{\mathscr{E}_{\mathbf{k}+\mathbf{q}} - \mathscr{E}_\mathbf{q} + \Delta + \hbar\omega + i\epsilon}. \tag{9.130}$$

If we set $\hbar\omega = -\hbar\Omega_\mathbf{k}$, say, then for $\mathbf{k} = 0$ the secular equation (9.130) for the excitation energy Ω_0 is

$$1 = I\frac{1}{N}\sum_\mathbf{q}\frac{f_{\mathbf{q}\downarrow} - f_{\mathbf{q}\uparrow}}{\Delta - \hbar\Omega_0} = \frac{I}{\Delta - \hbar\Omega_0}\,\frac{\Delta}{I},$$

which is satisfied if $\hbar\Omega_0 = 0$. Thus a solution of (9.130) exists which has the property $\lim_{\mathbf{k}\to 0}\hbar\Omega_\mathbf{k} = 0$ and this solution corresponds to a spin-wave excitation in the narrow energy band of electrons. If we plot the solutions of equation (9.130) we find they appear as in Fig. 9.7, i.e. in addition to solutions of the type shown in Fig. 9.6 (so-called Stoner excitations) we have now the spin-wave mode occurring in the gap around $\mathbf{k} = 0$.

† Calculations of $\chi_\mathbf{k}^{(\pm)}$, eqn (9.124), for a tight-binding model of Ni, at various temperatures, have been reported by Windsor *et al.*, *Phys. Rev. Lett.* **22**, 849 (1969).

For small values of $\mathbf{k}$ we expect the dispersion relation for the spin-wave mode in a cubic crystal to be of the form

$$\hbar\Omega_{\mathbf{k}} = Dk^2, \tag{9.131}$$

and we can determine the coefficient D as follows. We write

$$\frac{1}{N}\sum_{\mathbf{q}} \frac{f_{\mathbf{q}\downarrow}-f_{\mathbf{k}+\mathbf{q}\uparrow}}{\mathscr{E}_{\mathbf{k}+\mathbf{q}}-\mathscr{E}_{\mathbf{q}}+\Delta-\hbar\Omega_{\mathbf{k}}} = \frac{1}{N}\sum_{\mathbf{q}} \frac{f_{\mathbf{q}\downarrow}}{\mathscr{E}_{\mathbf{k}+\mathbf{q}}-\mathscr{E}_{\mathbf{q}}+\Delta-\hbar\Omega_{\mathbf{k}}} - \frac{1}{N}\sum_{\mathbf{q}} \frac{f_{\mathbf{k}+\mathbf{q}\uparrow}}{\mathscr{E}_{\mathbf{k}+\mathbf{q}}-\mathscr{E}_{\mathbf{q}}+\Delta-\hbar\Omega_{\mathbf{k}}},$$

and since

$$\frac{1}{N}\sum_{\mathbf{q}} \frac{f_{\mathbf{k}+\mathbf{q}\uparrow}}{\mathscr{E}_{\mathbf{k}+\mathbf{q}}-\mathscr{E}_{\mathbf{q}}+\Delta-\hbar\Omega_{\mathbf{k}}} = \frac{1}{N}\sum_{\mathbf{q}} \frac{f_{\mathbf{q}\uparrow}}{\mathscr{E}_{\mathbf{q}}-\mathscr{E}_{\mathbf{q}-\mathbf{k}}+\Delta-\hbar\Omega_{\mathbf{k}}},$$

we have, on expanding $\mathscr{E}_{\mathbf{q}+\mathbf{k}}$ and $\mathscr{E}_{\mathbf{q}-\mathbf{k}}$ in $\mathbf{k}$,

$$1 = I\frac{1}{N}\sum_{\mathbf{q}} \frac{f_{\mathbf{q}\downarrow}}{\mathbf{k}.\boldsymbol{\nabla}_{\mathbf{q}}\mathscr{E}_{\mathbf{q}}+\frac{1}{2}(\mathbf{k}.\boldsymbol{\nabla}_{\mathbf{q}})^2\mathscr{E}_{\mathbf{q}}-\hbar\Omega_{\mathbf{k}}+\Delta} + \frac{f_{\mathbf{q}\uparrow}}{-\mathbf{k}.\boldsymbol{\nabla}_{\mathbf{q}}\mathscr{E}_{\mathbf{q}}+\frac{1}{2}(\mathbf{k}.\boldsymbol{\nabla}_{\mathbf{q}})^2\mathscr{E}_{\mathbf{q}}+\hbar\Omega_{\mathbf{k}}-\Delta}. \tag{9.132}$$

Expanding in powers of $\mathbf{k}$ gives

$$1 = \frac{I}{\Delta N}\sum_{\mathbf{q}}(f_{\mathbf{q}\downarrow}-f_{\mathbf{q}\uparrow}) - \frac{I}{\Delta N}\sum_{\mathbf{q}}(f_{\mathbf{q}\downarrow}+f_{\mathbf{q}\uparrow})\frac{1}{2\Delta}(\mathbf{k}.\boldsymbol{\nabla}_{\mathbf{q}})^2\mathscr{E}_{\mathbf{q}} + \frac{I}{\Delta N}\sum_{\mathbf{q}}(f_{\mathbf{q}\downarrow}-f_{\mathbf{q}\uparrow})\frac{\hbar\Omega_{\mathbf{k}}}{\Delta} + \frac{I}{\Delta N}\sum_{\mathbf{q}}(f_{\mathbf{q}\downarrow}-f_{\mathbf{q}\uparrow})\frac{1}{\Delta^2}\{\mathbf{k}.\boldsymbol{\nabla}_{\mathbf{q}}\mathscr{E}_{\mathbf{q}}\}^2.$$

Now

$$\sum_{\mathbf{q}} f_{\mathbf{q}\downarrow} = N\langle\hat{n}_{\downarrow}\rangle$$

and, remembering that $\Delta = I(\langle\hat{n}_{\downarrow}\rangle-\langle\hat{n}_{\uparrow}\rangle)$, we get

$$\hbar\Omega_{\mathbf{k}} = \frac{I}{2\Delta}\frac{1}{N}\sum_{\mathbf{q}}\{(\mathbf{k}.\boldsymbol{\nabla}_{\mathbf{q}})^2\mathscr{E}_{\mathbf{q}}\}(f_{\mathbf{q}\uparrow}+f_{\mathbf{q}\downarrow}) + \frac{I}{\Delta^2}\frac{1}{N}\sum_{\mathbf{q}}(\mathbf{k}.\boldsymbol{\nabla}_{\mathbf{q}}\mathscr{E}_{\mathbf{q}})^2(f_{\mathbf{q}\uparrow}-f_{\mathbf{q}\downarrow}). \tag{9.133}$$

Thus for cubic lattices we find that D in eqn (9.131) is given by the expression†

$$D = \frac{I}{6\Delta}\frac{1}{N}\sum_{\mathbf{q}}\{\nabla_{\mathbf{q}}^2\mathscr{E}_{\mathbf{q}}\}(f_{\mathbf{q}\uparrow}+f_{\mathbf{q}\downarrow}) + \frac{I}{3\Delta^2}\frac{1}{N}\sum_{\mathbf{q}}(\nabla_{\mathbf{q}}\mathscr{E}_{\mathbf{q}})^2(f_{\mathbf{q}\uparrow}-f_{\mathbf{q}\downarrow}). \tag{9.134}$$

If we assume that $\mathscr{E}_{\mathbf{q}}$ can be represented by

$$\mathscr{E}_{\mathbf{q}} = \frac{\hbar^2}{2m^*}q^2, \tag{9.135}$$

† An exact expression for D has been derived by D. Edwards, *Proc. R. Soc.* **A300**, 373 (1967).

where m^* is the effective mass for the band, then it is straightforward to show that in this instance eqn (9.134) for D becomes

$$D = \frac{\hbar^2 In}{2m^*\Delta}\left\{1-\frac{4(\mathscr{E}_{f\downarrow}\langle\hat{n}_\downarrow\rangle-\mathscr{E}_{f\uparrow}\langle\hat{n}_\uparrow\rangle)}{5\Delta n}\right\}. \tag{9.136}$$

Turning now to the calculation of the cross-section (9.125), we obtain from eqn (9.124) the imaginary part of $\chi_{\mathbf{k}}^{(+)}[\omega]$, namely,

$$\operatorname{Im}\chi_{\mathbf{k}}^{(+)}[\omega] = \frac{\operatorname{Im}\chi_{\mathbf{k},0}^{(+)}[\omega]}{[1-\{I/(g\mu_{\mathrm{B}})^2\}\operatorname{Re}\chi_{\mathbf{k},0}^{(+)}[\omega]]^2+[\{I/(g\mu_{\mathrm{B}})^2\}\operatorname{Im}\chi_{\mathbf{k},0}^{(+)}[\omega]]^2}. \tag{9.137}$$

Now it follows from (9.127) that $\operatorname{Im}\chi_{\mathbf{k},0}^{(+)}[\omega]$ is non-zero only for values of $\hbar\omega$ equal to the energies of the Stoner excitations of Fig. 9.6. Hence it follows that when $\hbar\omega = \hbar\Omega_{\mathbf{k}}$, $\operatorname{Im}\chi_{\mathbf{k},0}^{(+)}$ is zero, except when $\Omega_{\mathbf{k}}$ merges into the continuum of the Stoner excitations. Leaving aside this particular case, we can in (9.137) take the limit $\operatorname{Im}\chi_{\mathbf{k},0}^{(+)} \to 0^-$ and, using the result (9.128), obtain

$$\frac{I}{(g\mu_{\mathrm{B}})^2}\operatorname{Im}\chi_{\mathbf{k}}^{(+)}[\omega] = -\pi\delta\left\{1-\frac{I}{(g\mu_{\mathrm{B}})^2}\operatorname{Re}\chi_{\mathbf{k},0}^{(+)}[\omega]\right\}. \tag{9.138}$$

But

$$1-\frac{I}{(g\mu_{\mathrm{B}})^2}\operatorname{Re}\chi_{\mathbf{k},0}^{(+)}[\omega] \simeq \frac{\hbar\omega+\hbar\Omega_{\mathbf{k}}}{\Delta}, \tag{9.139}$$

and thus

$$\operatorname{Im}\chi_{\mathbf{k}}^{(+)}[\omega] = -\pi(g\mu_{\mathrm{B}})^2\frac{\Delta}{I}\,\delta(\hbar\omega+\hbar\Omega_{\mathbf{k}}). \tag{9.140}$$

Using (9.140) with (9.126) and (9.125) leads to the expression†

$$\frac{\mathrm{d}^2\sigma}{\mathrm{d}\Omega\mathrm{d}E'} = \left(\frac{\gamma e^2}{m_{\mathrm{e}}c^2}\right)^2|F(\boldsymbol{\varkappa})|^2\frac{1}{4}\frac{k'}{k}(1+\tilde{\kappa}_z^2)N\Delta/I\times$$
$$\times\sum_{\mathbf{q}}[n(\Omega_{\mathbf{q}})\delta(\hbar\omega+\hbar\Omega_{\mathbf{q}})+\{1+n(\Omega_{\mathbf{q}})\}\delta(\hbar\omega-\hbar\Omega_{\mathbf{q}})] \tag{9.141}$$

for the transverse neutron cross-section. We immediately note the similarity of this expression for the scattering by the spin-wave mode in a narrow band of electrons with that for the scattering by spin waves in a Heisenberg ferromagnet, eqn (9.44). In fact, we can identify each factor in (9.141) with a corresponding factor in (9.44) when we recall that $\Delta/I = \langle\hat{n}_\downarrow\rangle-\langle\hat{n}_\uparrow\rangle$ is proportional to the spin moment per atom, and also different normalization constants arise in (9.141) because we use a Kronecker delta function to express the momentum condition.

The cross-section (9.141) does not include the contribution arising from the scattering by Stoner excitation, i.e. that determined by $\chi_{\mathbf{k},0}^{(+)}[\omega]$.

† (9.141) was derived first by Izuyama *et al.*, *J. phys. Soc. Japan* **18**, 1025 (1963).

However, it is clear that this scattering is nothing like that due to the spin-wave mode because it involves a continuum of states for a given wave vector and thus the geometrical conditions imposed by the conservation conditions are weak. Because the geometrical conditions are weak it also follows that the neutrons are scattered roughly in all directions, i.e. the cross-section is roughly isotropic and small. In contrast to this, the scattering from the spin-wave mode is strongly concentrated near the Bragg peaks.

To evaluate the temperature dependence of the coefficient D in (9.131) requires, for a detailed answer, the use of thermodynamic perturbation theory. Such an approach would lead us too far afield and we therefore content ourselves with an argument based on quite general considerations that shows what the form of the dependence on T must be.

In the absence of spin-wave interactions the total energy of the spin waves in the itinerant electron model is

$$\sum_{\mathbf{k}} \hat{n}_{\mathbf{k}} \hbar\Omega_{\mathbf{k}}.$$

If now we imagine the interaction terms to have been evaluated, the exact energy will involve terms linear in $\hat{n}_{\mathbf{k}}$ plus terms quadratic in $\hat{n}_{\mathbf{k}}$ plus higher-order terms. Hence

$$\mathscr{H} = \sum_{\mathbf{k}} \hat{n}_{\mathbf{k}} \hbar\Omega_{\mathbf{k}} - \sum_{\mathbf{k},\mathbf{q}} \hat{n}_{\mathbf{k}} \hat{n}_{\mathbf{q}} c_{\mathbf{k},\mathbf{q}}, \tag{9.142 a}$$

where $c_{\mathbf{k},\mathbf{q}}$ is a coefficient to be determined. The only assumption included in (9.142 a) is that the state of the crystal can be specified by just the occupation number operators $\hat{n}_{\mathbf{k}}$, i.e. the interactions of spin waves with other types of excitations in the crystal have been neglected.

If we now write for $\hat{n}_{\mathbf{k}} \hat{n}_{\mathbf{q}}$ in (9.142 a) $\hat{n}_{\mathbf{k}}\langle \hat{n}_{\mathbf{q}}\rangle + \langle \hat{n}_{\mathbf{k}}\rangle \hat{n}_{\mathbf{q}}$ then it is given by

$$\mathscr{H} = \sum_{\mathbf{k}} \Big\{\hbar\Omega_{\mathbf{k}} - 2\sum_{\mathbf{q}} c_{\mathbf{k},\mathbf{q}} \langle \hat{n}_{\mathbf{q}}\rangle\Big\} \hat{n}_{\mathbf{k}} = \sum_{\mathbf{k}} E_{\mathbf{k}} \hat{n}_{\mathbf{k}}. \tag{9.142 b}$$

In the interaction term in $E_{\mathbf{k}}$ the only dependence on $\mathbf{k}$ arises from the coefficient $c_{\mathbf{k},\mathbf{q}}$. This coefficient must go to zero as $\mathbf{k}$ tends to zero; otherwise $E_{\mathbf{k}}$ would possess an energy gap at $\mathbf{k} = 0$ and it is known that the uniform excitation mode ($\mathbf{k} = 0$) has zero energy. Because we are also confident that the energy–momentum relation is quadratic at small $\mathbf{k}$, it follows that the leading term in $c_{\mathbf{k},\mathbf{q}}$ must be of order k^2 at most at small $\mathbf{k}$. By definition $c_{\mathbf{k},\mathbf{q}}$ is symmetric in $\mathbf{k}$ and $\mathbf{q}$ so if it has a factor k^2 it must also have a factor q^2. Thus it follows that

$$Nc_{\mathbf{k},\mathbf{q}} = k^2 q^2 c. \tag{9.143}$$

On taking $\hbar\Omega_{\mathbf{k}} = Dk^2$ as in (9.131), the relation (9.143) shows, by arguments identical to those used in § 9.2, that

$$E_{\mathbf{k}} = k^2\left\{D - \frac{2c}{N}\sum_{\mathbf{q}} q^2\langle\hat{n}_{\mathbf{q}}\rangle\right\}$$

$$= k^2D\left\{1 - \frac{v_0\,4\pi c}{3\sqrt{3}D}\left(\frac{k_B\,T}{4\pi D}\right)^{\frac{5}{2}}\zeta(\tfrac{5}{2})\right\} \tag{9.144}$$

to first order in c.

Thus, we find the temperature dependence of D in the itinerant model to be the same as that in the Heisenberg model of ferromagnetism. On examining the argument leading to (9.144), we see that this conclusion is to be expected, since the assumptions made are equally valid for either model.

One of these assumptions was that the spin waves do not interact with any other excitation in the solid. Let us now consider the effect of the interaction between the spin waves and the electrons excited out of their zero-temperature configuration. If we denote the number operator for the thermally excited itinerant electrons by $\delta\hat{n}_{\mathbf{k}}$ then to (9.142a) we must add a term $\sum \mathscr{E}_{\mathbf{k}}\,\delta\hat{n}_{\mathbf{k}}$ corresponding to their kinetic energy, an interaction term of the form $-\sum \bar{c}_{\mathbf{k},\mathbf{q}}\,\delta\hat{n}_{\mathbf{k}}\,\delta\hat{n}_{\mathbf{q}}$ and a further term $-\sum c'_{\mathbf{k},\mathbf{q}}\,\hat{n}_{\mathbf{q}}\,\delta\hat{n}_{\mathbf{k}}$ representing the interaction of the electrons with the spin waves. Thus in place of the dispersion relation given in (9.142b) for the spin waves we now have the result

$$E_{\mathbf{k}} = \hbar\Omega_{\mathbf{k}} - 2\sum_{\mathbf{q}} c_{\mathbf{k},\mathbf{q}}\langle\hat{n}_{\mathbf{q}}\rangle - \sum_{\mathbf{q}} c'_{\mathbf{k},\mathbf{q}}\langle\delta\hat{n}_{\mathbf{q}}\rangle,$$

and it may be easily verified that the last term gives rise to a T^2 dependence.† Hence we expect for cubic metals that the coefficient D should have a temperature dependence of the form

$$D(T) = D_0 - D_1\,T^2 - D_2\,T^{\frac{5}{2}}. \tag{9.145}$$

The evaluation of the coefficient D_1 in simple cases is discussed by Herring.‡

In Fig. 9.3 we showed some measurements by Stringfellow of $D(T)$ for the two elements Fe and Ni, and noted that though $D(T)$ for Ni was in good agreement with a $T^{\frac{5}{2}}$ law there was a marked discrepancy between this law and the measurements on Fe at small T. Stringfellow analysed the Fe data on the basis of (9.145) and the result is shown in Fig. 9.8, where $D(T)$ is plotted against T^2 and the solid line is a function of the form (9.145). While it is obviously not a good policy to state that a fit

† T. Izuyama and R. Kubo, *J. appl. Phys.* **35**, 1074 (1964).
‡ op. cit.

to a few terms in a power-series expansion can discriminate between two slight modifications of the terms included, the agreement between (9.145) and the Fe data is very good and therefore strongly indicative of the correctness of a T^2 term in $D(T)$ for pure iron.

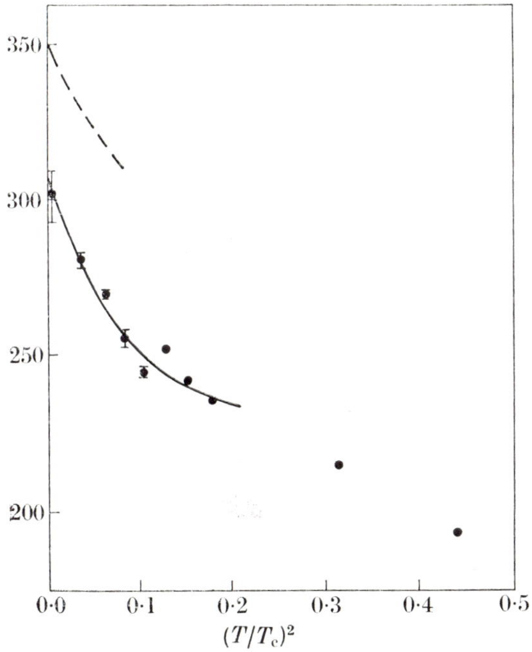

FIG. 9.8. The values of $D(T)$ for Fe given in Fig. 9.3 are here shown plotted against $(T/T_c)^2$ in the range $0 < T/T_c < 0{\cdot}42$. The solid line is the curve

$$D(T) = 314 \pm 10 - (1{\cdot}6 \pm 0{\cdot}5)10^{-3}T^2 + (5{\cdot}7 \pm 2{\cdot}2)10^{-5}T^{\frac{5}{2}} \text{ meV Å}^2.$$

Note that the value of D_0 is some 20–30 meV Å^2 greater than that proposed by Shirane *et al.* Fig. 9.2.

The broken curve in Fig. 9.8 represents the results of measurements by Phillips using spin-wave resonance (Phillips, T. G. (1966), *Proc. Roy. Soc.* A**292**, 224).

(e) *Longitudinal interacting susceptibility*

The longitudinal cross-section for the itinerant electron model behaves very differently to (9.141). In fact, just as we found for the Heisenberg ferromagnet, it does not yield spin-wave scattering. However, it would be important for a consideration of paramagnetic and critical scattering, and since its derivation is closely akin to that for the transverse cross-section it is logical for us to present the calculation at this juncture.

We follow the same scheme in calculating the longitudinal susceptibility $\chi_{\mathbf{k}}^{(0)}[\omega]$ as we adopted for the transverse susceptibility, namely, we calculate first the non-interacting value $\chi^{(0)}_{\mathbf{k},0}[\omega]$ and then its value in the interacting case. To the Hamiltonian (9.100) we therefore add the term

$$g\mu_{\mathrm{B}}\sum_{\mathbf{l}} H_{\mathbf{l}}(t)\hat{S}^z_{\mathbf{l}}(t) = \tfrac{1}{2}g\mu_{\mathrm{B}}\sum_{\mathbf{l},\sigma} H_{\mathbf{l}}(t)\sigma\hat{n}_{\mathbf{l}\sigma} \tag{9.146}$$

and hence the analogue of the Hamiltonian (9.107) is

$$\mathscr{H} = \sum_{\mathbf{l},\mathbf{l}'\sigma} T(\mathbf{l}-\mathbf{l}')\hat{c}^+_{\mathbf{l}\sigma}\hat{c}_{\mathbf{l}'\sigma}+I\sum_{\mathbf{l},\sigma}\langle\hat{n}_{-\sigma}\rangle\hat{n}_{\mathbf{l}\sigma}+\tfrac{1}{2}g\mu_{\mathrm{B}}\sum_{\mathbf{l},\sigma}H_{\mathbf{l}}(t)\sigma\hat{n}_{\mathbf{l}\sigma}. \tag{9.147}$$

If
$$H_{\mathbf{l}}(t) = \tfrac{1}{2}H\{\exp(\mathrm{i}\mathbf{k}.\mathbf{l}-\mathrm{i}\omega t)+\text{c.c.}\}\exp\epsilon t \tag{9.148}$$

then the Hamiltonian (9.147) in terms of Bloch state operators is

$$\mathscr{H} = \sum_{\mathbf{k}',\sigma}\{\mathscr{E}_{\mathbf{k}'}+I\langle\hat{n}_{-\sigma}\rangle\}\hat{c}^+_{\mathbf{k}'\sigma}\hat{c}_{\mathbf{k}'\sigma}+$$
$$+\tfrac{1}{4}g\mu_{\mathrm{B}}H\sum_{\mathbf{k}',\sigma}\sigma\{\exp(-\mathrm{i}\omega t)\hat{c}^+_{\mathbf{k}'\sigma}\hat{c}_{\mathbf{k}+\mathbf{k}'\sigma}+\exp(\mathrm{i}\omega t)\hat{c}^+_{\mathbf{k}+\mathbf{k}'\sigma}\hat{c}_{\mathbf{k}'\sigma}\}. \tag{9.149}$$

The equation of motion for $\hat{c}^+_{\mathbf{q}\sigma}$ is, therefore,

$$\mathrm{i}\hbar\dot{\hat{c}}^+_{\mathbf{q}\sigma} = -E_{\mathbf{q}\sigma}\hat{c}^+_{\mathbf{q}\sigma}-\tfrac{1}{4}g\mu_{\mathrm{B}}H\sigma\{\exp(-\mathrm{i}\omega t)\hat{c}^+_{\mathbf{q}-\mathbf{k}\sigma}+\exp(\mathrm{i}\omega t)\hat{c}^+_{\mathbf{q}+\mathbf{k}\sigma}\}\exp\epsilon t, \tag{9.150}$$

where we find it convenient to use the notation

$$E_{\mathbf{q}\sigma} = \mathscr{E}_{\mathbf{q}}+I\langle\hat{n}_{-\sigma}\rangle. \tag{9.151}$$

If $H = 0$, then the solution of (9.150) is

$$\hat{c}^+_{\mathbf{q}\sigma}(t) = \hat{c}^+_{\mathbf{q}\sigma}\exp(\mathrm{i}tE_{\mathbf{q}\sigma}/\hbar).$$

Because we require the solution of (9.150) only to first order in H we can use this result for the operators appearing in the coefficient of H in (9.150). Thus,

$$\mathrm{i}\hbar\frac{\partial}{\partial t}\exp(-\mathrm{i}tE_{\mathbf{q}\sigma}/\hbar)\hat{c}^+_{\mathbf{q}\sigma} = -\tfrac{1}{4}g\mu_{\mathrm{B}}H\sigma\{\exp(\mathrm{i}tE_{\mathbf{q}-\mathbf{k}\sigma}/\hbar-\mathrm{i}\omega t)\hat{c}^+_{\mathbf{q}-\mathbf{k}\sigma}+$$
$$+\exp(\mathrm{i}tE_{\mathbf{q}+\mathbf{k}\sigma}/\hbar+\mathrm{i}\omega t)\hat{c}^+_{\mathbf{q}+\mathbf{k}\sigma}\}\exp(-\mathrm{i}tE_{\mathbf{q}\sigma}/\hbar+\epsilon t),$$

which when integrated leads to the results

$$\hat{c}^+_{\mathbf{q}\sigma}(t) = \exp(\mathrm{i}tE_{\mathbf{q}\sigma}/\hbar)\hat{c}^+_{\mathbf{q}\sigma}-\tfrac{1}{4}g\mu_{\mathrm{B}}H\sigma\times$$
$$\times\left\{\frac{\exp(\mathrm{i}tE_{\mathbf{q}-\mathbf{k}\sigma}/\hbar-\mathrm{i}\omega t)\hat{c}^+_{\mathbf{q}-\mathbf{k}\sigma}}{\mathscr{E}_{\mathbf{q}}-\mathscr{E}_{\mathbf{q}-\mathbf{k}}+\hbar\omega+\mathrm{i}\epsilon}+\frac{\exp(\mathrm{i}tE_{\mathbf{q}+\mathbf{k}\sigma}/\hbar+\mathrm{i}\omega t)\hat{c}^+_{\mathbf{q}+\mathbf{k}\sigma}}{\mathscr{E}_{\mathbf{q}}-\mathscr{E}_{\mathbf{q}+\mathbf{k}}-\hbar\omega+\mathrm{i}\epsilon}\right\} \tag{9.152 a}$$

and

$$\hat{c}_{\mathbf{q}'\sigma}(t) = \exp(-\mathrm{i}tE_{\mathbf{q}'\sigma}/\hbar)\hat{c}_{\mathbf{q}'\sigma}-\tfrac{1}{4}g\mu_{\mathrm{B}}H\sigma\times$$
$$\times\left\{\frac{\exp(-\mathrm{i}tE_{\mathbf{q}'-\mathbf{k}\sigma}/\hbar+\mathrm{i}\omega t)\hat{c}_{\mathbf{q}'-\mathbf{k}\sigma}}{\mathscr{E}_{\mathbf{q}'}-\mathscr{E}_{\mathbf{q}'-\mathbf{k}}+\hbar\omega-\mathrm{i}\epsilon}+\frac{\exp(-\mathrm{i}tE_{\mathbf{q}'+\mathbf{k}\sigma}/\hbar-\mathrm{i}\omega t)\hat{c}_{\mathbf{q}'+\mathbf{k}\sigma}}{\mathscr{E}_{\mathbf{q}'}-\mathscr{E}_{\mathbf{q}'+\mathbf{k}}-\hbar\omega-\mathrm{i}\epsilon}\right\}. \tag{9.152 b}$$

Now
$$\langle\hat{S}^z_{\mathbf{l}}(t)\rangle = \frac{1}{N}\sum_{\mathbf{q},\mathbf{q}'\sigma}\tfrac{1}{2}\sigma\langle\hat{c}^+_{\mathbf{q}\sigma}(t)\hat{c}_{\mathbf{q}'\sigma}(t)\rangle\exp\{\mathrm{i}\mathbf{l}.(\mathbf{q}-\mathbf{q}')\}, \tag{9.153}$$

and we find from (9.152) by a straightforward calculation

$$\langle \hat{S}^z_{\mathbf{l}}(t)\rangle = \frac{1}{N}\sum_{\mathbf{q},\sigma}\tfrac{1}{2}\sigma\langle\hat{n}_{\mathbf{q}\sigma}\rangle - \tfrac{1}{4}g\mu_{\mathrm{B}}H\frac{1}{N}\sum_{\mathbf{q},\sigma}\{\langle\hat{n}_{\mathbf{q}\sigma}\rangle-\langle\hat{n}_{\mathbf{k}+\mathbf{q}\sigma}\rangle\}\times$$
$$\times\mathrm{Re}\left\{\frac{\exp(\mathrm{i}\mathbf{k}.\mathbf{l}-\mathrm{i}\omega t)}{\mathscr{E}_{\mathbf{k}+\mathbf{q}}-\mathscr{E}_{\mathbf{q}}+\hbar\omega+\mathrm{i}\epsilon}\right\}, \qquad (9.154)$$

where $$\langle\hat{n}_{\mathbf{k}\sigma}\rangle \equiv \langle\hat{c}^+_{\mathbf{k}\sigma}\hat{c}_{\mathbf{k}\sigma}\rangle = f_{\mathbf{k}\sigma}.$$

If the system were in a paramagnetic state the first term in (9.154) would be zero and the second independent of spin, the sum over σ giving just an additional factor of 2.

By definition the z component of magnetization $M^z_{\mathbf{l}}(t)$ is given in terms of $\chi^{(0)}_{\mathbf{k}}[\omega]$ through the relation

$$M^z_{\mathbf{l}}(t) = -\frac{1}{2}\frac{g\mu_{\mathrm{B}}}{N}\sum_{\mathbf{q},\sigma}\sigma f_{\mathbf{q}\sigma}+\frac{H}{N}\mathrm{Re}\{\exp(\mathrm{i}\mathbf{k}.\mathbf{l}-\mathrm{i}\omega t)\chi^{(0)}_{\mathbf{k}}[\omega]\}. \qquad (9.155)$$

Hence, by comparing (9.155) with $(-g\mu_{\mathrm{B}})$ times (9.154) we have the non-interacting longitudinal susceptibility

$$\chi^{(0)}_{\mathbf{k},0}[\omega] = -(\tfrac{1}{2}g\mu_{\mathrm{B}})^2\frac{1}{N}\sum_{\mathbf{q},\sigma}\frac{f_{\mathbf{k}+\mathbf{q}\sigma}-f_{\mathbf{q}\sigma}}{\mathscr{E}_{\mathbf{k}+\mathbf{q}}-\mathscr{E}_{\mathbf{q}}+\hbar\omega+\mathrm{i}\epsilon}. \qquad (9.156)$$

In order to calculate $\chi^{(0)}_{\mathbf{k}}[\omega]$, the longitudinal susceptibility in the interacting case, $\langle\hat{n}_{-\sigma}\rangle$ in (9.147) must be allowed to be a function of position, as it must be in the presence of the external field $H_{\mathbf{l}}(t)$. The result (9.152) tells us that to first order in H $\langle\hat{n}_{\mathbf{l}-\sigma}\rangle$ is given by

$$\langle\hat{n}_{\mathbf{l}-\sigma}\rangle = \frac{1}{N}\sum_{\mathbf{q}}f_{\mathbf{q}-\sigma}-\tfrac{1}{2}g\mu_{\mathrm{B}}H\sigma\frac{1}{N}\sum_{\mathbf{q}}(f_{\mathbf{k}+\mathbf{q}-\sigma}-f_{\mathbf{q}-\sigma})\times$$
$$\times\mathrm{Re}\left\{\frac{\exp(\mathrm{i}\mathbf{k}.\mathbf{l}-\mathrm{i}\omega t)}{\mathscr{E}_{\mathbf{k}+\mathbf{q}}-\mathscr{E}_{\mathbf{q}}+\hbar\omega+\mathrm{i}\epsilon}\right\}.$$

Hence its first-order correction *in the paramagnetic region* is

$$\frac{I}{g\mu_{\mathrm{B}}}H\sigma\,\mathrm{Re}\{\exp(\mathrm{i}\mathbf{k}.\mathbf{l}-\mathrm{i}\omega t)\chi^{(0)}_{\mathbf{k},0}[\omega]\}. \qquad (9.157)$$

In view of this result the perturbation term to be added to the unperturbed Hamiltonian is now

$$H\sum_{\mathbf{l},\sigma}\sigma\hat{n}_{\mathbf{l}\sigma}\mathrm{Re}\left\{\left(\tfrac{1}{2}g\mu_{\mathrm{B}}+\frac{I}{g\mu_{\mathrm{B}}}\chi^{(0)}_{\mathbf{k}}[\omega]\right)\exp(\mathrm{i}\mathbf{k}.\mathbf{l}-\mathrm{i}\omega t)\right\}\exp\epsilon t \qquad (9.158)$$

in place of (9.146). The calculation of $\hat{c}^+_{\mathbf{q}\sigma}(t)$ with (9.158) added to the unperturbed Hamiltonian proceeds just as in the non-interacting case.

It is readily shown that in the interacting case the susceptibility is

$$\chi_{\mathbf{k}}^{(0)}[\omega] = \frac{\chi_{\mathbf{k},0}^{(0)}[\omega]}{1-2/(g\mu_{\mathrm{B}})^2 I\chi_{\mathbf{k},0}^{(0)}[\omega]}. \tag{9.159}$$

It is clear from the structure of $\chi_{\mathbf{k},0}^{(0)}[\omega]$, eqn (9.156), that the paramagnetic longitudinal susceptibility (9.159) leads to scattering that is nothing like that arising from the transverse susceptibility. We shall not, therefore, dwell upon its behaviour or the scattering it describes, but refer the reader to Doniach's study.†

9.4. Heisenberg ferromagnet with dipole interaction

In many cases of practical importance the Heisenberg Hamiltonian (9.1) is inadequate to describe the properties of systems with localized magnetic moments because it does not include the dipole term ($\mathbf{r} \neq 0$)

$$\mathscr{H}_{\mathrm{d}} = \tfrac{1}{2}(g\mu_{\mathrm{B}})^2 \sum_{\mathbf{l},\mathbf{r}} \left\{\hat{\mathbf{S}}_{\mathbf{l}}\cdot\hat{\mathbf{S}}_{\mathbf{l}+\mathbf{r}} - 3\,\frac{(\hat{\mathbf{S}}_{\mathbf{l}}\cdot\mathbf{r})(\hat{\mathbf{S}}_{\mathbf{l}+\mathbf{r}}\cdot\mathbf{r})}{r^2}\right\}\frac{1}{r^3}. \tag{9.160}$$

To take this into account in the linear spin-wave approximation we first write

$$\mathbf{r} = \left\{\tfrac{1}{2}(r^+ + r^-),\ \frac{1}{2\mathrm{i}}(r^+ - r^-),\ z\right\}, \tag{9.161}$$

so that

$$\hat{\mathbf{S}}_{\mathbf{l}}\cdot\mathbf{r} = z\hat{S}_{\mathbf{l}}^z + \tfrac{1}{2}(r^+\hat{S}_{\mathbf{l}}^- + r^-\hat{S}_{\mathbf{l}}^+) \tag{9.162}$$

and similarly for $\hat{\mathbf{S}}_{\mathbf{l}+\mathbf{r}}\cdot\mathbf{r}$. We then find, for Bravais crystal structures,

$$\begin{aligned}\mathscr{H}_{\mathrm{d}} = \tfrac{1}{2}(g\mu_{\mathrm{B}})^2 \sum_{\mathbf{l},\mathbf{r}} \frac{1}{r^3}\{&\hat{S}_{\mathbf{l}}^z\hat{S}_{\mathbf{l}+\mathbf{r}}^z(1-3z^2/r^2) + \hat{S}_{\mathbf{l}}^+\hat{S}_{\mathbf{l}+\mathbf{r}}^-(1-3r^+r^-/2r^2) - \\ &-(3z/r^2)(r^+\hat{S}_{\mathbf{l}}^z\hat{S}_{\mathbf{l}+\mathbf{r}}^- + r^-\hat{S}_{\mathbf{l}}^z\hat{S}_{\mathbf{l}+\mathbf{r}}^+) - \\ &-(3/4r^2)(r^{+2}\hat{S}_{\mathbf{l}}^-\hat{S}_{\mathbf{l}+\mathbf{r}}^- + r^{-2}\hat{S}_{\mathbf{l}}^+\hat{S}_{\mathbf{l}+\mathbf{r}}^+)\}.\end{aligned} \tag{9.163}$$

If we use the representation (9.14) for $\hat{S}^z$, as is appropriate for linear spin-wave theory, it is clear that the third set of terms in (9.163) give rise only to terms independent of $\hat{S}^{\pm}$ and terms cubic in these operators. We may therefore neglect this term in a linear theory.

Using the commutation rule (9.13) we then obtain for the commutator $[\hat{S}_{\mathbf{l}}^+, \mathscr{H}_{\mathrm{d}}]$, which occurs in the equation of motion for $\hat{S}_{\mathbf{l}}^+$ when the Hamiltonian $\mathscr{H}_{\mathrm{d}}$ is added to the Heisenberg exchange Hamiltonian (9.1), the result

$$\begin{aligned}[\hat{S}_{\mathbf{l}}^+, \mathscr{H}_{\mathrm{d}}] = S(g\mu_{\mathrm{B}})^2 \sum_{\mathbf{r}} \frac{1}{r^3}\{&-\hat{S}_{\mathbf{l}}^+(1-3z^2/r^2) + \hat{S}_{\mathbf{l}+\mathbf{r}}^+(1-3r^+r^-/2r^2) - \\ &-(3/2r^2)(r^+)^2\hat{S}_{\mathbf{l}+\mathbf{r}}^-\}.\end{aligned} \tag{9.164}$$

† S. Doniach, *Proc. phys. Soc.* **91**, 86 (1967). See also Windsor *et al.*, *Phys. Rev. Lett.*, **22**, 849 (1969).

If we now write this commutator in terms of the operators $\hat{S}_{\mathbf{q}}^{\pm}$ we have, on adding the result to the right-hand side of (9.9), the equation of motion for $\hat{S}_{\mathbf{q}}^{+}$. The resulting equation of motion can be written

$$i\hbar\dot{\hat{S}}_{\mathbf{q}}^{+} = A_{\mathbf{q}}\,\hat{S}_{\mathbf{q}}^{+}+B_{\mathbf{q}}^{*}\hat{S}_{-\mathbf{q}}^{-}, \tag{9.165}$$

where

$$A_{\mathbf{q}} = g\mu_{\mathrm{B}}H+2S\{\mathscr{J}(0)-\mathscr{J}(\mathbf{q})\}+S(g\mu_{\mathrm{B}})^2\times$$
$$\times\left[\sum_{\mathbf{r}}\frac{\exp(\mathrm{i}\mathbf{q}.\mathbf{r})}{r^3}\{1-3(x^2+y^2)/2r^2\}-\sum_{\mathbf{r}}\frac{1}{r^3}(1-3z^2/r^2)\right] \tag{9.166 a}$$

and

$$B_{\mathbf{q}} = -\frac{3S}{2}(g\mu_{\mathrm{B}})^2\sum_{\mathbf{r}}\frac{\exp(-\mathrm{i}\mathbf{q}.\mathbf{r})}{r^5}(x-\mathrm{i}y)^2. \tag{9.166 b}$$

The Hamiltonian (9.160) is that derived from classical electromagnetics. However, as an indirect result of spin–orbit coupling there are usually other terms of the same structure as the classical dipole term.† These are usually referred to a 'pseudodipolar' terms and their inclusion in (9.160) would still result in an equation of motion of the form (9.165) but with modified values of the coefficients $A_{\mathbf{q}}$ and $B_{\mathbf{q}}$. We shall not explicitly consider these pseudodipolar terms. The evaluation of the coefficients $A_{\mathbf{q}}$ and $B_{\mathbf{q}}$, including the pseudodipolar terms, is given in full by Keffer.‡

We notice that

$$A_{-\mathbf{q}} = A_{\mathbf{q}} = A_{\mathbf{q}}^{*}$$

and

$$B_{-\mathbf{q}} = B_{\mathbf{q}} \neq B_{\mathbf{q}}^{*}. \tag{9.167}$$

We look for a solution of the form

$$\hat{S}_{\mathbf{q}}^{+} = u_{\mathbf{q}}\,\hat{\alpha}_{\mathbf{q}}+v_{\mathbf{q}}\,\hat{\alpha}_{-\mathbf{q}}^{+}, \tag{9.168 a}$$

and hence

$$\hat{S}_{\mathbf{q}}^{-} = u_{\mathbf{q}}^{*}\,\hat{\alpha}_{\mathbf{q}}^{+}+v_{\mathbf{q}}^{*}\,\hat{\alpha}_{-\mathbf{q}}, \tag{9.168 b}$$

and demand that the transformation diagonalizes the linear Hamiltonian and that $\hat{\alpha}_{\mathbf{q}}$, $\hat{\alpha}_{\mathbf{q}}^{+}$ satisfy the commutation rule

$$[\hat{\alpha}_{\mathbf{q}}, \hat{\alpha}_{\mathbf{q}'}^{+}] = \delta_{\mathbf{q},\mathbf{q}'}. \tag{9.169}$$

Inverting (9.168) we get

$$\hat{\alpha}_{\mathbf{q}} = (u_{-\mathbf{q}}^{*}\,\hat{S}_{\mathbf{q}}^{+}-v_{\mathbf{q}}\,\hat{S}_{-\mathbf{q}}^{-})/(u_{\mathbf{q}}\,u_{-\mathbf{q}}^{*}-v_{\mathbf{q}}\,v_{-\mathbf{q}}^{*}). \tag{9.170}$$

Then

$$\mathrm{i}\hbar\frac{\partial}{\partial t}\hat{\alpha}_{\mathbf{q}} = \hbar\omega_{\mathbf{q}}\,\hat{\alpha}_{\mathbf{q}}$$

† J. H. Van Vleck, *Phys. Rev.* **52,** 1195 (1937). ‡ Ibid.

gives $$u^*_{-\mathbf{q}}(A_{\mathbf{q}}-\hbar\omega_{\mathbf{q}})+v_{\mathbf{q}}B_{\mathbf{q}} = 0$$

and $$u^*_{-\mathbf{q}}B^*_{\mathbf{q}}+v_{\mathbf{q}}(A_{\mathbf{q}}+\hbar\omega_{\mathbf{q}}) = 0. \tag{9.171}$$

Hence $$\hbar\omega_{\mathbf{q}} = \{A^2_{\mathbf{q}}-|B_{\mathbf{q}}|^2\}^{\frac{1}{2}}. \tag{9.172}$$

We can choose to make

$$u_{\mathbf{q}} = u_{-\mathbf{q}} = u^*_{\mathbf{q}} = u^*_{-\mathbf{q}}. \tag{9.173}$$

Hence $$v_{\mathbf{q}} = v_{\ \mathbf{q}} = \frac{-u_{\mathbf{q}}B^*_{\mathbf{q}}}{(A_{\mathbf{q}}+\hbar\omega_{\mathbf{q}})}. \tag{9.174}$$

(9.169), (9.170), and (9.28) together give

$$u^2_{\mathbf{q}}-|v_{\mathbf{q}}|^2 = 2SN \tag{9.175}$$

and (9.174) and (9.175) together give

$$u^2_{\mathbf{q}} = 2SN\frac{(A_{\mathbf{q}}+\hbar\omega_{\mathbf{q}})}{2\hbar\omega_{\mathbf{q}}}$$

and $$2v_{\mathbf{q}}u_{\mathbf{q}} = \frac{-2SNB^*_{\mathbf{q}}}{\hbar\omega_{\mathbf{q}}}. \tag{9.176}$$

This completes the information we need on the transformation. We now note that

$$\begin{aligned}\hat{S}^x_{\mathbf{l}}(t) &= \tfrac{1}{2}\{\hat{S}^+_{\mathbf{l}}(t)+\hat{S}^-_{\mathbf{l}}(t)\}\\ &= \frac{1}{2N}\sum_{\mathbf{q}}\exp(\mathrm{i}\mathbf{q}.\mathbf{l})\{\hat{S}^+_{\mathbf{q}}(t)+\hat{S}^-_{-\mathbf{q}}(t)\}\\ &= \frac{1}{2N}\sum_{\mathbf{q}}\exp(\mathrm{i}\mathbf{q}.\mathbf{l})\{(u_{\mathbf{q}}+v^*_{\mathbf{q}})\hat{\alpha}_{\mathbf{q}}(t)+(u_{\mathbf{q}}+v_{\mathbf{q}})\hat{\alpha}^+_{-\mathbf{q}}(t)\}\\ &= \frac{1}{2N}\sum_{\mathbf{q}}\exp(\mathrm{i}\mathbf{q}.\mathbf{l})\{(u_{\mathbf{q}}+v^*_{\mathbf{q}})\exp(-\mathrm{i}\omega_{\mathbf{q}}t)\hat{\alpha}_{\mathbf{q}}+(u_{\mathbf{q}}+v_{\mathbf{q}})\exp(\mathrm{i}\omega_{\mathbf{q}}t)\hat{\alpha}^+_{-\mathbf{q}}\}\end{aligned} \tag{9.177}$$

and

$$\begin{aligned}\hat{S}^y_{\mathbf{l}}(t) &= \frac{1}{2\mathrm{i}}\{\hat{S}^+_{\mathbf{l}}(t)-\hat{S}_{\mathbf{l}}\ (t)\}\\ &= \frac{1}{2\mathrm{i}N}\sum_{\mathbf{q}}\exp(\mathrm{i}\mathbf{q}.\mathbf{l})\{\hat{S}^+_{\mathbf{q}}(t)-\hat{S}^-_{-\mathbf{q}}(t)\}\\ &= \frac{1}{2\mathrm{i}N}\sum_{\mathbf{q}}\exp(\mathrm{i}\mathbf{q}.\mathbf{l})\{(u_{\mathbf{q}}-v^*_{\mathbf{q}})\exp(-\mathrm{i}\omega_{\mathbf{q}}t)\hat{\alpha}_{\mathbf{q}}-\\ &\qquad -(u_{\mathbf{q}}-v_{\mathbf{q}})\exp(\mathrm{i}\omega_{\mathbf{q}}t)\hat{\alpha}^+_{-\mathbf{q}}\}.\end{aligned} \tag{9.178}$$

Hence

$$\langle \hat{S}^x_{\mathbf{l}}(0)\hat{S}^x_{\mathbf{l}'}(t)\rangle = \frac{1}{4N^2}\sum_{\mathbf{q}} \exp\{\mathrm{i}\mathbf{q}(\mathbf{l}-\mathbf{l}')\}|u_{\mathbf{q}}+v_{\mathbf{q}}|^2\{\exp(-\mathrm{i}\omega_{\mathbf{q}}t)n_{\mathbf{q}}+\exp(\mathrm{i}\omega_{\mathbf{q}}t)(n_{\mathbf{q}}+1)\}, \tag{9.179 a}$$

$$\langle \hat{S}^y_{\mathbf{l}}(0)\hat{S}^y_{\mathbf{l}'}(t)\rangle = \frac{1}{4N^2}\sum_{\mathbf{q}} \exp\{\mathrm{i}\mathbf{q}(\mathbf{l}-\mathbf{l}')\}|u_{\mathbf{q}}-v_{\mathbf{q}}|^2\{\exp(-\mathrm{i}\omega_{\mathbf{q}}t)n_{\mathbf{q}}+\exp(\mathrm{i}\omega_{\mathbf{q}}t)(n_{\mathbf{q}}+1)\}, \tag{9.179 b}$$

and

$$\langle \hat{S}^x_{\mathbf{l}}(0)\hat{S}^y_{\mathbf{l}'}(t)+\hat{S}^y_{\mathbf{l}}(0)\hat{S}^x_{\mathbf{l}'}(t)\rangle = \frac{1}{4\mathrm{i}N^2}\sum_{\mathbf{q}} \exp\{\mathrm{i}\mathbf{q}.(\mathbf{l}-\mathbf{l}')\}2u_{\mathbf{q}}(v_{\mathbf{q}}-v^*_{\mathbf{q}})\{\exp(\mathrm{i}\omega_{\mathbf{q}}t)(n_{\mathbf{q}}+1)+\exp(-\mathrm{i}\omega_{\mathbf{q}}t)n_{\mathbf{q}}\}. \tag{9.180}$$

Putting all these results together gives, finally,

$$\frac{\mathrm{d}^2\sigma}{\mathrm{d}\Omega\mathrm{d}E'} = \left(\frac{\gamma e^2}{m_e c^2}\right)^2\{\tfrac{1}{2}gF(\boldsymbol{\kappa})\}^2\frac{k'}{k}\exp\{-2W(\boldsymbol{\kappa})\}\frac{S}{2}\frac{(2\pi)^2}{v_0}\times$$
$$\times\sum_{\mathbf{q},\boldsymbol{\tau}}\left\{(1+\tilde{\kappa}^2_z)\frac{A_{\mathbf{q}}}{\hbar\omega_{\mathbf{q}}}+(1-\tilde{\kappa}^2_z)\frac{|B_{\mathbf{q}}|}{\hbar\omega_{\mathbf{q}}}\cos 2(\phi+\phi_{\mathbf{q}})\right\}\times$$
$$\times\{n_{\mathbf{q}}\delta(\hbar\omega+\hbar\omega_{\mathbf{q}})\delta(\boldsymbol{\kappa}+\mathbf{q}-\boldsymbol{\tau})+(n_{\mathbf{q}}+1)\delta(\hbar\omega-\hbar\omega_{\mathbf{q}})\delta(\boldsymbol{\kappa}-\mathbf{q}-\boldsymbol{\tau})\}, \tag{9.181}$$

where we have defined

$$B_{\mathbf{q}} = |B_{\mathbf{q}}|\exp(2\mathrm{i}\phi_{\mathbf{q}}),$$

and

$$\tilde{\boldsymbol{\kappa}} = (\sin\theta\cos\phi,\ \sin\theta\sin\phi,\ \cos\theta) \tag{9.182}$$

is the unit scattering vector.

If $|B_{\mathbf{q}}|$ is small then the cross-section (9.181) is seen to retain the same, simple dependence $1+\tilde{\kappa}^2_z$ as for the cross-section in the absence of dipolar forces. However, in the general case the dipole forces greatly complicate the angular dependence of the cross-section.

9.5. Rare earths

The discussion in § 9.2 of the scattering by spin waves in a Heisenberg ferromagnet was restricted to structures possessing a Bravais crystal lattice. We have previously remarked that the Heisenberg model ought to furnish a reasonable description of the rare-earth metals, and these commonly possess a hexagonal close-packed structure.† In this section we therefore consider the scattering by spin waves in a h.c.p. lattice,

† The physical properties of the rare earths are described by R. J. Elliott, *Magnetism*, vol. IIA, edited by G. T. Rado and H. Suhl (Academic Press, 1963). See also Cooper, *Solid St. Phys.* **21**, 393 (1968), edited by Seitz, Turnbull, and Ehrenreich.

taking into account the single-ion anisotropy terms appropriate to terbium. The analysis is, of course, also applicable to other h.c.p. ferromagnets (e.g. Co) in as far as they are adequately described by the Heisenberg model of ferromagnetism.

The h.c.p. lattice is constructed from two hexagonal Bravais lattices as described in § 2.1. We denote the spin operators on one sublattice by $\hat{S}$ and those associated with the ions on the other by $\hat{T}$. If the exchange interaction between the spins on different sublattices is denoted by J and that within a sublattice by J', then the exchange part of the Heisenberg Hamiltonian is

$$\mathscr{H}_{\text{ex}} = -\sum_{\mathbf{m},\mathbf{n}} 2J(\mathbf{m}-\mathbf{n})\hat{\mathbf{S}}_{\mathbf{m}}.\hat{\mathbf{T}}_{\mathbf{n}} - \sum_{\mathbf{m},\mathbf{m}'} J'(\mathbf{m}-\mathbf{m}')\hat{\mathbf{S}}_{\mathbf{m}}.\hat{\mathbf{S}}_{\mathbf{m}'} - \\ - \sum_{\mathbf{n},\mathbf{n}'} J'(\mathbf{n}-\mathbf{n}')\hat{\mathbf{T}}_{\mathbf{n}}.\hat{\mathbf{T}}_{\mathbf{n}'}. \quad (9.183)$$

In (9.183) we have adopted the notation of $\mathbf{m}$ for position vectors defining the sublattice with spin operators $\hat{\mathbf{S}}$ and $\mathbf{n}$ for the position vectors defining the sublattice with spin operators $\hat{\mathbf{T}}$.

The exchange interaction is not sufficient to describe the magnetic state of rare earths because they have, in general, unquenched orbital angular momentum. This means that the $4f$ electrons of terbium, for instance, with $S = L = 3$ and $J = 6$, do not form a spherical distribution but are expanded in a plane normal to the total angular momentum J.† This results in large single-ion anisotropy terms which have to be taken into account in order to describe adequately the spin-wave modes. We assume the anisotropy to give rise to the following additional terms in the total Hamiltonian,

$$\mathscr{H}_{\text{an}} = B\sum_{\mathbf{m}} (\hat{S}^z_{\mathbf{m}})^2 + B\sum_{\mathbf{n}} (\hat{T}^z_{\mathbf{n}})^2 - \tfrac{1}{2}G\sum_{\mathbf{m}} \{(\hat{S}^x_{\mathbf{m}} + i\hat{S}^y_{\mathbf{m}})^6 + (\hat{S}^x_{\mathbf{m}} - i\hat{S}^y_{\mathbf{m}})^6\} - \\ - \tfrac{1}{2}G\sum_{\mathbf{n}} \{(\hat{T}^x_{\mathbf{n}} + i\hat{T}^y_{\mathbf{n}})^6 + (\hat{T}^x_{\mathbf{n}} - i\hat{T}^y_{\mathbf{n}})^6\}. \quad (9.184)$$

In (9.184) $\hat{S}^z$ and $\hat{T}^z$ are the components of $\hat{\mathbf{S}}$ and $\hat{\mathbf{T}}$ in the z direction. The first terms on the right-hand side account for the tendency of the magnetic moments to lie in the hexagonal basal plane and the second set of terms is responsible for the preferred direction within the basal plane. With both B and G positive the crystal-field anisotropy energy

† The lattice parameters of Tb are $a = 3{\cdot}599$ Å and $c = 5{\cdot}696$ Å at room temperature. It is a ferromagnetic below 216 K, in the temperature range 216 K–226 K it forms an antiferromagnetic spiral structure, and is paramagnetic above 226 K. We also note that the mean radius of the $4f$ shell wave function is believed to be $0{\cdot}35$ Å, which is an order of magnitude less than the interatomic spacing.

has a minimum value when the moments are aligned in the x direction. In view of this latter fact,

$$\hat{S}^{\pm} = \hat{S}^{y} \pm \mathrm{i}\hat{S}^{z}. \tag{9.185}$$

In deriving the spin-wave modes in the system described by the Hamiltonian formed from the sum of (9.183) and (9.184) we shall use the modified commutation relation (9.13) to obtain the necessary linear equations of motion.

It follows from (9.185) that

$$(\hat{S}^{z})^{2} = -\tfrac{1}{4}\{(\hat{S}^{+})^{2}+(\hat{S}^{-})^{2}-2S-2\hat{S}^{-}\hat{S}^{+}\}. \tag{9.186}$$

To terms quadratic in $\hat{S}^{+}$ and $\hat{S}^{-}$ it is easy to show that

$$(\hat{S}^{x}+\mathrm{i}\hat{S}^{y})^{6}+(\hat{S}^{x}-\mathrm{i}\hat{S}^{y})^{6} \simeq S^{6}\left\{2-\frac{6}{S^{2}}\hat{S}^{-}\hat{S}^{+}-\frac{15}{2S^{2}}(\hat{S}^{+}+\hat{S}^{-})^{2}\right\} \tag{9.187}$$

on using the representation

$$\hat{S}^{x} = \hat{S}-\frac{1}{2S}\hat{S}^{-}\hat{S}^{+}$$

as appropriate for linear spin-wave theory (cf. eqn (9.14)).

With (9.186) and (9.187) the linear equation of motion for the operator $\hat{S}^{+}_{\mathbf{m}}$ is

$$\mathrm{i}\hbar\dot{\hat{S}}{}^{+}_{\mathbf{m}} = 2S\sum_{\mathbf{m}'} J'(\mathbf{m}-\mathbf{m}')(\hat{S}^{+}_{\mathbf{m}}-\hat{S}^{+}_{\mathbf{m}'})+2S\sum_{\mathbf{n}} J(\mathbf{m}-\mathbf{n})(\hat{S}^{+}_{\mathbf{m}}-\hat{T}^{+}_{\mathbf{n}})+$$
$$+\hat{S}^{+}_{\mathbf{m}}(SB+21S^{5}G)+\hat{S}^{-}_{\mathbf{m}}(15S^{5}G-SB) \tag{9.188}$$

and we shall henceforth denote the coefficient of $\hat{S}^{+}_{\mathbf{m}}$ in the third term of (9.188) by A and that of $\hat{S}^{-}_{\mathbf{m}}$ in the fourth by A'.

We now transform to operators $\hat{S}^{\pm}_{\mathbf{q}}$ and $\hat{T}^{\pm}_{\mathbf{q}}$ with the convention defined by eqn (9.8) to write (9.188) as

$$\mathrm{i}\hbar\dot{\hat{S}}{}^{+}_{\mathbf{q}} = \mathscr{A}_{\mathbf{q}}\hat{S}^{+}_{\mathbf{q}}+A'\hat{S}^{-}_{-\mathbf{q}}-2S\mathscr{J}(\mathbf{q})\hat{T}^{+}_{\mathbf{q}}, \tag{9.189}$$

where the coefficient $\mathscr{A}_{\mathbf{q}}$ is defined by

$$\mathscr{A}_{\mathbf{q}} = 2S\{\mathscr{J}'(0)-\mathscr{J}'(\mathbf{q})\}+A+2S\mathscr{J}(0). \tag{9.190}$$

In (9.189) and (9.190)

$$\mathscr{J}(\mathbf{q}) = \sum_{\mathbf{m}} J(\mathbf{m})\exp(-\mathrm{i}\mathbf{q}.\mathbf{m}) = \mathscr{J}^{*}(-\mathbf{q}) \tag{9.191 a}$$

and

$$\mathscr{J}'(\mathbf{q}) = \sum_{\mathbf{m}} J'(\mathbf{m})\exp(-\mathrm{i}\mathbf{q}.\mathbf{m}) = \mathscr{J}'(-\mathbf{q}). \tag{9.191 b}$$

$\mathscr{J}(\mathbf{q})$ is complex because, as we noted in § 2.1, the vector joining sites on different sublattices in h.c.p. is not a lattice vector, i.e. ions are not centres of symmetry with respect to ions on the opposite sublattice. Of course $\mathscr{J}'(\mathbf{q})$ is real because each sublattice is Bravais.

Since the equation of motion for $\hat{S}^+_{\mathbf{q}}$ contains in addition to $\hat{S}^+_{\mathbf{q}}$ the operators $\hat{S}^-_{-\mathbf{q}}$ and $\hat{T}^+_{\mathbf{q}}$, we have also to form their equations of motion. In fact we find that these equations of motion contain $\hat{T}^-_{-\mathbf{q}}$ so that all told we require four equations of motion including (9.189). The remaining three are readily shown to be

$$i\hbar\dot{\hat{S}}^-_{-\mathbf{q}} = -\mathscr{A}_{\mathbf{q}}\hat{S}^-_{-\mathbf{q}} - A'\hat{S}^+_{\mathbf{q}} + 2S\mathscr{J}(\mathbf{q})\hat{T}^-_{-\mathbf{q}}, \tag{9.192 a}$$

$$i\hbar\dot{\hat{T}}^+_{\mathbf{q}} = \mathscr{A}_{\mathbf{q}}\hat{T}^+_{\mathbf{q}} + A'\hat{T}^-_{-\mathbf{q}} - 2S\mathscr{J}^*(\mathbf{q})\hat{S}^+_{\mathbf{q}}, \tag{9.192 b}$$

and

$$i\hbar\dot{\hat{T}}^-_{-\mathbf{q}} = -\mathscr{A}_{\mathbf{q}}\hat{T}^-_{-\mathbf{q}} - A'\hat{T}^+_{\mathbf{q}} + 2S\mathscr{J}^*(\mathbf{q})\hat{S}^-_{-\mathbf{q}}. \tag{9.192 c}$$

The equations of motion (9.189) and (9.192) are derivable from the linearized Hamiltonian

$$\begin{aligned}\mathscr{H} = \frac{1}{2SN}\sum_{\mathbf{q}} \tfrac{1}{2}\mathscr{A}_{\mathbf{q}}(\hat{S}^-_{\mathbf{q}}\hat{S}^+_{\mathbf{q}} + \hat{S}^+_{\mathbf{q}}\hat{S}^-_{\mathbf{q}} + \hat{T}^-_{\mathbf{q}}\hat{T}^+_{\mathbf{q}} + \hat{T}^+_{\mathbf{q}}\hat{T}^-_{\mathbf{q}}) + \\ + \tfrac{1}{2}A'(\hat{S}^-_{\mathbf{q}}\hat{S}^-_{-\mathbf{q}} + \hat{S}^+_{\mathbf{q}}\hat{S}^+_{-\mathbf{q}} + \hat{T}^-_{\mathbf{q}}\hat{T}^-_{-\mathbf{q}} + \hat{T}^+_{\mathbf{q}}\hat{T}^+_{-\mathbf{q}}) - \\ - S[\mathscr{J}^*(\mathbf{q})\{\hat{S}^+_{\mathbf{q}}\hat{T}^-_{\mathbf{q}} + \hat{S}^-_{-\mathbf{q}}\hat{T}^+_{-\mathbf{q}}\} + \mathscr{J}(\mathbf{q})\{\hat{S}^-_{\mathbf{q}}\hat{T}^+_{\mathbf{q}} + \hat{S}^+_{-\mathbf{q}}\hat{T}^-_{-\mathbf{q}}\}].\end{aligned} \tag{9.193}$$

This may easily be verified with the aid of the commutation rule (9.28).

We can proceed in two completely equivalent ways; we may either solve the four coupled equations of motion formed by (9.189) and (9.192) or seek the transformation that diagonalizes the Hamiltonian (9.193). Since in § 9.1 we have already given an example of the equation of motion method we here opt for the second alternative and find it convenient to use a matrix notation.†

We write the Hamiltonian (9.193) as

$$\mathscr{H} = \frac{1}{2SN}\sum_{\mathbf{q}} \mathscr{H}_{\mathbf{q}}, \tag{9.194}$$

where $\mathscr{H}_{\mathbf{q}}$ is given in terms of the matrices $\hat{X}$ and $\mathscr{H}$. The column matrix operator $\hat{X}$ is defined to be

$$\hat{X} = \begin{pmatrix} \hat{S}^+_{\mathbf{q}} \\ \hat{S}^-_{-\mathbf{q}} \\ \hat{T}^+_{\mathbf{q}} \\ \hat{T}^-_{-\mathbf{q}} \end{pmatrix}, \tag{9.195}$$

and

$$\mathscr{H} = \begin{pmatrix} h & h_1^* \\ h_1 & h \end{pmatrix}, \tag{9.196 a}$$

where

$$h = \begin{pmatrix} \mathscr{A}_{\mathbf{q}} & A' \\ A' & \mathscr{A}_{\mathbf{q}} \end{pmatrix} \quad \text{and} \quad h_1 = \begin{pmatrix} C_{\mathbf{q}} & 0 \\ 0 & C_{\mathbf{q}} \end{pmatrix} \tag{9.196 b}$$

with

$$C_{\mathbf{q}} = -2S\mathscr{J}^*(\mathbf{q}). \tag{9.197}$$

† P. A. Lindgard *et al.* *Physics Chem. Solids* **28**, 1357 (1966).

In terms of $\hat{X}$ and $\mathscr{H}$,
$$\mathscr{H}_{\mathbf{q}} = \tfrac{1}{2}\hat{X}^{+}\mathscr{H}\hat{X}. \tag{9.198}$$
If
$$\hat{X} = T\hat{Y}, \tag{9.199}$$
where
$$\hat{Y} = \begin{pmatrix} \hat{\alpha}_{\mathbf{q}} \\ \hat{\alpha}^{+}_{-\mathbf{q}} \\ \hat{\beta}_{\mathbf{q}} \\ \hat{\beta}^{+}_{-\mathbf{q}} \end{pmatrix}, \tag{9.200}$$
then we can find the matrix T such that $\mathscr{H}_{\mathbf{q}}$ is diagonal in the operators $\hat{\alpha}$ and $\hat{\beta}$, i.e. such that
$$\mathscr{H}_{\mathbf{q}} = \tfrac{1}{2}\hat{Y}^{+}E\hat{Y}, \tag{9.201}$$
where E is a diagonal matrix.

As may be verified by straightforward, though perhaps tedious, matrix algebra the matrix T is given by
$$T = \begin{pmatrix} t_0 & t_1 \\ ct_0 & -ct_1 \end{pmatrix}, \tag{9.202 a}$$
where
$$t_a = \begin{pmatrix} p_a & -m_a \\ -m_a & p_a \end{pmatrix}, \quad \text{with } a = 0, 1, \tag{9.202 b}$$
and
$$c = C_{\mathbf{q}}/|C_{\mathbf{q}}| = -\mathscr{J}^{*}(\mathbf{q})/|\mathscr{J}(\mathbf{q})|. \tag{9.203}$$
The elements p_a and m_a of the matrix t_a are
$$m_a^2 = (2SN)\frac{1}{4E_a}\{(E_a^2+A'^2)^{\frac{1}{2}}-E_a\}, \tag{9.204}$$
and
$$m_a p_a = (2SN)A'/4E_a, \tag{9.205}$$
where the energies E_0 and E_1 are given by
$$E_{a=0}^2(\mathbf{q}) = -A'^2+(\mathscr{A}_{\mathbf{q}}+|C_{\mathbf{q}}|)^2\text{: optical}, \tag{9.206}$$
$$E_{a=1}^2(\mathbf{q}) = -A'^2+(\mathscr{A}_{\mathbf{q}}-|C_{\mathbf{q}}|)^2\text{: acoustic}. \tag{9.207}$$
We choose to call E_0 the optical magnon branch and E_1 the acoustic magnon branch by analogy with the terminology used in phonon theory.

Thus
$$\mathscr{H}_{\mathbf{q}} = \tfrac{1}{2}\hat{X}^{+}\mathscr{H}\hat{X} = \tfrac{1}{2}(T\hat{Y})^{+}\mathscr{H}T\hat{Y} = \tfrac{1}{2}\hat{Y}^{+}(T^{+}\mathscr{H}T)\hat{Y},$$
where
$$T^{+}\mathscr{H}T = E = (2SN)\begin{pmatrix} E_0(\mathbf{q}) & 0 & 0 & 0 \\ 0 & E_0(\mathbf{q}) & 0 & 0 \\ 0 & 0 & E_1(\mathbf{q}) & 0 \\ 0 & 0 & 0 & E_1(\mathbf{q}) \end{pmatrix} \tag{9.208}$$
and
$$\left(\frac{1}{2SN}\right)\tfrac{1}{2}\hat{Y}^{+}E\hat{Y} = E_0(\mathbf{q})(\hat{\alpha}_{\mathbf{q}}^{+}\hat{\alpha}_{\mathbf{q}}+\tfrac{1}{2})+E_1(\mathbf{q})(\hat{\beta}_{\mathbf{q}}^{+}\hat{\beta}_{\mathbf{q}}+\tfrac{1}{2}),$$
i.e.
$$\mathscr{H} = \sum_{\mathbf{q}}\{E_0(\mathbf{q})(\hat{\alpha}_{\mathbf{q}}^{+}\hat{\alpha}_{\mathbf{q}}+\tfrac{1}{2})+E_1(\mathbf{q})(\hat{\beta}_{\mathbf{q}}^{+}\hat{\beta}_{\mathbf{q}}+\tfrac{1}{2})\}. \tag{9.209}$$

The total x component of spin $\sum \hat{S}^x + \sum \hat{T}^x$ does not commute with the Hamiltonian (9.193). Thus the cross-section for the scattering by the optical and acoustic magnons will contain terms other than those occurring in the transverse inelastic cross-section (9.42), i.e. the correlation functions

$$\langle \hat{S}^+_{\mathbf{m}}(0)\hat{S}^+_{\mathbf{m}'}(t)\rangle \quad \text{and} \quad \langle \hat{S}^-_{\mathbf{m}}(0)\hat{S}^-_{\mathbf{m}'}(t)\rangle \tag{9.210}$$

and the equivalent ones in $\hat{T}^\pm_{\mathbf{n}}$, are non-zero.

If we denote the position of any site in the h.c.p. structure by $\mathbf{R}$, so that $\mathbf{R}$ takes $2N$ values, where N is the number of ions on one hexagonal sublattice, the correlation function required to calculate the cross-section for the creation or annihilation of a magnon is

$$\sum_{\alpha,\beta} (\delta_{\alpha\beta} - \tilde{\kappa}_\alpha \tilde{\kappa}_\beta) \sum_{\mathbf{R},\mathbf{R}'} \exp\{i\boldsymbol{\kappa}.(\mathbf{R}'-\mathbf{R})\}\langle \hat{S}^\alpha_{\mathbf{R}}(0)\hat{S}^\beta_{\mathbf{R}'}(t)\rangle, \tag{9.211}$$

where α, $\beta = y$ and z. Because

$$\langle \hat{S}^+_{\mathbf{R}}(0)\hat{S}^+_{\mathbf{R}'}(t) - \hat{S}^-_{\mathbf{R}}(0)\hat{S}^-_{\mathbf{R}'}(t)\rangle = 0$$

(9.211) contains no terms involving $\tilde{\kappa}_y \tilde{\kappa}_z$. Thus in (9.211)

$$\begin{aligned}\sum_{\alpha,\beta} &(\delta_{\alpha\beta} - \tilde{\kappa}_\alpha \tilde{\kappa}_\beta)\langle \hat{S}^\alpha_{\mathbf{R}}(0)\hat{S}^\beta_{\mathbf{R}'}(t)\rangle \\ &= \tfrac{1}{4}(1+\tilde{\kappa}_x^2)\langle \hat{S}^+_{\mathbf{R}}(0)\hat{S}^-_{\mathbf{R}'}(t) + \hat{S}^-_{\mathbf{R}}(0)\hat{S}^+_{\mathbf{R}'}(t)\rangle + \\ &\quad + \tfrac{1}{4}(\tilde{\kappa}_z^2 - \tilde{\kappa}_y^2)\langle \hat{S}^+_{\mathbf{R}}(0)\hat{S}^+_{\mathbf{R}'}(t) + \hat{S}^-_{\mathbf{R}}(0)\hat{S}^-_{\mathbf{R}'}(t)\rangle. \end{aligned} \tag{9.212}$$

The sums over $\mathbf{R}$ and $\mathbf{R}'$ in (9.211) must each be split into a sum over $\mathbf{m}$ and a sum over $\mathbf{n}$ and so mixed correlation functions of the type $\langle \hat{S}^+_{\mathbf{m}}(0)\hat{T}^+_{\mathbf{n}}(t)\rangle$ also occur in the evaluation of the cross-section, in addition to those of the type (9.210) and $\langle \hat{S}^\pm_{\mathbf{m}}(0)\hat{S}^\mp_{\mathbf{m}'}(t)\rangle$ and $\langle \hat{T}^\pm_{\mathbf{n}}(0)\hat{T}^\mp_{\mathbf{n}'}(t)\rangle$. Each of these correlation functions can be calculated in terms of the operators $\hat{\alpha}$, $\hat{\beta}$ via the transformation (9.199), and since these diagonalize $\mathscr{H}$ it follows that

$$\langle \hat{\alpha}\hat{\beta}\rangle = \langle \hat{\alpha}^+\hat{\beta}^+\rangle = \langle \hat{\alpha}^+\hat{\alpha}^+\rangle = \langle \hat{\alpha}\hat{\alpha}\rangle = \langle \hat{\beta}\hat{\beta}\rangle = \langle \hat{\beta}^+\hat{\beta}^+\rangle = 0.$$

The result of the calculation is

$$\begin{aligned} &\tfrac{1}{4}(1+\tilde{\kappa}_x^2)\sum_{\mathbf{R},\mathbf{R}'} \exp\{i\boldsymbol{\kappa}.(\mathbf{R}'-\mathbf{R})\}\langle \hat{S}^+_{\mathbf{R}}(0)\hat{S}^-_{\mathbf{R}'}(t) + \hat{S}^-_{\mathbf{R}}(0)\hat{S}^+_{\mathbf{R}'}(t)\rangle + \\ &\quad + \tfrac{1}{4}(\tilde{\kappa}_z^2 - \tilde{\kappa}_y^2)\sum_{\mathbf{R},\mathbf{R}'} \exp\{i\boldsymbol{\kappa}.(\mathbf{R}'-\mathbf{R})\}\langle \hat{S}^+_{\mathbf{R}}(0)\hat{S}^+_{\mathbf{R}'}(t) + \hat{S}^-_{\mathbf{R}}(0)\hat{S}^-_{\mathbf{R}'}(t)\rangle \qquad (9.213) \\ &= \frac{1}{N^2}\sum_{\mathbf{q},a} \{(n_{\mathbf{q},a}+1)\exp(i\omega_{\mathbf{q},a}t) + n_{\mathbf{q},a}\exp(-i\omega_{\mathbf{q},a}t\}\times \\ &\qquad \times\{\tfrac{1}{4}(1+\tilde{\kappa}_x^2)(m_a^2+p_a^2) + \tfrac{1}{2}(\tilde{\kappa}_y^2 - \tilde{\kappa}_z^2)m_a p_a\}\times \\ &\qquad \times[2\sum_{\mathbf{m},\mathbf{m}'} \exp\{i(\boldsymbol{\kappa}-\mathbf{q}).(\mathbf{m}'-\mathbf{m})\} + (-1)^a \sum_{\mathbf{m},\mathbf{n}} c\exp\{i(\boldsymbol{\kappa}-\mathbf{q}).(\mathbf{m}-\mathbf{n})\} + \\ &\qquad\qquad + c^*\exp\{-i(\boldsymbol{\kappa}-\mathbf{q}).(\mathbf{m}-\mathbf{n})\}]. \end{aligned}$$

In (9.213), the integer a takes the two values 0 ($\equiv$ optical) and 1 ($\equiv$ acoustic) and $n_{\mathbf{q},a}$ is the Bose factor

$$n_{\mathbf{q},a} = \{\exp(\hbar\omega_{\mathbf{q},a}\beta)-1\}^{-1}, \tag{9.214}$$

where
$$\hbar\omega_{\mathbf{q},a} = E_a(\mathbf{q}).$$

Consider the sum

$$\sum_{\mathbf{m},\mathbf{n}} \exp\{\mathrm{i}(\boldsymbol{\varkappa}-\mathbf{q}).(\mathbf{m}-\mathbf{n})\} \tag{9.215}$$

that occurs in (9.213). If we denote the vector joining an ion on one sublattice to its nearest neighbour on the other sublattice by $\boldsymbol{\rho}$, (9.215) can be written

$$\exp\{\mathrm{i}\boldsymbol{\rho}.(\boldsymbol{\varkappa}-\mathbf{q})\}\sum_{\mathbf{m},\mathbf{m}'}\exp\{\mathrm{i}(\boldsymbol{\varkappa}-\mathbf{q}).(\mathbf{m}-\mathbf{m}')\}$$
$$= \exp\{\mathrm{i}\boldsymbol{\rho}.(\boldsymbol{\varkappa}-\mathbf{q})\}N\frac{(2\pi)^3}{v_0}\sum_{\boldsymbol{\tau}}\delta(\boldsymbol{\varkappa}-\mathbf{q}-\boldsymbol{\tau}), \tag{9.216}$$

where $\boldsymbol{\tau}$ are the reciprocal lattice vectors of a *sublattice*. For the particular case of a h.c.p. structure

$$\boldsymbol{\rho} = \tfrac{1}{3}\mathbf{a}_1+\tfrac{1}{3}\mathbf{a}_2+\tfrac{1}{2}\mathbf{a}_3 \tag{9.217}$$

and the basic vectors $\mathbf{a}_j$, and also the reciprocal lattice vectors $\boldsymbol{\tau}$, are given in § 2.1.

Because $|c| = 1$ we may write in (9.213)

$$c = \exp \mathrm{i}\phi. \tag{9.218}$$

Thus, using (9.216) and (9.218) in (9.213), the latter becomes

$$\tfrac{1}{2}S\frac{(2\pi)^3}{v_0}\sum_{\boldsymbol{\tau},\mathbf{q}}\delta(\boldsymbol{\varkappa}-\mathbf{q}-\boldsymbol{\tau})\sum_a\{(n_{\mathbf{q},a}+1)\exp(\mathrm{i}\omega_{\mathbf{q},a}t)+n_{\mathbf{q},a}\exp(-\mathrm{i}\omega_{\mathbf{q},a}t)\}\times$$
$$\times\{1+(-1)^a\cos(\boldsymbol{\tau}.\boldsymbol{\rho}+\phi)\}\left\{(1+\tilde{\kappa}_x^2)\frac{(E_a^2+A'^2)^{\frac{1}{2}}}{E_a}+(\tilde{\kappa}_y^2-\tilde{\kappa}_z^2)\frac{A'}{E_a}\right\}. \tag{9.219}$$

From (9.219) we can immediately calculate the neutron cross-section. This is

$$\frac{\mathrm{d}^2\sigma}{\mathrm{d}\Omega\mathrm{d}E'} = \left(\frac{\gamma e^2}{m_\mathrm{e}c^2}\right)^2\{\tfrac{1}{2}gF(\boldsymbol{\varkappa})\}^2\frac{k'}{k}\exp\{-2W(\boldsymbol{\varkappa})\}\tfrac{1}{2}S\frac{(2\pi)^3}{v_0}\sum_{\tau,\mathbf{q}}\delta(\boldsymbol{\varkappa}-\mathbf{q}-\boldsymbol{\tau})\times$$
$$\times\sum_a[(n_{\mathbf{q},a}+1)\delta\{\hbar\omega-E_a(\mathbf{q})\}+n_{\mathbf{q},a}\delta\{\hbar\omega+E_a(\mathbf{q})\}]\times$$
$$\times\{1+(-1)^a\cos(\boldsymbol{\tau}.\boldsymbol{\rho}+\phi)\}\left\{(1+\tilde{\kappa}_x^2)\frac{(E_a^2+A'^2)^{\frac{1}{2}}}{E_a}+(\tilde{\kappa}_y^2-\tilde{\kappa}_z^2)\frac{A'}{E_a}\right\}. \tag{9.220}$$

The cross-section (9.220) is that for the inelastic scattering by the spin-wave modes of the Hamiltonian (9.193) and consists of the sum of two terms, one corresponding to magnon creation and the other to magnon annihilation of the optical and acoustic magnons with energies $E_0(\mathbf{q})$ and $E_1(\mathbf{q})$, respectively. If we set $A = A' = C_\mathbf{q} = 0$, the Hamiltonian (9.193) is just twice that for linear spin waves in an isotropic

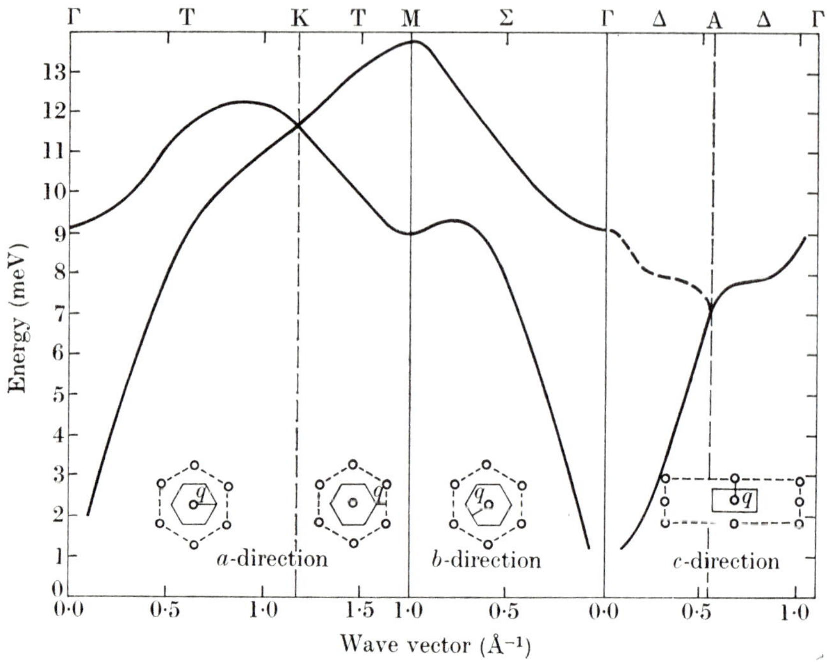

FIG. 9.9. Experimental magnon dispersion relations in Tb along symmetry lines in the Brillouin zone at 90 K.

Heisenberg ferromagnet and (9.220) becomes identical with the cross-section for this system, eqn (9.44).

The factor $\{1+(-1)^a \cos(\boldsymbol{\tau}.\boldsymbol{\rho}+\phi)\}$ in (9.220) is usually referred to as the magnetic structure factor since it originates from the two atoms per unit cell of the h.c.p. structure. In general ϕ is not zero nor a multiple of π, except in special symmetry directions.

In the rare earths the exchange interactions arise from an indirect interaction between the highly localized $4f$ electrons via the conduction electrons. This interaction has the characteristic of being long ranged and oscillatory and for this reason alone the evaluation of the cross-section for comparison with a measured cross-section is complicated. Fig. 9.9 shows the dispersion relations in Tb at 90 K as measured by

Møller and Houmann.† The curves show the dispersion curves along the symmetry lines in the Brillouin zone as indicated on the actual figures (the b direction is that of easy magnetization). From their measurements Møller and Houmann were able to extract the numerous exchange parameters that appeared to be significant in determining the dispersion curves.

The inelastic one-magnon cross-sections, and also the created polarization, for a ferromagnetic spiral system have been evaluated by Bar'yakhtar and Maleev.‡

9.6. Heisenberg antiferromagnets

While there are very few known non-metallic ferromagnets,§ there exist several simple ionic antiferromagnets that are well described by a Heisenberg exchange Hamiltonian with the addition of some single-ion anisotropy terms. In this section we consider the magnetic properties of Heisenberg antiferromagnets on much the same lines as the discussion of the Heisenberg ferromagnets in § 9.2. We shall, however, find it convenient to work with a more general Hamiltonian than that required to describe antiferromagnets, the Hamiltonian being that also appropriate for the description of ferrimagnets. The antiferromagnet is obtained as a limiting case; namely, when all the ions are identical. We shall discuss ferrimagnets in the following section, § 9.7.

We limit our discussion to those antiferromagnets and ferrimagnets that can be described in terms of two identical interpenetrating sublattices, with lattice sites given by the lattice vectors **m** and **n**. Often we distinguish between the two sublattices by referring to one as the 'up' sublattice (lattice vectors **m**) and the other as the 'down' sublattice (**n**).

(*a*) *Linear spin-wave theory*

For an antiferromagnet the Heisenberg Hamiltonian is of the form

$$\mathscr{H} = 2J \sum_{\mathbf{m},\rho} \hat{\mathbf{S}}_{\mathbf{m}} \cdot \hat{\mathbf{S}}_{\mathbf{m}+\rho}. \tag{9.221}$$

In (9.221) only a nearest-neighbour exchange is included for the present; we shall subsequently allow the exchange to have an arbitrary range.

† H. B. Møller and J. C. G. Houmann, *Phys. Rev. Lett.* **16**, 737 (1966) and J. C. G. Houmann, *Solid St. Commun.* **6**, 479 (1968).

‡ V. G. Bar'yakhtar and S. V. Maleev, *Soviet Phys. solid St.* **5**, 858 (1963).

§ EuO is perhaps the best known non-metallic ferromagnet, but the high neutron absorption cross-section of natural Eu excludes it from use in neutron studies.

There is a fundamental difference between calculating the spin-wave spectrum of (9.221) as compared to that for the Heisenberg ferromagnet, namely, that the exact ground state of the Hamiltonian is not known. This means that spin-wave theory is constructed from the approximate ground state, this state being that of lowest classical energy, or Néel state, in which the spins on one sublattice are oppositely directed to those on the other. As may be easily verified, this state is not, in fact, even an eigenstate of the Hamiltonian (9.221).

The spin-wave theory constructed from the Néel state predicts a correction to the latter which is manifest in a zero-point spin deviation. This situation is the same as that for the Heisenberg ferromagnet in the presence of dipolar terms, because in this case the fully aligned state, from which the spin-wave theory is constructed, is no longer the ground state and there are zero-point spin fluctuations.

We are not concerned here with the ground-state properties of the antiferromagnet, or those of the ferrimagnet, but are concerned with the excitations that occur as the spins deviate from their assumed ground-state values. Experiments show that spin-wave theory gives an extremely good description of these states. Also, with essentially the same, simple treatment of two-spin-wave interactions as we gave for the ferromagnet, the theory is found to apply remarkably well over a large temperature range.

Consider the Hamiltonian

$$\begin{aligned}\mathscr{H} = \sum_{\mathbf{m},\mathbf{r}} J(\mathbf{r})\hat{\mathbf{S}}_{\mathbf{m}}\cdot\hat{\mathbf{S}}_{\mathbf{m}+\mathbf{r}} + \sum_{\mathbf{n},\mathbf{r}} J(\mathbf{r})\hat{\mathbf{S}}_{\mathbf{n}}\cdot\hat{\mathbf{S}}_{\mathbf{n}+\mathbf{r}} + \\ + \sum_{\mathbf{m},\mathbf{R}} J_1(\mathbf{R})\hat{\mathbf{S}}_{\mathbf{m}}\cdot\hat{\mathbf{S}}_{\mathbf{m}+\mathbf{R}} + \sum_{\mathbf{n},\mathbf{R}} J_2(\mathbf{R})\hat{\mathbf{S}}_{\mathbf{n}}\cdot\hat{\mathbf{S}}_{\mathbf{n}+\mathbf{R}} - \\ - g_1\mu_{\mathrm{B}}(H+H_{A,1})\sum_{\mathbf{m}} \hat{S}^z_{\mathbf{m}} - g_2\mu_{\mathrm{B}}(H-H_{A,2})\sum_{\mathbf{n}} \hat{S}^z_{\mathbf{n}},\end{aligned} \quad (9.222)$$

where $\mathbf{r}$ connects sites on opposite sublattices and $\mathbf{R}$ those on the same sublattice. In (9.222) the spins on the $\mathbf{m}$ sublattice have magnitude S_1 and those on $\mathbf{n}$ magnitude S_2, and the exchange parameters within each sublattice are denoted by J_1 and J_2 respectively. A uniaxial anisotropy has been incorporated in (9.222) via effective magnetic fields $H_{A,1}$ and $H_{A,2}$, the gyromagnetic ratios for the two types of ion being denoted by g_1 and g_2.

In calculating the spin-wave modes of the Hamiltonian (9.222) it is convenient to apply a canonical transformation to the spin operators associated with ions on the $\mathbf{n}$ sublattice; we rotate the axes of the $\mathbf{n}$

sublattice through π about the x axes. Thus

$$\hat{S}^x_{\mathbf{n}} = \hat{T}^x_{\mathbf{n}}, \qquad \hat{S}^+_{\mathbf{n}} = \hat{T}^-_{\mathbf{n}},$$
$$\hat{S}^y_{\mathbf{n}} = -\hat{T}^y_{\mathbf{n}}, \qquad \hat{S}^-_{\mathbf{n}} = \hat{T}^+_{\mathbf{n}},$$
$$\hat{S}^z_{\mathbf{n}} = -\hat{T}^z_{\mathbf{n}}. \tag{9.223}$$

Also the Fourier transforms are defined

$$\hat{S}^+_{\mathbf{m}} = \frac{1}{N}\sum_{\mathbf{q}} \exp(\mathrm{i}\mathbf{q}\,.\,\mathbf{m})\hat{S}^+_{\mathbf{q}},$$

$$\hat{S}^-_{\mathbf{m}} = \frac{1}{N}\sum_{\mathbf{q}} \exp(-\mathrm{i}\mathbf{q}\,.\,\mathbf{m})\hat{S}^-_{\mathbf{q}},$$

$$\hat{T}^+_{\mathbf{n}} = \frac{1}{N}\sum_{\mathbf{q}} \exp(-\mathrm{i}\mathbf{q}\,.\,\mathbf{n})\hat{T}^+_{\mathbf{q}},$$

$$\hat{T}^-_{\mathbf{n}} = \frac{1}{N}\sum_{\mathbf{q}} \exp(\mathrm{i}\mathbf{q}\,.\,\mathbf{n})\hat{T}^-_{\mathbf{q}}, \tag{9.224}$$

where $\mathbf{q}$ are restricted to the first Brillouin zone of one sublattice, each sublattice having N lattice sites.

Using the commutation rule (9.13) and neglecting terms involving three raising and lowering operators, e.g. $\hat{S}^-_{\mathbf{m}}\hat{S}^+_{\mathbf{m}}\hat{T}^-_{\mathbf{m}+\mathbf{r}}$, the linear equation of motion for $\hat{S}^+_{\mathbf{m}}$ is found to be

$$\mathrm{i}\hbar\dot{\hat{S}}^+_{\mathbf{m}} = 2S_1\sum_{\mathbf{r}} J(\mathbf{r})\{\sigma\hat{S}^+_{\mathbf{m}}+\hat{T}^-_{\mathbf{m}+\mathbf{r}}\}+$$
$$+2S_1\sum_{\mathbf{R}} J_1(\mathbf{R})\{-\hat{S}^+_{\mathbf{m}}+\hat{S}^+_{\mathbf{m}+\mathbf{R}}\}+g_1\mu_{\mathrm{B}}(H+H_{A,1})\hat{S}^+_{\mathbf{m}}. \tag{9.225}$$

In this equation the parameter σ is

$$\sigma = S_2/S_1 \tag{9.226}$$

and it arises in (9.225) because the term

$$\hat{S}^+_{\mathbf{m}}\hat{T}^z_{\mathbf{m}+\mathbf{r}} = \hat{S}^+_{\mathbf{m}}\left(S_2-\frac{1}{2S_2}\hat{T}^-_{\mathbf{m}+\mathbf{r}}\hat{T}^+_{\mathbf{m}+\mathbf{r}}\right)$$

in the equation of motion for $\hat{S}^+_{\mathbf{m}}$ is replaced by

$$S_2\,\hat{S}^+_{\mathbf{m}}$$

in the linear approximation.

In terms of $S^+_{\mathbf{q}}$ and $\hat{T}^-_{\mathbf{q}}$, (9.225) reads

$$\mathrm{i}\hbar\dot{\hat{S}}^+_{\mathbf{q}} = [2S_1\sigma\mathscr{J}(0)-2S_1\{\mathscr{J}_1(0)-\mathscr{J}_1(\mathbf{q})\}+g_1\mu_{\mathrm{B}}(H+H_{A,1})]\hat{S}^+_{\mathbf{q}}+$$
$$+2S_1\mathscr{J}(\mathbf{q})\hat{T}^-_{\mathbf{q}}. \tag{9.227}$$

The equation of motion for $\hat{T}^-_{\mathbf{q}}$ is

$$i\hbar\dot{\hat{T}}^-_{\mathbf{q}} = -[2S_1\mathscr{J}(0)-2S_1\sigma\{\mathscr{J}_2(0)-\mathscr{J}_2(\mathbf{q})\}-g_2\mu_B(H-H_{A,2})]\hat{T}^-_{\mathbf{q}}-$$
$$-2S_1\sigma\mathscr{J}(\mathbf{q})\hat{S}^+_{\mathbf{q}}. \quad (9.228)$$

The solution of (9.227) and (9.228) for $\hat{S}^+_{\mathbf{q}}$ and $\hat{T}^-_{\mathbf{q}}$ parallels that of the equations (9.165).

Let
$$\hat{S}^+_{\mathbf{q}} = u_{\mathbf{q}}\hat{\alpha}_{\mathbf{q}}+v_{\mathbf{q}}\hat{\beta}^+_{\mathbf{q}}$$
and
$$\hat{T}^-_{\mathbf{q}} = \sigma^{\frac{1}{2}}\{u_{\mathbf{q}}\hat{\beta}^+_{\mathbf{q}}+v_{\mathbf{q}}\hat{\alpha}_{\mathbf{q}}\}. \quad (9.229)$$

For this case we may assume $u_{\mathbf{q}}$ and $v_{\mathbf{q}}$ are real. Then we have the relation
$$u^2_{\mathbf{q}}-v^2_{\mathbf{q}} = 2S_1N \quad (9.230)$$
in order to preserve the commutation relations
$$[\hat{S}^+_{\mathbf{m}}, \hat{S}^-_{\mathbf{m}'}] = 2S_1\delta_{\mathbf{m},\mathbf{m}'}$$
and
$$[\hat{T}^+_{\mathbf{n}}, \hat{T}^-_{\mathbf{n}'}] = 2S_2\delta_{\mathbf{n},\mathbf{n}'}. \quad (9.231)$$

We demand that $\hat{\alpha}_{\mathbf{q}}$ satisfies
$$\hbar\omega_{\mathbf{q},0}\frac{\partial}{\partial t}\hat{\alpha}_{\mathbf{q}} = [\hat{\alpha}_{\mathbf{q}},\mathscr{H}]$$
and that
$$\hbar\omega_{\mathbf{q},1}\frac{\partial}{\partial t}\hat{\beta}_{\mathbf{q}} = [\hat{\beta}_{\mathbf{q}},\mathscr{H}]. \quad (9.232)$$

From the former we immediately obtain, using (9.227), (9.228), and (9.229), the following pair of homogeneous equations for the coefficients $u_{\mathbf{q}}$ and $v_{\mathbf{q}}$;
$$(\hbar\omega_{\mathbf{q},0}-a_1-b_1)u_{\mathbf{q}}-2S_1\sigma^{\frac{1}{2}}\mathscr{J}(\mathbf{q})v_{\mathbf{q}} = 0,$$
$$2S_1\sigma^{\frac{1}{2}}\mathscr{J}(\mathbf{q})u_{\mathbf{q}}+(\hbar\omega_{\mathbf{q},0}-a_2+b_2)v_{\mathbf{q}} = 0. \quad (9.233)$$

The functions a_1, a_2, b_1, and b_2 are given by
$$a_1 = g_1\mu_B H, \qquad a_2 = g_2\mu_B H$$
and
$$b_1 = 2S_1\sigma\mathscr{J}(0)-2S_1\{\mathscr{J}_1(0)-\mathscr{J}_1(\mathbf{q})\}+g_1\mu_B H_{A,1},$$
$$b_2 = 2S_1\mathscr{J}(0)-2S_1\sigma\{\mathscr{J}_2(0)-\mathscr{J}_2(\mathbf{q})\}+g_2\mu_B H_{A,2}. \quad (9.234)$$

(9.233) possess a solution for
$$2\hbar\omega_{\mathbf{q},0} = (a_1+a_2+b_1-b_2)+2\Omega(\mathbf{q}),$$
where
$$2\Omega(\mathbf{q}) = [(a_1-a_2+b_1+b_2)^2-S_1S_2\{4\mathscr{J}(\mathbf{q})\}^2]^{\frac{1}{2}}. \quad (9.235\,\text{a})$$

Further, the second equation in (9.232) gives, in the same way,
$$2\hbar\omega_{\mathbf{q},1} = -(a_1+a_2+b_1-b_2)+2\Omega(\mathbf{q}). \quad (9.235\,\text{b})$$

These equations mean that the Hamiltonian (9.222) possesses two linear spin-wave modes, one associated with the operators $\hat{\alpha}^+_{\mathbf{q}}$, $\hat{\alpha}_{\mathbf{q}}$ of energy $\hbar\omega_{\mathbf{q},0}$ and another associated with the operators $\hat{\beta}^+_{\mathbf{q}}$, $\hat{\beta}_{\mathbf{q}}$ of energy $\hbar\omega_{\mathbf{q},1}$. In general these two modes are not degenerate.

The transition coefficients $u_{\mathbf{q}}$ and $v_{\mathbf{q}}$, both of which are purely real, are easily obtained from the pair of equations (9.233). It is found that $u_{\mathbf{q}}$ and $v_{\mathbf{q}}$ satisfy

$$u_{\mathbf{q}}^2 = (2S_1 N)(\hbar\omega_{\mathbf{q},0} - a_2 + b_2)/2\Omega(\mathbf{q})$$

and

$$u_{\mathbf{q}} v_{\mathbf{q}} = -(2S_1 N)S_1 \sigma^{\frac{1}{2}} \mathscr{J}(\mathbf{q})/\Omega(\mathbf{q}). \tag{9.236}$$

It is instructive to show that the transformation (9.229), with the coefficients $u_{\mathbf{q}}$ and $v_{\mathbf{q}}$ satisfying (9.236), does indeed diagonalize the linearized Hamiltonian that corresponds to the equations of motion (9.227) and (9.228); i.e. to show that the Hamiltonian

$$\mathscr{H} = \frac{1}{N}\sum_{\mathbf{q}} \left(\frac{a_1+b_1}{2S_1}\right)\hat{S}_{\mathbf{q}}^- \hat{S}_{\mathbf{q}}^+ - \left(\frac{a_2-b_2}{2S_2}\right)\hat{T}_{\mathbf{q}}^- \hat{T}_{\mathbf{q}}^+ + \mathscr{J}(\mathbf{q})(\hat{S}_{\mathbf{q}}^- \hat{T}_{\mathbf{q}}^- + \hat{S}_{\mathbf{q}}^+ \hat{T}_{\mathbf{q}}^+) \tag{9.237 a}$$

from which (9.227) and (9.228) can be derived, is equivalent to

$$\mathscr{H} = \sum_{\mathbf{q}} \hbar\omega_{\mathbf{q},0}\, \hat{\alpha}_{\mathbf{q}}^+ \hat{\alpha}_{\mathbf{q}} + \hbar\omega_{\mathbf{q},1}\, \hat{\beta}_{\mathbf{q}}^+ \hat{\beta}_{\mathbf{q}}. \tag{9.237 b}$$

We now specialize to the case of an antiferromagnet, returning to the ferrimagnet in § 9.7.

(b) Antiferromagnets

For an antiferromagnet $g_1 = g_2 = g$, $H_{A,1} = H_{A,2} = H_A$, $S_1 = S_2 = S$, and $\mathscr{J}_1 = \mathscr{J}_2 = \mathscr{J}'$. This means that the functions a_1, a_2, b_1, and b_2 satisfy $a_1 = a_2 = g\mu_{\mathrm{B}} H$ and

$$b_1 = b_2 = 2S\mathscr{J}(0) - 2S\{\mathscr{J}'(0) - \mathscr{J}'(\mathbf{q})\} + g\mu_{\mathrm{B}} H_A = b, \text{ say,} \tag{9.238}$$

so

$$\Omega(\mathbf{q}) \equiv [b^2 - \{2S\mathscr{J}(\mathbf{q})\}^2]^{\frac{1}{2}}$$

with

$$\hbar\omega_{\mathbf{q},a} = (-1)^a g\mu_{\mathrm{B}} H + \Omega(\mathbf{q}) \quad (a = 0, 1). \tag{9.239}$$

For small values of $\mathbf{q}$ and cubic crystals

$$\mathscr{J}(\mathbf{q}) = \sum_{\mathbf{r}} \exp(-\mathrm{i}\mathbf{q}.\mathbf{r})J(\mathbf{r}) \simeq \mathscr{J}(0) - \tfrac{1}{6}q^2 \sum_{\mathbf{r}} r^2 J(\mathbf{r})$$

and we define

$$J^{(2)} = \sum_{\mathbf{r}} r^2 J(\mathbf{r}), \tag{9.240}$$

i.e.

$$\mathscr{J}(0) - \mathscr{J}(\mathbf{q}) \simeq \tfrac{1}{6}q^2 J^{(2)}. \tag{9.241}$$

Similarly,

$$\mathscr{J}'(0) - \mathscr{J}'(\mathbf{q}) \simeq \tfrac{1}{6}q^2 J'^{(2)}. \tag{9.242}$$

(In (9.22) we defined the moments $\overline{l^n}$ through $\overline{l^n} = \Sigma_{\mathbf{l}}\, l^{n+2}J(\mathbf{l})/\Sigma_{\mathbf{l}}\, l^2 J(\mathbf{l})$. In the notation of (9.240) this would read $J^{(n+2)}/J^{(2)}$.)

Though (9.241) and (9.242) hold rigorously at small $\mathbf{q}$ only for cubic crystals, it is often a good approximation to use them in other cases.

The use of (9.241) and (9.242) in (9.239) yields

$$\hbar\omega_{\mathbf{q},a} = [g\mu_B H_A\{4S\mathscr{J}(0)+g\mu_B H_A\}+ \\ +\tfrac{1}{3}q^2 S\{4S\mathscr{J}(0)(J^{(2)}-J'^{(2)})-2g\mu_B H_A J'^{(2)}\}]^{\frac{1}{2}}+(-1)^a g\mu_B H \tag{9.243}$$

and this simplifies to

$$\hbar\omega_{\mathbf{q},a} = q\left\{\frac{4S^2}{3}\mathscr{J}(0)(J^{(2)}-J'^{(2)})\right\}^{\frac{1}{2}}+(-1)^a g\mu_B H \tag{9.244}$$

when $H_A = 0$.

In many antiferromagnets it is found that the nearest-neighbour exchange coupling is the dominant one. If this is denoted by J then from (9.239) with $H = 0$ the spin-wave spectrum is degenerate with

$$\hbar\omega_{\mathbf{q}} = 2rJS\{(1+h_A)^2-\gamma_{\mathbf{q}}^2\}^{\frac{1}{2}} = 2rJS\mathscr{E}_{\mathbf{q}}, \tag{9.245}$$

where we have defined a reduced anisotropy field $h_A = g\mu_B H_A/2rJS$.

It is seen that the anisotropy field H_A creates a gap in the spin-wave spectrum at $\mathbf{q} = 0$, i.e. it suppresses the low-energy excitations. One important consequence of this is that, particularly for large h_A, when the temperature is high enough to excite spin waves, there is a significant fraction with large values of $\mathbf{q}$.

Note that for the particular case when only nearest-neighbour coupling is taken into account the transformation coefficients have the particularly simple form

$$\left(\frac{1}{2SN}\right)u_{\mathbf{q}}^2 = (1+h_A+\mathscr{E}_{\mathbf{q}})/2\mathscr{E}_{\mathbf{q}},$$

$$\left(\frac{1}{2SN}\right)v_{\mathbf{q}}^2 = (1+h_A-\mathscr{E}_{\mathbf{q}})/2\mathscr{E}_{\mathbf{q}},$$

and

$$\left(\frac{1}{2SN}\right)u_{\mathbf{q}}v_{\mathbf{q}} = -\gamma_{\mathbf{q}}/2\mathscr{E}_{\mathbf{q}}. \tag{9.246}$$

Here, $\mathscr{E}_{\mathbf{q}}$ is defined by (9.245) and the expressions for $u_{\mathbf{q}}^2$, $v_{\mathbf{q}}^2$, and $u_{\mathbf{q}}v_{\mathbf{q}}$ follow directly from (9.236).

Let us now calculate some of the thermodynamic properties of the simple antiferromagnet with the spin-wave spectrum (9.245).

First, there are two sets of spin waves, one set corresponding to the operators $\hat{\alpha}_{\mathbf{q}}^+$, $\hat{\alpha}_{\mathbf{q}}$ the other to the operators $\hat{\beta}_{\mathbf{q}}^+$, $\hat{\beta}_{\mathbf{q}}$, and these are degenerate in the absence of a magnetic field. These spin waves can be considered as states in which the spins on the two sublattices precess in the same sense but with unequal amplitude. The doubly degenerate spectrum then corresponds to clockwise and anticlockwise precession of the spins about the anisotropy field direction; in our case this direction coincides with the z axes.

The sublattice magnetization can be easily calculated in terms of the operators $\hat{\alpha}_{\mathbf{q}}$ and $\hat{\beta}_{\mathbf{q}}$. For it follows from (9.229) that

$$\langle \hat{S}_{\mathbf{q}}^{-} \hat{S}_{\mathbf{q}}^{+} \rangle = \langle (u_{\mathbf{q}}\hat{\alpha}_{\mathbf{q}}^{+} + v_{\mathbf{q}}\hat{\beta}_{\mathbf{q}})(u_{\mathbf{q}}\hat{\alpha}_{\mathbf{q}} + v_{\mathbf{q}}\hat{\beta}_{\mathbf{q}}^{+}) \rangle$$
$$= u_{\mathbf{q}}^2 n_{\mathbf{q}} + v_{\mathbf{q}}^2 (n_{\mathbf{q}}+1) \tag{9.247}$$

and

$$\langle \hat{S}_{\mathbf{m}}^{z} \rangle = S - \frac{1}{2S}\frac{1}{N^2}\sum_{\mathbf{q}} \langle \hat{S}_{\mathbf{q}}^{-} \hat{S}_{\mathbf{q}}^{+} \rangle$$
$$= S - \frac{1}{2N}\sum_{\mathbf{q}} \left\{ \left(\frac{1+h_{\mathrm{A}}}{\mathscr{E}_{\mathbf{q}}} \right) \coth(\tfrac{1}{2}\hbar\omega_{\mathbf{q}}\beta) - 1 \right\}. \tag{9.248}$$

The last line of (9.248) follows immediately from (9.247) on using the expressions (9.246) for $u_{\mathbf{q}}^2$ and $v_{\mathbf{q}}^2$. It is worth while to note that the $\hat{\alpha}_{\mathbf{q}}^{+}$ spin-wave contribution to (9.247) contains $n_{\mathbf{q}}$, the $\hat{\beta}_{\mathbf{q}}^{+}$ spin-wave contribution the factor $1+n_{\mathbf{q}}$; at zero temperature only the latter is finite, indicating that the $\hat{\beta}_{\mathbf{q}}^{+}$ mode decreases the sublattice magnetization. The reverse is true for the down sublattice.

Taking the limit $\beta \to \infty$ ($T = 0$) in (9.248) we see that the average value of $\hat{S}^z$ is less than its value in the Néel ground state, S, by an amount

$$\delta\langle \hat{S}^z \rangle_0 = \frac{1}{2N}\sum_{\mathbf{q}} \left(\frac{1+h_{\mathrm{A}}}{\mathscr{E}_{\mathbf{q}}} \right) - 1. \tag{9.249}$$

$\delta\langle S^z \rangle_0$, the zero-point spin deviation, has a value $\simeq 0{\cdot}06$ for a body-centred antiferromagnet with $h_{\mathrm{A}} = 0$, and increases in magnitude as the number of neighbours decreases, e.g. for a quadratic layer ($r = 4$) it is $\simeq 0{\cdot}2$.

At a finite temperature we write for $\langle \hat{S}^z \rangle$

$$\langle \hat{S}^z \rangle = S - \delta\langle \hat{S}^z \rangle_0 - \frac{1}{N}\sum_{\mathbf{q}} \left(\frac{1+h_{\mathrm{A}}}{\mathscr{E}_{\mathbf{q}}} \right) \frac{1}{\exp(\hbar\omega_{\mathbf{q}}\beta)-1}. \tag{9.250}$$

For small q and $h_{\mathrm{A}} = 0$ (see eqn (9.17))

$$\mathscr{E}_{\mathbf{q}} = q(\rho^2/3)^{\frac{1}{2}} = Dq/2rJS, \tag{9.251}$$

which defines the parameter D for an antiferromagnet. Hence

$$\frac{1}{N}\sum_{\mathbf{q}} \frac{1}{(1-\gamma_{\mathbf{q}}^2)^{\frac{1}{2}}} \frac{1}{\exp(\hbar\omega_{\mathbf{q}}\beta)-1}$$
$$\simeq \frac{1}{N}\frac{V}{(2\pi)^3}\int_0^\infty \mathrm{d}q\, 4\pi q^2 \frac{\sqrt{3}}{\rho q}\frac{1}{\exp(\beta Dq)-1} = v_0 \frac{\sqrt{3}}{2\rho}\left(\frac{k_{\mathrm{B}}T}{\pi D}\right)^2 \zeta(2), \tag{9.252}$$

where v_0 is the volume per ion. We therefore see that the sublattice magnetization of an antiferromagnet decreases as T^2, in contrast to the $T^{\frac{3}{2}}$ dependence of the magnetization in a ferromagnet.

It is also of interest to calculate the exchange energy. A straightforward calculation shows that

$$E_{\text{exch}} = \langle \mathscr{H} \rangle - \langle \text{anisotropy energy} \rangle$$
$$= -2rJS(S+1)N + 2rJS \sum_{\mathbf{q}} \left\{ \frac{1+h_{\text{A}}-\gamma_{\mathbf{q}}^2}{\mathscr{E}_{\mathbf{q}}} \right\} \coth(\tfrac{1}{2}\hbar\omega_{\mathbf{q}}\beta) \quad (9.253)$$

for a nearest-neighbour model.

In the limit $\beta \to \infty$, E_{exch} can be written

$$E_{\text{exch}} = -2rJS^2N\{1+e_0/S\}: \quad T = 0, \quad (9.254)$$

where

$$e_0 = \frac{1}{N} \sum_{\mathbf{q}} \{1-(1+h_{\text{A}}-\gamma_{\mathbf{q}}^2)/\mathscr{E}_{\mathbf{q}}\}. \quad (9.255)$$

Note that increasing the anisotropy field *increases* ground-state energy, tending in the limit $h_{\text{A}} \to \infty$ to the ground-state energy for the corresponding Ising model, $-2rJS^2N$. For $h_{\text{A}} \gg 0$,

$$E_{\text{exch}} \simeq -2rJS^2N\{1+1/rS(1+h_{\text{A}})\}: \quad T = 0 \quad (9.256)$$

and for $h_{\text{A}} = 0$ and a body-centred orthorhombic crystal,

$$E_{\text{exch}} = -2rJS^2N\{1+0{\cdot}582/rS\}: \quad T = 0. \quad (9.257)$$

As one would anticipate, e_0 for a given value of h_{A} increases as the number of nearest neighbours decreases.

(c) *One-magnon cross-sections*

Now that we have familiarized ourselves with the fundamental properties of the antiferromagnet, we turn to the calculation of the neutron cross-section for inelastic scattering involving the creation or annihilation of single magnons. Because the *total* z component of spin commutes with the Hamiltonian (9.222) the cross-section contains only the terms $\langle \hat{S}^-\hat{S}^+ \rangle$ and $\langle \hat{S}^+\hat{S}^- \rangle$.

If by $\mathbf{R}$ we mean the position vector of *any* ion in the antiferromagnet, a straightforward calculation gives

$$\sum_{\mathbf{R},\mathbf{R}'} \exp\{i\boldsymbol{\kappa}.(\mathbf{R}'-\mathbf{R})\}\langle \hat{S}_{\mathbf{R}}^+(0)\hat{S}_{\mathbf{R}'}^-(t)\rangle$$
$$= \frac{1}{N^2} \sum_{\mathbf{q}} \{(1+n_{\mathbf{q},0})\exp(i\omega_{\mathbf{q},0}t)+n_{\mathbf{q},1}\exp(-i\omega_{\mathbf{q},1}t)\} \times$$
$$\times \left\{ (u_{\mathbf{q}}^2+v_{\mathbf{q}}^2) \left| \sum_{\mathbf{m}} \exp\{i(\boldsymbol{\kappa}-\mathbf{q}).\mathbf{m}\} \right|^2 + \right.$$
$$\left. + u_{\mathbf{q}}v_{\mathbf{q}} \sum_{\mathbf{m},\mathbf{n}} [\exp\{i(\boldsymbol{\kappa}-\mathbf{q}).(\mathbf{m}-\mathbf{n})\}+\text{c.c.}] \right\}, \quad (9.258)$$

where

$$n_{\mathbf{q},a} = \{\exp(\hbar\omega_{\mathbf{q},a}\beta)-1\}^{-1} \quad (a = 0, 1).$$

We now utilize the result (cf. § 9.5),

$$\sum_{\mathbf{m},\mathbf{n}} \exp\{\mathrm{i}(\boldsymbol{\varkappa}-\mathbf{q}).(\mathbf{m}-\mathbf{n})\} = \exp\{\mathrm{i}\boldsymbol{\rho}.(\boldsymbol{\varkappa}-\mathbf{q})\}\left|\sum_{\mathbf{m}} \exp\{\mathrm{i}(\boldsymbol{\varkappa}-\mathbf{q}).\mathbf{m}\}\right|^2, \tag{9.259}$$

where $\boldsymbol{\rho}$ is the vector joining an ion on one sublattice with its nearest neighbour on the opposite sublattice, to write (9.258) as

$$\frac{1}{N^2}\sum_{\mathbf{q}}\left|\sum_{\mathbf{m}} \exp\{\mathrm{i}(\boldsymbol{\varkappa}-\mathbf{q}).\mathbf{m}\}\right|^2\{(1+n_{\mathbf{q},0})\exp(\mathrm{i}\omega_{\mathbf{q},0}t)+n_{\mathbf{q},1}\exp(-\mathrm{i}\omega_{\mathbf{q},1}t)\}\times$$
$$\times\{u_{\mathbf{q}}^2+v_{\mathbf{q}}^2+2u_{\mathbf{q}}v_{\mathbf{q}}\cos\boldsymbol{\rho}.(\boldsymbol{\varkappa}-\mathbf{q})\}. \tag{9.260}$$

Since

$$\langle\hat{S}_{\mathbf{R}}^{-}(0)\hat{S}_{\mathbf{R}'}^{+}(t)\rangle = \langle\hat{S}_{\mathbf{R}'}^{+}(0)\hat{S}_{\mathbf{R}}^{-}(-t+\mathrm{i}\hbar\beta)\rangle, \tag{9.261}$$

the function

$$\sum_{\mathbf{R},\mathbf{R}'} \exp\{\mathrm{i}(\mathbf{R}'-\mathbf{R}).\boldsymbol{\varkappa}\}\langle\hat{S}_{\mathbf{R}}^{-}(0)\hat{S}_{\mathbf{R}'}^{+}(t)\rangle$$

is easily calculated from (9.260). Hence, the cross-sections for creation and annihilation of a single magnon are

$$\left(\frac{\mathrm{d}^2\sigma}{\mathrm{d}\Omega\mathrm{d}E'}\right)^{(\pm)} = \left(\frac{\gamma e^2}{m_{\mathrm{e}}c^2}\right)^2\{\tfrac{1}{2}gF(\boldsymbol{\varkappa})\}^2\frac{1}{4}\frac{k'}{k}(1+\tilde{\kappa}_z^2)\exp\{-2W(\boldsymbol{\varkappa})\}\times$$
$$\times\frac{(2\pi)^3}{v_0}\sum_{a=0,1}\sum_{\mathbf{q},\boldsymbol{\tau}}(n_{\mathbf{q},a}+\tfrac{1}{2}\pm\tfrac{1}{2})\delta(\hbar\omega_{\mathbf{q},a}\mp\hbar\omega)\delta(\boldsymbol{\varkappa}\mp\mathbf{q}-\boldsymbol{\tau})\times$$
$$\times\{u_{\mathbf{q}}^2+v_{\mathbf{q}}^2+2u_{\mathbf{q}}v_{\mathbf{q}}\cos\boldsymbol{\rho}.\boldsymbol{\tau}\}. \tag{9.262}$$

The reciprocal lattice vectors $\boldsymbol{\tau}$ are those of a sublattice. The cross-section (9.262) applies for any two-sublattice antiferromagnet; the coefficients $u_{\mathbf{q}}$ and $v_{\mathbf{q}}$ are given by (9.230) and (9.236) with $g_1 = g_2$, $H_{\mathrm{A},1} = H_{\mathrm{A},2}$, $\mathscr{J}_1 = \mathscr{J}_2$, and $S_1 = S_2$.

If $2\boldsymbol{\rho} = \mathbf{m}$, where $\mathbf{m}$ is some lattice vector of a sublattice, it follows by definition of the reciprocal lattice vectors $\boldsymbol{\tau}$ that

$$\boldsymbol{\rho}.\boldsymbol{\tau} = \tfrac{1}{2}\mathbf{m}.\boldsymbol{\tau} = \pi\times\text{integer}. \tag{9.263}$$

Thus

$$\cos\boldsymbol{\tau}.\boldsymbol{\rho} = \pm 1 \tag{9.264}$$

according to whether the integer is even (+) or odd (−).

For instance, consider a b.c.c. lattice. In this case the sublattices are simple cubic lattices, $\boldsymbol{\rho} = \frac{1}{2}a(1,1,1)$ and the reciprocal lattice vectors $\boldsymbol{\tau}$ are given by

$$\boldsymbol{\tau} = \frac{2\pi}{a}(\tau_1, \tau_2, \tau_3),$$

where τ_1, τ_2, τ_3 are arbitrary integers. The reciprocal lattice vectors for a b.c.c. structure, however, are given by

$$\boldsymbol{\tau}_n = \frac{2\pi}{a}(\tau_1, \tau_2, \tau_3),$$

where $\tau_1+\tau_2+\tau_3$ is an even integer. Hence

$$\boldsymbol{\rho}.\boldsymbol{\tau}_n = \pi \times \text{even integer} \tag{9.265}$$

and

$$\boldsymbol{\rho}.(\boldsymbol{\tau}_n+\mathbf{w}) = \pi \times \text{odd integer} \tag{9.266}$$

if

$$\mathbf{w} = \frac{2\pi}{a}(w_1, w_2, w_3),$$

where $w_1+w_2+w_3$ is an odd integer. Thus we may write

$$\sum_{\boldsymbol{\tau}} \delta(\boldsymbol{\varkappa}-\mathbf{q}-\boldsymbol{\tau})(u_{\mathbf{q}}^2+v_{\mathbf{q}}^2+2u_{\mathbf{q}}v_{\mathbf{q}}\cos\boldsymbol{\rho}.\boldsymbol{\tau}) \equiv \sum_{\boldsymbol{\tau}_n} (u_{\mathbf{q}} \pm v_{\mathbf{q}})^2 \begin{cases}\delta(\boldsymbol{\varkappa}-\mathbf{q}-\boldsymbol{\tau}_n)\\ \delta(\boldsymbol{\varkappa}-\mathbf{q}-\boldsymbol{\tau}_n-\mathbf{w}),\end{cases} \tag{9.267}$$

that is to say, the scattering occurs at all the reciprocal lattice vectors; but near the peaks that coincide with the nuclear reflections the intensity of scattering is small $[\sim (u_{\mathbf{q}}+v_{\mathbf{q}})^2]$ whereas near the *superlattice* reflections the intensity of scattering is large $[\sim (u_{\mathbf{q}}-v_{\mathbf{q}})^2]$.

Also note that for a b.c.c. lattice the vectors $\mathbf{R}$ are given by

$$\mathbf{R} = \tfrac{1}{2}a(R_1, R_2, R_3),$$

where the integers R_1, R_2, R_3 are either all even or all odd. Now $\exp(i\mathbf{w}.\mathbf{R}) = 1$ when all R_i are even and equals -1 when all R_i are odd, i.e. $\exp(i\mathbf{w}.\mathbf{R})$ alternates in sign as $\mathbf{R}$ goes from one sublattice to another.

Similarly, a s.c. lattice can be constructed from two sublattices corresponding to the two cases $\mathbf{m} = a(m_1, m_2, m_3)$ with $m_1+m_2+m_3$ an even integer and $\mathbf{n} = a(n_1, n_2, n_3)$ with $n_1+n_2+n_3$ an odd integer. For this lattice

$$\boldsymbol{\tau}_n = \frac{2\pi}{a}(\tau_1, \tau_2, \tau_3)$$

and we can choose $\mathbf{w} = (\pi/a)(1, 1, 1)$.

For the case when only a nearest-neighbour exchange is important, (9.246) shows that

$$(u_{\mathbf{q}} \pm v_{\mathbf{q}})^2 = 2SN(1+h_{\mathrm{A}} \mp \gamma_{\mathbf{q}})/\mathscr{E}_{\mathbf{q}}$$
$$\equiv 2SN(1+h_{\mathrm{A}} \mp \gamma_{\mathbf{q}})/\{(1+h_{\mathrm{A}})^2-\gamma_{\mathbf{q}}^2\}^{\frac{1}{2}}.$$

As we mentioned at the beginning of this section, the dispersion curves for several antiferromagnets have been measured by inelastic neutron scattering, but it is probably true that MnF_2 has received most attention, both from an experimental point of view and a theoretical one. This has a rutile crystal structure, the lattice parameters being such that nearest-neighbour ions reside on the same sublattice. However, the predominant exchange interaction is that between the nearest-neighbour ions on different sublattices. The magnitude of this exchange exceeds that between the (geometrically) nearest-neighbour ions by about an order of magnitude, and this makes MnF_2 an attractive crystal to study theoretically. From an experimental point of view MnF_2 is favoured because Mn^{2+} is an s-state ion and it can be grown in large single crystals.

The dispersion curves for MnF_2 at 4·2 K, as measured by neutron scattering, are shown in Fig. 9.10. The measurements are shown for two symmetry directions, (001) and (100).

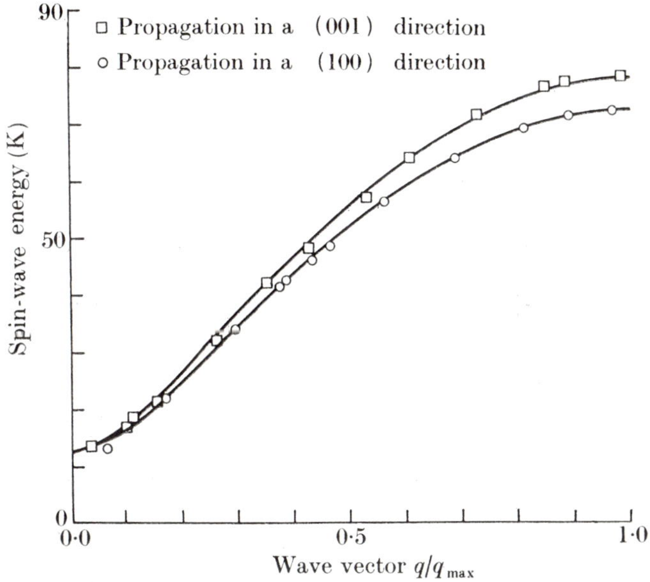

FIG. 9.10. Spin-wave dispersion curve for MnF_2 (A. Okazaki, K. C. Turberfield, and R. W. H. Stevenson, *Phys. Lett.* **8**, 9 (1964)).

The solid curves are from spin-wave theory. In the experiment the only adjustable parameter was the exchange parameter for geometrically nearest-neighbour Mn ions.

(*d*) *Two-magnon interactions in Heisenberg antiferromagnets: Hartree–Fock approximation*

To study the effects of the two-spin-wave dynamical interaction on the spin-wave dispersion relation of an antiferromagnet we use a slight extension of the method of § 9.2 (*c*) and for brevity we consider from the outset only a nearest-neighbour exchange coupling. The discussion can easily be extended to cover the case of a longer-range coupling.

For the spins on the 'up' sublattice

$$\hat{S}^z_{\mathbf{m}} = S-\hat{a}^+_{\mathbf{m}}\hat{a}_{\mathbf{m}}, \qquad \hat{S}^+_{\mathbf{m}} = (2S)^{\frac{1}{2}}\hat{a}_{\mathbf{m}}, \qquad \hat{S}^-_{\mathbf{m}} = (2S)^{\frac{1}{2}}\hat{a}^+_{\mathbf{m}}(1-\hat{a}^+_{\mathbf{m}}\hat{a}_{\mathbf{m}}/2S), \tag{9.268}$$

and, in view of (9.223), for the spins on the 'down' sublattice

$$\hat{S}^z_{\mathbf{n}} = -S+\hat{b}^+_{\mathbf{n}}\hat{b}_{\mathbf{n}}, \qquad \hat{S}^+_{\mathbf{n}} = (2S)^{\frac{1}{2}}\hat{b}^+_{\mathbf{n}}, \qquad \hat{S}^-_{\mathbf{n}} = (2S)^{\frac{1}{2}}(1-\hat{b}^+_{\mathbf{n}}\hat{b}_{\mathbf{n}}/2S)\hat{b}_{\mathbf{n}}. \tag{9.269}$$

The Hamiltonian we use is obtained from (9.221) with the addition of a uniaxial anisotropy;

$$\mathscr{H} = 2J\sum_{\mathbf{m},\boldsymbol{\rho}} \hat{\mathbf{S}}_{\mathbf{m}}\cdot\hat{\mathbf{S}}_{\mathbf{m}+\boldsymbol{\rho}} - g\mu_{\mathrm{B}}H_{\mathrm{A}}\sum_{\mathbf{m}}\hat{S}^z_{\mathbf{m}} + g\mu_{\mathrm{B}}H_{\mathrm{A}}\sum_{\mathbf{n}}\hat{S}^z_{\mathbf{n}}. \tag{9.270}$$

Inserting the transformations (9.268) and (9.269), and defining

$$\hat{a}_{\mathbf{m}} = \frac{1}{\sqrt{N}}\sum_{\mathbf{q}} \exp(\mathrm{i}\mathbf{q}\cdot\mathbf{m})\hat{a}_{\mathbf{q}}$$

and

$$\hat{b}^+_{\mathbf{n}} = \frac{1}{\sqrt{N}}\sum_{\mathbf{q}} \exp(\mathrm{i}\mathbf{q}\cdot\mathbf{n})\hat{b}^+_{\mathbf{q}}, \tag{9.271}$$

the Hamiltonian (9.270) becomes, apart from constant terms,

$$\frac{1}{2rJS}\mathscr{H} = \sum_{\mathbf{q}}(1+h_{\mathrm{A}})(\hat{a}^+_{\mathbf{q}}\hat{a}_{\mathbf{q}}+\hat{b}^+_{\mathbf{q}}\hat{b}_{\mathbf{q}})+\gamma_{\mathbf{q}}(\hat{a}_{\mathbf{q}}\hat{b}_{\mathbf{q}}+\hat{a}^+_{\mathbf{q}}\hat{b}^+_{\mathbf{q}})-$$
$$-\frac{1}{NS}\sum_{\mathbf{1,2,3,4}}\delta_{\mathbf{1+4,2+3}}(\gamma_{\mathbf{3-4}}\hat{a}^+_{\mathbf{1}}\hat{a}_{\mathbf{2}}\hat{b}^+_{\mathbf{3}}\hat{b}_{\mathbf{4}}+\tfrac{1}{2}\gamma_{\mathbf{1}}\hat{a}_{-\mathbf{1}}\hat{b}^+_{\mathbf{2}}\hat{b}_{-\mathbf{3}}\hat{b}_{\mathbf{4}}+$$
$$+\tfrac{1}{2}\gamma_{\mathbf{2}}\hat{a}^+_{\mathbf{1}}\hat{b}^+_{\mathbf{2}}\hat{a}^+_{-\mathbf{3}}\hat{a}_{-\mathbf{4}}). \tag{9.272}$$

In the second term in (9.272) we have used the shorthand notation $\mathbf{q}_1 = \mathbf{1}$, $\mathbf{q}_2 = \mathbf{2}$, etc.

The first part of (9.272) represents the non-interacting spin waves and is brought into diagonal form by using the transformation (cf. (9.229))

$$\hat{a}_{\mathbf{q}} = \bar{u}_{\mathbf{q}}\hat{\alpha}_{\mathbf{q}}+\bar{v}_{\mathbf{q}}\hat{\beta}^+_{\mathbf{q}}, \qquad \hat{b}^+_{\mathbf{q}} = \bar{u}_{\mathbf{q}}\hat{\beta}^+_{\mathbf{q}}+\bar{v}_{\mathbf{q}}\hat{\alpha}_{\mathbf{q}}. \tag{9.273}$$

The coefficients $\bar{u}_{\mathbf{q}}$ and $\bar{v}_{\mathbf{q}}$ differ from those given by eqn (9.246) only in the factor $(2SN)^{-1}$. This difference arises because of the different normalization adopted for the Bose operators $\hat{a}_{\mathbf{q}}$, $\hat{a}^+_{\mathbf{q}}$ as compared to that used for the operators $\hat{S}^{\pm}_{\mathbf{q}}$. We recall that the non-interacting spin waves have the energy spectrum

$$\hbar\omega_{\mathbf{q}} = 2rJS\{(1+h_{\mathrm{A}})^2-\gamma^2_{\mathbf{q}}\}^{\frac{1}{2}} \equiv 2rJS\mathscr{E}_{\mathbf{q}}. \tag{9.274}$$

The procedure we adopt to approximate the products of four Bose operators in the second part of (9.272) by products of two operators is just the same as that used in § 9.2 (*c*). Take first the term $\hat{a}^+_{\mathbf{1}}\hat{a}_{\mathbf{2}}\hat{b}^+_{\mathbf{3}}\hat{b}_{\mathbf{4}}$. Because the terms $\langle\hat{a}^+\hat{b}\rangle$, $\langle\hat{a}\hat{b}^+\rangle$, etc. are zero,

$$\hat{a}^+_{\mathbf{1}}\hat{a}_{\mathbf{2}}\hat{b}^+_{\mathbf{3}}\hat{b}_{\mathbf{4}} \to \delta_{\mathbf{1,2}}\langle\hat{a}^+_{\mathbf{1}}\hat{a}_{\mathbf{1}}\rangle\hat{b}^+_{\mathbf{3}}\hat{b}_{\mathbf{4}}+\delta_{\mathbf{3,4}}\hat{a}^+_{\mathbf{1}}\hat{a}_{\mathbf{2}}\langle\hat{b}^+_{\mathbf{3}}\hat{b}_{\mathbf{3}}\rangle+$$
$$+\delta_{\mathbf{1,3}}\langle\hat{a}^+_{\mathbf{1}}\hat{b}^+_{\mathbf{1}}\rangle\hat{a}_{\mathbf{2}}\hat{b}_{\mathbf{4}}+\delta_{\mathbf{2,4}}\hat{a}^+_{\mathbf{1}}\hat{b}^+_{\mathbf{3}}\langle\hat{a}_{\mathbf{2}}\hat{b}_{\mathbf{2}}\rangle. \tag{9.275}$$

In (9.275) we have dropped the constant terms. We denote the thermal averages

$$\langle \hat{a}_1^+ \hat{a}_1 \rangle = \langle \hat{b}_1^+ \hat{b}_1 \rangle = \eta_1$$

and

$$\langle \hat{a}_1^+ \hat{b}_1^+ \rangle = \langle \hat{a}_1 \hat{b}_1 \rangle = \xi_1. \tag{9.276}$$

Thus, from (9.275) and (9.276),

$$\hat{a}_1^+ \hat{a}_2 \hat{b}_3^+ \hat{b}_4 \delta_{1+4,2+3} \gamma_{3-4} \to \delta_{1,2} \delta_{3,4}(\eta_1 \hat{b}_3^+ \hat{b}_3 + \eta_3 \hat{a}_1^+ \hat{a}_1) + \\ + \delta_{1,3} \delta_{2,4}(\xi_1 \hat{a}_2 \hat{b}_2 + \xi_2 \hat{a}_1^+ \hat{b}_1^+)\gamma_{1-2}. \tag{9.277}$$

Similarly,

$$\hat{a}_{-1} \hat{b}_2^+ \hat{b}_{-3} \hat{b}_4 \delta_{1+4,2+3} \gamma_1 \\ \to (\xi_1 \hat{b}_2^+ \hat{b}_2 + \eta_2 \hat{a}_{-1} \hat{b}_{-1})\gamma_1(\delta_{1,3} \delta_{2,4} + \delta_{2,-3} \delta_{1,-4}) \tag{9.278}$$

and

$$\hat{a}_1^+ \hat{b}_2^+ \hat{a}_{-3}^+ \hat{a}_{-4} \delta_{1+4,2+3} \gamma_2 \\ \to \delta_{1,2} \delta_{3,4}(\xi_1 \hat{a}_{-3}^+ \hat{a}_{-3} + \eta_3 \hat{a}_1^+ \hat{b}_1^+)\gamma_1 + \delta_{1,-4} \delta_{2,-3}(\eta_1 \hat{b}_2^+ \hat{a}_2^+ + \xi_2 \hat{a}_1^+ \hat{a}_1). \tag{9.279}$$

If (9.277), (9.278), and (9.279) are inserted into the second part of the Hamiltonian (9.272) it reduces to

$$-\frac{1}{SN} \sum_{1,2} (\hat{a}_1^+ \hat{a}_1 + \hat{b}_1^+ \hat{b}_1)(\eta_2 + \gamma_2 \xi_2) + (\hat{a}_1 \hat{b}_1 + \hat{a}_1^+ \hat{b}_1^+)(\gamma_{1-2} \xi_2 + \gamma_1 \eta_2),$$

which when added to the first part of (9.272), the Hamiltonian of the non-interacting spin waves, yields

$$\frac{1}{2rJS} \mathscr{H} = \sum_{\mathbf{q}} (\hat{a}_{\mathbf{q}}^+ \hat{a}_{\mathbf{q}} + \hat{b}_{\mathbf{q}}^+ \hat{b}_{\mathbf{q}})\{1 + h_{\mathrm{A}} - C(T)\} + (\hat{a}_{\mathbf{q}} \hat{b}_{\mathbf{q}} + \hat{a}_{\mathbf{q}}^+ \hat{b}_{\mathbf{q}}^+)\gamma_{\mathbf{q}}\{1 - C(T)\}. \tag{9.280}$$

Here

$$C(T) = \frac{1}{NS} \sum_{\mathbf{q}} (\eta_{\mathbf{q}} + \gamma_{\mathbf{q}} \xi_{\mathbf{q}}). \tag{9.281}$$

In arriving at (9.280) we have made use of (9.61) to write

$$\sum_2 \gamma_{1-2} \xi_2 = \gamma_1 \sum_2 \gamma_2 \xi_2. \tag{9.282}$$

The Hamiltonian (9.280) has exactly the same form as for non-interacting spin waves except that $1+h_{\mathrm{A}}$ has been replaced by $1+h_{\mathrm{A}}-C(T)$ and $\gamma_{\mathbf{q}}$ by $\gamma_{\mathbf{q}}\{1-C(T)\}$, and hence we can immediately write down the quantities of interest. The transformation coefficients $\bar{u}_{\mathbf{q}}$ and $\bar{v}_{\mathbf{q}}$ in (9.273) for the diagonalization of (9.280) satisfy

$$\bar{u}_{\mathbf{q}}^2 - \bar{v}_{\mathbf{q}}^2 = 1, \qquad \bar{u}_{\mathbf{q}}^2 = \frac{1 + h_{\mathrm{A}} - C(T) + \mathscr{E}_{\mathbf{q}}(T)}{2\mathscr{E}_{\mathbf{q}}(T)}, \tag{9.283}$$

and

$$2\bar{u}_{\mathbf{q}} \bar{v}_{\mathbf{q}} = -\gamma_{\mathbf{q}}\{1 - C(T)\}/\mathscr{E}_{\mathbf{q}}(T).$$

In (9.283)

$$\mathscr{E}_{\mathbf{q}}(T) = [\{1+h_{\mathrm{A}}-C(T)\}^2-\gamma_{\mathbf{q}}^2\{1-C(T)\}^2]^{\frac{1}{2}}$$
$$= \{1-C(T)\}[\{1+h_{\mathrm{A}}(T)\}^2-\gamma_{\mathbf{q}}^2]^{\frac{1}{2}}, \quad (9.284)$$

where
$$h_{\mathrm{A}}(T) = \frac{h_{\mathrm{A}}}{1-C(T)} \simeq h_{\mathrm{A}}\{1+C(T)\}, \quad (9.285)$$

i.e. the dispersion relation is

$$\hbar\omega_{\mathbf{q}}'(T) = 2rJS\{1-C(T)\}[\{1+h_{\mathrm{A}}(T)\}^2-\gamma_{\mathbf{q}}^2]^{\frac{1}{2}}. \quad (9.286)$$

Thus, as we found in the ferromagnet, the dynamical interaction has led to a renormalization of the dispersion relation, but for the antiferromagnet there is the additional feature of renormalizing the anisotropy field.†

For the function $C(T)$

$$C(T) = \frac{1}{NS}\sum_{\mathbf{q}}\{\langle\hat{a}_{\mathbf{q}}^{+}\hat{a}_{\mathbf{q}}\rangle+\gamma_{\mathbf{q}}\langle\hat{a}_{\mathbf{q}}\hat{b}_{\mathbf{q}}\rangle\}$$

we eventually obtain the integral equation

$$C(T) = \frac{1}{2NS}\sum_{\mathbf{q}}\left[\frac{\{1-C(T)\}(1-\gamma_{\mathbf{q}}^2)+h_{\mathrm{A}}}{\mathscr{E}_{\mathbf{q}}(T)}\right]\coth\{\tfrac{1}{2}\hbar\omega_{\mathbf{q}}(T)\beta\}-1. \quad (9.287)$$

And, for $\langle\hat{S}^z\rangle$,

$$\langle\hat{S}^z\rangle = S-\frac{1}{N}\sum_{\mathbf{q}}\langle\hat{a}_{\mathbf{q}}^{+}\hat{a}_{\mathbf{q}}\rangle$$
$$= S-\frac{1}{2N}\sum_{\mathbf{q}}\left\{\frac{1+h_{\mathrm{A}}-C(T)}{\mathscr{E}_{\mathbf{q}}(T)}\right\}\coth\{\tfrac{1}{2}\hbar\omega_{\mathbf{q}}(T)\beta\}-1. \quad (9.288)$$

To calculate the dispersion relation (9.286) as a function of temperature or to compute the sublattice magnetization (9.288), we have to solve a complicated integral equation and this in general is a numerical problem. However, the equations undergo a great simplification for $h_{\mathrm{A}} = 0$ and from these it is relatively straightforward to calculate the leading order corrections introduced by the dynamical interaction.

Setting $h_{\mathrm{A}} = 0$ in (9.287), $C(T)$ is given by

$$C(T) = \frac{1}{2NS}\sum_{\mathbf{q}}\{(1-\gamma_{\mathbf{q}}^2)^{\frac{1}{2}}-1\}+\frac{1}{NS}\sum_{\mathbf{q}}(1-\gamma_{\mathbf{q}}^2)^{\frac{1}{2}}\frac{1}{\exp\{\hbar\omega_{\mathbf{q}}(T)\beta\}-1}, \quad (9.289)$$

† Eqn (9.286) for $h_{\mathrm{A}} = 0$ was first obtained by S. Liu, *Phys. Rev.* **142**, 267 (1966). This paper contains a detailed numerical analysis of the renormalized theory in the limit of zero anisotropy.

where now $$\hbar\omega_{\mathbf{q}}(T) = 2rJS\{1-C(T)\}(1-\gamma_{\mathbf{q}}^2)^{\frac{1}{2}}. \tag{9.290}$$

For small values of $|\mathbf{q}|$

$$\hbar\omega_{\mathbf{q}}(T) = D(T)q, \tag{9.291}$$

where $$D(T) = D\{1-C(T)\} \tag{9.292}$$

and D is given by eqn (9.251). Using the relation (9.291) and approximating $(1-\gamma_{\mathbf{q}}^2)^{\frac{1}{2}}$ by $\rho q/\sqrt{3}$ we obtain for

$$\frac{1}{NS}\sum_{\mathbf{q}}(1-\gamma_{\mathbf{q}}^2)^{\frac{1}{2}}\frac{1}{\exp\{\hbar\omega_{\mathbf{q}}(T)\beta\}-1}$$

the result $$\frac{v_0}{2\pi^2 S}\frac{\rho}{\sqrt{3}}\left\{\frac{k_{\mathrm{B}}T}{D(T)}\right\}^4\Gamma(4)\zeta(4) \tag{9.293}$$

and $\Gamma(4) = 3!$

If we denote the constant†

$$C_0 = \frac{1}{N}\sum_{\mathbf{q}}\{(1-\gamma_{\mathbf{q}}^2)^{\frac{1}{2}}-1\} \tag{9.294}$$

and treat $C(T)-C_0/2S$ as a small quantity, (9.293) leads to the result

$$C(T)-C_0/2S = \frac{3v_0}{\pi^2 S(1-C_0/2S)^4}\left(\frac{\sqrt{3}}{\rho}\right)^3\left(\frac{k_{\mathrm{B}}T}{2rJS}\right)^4\zeta(4)+ \\ +\frac{4}{(1-C_0/2S)^9}\left\{\frac{3v_0}{\pi^2 S}\left(\frac{\sqrt{3}}{\rho}\right)^3\zeta(4)\right\}^2\left(\frac{k_{\mathrm{B}}T}{2rJS}\right)^8. \tag{9.295}$$

From this we conclude that $D(T)$ for an antiferromagnet decreases with temperature, because of the dynamical interaction of the spin waves, as T^4. The renormalized anisotropy field, $h_{\mathrm{A}}(T) \simeq h_{\mathrm{A}}\{1+C(T)\}$, *increases* as T^4. The effect on the dispersion curve is to decrease the overall bandwidth, the anisotropy field gap

$$(h_{\mathrm{A}}^2+2h_{\mathrm{A}})^{\frac{1}{2}}-\frac{C(T)}{2(h_{\mathrm{A}}^2+2h_{\mathrm{A}})^{\frac{1}{2}}} \tag{9.296}$$

decreasing with temperature like T^4.

A detailed investigation of the validity of renormalized antiferromagnetic spin-wave theory was made by Low‡ for MnF_2. By far the most sensitive test that can be applied to any such theory is to compare its predictions with measured spin-wave dispersion curves. The reason that this is more sensitive than comparing macroscopic quantities is that the latter involve an integration over the whole spin-wave band and therefore smear out any of the finer points in the details of the theory. Fig. 9.11 shows the result of Low's calculation. It is seen that the

† Values of C_0 are given by Keffer, ibid.

‡ G. G. Low, *Proc. symposium on inelastic scattering of neutrons in solids and liquids*, 1964 (IAEA, 1965).

agreement between the measured dispersion curves and that given by the renormalized theory is extremely good; the agreement being almost remarkable at $T = 62$ K since $T_N \simeq 69$ K. In these calculations three sets of exchange integrals were taken into account, and the various sets of curves for $T = 49{\cdot}5$ K and 62 K are for calculations in which the renormalization factor in $\hbar\omega_{\mathbf{q}}$ is multiplied by the figure shown.

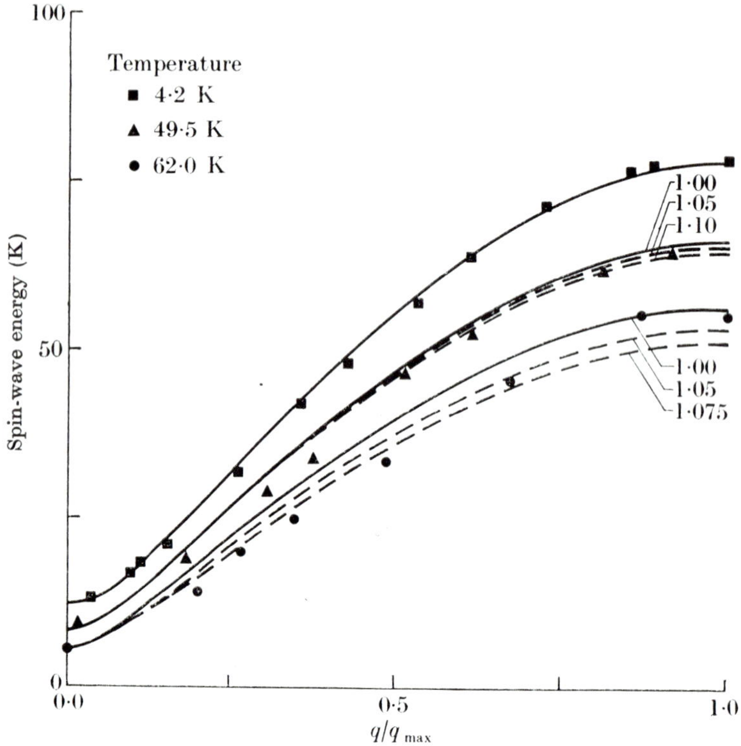

FIG. 9.11. The dispersion curves for MnF_2 as measured by Okazaki *et al.* (loc. cit.) compared to a renormalized spin-wave theory. The two sets of curves for $T = 49{\cdot}5$ K and 62·0 K correspond to calculations with the renormalization factor in $\hbar\omega_{\mathbf{q}}$ multiplied by the given figure (for MnF_2 $T_N = 69$ K).

Finally, we examine the sublattice magnetization to find the leading order correction arising from the two-spin-wave dynamical interaction. For $h_A = 0$

$$\frac{1}{N}\sum_{\mathbf{q}}\langle\hat{a}_{\mathbf{q}}^{+}\hat{a}_{\mathbf{q}}\rangle = \delta\langle\hat{S}^z\rangle_0 + \frac{1}{N}\sum_{\mathbf{q}}\frac{1}{(1-\gamma_{\mathbf{q}}^2)^{\frac{1}{2}}}\,\frac{1}{\exp\{\hbar\omega_{\mathbf{q}}(T)\beta\}-1}, \tag{9.297}$$

which for small $\mathbf{q}$ becomes

$$\delta\langle\hat{S}^z\rangle_0 + \frac{v_0}{2\pi^2}\left(\frac{\surd 3}{\rho}\right)^3\left(\frac{k_B T}{2rJS}\right)^2\frac{\zeta(2)}{\{1-C(T)\}^2}.$$

This evaluated to the first-order correction yields†

$$\langle \hat{S}^z \rangle = S - \delta \langle \hat{S}^z \rangle_0 - \frac{v_0}{2\pi^2(1-C_0/2S)^2}\left(\frac{\sqrt{3}}{\rho}\right)^3 \left(\frac{k_B T}{2rJS}\right)^2 \zeta(2) - \\ - \frac{3}{S(1-C_0/2S)^7}\left\{\frac{v_0}{\pi^2}\left(\frac{\sqrt{3}}{\rho}\right)^3\right\}^2 \left(\frac{k_B T}{2rJS}\right)^6 \zeta(2)\zeta(4). \quad (9.298)$$

From (9.298) we see that the dynamical interaction enters the sublattice magnetization as T^6, and modifies the non-interacting term by a factor $(1-C_0/2S)^{-2}$. The renormalized sublattice magnetization has been shown to be in remarkably good agreement with experiments, the simple treatment of the dynamical interaction leading to expressions that hold good over a remarkably large temperature range. Further details can be found in the article by Keffer cited at the beginning of the chapter.

9.7. Heisenberg ferrimagnets

Our discussion of the properties of ferrimagnets will be less detailed than that given for ferromagnets and antiferromagnets.‡ In part this is due to the complexity of the attendant algebra; a detailed discussion would involve numerical computations and the number of parameters makes this unreasonable. In addition, very little neutron work has been done with ferrimagnets to date. We shall in fact, content ourselves with writing down the cross-section in general form and then briefly survey the fundamental properties in simple cases. The analysis set up in § 9.6 is sufficient to treat a particular case in detail except, perhaps, in respect of the anisotropy terms considered, since these were taken to be uniaxial.

As we stated in § 9.6, the Hamiltonian (9.222) commutes with the *total* z component of spin; thus we have only to calculate $\langle \hat{S}^{\pm}\hat{S}^{\mp} \rangle$ to evaluate the neutron cross-sections for the processes involving the creation or annihilation of a single magnon. These correlation functions are simply calculated with the transformation (9.229).

If we denote a general lattice site by $\mathbf{R}$, as in § 9.6, a straightforward calculation gives

$$\sum_{\mathbf{R},\mathbf{R}'} \exp\{i\boldsymbol{\kappa}.(\mathbf{R}'-\mathbf{R})\}\langle \hat{S}^+_{\mathbf{R}}(0)\hat{S}^-_{\mathbf{R}'}(t)\rangle \\ = \frac{1}{N^2}\sum_{\mathbf{q},\boldsymbol{\tau}} N\frac{(2\pi)^3}{v_0}\delta(\boldsymbol{\kappa}-\mathbf{q}-\boldsymbol{\tau})[(1+n_{\mathbf{q},0})\exp(i\omega_{\mathbf{q},0}t)(u_{\mathbf{q}}^2+\sigma v_{\mathbf{q}}^2) + \\ + n_{\mathbf{q},1}\exp(-i\omega_{\mathbf{q},1}t)(\sigma u_{\mathbf{q}}^2+v_{\mathbf{q}}^2) + \\ + 2u_{\mathbf{q}}v_{\mathbf{q}}\sigma^{\frac{1}{2}}\cos\boldsymbol{\rho}.\boldsymbol{\tau}\{(1+n_{\mathbf{q},0})\exp(i\omega_{\mathbf{q},0}t)+n_{\mathbf{q},1}\exp(-i\omega_{\mathbf{q},1}t)\}]. \quad (9.299)$$

† Values of $\delta\langle \hat{S}^z \rangle_0$ can be found in the article by Keffer, op. cit.

‡ A detailed account of the thermodynamic properties of ferrimagnets is given by T. Nakamura and M. Bloch, *Phys. Rev.* **132**, 2528 (1963).

8512546 M

We remind the reader that the reciprocal lattice vectors $\boldsymbol{\tau}$ in (9.299) are those belonging to a sublattice and N is the number of ions situated on one sublattice. Also, the suffixes 0 and 1 refer to the two spin-wave modes in the ferrimagnet; these have energy spectra given by (9.235 a) (corresponding to $a = 0$) and (9.235 b) (corresponding to $a = 1$).

We can again use (9.261) to obtain

$$\sum_{\mathbf{R},\mathbf{R}'} \exp\{i\boldsymbol{\kappa}.(\mathbf{R}'-\mathbf{R})\}\langle \hat{S}^{+}_{\mathbf{R}}(0)\hat{S}^{-}_{\mathbf{R}'}(t)+\hat{S}^{-}_{\mathbf{R}}(0)\hat{S}^{+}_{\mathbf{R}'}(t)\rangle$$
$$= \frac{(2\pi)^3}{Nv_0}\sum_{\mathbf{q},\boldsymbol{\tau}} \delta(\boldsymbol{\kappa}-\mathbf{q}-\boldsymbol{\tau})[\{(1+n_{\mathbf{q},0})\exp(i\omega_{\mathbf{q},0}t)+n_{\mathbf{q},0}\exp(-i\omega_{\mathbf{q},0}t)\}\times$$
$$\times\{u_{\mathbf{q}}^2+\sigma v_{\mathbf{q}}^2+2u_{\mathbf{q}}v_{\mathbf{q}}\sigma^{\frac{1}{2}}\cos\boldsymbol{\rho}.\boldsymbol{\tau}\}+$$
$$+\{(1+n_{\mathbf{q},1})\exp(i\omega_{\mathbf{q},1}t)+n_{\mathbf{q},1}\exp(-i\omega_{\mathbf{q},1}t)\}\{\sigma u_{\mathbf{q}}^2+v_{\mathbf{q}}^2+2u_{\mathbf{q}}v_{\mathbf{q}}\sigma^{\frac{1}{2}}\cos\boldsymbol{\rho}.\boldsymbol{\tau}\}]. \tag{9.300}$$

Note that the result (9.300) shows that the two spin-wave modes have different structure factors associated with them.

We can immediately calculate from (9.300) the cross-sections in which a magnon of type $a = 0$ is created or annihilated and similarly for a magnon of type $a = 1$. If we neglect the difference in both the Debye–Waller factors for the two types of ion in the ferrimagnet and their atomic form factors, these cross-sections are

$$\left(\frac{d^2\sigma}{d\Omega dE'}\right)^{(\pm)}_{a=0} = \left(\frac{\gamma e^2}{m_e c^2}\right)^2\{\tfrac{1}{2}gF(\boldsymbol{\kappa})\}^2\frac{1}{4}\frac{k'}{k}(1+\tilde{\kappa}_z^2)\exp\{-2W(\boldsymbol{\kappa})\}\times$$
$$\times\frac{(2\pi)^3}{Nv_0}\sum_{\mathbf{q},\boldsymbol{\tau}}(n_{\mathbf{q},0}+\tfrac{1}{2}\pm\tfrac{1}{2})\delta(\hbar\omega_{\mathbf{q},0}\mp\hbar\omega)\delta(\boldsymbol{\kappa}\mp\mathbf{q}-\boldsymbol{\tau})\times$$
$$\times(u_{\mathbf{q}}^2+\sigma v_{\mathbf{q}}^2+2u_{\mathbf{q}}v_{\mathbf{q}}\sigma^{\frac{1}{2}}\cos\boldsymbol{\rho}.\boldsymbol{\tau}) \tag{9.301}$$

and

$$\left(\frac{d^2\sigma}{d\Omega dE'}\right)^{(\pm)}_{a=1} = \left(\frac{\gamma e^2}{m_e c^2}\right)^2\{\tfrac{1}{2}gF(\boldsymbol{\kappa})\}^2\frac{1}{4}\frac{k'}{k}(1+\tilde{\kappa}_z^2)\exp\{-2W(\boldsymbol{\kappa})\}\times$$
$$\times\frac{(2\pi)^3}{Nv_0}\sum_{\mathbf{q},\boldsymbol{\tau}}(n_{\mathbf{q},1}+\tfrac{1}{2}\pm\tfrac{1}{2})\delta(\hbar\omega_{\mathbf{q},1}\mp\hbar\omega)\delta(\boldsymbol{\kappa}\mp\mathbf{q}-\boldsymbol{\tau})\times$$
$$\times(\sigma u_{\mathbf{q}}^2+v_{\mathbf{q}}^2+2u_{\mathbf{q}}v_{\mathbf{q}}\sigma^{\frac{1}{2}}\cos\boldsymbol{\rho}.\boldsymbol{\tau}). \tag{9.302}$$

In general we might expect $\hbar\omega_{\mathbf{q},1}$ to be greater than $\hbar\omega_{\mathbf{q},0}$ in the limit $\mathbf{q}\to 0$, so by analogy with the phonon problem, it is usual to refer to (9.301) as giving the cross-sections for the creation and annihilation of the acoustic magnons and (9.302) the corresponding cross-sections for the optical magnons.

Under certain conditions, a ferrimagnet possesses thermodynamic properties akin to the ferromagnet. To see this set $H_{A,1} = H_{A,2} = 0$, $H = 0$ and in addition $\mathscr{J}_1 = \mathscr{J}_2 = 0$. The dispersion relations for the two spin-wave modes are in this case

$$\hbar\omega_{\mathbf{q},a} = (-1)^a \mathscr{J}(0)(S_2-S_1)+[\{\mathscr{J}(0)(S_1+S_2)\}^2-4S_1 S_2 \mathscr{J}^2(\mathbf{q})]^{\frac{1}{2}}. \quad (9.303)$$

This result follows immediately from (9.235).

If we now write (cf. eqn (9.240))

$$\mathscr{J}(0)-\mathscr{J}(\mathbf{q}) \simeq \tfrac{1}{6}q^2 J^{(2)}, \quad (9.304)$$

then

$$\mathscr{J}^2(\mathbf{q}) \simeq \mathscr{J}^2(0)-\tfrac{1}{3}q^2\mathscr{J}(0)J^{(2)}.$$

From this it follows that

$$\hbar\omega_{\mathbf{q},a} \simeq -\mathscr{J}(0)(S_1-S_2)(-1)^a+[\{\mathscr{J}(0)(S_1-S_2)\}^2+\tfrac{16}{3}S_1 S_2 q^2 \mathscr{J}(0)J^{(2)}]^{\frac{1}{2}}$$

$$\simeq \mathscr{J}(0)(S_1-S_2)\{1+(-1)^{a+1}\}+q^2\frac{\frac{2}{3}S_1 S_2 J^{(2)}}{(S_1-S_2)}$$

and since at low temperatures only the mode for which $a = 0$ (acoustic mode) is of importance we have

$$\hbar\omega_{\mathbf{q},0} \simeq q^2\left\{\frac{2}{3}\frac{S_1 S_2 J^{(2)}}{S_1-S_2}\right\}. \quad (9.305)$$

Thus, because of the q^2 dependence at small $\mathbf{q}$, the thermodynamic properties of a ferrimagnet (under the conditions specified above) are analogous to those of a ferromagnet, e.g. the magnetization decreases with temperature $T^{\frac{3}{2}}$. Note, however, that if $S_1 \sim S_2$ the effective D in the ferrimagnet can be quite large. For nearest-neighbour coupling only we have from (9.305) that D for a ferrimagnet is given by

$$D = \frac{4S_1 S_2 a^2 J}{S_1-S_2}. \quad (9.306)$$

10

POLARIZATION ANALYSIS†

To specify completely the physical state of a beam of neutrons we must give both its momentum and spin state. Up to now we have always taken the spins of the incident neutrons to be randomly orientated, i.e. unpolarized; neither have we asked for the spin state of the scattered beam. Clearly, however, if we use incident neutrons whose spins have some preferred axis, i.e. the beam is, to some degree, polarized, and in addition analyse not only the spatial and energy distribution of the scattered neutrons but also their spin states, we must gain more information about the nature of the scattering process than if we did neither. In such a scheme we are introducing two new coordinates into the scattering process, the polarization of the incident beam and the polarization of the scattered beam. We must, therefore, first ask how the cross-section for the scattering process depends on the polarization of the incident neutrons, and secondly, relate the polarization of the scattered neutrons to the properties of the target system. In some instances the scattered neutrons may be polarized even though the incident neutrons are unpolarized. Thus, in this instance, polarization is created in the scattering process and this polarization must be intimately connected with the physical properties of the target system sensed by the neutrons. The use of polarized incident neutrons, or just the analysis of the polarization of the scattered beam, results in a marked decrease in the measured intensity, usually by at *least* an order of magnitude. Until now, polarized neutrons have mainly been used in elastic scattering experiments, but high-flux reactors make inelastic magnetic scattering experiments possible with polarized neutrons.

Quite generally we can see that the cross-section for the scattering of polarized neutrons will be independent of the polarization if there is no preferred axis in the target system itself. Thus, pure nuclear scattering is independent of polarization because nuclear spins are always randomly orientated at temperatures of interest to us (at very low temperatures the nuclei may be polarized, see R. I. Schermer and

† A theoretical survey is given by Williams, *Introduction to elementary particles*, chap. 8 (Academic Press, 1961).

M. Blume, *Phys. Rev.* **166,** 554 (1968)). Even so, it is of interest to ask how the scattering affects the polarization if it is initially present, and we discuss this in detail. Because a ferromagnet does have a preferred axis (when the domains are aligned by an external magnetic field) the cross-section for scattering from a ferromagnet depends on the polarization. We find the purely nuclear and purely magnetic cross-sections are as before, but a new term due to interference between nuclear and magnetic scattering appears. Also, in the polarization of neutrons inelastically scattered by magnetic target systems we find created polarization, the orientation of which depends on whether it is associated with a scattering process in which an elementary excitation is created or annihilated.

10.1. Description of a polarized beam of neutrons

First, what do we mean when we say a beam of neutrons possesses a certain polarization? We define the polarization of a beam of neutrons as twice the average value of the spin of the neutrons in the beam; viz.

$$\mathbf{P} = 2\langle \hat{\mathbf{s}} \rangle = \langle \hat{\boldsymbol{\sigma}} \rangle. \tag{10.1}$$

In (10.1), $\hat{\sigma}^{\alpha}$ are the Pauli matrices

$$\hat{\sigma}^{x} = \begin{pmatrix} 0 & 1 \\ 1 & 0 \end{pmatrix}, \quad \hat{\sigma}^{y} = \begin{pmatrix} 0 & -\mathrm{i} \\ \mathrm{i} & 0 \end{pmatrix}, \quad \hat{\sigma}^{z} = \begin{pmatrix} 1 & 0 \\ 0 & -1 \end{pmatrix}. \tag{10.2}$$

For an *unpolarized beam*, $\mathbf{P} = 0$; for a *completely polarized beam* $|\mathbf{P}| = 1$. By a *partially polarized beam* we mean that $0 < |\mathbf{P}| < 1$.

If a beam of neutrons is partially polarized we cannot have sufficient information about it to give a complete quantum mechanical description, i.e. we cannot assign a wave function to the spin state of the beam. For if the spin state of a neutron is described by a wave function χ, which for a spin $\frac{1}{2}$ particle is a two-component spinor, it follows that in some direction in space the neutron has a definite spin value, i.e. it is either $\pm\frac{1}{2}$, and is therefore completely polarized. Hence, a partially polarized beam of neutrons can only be defined by some probability distribution. The physical properties of such a system, usually called a 'mixed' system, can be described in terms of a density matrix operator $\hat{\rho}$. We choose to introduce the density matrix operator by first considering a single neutron with a spin-wave χ and then construct the density matrix for the beam of incident neutrons by averaging over the density matrices of the individual neutrons.

The most general form of the spin wave function χ for a neutron is

$$\chi = a\chi_{\uparrow} + b\chi_{\downarrow}, \tag{10.3}$$

where $\chi_\uparrow$ and $\chi_\downarrow$ are the eigenfunctions of σ^z. a and b must satisfy

$$|a|^2+|b|^2 = 1 \tag{10.4}$$

for χ to be normalized to unity.

Clearly the quantities $|a|^2$ and $|b|^2$ are, respectively, the probabilities that a measurement of the z component of spin will show that the spin is parallel ($\uparrow$) or antiparallel ($\downarrow$) to the z direction. There will always exist a direction defined by the unit vector ξ such that

$$(\hat{\boldsymbol{\sigma}}.\xi)\chi = \chi,$$

i.e. the spin is *fully* aligned in the direction ξ.

We define the density matrix operator $\hat{\rho}$ as

$$\hat{\rho} = \chi\chi^+ = \begin{pmatrix} |a|^2 & ab^* \\ ba^* & |b|^2 \end{pmatrix}. \tag{10.5}$$

$\hat{\rho}$ is Hermitian and has unit trace, i.e.

$$\operatorname{Tr}\hat{\rho} = 1. \tag{10.6}$$

The average value of an arbitrary operator $\hat{O}$ in the state described by χ is $\chi^+\hat{O}\chi$. From (10.3) and (10.5) we see that this may be written in terms of $\hat{\rho}$, since

$$\langle\hat{O}\rangle = \chi^+\hat{O}\chi = \operatorname{Tr}\hat{O}\hat{\rho} = \operatorname{Tr}\hat{\rho}\hat{O}. \tag{10.7}$$

The last equality follows from the invariance of the trace to a cyclic permutation of the operators.

Because $\hat{\rho}$ is a 2×2 Hermitian matrix it can be expanded in terms of the unit matrix I and the Pauli matrices σ^α with real expansion coefficients. In fact

$$\hat{\rho} = \tfrac{1}{2}(I+\mathbf{P}.\hat{\boldsymbol{\sigma}}), \tag{10.8}$$

where we have defined

$$P_x = 2\operatorname{Re}(a^*b),$$
$$P_y = 2\operatorname{Im}(a^*b),$$
$$P_z = |a|^2-|b|^2. \tag{10.9}$$

That the vector $\mathbf{P}$ in (10.8) coincides with the definition of the polarization vector given in (10.1) follows immediately from (10.7) if we make use of

$$\operatorname{Tr}\hat{\sigma}^\alpha = 0 \quad \text{for all } \alpha, \tag{10.10}$$

$$\operatorname{Tr} I = 2, \tag{10.11}$$

and, also, the identity

$$\operatorname{Tr}\hat{\sigma}^\alpha\hat{\sigma}^\beta = 2\delta_{\alpha,\beta}. \tag{10.12}$$

Of course, in the present case $\mathbf{P}$ has unit magnitude because the neutron is fully polarized. This follows directly from (10.9) or by noting that from (10.5) we must have

$$\hat{\rho}^2 = \chi\chi^+\chi\chi^+ = \chi\chi^+ = \hat{\rho}. \tag{10.13}$$

Consider now a beam of neutrons. Each individual neutron will have its own spin state described by a density matrix. However, in general, the beam is not in a definite spin state. We therefore define the polarization of the beam as the average over the polarizations of each neutron. Thus if the jth neutron has a polarization $\mathbf{P}_j$ and there are $\mathcal{N}$ neutrons

$$\mathbf{P} = \frac{1}{\mathcal{N}} \sum_j \mathbf{P}_j. \tag{10.14}$$

The magnitude of $\mathbf{P}$ must lie between zero and unity, and the density matrix of the beam is, clearly,

$$\hat{\rho} = \tfrac{1}{2}(I + \mathbf{P}.\hat{\boldsymbol{\sigma}}) \quad (0 \leqslant |\mathbf{P}| \leqslant 1). \tag{10.15}$$

We note that $$\langle\hat{\rho}\rangle = \mathrm{Tr}\,\hat{\rho}\hat{\rho} = \tfrac{1}{2}\{1 + |\mathbf{P}|^2\}, \tag{10.16}$$

from which it follows that $$\langle\hat{\rho}\rangle \geqslant \tfrac{1}{2}, \tag{10.17}$$

and, since $|\mathbf{P}|^2 \leqslant 1$, $$\langle\hat{\rho}\rangle \leqslant 1. \tag{10.18}$$

The equality in (10.17) is satisfied when the neutron beam is unpolarized ($\mathbf{P} = 0$), whereas the equality in (10.18) is satisfied when the beam is completely polarized.

It is also worth while to note that an unpolarized beam can be regarded as the sum of two completely polarized beams that possess opposite polarizations. Let the two beams have polarizations in the β direction, then the above statement merely says that the density matrix

$$\{\tfrac{1}{2}(I + P_\beta\hat{\sigma}^\beta) + \tfrac{1}{2}(I - P_\beta\hat{\sigma}^\beta)\}/\mathrm{Tr}\{\tfrac{1}{2}(I + P_\beta\hat{\sigma}^\beta) + \tfrac{1}{2}(I - P_\beta\hat{\sigma}^\beta)\} = \tfrac{1}{2}I, \tag{10.19}$$

as required for an unpolarized beam.

10.2. Cross-section and polarization of scattered beam

The partial differential cross-section derived in Chapter 1 is given by

$$\frac{d^2\sigma}{d\Omega dE'} = \left(\frac{m}{2\pi\hbar^2}\right)^2 \frac{k'}{k} \sum_{\lambda,\sigma} p_\lambda p_\sigma \sum_{\lambda',\sigma'} \langle\lambda,\sigma|\hat{V}^+(\boldsymbol{\kappa})|\lambda',\sigma'\rangle \times$$
$$\times \langle\lambda',\sigma'|\hat{V}(\boldsymbol{\kappa})|\lambda,\sigma\rangle\delta(\hbar\omega + E_\lambda - E_{\lambda'}). \tag{10.20}$$

In (10.20), $\hat{V}(\boldsymbol{\kappa})$ is the Fourier transform of the interaction potential between the incident neutron and the target system.

For purely nuclear scattering from an array of rigid nuclei,

$$\hat{V}_{\mathrm{N}}(\boldsymbol{\kappa}) = \frac{2\pi\hbar^2}{m} \sum_{\mathbf{l},\mathbf{d}} \exp(\mathrm{i}\boldsymbol{\kappa}.\mathbf{R}_{ld})\hat{b}_{ld}. \tag{10.21}$$

Here $\hat{b}_{ld}$ is the scattering amplitude operator (cf. (1.26))

$$\hat{b}_{ld} = A_{ld} + \tfrac{1}{2}B_{ld}\hat{\boldsymbol{\sigma}}.\hat{\mathbf{I}}_{ld}, \tag{10.22}$$

where $\hat{\mathbf{i}}_{ld}$ is the angular momentum operator for the nucleus at lattice site $\mathbf{R}_{ld} = \mathbf{l}+\mathbf{d}$.

For purely magnetic scattering (cf. (5.10))

$$\hat{V}_{\mathrm{M}}(\boldsymbol{\varkappa}) = \left(\frac{2\pi\hbar^2}{m}\right)\left(\frac{\gamma e^2}{m_{\mathrm{e}}c^2}\right)\hat{\boldsymbol{\sigma}}.\sum_i \exp(\mathrm{i}\boldsymbol{\varkappa}.\mathbf{r}_i)\left\{\tilde{\boldsymbol{\varkappa}}\times(\hat{\mathbf{s}}_i\times\tilde{\boldsymbol{\varkappa}})-\frac{\mathrm{i}}{\hbar|\boldsymbol{\varkappa}|}\tilde{\boldsymbol{\varkappa}}\times\hat{\mathbf{p}}_i\right\}. \quad (10.23)$$

In a later section of this chapter we calculate the cross-section for the scattering of neutrons by the electric field produced by the nuclei and atomic electrons in a solid. The interaction potential for this process is there shown to be

$$\hat{V}_{\mathrm{E}}(\boldsymbol{\varkappa}) = \left(\frac{2\pi\hbar^2}{m}\right)\left(\frac{m_{\mathrm{e}}}{2m}\right)\left(\frac{\gamma e^2}{m_{\mathrm{e}}c^2}\right)\{\mathrm{i}\cot(\tfrac{1}{2}\theta)\tilde{\mathbf{n}}.\hat{\boldsymbol{\sigma}}-1\}\times$$

$$\times\left\{\sum_{\mathbf{l},\mathbf{d}}\exp(\mathrm{i}\boldsymbol{\varkappa}.\mathbf{R}_{ld})Z_d-\sum_i\exp(\mathrm{i}\boldsymbol{\varkappa}.\mathbf{r}_i)\right\}, \quad (10.24)$$

where

$$\mathbf{k}'\times\mathbf{k} = k^2\sin\theta\,\tilde{\mathbf{n}} \quad (10.25)$$

defines the unit vector $\tilde{\mathbf{n}}$ and eZ_d is the charge associated with the nucleus of site $\mathbf{d}$ within a unit cell.

An examination of the three types of interaction potential given above shows that they all have the form

$$\hat{v} = \hat{\beta}+\hat{\boldsymbol{\alpha}}.\hat{\boldsymbol{\sigma}}, \quad (10.26)$$

where the operators $\hat{\beta}$ and $\hat{\boldsymbol{\alpha}}$ refer to the target system. It is therefore sufficient in studying the modification to the cross-section due to polarization, to consider just the general form of interaction potential (10.26). Let us also, for the moment, consider that part of the cross-section (10.20) that depends on the neutron spin, namely,

$$\sum_{\sigma,\sigma'} p_\sigma\langle\sigma|\hat{v}^+|\sigma'\rangle\langle\sigma'|\hat{v}|\sigma\rangle. \quad (10.27)$$

The sum over σ' in (10.27) can be done by closure, to give

$$\sum_\sigma p_\sigma\langle\sigma|\hat{v}^+\hat{v}|\sigma\rangle. \quad (10.28)$$

This is valid only if there is no phase correlation between the states labelled with the quantum number σ, i.e. only if with respect to these states σ the density matrix is diagonal. But in this case the probability p_σ is just the diagonal element $\langle\sigma|\hat{\rho}|\sigma\rangle$, so (10.28) can be rewritten

$$\sum_\sigma \langle\sigma|\hat{\rho}|\sigma\rangle\langle\sigma|\hat{v}^+\hat{v}|\sigma\rangle. \quad (10.29)$$

Furthermore, if $\hat{\rho}$ is diagonal,

$$\langle\sigma'|\hat{\rho}|\sigma\rangle = \delta_{\sigma,\sigma'}\langle\sigma|\hat{\rho}|\sigma\rangle$$

so that (10.29) becomes

$$\sum_{\sigma,\sigma'} \langle\sigma|\,\hat{v}^{+}\hat{v}\,|\sigma'\rangle\langle\sigma'|\,\hat{\rho}\,|\sigma\rangle$$

and the sum over σ' can be done by closure to give for (10.27)

$$\sum_{\sigma,\sigma'} p_{\sigma}\langle\sigma|\,\hat{v}^{+}\,|\sigma'\rangle\langle\sigma'|\,\hat{v}\,|\sigma\rangle = \sum_{\sigma} \langle\sigma|\,\hat{v}^{+}\hat{v}\hat{\rho}\,|\sigma\rangle$$

$$\equiv \mathrm{Tr}\,\hat{v}^{+}\hat{v}\hat{\rho} \equiv \mathrm{Tr}\,\hat{\rho}\hat{v}^{+}\hat{v}. \qquad (10.30)$$

This final form is independent of the representation that is chosen to label the states and hence for this last form it does not matter whether or not $\hat{\rho}$ is diagonal. We conclude therefore that a formula which is much more general than (10.19) is

$$\frac{\mathrm{d}^{2}\sigma}{\mathrm{d}\Omega\mathrm{d}E'} = \left(\frac{m}{2\pi\hbar^{2}}\right)^{2}\frac{k'}{k}\sum_{\lambda,\lambda'} p_{\lambda}\,\mathrm{Tr}\,\hat{\rho}\langle\lambda|\,\hat{V}^{+}(\boldsymbol{\varkappa})\,|\lambda'\rangle\langle\lambda'|\,\hat{V}(\boldsymbol{\varkappa})\,|\lambda\rangle\delta(\hbar\omega+E_{\lambda}-E_{\lambda'}), \qquad (10.31)$$

where it is understood that the trace is to be taken with respect only to the neutron spin coordinates. Before we examine the structure of the spin-dependent part of this cross-section we derive an expression for the polarization of the scattered beam, $\mathbf{P}'$. The formula for $\mathbf{P}'$ must represent the transformation of the spin state of the incident neutron beam, defined by $\mathbf{P}$, due to the interaction with the target system; we must average the initial spin state of the beam over all possible scattering processes and sum over all possible final states. Thus†

$$\mathbf{P}' \propto \mathrm{Tr}\,\hat{\rho}\hat{v}^{+}\hat{\boldsymbol{\sigma}}\hat{v}.$$

The constant of proportionality is determined by normalization; viz.

$$\mathbf{P}' = \mathrm{Tr}\,\hat{\rho}\hat{v}^{+}\hat{\boldsymbol{\sigma}}\hat{v}/\mathrm{Tr}\,\hat{\rho}\hat{v}^{+}\hat{v}. \qquad (10.32)$$

In full we have

$$\mathbf{P}'\left(\frac{\mathrm{d}^{2}\sigma}{\mathrm{d}\Omega\mathrm{d}E'}\right) = \left(\frac{m}{2\pi\hbar^{2}}\right)^{2}\frac{k'}{k}\sum_{\lambda,\lambda'} p_{\lambda}\,\mathrm{Tr}\,\hat{\rho}\langle\lambda|\,\hat{V}^{+}(\boldsymbol{\varkappa})\,|\lambda'\rangle\hat{\boldsymbol{\sigma}}\times$$

$$\times\langle\lambda'|\,\hat{V}(\boldsymbol{\varkappa})\,|\lambda\rangle\delta(\hbar\omega+E_{\lambda}-E_{\lambda'}). \qquad (10.33)$$

Let us examine the structure of eqns (10.30) and (10.32) with the general form of $\hat{v}$ given by eqn (10.26). For the cross-section we need to evaluate (10.30),

$$\mathrm{Tr}\,\hat{\rho}\hat{v}^{+}\hat{v} = \tfrac{1}{2}\mathrm{Tr}(I+\mathbf{P}.\hat{\boldsymbol{\sigma}})(\hat{\beta}^{+}+\hat{\boldsymbol{\alpha}}^{+}.\hat{\boldsymbol{\sigma}})(\hat{\beta}+\hat{\boldsymbol{\alpha}}.\hat{\boldsymbol{\sigma}})$$

$$= \tfrac{1}{2}\mathrm{Tr}(I+\mathbf{P}.\hat{\boldsymbol{\sigma}})(\hat{\beta}^{+}\hat{\beta}+\hat{\beta}^{+}\hat{\boldsymbol{\alpha}}.\hat{\boldsymbol{\sigma}}+\hat{\boldsymbol{\alpha}}^{+}.\hat{\boldsymbol{\sigma}}\hat{\beta}+\hat{\boldsymbol{\alpha}}^{+}.\hat{\boldsymbol{\sigma}}\hat{\boldsymbol{\alpha}}.\hat{\boldsymbol{\sigma}}). \qquad (10.34)$$

† Compare Williams, loc. cit., p. 234.

The evaluation of the trace of products of Pauli operators is facilitated by making use of the identity

$$\hat{\sigma}^{\alpha}\hat{\sigma}^{\beta} = \delta_{\alpha,\beta} I + \mathrm{i} \sum_{\gamma} \epsilon^{\alpha\beta\gamma}\hat{\sigma}^{\gamma}, \tag{10.35}$$

where $\epsilon^{\alpha\beta\gamma}$ is the completely antisymmetrical tensor with three indices, i.e.

$$\begin{aligned} \epsilon^{\alpha\beta\gamma} &= +1 \quad \text{if } \alpha, \beta, \text{ and } \gamma \text{ are in cyclic order,} \\ &= -1 \quad \text{if } \alpha, \beta, \text{ and } \gamma \text{ are not in cyclic order,} \\ &= 0 \quad \text{otherwise.} \end{aligned}$$

This identity is derived by using the commutation relation for angular momentum operators,

$$[\hat{L}^{\alpha}, \hat{L}^{\beta}] = \mathrm{i} \sum_{\gamma} \epsilon^{\alpha\beta\gamma}\hat{L}^{\gamma}$$

and the fact that Pauli operators anticommute,

$$\hat{\sigma}^{\alpha}\hat{\sigma}^{\beta} + \hat{\sigma}^{\beta}\hat{\sigma}^{\alpha} = 2\delta_{\alpha\beta} I.$$

Because Tr $\hat{\sigma}^{\gamma}$ is zero, we obtain from (10.35)

$$\mathrm{Tr}\, \hat{\sigma}^{\alpha}\hat{\sigma}^{\beta} = 2\delta_{\alpha,\beta}. \tag{10.11}$$

Also, with the aid of this result,

$$\begin{aligned} \mathrm{Tr}\, \hat{\sigma}^{\alpha}\hat{\sigma}^{\beta}\hat{\sigma}^{\gamma} &= \mathrm{i} \sum_{\gamma'} \epsilon^{\alpha\beta\gamma'}\, \mathrm{Tr}\, \hat{\sigma}^{\gamma'}\hat{\sigma}^{\gamma} \\ &= 2\mathrm{i}\epsilon^{\alpha\beta\gamma}. \end{aligned} \tag{10.36}$$

With the aid of these formulae we find for (10.34) the result,

$$\mathrm{Tr}\, \hat{\rho}\hat{v}^{+}\hat{v} = \hat{\boldsymbol{\alpha}}^{+}.\hat{\boldsymbol{\alpha}} + \hat{\beta}^{+}\hat{\beta} + \hat{\beta}^{+}(\hat{\boldsymbol{\alpha}}.\mathbf{P}) + (\hat{\boldsymbol{\alpha}}^{+}.\mathbf{P})\hat{\beta} + \mathrm{i}\mathbf{P}.(\hat{\boldsymbol{\alpha}}^{+} \times \hat{\boldsymbol{\alpha}}). \tag{10.37}$$

When it is adequate to consider only nuclear and magnetic scattering, an examination of the corresponding interaction potentials, (10.21) and (10.23), shows that $\hat{\beta}$ contains only nuclear terms and $\hat{\boldsymbol{\alpha}}$ both nuclear and magnetic scattering. Thus the third and fourth terms in (10.37) result in interference between nuclear and magnetic scattering. This interference has been utilized to perform very accurate measurements of magnetic structure factors. Bearing in mind that the cross-section must be averaged over the orientations of the nuclei, it can be seen that for randomly orientated nuclei the last term on the right-hand side of (10.37) is purely magnetic.

In order to evaluate

$$\mathrm{Tr}\, \hat{\rho}\hat{v}^{+}\hat{\boldsymbol{\sigma}}\hat{v} \tag{10.38}$$

as required in the formula for $\mathbf{P}'$, eqn (10.32), we need, in addition to (10.12) and (10.36) the identity

$$\mathrm{Tr}\, \hat{\sigma}^{\alpha}\hat{\sigma}^{\beta}\hat{\sigma}^{\gamma}\hat{\sigma}^{\delta} = 2(\delta_{\alpha\beta}\delta_{\gamma\delta} - \delta_{\alpha\gamma}\delta_{\beta\delta} + \delta_{\alpha\delta}\delta_{\beta\gamma}), \tag{10.39}$$

which can be derived from (10.35). Also note that, if **A** and **B** are arbitrary vectors

$$(\mathbf{A}\times\mathbf{B})_\alpha = \sum_{\beta,\gamma} \epsilon^{\alpha\beta\gamma} A_\beta B_\gamma. \tag{10.40}$$

With these results,

$$\begin{aligned}\operatorname{Tr}\hat{\rho}\hat{v}^+\hat{\boldsymbol{\sigma}}\hat{v} = {} & \hat{\beta}^+\hat{\boldsymbol{\alpha}}+\hat{\boldsymbol{\alpha}}^+\hat{\beta}+\hat{\beta}^+\hat{\beta}\mathbf{P}+\hat{\boldsymbol{\alpha}}^+(\hat{\boldsymbol{\alpha}}.\mathbf{P})+(\hat{\boldsymbol{\alpha}}^+.\mathbf{P})\hat{\boldsymbol{\alpha}}- \\ & -\mathbf{P}(\hat{\boldsymbol{\alpha}}^+.\hat{\boldsymbol{\alpha}})-\mathrm{i}\hat{\boldsymbol{\alpha}}^+\times\hat{\boldsymbol{\alpha}}+\mathrm{i}\hat{\beta}^+(\hat{\boldsymbol{\alpha}}\times\mathbf{P})+\mathrm{i}(\mathbf{P}\times\hat{\boldsymbol{\alpha}}^+)\hat{\beta}.\end{aligned} \tag{10.41}$$

If we again consider just nuclear and magnetic scattering, we see that in the limit of an unpolarized incident beam of neutrons there is a creation of polarization by a purely magnetic term (for randomly orientated nuclei), $-\mathrm{i}\hat{\boldsymbol{\alpha}}^+\times\hat{\boldsymbol{\alpha}}$.

To sum up, we use (10.37) in conjunction with (10.31) to give the cross-sections and we use (10.41) in conjunction with (10.33) to give the polarization $\mathbf{P}'$ of the scattered neutrons. We shall discuss how to do this first for elastic scattering and then for inelastic magnetic scattering. We begin our discussion of elastic scattering by examining the scattering of neutrons by the electrostatic fields created by the nuclei and atomic electrons in a solid.

10.3. Neutron scattering by the Coulomb field of the nuclei and atomic electrons in a solid

If the nucleus at the site **d** in a unit cell has a charge eZ_d then the total charge density at position **r** due to all the nuclei in the solid is

$$\sum_{\mathbf{l},\mathbf{d}} eZ_d\,\delta(\mathbf{r}-\mathbf{R}_{ld}). \tag{10.42}$$

Since the charge density at **r** due to the atomic electrons is simply

$$-e\sum_i \delta(\mathbf{r}-\mathbf{r}_i),$$

the total charge density $n(\mathbf{r})$ is

$$n(\mathbf{r}) = e\Big\{\sum_{\mathbf{l},\mathbf{d}} Z_d\,\delta(\mathbf{r}-\mathbf{R}_{ld})-\sum_i \delta(\mathbf{r}-\mathbf{r}_i)\Big\}. \tag{10.43}$$

Because the crystal is electrostatically neutral we must have

$$\int \mathrm{d}\mathbf{r}\; n(\mathbf{r}) = 0. \tag{10.44}$$

The electrostatic potential $\phi(\mathbf{r})$ of the charge density $n(\mathbf{r})$ is determined by†

$$\nabla^2\phi(\mathbf{r}) = -4\pi n(\mathbf{r}) \tag{10.45}$$

and the electric field **E** satisfies

$$\mathbf{E} = -\boldsymbol{\nabla}\phi(\mathbf{r}). \tag{10.46}$$

† L. D. Landau and E. M. Lifshitz, *Electrodynamics of continuous media.*

The interaction potential between a neutron and the charge distribution $n(\mathbf{r})$ consists of two terms (cf. Chapter 5), namely

$$\hat{V}_{\mathrm{E}}(\mathbf{r}) = (-\gamma\mu_{\mathrm{N}}/mc)\hat{\boldsymbol{\sigma}}.\mathbf{E}\times\hat{\mathbf{p}}+(-\hbar\gamma\mu_{\mathrm{N}}/2mc)\boldsymbol{\nabla}.\mathbf{E}. \tag{10.47}$$

The first term on the right-hand side is usually referred to as the spin–orbit interaction while the second term is the neutron analogue of the Darwin interaction for electrons. The existence of this term for neutrons in an electrostatic field was first pointed out by Foldy.

For the contribution to the cross-section from the spin–orbit term we need

$$\int \mathrm{d}\mathbf{r}\,\exp(-\mathrm{i}\mathbf{k}'.\mathbf{r})(-\gamma\mu_{\mathrm{N}}/mc)\hat{\boldsymbol{\sigma}}.\mathbf{E}\times\hat{\mathbf{p}}\exp(\mathrm{i}\mathbf{k}.\mathbf{r})$$
$$=(-\gamma\mu_{\mathrm{N}}/mc)\hat{\boldsymbol{\sigma}}.\int \mathrm{d}\mathbf{r}\,\exp(\mathrm{i}\boldsymbol{\varkappa}.\mathbf{r})\mathbf{E}\times\hbar\mathbf{k}. \tag{10.48}$$

The Fourier integral of the electric field in (10.48) is readily effected by the following argument. From (10.45) the potential $\phi(\mathbf{r})$ is given by

$$\phi(\mathbf{r}) = \int \mathrm{d}\mathbf{r}'\,\frac{n(\mathbf{r}')}{|\mathbf{r}-\mathbf{r}'|}. \tag{10.49}$$

If $n(\boldsymbol{\varkappa})$ is the Fourier transform of the charge density, i.e.

$$n(\boldsymbol{\varkappa}) = \int \mathrm{d}\mathbf{r}\,\exp(-\mathrm{i}\boldsymbol{\varkappa}.\mathbf{r})n(\mathbf{r}) \tag{10.50}$$

then (10.49) shows that

$$\phi(\mathbf{r}) = \frac{1}{2\pi^2}\int \mathrm{d}\boldsymbol{\varkappa}\,\exp(\mathrm{i}\boldsymbol{\varkappa}.\mathbf{r})\frac{n(\boldsymbol{\varkappa})}{|\boldsymbol{\varkappa}|^2}. \tag{10.51}$$

In obtaining (10.51) from (10.49) it is convenient to use the integral

$$\int \mathrm{d}\mathbf{r}'\,\frac{\exp(\mathrm{i}\boldsymbol{\varkappa}.\mathbf{r}')}{|\mathbf{r}-\mathbf{r}'|} = \frac{4\pi}{|\boldsymbol{\varkappa}|^2}\exp(\mathrm{i}\boldsymbol{\varkappa}.\mathbf{r}).$$

Hence we have now

$$\int \mathrm{d}\mathbf{r}\,\exp(\mathrm{i}\boldsymbol{\varkappa}.\mathbf{r})\mathbf{E}\times\hbar\mathbf{k} = \int \mathrm{d}\mathbf{r}\,\exp(\mathrm{i}\boldsymbol{\varkappa}.\mathbf{r})\{-\boldsymbol{\nabla}\phi(\mathbf{r})\}\times\hbar\mathbf{k}$$
$$= \frac{\mathrm{i}(2\pi)^3}{2\pi^2}\frac{n(-\boldsymbol{\varkappa})}{|\boldsymbol{\varkappa}|^2}(\boldsymbol{\varkappa}\times\hbar\mathbf{k})$$
$$= -\frac{4\pi\mathrm{i}\hbar}{|\boldsymbol{\varkappa}|^2}n(-\boldsymbol{\varkappa})(\mathbf{k}'\times\mathbf{k}). \tag{10.52}$$

Noting that for elastic scattering $|\boldsymbol{\varkappa}| = 2k\sin\frac{1}{2}\theta$, (10.48) reads

$$\left(\frac{2\pi\hbar^2}{m}\right)\left(\frac{\gamma e}{2mc^2}\right)n(-\boldsymbol{\varkappa})\mathrm{i}\cot(\tfrac{1}{2}\theta)\tilde{\mathbf{n}}.\hat{\boldsymbol{\sigma}}, \tag{10.53}$$

where the unit vector $\tilde{\mathbf{n}}$ perpendicular to the scattering plane is defined in (10.25).

The evaluation of the contribution to the cross-section from the Foldy interaction is straightforward. From the definition of $n(\boldsymbol{\varkappa})$,

$$\int \mathrm{d}\mathbf{r}\,\exp(\mathrm{i}\boldsymbol{\varkappa}.\mathbf{r})\boldsymbol{\nabla}.\mathbf{E} = \int \mathrm{d}\mathbf{r}\,\exp(\mathrm{i}\boldsymbol{\varkappa}.\mathbf{r})\{4\pi n(\mathbf{r})\}$$
$$= 4\pi n(-\boldsymbol{\varkappa}). \qquad (10.54)$$

Therefore from (10.53) and (10.54)

$$\hat{V}_{\mathrm{E}}(\boldsymbol{\varkappa}) = \int \exp(-\mathrm{i}\mathbf{k}'.\mathbf{r})\hat{V}_{\mathrm{E}}(\mathbf{r})\exp(\mathrm{i}\mathbf{k}.\mathbf{r})\,\mathrm{d}\mathbf{r}$$
$$= \left(\frac{2\pi\hbar^2}{m}\right)\left(\frac{m_{\mathrm{e}}}{2m}\right)\left(\frac{\gamma e}{m_{\mathrm{e}}c^2}\right)n(-\boldsymbol{\varkappa})\{\mathrm{i}\cot(\tfrac{1}{2}\theta)\tilde{\mathbf{n}}.\hat{\boldsymbol{\sigma}}-1\}. \qquad (10.55)$$

By analogy with the scattering of neutrons by the unpaired electrons of an ion, we write for the matrix element

$$\langle\lambda|\,n(-\boldsymbol{\varkappa})\,|\lambda\rangle = e\sum_{\mathbf{l},d}\exp(\mathrm{i}\boldsymbol{\varkappa}.\mathbf{R}_{ld})\{Z_d - f_d(\boldsymbol{\varkappa})\}$$
$$= e\sum_{\mathbf{l}}\exp(\mathrm{i}\boldsymbol{\varkappa}.\mathbf{l})F_{\mathrm{E}}(\boldsymbol{\varkappa}), \qquad (10.56)$$

where $f_d(\boldsymbol{\varkappa})$ is the atomic form factor of the *charge density* associated with the atom at $\mathbf{R}_{ld}$, and $F_{\mathrm{E}}(\boldsymbol{\varkappa})$ is the *electrostatic unit-cell structure factor*.

The form factor entering in (10.56) is that encountered in X-ray scattering. Note that in view of (10.44) we have $f_d(0) = Z_d$.

In the notation of § 10.2, we see from (10.55) that

$$\langle\lambda|\,\hat{\boldsymbol{\alpha}}\,|\lambda\rangle \equiv \tilde{\mathbf{n}}\left(\frac{2\pi\hbar^2}{m}\right)\left(\frac{m_{\mathrm{e}}}{2m}\right)\left(\frac{\gamma e^2}{m_{\mathrm{e}}c^2}\right)\mathrm{i}\cot(\tfrac{1}{2}\theta)\sum_{\mathbf{l}}\exp(\mathrm{i}\boldsymbol{\varkappa}.\mathbf{l})F_{\mathrm{E}}(\boldsymbol{\varkappa})$$

and
$$\langle\lambda|\,\hat{\beta}\,|\lambda\rangle \equiv -\left(\frac{2\pi\hbar^2}{m}\right)\left(\frac{m_{\mathrm{e}}}{2m}\right)\left(\frac{\gamma e^2}{m_{\mathrm{e}}c^2}\right)\sum_{\mathbf{l}}\exp(\mathrm{i}\boldsymbol{\varkappa}.\mathbf{l})F_{\mathrm{E}}(\boldsymbol{\varkappa}).$$

By using these results in our general expression for the neutron spin-dependent part of the elastic cross-section, eqn (10.37), we obtain for the cross-section for the elastic scattering of polarized neutrons by the Coulomb field of the nuclei and atomic electrons in a solid the result

$$\left(\frac{\mathrm{d}\sigma}{\mathrm{d}\Omega}\right)_{\mathrm{EE}} = \left|\sum_{\mathbf{l}}\exp(\mathrm{i}\boldsymbol{\varkappa}.\mathbf{l})\right|^2\left(\frac{m_{\mathrm{e}}}{2m}\right)^2\left(\frac{\gamma e^2}{m_{\mathrm{e}}c^2}\right)^2(\sin\tfrac{1}{2}\theta)^{-2}|F_{\mathrm{E}}(\boldsymbol{\varkappa})|^2. \qquad (10.57)$$

The scattering is therefore confined to nuclear Bragg peaks, a result that is simply a consequence of the fact that the neutrons are scattered by a (rigid) periodic array of potentials. That $(\mathrm{d}\sigma/\mathrm{d}\Omega)_{\mathrm{EE}}$ is independent of the polarization $\mathbf{P}$ of the incident beam is an example of the general principle stated at the beginning of the chapter, namely, that the cross-section is independent of $\mathbf{P}$ if the target system does not possess a preferred axis. The cross-section for the electrostatic scattering of neutrons is seen from (10.57) to be very small. We recall that

$$(\gamma e^2/m_e c^2)^2 = 0{\cdot}292\ \text{b};$$

hence (10.57) is of order $(m_e/m)^2 = (1/1839)^2$

smaller than the cross-section for purely magnetic scattering.

We do not pause to calculate the polarization of the scattered beam but go on to consider the scattering of polarized neutrons through their nuclear interaction with a crystal.

10.4. Nuclear scattering

The cross-section for purely nuclear scattering of a beam of polarized neutrons is independent of the polarization **P** because unless the temperature is exceptionally low the nuclei are randomly oriented and thus the crystal does not possess a preferred axis. However, to gain some familiarity with the formulae it is worth while to demonstrate this and to ask for the polarization of the scattered beam.

From § 10.2 we know that

$$\hat{V}_{\mathrm{N}}(\boldsymbol{\kappa}) = \left(\frac{2\pi\hbar^2}{m}\right)\sum_{\mathbf{l},\mathbf{d}} \exp(\mathrm{i}\boldsymbol{\kappa}.\mathbf{R}_{ld})(A_{ld}+\tfrac{1}{2}B_{ld}\,\hat{\boldsymbol{\sigma}}.\hat{\mathbf{I}}_{ld}), \qquad (10.58)$$

so, from (10.26),

$$\hat{\boldsymbol{\alpha}} \equiv \left(\frac{2\pi\hbar^2}{m}\right)\sum_{\mathbf{l},\mathbf{d}} \exp(\mathrm{i}\boldsymbol{\kappa}.\mathbf{R}_{ld})\tfrac{1}{2}B_{ld}\,\hat{\mathbf{I}}_{ld}$$

and

$$\hat{\beta} \equiv \left(\frac{2\pi\hbar^2}{m}\right)\sum_{\mathbf{l},\mathbf{d}} \exp(\mathrm{i}\boldsymbol{\kappa}\cdot\mathbf{R}_{ld})A_{ld}. \qquad (10.59)$$

Now suppose (10.59) is inserted into (10.37) and (10.37) in turn inserted into (10.31). Then in (10.31) one operation we must perform is to average over nuclear spin orientations and isotope distributions. Because we assume the energy and all other properties of the target are independent of these factors we can do this immediately and it is convenient to do this step on (10.37) directly (rather than inserting it into (10.31) and then averaging).

For a discussion of polarization effects it is necessary to have an explicit notation for averaging over nuclear spin orientations separately from averaging over both orientations and isotope distributions. We shall denote $\mathcal{O}$ = *average over nuclear spin orientations*.

Now consider this averaging process over (10.37), i.e. consider

$$\mathcal{O}\,\mathrm{Tr}\{\hat{\rho}\hat{v}^{+}\hat{v}\}.$$

We notice that (10.37) contains a term $\hat{\beta}^{+}(\hat{\boldsymbol{\alpha}}.\mathbf{P})$ and from (10.59) this is linear in the nuclear spins $\hat{\mathbf{I}}_{ld}$. Hence

$$\mathcal{O}\{\hat{\beta}^{+}(\hat{\boldsymbol{\alpha}}.\mathbf{P})\} = 0. \tag{10.60 a}$$

Similarly

$$\mathcal{O}\{(\hat{\boldsymbol{\alpha}}^{+}.\mathbf{P})\hat{\beta}\} = 0. \tag{10.60 b}$$

We also notice that (10.37) contains a term $\mathrm{i}\mathbf{P}.\hat{\boldsymbol{\alpha}}^{+}\times\hat{\boldsymbol{\alpha}}$ and from (10.59) we see this involves expressions like

$$\hat{\mathbf{I}}_{ld}\times\hat{\mathbf{I}}_{l'd'} = \delta_{\mathbf{l},\mathbf{l}'}\delta_{\mathbf{d},\mathbf{d}'}\,\mathrm{i}\,\hat{\mathbf{I}}_{ld}.$$

Since this also is linear in the nuclear spins

$$\mathcal{O}\{\mathrm{i}\mathbf{P}.\hat{\boldsymbol{\alpha}}^{+}\times\hat{\boldsymbol{\alpha}}\} = 0. \tag{10.60 c}$$

We therefore see that the three terms in (10.37) which involve the polarization all vanish. It follows that the cross-section itself is independent of $\mathbf{P}$, as we had anticipated. We are left only with the terms $\hat{\boldsymbol{\alpha}}^{+}.\hat{\boldsymbol{\alpha}}$ and $\hat{\beta}^{+}\hat{\beta}$. The latter is independent of the $\mathcal{O}$ operation, hence

$$\begin{aligned}\mathcal{O}\,\mathrm{Tr}\{\hat{\rho}\hat{v}^{+}\hat{v}\} &= \hat{\beta}^{+}\hat{\beta}+\mathcal{O}\{\hat{\boldsymbol{\alpha}}^{+}.\hat{\boldsymbol{\alpha}}\}\\ &= \left(\frac{2\pi\hbar^{2}}{m}\right)^{2}\sum_{\mathbf{l},\mathbf{d},\mathbf{l}',\mathbf{d}'}\exp\{\mathrm{i}\boldsymbol{\kappa}(\mathbf{R}_{ld}-\mathbf{R}_{l'd'})\}\times\\ &\qquad\times\{A^{*}_{l'd'}A_{ld}+\tfrac{1}{4}\mathcal{O}(B^{*}_{l'd'}B_{ld}\hat{\mathbf{I}}_{l'd'}.\hat{\mathbf{I}}_{ld})\}\\ &= \left(\frac{2\pi\hbar^{2}}{m}\right)^{2}\sum_{\mathbf{l},\mathbf{d},\mathbf{l}',\mathbf{d}'}\exp(-\mathrm{i}\boldsymbol{\kappa}.\mathbf{R}_{l'd'})\exp(\mathrm{i}\boldsymbol{\kappa}.\mathbf{R}_{ld})\times\\ &\qquad\times[|\bar{A}_{d}|^{2}+\delta_{\mathbf{l},\mathbf{l}'}\delta_{\mathbf{d},\mathbf{d}'}\{\overline{|A_{d}|^{2}}-|\bar{A}_{d}|^{2}+\tfrac{1}{4}\overline{|B_{d}|^{2}i_{d}(i_{d}+1)}\}],\end{aligned} \tag{10.61}$$

where in the last step we have also averaged over isotope distributions. It is easily shown that (10.61) reproduces the results of Chapter 1.

We now suppose the incident beam is polarized and calculate the final polarization using (10.33) and (10.41). We first evaluate (10.41),

$$\mathcal{O}\{\mathrm{Tr}\,\hat{\rho}\hat{v}^{+}\hat{\boldsymbol{\sigma}}\hat{v}\}$$

and, as in the previous example, note that all the terms linear in $\hat{\boldsymbol{\alpha}}$ vanish and so also does the term $\mathrm{i}\hat{\boldsymbol{\alpha}}^{+}\times\hat{\boldsymbol{\alpha}}$. Furthermore,

$$\mathcal{O}\{\hat{\boldsymbol{\alpha}}^{+}(\hat{\boldsymbol{\alpha}}.\mathbf{P})\} = \mathcal{O}\{(\hat{\boldsymbol{\alpha}}^{+}.\mathbf{P})\hat{\boldsymbol{\alpha}}\} = \tfrac{1}{3}\mathbf{P}\mathcal{O}\{\hat{\boldsymbol{\alpha}}^{+}.\hat{\boldsymbol{\alpha}}\}.$$

Hence

$$\mathcal{O}\{\mathrm{Tr}\,\hat{\rho}\hat{v}^{+}\hat{\boldsymbol{\sigma}}\hat{v}\} = \mathbf{P}\{\hat{\beta}^{+}\hat{\beta}-\tfrac{1}{3}\mathcal{O}(\hat{\boldsymbol{\alpha}}^{+}.\hat{\boldsymbol{\alpha}})\}$$

and by comparison with the derivation of (10.61) this gives, after averaging over isotope distributions,

$$\begin{aligned}&\left(\frac{2\pi\hbar^{2}}{m}\right)^{2}\sum_{\mathbf{l},\mathbf{d},\mathbf{l}',\mathbf{d}'}\exp\{\mathrm{i}\boldsymbol{\kappa}(\mathbf{R}_{ld}-\mathbf{R}_{l'd'})\}\times\\ &\qquad\times\mathbf{P}[|\bar{A}_{d}|^{2}+\delta_{\mathbf{l},\mathbf{l}'}\delta_{\mathbf{d},\mathbf{d}'}\{\overline{|A_{d}|^{2}}-|\bar{A}_{d}|^{2}-\tfrac{1}{12}\overline{|B_{d}|^{2}i_{d}(i_{d}+1)}\}].\end{aligned} \tag{10.62}$$

Comparing (10.62) with (10.61) term by term we note that: (1) the nuclear coherent scattering has a final polarization equal to the initial polarization **P**; (2) the nuclear incoherent scattering due to the random isotope distributions also has a final polarization **P**; (3) the nuclear incoherent scattering due to the random nuclear spin orientations has a polarization $-\frac{1}{3}\mathbf{P}$. Physically these results are easy to understand. Neither (1) nor (2) can possibly involve a neutron spin-flip and therefore the polarization is unchanged. However, for (3) one-third of the interaction, namely $\sigma_z\, \hat{i}_z$, scatters the neutron without spin-flip whereas two-thirds, namely $\hat{\sigma}_x \hat{i}_x + \hat{\sigma}_y \hat{i}_y$, scatters with spin-flip and hence gives a final polarization of opposite sign and one-third the magnitude of the initial polarization.

In general we have incoherent scattering due to both random isotope distributions and nuclear spin orientations and therefore the final polarization of the incoherent scattering lies between the limits of **P** and $-\frac{1}{3}\mathbf{P}$. From (10.61) and (10.62) we see that

$$\mathbf{P}'_{\text{coh}} = \mathbf{P} \tag{10.63}$$

and

$$\mathbf{P}'_{\text{incoh}} = \mathbf{P}\,\frac{\sum_{\mathbf{d}}\{\overline{|A_d|^2} - |\bar{A}_d|^2 - \frac{1}{12}\overline{|B_d|^2 i_d(i_d+1)}\}}{\sum_{\mathbf{d}}\{\overline{|A_d|^2} - |\bar{A}_d|^2 + \frac{1}{4}\overline{|B_d|^2 i_d(i_d+1)}\}}. \tag{10.64}$$

We now remember, from Chapter 1, that

$$\bar{A}_d = \bar{b}_d \tag{10.65}$$

and

$$\overline{|A_d|^2} + \tfrac{1}{4}\overline{|B_d|^2 i_d(i_d+1)} = \overline{|b_d|^2} \tag{10.66}$$

and rewrite (10.64) as

$$\mathbf{P}'_{\text{incoh}} = \mathbf{P}\,\frac{\sum_{\mathbf{d}}\{\frac{4}{3}\overline{|A_d|^2} - |\bar{b}_d|^2 - \frac{1}{3}\overline{|b_d|^2}\}}{\sum_{\mathbf{d}}\{\overline{|b_d|^2} - |\bar{b}_d|^2\}}, \tag{10.67}$$

where the quantities $\bar{b}_d$ and $\overline{|b_d|^2}$ are defined in Chapter 1 and from the definition in this section

$$\overline{|A|^2} = \sum_{\xi} c_{\xi} |A_{\xi}|^2. \tag{10.68}$$

The experimental arrangement used in polarization analysis work is illustrated in Fig. 10.1. The actual layout shown is that used with the High Flux Isotope Reactor at Oak Ridge National Laboratory. It consists of a triple axis spectrometer with a polarizer and an analyser made from Co–Fe crystals. With this scheme an energy analysis of the scattered neutrons can be made, as with any triple-axis instrument, and, in addition, production of polarized neutrons and selection of a particular polarization state of the scattered beam. The polarization of the incident and scattered beams can be reversed by activating the

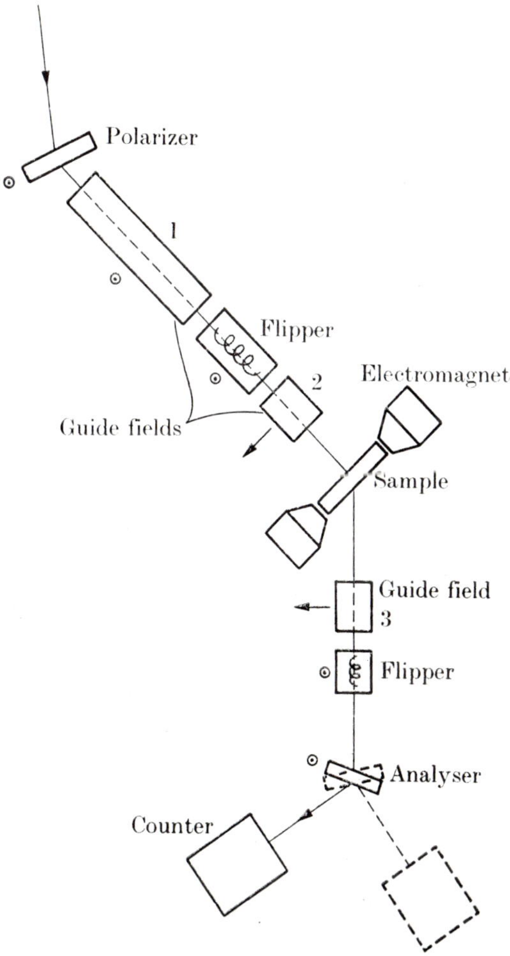

FIG. 10.1. Experimental layout used in polarization analysis experiments with HFIR at Oak Ridge National Laboratory.

(Figs. 10.1, 2, 3, 5, 6, 7, 8, and 9 are from lecture by R. M. Moon at Kjeller, Netherlands–Norwegian Reactor School and R. M. Moon, T. Riste, and W. C. Koehler, *Phys. Rev.* **181**, 920.)

r.f. coils, or flippers. The electromagnet around the target and the guide fields 2 and 3 can be rotated so that the polarization is either perpendicular or parallel to the scattering vector $\boldsymbol{\varkappa}$. In detail the experimental scheme is as follows.

With both flippers off, the beam of neutrons incident on the target system is polarized in a direction perpendicular to the scattering plane ($\uparrow$) and only those scattered neutrons in the same polarization state are reflected by the analysing crystal. With the first flipper on the neutrons incident on the target are in the $\downarrow$ state and the analyser affects only those scattered neutrons whose polarization is oppositely orientated to

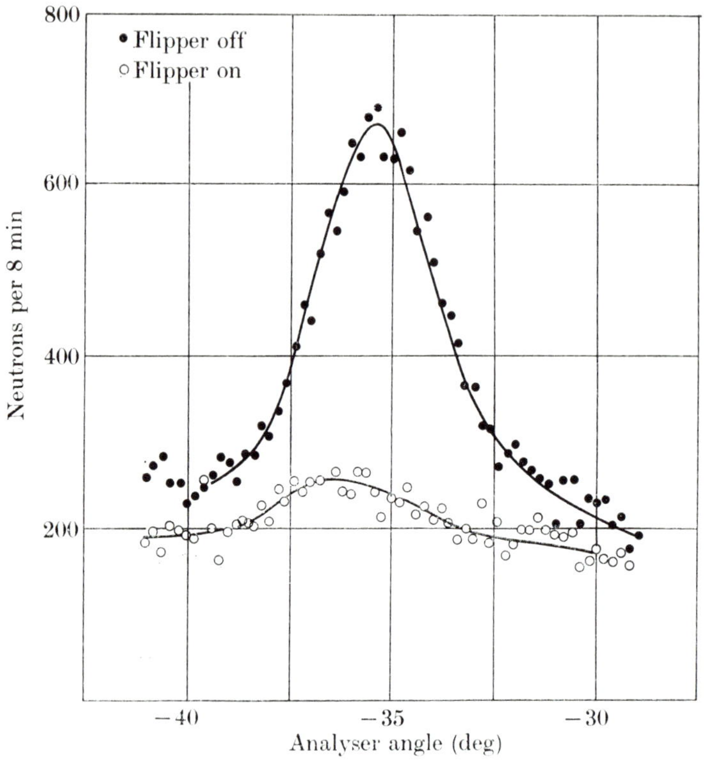

FIG. 10.2. Isotopic incoherent scattering from polycrystalline Ni.

that of the incident neutrons. Similarly, if just the second flipper is on, the measured cross-section is that for the $\uparrow$–$\downarrow$ transition. With both flippers on the measured cross-section corresponds to $\downarrow$–$\downarrow$ transitions.

An example of isotopic incoherent scattering is shown in Fig. 10.2. This gives the number of neutrons scattered from polycrystalline nickel for the two cases where neither flipper was activated and where one flipper was activated. All nickel isotopes have zero spin so our discussion given above tells us that the incoherent nuclear scattering has the same magnitude as for the scattering of unpolarized neutrons and the polarization of the scattered beam is parallel to that of the incident beam. These facts are clearly borne out by the experiment. The small peak in the flipper on measurements is magnon scattering.

Because there is only one vanadium isotope we expect $\frac{1}{3}$ of the incoherent scattering to be without spin-flip and $\frac{2}{3}$ with spin-flip. The measurements shown in Fig. 10.3 verify that this is so, independent of whether the incident beam is polarized parallel or perpendicular to $\boldsymbol{\varkappa}$.

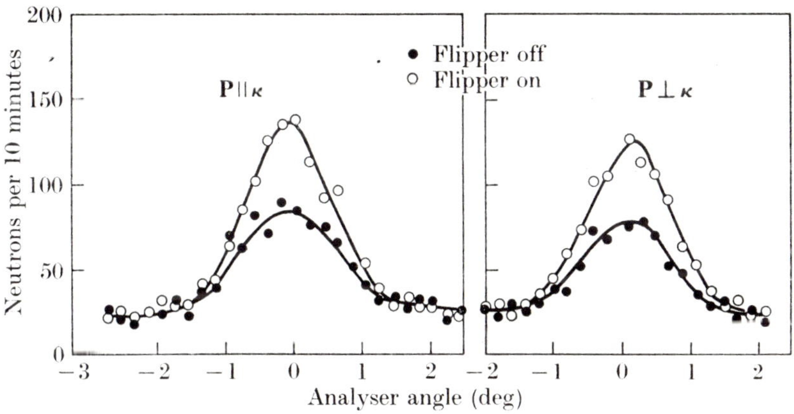

FIG. 10.3. Nuclear spin incoherent scattering from vanadium.

In the previous section we considered the electrostatic scattering of polarized neutrons, so let us now ask for the cross-section when we have both electrostatic and nuclear scattering. For this case

$$\langle\lambda|\,\hat{V}(\boldsymbol{\varkappa})\,|\lambda\rangle = \langle\lambda|\,\hat{V}_{\mathrm{E}}(\boldsymbol{\varkappa})+\hat{V}_{\mathrm{N}}(\boldsymbol{\varkappa})\,|\lambda\rangle \tag{10.69}$$

and so from (10.55) and (10.58)

$$\langle\lambda|\,\hat{\boldsymbol{\alpha}}\,|\lambda\rangle \equiv \left(\frac{2\pi\hbar^2}{m}\right)\sum_{\mathbf{l}}\exp(\mathrm{i}\boldsymbol{\varkappa}.\mathbf{l})\left\{\sum_{\mathbf{d}}\exp(\mathrm{i}\boldsymbol{\varkappa}.\mathbf{d})\tfrac{1}{2}B_{ld}\,\hat{\mathbf{I}}_{ld}+\right.$$
$$\left.+\left(\frac{\gamma e^2}{2mc^2}\right)\mathrm{i}\cot(\tfrac{1}{2}\theta)F_{\mathrm{E}}\,(\boldsymbol{\varkappa})\tilde{\mathbf{n}}\right\} \tag{10.70}$$

and

$$\langle\lambda|\hat{\beta}\,|\lambda\rangle \equiv \left(\frac{2\pi\hbar^2}{m}\right)\sum_{\mathbf{l}}\exp(\mathrm{i}\boldsymbol{\varkappa}.\mathbf{l})\left\{\sum_{\mathbf{d}}\exp(\mathrm{i}\boldsymbol{\varkappa}.\mathbf{d})A_{ld}-\left(\frac{\gamma e^2}{2mc^2}\right)F_{\mathrm{E}}\,(\boldsymbol{\varkappa})\right\}.$$

When we use the result (10.70) in (10.37) we shall, of course, recover the nuclear coherent, nuclear incoherent, and Coulomb cross-sections separately. In addition we now look for interference effects. No interference effects can come from the term $\hat{\boldsymbol{\alpha}}^{+}.\hat{\boldsymbol{\alpha}}$ because any such term would be linear in the nuclear spins $\hat{\mathbf{i}}_{ld}$ and therefore vanish with the averaging operating $\mathscr{O}$. For a similar reason no contribution comes from the term $\mathrm{i}\mathbf{P}.\hat{\boldsymbol{\alpha}}^{+}\times\hat{\boldsymbol{\alpha}}$. But the other terms in (10.37) do give an interference effect which, remembering to apply the averaging operation $\mathscr{O}$, adds to the previous results to give in total

$$\frac{\mathrm{d}\sigma}{\mathrm{d}\Omega} = \left(\frac{\mathrm{d}\sigma}{\mathrm{d}\Omega}\right)_{\mathrm{NN}} + \left(\frac{\mathrm{d}\sigma}{\mathrm{d}\Omega}\right)_{\mathrm{EE}} + \left|\sum_{\mathbf{l}} \exp(\mathrm{i}\boldsymbol{\varkappa}.\mathbf{l})\right|^2 \left(\frac{m_{\mathrm{e}}}{m}\right)\left(\frac{\gamma e^2}{m_{\mathrm{e}} c^2}\right) \times$$
$$\times\{\cot(\tfrac{1}{2}\theta)\mathbf{P}.\tilde{\mathbf{n}}\,\mathrm{Im}\,F_{\mathrm{N}}(\boldsymbol{\varkappa})F_{\mathrm{E}}^{*}(\boldsymbol{\varkappa}) - \mathrm{Re}\,F_{\mathrm{N}}(\boldsymbol{\varkappa})F_{\mathrm{E}}^{*}(\boldsymbol{\varkappa})\}. \qquad (10.71)$$

Thus we have in addition to the pure nuclear and pure electrostatic cross-sections two interference terms that contribute at Bragg reflections, one of which depends on the polarizations. The polarization-independent term arises from the Foldy part of the electrostatic interaction potential while the other term arises from the spin–orbit part of the electrostatic interaction. As a consequence of this latter term, containing the factor $\cot\frac{1}{2}\theta$, the Bragg cross-section is asymmetric, i.e. the cross-section differs for scattering to the right and to the left of the incident beam. The presence of the asymmetric term in the scattering from vanadium crystals has been demonstrated by Shull.† He measured the flipping ratios for the (110) vanadium reflection for θ and $-\theta$ and found that for the former setting the ratio was greater than unity and for the latter less than unity, the magnitude of the deviation from unity being the same in the two cases. Shull's results are given in Fig. 10.4.

Because it depends upon the real part of the scattering length, the polarization-independent interference term in (10.71) may well be an order of magnitude greater than the asymmetric term; nevertheless its presence need not be significant in an experiment undertaken to observe the asymmetric contribution. If the coherent nuclear scattering gives the major part of the measured intensity the flipping ratio can be written

$$R = \frac{1-\delta+\delta'}{1-\delta-\delta'},$$

where $$\delta = (\gamma e^2/mc^2)\mathrm{Re}\{F_{\mathrm{N}}(\boldsymbol{\tau})F_{\mathrm{E}}^{*}(\boldsymbol{\tau})\}/|F_{\mathrm{N}}(\boldsymbol{\tau})|^2$$

and $$\delta' = (\gamma e^2/mc^2)\cot(\tfrac{1}{2}\theta)\mathrm{Im}\{F_{\mathrm{N}}(\boldsymbol{\tau})F_{\mathrm{E}}^{*}(\boldsymbol{\tau})\}/|F_{\mathrm{N}}(\boldsymbol{\tau})|^2.$$

Assuming that both δ and δ' are much less than unity,

$$R \simeq 1+2\delta'-\delta^2.$$

Thus if $2\delta'$ is about the limit of what can be measured with any certainty the contribution δ^2 can certainly be neglected.

† C. G. Shull, *Phys. Rev. Lett.* **10**, 297 (1963).

In most cases the contributions to the cross-section coming from the nuclear–electrostatic interference terms are completely negligible. However, Shull pointed out that the presence of the asymmetric term means that all polarized-beam spectrometers are inherently left- and right-handed.

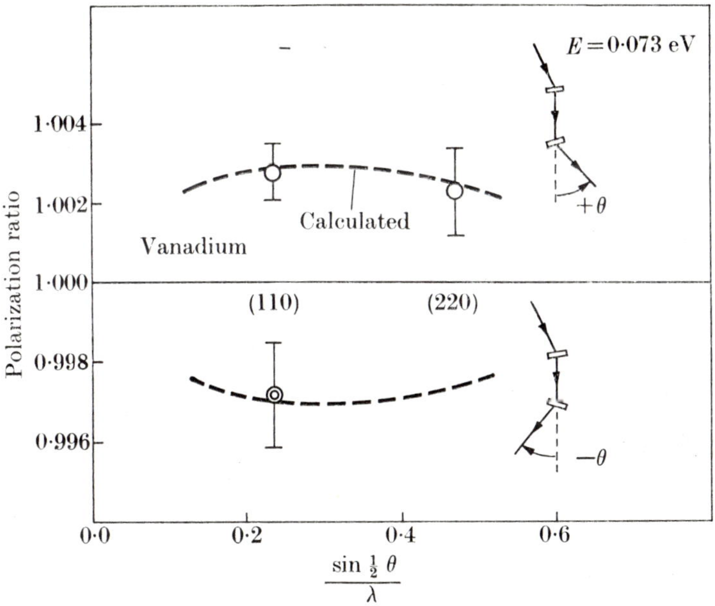

FIG. 10.4. Slow neutron spin–orbit asymmetry in the scattering by vanadium with zero magnetic field.
C. G. Shull, *Phys. Rev. Lett.* **10**, 297 (1963).

If the incident neutrons are unpolarized it is very easy to calculate the polarization of the scattered neutrons, for we have only to consider the terms

$$\hat{\beta}^{+}\hat{\boldsymbol{\alpha}}+\hat{\boldsymbol{\alpha}}^{+}\hat{\beta}-\mathrm{i}\hat{\boldsymbol{\alpha}}^{+}\times\hat{\boldsymbol{\alpha}} \tag{10.72}$$

in (10.41), and all these occur in the cross-section (10.71). Thus we have immediately (for a Bragg reflection)

$$\mathbf{P}'_{\mathrm{NE}} = \frac{(m_{\mathrm{e}}/m)(\gamma e^2/m_{\mathrm{e}}c^2)\cot(\tfrac{1}{2}\theta)\tilde{\mathbf{n}}\,\mathrm{Im}\,F_{\mathrm{N}}(\boldsymbol{\tau})F^{*}_{\mathrm{E}}(\boldsymbol{\tau})}{\{|F_{\mathrm{N}}(\boldsymbol{\tau})|^2+(\gamma e^2/2mc^2)^2(\sin\tfrac{1}{2}\theta)^{-2}|F_{\mathrm{E}}(\boldsymbol{\tau})|^2-(\gamma e^2/mc^2)\mathrm{Re}\,F_{\mathrm{N}}(\boldsymbol{\tau})F^{*}_{\mathrm{E}}(\boldsymbol{\tau})\}}. \tag{10.73}$$

Thus the polarization arising from the interference of the nuclear electrostatic scattering $\mathbf{P}'_{\mathrm{NE}}$ is perpendicular to the scattering plane. The magnitude of $\mathbf{P}'_{\mathrm{NE}}$ is probably largest at the first Bragg peak owing to the factor $\cot(\frac{1}{2}\theta)$.

10.5. Magnetic elastic scattering

(a) Perfect paramagnet

Because such a system does not possess a preferred axis the cross-section for scattering of polarized neutrons is independent of **P** and is given by (5.33). The polarization of the scattered beam **P**′ is of interest and readily calculated.

For purely magnetic scattering the interaction potential $\hat{V}_{\mathrm{M}}(\boldsymbol{\kappa})$ is given by eqn (10.23). From this equation we see that $\hat{\beta} \equiv 0$. Hence, from the general expression (10.41) for **P**′ we have only to calculate the terms corresponding to

$$\hat{\boldsymbol{\alpha}}^{+}(\hat{\boldsymbol{\alpha}}.\mathbf{P})+(\hat{\boldsymbol{\alpha}}^{+}.\mathbf{P})\hat{\boldsymbol{\alpha}}-\mathbf{P}(\hat{\boldsymbol{\alpha}}^{+}.\hat{\boldsymbol{\alpha}})-\mathrm{i}\hat{\boldsymbol{\alpha}}^{+}\times\hat{\boldsymbol{\alpha}}. \tag{10.74}$$

In the dipole approximation (cf. § 5.3)

$$\hat{\boldsymbol{\alpha}} \equiv \left(\frac{2\pi\hbar^2}{m}\right)\left(\frac{\gamma e^2}{m_{\mathrm{e}}c^2}\right)\sum_{\mathbf{l},\mathbf{d}} \exp(\mathrm{i}\boldsymbol{\kappa}.\mathbf{R}_{ld})\tfrac{1}{2}g_d F_d(\boldsymbol{\kappa})\{\tilde{\boldsymbol{\kappa}}\times(\hat{\mathbf{S}}_{ld}\times\tilde{\boldsymbol{\kappa}})\}. \tag{10.75}$$

To calculate the contribution to the cross-section coming from the first term in (10.74) we need to evaluate

$$\sum_{\mathbf{l},\mathbf{d}}\sum_{\mathbf{l}',\mathbf{d}'} \exp\{\mathrm{i}\boldsymbol{\kappa}.(\mathbf{R}_{ld}-\mathbf{R}_{l'd'})\}g_d g_{d'} F_d(\boldsymbol{\kappa})F^*_{d'}(\boldsymbol{\kappa})\times$$
$$\times\sum_{\lambda} p_\lambda\langle\lambda|\{\tilde{\boldsymbol{\kappa}}\times(\hat{\mathbf{S}}_{l'd'}\times\tilde{\boldsymbol{\kappa}})\}\{\mathbf{P}.\tilde{\boldsymbol{\kappa}}\times(\hat{\mathbf{S}}_{ld}\times\tilde{\boldsymbol{\kappa}})\}|\lambda\rangle. \tag{10.76}$$

Now

$$\sum_{\lambda} p_\lambda\langle\lambda|\{\tilde{\boldsymbol{\kappa}}\times(\hat{\mathbf{S}}_{l'd'}\times\tilde{\boldsymbol{\kappa}})\}\{\mathbf{P}.\tilde{\boldsymbol{\kappa}}\times(\hat{\mathbf{S}}_{ld}\times\tilde{\boldsymbol{\kappa}})\}|\lambda\rangle$$
$$= \sum_{\alpha,\beta}(\tilde{\boldsymbol{\alpha}}-\tilde{\boldsymbol{\kappa}}\tilde{\kappa}_\alpha)\{P^\beta-(\tilde{\boldsymbol{\kappa}}.\mathbf{P})\tilde{\kappa}_\beta\}\langle\hat{S}^\alpha_{l'd'}\hat{S}^\beta_{ld}\rangle$$
$$= \sum_{\alpha,\beta}(\tilde{\boldsymbol{\alpha}}-\tilde{\boldsymbol{\kappa}}\tilde{\kappa}_\alpha)\{P^\beta-(\tilde{\boldsymbol{\kappa}}.\mathbf{P})\tilde{\kappa}_\beta\}\delta_{\mathbf{l},\mathbf{l}'}\delta_{\mathbf{d},\mathbf{d}'}\delta_{\alpha,\beta}\tfrac{1}{3}S_d(S_d+1), \tag{10.77}$$

where to obtain the last line we have used (5.32). We therefore obtain for (10.76)

$$\tfrac{1}{3}N\sum_{\mathbf{d}}|g_d F_d(\boldsymbol{\kappa})|^2 S_d(S_d+1)\mathbf{P}_\perp, \tag{10.78}$$

where

$$\mathbf{P}_\perp = \tilde{\boldsymbol{\kappa}}\times(\mathbf{P}\times\tilde{\boldsymbol{\kappa}}) \tag{10.79}$$

is the component of **P** perpendicular to $\boldsymbol{\kappa}$.

The second term of (10.74) gives an identical result and the third term is easily evaluated as (10.78) with $-2\mathbf{P}$ replacing $\mathbf{P}_\perp$. Hence the first three terms of (10.74) give

$$\left(\frac{2\pi\hbar^2}{m}\right)^2\left(\frac{\gamma e^2}{m_{\mathrm{e}}c^2}\right)^2\tfrac{2}{3}N\sum_{\mathbf{d}}|\tfrac{1}{2}g_d F_d(\boldsymbol{\kappa})|^2 S_d(S_d+1)\{\mathbf{P}_\perp-\mathbf{P}\}$$
$$= \left(\frac{2\pi\hbar^2}{m}\right)^2\left(\frac{\gamma e^2}{m_{\mathrm{e}}c^2}\right)^2\tfrac{2}{3}N\sum_{\mathbf{d}}|\tfrac{1}{2}g_d F_d(\boldsymbol{\kappa})|^2 S_d(S_d+1)\{-\tilde{\boldsymbol{\kappa}}(\tilde{\boldsymbol{\kappa}}.\mathbf{P})\}. \tag{10.80}$$

We have now to evaluate the $\hat{\alpha}^+ \times \hat{\alpha}$ term in (10.74). This requires

$$\sum_\lambda p_\lambda \langle\lambda|\{\tilde{\varkappa}\times(\hat{\mathbf{S}}_{l'd'}\times\tilde{\varkappa})\}\times\{\tilde{\varkappa}\times(\hat{\mathbf{S}}_{ld}\times\tilde{\varkappa})\}|\lambda\rangle$$
$$= \langle\hat{\mathbf{S}}_{l'd'}\times\hat{\mathbf{S}}_{ld}\rangle + \langle(\tilde{\varkappa}\times\hat{\mathbf{S}}_{l'd'})(\tilde{\varkappa}.\hat{\mathbf{S}}_{ld}) - (\tilde{\varkappa}.\hat{\mathbf{S}}_{l'd'})(\tilde{\varkappa}\times\hat{\mathbf{S}}_{ld})\rangle. \quad (10.81)$$

The first term in (10.81) is zero by an argument analogous to that used in deriving (10.60 c). Further, each of the remaining two terms is zero.

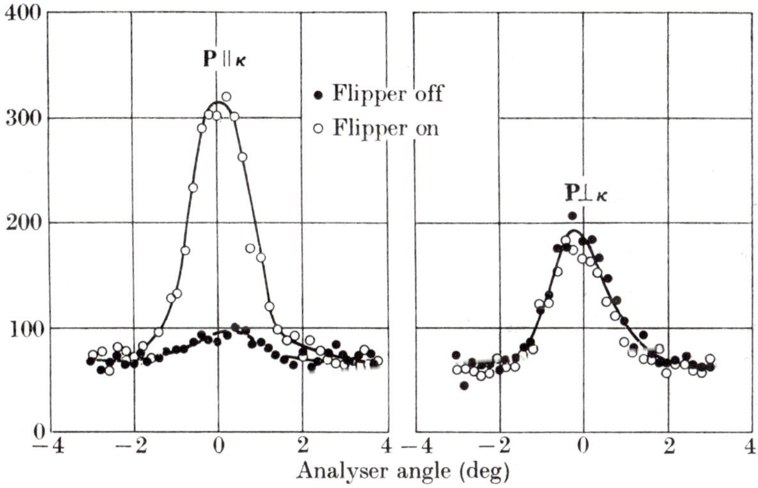

FIG. 10.5. Paramagnetic scattering from MnF_2.

For

$$\langle(\tilde{\varkappa}\times\hat{\mathbf{S}}_{l'd'})_\alpha(\tilde{\varkappa}.\hat{\mathbf{S}}_{ld})\rangle = \sum_{\beta,\gamma,\mu} \epsilon^{\alpha\beta\gamma}\tilde{\kappa}_\beta\tilde{\kappa}_\mu\langle\hat{S}^\gamma_{l'd'}\hat{S}^\mu_{ld}\rangle$$
$$= \sum_{\beta,\gamma,\mu} \epsilon^{\alpha\beta\gamma}\tilde{\kappa}_\beta\tilde{\kappa}_\mu\,\delta_{\gamma,\mu}\,\delta_{\mathbf{l},\mathbf{l}'}\,\delta_{\mathbf{d},\mathbf{d}'}\,\tfrac{1}{3}S_d(S_d+1) \quad (10.82)$$

and

$$\sum_{\beta,\gamma,\mu} \epsilon^{\alpha\beta\gamma}\tilde{\kappa}_\beta\tilde{\kappa}_\mu\,\delta_{\gamma,\mu} = \sum_{\beta,\gamma} \epsilon^{\alpha\beta\gamma}\tilde{\kappa}_\beta\tilde{\kappa}_\gamma = 0. \quad (10.83)$$

Thus (10.80) represents the only non-zero part of (10.74). The final result for $\mathbf{P}'$ is, therefore,

$$\mathbf{P}' = -\tilde{\varkappa}(\tilde{\varkappa}.\mathbf{P}). \quad (10.84)$$

The polarization of the scattered beam is seen to be zero when the incident beam is unpolarized, and when the incident beam is polarized the scattered beam is polarized in the direction of the scattering vector.

Result (10.84) for the polarization of neutrons scattered by a paramagnet is illustrated in Fig. 10.5. This shows the paramagnetic scattering from MnF_2 with **P** parallel to the scattering vector and with **P** perpendicular to the scattering vector. In the latter case no difference is observed in the scattered neutrons with flipper on or off, as expected from the above result. For **P** parallel to the scattering vector, the scattered neutrons are only reflected by the analyser when one of the flippers is activated, in accord with the negative sign in the result (10.84). The small peak observed in the **P** parallel to $\boldsymbol{\kappa}$ case with the flipper off, is due to multiple Bragg scattering.

Paramagnetic scattering studies afford an area of research where polarization analysis can be of great value. For in the usual diffuse scattering experiment where the total cross-section is measured several sources of scattering additional to the paramagnetic scattering can be present; e.g. isotopic disorder, multiple Bragg, Bragg, and nuclear spin coherent. If the spin-flip scattering is measured only the nuclear spin and paramagnetic scattering processes contribute, thus greatly simplifying the task of extracting the purely paramagnetic scattering.

(*b*) *Cross-section for ordered structures*

One of the most rewarding applications of polarized neutrons to date has been in the field of magnetic spin density and form factor measurements in ferromagnetic metals, a subject discussed in Chapter 7. These experiments utilize the polarization-dependent interference term between nuclear and magnetic scattering to obtain accurate measurements of the magnetic scattering amplitude. The actual quantity measured is the ratio of the scattered intensity (at a Bragg reflection) for one setting of the polarization to the intensity with polarization direction reversed. Since the interference term is linear in the magnetic scattering amplitude this technique enables the sign of the magnetic amplitude to be determined as well. We do not discuss elastic scattering of polarized neutrons specifically for ferromagnets but choose to discuss a far more general case. First we calculate the cross-section for the scattering of polarized neutrons though their nuclear and magnetic interactions with a solid, and then go on to evaluate the polarization of the scattered beam. Because of the importance of the polarized-beam technique in gaining detailed information on the magnetic properties of crystals we present the calculations in detail.

For elastic scattering by ions with only spin angular momentum,

$$\langle\lambda|\,\hat{\boldsymbol{\alpha}}\,|\lambda\rangle \equiv \left(\frac{2\pi\hbar^2}{m}\right)\left(\frac{\gamma e^2}{m_e c^2}\right)\sum_{\mathbf{l,d}} \exp(\mathrm{i}\boldsymbol{\kappa}\,.\,\mathbf{R}_{ld})\tfrac{1}{2}g_d\,F_d(\boldsymbol{\kappa})\langle\hat{S}_{ld}\rangle\times \\ \times[\tilde{\boldsymbol{\kappa}}\times\{\tilde{\boldsymbol{\eta}}_{ld}(\boldsymbol{\kappa})\times\tilde{\boldsymbol{\kappa}}\}], \quad (10.85)$$

where $\tilde{\boldsymbol{\eta}}_{ld}(\boldsymbol{\kappa})$ is a unit vector in the direction of the spin at site $\mathbf{R}_{ld}$ and for non-collinear spin systems it is a function of the scattering vector.†

† M. Blume, *Phys. Rev. Lett.* **10**, 489 (1963).

For brevity of notation we do not show this dependence in the course of the following calculations.

We must add to (10.85) the expression for $\hat{\boldsymbol{\alpha}}$ for purely nuclear scattering. This is given in (10.59) together with the required expression for $\hat{\beta}$. Because we assume the nuclear spins to be randomly orientated any term linear in $\hat{\mathbf{I}}_{ld}$ can be dropped. Similarly terms containing $\hat{\mathbf{I}}_{ld}\times\hat{\mathbf{I}}_{l'd'}$ can be dropped. Consider the term corresponding to $\hat{\boldsymbol{\alpha}}^{+}.\hat{\boldsymbol{\alpha}}$ in expression (10.37). We require

$$\left(\frac{2\pi\hbar^2}{m}\right)^2 \sum_{\mathbf{l,d}} \sum_{\mathbf{l',d'}} \exp\{\mathrm{i}\boldsymbol{\varkappa}.(\mathbf{R}_{ld}-\mathbf{R}_{l'd'})\}\times$$
$$\times\,\mathscr{O}\left[\left\{\tfrac{1}{2}B^*_{l'd'}\hat{\mathbf{I}}_{l'd'}+\left(\frac{\gamma e^2}{m_\mathrm{e}c^2}\right)\tfrac{1}{2}g_{d'}F^*_{d'}\langle\hat{S}_{l'd'}\rangle[\tilde{\boldsymbol{\varkappa}}\times(\tilde{\boldsymbol{\eta}}_{l'd'}\times\tilde{\boldsymbol{\varkappa}})]\right\}.\right.$$
$$\left.\cdot\left\{\tfrac{1}{2}B_{ld}\hat{\mathbf{I}}_{ld}+\left(\frac{\gamma e^2}{m_\mathrm{e}c^2}\right)\tfrac{1}{2}g_dF_d\langle\hat{S}_{ld}\rangle[\tilde{\boldsymbol{\varkappa}}\times(\tilde{\boldsymbol{\eta}}_{ld}\times\tilde{\boldsymbol{\varkappa}})]\right\}\right]$$
$$=\left(\frac{2\pi\hbar^2}{m}\right)^2 \sum_{\mathbf{l,d}} \sum_{\mathbf{l',d'}} \exp\{\mathrm{i}\boldsymbol{\varkappa}.(\mathbf{R}_{ld}-\mathbf{R}_{l'd'})\}\left[\tfrac{1}{4}\delta_{\mathbf{l,l'}}\delta_{\mathbf{d,d'}}\{|B_{ld}|^2i_d(i_d+1)\}+\right.$$
$$\left.+\left(\frac{\gamma e^2}{m_\mathrm{e}c^2}\right)^2\tfrac{1}{2}g_dF_d\langle\hat{S}_{ld}\rangle\tfrac{1}{2}g_{d'}F^*_{d'}\langle\hat{S}_{l'd'}\rangle[\tilde{\boldsymbol{\varkappa}}\times(\tilde{\boldsymbol{\eta}}_{ld}\times\tilde{\boldsymbol{\varkappa}})].[\tilde{\boldsymbol{\varkappa}}\times(\tilde{\boldsymbol{\eta}}_{l'd'}\times\tilde{\boldsymbol{\varkappa}})]\right]. \tag{10.86}$$

It remains to average the first term on the right-hand side over the (random) distribution of nuclear isotopes. This same result was given in eqn (10.61).

The calculation of the $\hat{\beta}^{+}\hat{\beta}$ term in (10.37) is straightforward. We have only to average

$$\left(\frac{2\pi\hbar^2}{m}\right)^2 \sum_{\mathbf{l,d}} \sum_{\mathbf{l',d'}} \exp\{\mathrm{i}\boldsymbol{\varkappa}.(\mathbf{R}_{ld}-\mathbf{R}_{l'd'})\}A_{ld}A^*_{l'd'} \tag{10.87}$$

over the distribution of isotopes, and find it to be

$$\left(\frac{2\pi\hbar^2}{m}\right)^2\left|\sum_{\mathbf{l}}\exp(\mathrm{i}\boldsymbol{\varkappa}.\mathbf{l})\right|^2|F_\mathrm{N}(\boldsymbol{\varkappa})|^2+N\sum_{\mathbf{d}}\{\overline{|A_d|^2}-|\bar{A}_d|^2\}. \tag{10.88}$$

Combining this with the first part of (10.86), and using (10.65) and (10.66), the total purely nuclear contribution to the cross-section is, apart from the constant factor,

$$\left(\frac{2\pi\hbar^2}{m}\right)^2\left\{N\sum_{\mathbf{d}}\{\overline{|b_d|^2}-|\bar{b}_d|^2\}+\left|\sum_{\mathbf{l}}\exp(\mathrm{i}\boldsymbol{\varkappa}.\mathbf{l})\right|^2|F_\mathrm{N}(\boldsymbol{\varkappa})|^2\right\}$$
$$=\left(\frac{2\pi\hbar^2}{m}\right)^2\left(\frac{\mathrm{d}\sigma}{\mathrm{d}\Omega}\right)_\mathrm{NN}. \tag{10.89}$$

This result is, of course, identical to that derived in Chapter 2; the first term is proportional to the incoherent nuclear cross-section and the second the coherent nuclear cross-section.

The evaluation of the terms $\hat{\beta}^+(\hat{\boldsymbol{\alpha}}.\mathbf{P})$ and $(\hat{\boldsymbol{\alpha}}^+.\mathbf{P})\hat{\beta}$ in (10.37) proceeds as follows. Because we need take only the magnetic part of $\hat{\boldsymbol{\alpha}}$, we have for the first of the two terms,

$$\left(\frac{2\pi\hbar}{m}\right)^2\left(\frac{\gamma e^2}{m_e c^2}\right)^2 \sum_{l,d}\sum_{l',d'} \exp\{-i\boldsymbol{\kappa}.(\mathbf{R}_{ld}-\mathbf{R}_{l'd'})\}\bar{b}_d^*\tfrac{1}{2}g_{d'}F_{d'}(\boldsymbol{\kappa})\langle\hat{S}_{l'd'}\rangle\times$$
$$\times\mathbf{P}.[\tilde{\boldsymbol{\kappa}}\times(\tilde{\boldsymbol{\eta}}_{l'd'}\times\tilde{\boldsymbol{\kappa}})], \quad (10.90)$$

where we have averaged over the isotope distribution. The term $(\hat{\boldsymbol{\alpha}}^+.\mathbf{P})\hat{\beta}$ is similar, the difference being that we take the complex conjugate of the magnetic part instead of the nuclear. Note that

$$\mathbf{P}.\tilde{\boldsymbol{\kappa}}\times(\tilde{\boldsymbol{\eta}}_{ld}\times\tilde{\boldsymbol{\kappa}}) = \tilde{\boldsymbol{\eta}}_{ld}.\tilde{\boldsymbol{\kappa}}\times(\mathbf{P}\times\tilde{\boldsymbol{\kappa}}) \equiv \tilde{\boldsymbol{\eta}}_{ld}.\mathbf{P}_\perp, \quad (10.91)$$

which shows that the interference term is zero for any spins that are orientated parallel to the scattering vector.

We have now only to calculate the $\hat{\boldsymbol{\alpha}}^+\times\hat{\boldsymbol{\alpha}}$ term in (10.37). For randomly orientated nuclear spins this is seen to be purely magnetic:

$$\left(\frac{2\pi\hbar^2}{m}\right)^2\left(\frac{\gamma e^2}{m_e c^2}\right)^2 \sum_{l,d}\sum_{l',d'} \exp\{i\boldsymbol{\kappa}(\mathbf{R}_{l'd'}-\mathbf{R}_{ld})\}\tfrac{1}{2}g_{d'}F_{d'}(\boldsymbol{\kappa})\langle\hat{S}_{l'd'}\rangle\times$$
$$\times\tfrac{1}{2}g_d F_d^*(\boldsymbol{\kappa})\langle\hat{S}_{ld}\rangle[\tilde{\boldsymbol{\kappa}}\times(\tilde{\boldsymbol{\eta}}_{l'd'}\times\tilde{\boldsymbol{\kappa}})]\times[\tilde{\boldsymbol{\kappa}}\times(\tilde{\boldsymbol{\eta}}_{ld}\times\tilde{\boldsymbol{\kappa}})]. \quad (10.92)$$

We find for the cross product in (10.92)

$$[\tilde{\boldsymbol{\kappa}}\times(\tilde{\boldsymbol{\eta}}_{l'd'}\times\tilde{\boldsymbol{\kappa}})]\times[\tilde{\boldsymbol{\kappa}}\times(\tilde{\boldsymbol{\eta}}_{ld}\times\tilde{\boldsymbol{\kappa}})]$$
$$= \tilde{\boldsymbol{\eta}}_{l'd'}\times\tilde{\boldsymbol{\eta}}_{ld}+(\tilde{\boldsymbol{\kappa}}\times\tilde{\boldsymbol{\eta}}_{l'd'})(\tilde{\boldsymbol{\kappa}}.\tilde{\boldsymbol{\eta}}_{ld})-(\tilde{\boldsymbol{\kappa}}\times\tilde{\boldsymbol{\eta}}_{ld})(\tilde{\boldsymbol{\kappa}}.\tilde{\boldsymbol{\eta}}_{l'd'}). \quad (10.93)$$

For certain magnetic structures this term is zero. Two important cases are (a) a ferromagnet in which all the spins are aligned parallel to a single direction and (b) an antiferromagnet in which the spins are either all parallel or antiparallel to a single direction.

We have now all the terms in the cross-section for the elastic scattering of polarized neutrons from a magnetic crystal. Combining (10.86), (10.89), (10.90), and (10.92) the result is

$$\frac{d\sigma}{d\Omega} = \left(\frac{d\sigma}{d\Omega}\right)_{NN}+\left(\frac{d\sigma}{d\Omega}\right)_{MM}+\left(\frac{\gamma e^2}{m_e c^2}\right)\sum_{l,d}\sum_{l',d'}\exp\{i\boldsymbol{\kappa}.(\mathbf{R}_{ld}-\mathbf{R}_{l'd'})\}\times$$
$$\times\Big[\{\bar{b}_{d'}^*\tfrac{1}{2}g_d F_d(\boldsymbol{\kappa})\langle\hat{S}_{ld}\rangle\tilde{\boldsymbol{\eta}}_{ld}.\mathbf{P}_\perp+\bar{b}_d\tfrac{1}{2}g_{d'}F_{d'}^*(\boldsymbol{\kappa})\langle\hat{S}_{l'd'}\rangle\tilde{\boldsymbol{\eta}}_{l'd'}.\mathbf{P}_\perp\}$$
$$+\left(\frac{\gamma e^2}{m_e c^2}\right)\tfrac{1}{2}g_{d'}F_{d'}^*(\boldsymbol{\kappa})\langle\hat{S}_{l'd'}\rangle\tfrac{1}{2}g_d F_d(\boldsymbol{\kappa})\langle\hat{S}_{ld}\rangle\times$$
$$\times i\mathbf{P}.[\tilde{\boldsymbol{\kappa}}\times(\tilde{\boldsymbol{\eta}}_{l'd'}\times\tilde{\boldsymbol{\kappa}})]\times[\tilde{\boldsymbol{\kappa}}\times(\tilde{\boldsymbol{\eta}}_{ld}\times\tilde{\boldsymbol{\kappa}})]\Big]. \quad (10.94)$$

In this expression $(d\sigma/d\Omega)_{MM}$ is the purely magnetic cross-section given by $(m/2\pi\hbar^2)^2$ times the second part of the right-hand side of (10.86).

We see from the expression (10.94) that the use of polarized neutrons introduces into the cross-section two additional contributions. First there is a contribution arising from interference between nuclear and magnetic scattering. This term will be zero for a given Bragg reflection for two sets of conditions: (a) if the spins concerned in the magnetic scattering are aligned parallel to the scattering vector, and (b) if both nuclear and magnetic scattering do not occur at the Bragg reflection in question. The latter case is, of course, encountered in scattering from antiferromagnets. For instance, in a rutile antiferromagnet like MnF_2 the $(0, t_2, t_3)$ reflections with t_2+t_3 odd are purely magnetic; reflections with $t_1+t_2+t_3$ even are, in the absence of covalency, purely nuclear; while all remaining reflections with $t_1+t_2+t_3$ odd are mixed nuclear and magnetic with the nuclear scattering coming solely from the nuclei of the anions. Thus in this type of antiferromagnet only the reflections with $t_1+t_2+t_3$ odd can be polarization dependent (unless covalency effects are present).

Besides the nuclear and magnetic interference terms the cross-section (10.94) contains an additional, purely magnetic term. As we have seen in (10.93), this term vanishes for a simple magnetic system in which all the spins are either parallel or antiparallel to a single direction. However, it can contribute in scattering from complicated spin configurations. An example of this is given in § 10.5 (*c*) when we calculate the cross-section for scattering of polarized neutrons by the antiferromagnetic MnO_2, which possesses a helical spin configuration.

A particular case of (10.94) that is of interest is that where all the spins in the crystal are either parallel or antiparallel to a single direction $\tilde{\boldsymbol{\eta}}$. Taking the Bragg part and writing the purely nuclear and magnetic contributions explicitly gives

$$\frac{d\sigma}{d\Omega} = \Big|\sum_{\mathbf{l}} \exp(i\boldsymbol{\varkappa}.\mathbf{l})\Big|^2 \Big[|F_N(\boldsymbol{\varkappa})|^2 + \Big(\frac{\gamma e^2}{m_e c^2}\Big)^2 \{1-(\tilde{\boldsymbol{\varkappa}}.\tilde{\boldsymbol{\eta}})^2\}|F_M(\tilde{\boldsymbol{\varkappa}})|^2 + \\ + \Big(\frac{2\gamma e^2}{m_e c^2}\Big)\tilde{\boldsymbol{\eta}}.\mathbf{P}_{\perp} \operatorname{Re} F_N(\boldsymbol{\varkappa})F_M^*(\boldsymbol{\varkappa})\Big], \quad (10.95)$$

where, as usual,

$$F_M(\boldsymbol{\varkappa}) = \sum_{\mathbf{d}} \exp(i\boldsymbol{\varkappa}.\mathbf{d})\tfrac{1}{2}g_d F_d(\boldsymbol{\varkappa})\langle \hat{S}_d\rangle\sigma_d.$$

Consider now a simple ferromagnet with just one atom per unit cell, i.e. Bravais lattice structure. Because each ion in such a system is at a centre of symmetry the atomic magnetic form factor $F(\boldsymbol{\kappa})$ for the ions is purely real. The cross-section for Bragg scattering is therefore

$$\frac{d\sigma}{d\Omega} = \left|\sum_{\mathbf{l}} \exp(i\boldsymbol{\kappa}.\mathbf{l})\right|^2 \left[\bar{b}^2 + \left(\frac{2\gamma e^2}{m_e c^2}\right)\tfrac{1}{2}gF(\boldsymbol{\kappa})\langle\hat{S}\rangle\bar{b}\,\tilde{\boldsymbol{\eta}}.\mathbf{P}_\perp + \right.$$
$$\left. + \left(\frac{\gamma e^2}{m_e c^2}\right)^2 \{1-(\tilde{\boldsymbol{\kappa}}.\tilde{\boldsymbol{\eta}})^2\}\{\tfrac{1}{2}gF(\boldsymbol{\kappa})\langle\hat{S}\rangle\}^2\right]. \quad (10.96)$$

If the magnetization of the ferromagnet is aligned by an external field perpendicular to the scattering plane,

$$\tilde{\boldsymbol{\eta}}.\mathbf{P}_\perp = \tilde{\boldsymbol{\eta}}.\mathbf{P} \quad \text{and} \quad \tilde{\boldsymbol{\kappa}}.\tilde{\boldsymbol{\eta}} = 0.$$

If now the polarization $\mathbf{P}$ is made to coincide with $\tilde{\boldsymbol{\eta}}$,

$$\frac{d\sigma}{d\Omega} = \left|\sum_{\mathbf{l}} \exp(i\boldsymbol{\kappa}.\mathbf{l})\right|^2 \left\{\bar{b} \pm \left(\frac{\gamma e^2}{m_e c^2}\right)\tfrac{1}{2}gF(\boldsymbol{\kappa})\langle\hat{S}\rangle\right\}^2, \quad (10.97)$$

where the $\pm$ signs are taken according to

$$\tilde{\boldsymbol{\eta}}.\mathbf{P} = \pm 1. \quad (10.98)$$

The ratio of the Bragg intensities, or the flipping ratio, is therefore given by

$$R = \left\{\frac{1+(\gamma e^2/m_e c^2)\frac{1}{2}gF(\boldsymbol{\tau})\langle\hat{S}\rangle/\bar{b}}{1-(\gamma e^2/m_e c^2)\frac{1}{2}gF(\boldsymbol{\tau})\langle\hat{S}\rangle/\bar{b}}\right\}^2. \quad (10.99)$$

Let us consider the value of R for scattering from nickel. The first Bragg reflection is (111) and the corresponding value of $F(\boldsymbol{\tau})$ is 0·79. Thus with

$$\tfrac{1}{2}g\langle\hat{S}\rangle = 0{\cdot}3$$

at room temperature, the magnetic scattering amplitude is

$$\left(\frac{\gamma e^2}{m_e c^2}\right)\tfrac{1}{2}g\langle\hat{S}\rangle F(\boldsymbol{\tau}) = 0{\cdot}54\times0{\cdot}3\times0{\cdot}79\times10^{-12}\text{ cm} = 0{\cdot}128\times10^{-12}\text{ cm}.$$

For Ni, $\bar{b} = 1{\cdot}03\times10^{-12}$ cm, so the ratio of the magnetic to nuclear scattering at the (111) reflection is

$$\frac{0{\cdot}128}{1{\cdot}03} = 0{\cdot}124 \quad \text{and} \quad R = \left(\frac{1+0{\cdot}124}{1-0{\cdot}124}\right)^2 = 1{\cdot}646.$$

At the (440) reflection $F \simeq 0{\cdot}06$ and the ratio of the magnetic and nuclear scattering is 0·0097 with

$$R = 1{\cdot}028.$$

(c) *Cross-section for helical structures*

In recent years several examples of complicated helical spin ordering have been determined by neutron diffraction. The use of polarized neutrons enables more information to be gained about these structures than would otherwise be possible. Yoshimori was the first to point out the existence of helical spin orderings and demonstrated that such an ordering was consistent with the neutron diffraction pattern obtained by Erickson for MnO_2. This was discussed in § 7.6. Here we evaluate the cross-section (10.94) for a helical spin structure and choose MnO_2 as the example.

MnO_2 is a rutile structure for which Yoshimori proposed that the unit vector $\tilde{\boldsymbol{\eta}}_l$ is given by

$$\tilde{\boldsymbol{\eta}}_l = \exp(i\mathbf{w}.\mathbf{l})\{\tilde{\mathbf{x}}\cos\mathbf{Q}.\mathbf{l}+\tilde{\mathbf{y}}\sin\mathbf{Q}.\mathbf{l}\}. \tag{10.100}$$

The reciprocal vector $\mathbf{w}$ is such that the factor $\exp(i\mathbf{w}.\mathbf{l})$ has the value $+1$ when l_1, l_2, and l_3 are all even and -1 when they are all odd. The vector $\mathbf{Q}$ in (10.100) is

$$\mathbf{Q} = \frac{2\pi}{c}(0, 0, \tfrac{2}{7}).$$

Thus (10.100) describes an antiferromagnetic spin configuration in which the spins screw along the z axis with pitch $7c/2$.

In § 7.6 we showed that the elastic magnetic cross-section for the scattering of unpolarized neutrons by MnO_2 is

$$\frac{d\sigma}{d\Omega} = \left(\frac{\gamma e^2}{m_e c^2}\right)^2 |F(\boldsymbol{\varkappa})|^2 \tfrac{1}{4}\langle \hat{S}\rangle^2(1+\tilde{\kappa}_z^2)\times$$
$$\times\left\{\left|\sum_{\mathbf{l}} \exp\{i\mathbf{l}.(\boldsymbol{\varkappa}+\mathbf{w}+\mathbf{Q})\}\right|^2+\left|\sum_{\mathbf{l}} \exp\{i\mathbf{l}.(\boldsymbol{\varkappa}+\mathbf{w}-\mathbf{Q})\}\right|^2\right\}, \tag{10.101}$$

i.e. the Bragg scattering occurs in satellite reflections equally displaced from $\boldsymbol{\varkappa}+\mathbf{w} = \boldsymbol{\tau}$ positions. This means that the magnetic and nuclear scattering occur at different positions and cannot therefore interfere with one another. Hence it is sufficient to consider only the purely magnetic part of the cross-section (10.94). This entails the evaluation of the contribution

$$\left(\frac{\gamma e^2}{m_e c^2}\right)^2 |F(\boldsymbol{\varkappa})|^2\langle \hat{S}\rangle^2 \sum_{\mathbf{l},\mathbf{l}'} \exp\{i\boldsymbol{\varkappa}.(\mathbf{l}-\mathbf{l}')\}i\mathbf{P}.[\tilde{\boldsymbol{\varkappa}}\times(\tilde{\boldsymbol{\eta}}_{l'}\times\tilde{\boldsymbol{\varkappa}})]\times[\tilde{\boldsymbol{\varkappa}}\times(\tilde{\boldsymbol{\eta}}_l\times\tilde{\boldsymbol{\varkappa}})]. \tag{10.102}$$

By using (10.93),

$$\mathbf{P}.[\tilde{\boldsymbol{\varkappa}}\times(\tilde{\boldsymbol{\eta}}_{l'}\times\tilde{\boldsymbol{\varkappa}})]\times[\tilde{\boldsymbol{\varkappa}}\times(\tilde{\boldsymbol{\eta}}_l\times\tilde{\boldsymbol{\varkappa}})] = \mathbf{P}.\tilde{\boldsymbol{\eta}}_{l'}\times\tilde{\boldsymbol{\eta}}_l+\sum_{\alpha,\beta}(\mathbf{P}\times\tilde{\boldsymbol{\varkappa}})_\alpha\tilde{\kappa}_\beta(\tilde{\eta}_{l'}^\alpha\tilde{\eta}_l^\beta-\tilde{\eta}_{l'}^\beta\tilde{\eta}_l^\alpha)$$

and since we have only α, $\beta = x$ or y this reduces to

$$[P_z+(\mathbf{P}\times\tilde{\boldsymbol{\varkappa}})_x\tilde{\kappa}_y-(\mathbf{P}\times\tilde{\boldsymbol{\varkappa}})_y\tilde{\kappa}_x](\tilde{\eta}_{l'}^x\tilde{\eta}_l^y-\tilde{\eta}_{l'}^y\tilde{\eta}_l^x) = \tilde{\kappa}_z(\mathbf{P}.\tilde{\boldsymbol{\varkappa}})(\tilde{\eta}_{l'}^x\tilde{\eta}_l^y-\tilde{\eta}_{l'}^y\tilde{\eta}_l^x). \tag{10.103}$$

From (10.100)

$$\tilde{\eta}_{l'}^x\tilde{\eta}_l^y-\tilde{\eta}_{l'}^y\tilde{\eta}_l^x \equiv \exp\{i\mathbf{w}.(\mathbf{l}-\mathbf{l}')\}\sin\mathbf{Q}.(\mathbf{l}-\mathbf{l}'). \tag{10.104}$$

Thus with the results (10.103) and (10.104) we have for (10.102),

$$\left(\frac{\gamma e^2}{m_e c^2}\right)^2 |F(\varkappa)|^2 \tfrac{1}{2} \langle \hat{S} \rangle^2 \tilde{\kappa}_z (\mathbf{P}.\tilde{\boldsymbol{\varkappa}}) \times$$

$$\times \left\{ \left| \sum_{\mathbf{l}} \exp\{i\mathbf{l}.(\boldsymbol{\varkappa}+\mathbf{w}+\mathbf{Q})\}\right|^2 - \left| \sum_{\mathbf{l}} \exp\{i\mathbf{l}.(\boldsymbol{\varkappa}+\mathbf{w}-\mathbf{Q})\}\right|^2 \right\}.$$

Adding this result to (10.101) as required in the cross-section (10.94), we have for the elastic magnetic cross-section for the scattering of polarized neutrons by the helical spin structure MnO_2 the result

$$\frac{d\sigma}{d\Omega} = \left(\frac{\gamma e^2}{m_e c^2}\right)^2 |F(\varkappa)|^2 \tfrac{1}{4} \langle \hat{S} \rangle^2 \left[\left| \sum_{\mathbf{l}} \exp\{i\mathbf{l}.(\boldsymbol{\varkappa}+\mathbf{w}+\mathbf{Q})\}\right|^2 \{1+\tilde{\kappa}_z^2+2\tilde{\kappa}_z(\mathbf{P}.\tilde{\boldsymbol{\varkappa}})\} + \right.$$

$$\left. + \left| \sum_{\mathbf{l}} \exp\{i\mathbf{l}.(\boldsymbol{\varkappa}+\mathbf{w}-\mathbf{Q})\}\right|^2 \{1+\tilde{\kappa}_z^2-2\tilde{\kappa}_z(\mathbf{P}.\tilde{\boldsymbol{\varkappa}})\} \right]. \quad (10.105)$$

From this we see that the polarization dependence is different for the two satellites, the difference arising from the

$$i\mathbf{P}.[\tilde{\boldsymbol{\varkappa}} \times (\tilde{\boldsymbol{\eta}}_{l'} \times \tilde{\boldsymbol{\varkappa}})] \times [\tilde{\boldsymbol{\varkappa}} \times (\tilde{\boldsymbol{\eta}}_{l} \times \tilde{\boldsymbol{\varkappa}})]$$

term in the cross-section.†

If both the scattering vector $\boldsymbol{\varkappa}$ and the polarization $\mathbf{P}$ are made parallel to the z axis, the intensity of the reflection at $\boldsymbol{\varkappa} = \boldsymbol{\tau}+\mathbf{Q}-\mathbf{w}$ is seen to be reduced to zero and the total intensity concentrated in the

$$\boldsymbol{\varkappa} = \boldsymbol{\tau}-\mathbf{Q}-\mathbf{w}$$

reflection. If $\boldsymbol{\varkappa}$ is maintained in the same direction but $\mathbf{P}$ altered to make it antiparallel then the situation is reversed. Finally, when $\mathbf{P}$ is perpendicular to the scattering plane the cross-section is independent of the polarization and the cross-section (10.105) reduces to that for the scattering of unpolarized neutrons.

(*d*) *Cross-section for ions with non-zero orbital angular momentum*

The above discussion of the cross-section for scattering of polarized neutrons by magnetic crystals applies only when the ions can be assigned an (effective) spin $\langle \hat{S} \rangle$ (cf. eqn (10.85)). When the ions possess both spin and orbital angular momentum the interference term between the nuclear and magnetic scattering amplitudes is readily derived in terms of the formalism presented in Chapter 6. For a crystal that possesses one type of ion the required expression is

$$2\bar{b}\left(\frac{\gamma e^2}{m_e c^2}\right)\mathbf{P}.\mathbf{E}, \quad (10.106)$$

† Yu. A. Izyumov, *Soviet Phys. JETP*, 1162 (1962).

where the qth spherical component of $\mathbf{E}$ is given by

$$E_q = \langle\phi_e|\,\hat{Q}_{\perp,q}\,|\phi_e\rangle. \tag{10.107}$$

Here $|\phi_e\rangle$ denotes the wave function of the unpaired electrons and matrix elements of $\hat{Q}_{\perp,q}$ in the SLJ basis are given by eqn (6.87 a). If $|\phi_e\rangle \equiv |SM_S\,LM_L\rangle$ the required matrix elements are derived from eqn (6.87 a) by the transformation (6.97). Note that in general the vector $\mathbf{E}$ is not parallel to $\hat{\mathbf{Q}}_\perp$, though it may be in certain special cases.

For the particular case of zero orbital angular momentum (10.106) clearly reduces to the result given previously, eqn (10.95), since in this instance

$$\mathbf{E} \rightarrow \tfrac{1}{2}gF(\boldsymbol{\varkappa})\langle\hat{S}\rangle[\tilde{\boldsymbol{\varkappa}}\times(\tilde{\boldsymbol{\eta}}\times\tilde{\boldsymbol{\varkappa}})].$$

A more detailed discussion of the derivation of (10.106) is given by Lovesey and Rimmer.†

We consider the structure of the interference term (10.106) for two cases: (i) the rare-earth ion Tb^{3+} (cf. § 6.6) and (ii) the transition-metal ion Co^{2+}.

(i) The wave function of Tb^{3+} is taken to be $S = L = 3$, $J = M = 6$ and we choose $\mathbf{P}$ parallel to our axis of quantization z. The interference term can be calculated from the theory given in § 6.6 and is found to be proportional to

$$\begin{aligned} &Y_0^0(\tilde{\boldsymbol{\varkappa}})[3{\cdot}0\bar{j}_0+1{\cdot}11111\bar{j}_2]+\\ &\quad+Y_0^2(\tilde{\boldsymbol{\varkappa}})[-1{\cdot}34164\bar{j}_0-0{\cdot}49690\bar{j}_2+0{\cdot}16262\bar{j}_4]+\\ &\quad+Y_0^4(\boldsymbol{\varkappa})[-0{\cdot}14876\bar{j}_4+0{\cdot}01907\bar{j}_6]+\\ &\quad+Y_0^6(\tilde{\boldsymbol{\varkappa}})[0{\cdot}02292\bar{j}_4-0{\cdot}02190\bar{j}_6]+\\ &\quad+Y_0^8(\tilde{\boldsymbol{\varkappa}})[0{\cdot}00528\bar{j}_6]. \end{aligned} \tag{10.108}$$

Note that this expression includes an eighth-order spherical harmonic, i.e. the highest-order spherical harmonic in (10.106) for $|\phi_e\rangle = |SLJM\rangle$ is not equal to the value of J. A second general feature is the decrease in the magnitude of the numerical coefficients associated with products of higher-order spherical harmonics and radial integrals, $\bar{j}_K$.

For the particular choice of $\boldsymbol{\varkappa}$ perpendicular to the axis of quantization, and therefore to $\mathbf{P}$, (10.108) simplifies since, with $\theta = \frac{1}{2}\pi$ (we omit the factor $(4\pi)^{-\frac{1}{2}}$ contained in the definition of the spherical harmonics),

$$Y_0^0 = 1,\quad Y_0^2 = -\tfrac{1}{2}\sqrt{5},\quad Y_0^4 = \tfrac{9}{8},\quad Y_0^6 = -\tfrac{5}{16}\sqrt{13},\quad \text{and}\quad Y_0^8 = \frac{105}{2^7}.$$

† S. W. Lovesey and D. E. Rimmer, *Rep. Prog. Phys.* **32,** 333 (1969). For saturated rare-earth ions, it follows from eqn (6.125) and (6.128) that the interference term (10.106) is proportional to the atomic form factor $F(\kappa, \theta)$ defined by eqn (6.131). See also Balcar *et. al.*, *J. Phys.* C**3,** 1292 (1970).

We obtain for (10.108) (the result

$$\tfrac{9}{2}\bar{j}_0+\tfrac{5}{3}\bar{j}_2-\tfrac{3}{8}\bar{j}_4+\tfrac{5}{16}\bar{j}_6 \quad (\theta = \tfrac{1}{2}\pi). \tag{10.109}$$

It is instructive to compare this exact result with that obtained in the dipole approximation (6.121), which retains all terms involving $\bar{j}_0$ and the main terms in $\bar{j}_2$. In this approximation we have

$$\tfrac{9}{2}\bar{j}_0+\tfrac{3}{2}\bar{j}_2, \tag{10.110}$$

i.e. in this instance the dipole approximation neglects a term in $\bar{j}_2$ with coefficient $\frac{1}{6}$.

(ii) The wave function of the unpaired electrons associated with Co^{2+} in a cubic environment possesses both unquenched orbital angular momentum and in addition a contribution due to spin–orbit coupling. However, for the purpose of illustration we present calculations of the interference term (10.106) for five sets of wave functions of varying complexity. One of these has $S = 0$, $L = 3$ and is included to illustrate general features of the orbital contribution to the interference term.

TABLE 10.1

Coefficients of the interference term (10.106) *for the five different wave functions for* Co^{2+} *defined in the text*

	(a)	(b)	(c)	(d)	(e)
$\bar{j}_0\ Y_1^2$	0·5477	0·5477	1·0954	0·6822	0·6700
$\bar{j}_2\ Y_1^2$	0·0140	0·5030	0·4844	0·1779	0·1392
$\bar{j}_2\ Y_1^4$	0·0365	−0·0456	−0·0913	−0·0482	−0·0817
$\bar{j}_4\ Y_1^2$	−0·0894	−0·0447	−0·0894	−0·0537	−0·0818
$\bar{j}_4\ Y_1^4$	0·0245	−0·0456	−0·0332	−0·0003	−0·0295
$\bar{j}_4\ Y_3^4$	—	—	—	0·1393	0·1514
$\bar{j}_4\ Y_1^6$	0·1167	—	0·0584	0·0548	0·0542
$\bar{j}_4\ Y_3^6$	—	—	—	0·0017	−0·0003
$\bar{j}_4\ Y_5^6$	—	—	—	−0·0075	0·0012

It is postulated that the crystal is a simple antiferromagnet with two Co^{2+} ions per unit cell and that we observe those Bragg reflections for which these sites are exactly out of phase (i.e. a superlattice reflection for which $\exp(i\mathbf{d}_1.\boldsymbol{\tau}) = +1$ and $\exp(i\mathbf{d}_2.\boldsymbol{\tau}) = -1$) giving twice the amplitude for a single ion. The spins are aligned parallel and antiparallel to the z axis and the neutron beam is assumed polarized parallel to the x axis. The results are given in Table 10.1. Here a harmonic denoted by $Y_M^L(\tilde{\boldsymbol{\kappa}})$ $(M > 0)$ actually refers to the combination

$$[Y_M^L(\tilde{\boldsymbol{\kappa}})+(-)^M Y_{\bar{M}}^L(\tilde{\boldsymbol{\kappa}})]$$

and a factor $1/\sqrt{(4\pi)}$ from the spherical harmonics has also been omitted.

(a) *Spin only*. In order to confirm some of the most basic properties we consider first a spin-only case. Selecting the ^4F manifold of $(3d)^7$ the state is chosen to be $|M_S = \frac{3}{2}; M_L = 0\rangle$. The first observation is that on substituting cartesian coordinates for the spherical harmonics

$$\begin{aligned}\sqrt{(4\pi)}\{Y_1^2(\tilde{\boldsymbol{\kappa}})-Y_{-1}^2(\tilde{\boldsymbol{\kappa}})\} &= -\sqrt{(30)}\kappa_x\kappa_z/\kappa^2,\\ \sqrt{(4\pi)}\{Y_1^4(\tilde{\boldsymbol{\kappa}})-Y_{-1}^4(\tilde{\boldsymbol{\kappa}})\} &= -(3\sqrt{5}/2)\kappa_x\kappa_z(7\kappa_z^2-3\kappa^2)/\kappa^4,\\ \sqrt{(4\pi)}\{Y_1^6(\tilde{\boldsymbol{\kappa}})-Y_{-1}^6(\tilde{\boldsymbol{\kappa}})\} &= -\sqrt{(273/2)}\kappa_x\kappa_z(33\kappa_z^4-30\kappa_z^2+5\kappa^4)/\kappa^6\end{aligned} \tag{10.111}$$

there appears an overall factor $\kappa_x\kappa_z$. This may be thought of as originating from $\mathbf{P}.\{\tilde{\boldsymbol{\kappa}}\times(\tilde{\boldsymbol{\eta}}\times\tilde{\boldsymbol{\kappa}})\}$, since $\mathbf{P}$ is parallel to x and $\tilde{\boldsymbol{\eta}}$ is parallel to z. Cancelling out $(-\tilde{\kappa}_x\tilde{\kappa}_z)$ it is found that the remaining angular functions associated with a particular $\bar{j}_K$ condense into a single harmonic to give

$$3\bar{j}_0+(7/5\sqrt{30})Y_0^2\bar{j}_2+\tfrac{6}{7}Y_0^4\bar{j}_4. \tag{10.112}$$

The coupling of equivalent orders of spherical harmonics and spherical Bessel functions arises out of taking the transform of a spin distribution,

$$\begin{aligned}&\int \mathrm{d}\mathbf{r}\, \exp(\mathrm{i}\boldsymbol{\kappa}.\mathbf{r}) \sum_{KQ} a_{KQ}\, Y_Q^K(\tilde{\mathbf{r}}) f_{3\mathrm{d}}^2(r)\\ &\qquad = \int \mathrm{d}\mathbf{r}\; 4\pi \sum_{LM} \mathrm{i}^L j_L(\kappa r) Y_M^{*L}(\tilde{\mathbf{r}}) Y_M^L(\tilde{\boldsymbol{\kappa}}) \sum_{KQ} a_{KQ}\, Y_Q^K(\tilde{\mathbf{r}}) f_{3\mathrm{d}}^2(r)\\ &\qquad = 4\pi \sum_{KQ} \mathrm{i}^K a_{KQ}\, Y_Q^K(\tilde{\boldsymbol{\kappa}}) \bar{j}_K.\end{aligned}$$

The coefficient of $\bar{j}_0$ gives the total spin component parallel to z, summed with phase factors over the unit cell. For all five cases it will be observed that the coefficient of $\bar{j}_0 Y_1^2$ is proportional to $\langle M_s+\frac{1}{2}M_L\rangle$, as has been demonstrated in § 6.5.

(b) *Orbit only*. In order to demonstrate the structure of the orbit term we next consider a rather unphysical case involving orbit terms only. Again in the ^4F manifold, this can be achieved using the state

$$\frac{1}{\sqrt{2}}[|M_S = \tfrac{3}{2}; M_L = 3\rangle+|M_S = -\tfrac{3}{2}; M_L = 3\rangle]. \tag{10.113}$$

First it will be noted that the cross-section can be rewritten as

$$\{0{\cdot}5477(\bar{j}_0+\bar{j}_2)-0{\cdot}0447(\bar{j}_2+\bar{j}_4)\}(Y_1^2-Y_{-1}^2)-0{\cdot}0456(\bar{j}_2+\bar{j}_4)(Y_1^4-Y_{-1}^4) \tag{10.114}$$

as the structure of the orbit expression requires. The appearance of $(\bar{j}_0+\bar{j}_2)$ in place of $\bar{j}_0$ as the leading term is responsible for the observed expansion of the form factor in the presence of orbital moment. The overall factor $\kappa_x\kappa_z$ is present, as required by the ability to write the orbit operator as $\mathbf{P}.\{\tilde{\boldsymbol{\kappa}}\times(\hat{\mathbf{Q}}\times\tilde{\boldsymbol{\kappa}})\}$ where $\hat{\mathbf{Q}}$ is parallel to $\mathbf{L}$ if $\mathbf{S} = 0$.

Finally, however, if $\kappa_x \kappa_z$ is factored out it is not possible to condense the remaining terms into the form of eqn (10.112). Terms with unequal ranks appear, showing the effect of the radial dependence of $\hat{\mathbf{Q}}$.

(c) *Spin and orbit.* If both spin and orbit are present it is more difficult to unscramble the output. With full spin–orbit coupling the ground state would be $|M_{\rm S} = \frac{3}{2}, M_{\rm L} = 3\rangle$ and the third column of the table refers to this. Without evaluating the electron density distribution it is difficult to analyse these figures. However, the necessary symmetry is retained.

(d) ${}^4F$ *approximation to* CoO. Going to a more realistic wave function we have available the calculation of Kanamori† (1957). His wave function for CoO is a linear combination of states from $|{}^4F\rangle$ and $|{}^4P\rangle$; as a next level of refinement we use his $|{}^4F\rangle$ state, which is

$$0{\cdot}900[\sqrt{(\tfrac{3}{8})}|M_{\rm S} = \tfrac{3}{2};\, M_{\rm L} = -1\rangle + \sqrt{(\tfrac{5}{8})}|M_{\rm S} = \tfrac{3}{2};\, M_{\rm L} = 3\rangle] - \\ -0{\cdot}401|M_{\rm S} = \tfrac{1}{2};\, M_{\rm L} = 0\rangle + \\ +0{\cdot}169[\sqrt{(\tfrac{3}{8})}|M_{\rm S} = -\tfrac{1}{2};\, M_{\rm L} = 1\rangle + \sqrt{(\tfrac{5}{8})}|M_{\rm S} = -\tfrac{1}{2};\, M_{\rm L} = -3\rangle]. \tag{10.115}$$

With the deviation of the magnetic moment away from the z direction there now appear terms which do not involve $\kappa_x \kappa_z$, though since they are associated with $\bar{j}_4$ integrals they will only be observable at large values of κ. Nevertheless it may be possible to detect from them reflections that would otherwise be absent.

(e) *Full Kanamori wave function.* Finally we consider the full wave function derived by Kanamori.† This is of the form

$$\sqrt{(1-\beta^2)}\, |{}^4F\rangle + \beta|{}^4P\rangle, \tag{10.116 a}$$

where $\beta = 0{\cdot}185$, $|{}^4F\rangle$ is the state used in (d) and $|{}^4P\rangle$ is

$$0{\cdot}900|M_{\rm S} = \tfrac{3}{2};\, M_{\rm L} = -1\rangle - 0{\cdot}410|M_{\rm S} = \tfrac{1}{2};\, M_{\rm L} = 0\rangle + \\ +0{\cdot}169|M_{\rm S} = -\tfrac{1}{2};\, M_{\rm L} = 1\rangle. \tag{10.116 b}$$

β is sufficiently small to cause a reduction in the leading term by only about 2 per cent, but it changes the dominant distortion term $\{Y_3^4(\tilde{\boldsymbol{\kappa}})\bar{j}_4\}$ by nearly 10 per cent. This is presumably because the direction of the effective magnetic moment associated with the $|{}^4P\rangle$ component makes a relatively large angle with the axis of quantization.

(e) *Polarization of scattered beam*

We turn now to the calculation of the polarization of the scattered beam for elastic magnetic scattering in the dipole approximation.

† J. Kanamori, *Prog. theor. Phys. Osaka* **17**, 177 (1957). See also T. Nagamiya and K. Motizuki, *Rev. Mod. Phys.* **30**, 89 (1956).

First note that there are several terms in (10.41) that we evaluated above when calculating the cross-section: thus, the first two terms are given in (10.90), the third is (10.88), the sixth is (10.86), and the seventh is (10.92). Of the remaining terms to be calculated we start with $\hat{\boldsymbol{\alpha}}^{+}(\hat{\boldsymbol{\alpha}}.\mathbf{P})$ and $(\hat{\boldsymbol{\alpha}}^{+}.\mathbf{P})\hat{\boldsymbol{\alpha}}$.

For the first of these we need

$$\left(\frac{2\pi\hbar^2}{m}\right)^2 \sum_{\mathbf{l},\mathbf{d}} \sum_{\mathbf{l}',\mathbf{d}'} \exp\{\mathrm{i}\boldsymbol{\varkappa}.(\mathbf{R}_{ld}-\mathbf{R}_{l'd'})\}\times$$

$$\mathscr{O}[\{\tfrac{1}{2}B^{*}_{l'd'}\hat{\mathbf{i}}_{l'd'}+\tfrac{1}{2}g_{d'}F^{*}_{d'}\langle\hat{S}_{l'd'}\rangle[\tilde{\boldsymbol{\varkappa}}\times(\tilde{\boldsymbol{\eta}}_{l'd'}\times\tilde{\boldsymbol{\varkappa}})]\}\{\tfrac{1}{2}B_{ld}\mathbf{P}.\hat{\mathbf{i}}_{ld}+\tfrac{1}{2}g_{d}F_{d}\langle\hat{S}_{ld}\rangle\tilde{\boldsymbol{\eta}}_{ld}.\mathbf{P}_{\perp}\}]$$

$$=\left(\frac{2\pi\hbar^2}{m}\right)^2 \sum_{\mathbf{l},\mathbf{d}} \sum_{\mathbf{l}',\mathbf{d}'} \exp\{\mathrm{i}\boldsymbol{\varkappa}.(\mathbf{R}_{ld}-\mathbf{R}_{l'd'})\}[\mathbf{P}\tfrac{1}{4}\delta_{\mathbf{l},\mathbf{l}'}\delta_{\mathbf{d},\mathbf{d}'}\{i_d(i_d+1)\tfrac{1}{3}|B_d|^2\}+$$

$$+\tfrac{1}{2}g_{d'}F^{*}_{d'}\langle\hat{S}_{l'd'}\rangle\{\tilde{\boldsymbol{\varkappa}}\times(\tilde{\boldsymbol{\eta}}_{l'd'}\times\tilde{\boldsymbol{\varkappa}})\}\tfrac{1}{2}g_{d}F_{d}\langle\hat{S}_{ld}\rangle\tilde{\boldsymbol{\eta}}_{ld}.\mathbf{P}_{\perp}]. \quad (10.117)$$

Here we have employed the same argument as that following (10.61). The calculation of the term corresponding to $(\hat{\boldsymbol{\alpha}}^{+}.\mathbf{P})\hat{\boldsymbol{\alpha}}$ is similar to that just given for $\hat{\boldsymbol{\alpha}}^{+}(\hat{\boldsymbol{\alpha}}.\mathbf{P})$.

The only terms in (10.41) that remain to be evaluated are the last two. Consider $\mathrm{i}\hat{\beta}^{+}(\hat{\boldsymbol{\alpha}}\times\mathbf{P})$; we need

$$\mathrm{i}\left(\frac{2\pi\hbar^2}{m}\right)^2\left(\frac{\gamma e^2}{m_{\mathrm{e}}c^2}\right) \sum_{\mathbf{l},\mathbf{d}} \sum_{\mathbf{l}',\mathbf{d}'} \exp\{\mathrm{i}\boldsymbol{\varkappa}.(\mathbf{R}_{ld}-\mathbf{R}_{l'd'})\}A^{*}_{l'd'}\tfrac{1}{2}g_{d}F_{d}\langle\hat{S}_{ld}\rangle\times$$

$$\times\{\tilde{\boldsymbol{\varkappa}}\times(\tilde{\boldsymbol{\eta}}_{ld}\times\tilde{\boldsymbol{\varkappa}})\}\times\mathbf{P}, \quad (10.118)$$

the nuclear part of $\hat{\boldsymbol{\alpha}}$ not entering because the nuclear spins are randomly orientated. The term corresponding to $\mathrm{i}(\mathbf{P}\times\hat{\boldsymbol{\alpha}})\hat{\beta}$ follows immediately from (10.118).

Combining the results (10.117) and (10.118) with those evaluated previously for the cross-section, the polarization of the scattered beam is given by the expression

$$\mathbf{P}'\left(\frac{\mathrm{d}\sigma}{\mathrm{d}\Omega}\right)=\mathbf{P}\left|\sum_{\mathbf{l}}\exp(\mathrm{i}\boldsymbol{\varkappa}.\mathbf{l})\right|^2|F_{\mathrm{N}}(\boldsymbol{\varkappa})|^2-\mathbf{P}N\sum_{\mathbf{d}}\{|\bar{b}_d|^2+\tfrac{1}{3}\overline{|b_d|^2}-\tfrac{4}{3}\overline{|A_d|^2}\}+$$

$$+\left(\frac{\gamma e^2}{m_{\mathrm{e}}c^2}\right)^2\sum_{\mathbf{l},\mathbf{d}}\sum_{\mathbf{l}',\mathbf{d}'}\exp\{\mathrm{i}\boldsymbol{\varkappa}.(\mathbf{R}_{ld}-\mathbf{R}_{l'd'})\}\tfrac{1}{2}g_{d'}F^{*}_{d'}(\boldsymbol{\varkappa})\langle\hat{S}_{l'd'}\rangle\tfrac{1}{2}g_{d}F_{d}(\boldsymbol{\varkappa})\langle\hat{S}_{ld}\rangle\times$$

$$\times\{-\mathbf{P}[\tilde{\boldsymbol{\varkappa}}\times(\tilde{\boldsymbol{\eta}}_{l'd'}\times\tilde{\boldsymbol{\varkappa}})].[\tilde{\boldsymbol{\varkappa}}\times(\tilde{\boldsymbol{\eta}}_{ld}\times\tilde{\boldsymbol{\varkappa}})]+[\tilde{\boldsymbol{\varkappa}}\times(\tilde{\boldsymbol{\eta}}_{l'd'}\times\tilde{\boldsymbol{\varkappa}})]\tilde{\boldsymbol{\eta}}_{ld}.\mathbf{P}_{\perp}+$$

$$+\tilde{\boldsymbol{\eta}}_{l'd'}.\mathbf{P}_{\perp}[\tilde{\boldsymbol{\varkappa}}\times(\tilde{\boldsymbol{\eta}}_{ld}\times\tilde{\boldsymbol{\varkappa}})]-\mathrm{i}[\tilde{\boldsymbol{\varkappa}}\times(\tilde{\boldsymbol{\eta}}_{l'd'}\times\tilde{\boldsymbol{\varkappa}})]\times[\tilde{\boldsymbol{\varkappa}}\times(\tilde{\boldsymbol{\eta}}_{ld}\times\tilde{\boldsymbol{\varkappa}})]\}+$$

$$+\left(\frac{\gamma e^2}{m_{\mathrm{e}}c^2}\right)\sum_{\mathbf{l},\mathbf{d}}\sum_{\mathbf{l}',\mathbf{d}'}\exp\{\mathrm{i}\boldsymbol{\varkappa}.(\mathbf{R}_{ld}-\mathbf{R}_{l'd'})\}\{\bar{b}^{*}_{d'}\tfrac{1}{2}g_{d}F_{d}(\boldsymbol{\varkappa})\langle\hat{S}_{ld}\rangle[\tilde{\boldsymbol{\varkappa}}\times(\tilde{\boldsymbol{\eta}}_{ld}\times\tilde{\boldsymbol{\varkappa}})]+$$

$$+\bar{b}_{d}\tfrac{1}{2}g_{d'}F^{*}_{d'}(\boldsymbol{\varkappa})\langle\hat{S}_{l'd'}\rangle[\tilde{\boldsymbol{\varkappa}}\times(\tilde{\boldsymbol{\eta}}_{l'd'}\times\tilde{\boldsymbol{\varkappa}})]+$$

$$+\mathrm{i}\bar{b}^{*}_{d'}\tfrac{1}{2}g_{d}F_{d}(\boldsymbol{\varkappa})\langle\hat{S}_{ld}\rangle[\tilde{\boldsymbol{\varkappa}}\times(\tilde{\boldsymbol{\eta}}_{ld}\times\tilde{\boldsymbol{\varkappa}})]\times\mathbf{P}+$$

$$+\mathrm{i}\bar{b}^{*}_{d'}\tfrac{1}{2}g_{d'}F^{*}_{d'}(\boldsymbol{\varkappa})\langle\hat{S}_{l'd'}\rangle\mathbf{P}\times[\tilde{\boldsymbol{\varkappa}}\times(\tilde{\boldsymbol{\eta}}_{l'd'}\times\tilde{\boldsymbol{\varkappa}})]\}. \quad (10.119)$$

The first two terms in (10.119) are the purely nuclear contributions to $\mathbf{P}'$ and are identical to the results derived in § 10.4. The second set of terms in (10.119), with the coefficient $(\gamma e^2/m_e c^2)^2$, are the purely magnetic contributions to $\mathbf{P}'$ and the third set, with coefficient $(\gamma e^2/m_e c^2)$, are the contributions arising from the interference between the nuclear and magnetic scattering. When we evaluate $\mathbf{P}'$ for a Bragg reflection the nuclear incoherent contribution can be neglected. Note that two contributions in (10.119) create polarization, one of them being purely magnetic and the other arising from the interference between nuclear and magnetic scattering. If for the Bragg reflection being examined there is not both nuclear and magnetic scattering then the latter cannot contribute to the polarization of an initially unpolarized beam.

If all the spins in the crystal are parallel or antiparallel to a single direction $\tilde{\boldsymbol{\eta}}$ the expression (10.119) for $\mathbf{P}'$ simplifies considerably. Taking just the Bragg part,

$$\mathbf{P}'\left(\frac{\mathrm{d}\sigma}{\mathrm{d}\Omega}\right) = \left|\sum_{\mathbf{l}} \exp(\mathrm{i}\boldsymbol{\varkappa}.\mathbf{l})\right|^2 \Big\{\mathbf{P}|F_{\mathrm{N}}(\boldsymbol{\varkappa})|^2 + $$
$$+\left(\frac{\gamma e^2}{m_e c^2}\right)^2 [-\mathbf{P}\{1-(\tilde{\boldsymbol{\varkappa}}.\tilde{\boldsymbol{\eta}})^2\}+2\tilde{\boldsymbol{\eta}}.\mathbf{P}_{\perp}\{\tilde{\boldsymbol{\varkappa}}\times(\tilde{\boldsymbol{\eta}}\times\tilde{\boldsymbol{\varkappa}})\}]|F_{\mathrm{M}}(\boldsymbol{\varkappa})|^2 +$$
$$+\left(\frac{2\gamma e^2}{m_e c^2}\right)[\tilde{\boldsymbol{\varkappa}}\times(\tilde{\boldsymbol{\eta}}\times\tilde{\boldsymbol{\varkappa}})]\mathrm{Re}\, F_{\mathrm{N}}(\boldsymbol{\varkappa})F_{\mathrm{M}}^*(\boldsymbol{\varkappa}) -$$
$$-\left(\frac{2\gamma e^2}{m_e c^2}\right)\mathbf{P}\times[\tilde{\boldsymbol{\varkappa}}\times(\tilde{\boldsymbol{\eta}}\times\tilde{\boldsymbol{\varkappa}})]\mathrm{Im}\, F_{\mathrm{N}}(\boldsymbol{\varkappa})F_{\mathrm{M}}^*(\boldsymbol{\varkappa})\Big\}, \tag{10.120}$$

where $(\mathrm{d}\sigma/\mathrm{d}\Omega)$ appearing in this expression is given by (10.95). Notice that the last term in this expression for $\mathbf{P}'$ involves the imaginary part of the magnetic unit-cell structure factor.† If the magnetic crystal structure is Bravais then each ion is at a centre of symmetry and the atomic magnetic form factor purely real. However, in non-centrosymmetric magnetic structures $F(\boldsymbol{\varkappa})$ can possess an imaginary part and this may be detected by observing the polarization of the scattered beam. For if it is arranged to have $\mathbf{P}$ perpendicular to $\tilde{\boldsymbol{\varkappa}}\times(\tilde{\boldsymbol{\eta}}\times\tilde{\boldsymbol{\varkappa}})$ the term containing Im $F(\boldsymbol{\varkappa})$ contributes a component to $\mathbf{P}'$ that is perpendicular to both vectors whereas all other contributions coincide with either $\mathbf{P}$ or $\tilde{\boldsymbol{\varkappa}}\times(\tilde{\boldsymbol{\eta}}\times\tilde{\boldsymbol{\varkappa}})$.

Fig. 10.6 shows the various vectors involved in (10.120) with the purpose of clarifying the relations amongst them.

† M. Blume, *Phys. Rev.* **130**, 1670 (1963).

For the case of a ferromagnet with just one atom per unit cell the expression (10.120) takes the very simple form

$$\mathbf{P}'\left(\frac{d\sigma}{d\Omega}\right) = \left|\sum_{\mathbf{l}} \exp(i\boldsymbol{\varkappa}.\mathbf{l})\right|^2 \left\{\mathbf{P}\bar{b}^2 + \left(\frac{\gamma e^2}{m_e c^2}\right)^2 [-\mathbf{P}\{1-(\tilde{\boldsymbol{\varkappa}}.\tilde{\boldsymbol{\eta}})^2\} + \right.$$
$$+2(\tilde{\boldsymbol{\eta}}.\mathbf{P}_\perp)\{\tilde{\boldsymbol{\varkappa}}\times(\tilde{\boldsymbol{\eta}}\times\tilde{\boldsymbol{\varkappa}})\}]\{\tfrac{1}{2}gF(\boldsymbol{\varkappa})\langle\hat{S}\rangle\}^2 +$$
$$\left. + \left(\frac{2\gamma e^2}{m_e c^2}\right)[\tilde{\boldsymbol{\varkappa}}\times(\tilde{\boldsymbol{\eta}}\times\boldsymbol{\varkappa})]\tfrac{1}{2}gF(\boldsymbol{\varkappa})\langle\hat{S}\rangle\bar{b}\right\} \quad (10.121)$$

with $(d\sigma/d\Omega)$ given by (10.96).

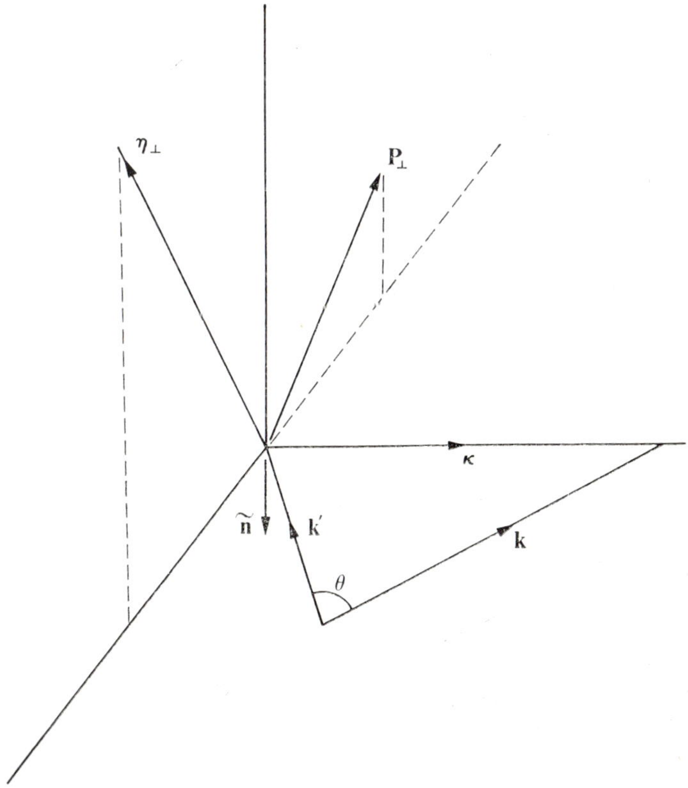

FIG. 10.6. Vectors in elastic neutron cross-section and expression for $\mathbf{P}_\perp = \tilde{\boldsymbol{\varkappa}}\times(\mathbf{P}'\times\tilde{\boldsymbol{\varkappa}})$ and similarly for $\boldsymbol{\eta}_\perp$. $\tilde{\mathbf{n}} = \mathbf{k}'\times\mathbf{k}/k^2 \sin\theta$.

A ferromagnet can be used to produce a polarized beam of neutrons. If the incident beam is unpolarized the expression (10.121) shows that the polarization of the scattered beam is given by

$$\mathbf{P}' = \frac{(2\gamma e^2/m_e c^2)\{\tilde{\boldsymbol{\tau}}\times(\tilde{\boldsymbol{\eta}}\times\tilde{\boldsymbol{\tau}})\}\tfrac{1}{2}gF(\boldsymbol{\tau})\langle\hat{S}\rangle\bar{b}}{\bar{b}^2+(\gamma e^2/m_e c^2)^2\{1-(\tilde{\boldsymbol{\tau}}.\tilde{\boldsymbol{\eta}})^2\}\{\tfrac{1}{2}gF(\boldsymbol{\tau})\langle\hat{S}\rangle\}^2}. \quad (10.122)$$

Hence if a reflection exists where

$$\frac{1}{b}\left(\frac{\gamma e^2}{m_e c^2}\right)\{1-(\tilde{\boldsymbol{\tau}}.\tilde{\boldsymbol{\eta}})^2\}^{\frac{1}{2}}\tfrac{1}{2}gF(\boldsymbol{\tau})\langle\hat{S}\rangle = \pm 1$$

the scattered Bragg beam is seen to be completely polarized. Most polarizers and analysers are made from crystals of cobalt–iron alloy with the arrangement $\tilde{\boldsymbol{\varkappa}}.\tilde{\boldsymbol{\eta}} = 0$. The above condition is then well fulfilled by the (111) reflection of the Co–Fe alloy containing 92 atomic per cent cobalt.

(f) Polarization of neutrons scattered from a helical structure

We calculate the polarization of neutrons scattered by a helical spin configuration only for the special case in which the incident beam is unpolarized. We choose as our example antiferromagnetic MnO_2. The spin configuration for this system is defined by $\tilde{\boldsymbol{\eta}}_l$ given in (10.100). Because the nuclear and magnetic scattering from MnO_2 do not coincide, the created polarization comes solely from the term

$$\left(\frac{\gamma e^2}{m_e c^2}\right)^2 |F(\boldsymbol{\varkappa})|^2\langle\hat{S}\rangle^2 \sum_{\mathbf{l},\mathbf{l}'} \exp\{\mathrm{i}\boldsymbol{\varkappa}.(\mathbf{l}-\mathbf{l}')\}(-\mathrm{i})[\tilde{\boldsymbol{\varkappa}}\times(\tilde{\boldsymbol{\eta}}_{l'}\times\tilde{\boldsymbol{\varkappa}})]\times[\tilde{\boldsymbol{\varkappa}}\times(\tilde{\boldsymbol{\eta}}_l\times\tilde{\boldsymbol{\varkappa}})] \tag{10.123}$$

in the general expression (10.119).

By using the results (10.103) and (10.104) we find that (10.123) is equal to

$$-\left(\frac{\gamma e^2}{m_e c^2}\right)^2 |F(\boldsymbol{\varkappa})|^2\tfrac{1}{2}\langle\hat{S}\rangle^2\tilde{\kappa}_z\tilde{\boldsymbol{\varkappa}}\left\{\left|\sum_{\mathbf{l}}\exp\{\mathrm{i}\mathbf{l}(\boldsymbol{\varkappa}+\mathbf{w}+\mathbf{Q})\}\right|^2 - \left|\sum_{\mathbf{l}}\exp\{\mathrm{i}\mathbf{l}(\boldsymbol{\varkappa}+\mathbf{w}-\mathbf{Q})\}\right|^2\right\}. \tag{10.124}$$

This result shows that the polarization created in the two Bragg peaks $\boldsymbol{\varkappa} = \boldsymbol{\tau}-\mathbf{w}\pm\mathbf{Q}$ is oppositely directed; in one it is parallel to the scattering vector $\boldsymbol{\varkappa}$ and in the other it is antiparallel to $\boldsymbol{\varkappa}$. We also note that if the scattering vector is arranged perpendicular to the screw axis, in this instance the z axis, the polarization is not observed.

(g) Cross-section and created polarization with nuclear, magnetic, and electrostatic scattering

As a final topic in elastic scattering we briefly discuss the cross-section and polarization of an initially unpolarized beam for scattering by nuclear, magnetic, and electrostatic interactions. Since we have considered pure nuclear, magnetic, and electrostatic scattering and also the nuclear–electrostatic and nuclear–magnetic interference terms, we have only to consider the magnetic and electrostatic interference terms. We simplify our discussion by restricting attention to magnetic systems where the spins are either parallel or antiparallel to the direction defined by the unit vector $\tilde{\boldsymbol{\eta}}$. The interaction potential for the magnetic and electrostatic scattering of neutrons is obtained from the sum of (10.55) and (10.85) and is, for the particular system under discussion,

$$\langle\lambda|\,\hat{V}(\boldsymbol{\varkappa})\,|\lambda\rangle = \left(\frac{2\pi\hbar^2}{m}\right)\left(\frac{\gamma e^2}{m_e c^2}\right)\sum_{\mathbf{l}} \exp(\mathrm{i}\boldsymbol{\varkappa}.\mathbf{l})\times$$

$$\times\left\{-\frac{1}{2}\frac{m_e}{m}F_E(\boldsymbol{\varkappa})+\hat{\boldsymbol{\sigma}}.\left[\tfrac{1}{2}\mathrm{i}\frac{m_e}{m}\cot(\tfrac{1}{2}\theta)F_E(\boldsymbol{\varkappa})\tilde{\mathbf{n}}+F_M(\boldsymbol{\varkappa})\{\tilde{\boldsymbol{\varkappa}}\times(\tilde{\boldsymbol{\eta}}\times\tilde{\boldsymbol{\varkappa}})\}\right]\right\}. \tag{10.125}$$

A straightforward calculation gives for the cross-section the result

$$\frac{\mathrm{d}\sigma}{\mathrm{d}\Omega} = \left(\frac{\mathrm{d}\sigma}{\mathrm{d}\Omega}\right)_{\mathrm{EE}}+\left(\frac{\mathrm{d}\sigma}{\mathrm{d}\Omega}\right)_{\mathrm{MM}}+\left|\sum_{\mathbf{l}}\exp(\mathrm{i}\boldsymbol{\varkappa}.\mathbf{l})\right|^2\frac{m_e}{m}\left(\frac{\gamma e^2}{m_e c^2}\right)^2\times$$

$$\times[\tilde{\boldsymbol{\eta}}.\tilde{\mathbf{n}}\cot(\tfrac{1}{2}\theta)\mathrm{Im}\,F_M(\boldsymbol{\varkappa})F_E^*(\boldsymbol{\varkappa})+\tilde{\boldsymbol{\eta}}.\{\cot(\tfrac{1}{2}\theta)(\tilde{\boldsymbol{\varkappa}}.\mathbf{P})(\tilde{\boldsymbol{\varkappa}}\times\tilde{\mathbf{n}})-\mathbf{P}_\perp\}\times$$

$$\times\mathrm{Re}\,F_M(\boldsymbol{\varkappa})F_E^*(\boldsymbol{\varkappa})]. \tag{10.126}$$

We see that there is a polarization-independent interference term between magnetic and electrostatic scattering. Since this term involves the imaginary parts of the magnetic and electrostatic unit-cell structure factors it can be dropped in the great majority of cases. For the other interference term we note that it involves the scalar product of $\tilde{\boldsymbol{\eta}}$ with two vectors both of which are perpendicular to the scattering vector. Thus this whole term is zero if $\tilde{\boldsymbol{\eta}}$ is parallel to $\boldsymbol{\varkappa}$, and the cross-section is then independent of the polarization **P**. That $\tilde{\boldsymbol{\eta}}$ be parallel to $\boldsymbol{\varkappa}$ is also the condition for the nuclear–magnetic interference term to be zero. In view of the identity

$$\tilde{\boldsymbol{\eta}}.\{\cot(\tfrac{1}{2}\theta)(\tilde{\boldsymbol{\varkappa}}.\mathbf{P})(\tilde{\boldsymbol{\varkappa}}\times\tilde{\mathbf{n}})-\mathbf{P}_\perp\} = \tilde{\boldsymbol{\eta}}.\left\{\frac{2\mathbf{k}(\tilde{\boldsymbol{\varkappa}}.\mathbf{P})}{|\boldsymbol{\varkappa}|}-\mathbf{P}\right\} \tag{10.127}$$

we observe that the polarization-dependent interference term has the dependence $-\tilde{\boldsymbol{\eta}}.\mathbf{P}$ if $\tilde{\boldsymbol{\eta}}$ is arranged to be perpendicular to the incident beam. Clearly the term is zero, if, in addition to this condition, we have also **P** perpendicular to the magnetization.

For the polarization of the scattered beam when the incident beam is unpolarized we find

$$\mathbf{P}'.\tilde{\mathbf{R}} \propto -\tilde{\mathbf{R}}[\{\tilde{\boldsymbol{\varkappa}}\times(\tilde{\boldsymbol{\eta}}\times\tilde{\boldsymbol{\varkappa}})\}+\cot(\tfrac{1}{2}\theta)\tilde{\boldsymbol{\varkappa}}\{\tilde{\boldsymbol{\eta}}.\tilde{\boldsymbol{\varkappa}}\times\tilde{\mathbf{n}}\}]\mathrm{Re}\,F_M(\boldsymbol{\varkappa})F_E^*(\boldsymbol{\varkappa}), \tag{10.128}$$

where $\tilde{\mathbf{R}}$ defines the direction relative to which the polarization is observed. Since

$$\tilde{\mathbf{R}}.[\{\tilde{\boldsymbol{\varkappa}}\times(\tilde{\boldsymbol{\eta}}\times\tilde{\boldsymbol{\varkappa}})\}+\cot(\tfrac{1}{2}\theta)\tilde{\boldsymbol{\varkappa}}\{\tilde{\boldsymbol{\eta}}.\tilde{\boldsymbol{\varkappa}}\times\tilde{\mathbf{n}}\}] = \tilde{\mathbf{R}}.\left\{\tilde{\boldsymbol{\eta}}+\frac{2\tilde{\boldsymbol{\varkappa}}(\mathbf{k}'.\tilde{\boldsymbol{\eta}})}{|\boldsymbol{\varkappa}|}\right\} \quad (10.129)$$

it is readily shown that the created polarization is perpendicular to the wave vector of the scattered beam $\mathbf{k}'$. Furthermore, (10.129) allows us to write the polarization in terms of the vectors $\tilde{\mathbf{n}}$ and $\tilde{\mathbf{n}}\times\mathbf{k}'$; namely

$$\mathbf{P}'.\tilde{\mathbf{R}} \propto \tilde{\mathbf{R}}.\left\{(\tilde{\boldsymbol{\eta}}.\tilde{\mathbf{n}})\tilde{\mathbf{n}}+\frac{2\tilde{\boldsymbol{\eta}}.(\tilde{\boldsymbol{\varkappa}}\times\tilde{\mathbf{n}})}{|\boldsymbol{\varkappa}|}(\tilde{\boldsymbol{\eta}}\times\mathbf{k}')\right\}\operatorname{Re} F_{\mathrm{M}}(\boldsymbol{\varkappa})F^{*}_{\mathrm{E}}(\boldsymbol{\varkappa}). \quad (10.130)$$

If we combine the results for magnetic and electrostatic scattering with those derived earlier for nuclear and electrostatic scattering, and nuclear and magnetic scattering, the polarization of an initially unpolarized beam of neutrons is given by

$$\begin{aligned}\mathbf{P}'.\tilde{\mathbf{R}} = \left(\frac{\gamma e^2}{m_{\mathrm{e}}c^2}\right)\Bigg[&\tilde{\mathbf{R}}.\{\tilde{\boldsymbol{\tau}}\times(\tilde{\boldsymbol{\eta}}\times\tilde{\boldsymbol{\tau}})\}2\operatorname{Re} F^{*}_{\mathrm{N}}(\boldsymbol{\tau})F_{\mathrm{M}}(\boldsymbol{\tau})+\\ &+\left(\frac{m_{\mathrm{e}}}{m}\right)\cot(\tfrac{1}{2}\theta)\tilde{\mathbf{R}}.\tilde{\mathbf{n}}\operatorname{Im} F_{\mathrm{N}}(\boldsymbol{\tau})F^{*}_{\mathrm{E}}(\boldsymbol{\tau})-\\ &-\left(\frac{m_{\mathrm{e}}}{m}\right)\left(\frac{\gamma e^2}{m_{\mathrm{e}}c^2}\right)\tilde{\mathbf{R}}.\left\{\tilde{\boldsymbol{\eta}}+\frac{2\tilde{\boldsymbol{\tau}}(\mathbf{k}'.\tilde{\boldsymbol{\eta}})}{|\boldsymbol{\tau}|}\right\}\operatorname{Re} F_{\mathrm{M}}(\boldsymbol{\tau})F^{*}_{\mathrm{E}}(\boldsymbol{\tau})\Bigg]\times\\ \times\Bigg[|F_{\mathrm{N}}(\boldsymbol{\tau})|^2+&\left(\frac{\gamma e^2}{m_{\mathrm{e}}c^2}\right)^2\{1-(\tilde{\boldsymbol{\tau}}.\tilde{\boldsymbol{\eta}})^2\}|F_{\mathrm{M}}(\boldsymbol{\tau})|^2+\\ &+\left(\frac{m_{\mathrm{e}}}{2m}\right)^2\left(\frac{\gamma e^2}{m_{\mathrm{e}}c^2}\right)^2(\sin\tfrac{1}{2}\theta)^{-2}|F_{\mathrm{E}}(\boldsymbol{\tau})|^2-\\ &-\left(\frac{m_{\mathrm{e}}}{m}\right)\left(\frac{\gamma e^2}{m_{\mathrm{e}}c^2}\right)\operatorname{Re} F_{\mathrm{N}}(\boldsymbol{\tau})F^{*}_{\mathrm{E}}(\boldsymbol{\tau})\Bigg]^{-1}. \quad (10.131)\end{aligned}$$

Only under special circumstances will the components in $\mathbf{P}'$ involving the electrostatic scattering be significant because normally the nuclear and magnetic contributions will be dominant.

10.6. Inelastic magnetic scattering

To calculate the cross-section for inelastic magnetic scattering of polarized neutrons and, also, the polarization of the scattered beam, we take in our general expressions (10.37) and (10.41), $\hat{\beta} \equiv 0$ and for $\hat{\boldsymbol{\alpha}}$

expression (10.75). Thus from (10.37) the full expression for the required cross-section is

$$\frac{d^2\sigma}{d\Omega dE'} = \left(\frac{m}{2\pi\hbar^2}\right)^2 \frac{k'}{k} \sum_{\lambda,\lambda'} p_\lambda \times$$

$$\times\{\langle\lambda|\,\hat{\boldsymbol{\alpha}}^+\,|\lambda'\rangle.\langle\lambda'|\,\hat{\boldsymbol{\alpha}}\,|\lambda\rangle + i\mathbf{P}.\langle\lambda|\,\hat{\boldsymbol{\alpha}}^+\,|\lambda'\rangle\times\langle\lambda'|\,\hat{\boldsymbol{\alpha}}\,|\lambda\rangle\}\delta(\hbar\omega+E_\lambda-E_{\lambda'}). \tag{10.132}$$

For the polarization of the scattered beam we have from (10.41)

$$\mathbf{P}'\left(\frac{d^2\sigma}{d\Omega dE'}\right) = \left(\frac{m}{2\pi\hbar^2}\right)^2 \frac{k'}{k} \sum_{\lambda,\lambda'} p_\lambda\{\langle\lambda|\,\hat{\boldsymbol{\alpha}}^+\,|\lambda'\rangle\langle\lambda'|\,\hat{\boldsymbol{\alpha}}.\mathbf{P}\,|\lambda\rangle +$$

$$+\langle\lambda|\,\hat{\boldsymbol{\alpha}}^+.\mathbf{P}\,|\lambda'\rangle\langle\lambda'|\,\hat{\boldsymbol{\alpha}}\,|\lambda\rangle - \mathbf{P}\langle\lambda|\,\hat{\boldsymbol{\alpha}}^+\,|\lambda'\rangle.\langle\lambda'|\,\hat{\boldsymbol{\alpha}}\,|\lambda\rangle -$$

$$-i\langle\lambda|\,\hat{\boldsymbol{\alpha}}^+\,|\lambda'\rangle\times\langle\lambda'|\,\hat{\boldsymbol{\alpha}}\,|\lambda\rangle\}\delta(\hbar\omega+E_\lambda-E_{\lambda'}). \tag{10.133}$$

We have previously shown that the cross-section for magnetic scattering of unpolarized neutrons can be expressed in terms of correlation functions of operators that belong to the target system (cf. Chapter 8). Such a programme can also be undertaken for the additional term in the cross-section (10.132) that arises because we now have polarized incident neutrons. Clearly then, each term in the expression on the right-hand side of (10.133) for $\mathbf{P}'$ can be written in terms of correlation functions. In carrying through this programme for the expressions (10.132) and (10.133) we need only look in detail at one term to see how the remaining terms go. We choose to examine the simplest term to occur in (10.132) and (10.133); namely

$$\left(\frac{m}{2\pi\hbar^2}\right)^2 \frac{k'}{k} \sum_{\lambda,\lambda'} p_\lambda\langle\lambda|\,\hat{\boldsymbol{\alpha}}^+\,|\lambda'\rangle.\langle\lambda'|\,\hat{\boldsymbol{\alpha}}\,|\lambda\rangle\delta(\hbar\omega+E_\lambda-E_{\lambda'}). \tag{10.134}$$

By using expression (10.75) for $\hat{\boldsymbol{\alpha}}$ and taking p_λ to be the Boltzmann distribution $p_\lambda = \exp(-\beta E_\lambda)/\mathrm{Tr}\exp(-\beta\mathscr{H})$ we have for (10.134)

$$\frac{k'}{k}\left(\frac{1}{\mathrm{Tr}\exp(-\beta\mathscr{H})}\right)\sum_{\lambda,\lambda'}\exp(-\beta E_\lambda)\sum_{\mathbf{l,d}}\sum_{\mathbf{l',d'}}\exp\{i\boldsymbol{\kappa}.(\mathbf{R}_{l'd'}-\mathbf{R}_{ld})\}\left(\frac{\gamma e^2}{m_e c^2}\right)^2\times$$

$$\times\tfrac{1}{2}g_d F_d^*(\boldsymbol{\kappa})\tfrac{1}{2}g_{d'}F_{d'}(\boldsymbol{\kappa})\langle\lambda|\,\tilde{\boldsymbol{\kappa}}\times(\hat{\mathbf{S}}_{ld}\times\tilde{\boldsymbol{\kappa}})\,|\lambda'\rangle.\langle\lambda'|\,\tilde{\boldsymbol{\kappa}}\times(\hat{\mathbf{S}}_{l'd'}\times\tilde{\boldsymbol{\kappa}})\,|\lambda\rangle\times$$

$$\times\frac{1}{2\pi\hbar}\int_{-\infty}^{\infty} dt\,\exp\{-i\omega t+it(E_{\lambda'}-E_\lambda)/\hbar\}. \tag{10.135}$$

Here we have used the integral representation of the delta function.

Now

$$\left(\frac{1}{\operatorname{Tr}\exp(-\beta\mathscr{H})}\right)\sum_{\lambda,\lambda'}\exp(-\beta E_\lambda)\langle\lambda|\,\tilde{\boldsymbol{\varkappa}}\times(\hat{\mathbf{S}}_{ld}\times\tilde{\boldsymbol{\varkappa}})|\lambda'\rangle.\langle\lambda'|\,\tilde{\boldsymbol{\varkappa}}\times(\hat{\mathbf{S}}_{l'd'}\times\tilde{\boldsymbol{\varkappa}})|\lambda\rangle\times$$

$$\times\frac{1}{2\pi\hbar}\int_{-\infty}^{\infty}\mathrm{d}t\,\exp\{-\mathrm{i}\omega t+\mathrm{i}t(E_{\lambda'}-E_\lambda)/\hbar\}$$

$$=\frac{1}{2\pi\hbar}\int_{-\infty}^{\infty}\mathrm{d}t\,\exp(-\mathrm{i}\omega t)\left(\frac{1}{\operatorname{Tr}\exp(-\beta\mathscr{H})}\right)\sum_{\lambda,\lambda'}\exp(-\beta E_\lambda)\times$$

$$\times\langle\lambda|\,\tilde{\boldsymbol{\varkappa}}\times\{\hat{\mathbf{S}}_{ld}(0)\times\tilde{\boldsymbol{\varkappa}}\}|\lambda'\rangle.\langle\lambda'|\,\tilde{\boldsymbol{\varkappa}}\times(\hat{\mathbf{S}}_{l'd'}(t)\times\tilde{\boldsymbol{\varkappa}})|\lambda\rangle$$

$$=\frac{1}{2\pi\hbar}\int_{-\infty}^{\infty}\mathrm{d}t\,\exp(-\mathrm{i}\omega t)\langle[\tilde{\boldsymbol{\varkappa}}\times\{\hat{\mathbf{S}}_{ld}(0)\times\tilde{\boldsymbol{\varkappa}}\}].[\tilde{\boldsymbol{\varkappa}}\times\{\hat{\mathbf{S}}_{l'd'}(t)\times\tilde{\boldsymbol{\varkappa}}\}]\rangle, \quad (10.136)$$

where $\hat{\mathbf{S}}_{ld}(t)=\exp(\mathrm{i}\mathscr{H}t/\hbar)\hat{\mathbf{S}}_{ld}\exp(-\mathrm{i}\mathscr{H}t/\hbar).$

Using (10.136) in (10.135) gives for the term (10.134)

$$\left(\frac{\gamma e^2}{m_\mathrm{e}c^2}\right)^2\frac{k'}{k}\sum_{\mathbf{l},\mathbf{d}}\sum_{\mathbf{l}',\mathbf{d}'}\exp\{\mathrm{i}\boldsymbol{\varkappa}.(\mathbf{R}_{l'd'}-\mathbf{R}_{ld})\}\tfrac{1}{2}g_d\,F_d^*(\boldsymbol{\varkappa})\tfrac{1}{2}g_{d'}\,F_{d'}(\boldsymbol{\varkappa})\times$$

$$\times\frac{1}{2\pi\hbar}\int_{-\infty}^{\infty}\mathrm{d}t\,\exp(-\mathrm{i}\omega t)\langle\hat{\mathbf{S}}_{ld}^{(\perp)}(0).\hat{\mathbf{S}}_{l'd'}^{(\perp)}(t)\rangle, \quad (10.137)$$

where for brevity we have defined

$$\hat{\mathbf{S}}_{ld}^{(\perp)}(t)=\tilde{\boldsymbol{\varkappa}}\times\{\hat{\mathbf{S}}_{ld}(t)\times\tilde{\boldsymbol{\varkappa}}\}. \quad (10.138)$$

In view of the result (10.137) for the term (10.134) in the cross-section it is clear that we can write the cross-section (10.132) for the magnetic scattering of polarized neutrons in the form

$$\frac{\mathrm{d}^2\sigma}{\mathrm{d}\Omega\mathrm{d}E'}=\left(\frac{\gamma e^2}{m_\mathrm{e}c^2}\right)^2\frac{k'}{k}\sum_{\mathbf{l},\mathbf{d}}\sum_{\mathbf{l}',\mathbf{d}'}\exp\{\mathrm{i}\boldsymbol{\varkappa}.(\mathbf{R}_{l'd'}-\mathbf{R}_{ld})\}\tfrac{1}{2}g_d\,F_d^*(\boldsymbol{\varkappa})\tfrac{1}{2}g_{d'}\,F_{d'}(\boldsymbol{\varkappa})\times$$

$$\times\frac{1}{2\pi\hbar}\int_{-\infty}^{\infty}\mathrm{d}t\,\exp(-\mathrm{i}\omega t)\{\langle\hat{\mathbf{S}}_{ld}^{(\perp)}(0).\hat{\mathbf{S}}_{l'd'}^{(\perp)}(t)\rangle+\mathrm{i}\mathbf{P}.\langle\hat{\mathbf{S}}_{ld}^{(\perp)}(0)\times\hat{\mathbf{S}}_{l'd'}^{(\perp)}(t)\rangle\}. \quad (10.139)$$

By inspecting (10.133) we see that the polarization of the scattered neutrons can be written

$$\mathbf{P}'\left(\frac{\mathrm{d}^2\sigma}{\mathrm{d}\Omega\mathrm{d}E'}\right)=\left(\frac{\gamma e^2}{m_\mathrm{e}c^2}\right)^2\frac{k'}{k}\sum_{\mathbf{l},\mathbf{d}}\sum_{\mathbf{l}',\mathbf{d}'}\exp\{\mathrm{i}\boldsymbol{\varkappa}(\mathbf{R}_{l'd'}-\mathbf{R}_{ld})\}\tfrac{1}{2}g_d\,F_d^*(\boldsymbol{\varkappa})\tfrac{1}{2}g_{d'}\,F_{d'}(\boldsymbol{\varkappa})\times$$

$$\times\frac{1}{2\pi\hbar}\int_{-\infty}^{\infty}\mathrm{d}t\,\exp(-\mathrm{i}\omega t)\{\langle\hat{\mathbf{S}}_{ld}^{(\perp)}(0)\{\mathbf{P}.\hat{\mathbf{S}}_{l'd'}^{(\perp)}(t)\}+\{\mathbf{P}.\hat{\mathbf{S}}_{ld}^{(\perp)}(0)\}\hat{\mathbf{S}}_{l'd'}^{(\perp)}(t)\rangle-$$

$$-\mathbf{P}\langle\hat{\mathbf{S}}_{ld}^{(\perp)}(0).\hat{\mathbf{S}}_{l'd'}^{(\perp)}(t)\rangle-\mathrm{i}\langle\hat{\mathbf{S}}_{ld}^{(\perp)}(0)\times\hat{\mathbf{S}}_{l'd'}^{(\perp)}(t)\rangle\}. \quad (10.140)$$

(10.139) and (10.140) are the key equations for discussing polarization effects in inelastic magnetic scattering.

In the particular case when the total z component of spin is a constant of motion equations (10.139) and (10.140) are greatly simplified. The reason for this is that if

$$[\hat{S}^z_{\text{tot}}, \mathscr{H}] = 0 \tag{10.141}$$

then we have the identity

$$\langle \hat{S}^{\pm}_{ld}(0)\hat{S}^{\pm}_{l'd'}(t)\rangle = 0 \tag{10.142}$$

and as a result many of the terms in (10.139) and (10.140) are identically zero.

In investigating the consequences of (10.141) in eqns (10.139) and (10.140) we assume a Bravais crystal structure and take $\tilde{\boldsymbol{\eta}}$ to be a unit vector in the direction of quantization, in this instance the z axis.

First look at

$$\langle \hat{\mathbf{S}}^{(\perp)}_l(0) . \hat{\mathbf{S}}^{(\perp)}_{l'}(t)\rangle = \sum_{\alpha,\beta} (\delta_{\alpha\beta} - \tilde{\kappa}_\alpha \tilde{\kappa}_\beta)\langle \hat{S}^\alpha_l(0)\hat{S}^\beta_{l'}(t)\rangle, \tag{10.143}$$

which appears in both (10.139) and (10.140).

The terms in this expression with coefficient $-\tilde{\kappa}_x\tilde{\kappa}_y$ are zero. For

$$\langle \hat{S}^x_l(0)\hat{S}^y_{l'}(t) + \hat{S}^y_l(0)\hat{S}^x_{l'}(t)\rangle = \frac{1}{2\mathrm{i}}\langle \hat{S}^+_l(0)\hat{S}^+_{l'}(t) - \hat{S}^-_l(0)\hat{S}^-_{l'}(t)\rangle = 0.$$

Furthermore, correlation functions of the form $\langle \hat{S}^z_l(0)\hat{S}^\alpha_{l'}(t)\rangle$ with $\alpha, \beta = x$ or y must be identically zero. In addition, $\langle \hat{S}^z_l(0)\hat{S}^z_{l'}(t)\rangle$ does not contribute in one-magnon processes.

As a consequence of these results we have for (10.143)

$$\begin{aligned}\langle \hat{\mathbf{S}}^{(\perp)}_l(0) . \hat{\mathbf{S}}^{(\perp)}_{l'}(t)\rangle &= (1-\tilde{\kappa}_x^2)\langle \hat{S}^x_l(0)\hat{S}^x_{l'}(t)\rangle + (1-\tilde{\kappa}_y^2)\langle \hat{S}^y_l(0)\hat{S}^y_{l'}(t)\rangle \\ &= \tfrac{1}{4}(1+\tilde{\kappa}_z^2)\langle \hat{S}^+_l(0)\hat{S}^-_{l'}(t) + \hat{S}^-_l(0)\hat{S}^+_{l'}(t)\rangle. \end{aligned} \tag{10.144}$$

The reduction of the second part of the cross-section (10.139) proceeds in much the same way. With the definition (10.138) of the operator $\hat{\mathbf{S}}^{(\perp)}_l(t)$

$$\begin{aligned}&\mathbf{P} . \langle \hat{\mathbf{S}}^{(\perp)}_l(0) \times \hat{\mathbf{S}}^{(\perp)}_{l'}(t)\rangle \\ &= \langle \mathbf{P} . \hat{\mathbf{S}}_l(0) \times \hat{\mathbf{S}}_{l'}(t) + \hat{\mathbf{S}}_l(0) . (\mathbf{P} \times \tilde{\boldsymbol{\kappa}})\{\tilde{\boldsymbol{\kappa}} . \hat{\mathbf{S}}_{l'}(t)\} - \{\tilde{\boldsymbol{\kappa}} . \hat{\mathbf{S}}_l(t)\}\hat{\mathbf{S}}_{l'}(t) . (\mathbf{P} \times \tilde{\boldsymbol{\kappa}})\rangle \\ &= \langle \mathbf{P} . \hat{\mathbf{S}}_l(0) \times \hat{\mathbf{S}}_{l'}(t) + \sum_{\alpha,\beta} (\mathbf{P} \times \tilde{\boldsymbol{\kappa}})_\alpha \tilde{\kappa}_\beta\{\hat{S}^\alpha_l(0)\hat{S}^\beta_{l'}(t) - \hat{S}^\beta_l(0)\hat{S}^\alpha_{l'}(t)\}\rangle.\end{aligned}$$

Because we only have to keep those terms that contain the products of $\hat{S}^x$ and $\hat{S}^y$ this reduces to

$$\begin{aligned}&\mathbf{P} . \langle \hat{\mathbf{S}}^{(\perp)}_l(0) \times \hat{\mathbf{S}}^{(\perp)}_{l'}(t)\rangle \\ &= \{P_z + (\mathbf{P} \times \tilde{\boldsymbol{\kappa}})_x \tilde{\kappa}_y - (\mathbf{P} \times \tilde{\boldsymbol{\kappa}})_y \tilde{\kappa}_x\}\langle \hat{S}^x_l(0)\hat{S}^y_{l'}(t) - \hat{S}^y_l(0)\hat{S}^x_{l'}(t)\rangle \\ &= (\mathbf{P} . \tilde{\boldsymbol{\kappa}})(\tilde{\boldsymbol{\kappa}} . \tilde{\boldsymbol{\eta}}) \frac{1}{2\mathrm{i}}\langle \hat{S}^-_l(0)\hat{S}^+_{l'}(t) - \hat{S}^+_l(0)\hat{S}^-_{l'}(t)\rangle.\end{aligned} \tag{10.145}$$

On combining this result with (10.144) we have the following expression for the cross-section for inelastic magnetic scattering of polarized neutrons from a target whose total component of spin in the direction defined by the unit vector $\tilde{\boldsymbol{\eta}}$ is a constant of motion:

$$\frac{\mathrm{d}^2\sigma}{\mathrm{d}\Omega\mathrm{d}E'} = \left(\frac{\gamma e^2}{m_\mathrm{e}c^2}\right)^2 \frac{k'}{k} |\tfrac{1}{2}gF(\boldsymbol{\kappa})|^2 \sum_{\mathbf{l},\mathbf{l}'} \exp\{\mathrm{i}\boldsymbol{\kappa}.(\mathbf{l}'-\mathbf{l})\} \frac{1}{2\pi\hbar} \int_{-\infty}^{\infty} \mathrm{d}t \exp(-\mathrm{i}\omega t) \times$$
$$\times \tfrac{1}{4}[\{1+(\tilde{\boldsymbol{\kappa}}.\tilde{\boldsymbol{\eta}})^2+2(\mathbf{P}.\tilde{\boldsymbol{\kappa}})(\tilde{\boldsymbol{\kappa}}.\tilde{\boldsymbol{\eta}})\}\langle\hat{S}_l^-(0)\hat{S}_{l'}^+(t)\rangle+$$
$$+\{1+(\tilde{\boldsymbol{\kappa}}.\tilde{\boldsymbol{\eta}})^2-2(\mathbf{P}.\tilde{\boldsymbol{\kappa}})(\tilde{\boldsymbol{\kappa}}.\tilde{\boldsymbol{\eta}})\}\langle\hat{S}_l^+(0)\hat{S}_{l'}^-(t)\rangle]. \quad (10.146)$$

Note that this expression consists of the sum of two terms, one containing the correlation function $\langle\hat{S}_l^-(0)\hat{S}_{l'}^+(t)\rangle$ and the other the correlation function $\langle\hat{S}_l^+(0)\hat{S}_{l'}^-(t)\rangle$. These correlation functions are related to each other via

$$\langle\hat{S}_l^+(0)\hat{S}_{l'}^-(t)\rangle = \langle\hat{S}_{l'}^-(0)\hat{S}_l^+(-t+\mathrm{i}\hbar\beta)\rangle. \quad (10.147)$$

If it is arranged to have the polarization and magnetization of the sample parallel to the scattering vector then, clearly, the cross-section contains a contribution from only one of the two correlation functions. At low temperatures there is a very small density of magnetic excitations in the target system, so the cross-section involving the creation of an excitation will be much greater in this instance than that involving the annihilation of an excitation.

The correlation functions $\langle\hat{S}_l^{\pm}(0)\hat{S}_{l'}^{\mp}(t)\rangle$ are evaluated for a Heisenberg ferromagnet and an antiferromagnet in §§ 9.2 (*b*) and 9.6 (*c*).

Using the results of this chapter, the cross-sections for magnon creation (upper sign) and magnon annihilation (lower sign) in the scattering of polarized neutrons from a Heisenberg ferromagnet are, from (10.146),

$$\left(\frac{\mathrm{d}^2\sigma}{\mathrm{d}\Omega\mathrm{d}E'}\right)^{(\pm)} = \left(\frac{\gamma e^2}{m_\mathrm{e}c^2}\right)^2 \frac{k'}{k} |\tfrac{1}{2}gF(\boldsymbol{\kappa})|^2 \tfrac{1}{2}S \frac{(2\pi)^3}{v_0} \sum_{\mathbf{q},\boldsymbol{\tau}} (n_\mathbf{q}+\tfrac{1}{2}\pm\tfrac{1}{2}) \times$$
$$\times \delta(\hbar\omega\mp\hbar\omega_\mathbf{q})\delta(\boldsymbol{\kappa}\mp\mathbf{q}-\boldsymbol{\tau})\{1+(\tilde{\boldsymbol{\kappa}}.\tilde{\boldsymbol{\eta}})^2\mp2(\mathbf{P}.\tilde{\boldsymbol{\kappa}})(\tilde{\boldsymbol{\kappa}}.\tilde{\boldsymbol{\eta}})\}, \quad (10.148)$$

where $\hbar\omega_\mathbf{q}$ is the energy of a magnon with wave vector $\mathbf{q}$ and

$$n_\mathbf{q} = \{\exp(\hbar\omega_\mathbf{q}\beta)-1\}^{-1}.$$

For a two-sublattice Heisenberg antiferromagnet in an external magnetic field H applied parallel to the axis of quantization, there are two spin-wave branches with energies

$$\hbar\omega_{\mathbf{q},a} = (-1)^a g\mu_\mathrm{B} H+\mathscr{E}_\mathbf{q} \quad (a = 0 \text{ or } 1). \quad (10.149)$$

The function $\mathscr{E}_\mathbf{q}$ depends upon the various exchange parameters and is given explicitly in § 9.6. The cross-sections for magnon creation and annihilation for the scattering of polarized neutrons by a Heisenberg two-sublattice antiferromagnet can be shown with the results of § 9.6 to be given by

$$\left(\frac{\mathrm{d}^2\sigma}{\mathrm{d}\Omega\mathrm{d}E'}\right)^{(\pm)} = \sum_{a=0,1}\left(\frac{\mathrm{d}^2\sigma}{\mathrm{d}\Omega\mathrm{d}E'}\right)_a^{(\pm)}, \tag{10.150}$$

where

$$\left(\frac{\mathrm{d}^2\sigma}{\mathrm{d}\Omega\mathrm{d}E'}\right)_a^{(\pm)} = \left(\frac{\gamma e^2}{m_e c^2}\right)^2 \frac{k'}{k}\,|\tfrac{1}{2}gF(\boldsymbol{\kappa})|^2 \frac{1}{4N}\frac{(2\pi)^3}{v_0}\sum_{\mathbf{q},\boldsymbol{\tau}}(n_{\mathbf{q},a}+\tfrac{1}{2}\pm\tfrac{1}{2})\times$$
$$\times\,\delta(\hbar\omega\mp\hbar\omega_{\mathbf{q},a})\delta(\boldsymbol{\kappa}\mp\mathbf{q}-\boldsymbol{\tau})(u_{\mathbf{q}}^2+v_{\mathbf{q}}^2+2u_{\mathbf{q}}v_{\mathbf{q}}\cos\boldsymbol{\tau}.\boldsymbol{\rho})\times$$
$$\times\{1+(\tilde{\boldsymbol{\kappa}}.\tilde{\boldsymbol{\eta}})^2\mp(-1)^a\,2(\mathbf{P}.\tilde{\boldsymbol{\kappa}})(\tilde{\boldsymbol{\kappa}}.\tilde{\boldsymbol{\eta}})\}. \tag{10.151}$$

The functions $u_{\mathbf{q}}$ and $v_{\mathbf{q}}$ appearing in (10.151) are given in § 9.6. $\boldsymbol{\rho}$ is the vector joining nearest-neighbour ions on opposite sublattices.

If no external field H is applied to the antiferromagnet, we see from (10.149) that the two spin-wave branches become degenerate, each having energy $\mathscr{E}_{\mathbf{q}}$. In this instance the cross-sections $(\mathrm{d}^2\sigma/\mathrm{d}\Omega\mathrm{d}E')^{(\pm)}$ become independent of the polarization; in (10.151) $n_{\mathbf{q},a} = n_{\mathbf{q}}$ and $\omega_{\mathbf{q},a} = \mathscr{E}_{\mathbf{q}}$. This result could have been anticipated because it is merely an example of the general principle stated in the introduction to this chapter, namely that the cross-section is independent of $\mathbf{P}$ if the target system has no preferred axis. In the present case, the two sublattices are identical in the absence of an external magnetic field and as a consequence the two spin-wave branches are degenerate.

Let us now seek the expression for the polarization of the scattered neutrons that corresponds to the cross-section (10.146). From the general expression for $\mathbf{P}'$, eqn (10.140), we see that we have only to consider the first correlation function appearing on the right-hand side because the remaining two have already been considered. For this correlation function we have

$$\langle\hat{\mathbf{S}}_l^{(\perp)}(0)\{\mathbf{P}.\hat{\mathbf{S}}_{l'}^{(\perp)}(t)\}+\{\mathbf{P}.\hat{\mathbf{S}}_l^{(\perp)}(0)\}\hat{\mathbf{S}}_{l'}^{(\perp)}(t)\rangle$$
$$=\sum_{\alpha\beta}\{\tilde{\boldsymbol{\alpha}}P_\beta+\tilde{\boldsymbol{\beta}}P_\alpha-\tilde{\boldsymbol{\kappa}}(\tilde{\kappa}_\alpha P_\beta+\tilde{\kappa}_\beta P_\alpha)-\mathbf{P}.\tilde{\boldsymbol{\kappa}}(\tilde{\boldsymbol{\alpha}}\tilde{\kappa}_\beta+\tilde{\boldsymbol{\beta}}\tilde{\kappa}_\alpha)+$$
$$+2\tilde{\boldsymbol{\kappa}}(\mathbf{P}.\tilde{\boldsymbol{\kappa}})(\tilde{\kappa}_\alpha\tilde{\kappa}_\beta)\}\langle\hat{S}_l^\alpha(0)\hat{S}_{l'}^\beta(t)\rangle.$$

Taking $\alpha = \beta = x$ or y, as required for the particular case under consideration, this reduces to

$$\tfrac{1}{2}\{\mathbf{P}_x+\mathbf{P}_y-\mathbf{P}.\tilde{\boldsymbol{\kappa}}(\tilde{\boldsymbol{\kappa}}_x+\tilde{\boldsymbol{\kappa}}_y)+\tilde{\boldsymbol{\kappa}}(\tilde{\boldsymbol{\kappa}}.\tilde{\boldsymbol{\eta}})\tilde{\boldsymbol{\eta}}.\mathbf{P}_\perp\}\langle\hat{S}_l^+(0)\hat{S}_{l'}^-(t)+\hat{S}_l^-(0)\hat{S}_{l'}^+(t)\rangle$$
$$=\tfrac{1}{2}\{\mathbf{P}_\perp-\boldsymbol{\eta}_\perp(\boldsymbol{\eta}_\perp.\mathbf{P}_\perp)\}\langle\hat{S}_l^+(0)\hat{S}_{l'}^-(t)+\hat{S}_l^-(0)\hat{S}_{l'}^+(t)\rangle,$$

where, in the usual notation, $\boldsymbol{\eta}_\perp = \tilde{\boldsymbol{\kappa}}\times(\tilde{\boldsymbol{\eta}}\times\tilde{\boldsymbol{\kappa}})$.

By combining this result with the results (10.144) and (10.145) we obtain from the general expression (10.140) for $\mathbf{P}'$ the following result, valid under the same conditions as applied to eqn (10.146) for the cross-section:

$$\mathbf{P}'\left(\frac{d^2\sigma}{d\Omega dE'}\right)$$
$$= \left(\frac{\gamma e^2}{m_e c^2}\right)^2 \frac{k'}{k} |\tfrac{1}{2}gF(\boldsymbol{\varkappa})|^2 \sum_{\mathbf{l},\mathbf{l}'} \exp\{i\boldsymbol{\varkappa}.(\mathbf{l}'-\mathbf{l})\} \frac{1}{2\pi\hbar} \int_{-\infty}^{\infty} dt \exp(-i\omega t) \times$$
$$\times \tfrac{1}{4}\{[2\boldsymbol{\eta}_\perp \times (\mathbf{P}_\perp \times \boldsymbol{\eta}_\perp) - \mathbf{P}\{1-(\tilde{\boldsymbol{\varkappa}}.\tilde{\boldsymbol{\eta}})^2\} -$$
$$-2\boldsymbol{\varkappa}(\tilde{\boldsymbol{\varkappa}}.\tilde{\boldsymbol{\eta}})\{(\tilde{\boldsymbol{\varkappa}}.\mathbf{P})(\tilde{\boldsymbol{\varkappa}}.\tilde{\boldsymbol{\eta}})+1\}]\langle \hat{S}_l^-(0)\hat{S}_{l'}^+(t)\rangle +$$
$$+[2\boldsymbol{\eta}_\perp \times (\mathbf{P}_\perp \times \boldsymbol{\eta}_\perp) - \mathbf{P}\{1-(\tilde{\boldsymbol{\varkappa}}.\tilde{\boldsymbol{\eta}})^2\} -$$
$$-2\tilde{\boldsymbol{\varkappa}}(\tilde{\boldsymbol{\varkappa}}.\tilde{\boldsymbol{\eta}})\{(\tilde{\boldsymbol{\varkappa}}.\mathbf{P})(\tilde{\boldsymbol{\varkappa}}.\tilde{\boldsymbol{\eta}})-1\}]\langle \hat{S}_l^+(0)\hat{S}_{l'}^-(t)\rangle\}. \quad (10.152)$$

In this expression the vector $\boldsymbol{\eta}_\perp \times (\mathbf{P}_\perp \times \boldsymbol{\eta}_\perp)$ is perpendicular to both the scattering vector and $\boldsymbol{\eta}_\perp$. If the axis of quantization and the scattering vector are parallel then, clearly, $\mathbf{P}'$ is parallel to the scattering vector

For the particular case of the Heisenberg ferromagnet considered before, the polarization of the scattered neutrons is

$$\mathbf{P}'_{(\pm)} = [-2\tilde{\boldsymbol{\varkappa}}(\tilde{\boldsymbol{\varkappa}}.\tilde{\boldsymbol{\eta}})\{(\tilde{\boldsymbol{\varkappa}}.\mathbf{P})(\tilde{\boldsymbol{\varkappa}}.\tilde{\boldsymbol{\eta}})\mp 1\} - \mathbf{P}\{1-(\tilde{\boldsymbol{\varkappa}}.\tilde{\boldsymbol{\eta}})^2\} + 2\boldsymbol{\eta}_\perp \times (\mathbf{P}_\perp \times \boldsymbol{\eta}_\perp)] \times$$
$$\times \{1+(\tilde{\boldsymbol{\varkappa}}.\tilde{\boldsymbol{\eta}})^2 \mp 2(\tilde{\boldsymbol{\varkappa}}.\mathbf{P})(\tilde{\boldsymbol{\varkappa}}.\tilde{\boldsymbol{\eta}})\}^{-1}, \quad (10.153)$$

where the upper sign refers to the creation process and the lower the annihilation process. If the initial beam is unpolarized ($\mathbf{P} = 0$) (10.153) reduces to

$$\mathbf{P}'_{(\pm)} = \frac{\pm 2\tilde{\boldsymbol{\varkappa}}(\tilde{\boldsymbol{\varkappa}}.\tilde{\boldsymbol{\eta}})}{1+(\tilde{\boldsymbol{\varkappa}}.\tilde{\boldsymbol{\eta}})^2}. \quad (10.154)$$

Thus the polarization created in the two scattering processes is oppositely aligned; for the creation of a single magnon the created polarization is parallel to the scattering vector and for the annihilation process the created polarization is antiparallel to the scattering vector. Complete polarization is achieved if the magnetization of the ferromagnet is aligned parallel to the scattering vector, i.e. if $-\tilde{\boldsymbol{\eta}}$ is made parallel to $\boldsymbol{\varkappa}$.

The various results derived so far in this section for ferromagnets are illustrated in Figs. 10.7, 10.8, and 10.9 for scattering of both polarized and unpolarized neutrons from lithium ferrite. If the polarization and magnetization are made to coincide with the scattering vector then either the creation or annihilation cross-sections are zero and correspondingly the polarization of the scattered neutrons is zero for one

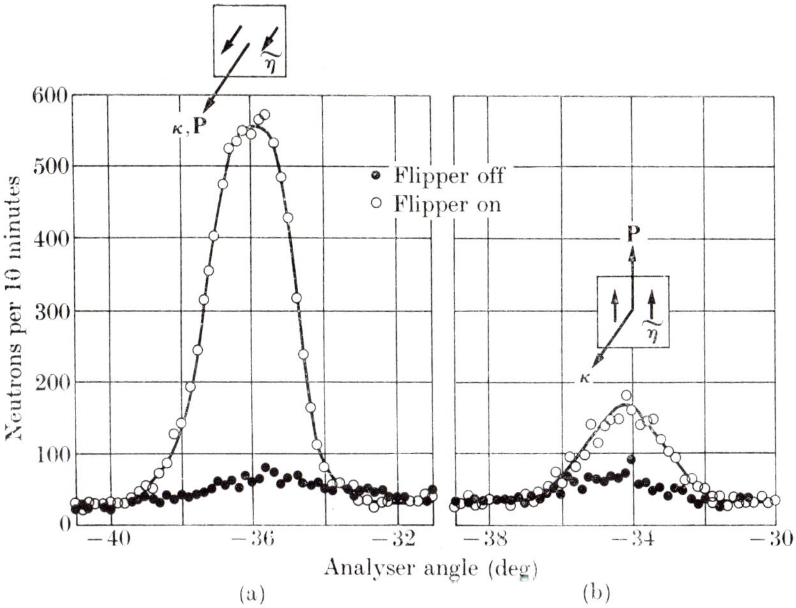

FIG. 10.7. Polarization reversal in magnon scattering from Li ferrite.

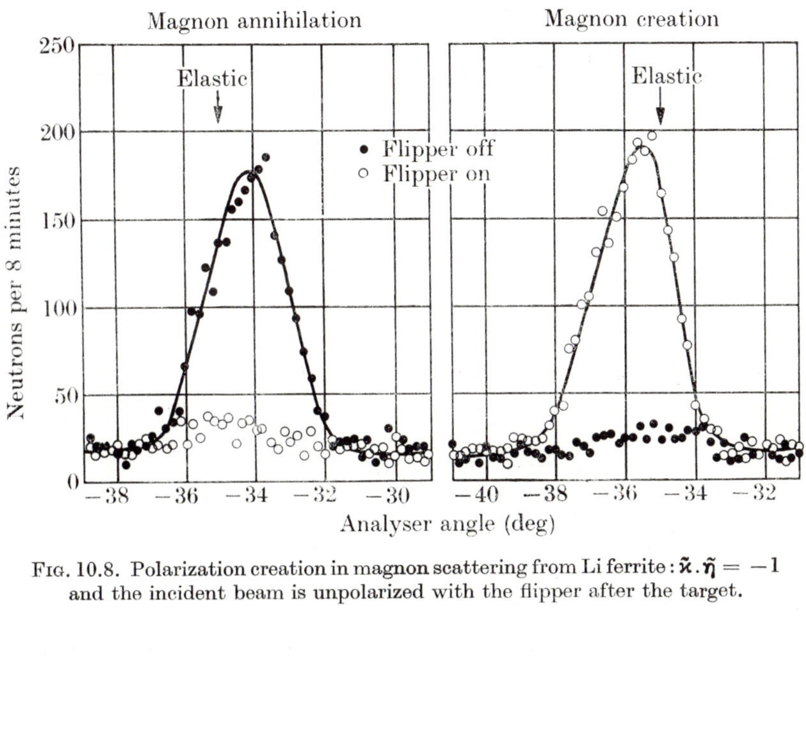

FIG. 10.8. Polarization creation in magnon scattering from Li ferrite: $\tilde{\boldsymbol{\kappa}}.\tilde{\boldsymbol{\eta}} = -1$ and the incident beam is unpolarized with the flipper after the target.

process. Fig. 10.7 (a) shows the number of neutrons scattered from Li ferrite with $\mathbf{P}$, $\tilde{\boldsymbol{\eta}}$, and $\boldsymbol{\varkappa}$ all parallel, and it is seen that neutrons are reflected by the analyser only if one flipper is activated. Thus the allowed scattering processes involve spin-flip, i.e. are either $\uparrow$–$\downarrow$ or $\downarrow$–$\uparrow$

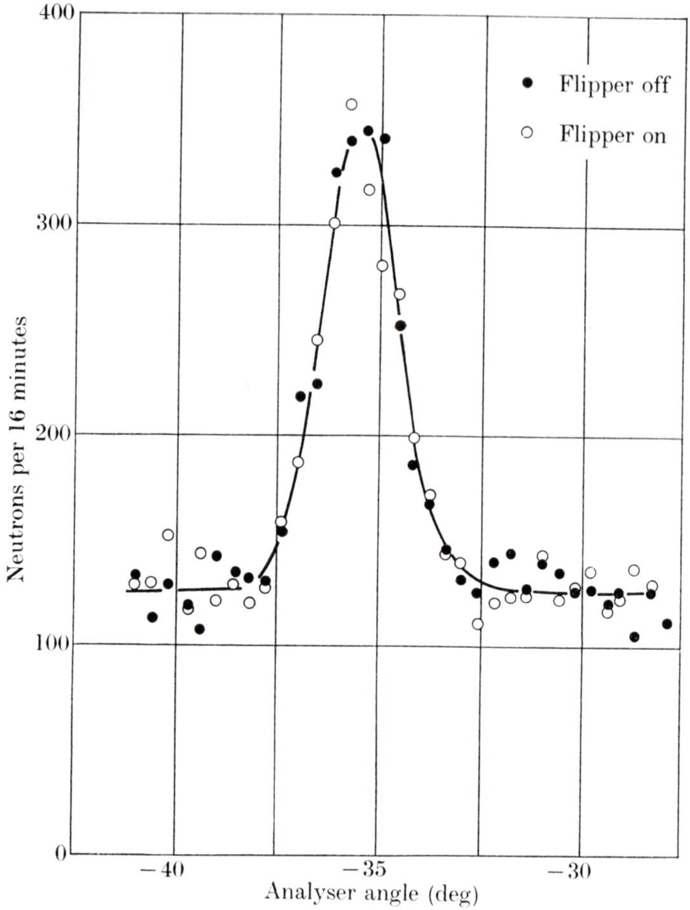

FIG. 10.9. Absence of polarization dependence in magnon scattering from Li ferrite with $\tilde{\boldsymbol{\varkappa}}.\tilde{\boldsymbol{\eta}} = 0$. Incident beam unpolarized and flipper after target.

in agreement with (10.153). The small peak in the measurements shown in Fig. 10.7 with the flipper off are due to nuclear disorder scattering. If the polarization and magnetization coincide and are perpendicular, the result (10.153) shows that neutrons will be reflected by the analyser only if one flipper is activated, in accord with the results shown in Fig. 10.7 (b). Also (10.156) shows that the cross-section of the scattered neutrons is only $\frac{1}{4}$ of its value in this instance compared to the previous case where $\mathbf{P}$, $\tilde{\boldsymbol{\eta}}$, and $\boldsymbol{\kappa}$ were all parallel. This too is borne out by Fig. 10.7 (a) and (b).

If the incident neutron beam is unpolarized, the polarization of the scattered beam is oppositely directed for magnon creation and magnon annihilation scattering. This point is illustrated in Fig. 10.8. The activated flipper was placed after the sample and the magnetization and the scattering vector made parallel, i.e. $\tilde{\boldsymbol{\kappa}}.\tilde{\boldsymbol{\eta}} = -1$ in eqn (10.154). Fig. 10.9 shows the number of neutrons inelastically scattered from Li ferrite with unpolarized incident neutrons and the magnetization perpendicular to the scattering vector. It is seen that there is no polarization dependence of the cross-section in accord with the theory given above.

The expression corresponding to (10.153) for a simple two-sublattice antiferromagnet is easily shown to be

$$\mathbf{P}'_{a,(\pm)} = [-2\boldsymbol{\kappa}(\tilde{\boldsymbol{\kappa}}.\tilde{\boldsymbol{\eta}})\{(\tilde{\boldsymbol{\kappa}}.\mathbf{P})(\tilde{\boldsymbol{\kappa}}.\tilde{\boldsymbol{\eta}})\mp(-1)^a\}-\mathbf{P}\{1-(\tilde{\boldsymbol{\kappa}}.\tilde{\boldsymbol{\eta}})^2\}+ \\ +2\boldsymbol{\eta}_\perp\times(\mathbf{P}_\perp\times\boldsymbol{\eta}_\perp)]\times\{1+(\tilde{\boldsymbol{\kappa}}.\tilde{\boldsymbol{\eta}})^2\mp(-1)^a(\tilde{\boldsymbol{\kappa}}.\mathbf{P})(\tilde{\boldsymbol{\kappa}}.\tilde{\boldsymbol{\eta}})\}^{-1}. \quad (10.155)$$

Thus the created polarization for a given scattering process is seen to lie in opposite directions for the two spin-wave modes.

If there is no external field applied to the antiferromagnet, so that the two sublattices are identical to one another, then in place of (10.155) we have the result

$$\mathbf{P}' = -\mathbf{P}+\frac{2\{\mathbf{P}_\perp-(\tilde{\boldsymbol{\eta}}.\mathbf{P}_\perp)\boldsymbol{\eta}_\perp\}}{1+(\tilde{\boldsymbol{\kappa}}.\tilde{\boldsymbol{\eta}})^2}. \quad (10.156)$$

From this we see that there is no created polarization nor any dependence of the polarization of the scattered beam on the scattering process involved. These conclusions are just what we expect for scattering from a target system with no preferred axis of quantization.

11

NEUTRON SCATTERING FROM LIQUIDS

THE theory of the dynamical properties of liquids is very much in its infancy. For this reason most neutron scattering measurements made on liquids have so far been interpreted in terms of plausible models. In fact, for each facet of research in the properties of liquids there exist several models, many of which can be regarded *a priori* as equally plausible.† We do not intend to review these models, instead in the main we choose to examine regions where the factors governing the behaviour of the cross-section are known on purely physical arguments, e.g. the coherent scattering from liquids at small $\boldsymbol{\varkappa}$ and ω is determined by the equations of hydrodynamics. Furthermore, we restrict our attention throughout to monatomic, non-molecular liquids.

11.1. Review of nuclear correlation functions

The partial differential cross-section for purely nuclear scattering is from Chapter 3

$$\frac{\mathrm{d}^2\sigma}{\mathrm{d}\Omega\mathrm{d}E'} = N\frac{k'}{k}\frac{1}{4\pi}\{\sigma_{\mathrm{c}}\,S(\boldsymbol{\varkappa},\omega)+\sigma_{\mathrm{i}}\,S_{\mathrm{i}}\,(\boldsymbol{\varkappa},\omega)\}. \tag{11.1}$$

Here σ_{c} and σ_{i} are the coherent and incoherent cross-sections for scattering by single bound nuclei. $S(\boldsymbol{\varkappa},\omega)$, the scattering law, was derived in various forms in Chapter 3 and for ease of reference we summarize the main results as

$$S(\boldsymbol{\varkappa},\omega)$$

$$= \frac{1}{2\pi\hbar}\int\limits_{-\infty}^{\infty}\mathrm{d}t\,\exp(-\mathrm{i}\omega t)\frac{1}{N}\sum_{j,j'}\langle\exp\{-\mathrm{i}\boldsymbol{\varkappa}\,.\,\hat{\mathbf{R}}_j(0)\}\exp\{\mathrm{i}\boldsymbol{\varkappa}\,.\,\hat{\mathbf{R}}_{j'}(t)\}\rangle$$

$$= \frac{1}{2\pi\hbar}\int\limits_{-\infty}^{\infty}\mathrm{d}t\,\exp(-\mathrm{i}\omega t)\int\mathrm{d}\mathbf{r}\,\exp(\mathrm{i}\boldsymbol{\varkappa}\,.\,\mathbf{r})G(\mathbf{r},t)$$

$$= \frac{1}{2\pi\hbar}\int\limits_{-\infty}^{\infty}\mathrm{d}t\,\exp(-\mathrm{i}\omega t)I(\boldsymbol{\varkappa},t)$$

† See, for instance, the review by A. Sjölander, *Thermal neutron scattering*, ed. by Egelstaff, Academic Press (1965).

$$= \frac{1}{2\pi\hbar}\int_{-\infty}^{\infty} \mathrm{d}t \exp(-\mathrm{i}\omega t)\int \mathrm{d}\mathbf{r} \exp(\mathrm{i}\boldsymbol{\varkappa}.\mathbf{r})\frac{1}{N}\int \mathrm{d}\mathbf{r}' \, \langle\hat{\rho}(\mathbf{r}'-\mathbf{r},0)\hat{\rho}(\mathbf{r}',t)\rangle$$

$$= \frac{1}{2\pi\hbar}\int_{-\infty}^{\infty} \mathrm{d}t \exp(-\mathrm{i}\omega t)\frac{1}{N}\langle\hat{\rho}_{\varkappa}(0)\hat{\rho}_{-\varkappa}(t)\rangle. \tag{11.2 a}$$

Similarly

$$S_{\mathrm{i}}(\boldsymbol{\varkappa},\omega) = \frac{1}{2\pi\hbar}\int_{-\infty}^{\infty} \mathrm{d}t \exp(-\mathrm{i}\omega t)\frac{1}{N}\sum_{j}\langle\exp\{-\mathrm{i}\boldsymbol{\varkappa}.\hat{\mathbf{R}}_j(0)\}\exp\{\mathrm{i}\boldsymbol{\varkappa}.\hat{\mathbf{R}}_j(t)\}\rangle$$

$$= \frac{1}{2\pi\hbar}\int_{-\infty}^{\infty} \mathrm{d}t \exp(-\mathrm{i}\omega t)\int \mathrm{d}\mathbf{r} \exp(\mathrm{i}\boldsymbol{\varkappa}.\mathbf{r})G_{\mathrm{s}}(\mathbf{r},t)$$

$$= \frac{1}{2\pi\hbar}\int_{-\infty}^{\infty} \mathrm{d}t \exp(-\mathrm{i}\omega t)I_{\mathrm{s}}(\boldsymbol{\varkappa},t). \tag{11.2 b}$$

These two equations (11.2 a) and (11.2 b) also serve to define the relations between the time-dependent quantities occurring inside the time integrals on the right-hand side.

By simple inspection we notice that

$$\int \mathrm{d}\mathbf{r}\; G_{\mathrm{s}}(\mathbf{r},t) = I_{\mathrm{s}}(0,t) = 1 \tag{11.3 a}$$

and

$$\int \mathrm{d}\mathbf{r}\; G(\mathbf{r},t) = I(0,t) = N. \tag{11.3 b}$$

Furthermore,

$$G_{\mathrm{s}}(\mathbf{r},0) = \delta(\mathbf{r}), \tag{11.3 c}$$

$$I_{\mathrm{s}}(\boldsymbol{\varkappa},0) = 1, \tag{11.3 d}$$

and so

$$\hbar\int_{-\infty}^{\infty} \mathrm{d}\omega\; S_{\mathrm{i}}(\boldsymbol{\varkappa},\omega) = 1. \tag{11.3 e}$$

Also

$$G(\mathbf{r},0) = \delta(\mathbf{r})+g(\mathbf{r}). \tag{11.3 f}$$

Hence

$$\begin{aligned}\hbar\int_{-\infty}^{\infty} \mathrm{d}\omega\; S(\boldsymbol{\varkappa},\omega) &= I(\boldsymbol{\varkappa},0)\\ &= \int \mathrm{d}\mathbf{r} \exp(\mathrm{i}\boldsymbol{\varkappa}.\mathbf{r})G(\mathbf{r},0)\\ &= 1+\gamma(\boldsymbol{\varkappa})\\ &= \frac{1}{N}\langle\hat{\rho}_{\varkappa}(0)\hat{\rho}_{-\varkappa}(0)\rangle. \end{aligned}\tag{11.3 g}$$

In this last equation we have introduced the Fourier transform

$$\gamma(\boldsymbol{\varkappa}) = \int \mathrm{d}\mathbf{r} \exp(\mathrm{i}\boldsymbol{\varkappa}.\mathbf{r})g(\mathbf{r}). \tag{11.4 a}$$

Sometimes the correlation function is separated into 'self' and 'distinct' parts, thus

$$G(\mathbf{r},t) = G_s(\mathbf{r},t)+G_d(\mathbf{r},t) \tag{11.4 b}$$

and then $G_d(\mathbf{r},0)$ is identical to $g(\mathbf{r})$.

For a homogeneous system with mean number density $\rho = N/V$ we can use (11.2 a) to express $G(\mathbf{r},t)$ as

$$G(\mathbf{r},t) = \frac{1}{N}\int d\mathbf{r}'\, \langle\hat{\rho}(\mathbf{r}'-\mathbf{r},0)\hat{\rho}(\mathbf{r}',t)\rangle$$

$$= \frac{1}{\rho}\langle\hat{\rho}(0,0)\hat{\rho}(\mathbf{r},t)\rangle \tag{11.4 c}$$

and hence it follows that

$$\lim_{|\mathbf{r}|\to\infty} G_s(\mathbf{r},t) = \lim_{t\to\infty} G_s(\mathbf{r},t) = 0 \tag{11.5 a}$$

and

$$\lim_{|\mathbf{r}|\to\infty} G(\mathbf{r},t) = \lim_{t\to\infty} G(\mathbf{r},t) = \rho. \tag{11.5 b}$$

Now in Chapter 3 we saw that elastic scattering, i.e. scattering with a cross-section involving $\delta(\omega)$, is given by the value of correlation functions at infinite time. But from (11.5 a) it follows that since $G_s(\mathbf{r},\infty)$ is zero there is no elastic incoherent scattering. Furthermore, if we substitute (11.5 b) into (11.2 a) we get

$$S(\boldsymbol{\kappa},\omega)_{\text{elas}} = \delta(\hbar\omega)N\delta_{\boldsymbol{\kappa},0}, \tag{11.5 c}$$

which does not correspond to real scattering because $\mathbf{k}'$ is fixed equal to $\mathbf{k}$. (Although if we were to depart from Born approximation this term would correspond to diffraction around the sample as a whole.) We conclude that there is no elastic scattering of any type from a liquid. The same conclusion applies equally well to X-ray scattering from a liquid and hence we see that the so-called 'elastic X-ray scattering' from liquids is really only quasi-elastic.

It is usually convenient, for formal reasons, to take account of (11.5 c) by concentrating on the true scattering cross-section. To obtain this we need to amend (11.2 a) to the form

$$S(\boldsymbol{\kappa},\omega)$$

$$= \frac{1}{2\pi\hbar}\int_{-\infty}^{\infty} dt\, \exp(-i\omega t)\,\frac{1}{N}\sum_{j,j'}\langle[\exp\{-i\boldsymbol{\kappa}.\hat{\mathbf{R}}_j(0)\}\exp\{i\boldsymbol{\kappa}.\hat{\mathbf{R}}_{j'}(t)\}-\delta_{\boldsymbol{\kappa},0}]\rangle$$

$$= \frac{1}{2\pi\hbar}\int_{-\infty}^{\infty} dt\, \exp(-i\omega t)\int d\mathbf{r}\, \exp(i\boldsymbol{\kappa}.\mathbf{r})\{G(\mathbf{r},t)-\rho\}$$

$$= \frac{1}{2\pi\hbar} \int_{-\infty}^{\infty} \mathrm{d}t \exp(-\mathrm{i}\omega t)\{I(\boldsymbol{\varkappa}, t) - N\delta_{\boldsymbol{\varkappa},0}\}$$

$$= \frac{1}{2\pi\hbar} \int_{-\infty}^{\infty} \mathrm{d}t \exp(-\mathrm{i}\omega t) \int \mathrm{d}\mathbf{r} \exp(\mathrm{i}\boldsymbol{\varkappa}.\mathbf{r}) \frac{1}{N} \int \mathrm{d}\mathbf{r}' \langle \delta\hat{\rho}(\mathbf{r}'-\mathbf{r}, 0)\, \delta\hat{\rho}(\mathbf{r}', t)\rangle$$

$$= \frac{1}{2\pi\hbar} \int_{-\infty}^{\infty} \mathrm{d}t \exp(-\mathrm{i}\omega t) \frac{1}{N} \langle \delta\hat{\rho}_{\boldsymbol{\varkappa}}(0)\, \delta\hat{\rho}_{-\boldsymbol{\varkappa}}(t)\rangle, \qquad (11.6\,\mathrm{a})$$

and then (11.3 g) becomes

$$\hbar \int_{-\infty}^{\infty} \mathrm{d}\omega\, S(\boldsymbol{\varkappa}, \omega) = I(\boldsymbol{\varkappa}, 0) - N\delta_{\boldsymbol{\varkappa},0}$$

$$= \int \mathrm{d}\mathbf{r} \exp(\mathrm{i}\boldsymbol{\varkappa}.\mathbf{r})\{G(\mathbf{r}, 0) - \rho\}$$

$$= 1 + \gamma(\boldsymbol{\varkappa}) - N\delta_{\boldsymbol{\varkappa},0}$$

$$= \frac{1}{N} \langle \delta\hat{\rho}_{\boldsymbol{\varkappa}}(0)\, \delta\hat{\rho}_{-\boldsymbol{\varkappa}}(0)\rangle. \qquad (11.6\,\mathrm{b})$$

The incoherent scattering is determined by the motion of individual nuclei in the liquid, whereas the coherent scattering depends upon their relative motion. It is a simpler theoretical task to seek the behaviour of a single nucleus moving in the field of the remaining nuclei in a liquid than to attempt to determine their relative motions. For this reason more work has been done on interpreting the incoherent scattering than the coherent. However, the coherent scattering is perhaps the more rewarding to understand because recent measurements have shown that collective excitations exist in liquid lead. From an experimental point of view it is not possible, at present, to study solely incoherent scattering because no suitable liquid is known. Both vanadium and hydrogen are almost totally incoherent scatterers but the melting-point of vanadium is too high to make it feasible to examine, and hydrogen forms a molecular system which complicates a theoretical analysis. There are, however, monatomic non-molecular liquids that are totally coherent scatterers, e.g. tin and lead, which have melting-points of 272 °C and 327 °C, respectively.

Since for most atomic species we know the nuclear spin and isotope dependence of the scattering length, it is possible in principle to measure the coherent and incoherent cross-sections separately, by using polarized neutrons and measuring the polarization of the scattered beam. The necessary formulae are given in Chapter 10. But experiments of this kind are difficult because of the present low intensities of polarized beams.

Even though it is not possible to isolate incoherent scattering experimentally it is convenient to discuss the incoherent and coherent cross-sections separately. The temperatures at which most solids become liquid is sufficiently high for them to be described in terms of classical statistical mechanics. However, we must be careful when calculating cross-sections for the scattering from classical systems to ensure that fundamental symmetry properties of the process are not violated. Thus it is not correct merely to replace either $S(\boldsymbol{\kappa},\omega)$ or $S_{\mathrm{i}}(\boldsymbol{\kappa},\omega)$ by their corresponding classical values because the resulting cross-sections would not satisfy the detailed balance condition (cf. Chapter 3).

Before discussing incoherent scattering it is convenient to gather together the fundamental properties of $S(\boldsymbol{\kappa},\omega)$ and $S_{\mathrm{i}}(\boldsymbol{\kappa},\omega)$ and related functions. In the following it is to be understood that the properties of $S_{\mathrm{i}}(\boldsymbol{\kappa},\omega)$ under discussion are the same as those of $S(\boldsymbol{\kappa},\omega)$.

First, detailed balance is expressed in the relation

$$S(\boldsymbol{\kappa},\omega) = \exp(\hbar\omega\beta)S(-\boldsymbol{\kappa},-\omega). \tag{11.7 a}$$

From this it follows that

$$\tilde{S}(\boldsymbol{\kappa},\omega) = \exp(-\tfrac{1}{2}\hbar\omega\beta)S(\boldsymbol{\kappa},\omega) \tag{11.7 b}$$

is an even function of both its arguments. Hence the function

$$\tilde{G}(\mathbf{r},t) = \frac{\hbar}{(2\pi)^3}\int \mathrm{d}\boldsymbol{\kappa}\int \mathrm{d}\omega\, \exp(-\mathrm{i}\boldsymbol{\kappa}.\mathbf{r}+\mathrm{i}\omega t)\tilde{S}(\boldsymbol{\kappa},\omega) \tag{11.8}$$

is even in $\mathbf{r}$ and t and purely real. In contrast $G(\mathbf{r},t)$ is a complex function which, because $S(\boldsymbol{\kappa},\omega)$ is real, satisfies

$$G^*(\mathbf{r},t) = G(-\mathbf{r},-t). \tag{11.9}$$

For isotropic systems we see from (11.9) that $G(\mathbf{r},0)$ is purely real (cf. (11.3 c) and (11.3 f)).

G and $\tilde{G}$ are related through

$$\tilde{G}(\mathbf{r},t) = G(\mathbf{r},t+\tfrac{1}{2}\mathrm{i}\hbar\beta). \tag{11.10}$$

The fact that $G(\mathbf{r},t)$ is a complex function is a consequence of its quantum mechanical aspects, as discussed in Chapter 3. Schofield† suggested that in calculating the classical approximation to the cross-section one should substitute the classical result G^{cl} for $\tilde{G}$ rather than G. This prescription leads to a cross-section that does not violate the detailed balance condition.

† P. Schofield, *Phys. Rev. Lett.* **4**, 239 (1960).

11.2. Incoherent scattering: classical theory

(a) Gaussian approximation

If we can calculate the classical function $G_s^{cl}(\mathbf{r},t)$ then the Schofield prescription is expressed in the relation

$$G_s(\mathbf{r},t) = \tilde{G}_s(\mathbf{r},t-\tfrac{1}{2}i\hbar\beta) \simeq G_s^{cl}(\mathbf{r},t-\tfrac{1}{2}i\hbar\beta). \quad (11.11)$$

For those simple problems where it is possible to evaluate $G_s^{cl}(\mathbf{r},t)$ exactly, i.e. for a perfect gas, a diffusing atom, and a harmonic oscillator, it is found that the function is a Gaussian in $\mathbf{r}$.

Hence we propose that in general the spatial dependence of $G_s^{cl}(\mathbf{r},t)$ is approximated by a Gaussian, i.e.

$$G_s^{cl}(\mathbf{r},t) \propto \exp\{-r^2/2\bar{w}(t)\}. \quad (11.12)$$

The normalization condition (eqn (11.3 a))

$$\int d\mathbf{r}\, G_s^{cl}(\mathbf{r},t) = 1 = 4\pi\int_0^\infty dr\, r^2 G_s(r,t) \quad (11.13)$$

determines the coefficient of proportionality in (11.12), viz.

$$G_s^{cl}(\mathbf{r},t) = \left\{\frac{1}{2\pi\bar{w}(t)}\right\}^{\frac{3}{2}} \exp\{-r^2/2\bar{w}(t)\}. \quad (11.14)$$

Now, because $G_s^{cl}(\mathbf{r},t)$ is both purely real and even in t, the function $\bar{w}(t)$ must also satisfy identical conditions.

From the representation (11.14)

$$I_s^{cl}(\boldsymbol{\varkappa},t) = \int d\mathbf{r}\, \exp(i\boldsymbol{\varkappa}.\mathbf{r}) G_s^{cl}(\mathbf{r},t) = \exp\{-\tfrac{1}{2}\kappa^2\bar{w}(t)\}. \quad (11.15)$$

Let us now seek the physical significance of $\bar{w}(t)$, which as a consequence of (11.15), is often referred to as the width function. The mean square displacement $\langle\{\mathbf{r}(t)-\mathbf{r}(0)\}^2\rangle$ is given by

$$\int d\mathbf{r}\, r^2 G_s^{cl}(\mathbf{r},t) = \frac{4\pi}{\{2\pi\bar{w}(t)\}^{\frac{3}{2}}}\int_0^\infty dr\, r^4 \exp\{-r^2/2\bar{w}(t)\} = 3\bar{w}(t), \quad (11.16)$$

and this gives one immediate physical significance to $\bar{w}(t)$; namely $3\bar{w}(t)$ is the mean square displacement after a time t.

Let us now pause to examine the various forms of $\bar{w}(t)$ taken for simple cases we have discussed elsewhere in this book. In Chapter 3 we calculated correlation functions for a single free nucleus and obtained, from (3.66),

$$I_s(\boldsymbol{\varkappa},t) = \exp\left[\frac{-\kappa^2}{2M\beta}t(t-i\hbar\beta)\right] \quad \text{(free nucleus)}, \quad (11.17)$$

from (3.67) $$G_s(\mathbf{r},t) = \left(\frac{\beta M}{2\pi t(t-i\hbar\beta)}\right)^{\frac{3}{2}} \exp\left[\frac{-\beta M r^2}{2t(t-i\hbar\beta)}\right], \tag{11.18}$$

and from (3.68)

$$S_s(\boldsymbol{\varkappa},\omega) = \left(\frac{\beta M}{2\pi\hbar^2\kappa^2}\right)^{\frac{1}{2}} \exp\left[\frac{-\beta M}{2\hbar^2\kappa^2}\left(\hbar\omega - \frac{\hbar^2\kappa^2}{2M}\right)^2\right]. \tag{11.19}$$

From these results and (11.11) we can immediately write down

$$\tilde{I}_s(\boldsymbol{\varkappa},t) = I_s(\boldsymbol{\varkappa},t+\tfrac{1}{2}i\hbar\beta) = \exp\left[-\frac{\kappa^2}{2M\beta}\{t^2+\tfrac{1}{4}\hbar^2\beta^2\}\right] \tag{11.20}$$

$$\tilde{G}_s(\mathbf{r},t) = G_s(\mathbf{r},t+\tfrac{1}{2}i\hbar\beta) = \left\{\frac{\beta M}{2\pi(t^2+\frac{1}{4}\hbar^2\beta^2)}\right\}^{\frac{3}{2}} \exp\left[\frac{-\beta M r^2}{2(t^2+\frac{1}{4}\hbar^2\beta^2)}\right] \tag{11.21}$$

and

$$\tilde{S}_s(\boldsymbol{\varkappa},\omega) = \exp(-\tfrac{1}{2}\hbar\omega\beta)S_s(\boldsymbol{\varkappa},\omega) = \left\{\frac{\beta M}{2\pi\hbar^2\kappa^2}\right\}^{\frac{1}{2}} \exp\left[\frac{-\beta M}{2\hbar^2\kappa^2}\left\{\hbar^2\omega^2+\left(\frac{\hbar^2\kappa^2}{2M}\right)^2\right\}\right]. \tag{11.22}$$

These results give us an indication of the meaning of the Schofield prescription (11.11). If we had used the classical result for $G_s(\mathbf{r},t)$ directly we would have had to replace $t(t-i\hbar\beta)$ in $G_s(\mathbf{r},t)$ and $I_s(\boldsymbol{\varkappa},t)$ by simply t^2 and this brings errors of order $t\hbar\beta$ which upset the condition of detailed balance. However, if we use the classical result for $\tilde{G}_s(\mathbf{r},t)$ and $\tilde{I}_s(\boldsymbol{\varkappa},t)$ then we are replacing $t^2+\frac{1}{4}\hbar^2\beta^2$ by t^2, and this gives the smaller error of order $\hbar^2\beta^2$. We notice that using the classical form for $\tilde{G}_s(\mathbf{r},t)$ and $\tilde{I}_s(\boldsymbol{\varkappa},t)$ is not exact and in the cross-section formulae we notice from (11.19) that it is equivalent to neglecting 'recoil effects'—i.e. neglecting terms of the type $\hbar^2\kappa^2/2M$ compared to $\hbar\omega$. This is quite a reasonable approximation when the mass of the target nuclei is large compared to the neutron mass. Comparing (11.15) with (11.20) we see that in this case

$$\bar{w}(t) = \frac{1}{M\beta}t^2 \quad \text{(free nucleus).} \tag{11.23}$$

Now let us examine a second example. In Chapter 4 we calculated the correlation functions associated with phonon scattering. Thus, for example, from (4.69) we deduce that for a cubic crystal, if $Z(\omega)$ is the density of phonon states

$$I_s(\boldsymbol{\varkappa},t) = \exp\{-2W(\boldsymbol{\varkappa})\}\exp\langle\boldsymbol{\varkappa}.\hat{\mathbf{u}}(l,0)\boldsymbol{\varkappa}.\hat{\mathbf{u}}(l,t)\rangle \tag{11.24}$$

$$= \exp\left[-\frac{\hbar\kappa^2}{2M}\int_0^\infty d\omega\, Z(\omega)\frac{1}{\omega}\{(1-\cos\omega t)\coth\tfrac{1}{2}\hbar\omega\beta - i\sin\omega t\}\right],$$

and hence

$$\tilde{I}_{\mathrm{s}}(\boldsymbol{\varkappa},t) = \exp\left[-\frac{\hbar\kappa^2}{2M}\int_0^\infty \mathrm{d}\omega\; Z(\omega)\frac{1}{\omega}\frac{\{\cosh\frac{1}{2}\hbar\omega\beta - \cos\omega t\}}{\sinh\frac{1}{2}\hbar\omega\beta}\right] \tag{11.25 a}$$

and
$$I_{\mathrm{s}}^{\mathrm{cl}}(\boldsymbol{\varkappa},t) = \exp\left[-\frac{\kappa^2}{M\beta}\int_0^\infty \mathrm{d}\omega\; Z(\omega)\frac{1}{\omega^2}\{1-\cos\omega t\}\right]. \tag{11.25 b}$$

Hence
$$\bar{w}(t) = \frac{2}{M\beta}\int_0^\infty \mathrm{d}\omega\; Z(\omega)\frac{1}{\omega^2}\{1-\cos\omega t\} \qquad \text{(crystal).} \tag{11.26}$$

From this last formula we notice that if we expand in powers of t we recover (11.23) for small t (since the integral over $Z(\omega)$ is normalized to unity by definition). This is as we expect; because at very short times each target nucleus moves as a free particle. But at long times (11.26) does not continuously increase but approaches a finite limit; it is this last feature which distinguishes a solid from all other forms of condensed matter. Again we notice that the approximation of replacing $\tilde{I}_{\mathrm{s}}(\boldsymbol{\varkappa},t)$ by $I_{\mathrm{s}}^{\mathrm{cl}}(\boldsymbol{\varkappa},t)$ means neglecting terms of order $(\hbar\omega\beta)^2$.

As a final example we anticipate the result to be derived later in this section, eqn (11.50), which says that for a diffusing nucleus

$$\bar{w}(t) = 2D\{|t| - D\beta M\} \qquad \text{(diffusion).}$$

The results for $\bar{w}(t)$ are shown schematically in Fig. 11.1. From these examples we see that, within the limitations of the Gaussian approximation and the neglect of recoil effects, the function $\bar{w}(t)$ gives us the complete statistical knowledge of the motion of a single nucleus in a liquid.

We have previously remarked, see (11.16), that $\bar{w}(t)$ was related to the mean square displacement after a time t. We now return to the general discussion and show how $\bar{w}(t)$ is related to velocity correlations. For a system in equilibrium

$$\frac{\partial^2}{\partial t^2}\langle\{\mathbf{r}(t)-\mathbf{r}(0)\}^2\rangle = 2\langle\mathbf{v}(0).\mathbf{v}(t)\rangle, \tag{11.27}$$

where $\mathbf{v}(t)$ is the velocity of a nucleus at time t. The equilibrium condition means that

$$\frac{\partial}{\partial t_0}\langle\mathbf{v}(t_0).\mathbf{v}(t_0+t)\rangle = 0. \tag{11.28}$$

By combining (11.16) and (11.27) we obtain the relationship

$$\ddot{\bar{w}}(t) \equiv \frac{\partial^2}{\partial t^2}\bar{w}(t) = \tfrac{2}{3}\langle\mathbf{v}(0).\mathbf{v}(t)\rangle \tag{11.29}$$

between the second differential of $\bar{w}(t)$ with respect to time and the velocity–velocity correlation function $\langle \mathbf{v}(0).\mathbf{v}(t)\rangle$.

Integrating (11.29) and noting that $\dot{\bar{w}}(0) = 0$, since $\bar{w}(t)$ is an even function of t,

$$\dot{\bar{w}}(t) = \tfrac{2}{3}\int\limits_0^t \mathrm{d}t'\ \langle \mathbf{v}(0).\mathbf{v}(t')\rangle. \tag{11.30}$$

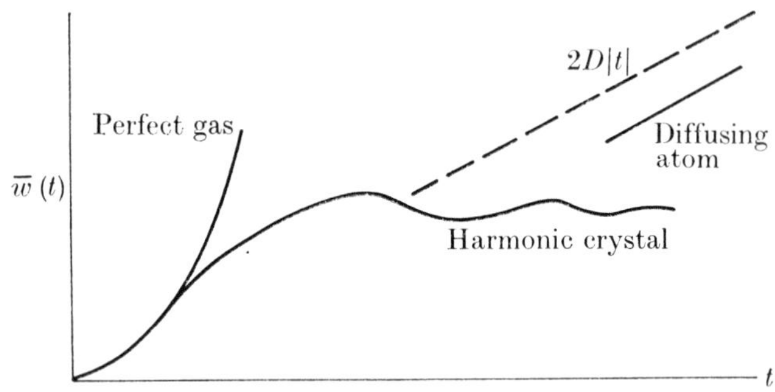

FIG. 11.1. Schematic diagram of $\bar{w}(t)$ for various simple models.

Furthermore, $\bar{w}(0) = 0$, for by definition

$$G_s^{\mathrm{cl}}(\mathbf{r}, 0) = \delta(\mathbf{r}) \tag{11.31}$$

(cf. eqn (11.3 c)), whence we obtain from (11.30) the desired relation, namely

$$\begin{aligned}\bar{w}(t) &= \tfrac{2}{3}\int\limits_0^t \mathrm{d}t' \int\limits_0^{t'} \mathrm{d}t''\ \langle \mathbf{v}(0).\mathbf{v}(t'')\rangle \\ &= \tfrac{2}{3}\int\limits_0^t \mathrm{d}t'\ (t-t')\langle \mathbf{v}(0).\mathbf{v}(t')\rangle.\end{aligned} \tag{11.32}$$

This second expression for $\bar{w}(t)$ gives it in terms of the velocity–velocity correlation function and is readily verified by differentiating it once with respect to t. From (11.29) we observe that since $\bar{w}(t)$ is real and even in t, so also is $\langle \mathbf{v}(0).\mathbf{v}(t)\rangle$.

At this stage we have shown that, in the approximation of assuming a Gaussian form for the spatial dependence of G_s^{cl}, the width function $\bar{w}(t)$ is intimately related to the velocity–velocity correlation function. We now introduce a velocity frequency function $p(\omega)$ through the relation

$$p(\omega) = \frac{\beta M}{3\pi}\int\limits_{-\infty}^{\infty} \mathrm{d}t\, \exp(\mathrm{i}\omega t)\langle \mathbf{v}(0).\mathbf{v}(t)\rangle, \tag{11.33}$$

where M is the mass of a nucleus and the coefficient is chosen such that

$$\int_0^\infty d\omega\, p(\omega) = 1. \tag{11.34}$$

Because $\langle \mathbf{v}(0).\mathbf{v}(t)\rangle$ is real and even in t, $p(\omega)$ is real and even in ω.

The velocity frequency function characterizes the dynamical properties of the liquid and can be extracted from the cross-section, as shown below. We also verify the normalization condition (11.34).

From (11.15)

$$I_s^{cl}(\boldsymbol{\kappa}, t) = 1 - \tfrac{1}{2}\kappa^2 \bar{w}(t) + O(\kappa^4).$$

Hence
$$\frac{\partial^2}{\partial t^2} I_s^{cl}(\boldsymbol{\kappa}, t) = -\tfrac{1}{2}\kappa^2 \ddot{\bar{w}}(t) + O(\kappa^4),$$

and therefore
$$\ddot{\bar{w}}(t) = \lim_{|\boldsymbol{\kappa}|\to 0}\left(\frac{-2}{\kappa^2}\right)\frac{\partial^2}{\partial t^2} I_s^{cl}(\boldsymbol{\kappa}, t).$$

But
$$I_s^{cl}(\boldsymbol{\kappa}, t) = \hbar \int_{-\infty}^{\infty} d\omega\, \exp(i\omega t)\tilde{S}_i(\boldsymbol{\kappa}, \omega),$$

and hence

$$\left(\frac{-2}{\kappa^2}\right)\frac{\partial^2}{\partial t^2} I_s^{cl}(\boldsymbol{\kappa}, t) = \frac{2\hbar}{\kappa^2}\int_{-\infty}^{\infty} d\omega\, \exp(i\omega t)\omega^2 \tilde{S}_i(\boldsymbol{\kappa}, \omega).$$

Now from (11.29) and (11.33)

$$\ddot{\bar{w}}(t) = \frac{1}{\beta M}\int_{-\infty}^{\infty} d\omega\, \exp(i\omega t) p(\omega).$$

Comparing these equations we conclude that

$$\frac{1}{2\hbar\beta M} p(\omega) = \lim_{|\boldsymbol{\kappa}|\to 0} \frac{\omega^2}{\kappa^2}\tilde{S}_i(\boldsymbol{\kappa}, \omega). \tag{11.35}$$

Eqn (11.35) gives the relation between the velocity frequency function $p(\omega)$ and the measured cross-section, namely, $p(\omega)$ is proportional to the cross-section extrapolated to the forward direction $\boldsymbol{\kappa} = 0$. Having obtained $p(\omega)$ it is then possible to calculate $\bar{w}(t)$ because, from (11.32) and (11.33),

$$\bar{w}(t) = \frac{2}{3}\int_0^t dt'\,(t-t')\frac{3}{2\beta M}\int_{-\infty}^{\infty} d\omega\, \exp(-i\omega t')p(\omega)$$

$$= \frac{2}{\beta M}\int_0^\infty d\omega \left\{\frac{1-\cos\omega t}{\omega^2}\right\} p(\omega). \tag{11.36}$$

(On comparing eqns (11.36) and (11.26) we see that $p(\omega)$ is the analogue of the density of phonon states $Z(\omega)$.)

To verify the normalization of $p(\omega)$ we proceed as follows. The first moment relation for $S(\boldsymbol{\kappa}, \omega)$ is

$$\int_{-\infty}^{\infty} \mathrm{d}\omega\, \omega S(\boldsymbol{\kappa}, \omega) = \frac{\kappa^2}{2M}. \tag{11.37}$$

Now in the classical limit, $\hbar\omega\beta \ll 1$ and

$$S(\boldsymbol{\kappa}, \omega) \simeq (1+\tfrac{1}{2}\hbar\omega\beta)\tilde{S}(\boldsymbol{\kappa}, \omega);$$

then (11.37) yields (note that $\tilde{S}$ is even in ω)

$$\int_{-\infty}^{\infty} \mathrm{d}\omega\, \omega^2 \tilde{S}(\boldsymbol{\kappa}, \omega) = \frac{\kappa^2}{\hbar\beta M}. \tag{11.38}$$

Combining this result with (11.35) confirms (11.34).

One further result to be noted from (11.33) and (11.30) is

$$\beta M \dot{\bar{w}}(t = \infty) = \pi p(0). \tag{11.39}$$

Let us now investigate the behaviour of $\bar{w}(t)$. In the limit of small t $G_{\mathrm{s}}^{\mathrm{cl}}(\mathbf{r}, t)$ must describe an ideal classical gas. For the limit of small t corresponds to high incident energies and in this instance the individual nuclei in the liquid must appear in the scattering process to be completely free. $G_{\mathrm{s}}^{\mathrm{cl}}(\mathbf{r}, t)$ for an ideal gas is obtained from the classical limit of (11.21), i.e.

$$G_{\mathrm{s}}^{\mathrm{cl}}(\mathbf{r}, t) = \left(\frac{\beta M}{2\pi t^2}\right)^{\frac{3}{2}} \exp\left\{-\frac{\beta M}{2t^2} r^2\right\} \quad \text{(ideal gas)}, \tag{11.40}$$

which gives (11.23) for $\bar{w}(t)$.

For this system the velocity frequency function $p(\omega)$ is proportional to $\delta(\omega)$. To see this we have from (11.29) and (11.33) the relationship

$$p(\omega) = \frac{\beta M}{2\pi} \int_{-\infty}^{\infty} \mathrm{d}t\, \exp(\mathrm{i}\omega t) \ddot{\bar{w}}(t). \tag{11.41}$$

Thus with the result (11.23) for $\bar{w}(t)$

$$\begin{aligned} p(\omega) &= \frac{\beta M}{2\pi} \frac{2}{\beta M} \int_{-\infty}^{\infty} \mathrm{d}t\, \exp(\mathrm{i}\omega t) \\ &= 2\delta(\omega) \qquad \text{(ideal gas)}. \end{aligned} \tag{11.42}$$

Consider now the opposite limit of long times. A given nucleus undergoes many collisions with the other nuclei in the liquid due to thermal agitation. When viewed over a time interval much greater than that between collisions it appears to move slowly, its velocity being determined by the equation of (Brownian) motion

$$M\dot{v}(t) = -\gamma M v(t) + R(t). \tag{11.43}$$

The frictional force exerted by the liquid is represented by the first term on the right-hand side of (11.43) and the second term, $R(t)$ represents the random force due to the random collisions with the other nuclei.

From the equation of motion (11.43) we obtain for the velocity–velocity correlation function the result

$$\langle \mathbf{v}(0).\mathbf{v}(t)\rangle = \langle \mathbf{v}^2(0)\rangle \exp(-\gamma|t|). \tag{11.44}$$

Note that this result cannot hold for small t because $\langle \mathbf{v}(0).\mathbf{v}(t)\rangle$ in reality cannot have a discontinuous gradient at $t = 0$ as (11.44) implies.

The diffusion constant D is defined to be

$$D = \tfrac{1}{3} \lim_{t\to\infty} \frac{1}{2t} \langle\{\mathbf{r}(0)-\mathbf{r}(t)\}^2\rangle. \tag{11.45}$$

Since
$$\mathbf{r}(t)-\mathbf{r}(0) = \int_0^t \mathrm{d}t'\ \mathbf{v}(t'),$$

we have
$$\begin{aligned}\langle\{\mathbf{r}(0)-\mathbf{r}(t)\}^2\rangle &= \left\langle\left\{\int_0^t \mathrm{d}t'\ \mathbf{v}(t')\right\}^2\right\rangle \\ &= 2\int_0^t \mathrm{d}t_1 \int_0^{t_1} \mathrm{d}t_2\ \langle \mathbf{v}(t_1).\mathbf{v}(t_2)\rangle \\ &= 2\int_0^t \mathrm{d}t'\ (t-t')\langle \mathbf{v}(0).\mathbf{v}(t')\rangle.\end{aligned}$$

Therefore D is given by

$$D = \tfrac{1}{3}\int_0^\infty \mathrm{d}t\ \langle \mathbf{v}(0).\mathbf{v}(t)\rangle. \tag{11.46}$$

In view of the representation (11.32) of $\bar{w}(t)$ in terms of $\langle \mathbf{v}(0).\mathbf{v}(t)\rangle$ and eqn (11.46) for D, it is evident that in the limit of large times

$$\bar{w}(t) \simeq 2D|t|. \tag{11.47}$$

Thus the diffusion constant determines the long-time behaviour of the width function, as we should expect.

We can use (11.46) to obtain the relation between the frictional coefficient γ and the diffusion constant D. Substituting the result

(11.44) in (11.46) the latter becomes

$$D = \tfrac{1}{3}\langle \mathbf{v}^2(0)\rangle \int_0^\infty dt \exp(-\gamma t) = \frac{1}{3\gamma}\langle \mathbf{v}^2(0)\rangle. \tag{11.48 a}$$

However, assuming the law of equipartition of energy,

$$\langle \mathbf{v}^2(0)\rangle = 3/\beta M, \tag{11.48 b}$$

(11.48) gives the result

$$\gamma D = 1/\beta M, \tag{11.49}$$

which is the familiar relation between the diffusion constant and frictional coefficient first derived in a study of Brownian motion by Einstein in 1905. Furthermore, the approximation (11.44) inserted into (11.32) gives a more precise form for $\bar{w}(t)$ at long t than that given by (11.47), namely

$$\bar{w}(t) \to 2D\{|t| - D\beta M\}. \tag{11.50}$$

By comparing the definition of the velocity frequency function, eqn (11.33), and expression (11.46) for D we see that the following relation holds true

$$D = \frac{\pi p(0)}{2\beta M}. \tag{11.51}$$

Let us now consider the explicit form of $p(\omega)$ in the diffusion limit. From the definition of $p(\omega)$ in (11.33), we have, using (11.44),

$$p(\omega) = \frac{\beta M}{3\pi}\langle \mathbf{v}^2(0)\rangle \int_{-\infty}^{\infty} dt \exp(i\omega t - \gamma|t|), \tag{11.52 a}$$

which, using (11.49), is

$$p(\omega) = \frac{1}{\pi}\frac{2\gamma}{\omega^2+\gamma^2} \qquad \text{(diffusion)}. \tag{11.52 b}$$

This is to be contrasted with the δ-function result for an ideal gas (11.42), which characterizes the short-time behaviour. As a consequence of diffusion $p(\omega)$ has a Lorentzian form. Note that in the limit of $\gamma \to 0$ the result (11.52 b) becomes identical with that for the ideal gas case. Furthermore, by taking $\omega = 0$ in (11.52 b) the result taken with the general relationship (11.51) is consistent with relation (11.49) between γ and D.

The effect of diffusion on the cross-section is readily shown. By making use of (11.15) the cross-section is proportional to

$$\begin{aligned} S_{\mathrm{i}}(\boldsymbol{\kappa}, \omega) &\simeq \exp(\tfrac{1}{2}\hbar\omega\beta)\frac{1}{2\pi\hbar}\int_{-\infty}^{\infty} dt \exp(-i\omega t)\exp\{-\tfrac{1}{2}\kappa^2(2D|t|)\} \\ &= \exp(\tfrac{1}{2}\hbar\omega\beta)\frac{1}{\pi\hbar}\frac{D\kappa^2}{\omega^2+(D\kappa^2)^2}. \end{aligned} \qquad \text{(diffusion)} \quad (11.53)$$

Thus far we have demonstrated that if the spatial dependence of the self-function $G_s(\mathbf{r}, t)$ is approximated by a Gaussian then the dynamical behaviour of the liquid, as contained in $\bar{w}(t)$, is determined by the velocity–velocity correlation function. In constructing a theory of incoherent scattering from liquids it is therefore sufficient to calculate only this correlation function, for we can then in principle calculate the cross-section. On the basis of purely physical arguments $\bar{w}(t)$ in the limit of small t is proportional to t^2 and at long times it is proportional to $|t|$.

(*b*) *Generalized Langevin equation*†

We noted above that the (Brownian) equation of motion (11.43) cannot describe the behaviour of a liquid for small times. To overcome this difficulty Kubo introduced a frequency-dependent frictional force, writing in place of (11.43) the generalized Langevin equation

$$\dot{v}(t) = -\int_{-\infty}^{\infty} \gamma(t-t')v(t')\,\mathrm{d}t' + \frac{1}{M}f(t). \tag{11.54}$$

In this equation $\gamma(t)$ represents the retarded effect of the frictional force, and $f(t)$ is an external force. A discussion of the properties of the generalized Langevin equation and the calculation of the velocity–velocity correlation function are most easily accomplished by using linear response theory (cf. Appendix B). The salient points of this theory are briefly reviewed prior to considering eqn (11.54) in detail.

Let A be an observable quantity and let an external potential with a time dependence $f(t)$ act on the system to which it belongs. The change in the average value of A due to the external potential is then given by

$$\overline{A(t)} - \langle A \rangle = \exp(-\mathrm{i}\omega t) f \chi[\omega], \tag{11.55}$$

where we have taken $f(t) = \exp(-\mathrm{i}\omega t) f$ and the constant f is defined to be sufficiently small for us to need only the change in A linear in f. The function $\chi[\omega]$ in eqn (11.55) is the generalized susceptibility function. We note that

$$\chi[\omega] = \chi'[\omega] + \mathrm{i}\chi''[\omega], \tag{11.56}$$

where the real part $\chi'[\omega]$ is even in ω and the imaginary part $\chi''[\omega]$ is odd in ω. The fluctuation-dissipation theorem tells us that if B is the physical quantity associated with $f(t)$ in the perturbed Hamiltonian then (for classical systems)

$$\langle B(0)A(t)\rangle = \langle A(t)B(0)\rangle = \frac{1}{\pi\beta}\int_{-\infty}^{\infty} \mathrm{d}\omega \exp(-\mathrm{i}\omega t)\frac{\chi''[\omega]}{\omega}. \tag{11.57}$$

† The reader who is unfamiliar with linear response theory and the fluctuation-dissipation theorem is urged to consult Appendix B and R. Kubo, *Rep. Prog. Phys.* **29**, 255 (1966). See also C. Murase, *J. phys. Soc. Japan* **29**, 549 (1970).

Eqn (11.57) is valid when A and B transform in the same way under time reversal.

By looking at the equation of motion (11.54) we see that the external potential must be $-xf(t)$, where $\dot{x}(t) = v(t)$. If we also take $A = x$, then the above equations read

$$\overline{x(t)} = \exp(-\mathrm{i}\omega t) f\chi[\omega] \tag{11.58}$$

and

$$\langle x(t)x(0)\rangle = \frac{1}{\pi\beta}\int_{-\infty}^{\infty} \mathrm{d}\omega \exp(-\mathrm{i}\omega t)\frac{\chi''[\omega]}{\omega}. \tag{11.59}$$

From the stationary condition

$$\langle x(0)\ddot{x}(t)\rangle = -\langle \dot{x}(0)\dot{x}(t)\rangle \equiv -\langle v(0)v(t)\rangle, \tag{11.60}$$

so that (11.59) yields the result

$$\langle v(0)v(t)\rangle = \tfrac{1}{3}\langle \mathbf{v}(0).\mathbf{v}(t)\rangle = \frac{1}{\pi\beta}\int_{-\infty}^{\infty} \mathrm{d}\omega\, \omega\chi''[\omega]\exp(-\mathrm{i}\omega t). \tag{11.61}$$

This is our key equation; it expresses the velocity–velocity correlation function in terms of the imaginary part of the generalized susceptibility $\chi[\omega]$.

Assuming

$$\langle \mathbf{v}^2(0)\rangle = \frac{3}{\beta M},$$

it follows that

$$\int_{-\infty}^{\infty} \mathrm{d}\omega\, \omega\chi''[\omega] = \frac{\pi}{M}. \tag{11.62}$$

The generalized susceptibility function $\chi[\omega]$ and the velocity frequency function $p(\omega)$ introduced in the previous section are intimately related. From the definition of $p(\omega)$, eqn (11.33), and the relation (11.62) we have

$$\begin{aligned} p(\omega) &= \frac{\beta M}{\pi}\frac{1}{\pi\beta}\int_{-\infty}^{\infty} \mathrm{d}t \exp(\mathrm{i}\omega t)\int_{-\infty}^{\infty} \mathrm{d}\omega'\, \omega'\chi''[\omega']\exp(-\mathrm{i}\omega' t) \\ &= \frac{2M}{\pi}\omega\chi''[\omega]. \end{aligned} \tag{11.63}$$

This relationship implies that

$$\omega\chi''[\omega] = -\omega\chi''[-\omega] \geqslant 0. \tag{11.64 a}$$

Hence from (11.61) and (11.64) we conclude that

$$\left[\frac{\partial}{\partial t}\langle v(0)v(t)\rangle\right]_{t=0} = 0 \tag{11.64 b}$$

and similarly for all other odd derivatives. Also the second derivative is negative and non-zero, i.e.

$$\left[\frac{\partial^2}{\partial t^2}\langle v(0)v(t)\rangle\right]_{t=0} < 0. \tag{11.64 c}$$

We notice that neither of these rigorous conditions are satisfied by the simple formula of (11.44).

By making use of the identity (11.62) it is a simple matter to verify that the expression (11.63) for $p(\omega)$ satisfies the normalization condition for $p(\omega)$, eqn (11.34).

We can also express the width function $\bar{w}(t)$ in terms of $\chi''[\omega]$. From the representation (11.32) of $\bar{w}(t)$,

$$\begin{aligned}\bar{w}(t) &= 2\int_0^t \mathrm{d}t'\,(t-t')\frac{1}{\pi\beta}\int_{-\infty}^{\infty}\mathrm{d}\omega\,\omega\chi''[\omega]\exp(-\mathrm{i}\omega t)\\ &= \frac{2}{\pi\beta}\int_{-\infty}^{\infty}\mathrm{d}\omega\,\chi''[\omega]\left\{\frac{1}{\omega}(1-\cos\omega t)-\mathrm{i}\left(t-\frac{1}{\omega}\sin\omega t\right)\right\}\\ &= \frac{2}{\pi\beta}\int_{-\infty}^{\infty}\mathrm{d}\omega\,\frac{\chi''[\omega]}{\omega}(1-\cos\omega t),\end{aligned} \tag{11.65}$$

where in deriving the last line we have used the fact that $\chi''[\omega]$ is an odd function of ω. The result (11.65) could have alternatively been obtained by combining (11.36) and (11.63).

One further relation of importance is that between the diffusion constant D and $\chi''[\omega]$. Proceeding from eqn (11.51) and making use of (11.63) we can derive the result

$$D = \frac{1}{\beta}\lim_{\omega\to 0^+}\omega\chi''[\omega]. \tag{11.66}$$

So far we have employed the fluctuation-dissipation theorem to express the quantities of interest in calculating the incoherent cross-section, namely $p(\omega)$ and $\bar{w}(t)$, in terms of the generalized susceptibility function. The theory is quite general, though our motivation in setting it up was to consider the properties of the dynamical motion of the nuclei as defined by eqn (11.54); this we do now.

In eqn (11.54) set $f(t) = f\exp(-\mathrm{i}\omega t)$, then $\overline{x(t)} = \exp(-\mathrm{i}\omega t)\overline{x(0)}$. We then find that

$$\overline{x(t)} = \frac{-f\exp(-\mathrm{i}\omega t)}{M\{\omega^2+\mathrm{i}\omega\gamma(\omega)\}}, \tag{11.67}$$

where
$$\gamma(\omega) = \int_{-\infty}^{\infty} \mathrm{d}t\, \gamma(t)\exp(\mathrm{i}\omega t). \tag{11.68}$$

From the definition of $\chi[\omega]$ we have the result

$$\chi[\omega] = \frac{-1}{M\{\omega^2+\mathrm{i}\omega\gamma(\omega)\}}, \tag{11.69}$$

i.e.
$$\chi''[\omega] = \frac{1}{M}\frac{\gamma'(\omega)}{\omega[\{\omega-\gamma''(\omega)\}^2+\{\gamma'(\omega)\}^2]}.$$

Because $\gamma(t)$ is a casual function it must possess the same analytic properties as the generalized susceptibility. In particular we have from Appendix B, eqns (B.11 a) and (B.11 b): $\gamma(\omega) = \gamma'(\omega)+\mathrm{i}\gamma''(\omega)$, where

$$\gamma''(\omega) = \frac{1}{\pi}\,\mathrm{P}\int_{-\infty}^{\infty} \frac{\mathrm{d}u\, \gamma'(u)}{\omega-u} \tag{11.70}$$

and
$$\gamma'(\omega) = -\frac{1}{\pi}\,\mathrm{P}\int_{-\infty}^{\infty} \frac{\mathrm{d}u\, \gamma''(u)}{\omega-u}. \tag{11.71}$$

In practical applications of the theory it is convenient to rewrite eqn (11.70) in the form

$$\gamma''(\omega) = \frac{2}{\pi}\int_0^{\infty} \mathrm{d}u\, \gamma'(u) \int_0^{\infty} \mathrm{d}s\, \sin(s\omega)\cos(su). \tag{11.72}$$

Also note from (11.68) that since $\gamma(t)$ is real, $\gamma'(\omega)$ must be an even function of ω and $\gamma''(\omega)$ must be an odd function of ω. Hence $\gamma''(0)$ is zero and in view of (11.66) and (11.69) we have the relationship

$$D = \frac{1}{\beta M}\frac{1}{\gamma'(0)}, \tag{11.73}$$

which implies that $\gamma'(0)$ is non-zero. From (11.71) we note that $\gamma'(0)$ is related to $\gamma''(\omega)$ through

$$\gamma'(0) = \frac{1}{\pi}\int_{-\infty}^{\infty} \mathrm{d}u\, \frac{\gamma''(u)}{u} = \frac{1}{D\beta M}. \tag{11.74}$$

It is worth while at this stage to verify that the theory developed so far is in agreement with the simple diffusion theory presented in § 11.2 (*a*). In this instance

$$\gamma(t) = \gamma\delta(t)$$

so that $\gamma(\omega) = \gamma$, $\gamma''(\omega)$ is zero, and (11.69) becomes

$$\omega\chi''[\omega] = \frac{\gamma}{M\{\omega^2+\gamma^2\}}.$$

Whence
$$\int_{-\infty}^{\infty} d\omega\, \omega\chi''[\omega] = \frac{\pi}{M}$$
in agreement with (11.62), and from (11.63)
$$p(\omega) = \frac{1}{\pi}\frac{2\gamma}{\omega^2+\gamma^2} \qquad \text{(diffusion)} \qquad (11.52\,\text{b})$$
as required. Furthermore, from (11.61) we get
$$\langle v(0)v(t)\rangle = \frac{1}{\beta M}\exp(-\gamma|t|).$$

Let us now examine an improved theory which assumes that $\gamma'(\omega)$ has the form
$$\gamma'(\omega) = \frac{\gamma}{1+(\omega\tau)^2}, \qquad (11.75\,\text{a})$$
where τ is a parameter. $\gamma''(\omega)$ is determined either by a contour integral directly from (11.70) or from (11.72); viz.
$$\gamma''(\omega) = \gamma\frac{2}{\pi}\int_0^{\infty} ds\, \sin(s\omega)\int_0^{\infty} du\, \frac{\cos(su)}{1+(u\tau)^2}$$
$$= \gamma\frac{2}{\pi}\frac{\pi}{2\tau}\int_0^{\infty} ds\, \sin(s\omega)\exp(-s/\tau)$$
$$= \frac{\gamma}{\tau}\frac{\omega}{\omega^2+\tau^{-2}},$$
i.e.
$$\gamma''(\omega) = \tau\gamma\frac{\omega}{1+(\omega\tau)^2}. \qquad (11.75\,\text{b})$$
Combining (11.75 a) and (11.75 b) we see that
$$\gamma(\omega) = \frac{\gamma}{1-i\omega\tau}. \qquad (11.75\,\text{c})$$
Furthermore, $\gamma(t)$ is given by
$$\gamma(t) = \frac{1}{2\pi}\int_{-\infty}^{\infty} d\omega\, \exp(-i\omega t)\gamma(\omega)$$
$$= \frac{\gamma}{\tau}\exp(-t/\tau) \quad (t>0)$$
$$= 0 \qquad (t<0). \qquad (11.76)$$

This verifies that our $\gamma(t)$ is indeed casual and in this instance it decays exponentially with lifetime τ.

We can guess immediately that, at small times, (11.76) is not a good representation of $\gamma(t)$: but because the assumption of exponential decay is now made for $\gamma(t)$ and not for $\langle v(0)v(t)\rangle$ directly, we expect this new theory to give a better representation of $\langle v(0)v(t)\rangle$. From (11.69) and (11.75) some simple algebra gives

$$\omega\chi''[\omega] = \frac{(\gamma/M)}{\omega^4\tau^2+\omega^2(1-2\tau\gamma)+\gamma^2}. \tag{11.77}$$

Putting (11.77) into (11.61) gives, after a contour integration,

$$\langle v(0)v(t)\rangle = \frac{1}{2M\beta R}[(1+R)\exp\{-|t|(1-R)/2\tau\}-$$
$$-(1-R)\exp\{-|t|(1+R)/2\tau\}], \tag{11.78 a}$$

where
$$R = (1-4\tau\gamma)^{\frac{1}{2}}. \tag{11.78 b}$$

We notice that (11.78 a) is an improvement on (11.44) because at small times

$$\langle v(0)v(t)\rangle = \frac{1}{M\beta}\left(1-\frac{t^2\gamma}{2\tau}+...\right)$$

and therefore does not have a discontinuous first derivative at $t = 0$. As we expect, (11.78 a) gives the equipartition of energy described by (11.49), and substituting (11.78 a) into (11.46) and performing the elementary integration immediately gives the diffusion constant D as in (11.50). At long times only the first term of (11.78 a) matters and we notice that in the limit $\tau \to 0$ we recover the simple result (11.44). We conclude that this is a better theory than the simple diffusion theory, although it is certainly not rigorous; for example (11.78 a) does not satisfy (11.64 b) applied to the third derivative.

From (11.32) and (11.78 a) we obtain for this model the result

$$\bar{w}(t) = \frac{2}{M\beta\gamma^2}\Big[-1+\tau\gamma+|t|\gamma+\frac{1}{8R}\exp(-|t|/2\tau)\{(1+R)^3\exp(|t|R/2\tau)-$$
$$-(1-R)^3\exp(-|t|R/2\tau)\}\Big]. \tag{11.78 c}$$

It is evident that this has the correct behaviour close to $t = 0$, and at large times gives the expected result, $2D|t|$.

(c) *Model of Egelstaff and Schofield*

The result (11.78 c) for $\bar{w}(t)$ is a reasonable approximation but it is inconvenient to use in calculations. Egelstaff and Schofield noted that it is possible to fabricate a simple $\bar{w}(t)$ that possesses the correct short-

and long-time behaviour. By correct we mean that for small t, $\bar{w}(t)$ is proportional to t^2 and for long times proportional to $|t|$. The function

$$\bar{w}(t) = 2D\{(t^2+c^2)^{\frac{1}{2}}-c\} \tag{11.79}$$

is readily shown to fulfil these conditions. c is a parameter characteristic of the model. The function (11.79) is simple enough to permit the cross-section to be calculated in closed form (unlike (11.78 c)).

From (11.2 b) and (11.15) it is evident that to calculate the cross-section we need to evaluate

$$\int_{-\infty}^{\infty} \mathrm{d}t \exp(-\mathrm{i}\omega t)\exp[-D\kappa^2\{(t^2+c^2)^{\frac{1}{2}}-c\}]. \tag{11.80}$$

This integral can be rewritten in terms of a modified Bessel function of the second kind $K_1(x)$. One definition of $K_1(x)$ is

$$2\left(\frac{a}{b}\right)^{\frac{1}{2}} K_1\{2\surd(ab)\} = \int_0^{\infty} \mathrm{d}t\, \frac{1}{t^2}\exp\left(-\frac{b}{t}-at\right). \tag{11.81}$$

By making use of the identity

$$\exp(-A\surd s) = \frac{A}{2\surd\pi}\int_0^{\infty} \mathrm{d}t'\, t'^{-\frac{3}{2}}\exp(-st'-A^2/4t')$$

we find from (11.80)

$$S_{\mathrm{i}}(\boldsymbol{\kappa}, \omega) \simeq \frac{1}{2\pi\hbar}\exp(\tfrac{1}{2}\hbar\omega\beta+c\kappa^2D)\frac{2c\kappa^2D}{\surd\{\omega^2+(\kappa^2D)^2\}}K_1[c\surd\{\omega^2+(\kappa^2D)^2\}]. \tag{11.82}$$

The asymptotic limits of $K_1(x)$ are

$$K_1(x) \simeq \frac{1}{x} \quad (x \to 0)$$

and

$$K_1(x) \simeq \sqrt{\left(\frac{\pi}{2x}\right)}\exp(-x) \quad (x \to \infty). \tag{11.83}$$

With the aid of these results we find that $S_{\mathrm{i}}(\boldsymbol{\kappa}, \omega)$ for the Egelstaff and Schofield model has the following limiting forms. For $c\kappa^2D \gg 1$

$$S_{\mathrm{i}}(\boldsymbol{\kappa}, \omega) \simeq \frac{1}{\hbar}\left\{\frac{c}{2\pi\kappa^2D}\right\}^{\frac{1}{2}}\exp\left\{\tfrac{1}{2}\hbar\omega\beta-\frac{1}{2}\frac{c\omega^2}{\kappa^2D}\right\}. \tag{11.84}$$

This corresponds to the behaviour of the cross-section at high energies.

In the opposite limit of $c\kappa^2D \ll 1$

$$S_{\mathrm{i}}(\boldsymbol{\kappa}, \omega) \simeq \exp(\tfrac{1}{2}\hbar\omega\beta)\,\frac{1}{\pi\hbar}\,\frac{\kappa^2D}{\omega^2+(\kappa^2D)^2}. \tag{11.85}$$

This agrees exactly with our previous result for simple diffusion theory, eqn (11.53).

The adequacy of the Egelstaff–Schofield model to account for the measured incoherent scattering from liquid sodium has been investigated in detail by Cocking. In his analysis Cocking multiplied the cross-section derived above by an 'effective Debye–Waller factor', $\exp\{-\frac{1}{2}\kappa^2 w_1(0)\}$ to

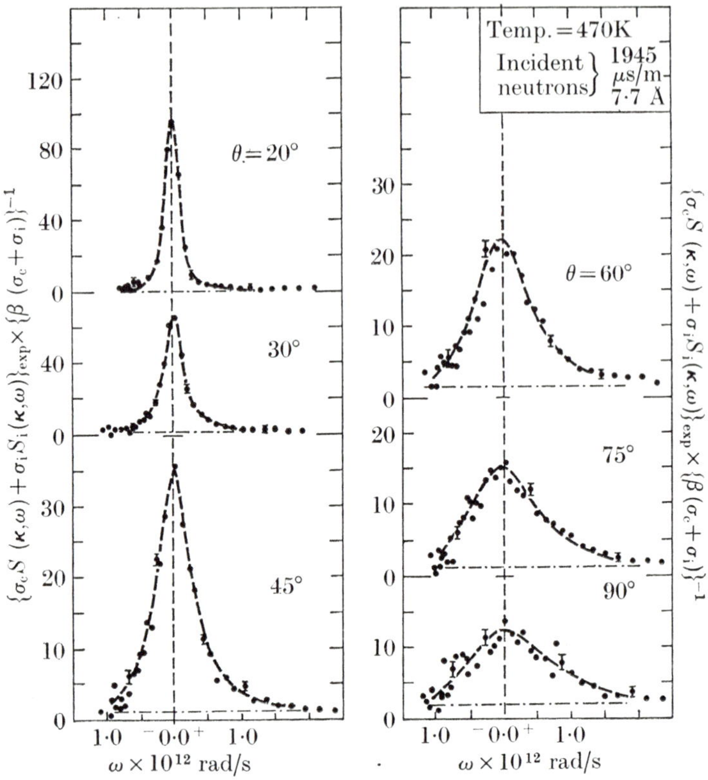

FIG. 11.2. Comparison between the Eglestaff–Schofield model of incoherent scattering from sodium (broken curve) and the experimental data of Cocking, showing that the coherent contribution is small and the model adequate. After S. J. Cocking, *J. Phys.* C **2**, 2047 (1969).

make some allowance for high-frequency modes. Thus the model studied by Cocking contained two parameters, c and $w_1(0)$. For a solid $w_1(0)$ can be calculated from the phonon density of states, as shown in Chapter 4. In the case of a liquid such a procedure cannot be adopted and therefore c and $w_1(0)$ were taken as parameters to be determined by fitting the measured quasi-elastic scattering. We do not pursue Cocking's analysis in great detail, but the comparison between the measured cross-section and the calculated cross-section in Fig. 11.2 shows clearly that the simple model gives a good account of the data.

The Egelstaff–Schofield model is valuable in analysing measurements first because the fabricated width function is based on fundamental physical arguments and secondly because it is sufficiently simple to permit the cross-section to be calculated in a convenient form. The analysis by Cocking has shown that, for sodium, it gives a good account of the measured quasi-elastic scattering (Fig 11.2).

11.3. Incoherent scattering: quantum theory

The definition of $I_s(\boldsymbol{\kappa}, t)$ is

$$I_s(\boldsymbol{\kappa}, t) = \frac{1}{N}\sum_j \langle \exp\{-i\boldsymbol{\kappa}.\hat{\mathbf{R}}_j(0)\}\exp\{i\boldsymbol{\kappa}.\hat{\mathbf{R}}_j(t)\}\rangle. \tag{11.86}$$

For a monatomic liquid the averaging over the atoms can be dropped.

Because $\exp\{i\boldsymbol{\kappa}.\hat{\mathbf{R}}(t)\}$ is an operator the two exponential factors in (11.86) cannot be written as simply $\exp[i\boldsymbol{\kappa}.\{\hat{\mathbf{R}}(t)-\hat{\mathbf{R}}(0)\}]$. However, we can define an operator $\hat{\Gamma}(t)$ such that

$$\exp\{-i\boldsymbol{\kappa}.\hat{\mathbf{R}}(0)\}\exp\{i\boldsymbol{\kappa}.\hat{\mathbf{R}}(t)\} = \exp[-i\boldsymbol{\kappa}.\{\hat{\mathbf{R}}(0)-\hat{\mathbf{R}}(t)\}+\tfrac{1}{2}i\hat{\Gamma}(t)]. \tag{11.87}$$

The formula to calculate $\hat{\Gamma}(t)$ is given in § 3.4. In the special case when $[\boldsymbol{\kappa}.\hat{\mathbf{R}}(0), \boldsymbol{\kappa}.\hat{\mathbf{R}}(t)]$ is a c-number, $\hat{\Gamma}(t)$ is also a c-number and has the value

$$\Gamma(t) = -i[\boldsymbol{\kappa}.\hat{\mathbf{R}}(0), \boldsymbol{\kappa}.\hat{\mathbf{R}}(t)]. \tag{11.88}$$

Note that $\Gamma(t)$ as given by (11.88) is purely real.

Let us make the assumption that $\hat{\Gamma}(t)$ is always a c-number. Then

$$\langle \exp\{-i\boldsymbol{\kappa}.\hat{\mathbf{R}}(0)\}\exp\{i\boldsymbol{\kappa}.\hat{\mathbf{R}}(t)\rangle = \exp\{\tfrac{1}{2}i\Gamma(t)\}\langle \exp[-i\boldsymbol{\kappa}.\{\hat{\mathbf{R}}(0)-\hat{\mathbf{R}}(t)\}]\rangle.$$

By analogy with the rigorous result for harmonic crystals let us write

$$\langle \exp[-i\boldsymbol{\kappa}.\{\hat{\mathbf{R}}(0)-\hat{\mathbf{R}}(t)\}]\rangle \equiv \exp\{-\tfrac{1}{2}\kappa^2 w(t)\}; \tag{11.89}$$

then $I_s(\boldsymbol{\kappa}, t)$ is

$$I_s(\boldsymbol{\kappa}, t) = \exp\{\tfrac{1}{2}i\Gamma(t)-\tfrac{1}{2}\kappa^2 w(t)\}, \tag{11.90}$$

i.e. the corresponding $G_s(\mathbf{r}, t)$ is a Gaussian in $\mathbf{r}$. The argument leading to (11.90) clarifies the assumptions made in assuming that $G_s(\mathbf{r}, t)$ has a Gaussian spatial dependence. We shall later indicate a more rigorous argument which allows us in principle to evaluate corrections to the Gaussian approximation.

For an ideal gas of nuclei $I_s(\boldsymbol{\varkappa}, t)$ can be evaluated exactly (cf. Chapter 3 and (11.17))

$$I_s(\boldsymbol{\varkappa}, t) = \exp\left\{\frac{-\kappa^2}{2M\beta}(t - \mathrm{i}\hbar\beta)t\right\} \qquad \text{(ideal gas).} \tag{11.91}$$

For small t, $I_s(\boldsymbol{\varkappa}, t)$ must coincide with the ideal gas result, eqn (11.91). Thus we can identify $\Gamma(t)$ in (11.90) with $\hbar\kappa^2 t/M$:

$$\Gamma(t) \equiv \hbar^2\kappa^2 t/M. \tag{11.92}$$

We now put the above discussion on a more rigorous basis. The required function is

$$\langle\exp\{-\mathrm{i}\boldsymbol{\varkappa}.\hat{\mathbf{R}}(0)\}\exp\{\mathrm{i}\boldsymbol{\varkappa}.\hat{\mathbf{R}}(t)\}\rangle$$
$$= \langle\exp(-\mathrm{i}\boldsymbol{\varkappa}.\mathbf{R})\exp(\mathrm{i}t\mathscr{H}/\hbar)\exp(\mathrm{i}\boldsymbol{\varkappa}.\mathbf{R})\exp(-\mathrm{i}t\mathscr{H}/\hbar)\rangle.$$

In view of the identity

$$\exp(-\mathrm{i}\boldsymbol{\varkappa}.\mathbf{r})f(\mathbf{r}, \hat{\mathbf{p}})\exp(\mathrm{i}\boldsymbol{\varkappa}.\mathbf{r}) = f(\mathbf{r}, \hat{\mathbf{p}}+\hbar\boldsymbol{\varkappa}),$$

where $\hat{\mathbf{p}}$ is the momentum operator conjugate to $\mathbf{r}$,

$$\exp(-\mathrm{i}\boldsymbol{\varkappa}.\mathbf{R}_j)\exp(\mathrm{i}t\mathscr{H}/\hbar)\exp(\mathrm{i}\boldsymbol{\varkappa}.\mathbf{R}_j)$$
$$= \exp(\tfrac{1}{2}\mathrm{i}\hbar\kappa^2 t/M)\exp(\mathrm{i}t\mathscr{H}/\hbar+\mathrm{i}t\boldsymbol{\varkappa}.\hat{\mathbf{v}}_j). \tag{11.93}$$

Here the label j has been re-introduced merely to clarify the result of the unitary transformation on the operator $\exp(\mathrm{i}t\mathscr{H}/\hbar)$.

With the result (11.93)

$$\langle\exp\{-\mathrm{i}\boldsymbol{\varkappa}.\hat{\mathbf{R}}(0)\}\exp\{\mathrm{i}\boldsymbol{\varkappa}.\hat{\mathbf{R}}(t)\}\rangle$$
$$= \exp(\tfrac{1}{2}\mathrm{i}\hbar\kappa^2 t/M)\langle\exp(\mathrm{i}t\mathscr{H}/\hbar+\mathrm{i}t\boldsymbol{\varkappa}.\hat{\mathbf{v}})\exp(-\mathrm{i}t\mathscr{H}/\hbar)\rangle$$
$$= \exp(\tfrac{1}{2}\mathrm{i}\hbar\kappa^2 t/M)\langle\hat{O}(t)\rangle. \tag{11.94}$$

To determine the operator $\hat{O}(t)$ defined in (11.94) we derive its equation of motion

$$\frac{\partial}{\partial t}\hat{O}(t) = \left(\frac{\mathrm{i}}{\hbar}\mathscr{H}+\mathrm{i}\boldsymbol{\varkappa}.\hat{\mathbf{v}}\right)\exp(\mathrm{i}t\mathscr{H}/\hbar+\mathrm{i}t\boldsymbol{\varkappa}.\hat{\mathbf{v}})\exp(-\mathrm{i}t\mathscr{H}/\hbar)+$$
$$+\exp(\mathrm{i}t\mathscr{H}/\hbar+\mathrm{i}t\boldsymbol{\varkappa}.\hat{\mathbf{v}})\left(-\frac{\mathrm{i}\mathscr{H}}{\hbar}\right)\exp(-\mathrm{i}t\mathscr{H}/\hbar)$$
$$= \exp(\mathrm{i}t\mathscr{H}/\hbar+\mathrm{i}t\boldsymbol{\varkappa}.\hat{\mathbf{v}})(\mathrm{i}\boldsymbol{\varkappa}.\hat{\mathbf{v}})\exp(-\mathrm{i}t\mathscr{H}/\hbar)$$
$$= \hat{O}(t)\exp(\mathrm{i}t\mathscr{H}/\hbar)(\mathrm{i}\boldsymbol{\varkappa}.\hat{\mathbf{v}})\exp(-\mathrm{i}t\mathscr{H}/\hbar).$$

Thus $\hat{O}(t)$ satisfies the equation

$$\frac{\partial\hat{O}(t)}{\partial t} = \hat{O}(t)\mathrm{i}\boldsymbol{\varkappa}.\hat{\mathbf{v}}(t) \tag{11.95}$$

or

$$\hat{O}(t) = 1+\int_0^t \mathrm{d}t'\ \hat{O}(t')\mathrm{i}\boldsymbol{\varkappa}.\hat{\mathbf{v}}(t').$$

Iterating the latter equation gives

$$\hat{O}(t) = 1 + \int_0^t \mathrm{d}t_1\, \mathrm{i}\boldsymbol{\kappa}.\hat{\mathbf{v}}(t_1) + \int_0^t \mathrm{d}t_1 \int_0^{t_1} \mathrm{d}t_2\, \mathrm{i}\boldsymbol{\kappa}.\hat{\mathbf{v}}(t_2)\mathrm{i}\boldsymbol{\kappa}.\hat{\mathbf{v}}(t_1) +$$

$$+ \ldots + \int_0^t \mathrm{d}t_1 \int_0^{t_1} \mathrm{d}t_2 \ldots \int_0^{t_{n-1}} \mathrm{d}t_n\, \mathrm{i}\boldsymbol{\kappa}.\hat{\mathbf{v}}(t_n)\mathrm{i}\boldsymbol{\kappa}.\hat{\mathbf{v}}(t_{n-1}) \ldots \mathrm{i}\boldsymbol{\kappa}.\hat{\mathbf{v}}(t_1) + \ldots, \tag{11.96}$$

where $t_1 > t_2 > t_3 > \ldots > t_n$.

We actually need the thermal average of $\hat{O}(t)$ in (11.94). Those terms in the expansion (11.96) containing an odd number of $\hat{\mathbf{v}}(t)$'s do not contribute to $\langle \hat{O}(t) \rangle$. Look at the second non-vanishing term in the expansion for $\langle \hat{O}(t) \rangle$ (note that the trace of a product of operators is invariant to a cyclic permutation of the operators; hence we can re-order the time in (11.96) after taking the thermal average); this is

$$(\mathrm{i})^2 \int_0^t \mathrm{d}t_1 \int_0^{t_1} \mathrm{d}t_2\, \langle \boldsymbol{\kappa}.\hat{\mathbf{v}}(t_1)\boldsymbol{\kappa}.\hat{\mathbf{v}}(t_2) \rangle \quad (t > t_1 > t_2) \tag{11.97}$$

and can be written

$$\tfrac{1}{3}(\mathrm{i})^2\kappa^2 \int_0^t \mathrm{d}t'\, (t-t')\langle \hat{\mathbf{v}}(0).\hat{\mathbf{v}}(t') \rangle$$

because we are concerned with an isotropic system. We define

$$w(t) = \tfrac{2}{3} \int_0^t \mathrm{d}t'\, (t-t')\langle \hat{\mathbf{v}}(0).\hat{\mathbf{v}}(t') \rangle \tag{11.98}$$

so that the second term in the expansion of $\langle \hat{O}(t) \rangle$, eqn (11.96), reads

$$-\tfrac{1}{2}\kappa^2 w(t). \tag{11.99}$$

Now let us consider the third non-zero term in the expansion of $\langle \hat{O}(t) \rangle$;

$$(\mathrm{i})^4 \int_0^t \mathrm{d}t_1 \int_0^{t_1} \mathrm{d}t_2 \int_0^{t_2} \mathrm{d}t_3 \int_0^{t_3} \mathrm{d}t_4\, \langle \boldsymbol{\kappa}.\hat{\mathbf{v}}(t_1)\boldsymbol{\kappa}.\hat{\mathbf{v}}(t_2)\boldsymbol{\kappa}.\hat{\mathbf{v}}(t_3)\boldsymbol{\kappa}.\hat{\mathbf{v}}(t_4) \rangle. \tag{11.100}$$

We treat this term by approximating the four-fold correlation function by the products of velocity–velocity correlation functions. Schematically the approximation reads

$$\langle 1234 \rangle \Rightarrow \langle 12 \rangle\langle 34 \rangle + \langle 13 \rangle\langle 24 \rangle + \langle 14 \rangle\langle 23 \rangle.$$

On performing this 'decoupling' in (11.100) the latter becomes

$$\frac{1}{2!}\{\tfrac{1}{2}\kappa^2 w(t)\}^2. \tag{11.101}$$

An analogous procedure applied to the remaining terms enables us to write

$$\langle \hat{O}(t) \rangle \simeq \exp\{-\tfrac{1}{2}\kappa^2 w(t)\} \tag{11.102}$$

with $w(t)$ given by (11.98). Thus we regain the result derived earlier, namely, within the above approximations,

$$I_s(\boldsymbol{\kappa}, t) = \exp\left\{\frac{\mathrm{i}}{2}\frac{\hbar\kappa^2 t}{M} - \tfrac{1}{2}\kappa^2 w(t)\right\}. \tag{11.103}$$

The derivation of approximation (11.102) for $\langle\hat{O}(t)\rangle$ was first made by Rahman, Singwi, and Sjölander.† These authors also showed how to evaluate corrections to the Gaussian (i.e. the result (11.103) for $I_s(\boldsymbol{\kappa}, t)$) approximation and we refer the reader to their paper for details.

In contrast to the classical width function $\bar{w}(t)$, the width function $w(t)$ is complex. From the requirement that the time Fourier transform of $I_s(\boldsymbol{\kappa}, t)$ be real, $w(t)$ must satisfy

$$w^*(t) = w(-t). \tag{11.104}$$

We can, with the use of the fluctuation-dissipation theorem, relate the real and imaginary parts of the correlation function $\langle\hat{\mathbf{v}}(0).\hat{\mathbf{v}}(t)\rangle$ and thence, via the definition of $w(t) = w'(t)+\mathrm{i}w''(t)$, eqn (11.98), relate w' and w''. The fluctuation-dissipation theorem states (Appendix B, eqn (B.56)) that if

$$\phi(t) = \frac{\mathrm{i}}{\hbar}\langle[\hat{A}(t), \hat{B}(0)]\rangle \tag{11.105}$$

is an odd function of t, then

$$\tfrac{1}{2}\int\limits_{-\infty}^{\infty} \mathrm{d}t\; \mathrm{e}^{\mathrm{i}\omega t}\langle\hat{A}(t)\hat{B}(0)+\hat{B}(0)\hat{A}(t)\rangle$$

$$= \hbar\coth(\tfrac{1}{2}\hbar\omega\beta)\,\mathrm{Im}\int\limits_{0}^{\infty} \mathrm{d}t\; \mathrm{e}^{\mathrm{i}\omega t}\phi(t). \tag{11.106}$$

With $\hat{A} = \hat{B} = \hat{v}$, $\phi(t)$ is an odd function of t, as required, and further

$$\phi(t) = \frac{2}{\hbar}\,\mathrm{Im}\langle\hat{v}(0)\hat{v}(t)\rangle. \tag{11.107}$$

Also

$$\langle\hat{v}(0)\hat{v}(t)+\hat{v}(t)\hat{v}(0)\rangle = 2\,\mathrm{Re}\langle\hat{v}(0)\hat{v}(t)\rangle. \tag{11.108}$$

Substituting (11.107) and (11.108) in (11.106) yields the result

$$\int\limits_{-\infty}^{\infty} \mathrm{d}t\; \mathrm{e}^{\mathrm{i}\omega t}\,\mathrm{Re}\langle\hat{v}(0)\hat{v}(t)\rangle = 2\coth(\tfrac{1}{2}\hbar\omega\beta)\,\mathrm{Im}\int\limits_{0}^{\infty} \mathrm{d}t\; \mathrm{e}^{\mathrm{i}\omega t}\,\mathrm{Im}\langle\hat{v}(0)\hat{v}(t)\rangle$$

$$= -\mathrm{i}\coth(\tfrac{1}{2}\hbar\omega\beta)\int\limits_{-\infty}^{\infty} \mathrm{d}t\; \mathrm{e}^{\mathrm{i}\omega t}\,\mathrm{Im}\langle\hat{v}(0)\hat{v}(t)\rangle \tag{11.109}$$

or

$$\mathrm{Re}\langle\hat{v}(0)\hat{v}(t)\rangle = -\frac{\mathrm{i}}{2\pi}\int\limits_{-\infty}^{\infty} \mathrm{d}\omega\; \mathrm{e}^{-\mathrm{i}\omega t}\coth(\tfrac{1}{2}\hbar\omega\beta)\int\limits_{-\infty}^{\infty} \mathrm{d}t'\; \mathrm{e}^{\mathrm{i}\omega t'}\,\mathrm{Im}\langle\hat{v}(0)v(\hat{t}')\rangle.$$

† A. Rahman, K. S. Singwi, and A. Sjölander, *Phys. Rev.* **126**, 986 (1962).

By making use of (11.98) and (11.109) we obtain for $w(t)$

$$w(t) = -\frac{2\mathrm{i}}{3\pi}\int_{-\infty}^{\infty} \mathrm{d}\omega \, \frac{1}{\mathrm{e}^{\hbar\omega\beta}-1}\left\{\frac{1}{\omega^2}(1-\cos\omega t)-\frac{\mathrm{i}}{\omega}\left(t-\frac{1}{\omega}\sin\omega t\right)\right\}\times$$
$$\times\int_{-\infty}^{\infty} \mathrm{d}t' \, \mathrm{e}^{\mathrm{i}\omega t'}\,\mathrm{Im}\langle\hat{\mathbf{v}}(0)\,.\,\hat{\mathbf{v}}(t')\rangle. \quad (11.110)$$

In discussing the dynamical behaviour of classical liquids we found it convenient to define a velocity frequency function $p(\omega)$. By analogy we introduce in the present quantum case a velocity frequency function $Z(\omega)$ such that

$$\mathrm{Im}\langle\hat{\mathbf{v}}(0)\,.\,\hat{\mathbf{v}}(t)\rangle = \frac{3\hbar}{2M}\int_{0}^{\infty} \mathrm{d}\omega \, \omega Z(\omega)\sin\omega t. \quad (11.111)$$

Note that on taking the limit $\hbar \to 0$, $\mathrm{Im}\langle\hat{\mathbf{v}}(0)\,.\,\hat{\mathbf{v}}(t)\rangle \to 0$, i.e. the imaginary part of the velocity–velocity correlation function is zero in the classical limit as, of course, it must be. From (11.109) the real part of the velocity–velocity correlation function is

$$\mathrm{Re}\langle\hat{\mathbf{v}}(0)\,.\,\hat{\mathbf{v}}(t)\rangle = \frac{3\hbar}{2M}\int_{0}^{\infty} \mathrm{d}\omega \, \omega Z(\omega)\cos(\omega t)\coth(\tfrac{1}{2}\hbar\omega\beta). \quad (11.112)$$

Taking $\hbar \to 0$ in this expression gives

$$\lim_{\hbar\to 0}\mathrm{Re}\langle\hat{\mathbf{v}}(0)\,.\,\hat{\mathbf{v}}(t)\rangle = \frac{3\hbar}{2M}\int_{0}^{\infty} \mathrm{d}\omega \, \omega Z(\omega)\cos(\omega t)\frac{2}{\hbar\omega\beta}$$
$$= \frac{3}{\beta M}\int_{0}^{\infty} \mathrm{d}\omega \, Z(\omega)\cos(\omega t) \quad (11.113)$$

and this agrees with the classical result if we identify $Z(\omega)$ with $p(\omega)$.

For the width function $w(t)$ we have from (11.110) and (11.111) the result

$$w(t) = \frac{\hbar}{M}\int_{-\infty}^{\infty} \mathrm{d}\omega \, \frac{Z(\omega)}{\mathrm{e}^{\hbar\omega\beta}-1}\left\{\frac{1}{\omega}(1-\cos\omega t)-\mathrm{i}\left(t-\frac{1}{\omega}\sin\omega t\right)\right\}. \quad (11.114\,\mathrm{a})$$

If we now utilize the fact that $Z(\omega) = Z(-\omega)$ in conjunction with the normalization condition

$$\int_{0}^{\infty} \mathrm{d}\omega \, Z(\omega) = 1,$$

the above expression for $w(t)$ can be transformed into

$$-\frac{\mathrm{i}\hbar t}{M}+w(t) = \frac{\hbar}{M}\int_0^\infty \mathrm{d}\omega\,\frac{Z(\omega)}{\omega}\{(1-\cos\omega t)\coth(\tfrac{1}{2}\hbar\omega\beta)-\mathrm{i}\sin\omega t\}. \tag{11.114 b}$$

Two points should be made about this expression for $w(t)$ in terms of $Z(\omega)$. First, if we take the limit $\hbar \to 0$ we regain result (11.36) for the classical width function $\bar{w}(t)$. The second point is that (11.114 b) inserted into (11.102) and (11.94) is the exact analogue of the corresponding expression derived for incoherent scattering from a harmonic crystal (see (11.24)). The latter result we point out is exact, and does not rest on the approximation of assuming a Gaussian form for $G_s(\mathbf{r}, t)$.

For free nuclei $Z(\omega) = 2\delta(\omega)$. This result when substituted in (11.114 a) gives $w(t) = t^2/M\beta$, in agreement with the exact answer, eqn (11.91).

11.4. Coherent scattering: long-wavelength fluctuations

So far in this chapter we have been concerned with incoherent scattering and the self correlation function $G_s(\mathbf{r}, t)$. We now turn to a consideration of coherent scattering and the full correlation function $G(\mathbf{r}, t)$, which, as (11.2 a) demonstrates, is intimately connected with the density of fluctuations in the liquid. At the present time there is no really satisfactory theory of these effects:† the discussion is therefore more difficult but potentially more rewarding than the discussion of incoherent effects.

We shall give no discussion of the various theories that, starting from a given interatomic potential, attempt to calculate $G(\mathbf{r}, 0)$ from first principles. In effect we shall take

$$G(\mathbf{r}, 0) = \delta(\mathbf{r})+g(\mathbf{r}) = \frac{1}{\rho}\langle\hat{\rho}(0,0)\hat{\rho}(\mathbf{r},0)\rangle$$

as already given, by an X-ray experiment for example, in most of what we do. However, in the long wavelength limit we can calculate the density fluctuations from first principles. It is convenient to give this calculation in § 11.4 (*a*), but we can quote one of the results of that section immediately. In the nomenclature of (11.6 b),

$$\begin{aligned}\lim_{\varkappa\to 0}\{I(\varkappa, 0)-N\delta_{\varkappa,0}\} &= \frac{1}{N}\langle\delta\hat{\rho}_0(0)\delta\hat{\rho}_0(0)\rangle\\ &= 1+\int \mathrm{d}\mathbf{r}\,\{g(\mathbf{r})-\rho\}\\ &= \rho k_\mathrm{B}\,T\kappa_\mathrm{T},\end{aligned}$$

† Hubbard and Beeby, *J. Phys.* C **2**, 556 (1969); Singwi *et. al.*, *Phys. Rev.* **1**, 454 (1970); and Musare, loc. cit.

where κ_T is the isothermal compressibility. For long wavelengths this is normally all we need to know about $I(\boldsymbol{\kappa}, 0)$. However, near the critical point it is essential to take account of the variation of $I(\boldsymbol{\kappa}, 0)$ as κ departs from zero because, at a critical point, κ_T is large and $I(\boldsymbol{\kappa}, 0)$ varies very rapidly with $\boldsymbol{\kappa}$. The discussion of critical phenomena is given in § 11.4 (*c*).

Irrespective of whether or not the liquid is in the critical region we shall assume that the time dependence of $I(\boldsymbol{\kappa}, t)$, and hence the ω-dependence of the coherent scattering, is given by the theory of the time dependence of thermodynamic fluctuations as applied to hydrodynamics. To introduce this theory we first give in § 11.4 (*c*) a discussion of a simple Langevin equation before applying it to the full hydrodynamic equations.

In the next sub-section, § 11.4 (*a*), we shall give some initial consideration to the type of density fluctuations to be expected and we postpone to § 11.5 the discussion of coherent scattering at general wavelengths, because this general case must necessarily be qualitative at the present time.

(*a*) *Initial considerations*†

The nuclei in a solid at a finite temperature undergo oscillations about their equilibrium lattice positions. In the lowest order approximation (the harmonic approximation) these vibrations are described in terms of a set of independent travelling waves each with a wave vector and a polarization vector. In a quantum mechanical treatment of the theory of vibrations these independent travelling waves correspond to collective excitations of the nuclei, or phonons. Thus in the scattering of neutrons by the nuclei of the solid it is possible for the neutrons to lose energy through the creation of a phonon. Alternatively, the scattering process in which a phonon is destroyed and the neutrons gain energy can occur, though at low temperatures this process is less probable than that in which a phonon is created.

As the temperature of the solid is raised the number of phonons increases and with it the probability of two-phonon collision processes. This results in the phonons acquiring a finite lifetime. Experimental evidence exists to show that even when the temperature of a solid is raised to the point at which it becomes a liquid it remains meaningful to speak of the existence of phonons, i.e. liquids can support collective oscillations of their constituent nuclei. Of course, because of the inherent fluidity the phonons are not so well defined as in the solid state.

We can therefore conclude that when a density fluctuation is created

† For the corresponding discussion for the scattering of light by a liquid, see, for example, L. D. Landau and E. M. Lifshitz, *Electrodynamics of continuous media*, § 94.

in a liquid as a consequence of the scattering of the incident neutrons, the fluctuation can be relieved through a mechanical process that involves the creation (annihilation) of phonons. For low energies and long wavelengths the neutrons cannot sense the detailed structure of the liquid and the density fluctuation will be governed by the hydrodynamic equations for the description of a viscous fluid. Thus we expect the coherent cross-section for the scattering from a liquid to possess two peaks centred at energies $\hbar\omega = \pm\hbar\kappa c_0$, where c_0 is the velocity of sound in the liquid, corresponding to phonon scattering processes. These two peaks are usually referred to as the Brillouin doublet. However, this is not the full story.

The thermodynamic state of a system made up of identical particles is determined by the values of any two quantities such as pressure, volume, temperature, entropy, etc. We choose the *statistically independent* variables of pressure p and entropy S; the pressure can be regarded as a 'mechanical' quantity and the entropy a 'thermal' quantity. In terms of these variables the fluctuation in the particle density is

$$\delta\rho = \left(\frac{\partial\rho}{\partial p}\right)_S \delta p + \left(\frac{\partial\rho}{\partial S}\right)_p \delta S.$$

If we Fourier analyse the density fluctuation $\delta\rho$ each Fourier component consists of two parts, the first part being mechanical and the second thermal. The mechanical part corresponds to the collective oscillations of the nuclei considered above. The thermal part is non-oscillatory and results in a peak in the cross-section centred about $\hbar\omega = 0$. The importance of this peak in the theory of light scattering from a liquid was first pointed out by Landau and Placzek.†

Let us now evaluate the intensity of both the thermal and mechanical parts of the density fluctuations and relate these to the neutron scattering cross-section in the forward direction, i.e. in the limit as $\boldsymbol{\varkappa} \to 0$. We can anticipate that, as $\boldsymbol{\varkappa} \to 0$, the time dependence of the fluctuations becomes slow and hence the scattering may be treated as quasi-elastic. Hence from (11.1) and (11.6 b) we have

$$\begin{aligned}\frac{d\sigma}{d\Omega} &= \frac{N\sigma_c}{4\pi} \lim_{\boldsymbol{\varkappa}\to 0} \hbar \int_{-\infty}^{\infty} d\omega\, S(\boldsymbol{\varkappa}, \omega) \\ &= \frac{N\sigma_c}{4\pi} \lim_{\boldsymbol{\varkappa}\to 0} \{I(\boldsymbol{\varkappa}, 0) - N\delta_{\boldsymbol{\varkappa},0}\} \\ &= \frac{\sigma_c}{4\pi} \langle \delta\rho_0(0)\delta\rho_0(0)\rangle.\end{aligned}$$

† *Phys. Z. SowjUn.* **5**, 172 (1934).

Now
$$\delta\rho_0 = \lim_{\varkappa\to 0}\int \mathrm{d}\mathbf{r}\,\exp(\mathrm{i}\boldsymbol{\varkappa}.\mathbf{r})\delta\rho = V\delta\rho.$$

Hence
$$\frac{\mathrm{d}\sigma}{\mathrm{d}\Omega} = \frac{\sigma_\mathrm{c}}{4\pi}V^2\langle\delta\rho^2\rangle.$$

But
$$\langle\delta\rho^2\rangle = \left(\frac{\partial\rho}{\partial p}\right)_S^2\langle(\delta p)^2\rangle+\left(\frac{\partial\rho}{\partial S}\right)_p^2\langle(\delta S)^2\rangle+2\left(\frac{\partial\rho}{\partial p}\right)_S\left(\frac{\partial\rho}{\partial S}\right)_p\langle\delta p\delta S\rangle,$$
where $\langle(\delta p)^2\rangle$ is the mean square thermodynamic fluctuation in the pressure, $\langle(\delta S)^2\rangle$ is the mean square thermodynamic fluctuation in the entropy, and $\langle\delta p\delta S\rangle$ the correlation between them. These have the values†
$$\langle(\delta p)^2\rangle = -k_\mathrm{B}T\Big/\left(\frac{\partial V}{\partial p}\right)_S, \qquad \langle(\delta S)^2\rangle = k_\mathrm{B}C_p, \qquad \langle\delta p\delta S\rangle = 0,$$
where the specific heat at constant pressure is
$$C_p = T\left(\frac{\partial S}{\partial T}\right)_p$$
and
$$\left(\frac{\partial V}{\partial p}\right)_S = -V\kappa_s = -V\frac{C_v}{C_p}\kappa_\mathrm{T},$$
where κ_s is the adiabatic compressibility and κ_T is the isothermal compressibility.

Now
$$\left(\frac{\partial\rho}{\partial p}\right)_S = -\frac{N}{V^2}\left(\frac{\partial V}{\partial p}\right)_S = \rho\kappa_s$$
and
$$\left(\frac{\partial\rho}{\partial S}\right)_p = -\frac{N}{V^2}\left(\frac{\partial V}{\partial S}\right)_p = -\frac{N}{V^2}\left(\frac{\partial V}{\partial T}\right)_p\left(\frac{\partial T}{\partial S}\right)_p = \frac{\rho\alpha T}{C_p},$$
where the thermal expansion coefficient is
$$\alpha = -\frac{1}{V}\left(\frac{\partial V}{\partial T}\right)_p.$$

Putting all these results together,
$$\langle\delta\rho^2\rangle = \frac{\rho^2k_\mathrm{B}T}{V}\left(\kappa_s+\frac{TV\alpha^2}{C_p}\right).$$

We now use the thermodynamic identity
$$C_p-C_v = \frac{TV\alpha^2}{\kappa_T}$$
to obtain the result
$$\langle\delta\rho^2\rangle = \frac{\rho^2k_\mathrm{B}T}{V}\left\{\kappa_s+\kappa_T\frac{C_p-C_v}{C_p}\right\}.$$

† L. D. Landau and E. M. Lifshitz, *Statistical physics*.

From this we can immediately deduce that the ratio of intensities between thermal and mechanical fluctuations is

$$\frac{\kappa_T}{\kappa_s}\frac{C_p-C_v}{C_p} = \frac{C_p-C_v}{C_v}$$

and furthermore the total fluctuation is given by

$$\langle\delta\rho^2\rangle = \frac{\rho^2 k_B T}{V}\kappa_T.$$

For a solid $(C_p-C_v)/C_v$ is small and therefore it is usually reasonable to neglect the thermal fluctuations in the density. But for a liquid $(C_p-C_v)/C_v$ is larger and this can no longer be done. Furthermore, at the liquid–gas critical point, C_p diverges while C_v remains finite (cf. § 11.4 (*c*)), and hence in the critical region the thermal fluctuations dominate the mechanical fluctuations. Since κ_T diverges at the critical point, the neutron cross-section will be large in the critical region.

(*b*) *Hydrodynamic limit*

This sub-section is concerned with the time dependence of fluctuations. The full theory of the density fluctuations, and in particular their fluctuation in time, in the hydrodynamic limit gives a calculation that is inevitably a lengthy one and therefore in this sub-section we first illustrate the fundamental ideas of the theory with a simple problem that we have previously mentioned briefly, namely the Brownian motion of a particle in one dimension. The Langevin equation describing the particle motion is

$$\dot{v}(t) = -\gamma v(t)+\frac{1}{M}R(t), \tag{11.115}$$

where M is the mass of the particle, the first term on the right-hand side represents the frictional force exerted by the medium in which the particle is embedded and $R(t)$ is the random force due to random collisions with the surrounding particles. The random force is assumed to possess the property that its values at different times are completely uncorrelated, i.e.

$$\langle R(t)R(t')\rangle \propto \delta(t-t'). \tag{11.116}$$

Writing
$$\langle R(t)R(t')\rangle = \frac{1}{2\pi}\int\limits_{-\infty}^{\infty} d\omega\,(R^2)_\omega \exp\{-i\omega(t-t')\}, \tag{11.117}$$

we see that the constant of proportionality in (11.116) is $(R^2)_\omega = (R^2)$.

If we also write

$$R(t) = \frac{1}{2\pi}\int\limits_{-\infty}^{\infty} \mathrm{d}\omega\; R(\omega)\exp(-\mathrm{i}\omega t), \tag{11.118}$$

then

$$\langle R(t)R(t')\rangle = \left(\frac{1}{2\pi}\right)^2 \int\limits_{-\infty}^{\infty} \mathrm{d}\omega \int\limits_{-\infty}^{\infty} \mathrm{d}\omega'\; \langle R(\omega)R(\omega')\rangle \exp(-\mathrm{i}\omega t - \mathrm{i}\omega' t'),$$

and so

$$\langle R(\omega)R(\omega')\rangle = 2\pi(R^2)\delta(\omega+\omega'). \tag{11.119}$$

With

$$v(t) = \frac{1}{2\pi}\int \mathrm{d}\omega\; v(\omega)\exp(-\mathrm{i}\omega t)$$

where the ω integration is along a contour to be determined later, we obtain from the equation of motion (11.115)

$$v(\omega) = \frac{R(\omega)/M}{\gamma - \mathrm{i}\omega}. \tag{11.120}$$

Hence

$$M^2\langle v(0)v(t)\rangle = \left(\frac{1}{2\pi}\right)^2 \int \mathrm{d}\omega' \int \mathrm{d}\omega\; \mathrm{e}^{-\mathrm{i}\omega t} \frac{\langle R(\omega')R(\omega)\rangle}{(\gamma\;\; \mathrm{i}\omega')(\gamma-\mathrm{i}\omega)}$$

$$= \frac{(R^2)}{2\pi}\int \mathrm{d}\omega\; \frac{\mathrm{e}^{\mathrm{i}\omega t}}{\gamma^2+\omega^2}. \tag{11.121 a}$$

We have not yet defined the contour for the ω integration. To do this we note that we want $\langle v(0)v(t)\rangle$ to be even in t. It follows that the ω contour runs from $-\infty$ to ∞ between the two poles at $\pm\mathrm{i}\gamma$. Hence

$$M^2\langle v(0)v(t)\rangle = \frac{(R^2)}{2\pi}\frac{\pi}{\gamma}\mathrm{e}^{-\gamma|t|}, \tag{11.121 b}$$

and we must therefore have

$$(R^2) = 2\gamma M/\beta, \tag{11.122 a}$$

because $\langle v^2(0)\rangle = (\beta M)^{-1}$ if we assume the law of equipartition of energy. Whence we obtain from (11.121 b) the result

$$\langle v(0)v(t)\rangle = \langle v^2(0)\rangle \mathrm{e}^{-\gamma|t|}, \tag{11.122 b}$$

in agreement with the result given in § 11.2.

We notice that, though the correlation $\langle v(0)v(t)\rangle$ decays exponentially with time, the instantaneous values of $v(t)$ do not decay because, of course, $\langle v(t)v(t)\rangle$ is the same as $\langle v(0)v(0)\rangle$. We must think of the random force $(1/M)R(t)$ constantly making a contribution to $v(t)$, which then decays with the characteristic time $1/\gamma$. To see this point more directly we can solve (11.115) without Fourier transforming. Multiplying

(11.115) by $e^{\gamma t}$ we find it can be integrated directly to give

$$v(t) = e^{-\gamma t}\left\{v(0)+\int\limits_0^t dt'\, e^{\gamma t'}\frac{1}{M}\,R(t')\right\};$$

then, since $\langle v(0)R(t')\rangle$ is zero we immediately obtain an exponential decay for $\langle v(t)v(0)\rangle$, although

$$\langle v^2(t)\rangle = e^{-2\gamma t}\left\{\langle v^2(0)\rangle+\int\limits_0^t dt'\int\limits_0^t dt''\, e^{\gamma(t'+t'')}\frac{1}{M^2}\langle R(t')R(t'')\rangle\right\}$$

$$= e^{-2\gamma t}\left\{\langle v^2(0)\rangle+\frac{(R^2)}{M^2}\frac{1}{2\gamma}(e^{2\gamma t}-1)\right\}$$

$$= \langle v^2(0)\rangle.$$

In the more general case where $v(t)$ stands for an arbitrary quantity we do not know that $\langle v^2(0)\rangle$ is $1/\beta M$. Landau and Lifshitz show that then we need to know how the rate of production of entropy is related to $v(t)$. If we are given this then the theory as sketched out above follows as before and we determine $\langle v^2(0)\rangle$ by the following argument.

In the more general case of several fluctuating quantities x_1, x_2,... Landau and Lifshitz† show that the set of equations corresponding to (11.115) are to be written

$$\dot{x}_a = -\sum_b \gamma_{ab} X_b + R_a \tag{11.123}$$

and, in place of (11.116),

$$\langle R_a(t)R_b(t')\rangle = (\gamma_{ab}+\gamma_{ba})\delta(t-t'). \tag{11.124}$$

We now turn to the full hydrodynamic problem. To calculate the cross-section we need to evaluate the function

$$\langle\hat{\rho}_{\varkappa}(0)\hat{\rho}_{-\varkappa}(t)\rangle. \tag{11.125}$$

We make two assumptions in performing the calculation. First, we calculate the classical quantity corresponding to the function (11.125) and then use the Schofield prescription to evaluate the cross-section. Secondly, we write

$$\langle\rho_{\varkappa}(0)\rho_{-\varkappa}(t)\rangle = \langle\rho_{\varkappa}(0)\rho_{-\varkappa}(0)\rangle f_{\varkappa}(t),$$

where
$$f_{\varkappa}(t) = \frac{\langle\rho_{\varkappa}(0)\rho_{-\varkappa}(t)\rangle}{\langle\rho_{\varkappa}(0)\rho_{-\varkappa}(0)\rangle}, \tag{11.126}$$

and make the assumption that $\langle\rho_{\varkappa}(0)\rho'_{-\varkappa}(0)\rangle$ is known, either from an experimental determination of $\gamma(\boldsymbol{\varkappa})$ and the use of (11.6b), or from a theory of fluctuations as described in § 11.4 (*c*). Furthermore, we assume

† *Statistical physics*, §§ 121, 124.

that $f_\varkappa(t)$ can be calculated as the time dependence of any fluctuation of wave vector $\boldsymbol{\varkappa}$ and is independent of the initial value $\rho_\varkappa(0)$. To do this we use the linearized equations of hydrodynamics. We start by writing down the three equations that express conservation of mass, momentum, and energy of a viscous fluid.

The velocity of a mass point (i.e. an infinitesimal volume) at position $\mathbf{r}$ at time t is $\mathbf{v} = \mathbf{v}(\mathbf{r},t)$. The *conservation of mass* is expressed by the equation of continuity

$$\frac{\mathrm{D}\rho}{\mathrm{D}t}+\rho\boldsymbol{\nabla}.\mathbf{v} = 0, \tag{11.127}$$

where

$$\frac{\mathrm{D}(\;)}{\mathrm{D}t} = \frac{\partial(\;)}{\partial t}+\mathbf{v}.\boldsymbol{\nabla}(\;). \tag{11.128}$$

$$\left(\boldsymbol{\nabla} \equiv \sum_\alpha \tilde{\boldsymbol{\alpha}}\frac{\partial}{\partial r_\alpha} = \sum_\alpha \tilde{\boldsymbol{\alpha}}\nabla_\alpha \quad \text{and} \quad \nabla^2 \equiv \sum_\alpha \frac{\partial^2}{\partial r_\alpha^2} = \sum_\alpha \nabla_\alpha^2.\right)$$

The *conservation of momentum* is expressed by the force equation

$$M\rho\frac{\mathrm{D}v_\alpha}{\mathrm{D}t} = -\nabla_\alpha p+\sum_\beta \nabla_\beta \sigma'_{\alpha\beta}, \tag{11.129}$$

where p is the thermodynamic pressure and $\sigma'_{\alpha\beta}$ the *stress tensor* defined by

$$\sigma'_{\alpha\beta} = \eta(\nabla_\beta v_\alpha+\nabla_\alpha v_\beta-\tfrac{2}{3}\delta_{\alpha\beta}\boldsymbol{\nabla}.\mathbf{v})+\eta_\mathrm{B}\,\delta_{\alpha\beta}\boldsymbol{\nabla}.\mathbf{v}. \tag{11.130}$$

In (11.130), η is the 'first' shear viscosity coefficient and η_B the bulk viscosity coefficient which we also write in terms of η' (the 'second' viscosity coefficient) defining the latter through $\eta_\mathrm{B} = \eta'+\frac{2}{3}\eta$. Note that the force equation can be written

$$M\rho\frac{\mathrm{D}\mathbf{v}}{\mathrm{D}t} = -\boldsymbol{\nabla}p+\eta\nabla^2\mathbf{v}+(\eta+\eta')\boldsymbol{\nabla}(\boldsymbol{\nabla}.\mathbf{v}). \tag{11.131}$$

Finally we have the equation that expresses the *conservation of energy*, namely

$$M\rho T\frac{\mathrm{D}s}{\mathrm{D}t} = \tfrac{1}{2}\sum_{\alpha,\beta}\sigma'_{\alpha\beta}(\nabla_\beta v_\alpha+\nabla_\alpha v_\beta)-\boldsymbol{\nabla}.\boldsymbol{\theta}. \tag{11.132}$$

Here s is the entropy per unit mass and $\boldsymbol{\theta}$ the heat flux vector, which is given by

$$\boldsymbol{\theta} = -\lambda\boldsymbol{\nabla}T, \tag{11.133}$$

where λ is the *thermal conductivity* and T the temperature.

If $\mathscr{E}$ is the internal energy per unit mass and $v = V/M$, the pressure and temperature are defined here through the relations

$$p = -\left(\frac{\partial\mathscr{E}}{\partial v}\right)_s, \qquad T = \left(\frac{\partial\mathscr{E}}{\partial s}\right)_v. \tag{11.134}$$

The remaining thermodynamic function that will concern us are:

$$\text{specific heat at constant volume, } C_v = T\left(\frac{\partial s}{\partial T}\right)_v,$$

$$\text{specific heat at constant pressure, } C_p = T\left(\frac{\partial s}{\partial T}\right)_p,$$

$$\text{coefficient of thermal expansion, } \alpha = -\frac{1}{\rho}\left(\frac{\partial \rho}{\partial T}\right)_p.$$

$$\text{isothermal compressibility, } \kappa_T = \frac{1}{\rho}\left(\frac{\partial \rho}{\partial p}\right)_T. \tag{11.135}$$

We shall also make use of the identity

$$T\frac{\mathrm{D}s}{\mathrm{D}t} = \frac{(C_p - C_v)}{\alpha}\boldsymbol{\nabla}.\mathbf{v} + C_v\frac{\mathrm{D}T}{\mathrm{D}t}. \tag{11.136}$$

Eqns (11.130) and (11.133) for the stress tensor $\sigma'_{\alpha\beta}$ and heat flux $\boldsymbol{\theta}$ express these quantities in terms of the velocity and temperature gradients, respectively. When we have fluctuations in the liquid there are in addition spontaneous local stresses and heat fluxes in the liquid, which are unrelated to the values of $\mathbf{v}$ or $\boldsymbol{\nabla}T$ averaged over these fluctuations: these are the random quantities corresponding to $R_a(t)$ in eqn (11.123). We must therefore replace $\sigma'_{\alpha\beta}$ and $\boldsymbol{\theta}$ in (11.129) and (11.132) by

$$\sigma'_{\alpha\beta} + s_{\alpha\beta} \quad \text{and} \quad \boldsymbol{\theta} + \mathbf{g},$$

where $s_{\alpha\beta}$ and $\mathbf{g}$ denote the random quantities. The quantities $\dot{x}_a$ (11.123) are the components of the tensor $\sigma'_{\alpha\beta}$ and the vector $\boldsymbol{\theta}$. We do not present here the calculation of the correlation functions

$$\langle s_{\alpha\beta}(\mathbf{r},t)g_\gamma(\mathbf{r}',t')\rangle, \quad \langle s_{\alpha\beta}(\mathbf{r},t)s_{\alpha'\beta'}(\mathbf{r}',t')\rangle, \quad \text{and} \quad \langle g_\alpha(\mathbf{r},t)g_\beta(\mathbf{r}',t')\rangle$$

for it is given in detail by Landau and Lifshitz, *Fluid mechanics*, § 132. They find

$$\langle s_{\alpha\beta}(\mathbf{r},t)g_\gamma(\mathbf{r}',t')\rangle = 0,$$

$$\langle s_{\alpha\beta}(\mathbf{r},t)s_{\alpha'\beta'}(\mathbf{r}',t')\rangle = \frac{2}{\beta}\{\eta(\delta_{\alpha\alpha'}\delta_{\beta\beta'} + \delta_{\alpha\beta'}\delta_{\alpha'\beta}) + \eta'\delta_{\alpha\beta}\delta_{\alpha'\beta'}\}\delta(\mathbf{r}-\mathbf{r}')\delta(t-t') \tag{11.137}$$

and

$$\langle g_\alpha(\mathbf{r},t)g_\beta(\mathbf{r}',t')\rangle = \frac{2\lambda}{\beta}\delta_{\alpha\beta}\delta(\mathbf{r}-\mathbf{r}')\delta(t-t').$$

From these results we observe that $s_{\alpha\beta}$ and g_γ are entirely uncorrelated, and, as we should expect, $\langle s_{\alpha\beta}(\mathbf{r},t)s_{\alpha'\beta'}(\mathbf{r}',t')\rangle$ represents the mechanical random fluctuation in the liquid and $\langle g_\alpha(\mathbf{r},t)g_\beta(\mathbf{r}',t')\rangle$ the purely thermal random fluctuation.

The density of the liquid before the application of the external perturbation (i.e. the incident neutrons) is a constant ρ_0. The fluctuation from this value that occurs as a consequence of the external perturbation we denote by $\delta\rho(\mathbf{r},t)$. Similarly we write for the temperature, pressure, and velocity, $T = T_0+\delta T$, $p = p_0+\delta p$, and $\mathbf{v} = 0+\delta\mathbf{v}$. Because $\delta\rho$, δT, δp, and $\delta\mathbf{v}$ are small we need retain in calculating $\delta\rho$ from the hydrodynamic equations only those terms that are linear in these small quantities. Thus, for example, we write

$$\rho\mathbf{v} \simeq \rho_0\,\delta\mathbf{v}.$$

Furthermore, we can express δp in terms of the temperature and density. For

$$\delta p = \left[\left(\frac{\partial p}{\partial\rho}\right)_T\right]_0(\rho-\rho_0)+\left[\left(\frac{\partial p}{\partial T}\right)_\rho\right]_0(T-T_0)+\dots$$

and so
$$\delta p = \frac{c_0^2 C_v}{C_p}(\delta\rho+\alpha\rho_0\,\delta T), \tag{11.138}$$

where $c_0^2 = \left[\left(\frac{\partial p}{\partial\rho}\right)_s\right]_0 = \frac{1}{\rho_0}\frac{C_p}{C_v}\kappa_T^{-1}$ is the square of the velocity of sound in the liquid.

By making use of this result and the identity (11.136) we find the linearized hydrodynamic equations are ($\delta\dot\rho \equiv (\partial/\partial t)\delta\rho$, etc.)

$$\delta\dot\rho+\rho_0\,\boldsymbol{\nabla}.\delta\mathbf{v} = 0,$$

$$\rho_0\,\delta\dot{\mathbf{v}}+\frac{c_0^2C_v}{C_p}(\boldsymbol{\nabla}\,\delta\rho+\alpha\rho_0\,\boldsymbol{\nabla}\,\delta T)-(\eta+\eta')\boldsymbol{\nabla}(\boldsymbol{\nabla}.\delta\mathbf{v})-\eta\nabla^2\delta\mathbf{v}-\sum_{\alpha,\beta}\tilde{\boldsymbol{\alpha}}\nabla_\beta s_{\alpha\beta} = 0. \tag{11.139}$$

To proceed further we write

$$\delta\rho(\mathbf{r},t) = \left(\frac{1}{2\pi}\right)^4\int d\mathbf{q}\int d\omega\,\exp(i\mathbf{q}.\mathbf{r}-i\omega t)\delta\rho_{\mathbf{q}}(\omega) \tag{11.140 a}$$

and similarly for all other quantities, and thereby obtain from eqn (11.139) the relations

$$-\omega\delta\rho_{\mathbf{q}}(\omega)+\rho_0\,\mathbf{q}.\delta\mathbf{v}_{\mathbf{q}}(\omega) = 0,$$

$$-\omega\rho_0\,\delta\mathbf{v}_{\mathbf{q}}(\omega)+\mathbf{q}\frac{c_0^2C_v}{C_p}\{\delta\rho_{\mathbf{q}}(\omega)+\alpha\rho_0\,\delta T_{\mathbf{q}}(\omega)\}-$$

$$-i\eta q^2\delta\mathbf{v}_{\mathbf{q}}(\omega)-i\mathbf{q}(\eta+\eta')\mathbf{q}.\delta\mathbf{v}_{\mathbf{q}}(\omega)-\sum_{\alpha,\beta}\tilde{\boldsymbol{\alpha}}q_\beta\,s_{\alpha\beta,\mathbf{q}}(\omega) = 0,$$

and

$$-C_v\,\omega\rho_0\,\delta T_{\mathbf{q}}(\omega)+\alpha^{-1}(C_p-C_v)\omega\delta\rho_{\mathbf{q}}(\omega)-i\lambda q^2\delta T_{\mathbf{q}}(\omega)+\mathbf{q}.\mathbf{g}_{\mathbf{q}}(\omega) = 0. \tag{11.140 b}$$

It is evident from the first equation of (11.140b) that the velocity fluctuation $\delta \mathbf{v}_q(\omega)$ is coupled to the density fluctuation only if $\delta \mathbf{v}_{\mathbf{q}}(\omega)$ is parallel to $\mathbf{q}$. We therefore assume this is the case from now on.

Taking the scalar product of the second equation in (11.140b) with $\mathbf{q}$ and utilizing the first equation to eliminate $\mathbf{q}.\delta\mathbf{v}_{\mathbf{q}}(\omega)$ we can readily solve for $\delta\rho_{\mathbf{q}}(\omega)$ and find

$$\delta\rho_{\mathbf{q}}(\omega) = \frac{(C_v\omega\rho_0+\mathrm{i}\lambda q^2)\Big\{\sum_{\alpha,\beta} q_\alpha q_\beta s_{\alpha\beta,\mathbf{q}}(\omega)\Big\}-(C_v/C_p)c_0^2\alpha\rho_0 q^2\mathbf{q}\,.\,\mathbf{g}_{\mathbf{q}}(\omega)}{\rho_0 C_v[c_0^2q^2\{\omega+(\mathrm{i}\lambda q^2/C_p\rho_0)\}-\omega\{\omega+(\mathrm{i}\lambda q^2/C_v\rho_0)\}\{\omega+\mathrm{i}(2\eta+\eta')q^2/\rho_0\}]}. \tag{11.141}$$

Now from the results (11.137) it follows that

$$\langle s_{\alpha\beta,\mathbf{q}}(\omega)s_{\alpha'\beta',\mathbf{q}'}(\omega')\rangle = (2\pi V)\frac{2}{\beta}\{\eta(\delta_{\alpha\alpha'}\delta_{\beta\beta'}+\delta_{\alpha\beta'}\delta_{\alpha'\beta})+\eta'\delta_{\alpha\beta}\delta_{\alpha'\beta'}\}\delta_{\mathbf{q},-\mathbf{q}'}\delta(\omega+\omega') \tag{11.142}$$

and
$$\langle g_{\alpha,\mathbf{q}}(\omega)g_{\beta,\mathbf{q}'}(\omega')\rangle = (2\pi V)\frac{2}{\beta}\lambda\delta_{\alpha\beta}\delta_{\mathbf{q},-\mathbf{q}'}\delta(\omega+\omega').$$

Hence
$$\langle\delta\rho_{\mathbf{q}}(\omega)\delta\rho_{\mathbf{q}'}(\omega')\rangle = (2\pi V)2k_{\mathrm{B}}\,\delta_{\mathbf{q},-\mathbf{q}'}\delta(\omega+\omega')\Psi(\mathbf{q},\omega), \tag{11.143}$$

where

$$\Psi(\mathbf{q},\omega) = \frac{\lambda q^6T^2+(2\eta+\eta')q^4T(C_p/c_0^2\alpha)^2\left|\omega+\dfrac{\mathrm{i}\lambda q^2}{C_v\rho_0}\right|^2}{\left|\alpha^{-1}C_pq^2\left(\omega+\dfrac{\mathrm{i}\lambda q^2}{C_p\rho_0}\right)-\dfrac{C_p\omega}{c_0^2\alpha}\left(\omega+\dfrac{\mathrm{i}\lambda q^2}{C_v\rho_0}\right)\left(\omega+\dfrac{\mathrm{i}(2\eta+\eta')q^2}{\rho_0}\right)\right|^2}. \tag{11.144}$$

Some tedious algebra verifies

$$\int_{-\infty}^{\infty}\mathrm{d}\omega\;\Psi(\mathbf{q},\omega) = \pi\rho_0^2T\kappa_T. \tag{11.145}$$

We now have all the results we need to calculate the cross-section. From (11.2a)

$$\tilde{S}(\boldsymbol{\varkappa},\omega) = \frac{1}{2\pi\hbar N}\int_{-\infty}^{\infty}\mathrm{d}t\;\mathrm{e}^{-\mathrm{i}\omega t}\langle\rho_{\boldsymbol{\varkappa}}(0)\rho_{-\boldsymbol{\varkappa}}(t)\rangle,$$

where it is now understood that the correlation function on the right-hand side is to be evaluated classically. Now, from (11.140a),

$$\langle\rho_{\boldsymbol{\varkappa}}(0)\rho_{-\boldsymbol{\varkappa}}(t)\rangle = \frac{1}{(2\pi)^2}\int_{-\infty}^{\infty}\mathrm{d}\omega\int_{-\infty}^{\infty}\mathrm{d}\omega'\;\mathrm{e}^{-\mathrm{i}\omega't}\langle\delta\rho_{\boldsymbol{\varkappa}}(\omega)\delta\rho_{-\boldsymbol{\varkappa}}(\omega')\rangle,$$

and, using (11.143), this is

$$\langle \rho_{\varkappa}(0)\rho_{-\varkappa}(t)\rangle = \frac{V}{2\pi}2k_{\mathrm{B}}\int_{-\infty}^{\infty}\mathrm{d}\omega\, \mathrm{e}^{\mathrm{i}\omega t}\Psi(\varkappa,\omega).$$

Hence
$$\tilde{S}(\varkappa,\omega) = \frac{Vk_{\mathrm{B}}}{\pi\hbar N}\Psi(\varkappa,\omega) = \frac{k_{\mathrm{B}}}{\pi\hbar\rho_0}\Psi(\varkappa,\omega),$$

and
$$\int_{-\infty}^{\infty}\mathrm{d}\omega\, \tilde{S}(\varkappa,\omega) = \frac{\rho_0\kappa_T}{\hbar\beta}.$$

In terms of the coherent cross-section itself

$$\left(\frac{\mathrm{d}^2\sigma}{\mathrm{d}\Omega\mathrm{d}E'}\right)_{\mathrm{coh}} = N\frac{k'}{k}\frac{\sigma_{\mathrm{c}}}{4\pi}\mathrm{e}^{\frac{1}{2}\hbar\omega\beta}\frac{k_{\mathrm{B}}}{\pi\hbar\rho_0}\Psi(\varkappa,\omega).$$

Hence
$$\int_{-\infty}^{\infty}\mathrm{d}E'\left(\frac{\mathrm{d}^2\sigma}{\mathrm{d}\Omega\mathrm{d}E'}\right)_{\mathrm{coh}}\frac{k}{k'}\mathrm{e}^{-\frac{1}{2}\hbar\omega\beta} = N\frac{\sigma_{\mathrm{c}}}{4\pi}\frac{\rho_0\kappa_T}{\beta}.$$

Alternatively we note that in the quasi-elastic approximation

$$\left(\frac{\mathrm{d}\sigma}{\mathrm{d}\Omega}\right)_{\mathrm{coh}} = N\frac{\sigma_{\mathrm{c}}}{4\pi}\frac{\rho_0\kappa_T}{\beta}$$

These results show again that in the forward direction, where $\varkappa$ is small enough for the hydrodynamic limit to hold, the cross-section is proportional to the isothermal compressibility. This result was derived in a simpler manner in § 11.4 (*a*).

This result is independent of $\varkappa$ and, to some extent, is therefore unrealistic (except in the limit $\varkappa \to 0$). We know that in practice the cross-section integrated over ω has a significant $\varkappa$-dependence (the liquid structure factor, see (11.3g)) and also we know that the cross-section has a strong $\varkappa$-dependence near a critical point. However, we recall the discussion centred about (11.125) and (11.126). Our purpose in deriving $\Psi(\varkappa,\omega)$ is to describe the ω-dependence of fluctuations of given wave vector $\varkappa$. We therefore assume that $\Psi(\varkappa,\omega)$, properly normalized and together with the Schofield prescription, gives the ω-dependence of the cross-section and we look for other techniques of examining the wave vector dependence. Hence we write

$$\frac{\mathrm{d}^2\sigma}{\mathrm{d}\Omega\mathrm{d}E'} = \frac{\sigma_{\mathrm{c}}}{4\pi}N\frac{k'}{k}\langle\rho_{\varkappa}(0)\rho_{-\varkappa}(0)\rangle\mathrm{e}^{\frac{1}{2}\hbar\omega\beta}\frac{\Psi(\varkappa,\omega)}{\pi\rho_0^2T\kappa_T}. \qquad (11.146)$$

Let us now examine the nature of our result (11.146). $\Psi(\varkappa,\omega)$ is, apart from constants, the (linear) density response of a viscous fluid to an external perturbation with time dependence $\exp(\mathrm{i}\omega t)$ and a spatial dependence $\exp(\mathrm{i}\varkappa.\mathbf{r})$. Thus if $\hbar\omega$ coincides with the energy of a collective excitation in the liquid, $\Psi(\boldsymbol{\kappa},\omega)$ must peak at this particular energy. From eqn (11.144) we see that the poles of $\Psi(\boldsymbol{\kappa},\omega)$ occur at frequencies ω that are solutions of the cubic equation:

$$\omega^3-\omega^2\left(2\eta+\eta'+\frac{\lambda}{C_v}\right)(\mathrm{i}q^2/\rho_0)-\omega\left\{\frac{\lambda q^4(2\eta+\eta')}{C_v\rho_0^2}+q^2c_0^2\right\}+\mathrm{i}q^4\lambda c_0^2/C_p\,\rho_0=0. \tag{11.147}$$

Though, of course, we could solve exactly for the three roots of this equation, it involves heavy algebra for general $\mathbf{q}$ and ω and the theory is anyway only applicable for small ω and $\mathbf{q}$. We therefore examine only the long wavelength limit. To order q^2 we find the roots of (11.147) are

$$\omega_1=\frac{iq^2\lambda}{C_p\,\rho_0} \tag{11.148 a}$$

and

$$\omega_{2,3}=\pm qc_0+\frac{iq^2}{2\rho_0}\left\{2\eta+\eta'+\frac{\lambda(C_p-C_v)}{C_p\,C_v}\right\}. \tag{11.148 b}$$

The solutions (11.148 b) represent sound waves which are slowly damped, and this damping does not change radically near the liquid–gas critical point. These are the longitudinal phonon modes. The solution (11.148 a) represents a steadily decaying mode. The rate of decay is slow because λ is small; the rate is proportional to q^2 and inversely proportional to C_p, which diverges at the critical point (cf. § 11.4 (*c*)).

If $\lambda=0$ in (11.147) then the mode centred about $\omega=0$ is zero, i.e. the central peak in $\Psi(\boldsymbol{\kappa},\omega)$ is due to the thermal decay of the density fluctuation. Also note that

$$\int_{-\infty}^{\infty}\mathrm{d}\omega\lim_{\lambda\to 0}\Psi(\boldsymbol{\kappa},\omega)=\pi\rho_0^2\,T\frac{C_v}{C_p}\kappa_T, \tag{11.149}$$

which does not diverge at the critical point. Thus the divergence must arise from the central peak and the ratio of this peak to the phonon peaks is of order (cf. § 11.4 (*a*)):

$$\frac{C_p-C_v}{C_v}. \tag{11.150}$$

(*c*) *Critical scattering from liquids*†

Considering the thermodynamic state of a system in terms of the two variables pressure p and temperature T, the phase equilibrium curve in the p, T plane may terminate at some point (cf. Fig. 11.3). Such a point is called critical and the values of the variables p and T that

† For reviews of theoretical and experimental investigations of static critical phenomena see M. E. Fisher, *Rep. Prog. Phys.* **30**, 615 (1967) and P. Heller, ibid. 731.

specify it are called the critical pressure p_c and critical temperature T_c. At the critical point the two phases liquid and gas are indistinguishable from one another.

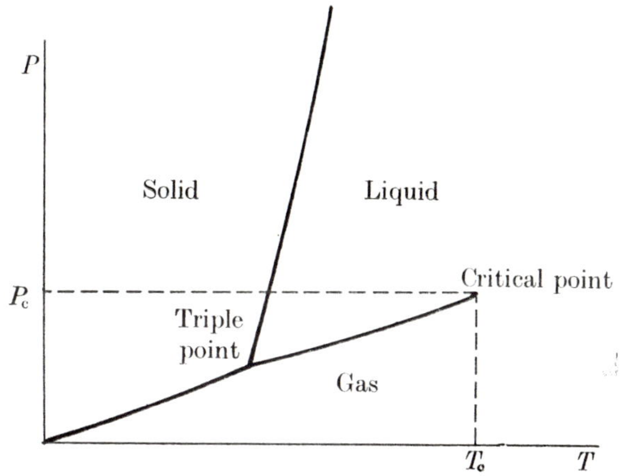

FIG. 11.3. Phase equilibrium curves in the P–T plane.

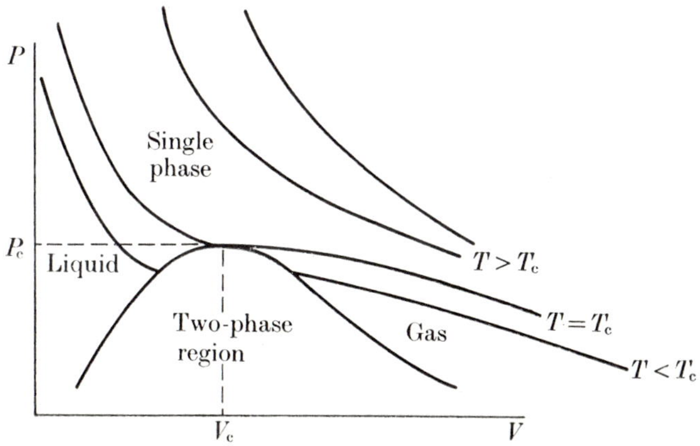

FIG. 11.4. Phase equilibrium curves in the P–V plane.

The equilibrium states as a function of the variables p and V are shown schematically in Fig. 11.4. At the critical point it is evident that

$$\left(\frac{\partial p}{\partial V}\right)_T = 0. \tag{11.151}$$

In view of the thermodynamic identity

$$C_p - C_v = -T\left(\frac{\partial p}{\partial T}\right)_V^2 \bigg/ \left(\frac{\partial p}{\partial V}\right)_T, \tag{11.152}$$

(11.151) implies that the specific heat at constant pressure C_p diverges at the critical point, the specific heat at constant volume C_v usually being assumed finite. (Because of the fundamental thermodynamic inequality $(\partial p/\partial V)_T < 0$ we *always* have, in view of (11.152), the inequality $C_p > C_v$.) Recent measurements have indicated that, for $V = V_c$, C_v tends to infinity as T_c is approached from either side, the singularity being much weaker than that in C_p.

We also have from (11.151) the result

$$-\kappa_T = \frac{1}{V}\left(\frac{\partial V}{\partial p}\right)_T \to \infty \text{ at the critical point.} \tag{11.153}$$

The identity

$$\frac{\kappa_s}{\kappa_T} = \frac{C_v}{C_p} < 1 \tag{11.154}$$

tells us that the adiabatic compressibility κ_s behaves like C_v.

To gain some understanding of the behaviour of the thermodynamic quantities in the vicinity of T_c we develop a simple theory for them, which, though incorrect in detail, possesses the gross features of their observed behaviour.

Near the critical point we can expand the pressure p in a power series in $V-V_c$ to get

$$p-p_c = A(T)-B(T)(V-V_c)+C(T)(V-V_c)^2-D(T)(V-V_c)^3+\ldots, \tag{11.155}$$

where A, B, C, and D are functions of the temperature T alone. By definition $A(T)$ is zero at T_c; by (11.153) it follows that $B(T)$ must also be zero. We make the assumption that both $A(T)$ and $B(T)$ behave like $T-T_c$ in the vicinity of the critical point. At T_c the first two terms in the expansion (11.155) are therefore zero and to reproduce the curve of Fig. 11.3 we must have $C(T)$ vanish as well. It follows that the third term of (11.155) is always small and can be neglected. Hence we have

$$p-p_c = A(T-T_c)-B(T-T_c)(V-V_c)-D(V-V_c)^3 \tag{11.156}$$

on taking $D(T)$ to be a constant independent of T.

From (11.156) it follows that

$$\left(\frac{\partial p}{\partial T}\right)_V = A-B(V-V_c),$$

$$-\left(\frac{\partial p}{\partial V}\right)_T = B(T-T_c)+3D(V-V_c)^2,$$

and

$$\left(\frac{\partial V}{\partial T}\right)_p = -\left(\frac{\partial p}{\partial T}\right)_V \bigg/ \left(\frac{\partial p}{\partial V}\right)_T.$$

Thus
$$\kappa_T = -\frac{1}{V}\left(\frac{\partial V}{\partial p}\right)_T = \frac{1}{V\{B(T-T_c)+3D(V-V_c)^2\}},$$
$$\alpha = \frac{1}{V}\left(\frac{\partial V}{\partial T}\right)_p = \frac{A-B(V-V_c)}{V\{B(T-T_c)+3D(V-V_c)^2\}}, \tag{11.157}$$
and, finally,
$$C_p - C_v = -T\left(\frac{\partial V}{\partial T}\right)_p^2 \Big/ \left(\frac{\partial V}{\partial p}\right)_T = \frac{T\{A-B(V-V_c)\}^2}{B(T-T_c)+3D(V-V_c)^2}.$$

We note that for this simple theory κ_T, α, and C_p all diverge at the critical point at the same rate.

Very accurate measurements of the compressibility κ_T for the two fluids CO_2 and Xe indicate that κ_T does not diverge linearly with $T-T_c$ ($V = V_c$, $T > T_c$), as the simple theory (11.157) predicts, but rather as $(T-T_c)^\gamma$, where the exponent $\gamma \simeq 1{\cdot}2$–$1{\cdot}3$. Measurements of the velocity of sound $c = (\rho\kappa_s)^{-\frac{1}{2}}$ yield values of κ_s. As the temperature is raised toward the critical point the components of the Brillouin doublet broaden, the separation of the lines decreases, and they finally merge into the Rayleigh line at the critical point.

Let us now examine the behaviour of the coherent neutron cross-section for the scattering by a liquid in the vicinity of its critical point. Since
$$\delta\rho_{\mathbf{q}}(t) = \int \mathrm{d}\mathbf{r}\, \exp(-\mathrm{i}\mathbf{q}.\mathbf{r})\{\rho(\mathbf{r},t)-\rho\} = \rho_{\mathbf{q}}(t) - N\delta_{\mathbf{q},0}$$
and $\langle\delta\rho_{\mathbf{q}}(t)\rangle = 0$ because $\langle\rho(\mathbf{r},t)\rangle$ is a constant, ρ, it follows that
$$S(\boldsymbol{\varkappa},\omega) = \frac{1}{2\pi\hbar N}\int_{-\infty}^{\infty} \mathrm{d}t\, \mathrm{e}^{-\mathrm{i}\omega t}\langle\{\delta\rho_{\boldsymbol{\varkappa}}(0)+N\delta_{\boldsymbol{\varkappa},0}\}\{\delta\rho_{-\boldsymbol{\varkappa}}(t)+N\delta_{\boldsymbol{\varkappa},0}\}\rangle$$
$$= N\delta(\hbar\omega)\delta_{\boldsymbol{\varkappa},0} + \frac{1}{2\pi\hbar N}\int_{-\infty}^{\infty} \mathrm{d}t\, \mathrm{e}^{-\mathrm{i}\omega t}\langle\delta\rho_{\boldsymbol{\varkappa}}(0)\delta\rho_{-\boldsymbol{\varkappa}}(t)\rangle.$$

The term containing the Kronecker delta $\delta_{\boldsymbol{\varkappa},0}$ corresponds to no scattering, so the cross-section of interest to us here is (cf. 11.6 a)
$$\frac{\mathrm{d}^2\sigma}{\mathrm{d}\Omega\mathrm{d}E'} = \frac{\sigma_c}{4\pi}\frac{k'}{k}\frac{1}{2\pi\hbar}\int_{-\infty}^{\infty} \mathrm{d}t\, \mathrm{e}^{-\mathrm{i}\omega t}\langle\delta\rho_{\boldsymbol{\varkappa}}(0)\delta\rho_{-\boldsymbol{\varkappa}}(t)\rangle. \tag{11.158}$$

If we are not interested in the energy spread of the scattered beam we can integrate over E' in (11.158) to give the differential cross-section as
$$\frac{\mathrm{d}\sigma}{\mathrm{d}\Omega} = \frac{\sigma_c}{4\pi}\int_{-\infty}^{E/\hbar} \mathrm{d}\omega\, \frac{k'}{k}\frac{1}{2\pi}\int_{-\infty}^{\infty} \mathrm{d}t\, \mathrm{e}^{-\mathrm{i}\omega t}\langle\delta\rho_{\boldsymbol{\varkappa}}(0)\delta\rho_{-\boldsymbol{\varkappa}}(t)\rangle. \tag{11.159}$$

When the incoming energy is large the integral over ω can be extended to $+\infty$ and, furthermore, the dependence of $\boldsymbol{\kappa}$ on ω can be neglected to give

$$\frac{\mathrm{d}\sigma}{\mathrm{d}\Omega} = \frac{\sigma_{\mathrm{c}}}{4\pi}\langle\delta\rho_{\boldsymbol{\kappa}}(0)\delta\rho_{-\boldsymbol{\kappa}}(0)\rangle \tag{11.160}$$

and the scattering is treated as though it were elastic. Note, however, that there is no purely elastic scattering from a liquid because purely elastic scattering is that which is proportional to $\delta(\omega)$ and in (11.159) the coefficient of $\delta(\omega)$ is proportional to

$$\langle\delta\rho_{\boldsymbol{\kappa}}(0)\delta\rho_{-\boldsymbol{\kappa}}(\infty)\rangle = \langle\delta\rho_{\boldsymbol{\kappa}}(0)\rangle\langle\delta\rho_{-\boldsymbol{\kappa}}(\infty)\rangle = 0. \tag{11.161}$$

The scattering predicted by (11.160) is therefore quasi-elastic. An equivalent formula is an excellent approximation for the scattering of light or X-rays but for neutrons, because of their low incident energy, (11.160) is not generally such a good approximation. We shall return to this point later in this section.

Assuming the validity of (11.160) we can recast it into a very useful form. First (see eqn (11.4 c)) we note that for a classical liquid

$$G(\mathbf{r},t) = \frac{1}{\rho}\langle\rho(0,0)\rho(\mathbf{r},t)\rangle. \tag{11.162}$$

(Strictly speaking we should attach a suffix 0 to the equilibrium values of the thermodynamic quantities as in § 11.4 (*b*). For convenience, however, we omit them except where it is likely to cause confusion.)

Writing
$$\rho(\mathbf{r},t) = \frac{1}{V}\sum_{\mathbf{q}}\rho_{\mathbf{q}}(t)\exp(\mathrm{i}\mathbf{q}.\mathbf{r}), \tag{11.163}$$

we obtain from (11.162) the relation

$$G(\mathbf{r},t) = \frac{1}{\rho V^2}\sum_{\mathbf{q}}\langle\rho_{\mathbf{q}}(0)\rho_{-\mathbf{q}}(t)\rangle\exp(-\mathrm{i}\mathbf{q}.\mathbf{r}),$$

or
$$\langle\rho_{\boldsymbol{\kappa}}(0)\rho_{-\boldsymbol{\kappa}}(t)\rangle = N\int\mathrm{d}\mathbf{r}\,\exp(\mathrm{i}\boldsymbol{\kappa}.\mathbf{r})G(\mathbf{r},t). \tag{11.164}$$

The static pair-distribution function $g(\mathbf{r})$ is related to $G(\mathbf{r},0)$ through (cf. eqns (3.23) and (11.3 f))

$$G(\mathbf{r},0) = \frac{1}{\rho}\langle\rho(0,0)\rho(\mathbf{r},0)\rangle = \delta(\mathbf{r})+g(\mathbf{r}). \tag{11.165}$$

Hence from (11.164)

$$\langle\delta\rho_{\boldsymbol{\kappa}}(0)\delta\rho_{-\boldsymbol{\kappa}}(0)\rangle = N[1+\int\mathrm{d}\mathbf{r}\,\exp(\mathrm{i}\boldsymbol{\kappa}.\mathbf{r})\{g(\mathbf{r})-\rho\}]. \tag{11.166}$$

From this we see that our quasi-elastic approximation to the differential cross-section, (11.160), is proportional to the Fourier integral of the static pair-distribution function.

Also note that from (11.165)

$$\langle\delta\rho(\mathbf{r}_1,0)\delta\rho(\mathbf{r}_2,0)\rangle = \langle\{\rho(\mathbf{r}_1,0)-\rho\}\{\rho(\mathbf{r}_2,0)-\rho\}\rangle$$
$$= \rho\delta(\mathbf{r}_1-\mathbf{r}_2)+\rho\{g(\mathbf{r}_1-\mathbf{r}_2)-\rho\}. \qquad (11.167)$$

We shall show later from thermodynamic arguments that the integral of $\{g(\mathbf{r})-\rho\}$ over the volume of the liquid is related to the isothermal compressibility κ_T through

$$\frac{\rho\kappa_T}{\beta} = 1+\int \mathrm{d}\mathbf{r}\,\{g(\mathbf{r})-\rho\}. \qquad (11.168)$$

By making use of this identity we have, in view of (11.166), the result that in the limit of $\boldsymbol{\varkappa} \to 0$ the quasi-elastic cross-section (11.160) tends to

$$\lim_{\boldsymbol{\varkappa}\to 0}\left(\frac{\mathrm{d}\sigma}{\mathrm{d}\Omega}\right) = \frac{\sigma_c}{4\pi}N\frac{\rho\kappa_T}{\beta}. \qquad (11.169)$$

This is an important result for it tells us that, within the quasi-elastic approximation, the scattering in the forward direction depends only on the isothermal compressibility κ_T and therefore diverges at the critical point.

We now verify (11.168).† One proof came as a by-product in § 11.4 (*b*), but that is a long and tedious calculation; we look for something more elegant. If we perturb a system in equilibrium by applying the interaction described by the Hamiltonian

$$\mathscr{H}_1 = -\int \mathrm{d}\mathbf{r}'\,A(\mathbf{r}')B(\mathbf{r}') \qquad (11.170)$$

the change in the thermal average of $A(\mathbf{r})$ to first order in B is (cf. Appendix B)

$$\delta\langle A(\mathbf{r})\rangle = \beta\int \mathrm{d}\mathbf{r}'\,B(\mathbf{r}')\{\langle A(\mathbf{r})A(\mathbf{r}')\rangle-\langle A(\mathbf{r})\rangle\langle A(\mathbf{r}')\rangle\}$$
$$\equiv \beta\int \mathrm{d}\mathbf{r}'\,B(\mathbf{r}')\langle\{A(\mathbf{r})-\langle A(\mathbf{r})\rangle\}\{A(\mathbf{r}')-\langle A(\mathbf{r}')\rangle\}\rangle. \qquad (11.171)$$

The thermal averages appearing on the right-hand side of (11.171) are taken with respect to the unperturbed system. If for $A(\mathbf{r})$ we choose the particle density $\rho(\mathbf{r},0)$ then the thermodynamic conjugate variable $B(\mathbf{r})$ is the change in the equilibrium value of the chemical potential $\delta\mu(\mathbf{r})$. The variables T and V are to be kept constant. Taking $\delta\mu(\mathbf{r}) = \delta\mu$, (11.171) yields

$$\delta\rho|_{T,V} = \beta\,\delta\mu\int \mathrm{d}\mathbf{r}'\,\langle\delta\rho(0,0)\delta\rho(\mathbf{r}',0)\rangle$$

from which it follows that

$$\left(\frac{\partial\rho}{\partial\mu}\right)_{T,V} = \beta\int \mathrm{d}\mathbf{r}'\,\langle\delta\rho(0,0)\delta\rho(\mathbf{r}',0)\rangle. \qquad (11.172)$$

† An alternative proof to that given here is given by L. D. Landau and E. M. Lifshitz, *Statistical physics*, § 115.

Now
$$\left(\frac{\partial\rho}{\partial\mu}\right)_{T,V} = \rho\left(\frac{\partial\rho}{\partial p}\right)_T = \rho^2\kappa_T \tag{11.173}$$
so making use of (11.167) we obtain the identity (11.168), as required.

Eqn (11.172) is important because it relates the thermodynamic derivative $(\partial\rho/\partial\mu)_{T,V}$ to the integral over all space of the density fluctuation correlation. As the critical point is approached the correlation range must increase in such a way that the integral finally diverges at $T = T_c$ $(V = V_c)$.

The mean square thermodynamic fluctuation in N is†
$$\langle(\delta N)^2\rangle = \frac{1}{\beta}\left(\frac{\partial N}{\partial\mu}\right)_{T,V}. \tag{11.174}$$
Hence from (11.168) and (11.173) it follows that
$$\rho\kappa_T = \frac{\beta}{N}\langle(\delta N)^2\rangle = \beta+\beta\int d\mathbf{r}\,\{g(\mathbf{r})-\rho\}. \tag{11.175}$$

The first equality in (11.175) elucidates the physical reason for the observed increase in the isothermal compressibility κ_T on approaching the liquid–gas transition; for in changing from a liquid (high density) to a gas (low density) the number of particles in a given volume of the fluid undergoes large fluctuations about its mean value and therefore $\langle(\delta N)^2\rangle$ increases markedly over its value far from the critical point. The Fermi pseudo-potential couples the incident neutrons to the density in the target system. We therefore anticipate that the large fluctuations in the particle density of the target (fluid) that occur as the critical point is approached will result in a marked increase in the scattering cross-section. Further, it is easy for us to see from this line of argument the general condition for the quasi-elastic approximation to be justified, namely, that the density fluctuations should appear static to the incident neutrons. Of course, as already noted, one way of satisfying this criterion is to use sufficiently high-energy incident neutrons. We find, however, that the density fluctuations slow down at the critical point; this phenomenon is usually called the *thermodynamic slowing down* of fluctuation. In consequence the quasi-elastic approximation becomes increasingly good in the vicinity of the liquid–gas phase transition.

We turn now to evaluating the quasi-elastic cross-section (11.160) when $\boldsymbol{\kappa}$ is small but not zero. For this we must derive an equation for the fluctuation in $\rho(\mathbf{r}, 0)$ due to the perturbation (11.170). The application of the perturbation described by the Hamiltonian (11.170), with $A(\mathbf{r}) = \rho(\mathbf{r}, 0)$ and $B(\mathbf{r}) = \delta\mu(\mathbf{r})$, will cause the free energy F to deviate

† Ibid. § 111.

from its value in the initial equilibrium state. Regarding the free energy as a function of $\rho(\mathbf{r}, 0)$ the deviation in F (at constant temperature and volume) is

$$\delta F = \int d\mathbf{r}\, [a_0\, \delta\rho(\mathbf{r},0) + \tfrac{1}{2}a_1\{\delta\rho(\mathbf{r},0)\}^2 + \ldots - \delta\mu(\mathbf{r})\delta\rho(\mathbf{r},0)].$$

Since $\int d\mathbf{r}\; \rho(\mathbf{r}, 0) = \int d\mathbf{r}\; \rho$, the first term in the Taylor expansion in $\delta\rho(\mathbf{r}, 0)$ is zero. The coefficient a_1 of the second term in the expansion is positive and vanishes at the critical point. For, by definition,

$$a_1 = \frac{1}{V}\left(\frac{\partial^2 F}{\partial \rho^2}\right)_{T,V},$$

and since

$$\left(\frac{\partial F}{\partial \rho}\right)_{T,V} = V\mu,$$

it follows from (11.173) that

$$a_1 = \left(\frac{\partial \mu}{\partial \rho}\right)_{T,V} = (\rho^2\kappa_T)^{-1}. \tag{11.176}$$

The terms included in the Taylor expansion of the free energy per unit volume in $\delta\rho(\mathbf{r}, 0)$ are not the only ones that contribute. As a consequence of the applied perturbation the fluid is not homogeneous and accordingly we must take into account the spatial derivatives of $\rho(\mathbf{r}, 0)$. Because the fluid is isotropic the lowest order term of this kind that can enter in the expansion of F is proportional to $\{\boldsymbol{\nabla}\rho(\mathbf{r}, 0)\}^2$. The coefficient of this expansion is taken to be a constant $\frac{1}{2}r_1^2$. We thus obtain

$$\delta F = \int d\mathbf{r}\, [\tfrac{1}{2}a_1\{\delta\rho(\mathbf{r},0)\}^2 + \tfrac{1}{2}r_1^2\{\boldsymbol{\nabla}\delta\rho(\mathbf{r},0)\}^2 - \delta\mu(\mathbf{r})\delta\rho(\mathbf{r},0)]. \tag{11.177}$$

In order that this should be a minimum with respect to $\delta\rho(\mathbf{r}, 0)$ the latter must satisfy

$$a_1\, \delta\rho(\mathbf{r},0) - r_1^2\nabla^2\delta\rho(\mathbf{r},0) - \delta\mu(\mathbf{r}) = 0. \tag{11.178}$$

This is the desired equation for the fluctuation $\delta\rho$ in the particle density due to the change in the chemical potential from μ to $\mu + \delta\mu(\mathbf{r})$.

Now (11.171) tells us that

$$\langle\rho(\mathbf{r},0)\rangle - \rho = \langle\delta\rho(\mathbf{r},0)\rangle = \beta \int d\mathbf{r}'\; \delta\mu(\mathbf{r}')\langle\delta\rho(\mathbf{r},0)\delta\rho(\mathbf{r}',0)\rangle. \tag{11.179}$$

Operating on (11.179) with $(a_1 - r_1^2\, \nabla_r^2)$ and then using (11.178) yields the equation

$$\beta \int d\mathbf{r}'\; \delta\mu(\mathbf{r}')(a_1 - r_1^2\, \nabla_r^2)\langle\delta\rho(\mathbf{r},0)\delta\rho(\mathbf{r}',0)\rangle = \delta\mu(\mathbf{r}),$$

which can be rewritten

$$\int d\mathbf{r}'\ \delta\mu(\mathbf{r}')\{\beta(a_1-r_1^2\nabla_r^2)\langle\delta\rho(\mathbf{r},0)\delta\rho(\mathbf{r}',0)\rangle-\delta(\mathbf{r}-\mathbf{r}')\} = 0. \tag{11.180}$$

Because $\delta\mu(\mathbf{r})$ is arbitrary,

$$\left(\nabla^2-\frac{a_1}{r_1^2}\right)\langle\delta\rho(\mathbf{r},0)\delta\rho(0,0)\rangle = -\frac{1}{r_1^2\beta}\delta(\mathbf{r}). \tag{11.181}$$

By making use of the identity

$$\nabla^2\left(\frac{1}{r}\right) = -4\pi\delta(\mathbf{r}) \tag{11.182}$$

it is readily verified that the solution of (11.181) that is zero for $r\to\infty$ is†

$$\langle\delta\rho(\mathbf{r},0)\delta\rho(0,0)\rangle = \frac{1}{4\pi r_1^2\beta}\frac{\exp(-\kappa_1 r)}{r}, \tag{11.183}$$

where $$\kappa_1^2 = a_1/r_1^2 = 1/(\rho^2 r_1^2\kappa_T). \tag{11.184}$$

Near the critical point κ_1 become very small, so

$$\langle\delta\rho(\mathbf{r},0)\delta\rho(0,0)\rangle \simeq \frac{1}{4\pi r_1^2\beta}\frac{1}{r}.$$

The correlation amongst pairs of nuclei is seen to decrease slowly with their separation; that is to say the correlation is much stronger than when the fluid is far from its critical point.

We must inquire after the conditions under which the solution (11.183) is valid. Because we retained in the Taylor expansion of the free energy (per unit volume) only the first non-zero term in $\delta\rho(\mathbf{r},0)$ and the lowest-order contribution involving the derivative of $\delta\rho(\mathbf{r},0)$, it is evident that the theory is limited to small values of $\delta\rho(\mathbf{r},0)$ which vary slowly with $\mathbf{r}$. In other words only those Fourier components of the disturbance that possess a long wavelength are taken into account. From (11.183) it is evident that these conditions are fulfilled for $r\gg\kappa_1^{-1}$.

It is straightforward to evaluate the quasi-elastic cross-section corresponding to the solution (11.183). $\langle\delta\rho_{\boldsymbol{\kappa}}(0)\delta\rho_{-\boldsymbol{\kappa}}(0)\rangle$ is given in terms of $\langle\delta\rho(\mathbf{r},0)\delta\rho(0,0)\rangle$ through

$$\langle\delta\rho_{\boldsymbol{\kappa}}(0)\delta\rho_{-\boldsymbol{\kappa}}(0)\rangle = V\int d\mathbf{r}\ \exp(i\boldsymbol{\kappa}.\mathbf{r})\langle\delta\rho(\mathbf{r},0)\delta\rho(0,0)\rangle, \tag{11.185}$$

and by making use of the result

$$\int d\mathbf{r}\ \frac{\exp(-\kappa_1 r+i\boldsymbol{\kappa}.\mathbf{r})}{r} = \frac{4\pi}{\kappa_1^2+\kappa^2} \tag{11.186}$$

† This form of the fluctuation density–density correlation function was first derived by L. S. Ornstein and F. Zernike, *Proc. Sect. Sci. K. ned. Akad. Wet.* **17,** 793 (1914); *Phys. Z.* **19,** 134 (1918).

we find, within the quasi-elastic approximation,

$$\frac{d\sigma}{d\Omega} = \frac{\sigma_c}{4\pi}\langle\delta\rho_{\varkappa}(0)\delta\rho_{-\varkappa}(0)\rangle$$

$$= \frac{\sigma_c}{4\pi}\frac{V}{r_1^2\beta}\frac{1}{\kappa^2+\kappa_1^2} \quad (|\varkappa| \ll \kappa_1). \tag{11.187}$$

Note that in the limit $\kappa \to 0$ we retrieve via (11.169) relation (11.184); viz. $(r_1\kappa_1)^2 = (\rho^2\kappa_T)^{-1}$.

For a given temperature and pressure the formula (11.187) describes the form of the quasi-elastic cross-section for scattering vectors $\varkappa$ that satisfy $|\varkappa| \ll \kappa_1$. Because κ_1 decreases as $T \to T_c$, $V \to V_c$, the range of $\varkappa$ for which (11.187) is valid decreases as the critical point is approached, tending to zero at the critical point itself.

The simple theory outlined at the beginning of this section predicts, with $V = V_c$, κ_1 tending to zero for T approaching T_c from above like $(T-T_c)^{\frac{1}{2}}$. As noted there, experiment suggests that the exponent is $\simeq$ 0·6–0·65.

We turn now to considering the time dependence of density fluctuations near the critical point. In general a density fluctuation can be relieved by either a mechanical or thermal fluctuation process. The mechanical process is the adiabatic fluctuation of pressure which is propagated as sound waves. The thermal process corresponds to fluctuations in entropy at constant pressure and are not propagated by the fluid but are slowly damped as a result of thermal conduction. As the fluid approaches its critical point the latter process dominates. In the preceding section we examined density fluctuations in a fluid within the context of the linearized hydrodynamic equations and derived the result (cf. 11.148 a)

$$q^2\Lambda = q^2\frac{\lambda}{\rho C_p} \tag{11.188}$$

for the thermal decay constant of the qth Fourier component of a density fluctuation. The time dependence of the Fourier components near the critical point is therefore $\exp(-q^2\Lambda|t|)$.

The result (11.183) for $\langle\delta\rho(\mathbf{r},0)\delta\rho(0,0)\rangle$ can be written as

$$\langle\delta\rho(\mathbf{r},0)\delta\rho(0,0)\rangle = \frac{1}{\beta r_1^2}\left(\frac{1}{2\pi}\right)^3\int d\mathbf{q}\,\frac{\exp(-i\mathbf{q}.\mathbf{r})}{\kappa_1^2+q^2},$$

and we propose that the time-dependent correlation function is

$$\langle\delta\rho(\mathbf{r},t)\delta\rho(0,0)\rangle = \frac{1}{\beta r_1^2}\left(\frac{1}{2\pi}\right)^3\int d\mathbf{q}\,\frac{\exp(-i\mathbf{q}.\mathbf{r}-q^2\Lambda|t|)}{\kappa_1^2+q^2}. \tag{11.189}$$

The partial differential cross-section of interest to us is given in terms of $\langle\delta\rho(\mathbf{r},t)\delta\rho(0,0)\rangle$ by

$$\frac{\mathrm{d}^2\sigma}{\mathrm{d}\Omega\mathrm{d}E'} = N\frac{\sigma_c}{4\pi}\frac{k'}{k}\frac{1}{2\pi\hbar}\int \mathrm{d}\mathbf{r}\int_{-\infty}^{\infty}\mathrm{d}t\,\exp(\mathrm{i}\boldsymbol{\kappa}.\mathbf{r}-\mathrm{i}\omega t)\{G(\mathbf{r},t-\tfrac{1}{2}\mathrm{i}\hbar\beta)-\rho\}$$

$$= N\frac{\sigma_c}{4\pi}\frac{k'}{k}\frac{1}{2\pi\hbar}\int \mathrm{d}\mathbf{r}\int_{-\infty}^{\infty}\mathrm{d}t\,\exp(\mathrm{i}\boldsymbol{\kappa}.\mathbf{r}-\mathrm{i}\omega t)\,\frac{\langle\delta\rho(\mathbf{r},t-\frac{1}{2}\mathrm{i}\hbar\beta)\delta\rho(0,0)\rangle}{\rho}. \tag{11.190}$$

From (11.189) we obtain the result

$$\frac{\mathrm{d}^2\sigma}{\mathrm{d}\Omega\mathrm{d}E'} = \frac{\sigma_c}{4\pi}\frac{k'}{k}\frac{V}{\pi\hbar\beta r_1^2}\frac{1}{\kappa^2+\kappa_1^2}\frac{\exp(\frac{1}{2}\hbar\omega\beta)(\kappa^2\Lambda)}{\omega^2+(\kappa^2\Lambda)^2}. \tag{11.191}$$

Note that in the limit $\Lambda \to 0$ this cross-section tends to

$$\frac{\sigma_c}{4\pi}\frac{V}{\beta r_1^2}\frac{1}{\kappa^2+\kappa_1^2}\delta(\hbar\omega)$$

in accord with (11.187) obtained in the quasi-elastic approximation.

Since Λ is inversely proportional to the specific heat at constant pressure its magnitude decreases rapidly as the critical point is approached, and therefore the inelasticity of the scattering decreases; this is a manifestation of the thermodynamic slowing down of the density fluctuations. From this we conclude that the use of the quasi-elastic approximation becomes more and more justified as the fluid approaches the liquid–gas phase transition.

There are several points to be noted about the above theory, which belongs to a wide class of theories of critical phenomena collectively referred to as *mean-field* theories. An example of another such theory is the Weiss molecular-field theory of magnetism discussed in Chapter 13. First, it would appear that the theory is not consistent. For in making the expansion of the free energy (11.177) it was assumed that the fluctuations in the density were small and slowly varying in space, yet we conclude, as a result of the ensuing calculation, that near the critical point the fluctuations are the dominant manifestation of criticality. The error in this argument lies in attempting to apply the theory outside its domain of validity. Clearly the latter is defined such that the theory is consistent; in other words the mean-field theory is not valid *close* to the critical point. Experiment indicates that, in general, it gives the correct behaviour on *approaching* the critical point. Putting this argument on a more quantitative basis it has been suggested† that the domain of validity is defined by the condition

$$\langle\delta\rho(\mathbf{r},0)\delta\rho(0,0)\rangle|_{r\kappa_1\sim 1} \ll \rho^2,$$

which must be examined for each particular system of interest.

As far as the results we have derived for the cross-section are concerned they will not hold unless $|\boldsymbol{\varkappa}| \ll \kappa_1$. Clearly this condition becomes increasingly more difficult to fulfil the closer the fluid is to its critical point.

11.5. Coherent scattering: short-wavelength fluctuations

When we consider the coherent scattering from liquids, when $\boldsymbol{\varkappa}$ in general lies outside the hydrodynamic limit, we have at present no really satisfactory theory to guide us. Therefore we cannot be sure which parts of the discussions in this section will survive careful analysis in future. We begin by reviewing the behaviour of the instantaneous correlation functions and then go on to consider various attempts to calculate the time-dependent correlation functions.

(a) *Instantaneous correlation functions*

Various results for these have been quoted previously in this chapter but for convenience we gather the results together here. From the definitions in (11.3), (11.4), and (11.6),

$$G(\mathbf{r},0) = \delta(\mathbf{r})+g(\mathbf{r}), \tag{11.192}$$

$$I(\boldsymbol{\varkappa},0) = 1+\gamma(\boldsymbol{\varkappa}), \tag{11.193}$$

and

$$\frac{d^2\sigma}{d\Omega dE'} = N\frac{k'}{k}\frac{\sigma_c}{4\pi}\frac{1}{2\pi\hbar}\int\limits_{-\infty}^{\infty} dt\, e^{-i\omega t}\{I(\boldsymbol{\varkappa},t)-N\delta_{\boldsymbol{\varkappa},0}\}. \tag{11.194}$$

Then rigorously

$$\begin{aligned}\hbar\int\limits_{-\infty}^{\infty} d\omega\, S(\boldsymbol{\varkappa},\omega) &= I(\boldsymbol{\varkappa},0)-N\delta_{\boldsymbol{\varkappa},0}\\ &= \int d\mathbf{r}\, \exp(i\boldsymbol{\varkappa}.\mathbf{r})\{G(\mathbf{r},0)-\rho\}\\ &= 1+\int d\mathbf{r}\, \exp(i\boldsymbol{\varkappa}.\mathbf{r})\{g(\mathbf{r})-\rho\},\end{aligned} \tag{11.195}$$

and in the quasi-elastic approximation

$$\frac{d\sigma}{d\Omega} \simeq N\frac{\sigma_c}{4\pi}\hbar\int\limits_{-\infty}^{\infty} d\omega\, S(\boldsymbol{\varkappa},\omega). \tag{11.196}$$

† V. L. Ginzburg, *Soviet Phys. solid St.* **2**, 1824 (1960).

The functions $g(\mathbf{r})$ and $1+\gamma(\boldsymbol{\varkappa})$ are plotted schematically in Figs. 11.5 and 11.6. Obviously the departure of $g(\mathbf{r})$ from the asymptotic value ρ is associated with the failure of hydrodynamics at small distance (corresponding to large $\boldsymbol{\varkappa}$). The oscillations in $g(\mathbf{r})$ represent the residual crystalline behaviour of the liquid. We note that $\gamma(\boldsymbol{\varkappa})$ and hence $g(\mathbf{r})$ can be determined by X-ray scattering because then the quasi-elastic

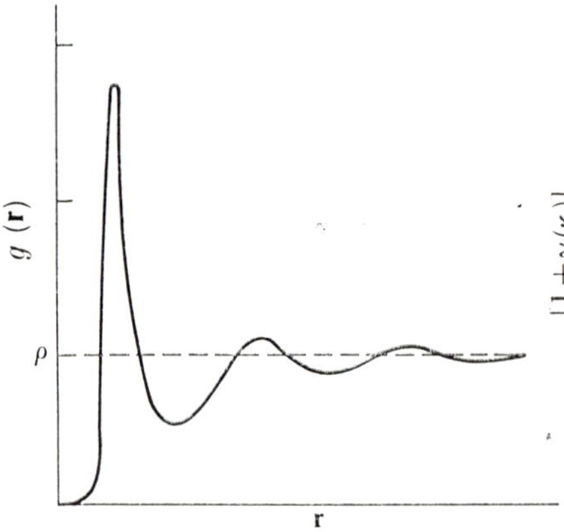

FIG. 11.5. Schematic diagram of $g(\mathbf{r})$.

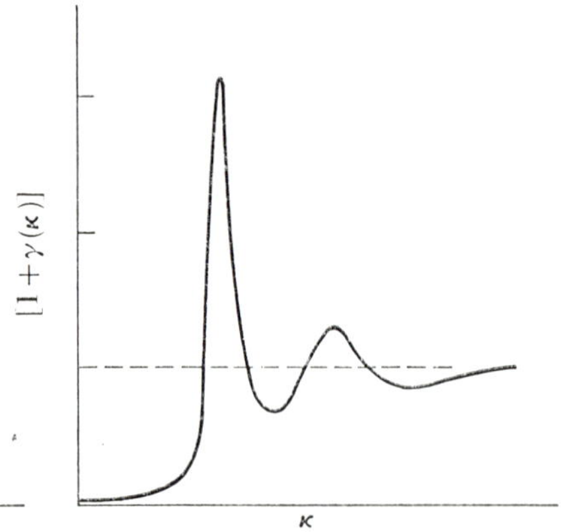

FIG. 11.6. Schematic diagram of $1+\gamma(\boldsymbol{\varkappa})$.

approximation is excellent. We note also that $\gamma(\boldsymbol{\varkappa})$ can be measured by neutron scattering, but only if high-energy neutrons are used or alternatively if the cross-section is measured at all $\boldsymbol{\varkappa}$ and ω so that an experimental result can be obtained for the integral in (11.195) which involves an integral over *all* ω at *fixed* κ.

(b) Convolution approximation

The calculation of $G(\mathbf{r},t)$ is not yet possible in any rigorous way. An early suggestion by Vineyard† was to approximate it as follows. We know that

$$G(\mathbf{r},t) = G_s(\mathbf{r},t)+\int d\mathbf{r}'\, g(\mathbf{r}')W(\mathbf{r},\mathbf{r}';t), \tag{11.197}$$

where $W(\mathbf{r},\mathbf{r}';t)$ is the probability, given that an atom is at the origin at $t = 0$ and a second atom is at $\mathbf{r}'$ at $t = 0$, that the second atom will move to $\mathbf{r}$ by the time t. Vineyard suggested that, in estimating $W(\mathbf{r},\mathbf{r}';t)$ the knowledge that an atom is at the origin at $t = 0$ might be ignored. If this is done $W(\mathbf{r},\mathbf{r}';t)$ becomes just the self-correlation function $G_s(\mathbf{r}-\mathbf{r}';t)$. Hence (11.197) becomes

$$G(\mathbf{r},t) = G_s(\mathbf{r},t)+\int d\mathbf{r}'\, g(\mathbf{r}')G_s(\mathbf{r}-\mathbf{r}';t), \tag{11.198}$$

† G. H. Vineyard, *Phys. Rev.* **110**, 999 (1958).

which in turn gives

$$I(\varkappa, t) = I_s(\varkappa, t)\{1+\gamma(\varkappa)\} \tag{11.199}$$

and

$$S(\varkappa, \omega) = S_s(\varkappa, \omega)\{1+\gamma(\varkappa)\}. \tag{11.200}$$

It follows that in this simple approximation the coherent cross-section may be discussed in as much detail as in § 11.2 and § 11.3 we discussed the incoherent cross-section. Unfortunately it turns out that this convolution approximation is not good enough; in particular it does not satisfy the rigorous moment relations we shall derive in § 11.5 (*d*), and therefore we shall not discuss it further.

(*c*) *Extrapolation of hydrodynamics*

Very careful experiments, for example by Cocking and Egelstaff on liquid lead and by Larsson and Dahlborg on liquid aluminium, show that for most wave vectors there is remarkably little difference in the cross-section for neutron scattering from liquids as compared to that from polycrystalline solids. This immediately suggests that a good theory might be constructed from some approximation to phonon theory, but no complete formulation along these lines has yet been worked out.† Therefore we shall content ourselves with a simple discussion based upon extending the hydrodynamic theory of § 11.4 (*b*).

The main result of § 11.4 (*b*) is given by (11.146), which we rewrite

$$S(\varkappa, \omega) = \{1+\gamma(\varkappa)\}e^{\frac{1}{2}\hbar\omega\beta}\frac{\Psi(\varkappa, \omega)}{\pi\rho_0^2 T\kappa_T}. \tag{11.201}$$

We recall that the poles of $\Psi(\varkappa, \omega)$ are given by (11.148) and conclude that, away from the critical point, there are two important peaks in $S(\varkappa, \omega)$ as a function of ω at $\omega = \pm\varkappa c_0$. Here c_0 is the velocity of sound and is given by

$$c_0^2 = \frac{1}{\rho_0}\frac{C_p}{C_v}\frac{1}{\kappa_T} = \frac{1}{\rho_0}\frac{C_p}{C_v}B_T = \frac{1}{\rho_0}B_s, \tag{11.202}$$

where we have introduced the isothermal bulk modulus $B_T = (\kappa_T)^{-1}$ and the adiabatic bulk modulus, $B_s = (\kappa_s)^{-1}$. We note immediately that (11.202) cannot be correct for general values of $\varkappa$ because, for an elastic solid, (11.202) would be replaced by

$$c_0^2 \simeq \frac{1}{\rho_0}(B_s+\tfrac{4}{3}G), \tag{11.203}$$

where G is the shear modulus. For high-frequency modes we expect the liquid to show a 'solid-like' behaviour and hence to show modes with velocities corresponding to (11.203) rather than (11.202). We therefore

† See, however, Hubbard and Beeby, loc. cit.

see that if the hydrodynamic theory of § 11.4 (*b*) is to be extended into higher values of the wave vector, then this is an important effect to put into the theory. The key point is concerned with the choice of the stress tensor (11.130). Let us put the bulk viscosity zero, as is appropriate for a simple monatomic liquid; then (11.130) becomes

$$\sigma'_{\alpha\beta} = \eta\{\nabla_\beta v_\alpha + \nabla_\alpha v_\beta - \tfrac{2}{3}\delta_{\alpha\beta}\boldsymbol{\nabla}.\mathbf{v}\}. \tag{11.204}$$

In contrast, in a solid we would work in terms of displacements, $\mathbf{r}$, of the particles and have

$$\sigma_{\alpha\beta} = G\{\nabla_\beta r_\alpha + \nabla_\alpha r_\beta - \tfrac{2}{3}\delta_{\alpha\beta}\boldsymbol{\nabla}.\mathbf{r}\}. \tag{11.205}$$

We need a theory which reproduces (11.204) at low frequencies and (11.205) at high frequencies. A simple interpolation formula first suggested by Frenkel is clearly

$$\left(\frac{1}{\eta}+\frac{1}{G}\frac{\mathrm{d}}{\mathrm{d}t}\right)\sigma_{\alpha\beta} = \nabla_\beta v_\alpha + \nabla_\alpha v_\beta - \tfrac{2}{3}\delta_{\alpha\beta}\boldsymbol{\nabla}.\mathbf{v}. \tag{11.206}$$

Fortunately we do not have to repeat all the calculations of § 11.4 (*b*). We first note that, since we have assumed the bulk viscosity to be zero, we should replace the second viscosity coefficient η' by $-\frac{2}{3}\eta$ everywhere it occurs. Next we note that when we introduce the $\mathbf{q}$ and ω dependence through equations like (11.140 a), then (11.206) becomes

$$\left(\frac{1}{\eta}-\frac{\mathrm{i}\omega}{G}\right)\sigma_{\alpha\beta,\mathbf{q}}(\omega) = \mathrm{i}\{q_\beta v^\alpha_\mathbf{q}(\omega) + q_\alpha v^\beta_\mathbf{q}(\omega) - \tfrac{2}{3}\delta_{\alpha\beta}\,\mathbf{q}.\mathbf{v}_\mathbf{q}(\omega)\}. \tag{11.207}$$

Hence the hydrodynamic results are modified by the replacement

$$\frac{1}{\eta} \to \frac{1}{\eta}\left(1-\frac{\mathrm{i}\omega\eta}{G}\right), \tag{11.208 a}$$

i.e.

$$\eta \to \frac{\eta}{(1-\mathrm{i}\omega\eta/G)}. \tag{11.208 b}$$

With these changes, in place of (11.140 b) we get

$$-\omega\delta\rho_\mathbf{q}(\omega) + \rho_0\,\mathbf{q}.\delta\mathbf{v}_\mathbf{q}(\omega) = 0, \tag{11.209}$$

$$-\omega\rho_0\,\delta\mathbf{v}_\mathbf{q}(\omega) + \mathbf{q}\frac{c_0^2 C_v}{C_p}\{\delta\rho_\mathbf{q}(\omega)+\alpha\rho_0\,\delta T_\mathbf{q}(\omega)\} -$$
$$-\frac{\mathrm{i}\eta q^2}{(1-\mathrm{i}\omega\eta/G)}\delta\mathbf{v}_\mathbf{q}(\omega) - \mathrm{i}\mathbf{q}\frac{1}{3}\frac{\eta}{(1-\mathrm{i}\omega\eta/G)}\mathbf{q}.\delta\mathbf{v}_\mathbf{q}(\omega)$$
$$= \sum_{\alpha,\beta}\tilde{\boldsymbol{\alpha}}q_\beta\, s_{\alpha\beta,\mathbf{q}}(\omega), \tag{11.210}$$

and

$$C_v\,\omega\rho_0\,\delta T_\mathbf{q}(\omega) - \frac{(C_p-C_v)}{\alpha}\omega\delta\rho_\mathbf{q}(\omega) + \mathrm{i}\lambda q^2\delta T_\mathbf{q}(\omega) = \mathbf{q}.\tilde{\mathbf{g}}_\mathbf{q}(\omega). \tag{11.211}$$

We now recognize that the change from simple hydrodynamics to a viscous–elastic theory affects the behaviour of the correlations in time but not the scale of the random fluctuations $s_{\alpha\beta,\mathbf{q}}(\omega)$ and $\mathbf{g}_{\mathbf{q}}(\omega)$. Hence (11.142) are simply

$$\langle s_{\alpha\beta,\mathbf{q}}(\omega)s_{\alpha'\beta',\mathbf{q}'}(\omega')\rangle = \frac{4\pi V\eta}{\beta}[\delta_{\alpha,\alpha'}\delta_{\beta,\beta'}+\delta_{\alpha,\beta'}\delta_{\alpha',\beta}-\tfrac{2}{3}\delta_{\alpha,\beta}\delta_{\alpha',\beta'}]\times \times\delta_{\mathbf{q},-\mathbf{q}'}\delta(\omega+\omega') \qquad (11.212)$$

and
$$\langle g_{\alpha,\mathbf{q}}(\omega)g_{\beta,\mathbf{q}'}(\omega')\rangle = \frac{4\pi V\lambda}{\beta}\delta_{\alpha,\beta}\delta_{\mathbf{q},-\mathbf{q}'}\delta(\omega+\omega'). \qquad (11.213)$$

We first notice that the transverse modes which are not connected with any density or temperature fluctuations have propagation characteristics described by

$$\omega = -\frac{\mathrm{i}\eta}{\rho_0}\frac{q^2}{(1-\mathrm{i}\omega\eta/G)}, \qquad (11.214)$$

i.e. by
$$\omega = -\frac{\mathrm{i}}{2\tau}\pm\frac{1}{2}\left(4c_T^2q^2-\frac{1}{\tau^2}\right)^{\frac{1}{2}}, \qquad (11.215)$$

where we have introduced the shorthand

$$c_T = (G/\rho)^{\frac{1}{2}} \qquad (11.216)$$

as the velocity of undamped transverse waves and

$$\tau = \eta/G \qquad (11.217)$$

as a relaxation time.

(11.215) tells us that for large enough $\mathbf{q}$ it is possible to propagate transverse modes, though they are always damped. No transverse waves can propagate for

$$q < \frac{1}{2c_T\tau}. \qquad (11.218)$$

We conclude that modes analogous to transverse phonons exist for high enough values of $\mathbf{q}$.

We turn next to the density fluctuations. From eqns (11.209) to (11.213) we can deduce immediately that in place of (11.144) we now obtain

$$\Psi(\mathbf{q},\omega) = \frac{\lambda q^6T^2+(4\eta/3)q^4T(C_p/c_0^2\alpha)^2|\omega-(\mathrm{i}\lambda q^2/C_v\rho_0)|^2}{\left|\frac{C_pq^2}{\alpha}\left(\omega+\frac{\mathrm{i}\lambda q^2}{C_p\rho_0}\right)-\frac{C_p\omega}{c_0^2\alpha}\left(\omega+\frac{\mathrm{i}\lambda q^2}{C_v\rho_0}\right)\left\{\omega+\frac{4\mathrm{i}\eta q^2}{3\rho_0(1-\mathrm{i}\omega\eta/G)}\right\}\right|^2}. \qquad (11.219)$$

It is easy to verify that this new form of $\Psi(\mathbf{q},\omega)$ has poles at points given by

$$\omega^2 = \frac{B_s}{\rho_0}q^2 - \frac{B_s q^2}{\rho_0}\left(\frac{C_p-C_v}{C_p}\right)\frac{\mathrm{i}\lambda q^2}{C_v\rho_0}\frac{1}{(\omega+\mathrm{i}\lambda q^2/C_v\rho_0)} - \frac{4q^2\mathrm{i}\eta\omega}{3\rho_0(1-\mathrm{i}\omega\eta/G)}. \tag{11.220}$$

At low frequencies the solution of (11.220) gives sound waves with

$$\omega \simeq q\left(\frac{B_s}{\rho_0}\right)^{\frac{1}{2}} \tag{11.221}$$

and the addition of imaginary terms to represent damping. This solution is therefore essentially the same as that obtained in § 11.4 (*b*). But, for frequencies above $1/\tau$, the solution in (11.220) is

$$\omega \simeq q\left(\frac{B_s+4G/3}{\rho_0}\right)^{\frac{1}{2}} \tag{11.222}$$

as we anticipated with (11.203). Notice that at all frequencies the longitudinal waves are damped but near the transition frequency from (11.221) to (11.222) the damping is particularly important.

11.6. Moment calculations for a classical liquid

In the preceding sections of this chapter we have examined $S(\boldsymbol{\varkappa},\omega)$ for several limiting values of its arguments. One method of gaining some insight into the general behaviour of $S(\boldsymbol{\varkappa},\omega)$ is to calculate its frequency moments, i.e. to examine the functions

$$\int\limits_{-\infty}^{\infty} \mathrm{d}\omega\ \omega^n S(\boldsymbol{\varkappa},\omega).$$

A prescription for calculating the moments has been discussed in detail in § 3.3 where we derived the result

$$\int\limits_{-\infty}^{\infty} \mathrm{d}\omega\ \omega S(\boldsymbol{\varkappa},\omega) = \frac{\kappa^2}{2M}. \tag{11.223}$$

This is also correct for $S_\mathrm{i}(\boldsymbol{\varkappa},\omega)$. We note that the first moment is independent of the detailed nature of the interaction between the atoms. This is no longer true for higher order moments, as can be readily verified on examining the formulae derived in § 3.3. Hence we choose to examine the higher order moments for a classical liquid, which means we ignore both quantum mechanical and recoil effects. We adhere closely to the discussion of this problem given by DeGennes.†

† P. G. DeGennes, *Physica 's Grav.* **25**, 825 (1959).

We define the moments $\overline{\omega^n_{\text{coh}}}$ and $\overline{\omega^n_{\text{incoh}}}$ by

$$\overline{\omega^n_{\text{coh}}} = \int \omega^n S(\boldsymbol{\varkappa}, \omega)\, d\omega \Big/ \int S(\boldsymbol{\varkappa}, \omega)\, d\omega \tag{11.224}$$

and

$$\overline{\omega^n_{\text{incoh}}} = \int \omega^n S_{\text{i}}(\boldsymbol{\varkappa}, \omega)\, d\omega \Big/ \int S_{\text{i}}(\boldsymbol{\varkappa}, \omega)\, d\omega. \tag{11.225}$$

From § 11.1 we know that

$$\hbar \int S(\boldsymbol{\varkappa}, \omega)\, d\omega = 1+\gamma(\boldsymbol{\varkappa}) \tag{11.226}$$

and

$$\hbar \int S_{\text{i}}(\boldsymbol{\varkappa}, \omega)\, d\omega = 1. \tag{11.227}$$

Consider first the calculation of $\overline{\omega^2_{\text{coh}}}$. This requires the evaluation

$$\int\limits_{-\infty}^{\infty} \omega^2 S(\boldsymbol{\varkappa}, \omega)\, d\omega = -\frac{1}{\hbar N}\langle \rho_{\boldsymbol{\varkappa}}(0)\ddot{\rho}_{-\boldsymbol{\varkappa}}(t)\rangle = \frac{1}{\hbar N}\langle |\dot{\rho}_{-\boldsymbol{\varkappa}}|^2\rangle, \tag{11.228}$$

where we recall $\rho_{\boldsymbol{\varkappa}} = \sum_j \exp(-\mathrm{i}\boldsymbol{\varkappa}.\mathbf{R}_j)$.

If we choose x_i to be parallel to $\boldsymbol{\varkappa}$ then $\boldsymbol{\varkappa}.\mathbf{R}_i = \kappa x_i$ and (11.228) becomes

$$\frac{\kappa^2}{\hbar N}\sum_{i,j}\langle \dot{x}_i \dot{x}_j \exp\{\mathrm{i}\kappa(x_i - x_j)\}\rangle. \tag{11.229}$$

Since there is no correlation between the velocities of the particles $\dot{x}_i$ and also $\langle \dot{x}_i\rangle = 0$ the terms with $i \neq j$ in (11.229) do not contribute. Hence

$$\int\limits_{-\infty}^{\infty} \omega^2 S(\boldsymbol{\varkappa}, \omega)\, d\omega = \frac{\kappa^2}{\hbar N}\sum_j \langle \dot{x}_j^2\rangle = \frac{\kappa^2}{\hbar}\langle \dot{x}^2\rangle \tag{11.230}$$

and to evaluate $\langle \dot{x}^2\rangle$ we recall that the Maxwell distribution is

$$\left(\frac{M\beta}{2\pi}\right)^{\frac{3}{2}} \exp\left\{-\frac{M\beta}{2}(\dot{x}^2+\dot{y}^2+\dot{z}^2)\right\} d\dot{x}d\dot{y}d\dot{z},$$

which gives (cf. 11.48 b) $\langle \dot{x}^2\rangle = \dfrac{1}{M\beta}$.

Combining this with (11.230) and (11.226) we obtain for $\overline{\omega^2_{\text{coh}}}$ the result

$$\overline{\omega^2_{\text{coh}}} = \frac{\kappa^2}{M\beta\{1+\gamma(\boldsymbol{\varkappa})\}}. \tag{11.231}$$

A calculation analogous to that leading to (11.231) gives for $\overline{\omega^2_{\text{incoh}}}$ the result

$$\overline{\omega^2_{\text{incoh}}} = \frac{\kappa^2}{M\beta}, \tag{11.232}$$

which has the same value in a real liquid as in a perfect gas for a given temperature and M.

The result (11.231) tells us that the energy spread in the coherent cross-section depends upon the scattering vector and will be narrowed at those $\boldsymbol{\varkappa}$ values at which $\gamma(\boldsymbol{\varkappa})$ has a maximum value (cf. Fig. 11.6).

In the present discussion only the even moments of $S(\boldsymbol{\varkappa},\omega)$ and $S_{\mathrm{i}}(\boldsymbol{\varkappa},\omega)$ are non-zero. We shall not present the details of the derivation of the fourth moments but quote the results of DeGennes. If the interaction potential for the liquid is of the form $\sum_{i\neq j} V(\mathbf{R}_i-\mathbf{R}_j)$ then

$$\overline{\omega^4_{\mathrm{coh}}} = \frac{\kappa^4}{\beta M^2\{1+\gamma(\boldsymbol{\varkappa})\}}\left[\frac{3}{\beta}+\int \mathrm{d}\mathbf{r}\; g(\mathbf{r})\left\{\frac{1-\cos(\kappa x)}{\kappa^2}\right\}\frac{\partial^2 V}{\partial x^2}\right] \tag{11.233}$$

and

$$\overline{\omega^4_{\mathrm{incoh}}} = \frac{\kappa^4}{\beta M^2}\left[\frac{3}{\beta}+\int \mathrm{d}\mathbf{r}\; g(\mathbf{r})\kappa^{-2}\frac{\partial^2 V}{\partial x^2}\right] \tag{11.234}$$

$$= 3[\overline{\omega^2_{\mathrm{incoh}}}]^2+\overline{\omega^2_{\mathrm{incoh}}}\,\frac{1}{M}\int \mathrm{d}\mathbf{r}\; g(\mathbf{r})\frac{\partial^2 V}{\partial x^2}.$$

Note that both $\overline{\omega^4_{\mathrm{coh}}}$ and $\overline{\omega^4_{\mathrm{incoh}}}$ are proportional to the fourth power of the scattering vector and involve an integration over the pair distribution function $g(\mathbf{r})$ and the second derivative of the pair interaction potential.

For incoherent scattering the results (11.232) and (11.234) for, respectively, $\overline{\omega^2_{\mathrm{incoh}}}$ and $\overline{\omega^4_{\mathrm{incoh}}}$ imply the following.

(a) For large values of the scattering vector the first term on the right-hand side of (11.234) for $\overline{\omega^4}$ will dominate so that $\overline{\omega^4_{\mathrm{incoh}}} \to 3[\overline{\omega^2_{\mathrm{incoh}}}]^2$. This means that in this limit the fourth moment assumes the same behaviour as in the Gaussian approximation.

(b) In the opposite limit of small $|\boldsymbol{\varkappa}|$ the fourth moment has a value too large to be compatible with a Gaussian. Hence it is more appropriate for small $|\boldsymbol{\varkappa}|$ to approximate $S_{\mathrm{i}}(\boldsymbol{\varkappa},\omega)$ by a Lorentz curve as is obtained from diffusion theory.

The results for $\overline{\omega^2_{\mathrm{coh}}}$ and $\overline{\omega^4_{\mathrm{coh}}}$, eqns (11.231) and (11.233), respectively, are more complicated than those for incoherent scattering, and this makes it more difficult to make deductions as to the general behaviour of $S(\boldsymbol{\varkappa},\omega)$ than for that of $S_{\mathrm{i}}(\boldsymbol{\varkappa},\omega)$. However, an important feature observed by DeGennes is that when $|\boldsymbol{\varkappa}|$ coincides with those values at which there is a diffraction peak in the X-ray pattern, $\overline{\omega^2_{\mathrm{coh}}}$ becomes small, as previously noted, and $\overline{\omega^4_{\mathrm{coh}}}/3[\overline{\omega^2_{\mathrm{coh}}}]^2$ becomes large. This means that for these values of $|\boldsymbol{\varkappa}|$ the behaviour of $S(\boldsymbol{\varkappa},\omega)$ is closer to that of a Lorentzian than to a Gaussian function.

12

SCATTERING BY GASES

12.1. Introduction

BY a free molecular gas we mean a gas whose temperature and density are such that we can neglect both the direct interaction between the molecules and also the purely quantum mechanical effects arising from their identity. The study of the cross-sections for the scattering of thermal and cold neutrons from such systems does not, in general, lead to any information about the dynamical behaviour of the internal motions of the molecules that could not be obtained with the same, or greater, accuracy from infra-red and Raman spectra. In fact, the various parameters that are required to evaluate the expressions for the cross-sections are obtained from infra-red and Raman spectra. In this sense the problem of calculating the neutron cross-section is rather an academic one, but one that has attracted much attention.

We assume that the molecules do not possess a magnetic moment and we will only concern ourselves with the scattering from gases consisting of diatomic, spherical, and symmetric top polyatomic molecules. Because the neutrons are principally scattered by the nuclei and since, furthermore, the energy required to excite a molecule out of its normal electron configuration is large compared to the energy of a thermal neutron, the atomic electrons do not give a direct contribution to the scattering. However, they are involved in so far as the dynamical states of the nuclei depend on the total symmetry of the molecule and, therefore, on the symmetry of the normal electron configuration. It is found that in the great majority of cases the wave function of the electrons in this state is completely symmetrical, i.e. the wave function is invariant with respect to all the elements of the symmetry group of the molecule. This empirical rule is found to apply to nearly all diatomic and polyatomic molecules.

Because the separation of the vibrational and rotational energy states of molecules can be of the same order of magnitude as the energy associated with thermal and cold neutrons we can meaningfully discuss the scattering from molecules only if we have some understanding of these dynamical states. Thus we will briefly discuss the molecular terms, first for diatomic molecules and then for polyatomic molecules, before considering the calculation of the neutron cross-sections.

The energy levels of a diatomic molecule can be shown to be represented by†

$$E = U_e + B_e K(K+1) + \hbar\omega_e(\nu+\tfrac{1}{2}). \tag{12.1}$$

In (12.1) U_e is the electron energy, including the energy of the Coulomb interaction of the nuclei at their equilibrium separation, the second term is the rotational energy from the rotation of the molecule, and the third the energy of the vibrations of the nuclei within the molecule. The parameters in the second two terms of (12.1) are defined as follows (for molecules with zero electron spin; it is found that the normal state of nearly all molecules is a singlet): $\hat{\mathbf{K}}$ is the sum of the orbital angular momentum of the electrons and the angular momentum of the nuclei, so that $\hat{\mathbf{K}}^2 = K(K+1)$ is a conserved quantity, $B_e = \hbar^2/(2M_{\text{mol}} r_e^2)$ (r_e is the equilibrium separation of the nuclei and M_{mol} the mass of the molecule) is called the rotational constant, ω_e is the frequency of the vibrations of the nuclei, and the number $\nu = 0, 1, 2,\ldots$, called the vibrational quantum number, denumerates the levels with a given K in order of increasing energy. This expression is a first approximation to the energy levels and is valid when the nuclei undergo small displacements from their equilibrium separation. In the next approximation the rotational and vibrational degrees of freedom are coupled. In all but one of the examples considered here we shall assume that the rotational and vibrational motions are independent of each other.

If the intervals between the vibrational, rotational, and electron levels are denoted by ΔE_ν, ΔE_r, and ΔE_{el}, respectively, then, quite generally,

$$\Delta E_{el} \gg \Delta E_\nu \gg \Delta E_r. \tag{12.2}$$

Hence, as is clear from the following table, the rotational levels must dominate the structure of the neutron cross-section.

(eV)	H_2	N_2
$-U_e$	4·7	7·5
$\hbar\omega_e$	0·54	0·29
$10^3 \times B_e$	7·6	0·25
r_e (Å)	0·742	1·09

† All the topics considered in this section are discussed fully by G. Herzberg in *Molecular spectra and molecular structure*, vols. 1, 2, and 3, Van Nostrand.

Let us now examine the restrictions imposed on the rotational levels of a *diatomic molecule* by the spatial symmetry of the nuclei. We assume the molecule has two identical nuclei at positions $\mathbf{R}_1$, $\mathbf{R}_2$ so that the total wave functions can be represented as

$$\Psi = \Phi_{\text{Elec}}\Phi_{\text{Vib}}(|\mathbf{R}_1-\mathbf{R}_2|)\Phi_{\text{Rot}}(\mathbf{R}_1-\mathbf{R}_2)\chi_{\text{Nuc}}, \tag{12.3}$$

where Φ_{Elec} is the electron wave function involving all the electron coordinates $\mathbf{r}_1, \mathbf{r}_2,..., \mathbf{r}_n$ and their spins $s_1, s_2,..., s_n$ with the nuclear positions $\mathbf{R}_1$ and $\mathbf{R}_2$ appearing as parameters. Φ_{Vib} is the vibrational wave function of the two nuclei and is even in their separation $\mathbf{R}_1-\mathbf{R}_2$. $\Phi_{\text{Rot}}(\mathbf{R}_1-\mathbf{R}_2)$ is the rotational wave function of the two nuclei and χ_{Nuc} is the total nuclear spin wave function of the molecule. We know that Ψ must be antisymmetric in the exchange of any pair of electrons, so that commonly Φ_{Elec} can in a first approximation be represented as a determinant wave function. Because the nuclei are identical Ψ must also be either antisymmetric or symmetric under interchange of the nuclei, and this introduces an important restriction on the allowed rotational wave functions. To understand this we must examine in outline the classification of molecular eigenstates.

We first consider the electronic wave function of a diatomic molecule with identical nuclei. Obviously the molecular axis is a symmetry axis and in almost every molecule (NO is a well-known exception) the total electronic wave function has the maximum possible symmetry in the ground state and is non-degenerate. It is customary to label such states as Σ-type. In addition we specify three other electronic quantum numbers; the spin multiplicity, $2S+1$; the symmetry on reflection (R) in a plane containing the molecular axis, even (+) or odd (−); the symmetry on inversion of coordinates of all the electrons ($\mathscr{I}$) about the middle of the molecular axis, even (*gerade*) or odd (*ungerade*).

To illustrate this notation consider Fig. 12.1, which shows a diatomic system and, schematically, an s orbital and two p orbitals on each atom. Then, for example, a single electron in the *s-bonding orbital*

$$S_{\text{b}} \simeq \{s_1(\mathbf{r})+s_2(\mathbf{r})\}$$

gives a wave function which is non-degenerate, therefore labelled Σ; has spin $\frac{1}{2}$ and therefore spin multiplicity 2, therefore labelled $^2\Sigma$; is even on reflection in the x–z plane, therefore labelled $^2\Sigma^+$; and even on inversion $\mathscr{I}$, therefore labelled $^2\Sigma_{\text{g}}^+$. A single electron in the *s-antibonding orbital*

$$S_{\text{ab}} \simeq \{s_1(\mathbf{r})-s_2(\mathbf{r})\}$$

would give a state labelled $^2\Sigma_{\text{u}}^+$. Two electrons in the s-bonding orbital would give an electronic wave function corresponding to the H_2 molecule and then

$$\Phi_{\text{Elec}} = S_{\text{b}}(\mathbf{r}_1)S_{\text{b}}(\mathbf{r}_2)\frac{1}{\sqrt{2}}(\alpha_1\beta_2-\beta_1\alpha_2)$$

is labelled ${}^1\Sigma_g^+$.

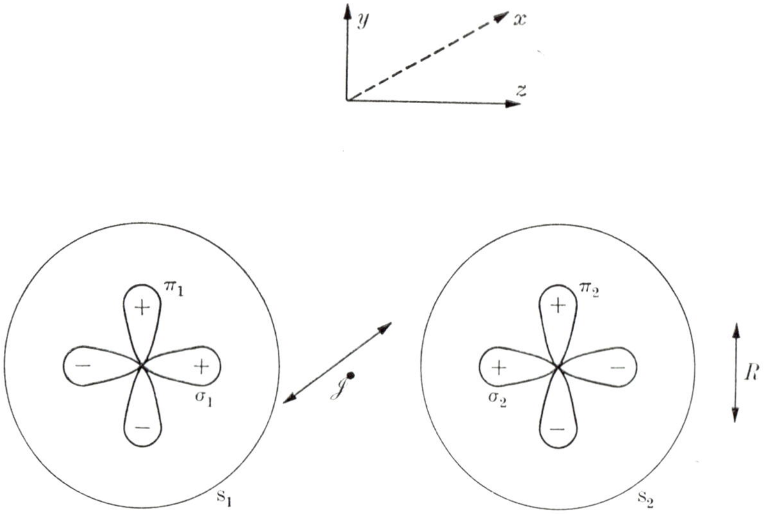

FIG. 12.1. Schematic diagram of s and p orbitals in a diatomic molecule. The other p orbitals, μ_1 and μ_2, are degenerate with π_1 and π_2 and point into the page, i.e. along the x axes.

As a final example, the ground state of O_2 has, apart from closed electronic shells, eight electrons to fit in the twelve possible p states. In a simple molecular orbital theory

$$\Phi_{\text{Elec}} = \text{Det}\{\sigma_{\text{b}}^{\alpha}(1)\sigma_{\text{b}}^{\beta}(2)\pi_{\text{b}}^{\alpha}(3)\pi_{\text{b}}^{\beta}(4)\mu_{\text{b}}^{\alpha}(5)\mu_{\text{b}}^{\beta}(6)\pi_{\text{ab}}^{\alpha}(7)\mu_{\text{ab}}^{\alpha}(8)\},$$

where Det{ } stands for the full determinant of the eight electrons. This state has $S = 1$, is odd because an odd number of π orbitals is involved, and is *gerade* because an even number of anti-bonding orbitals are occupied. Therefore the ground state of O_2 is labelled ${}^3\Sigma_g^-$.

Consider now the total wave function of the diatomic molecule. The Hamiltonian of the molecule is invariant with respect to a simultaneous change in sign of the coordinates of all the *electrons* and the *nuclei*. Hence such an operation either leaves the total wave function unchanged (*positive*) or multiplies it by -1 (*negative*). For Σ terms the sign depends only on K and is given by $(-1)^K$ for Σ^+ and $(-1)^{K+1}$ for Σ^- terms. These results can be understood as follows. From (12.1) the motion of the nuclei is seen to be equivalent to that of a single particle, of orbital angular momentum K, in a centrally symmetric field. Hence, when the signs of all the coordinates are changed the nuclear wave function is multiplied by $(-1)^K$. For the electron Σ term, inversion is equivalent to reflection in a plane passing through the molecular axis and we thus arrive at the above results for the signs of $\Sigma^{\pm}$ terms under inversion.

Lastly, consider the wave function of the two identical nuclei. If the nuclei have integral spin i the coordinate nuclear wave function must be *symmetric* under inversion whereas if i is half-integral the coordinate wave function must be *antisymmetric*. It follows that if the total nuclear spin I is even, then the coordinate nuclear wave function must be symmetric whereas if the total nuclear spin I is odd then the coordinate nuclear wave function must be antisymmetric.

We have now determined how both the total wave function of the electrons and nuclei transforms under inversion and also how the separate wave functions of the electrons and nuclei transform under inversion. Since the transformation property of the nuclear wave function under inversion depends on the total nuclear spin I and the transformation properties of the total wave function of the electrons and nuclei are determined by K, it follows that I and K are correlated for a given electron Σ-type level. As an example we examine a $\Sigma_g^{\pm}$ term. The sign of the total wave function of a Σ^+ term is given by $(-1)^K$; hence for Σ_g^+ the total wave function is positive for even K, and therefore symmetric in the nuclear wave function, while for odd K the total wave function is negative and consequently the nuclear wave function is antisymmetric. If $i = 0$ there are no antisymmetric states and thus no rotational states with odd K. An important example is that of the hydrogen molecule, where $i = \frac{1}{2}$. For $I = 0$ (para-hydrogen) there are no rotational states with odd K, and states with odd values of K only for $I = 1$ (ortho-hydrogen).

At high temperatures, when many rotational levels are excited, these rules are not of importance, but for low excitation numbers the restrictions are quite necessary.

For a polyatomic molecule containing N nuclei we are concerned with the rotational and vibrational motions of a system with $3N$ degrees of freedom. Of these, three correspond to translational and three to rotational motion of the system as a whole. (If the molecule is linear, the number of vibrational degrees of freedom is $3N-5$ since, in this case, it is not meaningful to speak of a rotation of a linear molecule about its axis.) The energy of a system of particles undergoing small displacements is

$$E = \sum_{\xi,\alpha} \tfrac{1}{2} m_\xi (\dot{u}_\xi^{(\alpha)})^2 + \sum_{\substack{\xi,\xi' \\ \alpha,\beta}} \tfrac{1}{2} u_\xi^{(\alpha)} k_{\xi,\xi'}^{(\alpha,\beta)} u_{\xi'}^{(\beta)}, \tag{12.4}$$

where $u_\xi^{(\alpha)}$ is the α component of the displacement of the ξth nucleus from its equilibrium position, m_ξ its mass, and $k_{\xi,\xi'}^{(\alpha,\beta)}$ a force constant. We can now make a linear transformation of the displacements that will eliminate the coordinates corresponding to the translational motion and rotation of the system in (12.4), and reduce the second term to the sum of squares. We will not pause here to show this for the classical case but we quantize the system and show how the Hamiltonian corresponding to (12.4) can be expressed as the sum of $3N-6$ independent harmonic oscillators.

Let $\phi_\xi^{(\alpha)}(\lambda)$ satisfy

$$\sum_{\xi,\alpha} m_\xi^{-\frac{1}{2}} k_{\xi,\xi'}^{(\alpha,\beta)} m_{\xi'}^{-\frac{1}{2}} \phi_\xi^{(\alpha)}(\lambda) = \omega_\lambda^2 \phi_{\xi'}^{(\beta)}(\lambda), \tag{12.5}$$

where, for a unitary transformation,

$$\sum_{\xi,\alpha} [\phi_\xi^{(\alpha)}(\lambda)]^* \phi_\xi^{(\alpha)}(\lambda') = \delta_{\lambda,\lambda'}, \tag{12.6}$$

where λ labels the $3N-6$ eigenmodes. We now define Bose operators $\hat{a}_\lambda$ and $\hat{a}_\lambda^+$ such that

$$\hat{u}_\xi^{(\alpha)} = \sum_\lambda \left(\frac{\hbar}{2m_\xi \omega_\lambda}\right)^{\frac{1}{2}} \phi_\xi^{(\alpha)}(\lambda)[\hat{a}_\lambda \exp(-\mathrm{i}\omega_\lambda t) + \hat{a}_{-\lambda}^+ \exp(\mathrm{i}\omega_\lambda t)], \tag{12.7}$$

so $\hat{u}_\xi^{(\alpha)}$ is real if $[\phi_\xi^{(\alpha)}(\lambda)]^* = \phi_\xi^{(\alpha)}(-\lambda)$. In terms of the Bose operators the Hamiltonian corresponding to (12.4) is simply

$$\mathscr{H} = \sum_\lambda \hbar\omega_\lambda (\hat{a}_\lambda^+ \hat{a}_\lambda + \tfrac{1}{2}). \tag{12.8}$$

The eigenmodes can, of course, be classified according to the symmetry elements of the equilibrium configuration of nuclei. For example, the spherical top molecule CH_4 has $3N-6 = 9$ normal vibrations: a non-degenerate one (belonging to the representation Γ_1) that involves only the hydrogen nuclei (along the C—H bond) with $\hbar\omega_1 = 0{\cdot}37$ eV; a two-fold degenerate one (Γ_3) involving only the hydrogen nuclei (perpendicular to the C—H bond), $\hbar\omega_2 = 0{\cdot}172$ eV; and two three-fold degenerate ones (Γ_4), $\hbar\omega_3 = 0{\cdot}168$ eV and $\hbar\omega_4 = 0{\cdot}391$ eV.

If we ignore the vibrational motion of the nuclei the rotational states of a polyatomic molecule are given by the eigenstates of the Hamiltonian of a rigid body, namely

$$\mathscr{H} = \tfrac{1}{2}\hbar^2 \left[\frac{\hat{J}_\xi^2}{I_\mathrm{A}} + \frac{\hat{J}_\eta^2}{I_\mathrm{B}} + \frac{\hat{J}_\zeta^2}{I_\mathrm{C}}\right], \tag{12.9}$$

where $\hat{J}_\xi$, $\hat{J}_\eta$, and $\hat{J}_\zeta$ are the components of the angular momentum for the three principal axes ξ, η, ζ of the molecule and I_A, I_B, and I_C the principal moments of inertia.

The simplest case of (12.9) is when all three principal moments of inertia are equal, $I_A = I_B = I_C = I$. For this case,

$$\mathscr{H} = \frac{\hbar^2}{2I}\hat{J}^2, \tag{12.10}$$

with eigenvalues
$$E_J = \frac{\hbar^2}{2I}J(J+1). \tag{12.11}$$

Each of these energy levels is degenerate with respect to the $2J+1$ directions of the angular momentum relative to the body itself, i.e. with respect to the values of $J_\zeta = K$. (We use K to denote both the total orbital a.m. of a diatomic molecule and also the eigenvalue of the ζ component of a.m. for polyatomic molecules. This should not lead to any confusion). The value of the rotational constant $\hbar^2/2I$ for methane is $\sim 6{\cdot}5\times 10^{-4}$ eV.

The symmetric top has two principal moments equal, I_A and I_B, say, and a corresponding Hamiltonian

$$\mathscr{H} = \frac{\hbar^2}{2I_A}\hat{J}^2+\tfrac{1}{2}\hbar^2\left(\frac{1}{I_C}-\frac{1}{I_A}\right)\hat{J}_\zeta^2. \tag{12.12}$$

For a given J and K the eigenvalues are

$$E_{JK} = \hbar^2 J(J+1)/2I_A+\tfrac{1}{2}\hbar^2\left(\frac{1}{I_C}-\frac{1}{I_A}\right)K^2 \tag{12.13}$$

The degeneracy with respect to values of K which occurred for a spherical top are here partly removed. The values of the energy are the same only for values of K differing in sign alone. It can be shown that the eigenfunctions of (12.12) can be expressed in terms of the rotation matrix† $\mathscr{D}^{(J)}_{MK}(\alpha\beta\gamma)$, where the Euler angles α, β, γ define the orientation of the axes fixed in the molecule with respect to the laboratory frame of reference. The eigenfunctions of (12.12), properly normalized, are

$$|JMK\rangle = \left(\frac{8\pi^2}{2J+1}\right)^{\frac{1}{2}}\mathscr{D}^{(J)}_{MK}(\alpha\beta\gamma). \tag{12.14}$$

$\mathscr{D}^{(J)}_{MK}$ is the eigenfunction of $\hat{J}^2$ with eigenvalue $\hbar^2 J(J+1)$ and the eigenfunction of $\hat{J}_z$, say, with eigenvalue $\hbar M$. It is simultaneously the eigenfunction of $\hat{J}_\zeta$ with an eigenvalue $\hbar K$.

† We use the same definitions as those employed by A. R. Edmonds, *Angular momentum in quantum mechanics*, Princeton University Press.

We cannot give a discussion of the classification of the molecular terms for a polyatomic molecule without the aid of group theory. Since this would lead us too far afield we shall not pursue this topic any further but merely note that the wave function of the normal electron state is completely symmetrical in nearly all stable polyatomic molecules.

12.2. Diatomic molecules—partial cross-section

The interaction between an incident neutron and the nuclei of a single molecule is

$$\hat{V}(\mathbf{r}) = \frac{2\pi\hbar^2}{m}\sum_{\xi} \hat{b}_{\xi}\,\delta(\mathbf{r}-\mathbf{R}_{\xi}), \tag{12.15}$$

where $\hat{b}_{\xi}$ is the scattering amplitude operator for the ξth nucleus and $\mathbf{R}_{\xi}$ its position vector.

The most convenient way to calculate the neutron cross-section for a diatomic molecule is to transform to the frame of reference in which the centre of mass of the system that consists of the incident neutron and the molecule is at rest.

The cross-section in this frame of reference for the process in which the nuclei change from the states λ to λ' is

$$\frac{\mathrm{d}\sigma_{\lambda\lambda'}}{\mathrm{d}\Omega} = \frac{k'}{k}\left(\frac{\mu}{2\pi\hbar^2}\right)^2 |\langle\lambda',\sigma'|\int \mathrm{d}\mathbf{r}\;\hat{V}(\mathbf{r})\exp(\mathrm{i}\boldsymbol{\kappa}.\mathbf{r})|\lambda,\sigma\rangle|^2. \tag{12.16}$$

In (12.16) $\mu = \dfrac{mM_{\text{mol}}}{m+M_{\text{mol}}}$ is the reduced mass and

$$\frac{\hbar^2k^2}{2\mu}+E_{\lambda} = \frac{\hbar^2k'^2}{2\mu}+E_{\lambda'}.$$

The labels λ and λ' denote both the spin states of the nuclei and their dynamical states, which, for the approximate energy levels given by (12.1), are specified by the quantum numbers K, M_K, and ν. For the interaction given by eqn (12.15) the cross-section reduces to

$$\frac{\mathrm{d}\sigma_{\lambda\lambda'}}{\mathrm{d}\Omega} = \frac{k'}{k}(\mu/m)^2|\langle\lambda',\sigma'|\sum_{\xi}\exp(\mathrm{i}\boldsymbol{\kappa}.\mathbf{r}_{\xi})\hat{b}_{\xi}|\lambda,\sigma\rangle|^2. \tag{12.17}$$

In the introduction to this chapter we noted that for a diatomic molecule, with identical nuclei (to which we shall restrict our discussion), in the state Σ_g^+ only even K rotational states are allowed when the total nuclear spin I is even, and odd K rotational states when I is odd. Hence for this state we must distinguish between the two cases: (i) when the nuclei make transitions between states with values of K of the same parity, which are possible only for the process in which I remains either even or odd, and (ii) when the nuclei make transitions between states with values of K of different parity, which are possible only when the total nuclear spin changes from an even number to an odd one, or vice versa.

In the former case,

$$\langle n|\exp(\mathrm{i}\boldsymbol{\kappa}.\mathbf{r}/2)|n'\rangle = \langle n|\exp(-\mathrm{i}\boldsymbol{\kappa}.\mathbf{r}/2)|n'\rangle$$
$$= \langle n|\cos\tfrac{1}{2}\boldsymbol{\kappa}.\mathbf{r}|n'\rangle$$

because the rotational levels are multiplied by $(-1)^K$ on changing the sign of $\mathbf{r}$, the position vector between the two nuclei, and the values of K in the dynamical states n and n' differ by an even integer.

Since (cf. Chapter 1)

$$\hat{b}_\xi = \frac{1}{2i_\xi+1}\{(i_\xi+1)b_\xi^{(+)}+i_\xi b^{(-)}\}+\frac{\hat{\boldsymbol{\sigma}}.\hat{\mathbf{i}}_\xi}{2i_\xi+1}\{b_\xi^{(+)}-b_\xi^{(-)}\} = A_\xi+\tfrac{1}{2}B_\xi\hat{\boldsymbol{\sigma}}.\hat{\mathbf{i}}_\xi,$$

we have for case (i)

$$\sum_\xi \langle\lambda'|\,\hat{b}_\xi\,|\lambda\rangle = \langle\lambda'|\cos(\tfrac{1}{2}\boldsymbol{\kappa}.\mathbf{r})(2A+\tfrac{1}{2}B\hat{\boldsymbol{\sigma}}.\hat{\mathbf{I}})|\lambda\rangle. \qquad (12.18)$$

For the second case, where the values of K in the states n and n' differ by an odd integer

$$\langle n|\exp(\mathrm{i}\boldsymbol{\kappa}.\mathbf{r}/2)|n'\rangle = -\langle n|\exp(-\mathrm{i}\boldsymbol{\kappa}.\mathbf{r}/2)|n'\rangle$$
$$= \mathrm{i}\langle n|\sin\tfrac{1}{2}\boldsymbol{\kappa}.\mathbf{r}|n'\rangle,$$

so that

$$\sum_\xi \langle\lambda'|\,\hat{b}_\xi\,|\lambda\rangle = \mathrm{i}\,\tfrac{1}{2}B\langle\lambda'|\sin(\tfrac{1}{2}\boldsymbol{\kappa}.\mathbf{r})\hat{\boldsymbol{\sigma}}.(\hat{\mathbf{i}}_1-\hat{\mathbf{i}}_2)\,|\lambda\rangle, \qquad (12.19)$$

and this will, clearly, result in scattering that depends solely on the incoherent scattering length B.

To obtain the total neutron cross-section we must, first of all, integrate the cross-section over all angles. We have, furthermore, to sum over all the final states of the molecule and over all the initial states of molecule, with their appropriate statistical weights. Finally we must transform from the centre of mass frame of reference to the laboratory frame of reference. If the total cross-section in the former is $\tilde{\sigma}$, then in the laboratory frame of reference the total cross-section σ is simply

$$\sigma = |1-V/\hbar\mathbf{k}|\tilde{\sigma},$$

where V is the velocity of the centre of mass.

We have not, so far, indicated how one calculates the matrix elements $(\sin\frac{1}{2}\boldsymbol{\kappa}.\mathbf{r})_{n'n}$ and $(\cos\frac{1}{2}\boldsymbol{\kappa}.\mathbf{r})_{n'n}$. On the basis of the approximate expression (12.1) for the energy levels of a diatomic molecule the wave functions of the rotational states are the spherical harmonics $Y^K_{M_K}$ and the wave functions of the vibrational states are proportional to Hermite polynomials of degree ν. Hence the calculation of the matrix elements, and the total neutron cross-section is, at least in principle, quite straightforward. For a detailed comparison of theory with experiment over a wide energy range, and the diatomic molecule is really the only molecule for which a 'first principles' calculation is at all feasible, this is not sufficient because the rotational and vibrational motions are, of course, coupled. More explicitly, if one writes down the Schrödinger equation for a particle in an isotropic, harmonic oscillator potential the equation can be separated into an equation for the rotational motion, which has solutions $Y^K_{M_K}$ and a radial equation. The latter contains not only a potential of the form $(r-r_e)^2$ but also a centrifugal potential $K(K+1)/r^2$ and this equation has to be solved numerically.

Brinberg† has made a detailed numerical calculation of the total neutron cross-section for the scattering of unpolarized thermal and cold neutrons by hydrogen, taking into account the coupling of the rotational and vibrational motions. In addition he has compared numerical calculations for a rigid rotator, and also an anharmonic potential (Morse potential) model with experimental results. Before discussing these results we consider the cross-sections for the various allowed transitions for the particular case of the scattering of unpolarized neutrons by hydrogen.

The total spin nuclear spin I for hydrogen can have the values $I = 0$ and $I = 1$, para- and ortho-hydrogen, respectively. For $I = 0$, K must be an even integer and for $I = 1$ it must be odd. The allowed transitions are para $\rightarrow$ para and ortho $\rightarrow$ ortho, for both of which $K'-K = \Delta K$ is an even integer, and the transitions ortho $\rightarrow$ para and para $\rightarrow$ ortho, in which ΔK is an odd integer. The squared modulus of (12.18) averaged over the spin directions of the incident neutron is

$$|(\cos\tfrac{1}{2}\boldsymbol{\varkappa}.\mathbf{r})_{n'n}|^2\{4A^2+B^2\tfrac{1}{4}I(I+1)\}$$
$$= |(\cos\tfrac{1}{2}\boldsymbol{\varkappa}.\mathbf{r})_{n'n}|^2\{(3a^{(+)}+a^{(-)})^2+I(I+1)(a^{(+)}-a^{(-)})^2\}, \quad (12.20)$$

where we have introduced the *free scattering lengths* $a^{(\pm)} = \frac{1}{2}b^{(\pm)}$. From (12.20) the cross-sections for the para $\rightarrow$ para and ortho $\rightarrow$ ortho transitions are, respectively,

$$\frac{d\sigma_{n,n'}}{d\Omega} = \frac{k'}{k}\frac{4}{9}|(\cos\tfrac{1}{2}\boldsymbol{\varkappa}.\mathbf{r})_{n'n}|^2(3a^{(+)}+a^{(-)})^2 \quad (12.21\,\text{a})$$

and

$$\frac{d\sigma_{n,n'}}{d\Omega} = \frac{k'}{k}\frac{4}{9}|(\cos\tfrac{1}{2}\boldsymbol{\varkappa}.\mathbf{r})_{n'n}|^2\{(3a^{(+)}+a^{(-)})^2+2(a^{(+)}-a^{(-)})^2\}. \quad (12.21\,\text{b})$$

† S. Brinberg, thesis, Department of Mathematics, Royal Institute of Technology, Stockholm (1956).

The cross-sections of the para → ortho and ortho → para transitions involve the squared modulus of (12.19) averaged over the spin directions of the incident neutron. Denoting this average by a bar, we have

$$\begin{aligned}\overline{\{\hat{\boldsymbol{\sigma}}.(\hat{\mathbf{i}}_1-\hat{\mathbf{i}}_2)\}^2} &= \overline{(\hat{\boldsymbol{\sigma}}.\hat{\mathbf{i}}_1)^2}+\overline{(\hat{\boldsymbol{\sigma}}.\hat{\mathbf{i}}_2)^2}-2\overline{(\hat{\boldsymbol{\sigma}}.\hat{\mathbf{i}}_1)(\hat{\boldsymbol{\sigma}}.\hat{\mathbf{i}}_2)}\\ &= \hat{\mathbf{i}}_1^2+\hat{\mathbf{i}}_2^2-2\hat{\mathbf{i}}_1.\hat{\mathbf{i}}_2\\ &= 2\hat{\mathbf{i}}_1^2+2\hat{\mathbf{i}}_2^2-\hat{\mathbf{I}}^2\\ &= 3-I(I+1)\end{aligned}$$

(cf. p. 12). Thus for the para → ortho transition the cross-section is

$$\frac{d\sigma_{n,n'}}{d\Omega} = \frac{k'}{k}\frac{4}{3}|(\sin\tfrac{1}{2}\boldsymbol{\kappa}.\mathbf{r})_{n'n}|^2(a^{(+)}-a^{(-)})^2, \qquad (12.22\,\text{a})$$

and for the ortho → para transition

$$\frac{d\sigma_{n,n'}}{d\Omega} = \frac{k'}{k}\frac{4}{9}|(\sin\tfrac{1}{2}\boldsymbol{\kappa}.\mathbf{r})_{n'n}|^2(a^{(+)}-a^{(-)})^2. \qquad (12.22\,\text{b})$$

The cross-sections (12.22 a) and (12.22 b) illustrate a rule which says that transitions between molecular states that have different symmetries occur only in those processes that involve the incoherent scattering length B. For a diatomic molecule this rule follows immediately from the expressions (12.18) and (12.19). If the initial and final nuclear states in the matrix element (12.18) have different total nuclear spins then, obviously, the term containing only A is zero. The other terms in the cross-section, which involve B, contain the operators

$$\overline{\hat{\boldsymbol{\sigma}}.\hat{\mathbf{I}}} \quad \text{and} \quad \overline{(\hat{\boldsymbol{\sigma}}.\hat{\mathbf{I}})^2} = \hat{\mathbf{I}}^2-\overline{\hat{\boldsymbol{\sigma}}.\hat{\mathbf{I}}},$$

neither of which has non-zero matrix elements between states which differ in the total nuclear spin by two units. Clearly this is true for both polarized and unpolarized incident neutrons. Thus we conclude that there are no contributions to the cross-section from transitions that involve a change in the nuclear spin by 2, 4,... units of angular momentum, i.e. (12.18) gives a cross-section that involves only those transitions in which the total nuclear spin is unchanged. (For example, the

$$I = 0 \leftrightarrow I = 2$$

transition does not contribute to the cross-section for the scattering of neutrons from molecular deuterium.)

The equilibrium separation of the nuclei in the hydrogen molecule is $r_e = 0{\cdot}742$ Å, so that for cold neutrons the condition $|\frac{1}{2}\boldsymbol{\kappa}.\mathbf{r}| \ll 1$ holds. In this case the para → ortho and the ortho → para scattering processes have a negligible amplitude. Furthermore, if we set $\cos(\frac{1}{2}\boldsymbol{\kappa}.\mathbf{r}) \sim 1$ in

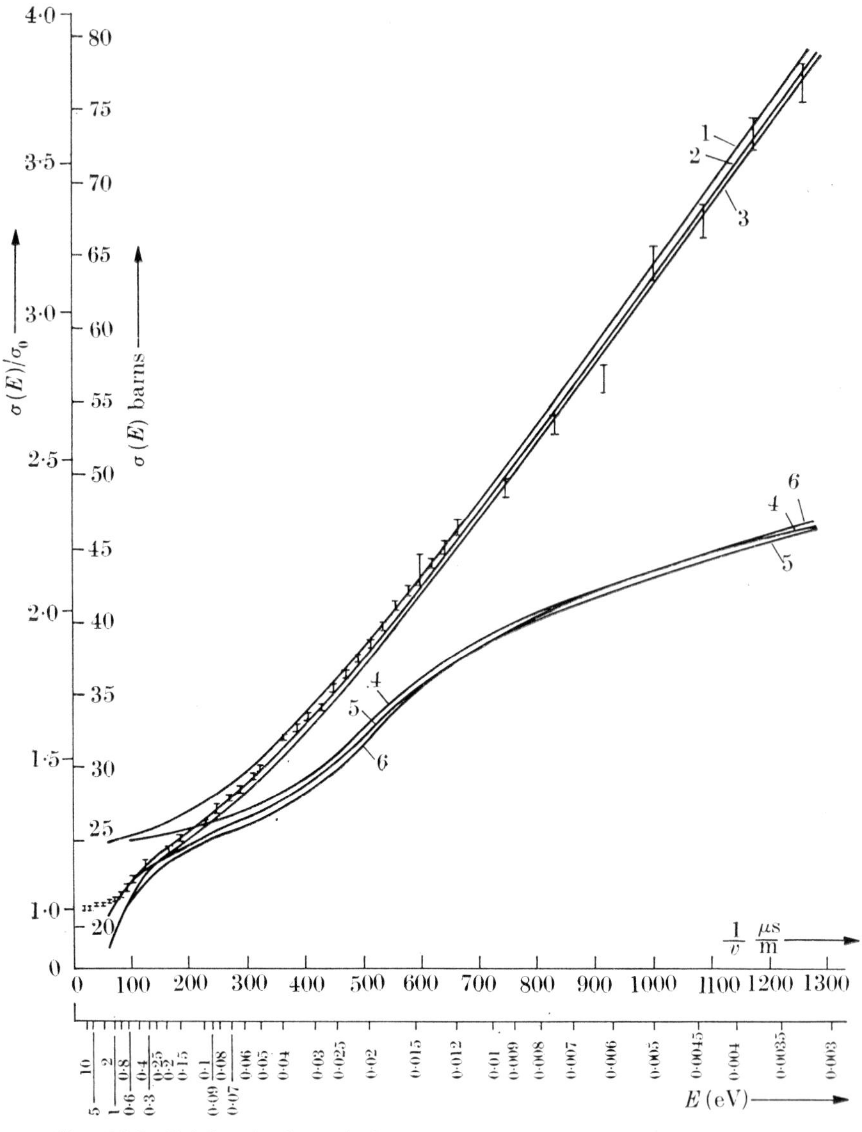

FIG. 12.2. Brinberg's theoretical values of the total scattering cross-section of slow neutrons by H_2 per hydrogen atom $\sigma_H(E_n)/\sigma_0$ or $\sigma_H(E_n)$ barns as a function of the neutron laboratory energy at the hydrogen gas temperatures $T = 300$ K (room temperature) and $T = 0$ K. The theoretical curves represent transitions in the vibrational ground state only. Free cross-section $\sigma_0 = 20{\cdot}36$ b.

1. The rigid rotator case: $T = 300$ K
2. The harmonic oscillator case: $T = 300$ K
3. The anharmonic oscillator case: $T = 300$ K
4. The rigid rotator case: $T = 0$ K
5. The harmonic oscillator case: $T = 0$ K
6. The anharmonic oscillator case: $T = 0$ K

I represent the experimental points of Melkonian (1949).

(12.21 a) and (12.21 b) then clearly the only non-zero matrix elements are those for which $n = n'$, i.e. the scattering is elastic in the centre of mass frame of reference.

In Fig. 12.2 we show the total neutron cross-section calculated by Brinberg for three models of the vibrational motion at both room temperature and absolute zero. Brinberg assumed that only the vibrational ground state need be considered and that the thermal average was to be taken only over the translational motion, i.e. the velocity $\mathbf{V}$ of the centre of mass is merely averaged over a Maxwell distribution $\exp(-\beta \frac{1}{2} M_{\text{mol}} V^2)$. The ultimate justification for these approximations is the agreement with the experimental results of Melkonian.† The agreement is clearly best for the isotropic, harmonic oscillator model. The deviation between the calculated curves and the measurements above $E = 0{\cdot}8$ eV is most probably due to the influence of the first vibrational state, since for hydrogen $\hbar\omega_e = 0{\cdot}54$ eV.

12.3. Perfect monatomic gas

The analysis given above for the scattering from a gas of diatomic molecules is not very suitable for a discussion of the cross-section for the scattering from other systems, particularly where thermal effects are to be taken into account. Also, experiments have been made to measure the partial differential cross-sections for several gases. For these reasons we examine in the remainder of the chapter the calculation of the cross-section for the scattering from molecular gases based on the correlation function approach introduced in Chapter 3.‡ In this section we calculate with this approach the cross-section for the simplest system, the perfect monatomic gas.

A perfect gas is one in which both the direct interaction between the constituent particles and the purely quantum mechanical effects which arise because all the particles are identical can be neglected. The direct interaction can be neglected if the mean occupation numbers of all the states are small compared to unity, i.e. the gas is sufficiently rarefied. This condition is satisfied if, for a given density ρ of the particles,

$$\rho(\hbar^2\beta/M)^{\frac{3}{2}} \ll 1.$$

If this is so, then at any moment of time no more than one particle is in fact in each quantum state. As a result one can also neglect the purely quantum mechanical exchange effects.

† E. Melkonian, *Phys. Rev.* **76**, 1750 (1949).
‡ A. C. Zemach and R. J. Glauber, *Phys. Rev.* **101**, 118 and 129 (1956).

For a system that can be regarded as a perfect gas of N nuclei† the interaction potential is

$$\frac{2\pi\hbar^2}{m}\, b \sum_l \delta(\mathbf{r}-\mathbf{R}_l)$$

and the Hamiltonian of the target system is simply

$$\mathscr{H}_{\mathrm{T}} = \frac{1}{2M}\sum_l \hat{\mathbf{p}}_l^2.$$

Hence, from Chapter 3,

$$\frac{\mathrm{d}^2\sigma}{\mathrm{d}\Omega \mathrm{d}E'} = N\frac{k'}{k}\, b^2 \frac{1}{2\pi\hbar}\int_{-\infty}^{\infty} \mathrm{d}t\, \mathrm{e}^{-\mathrm{i}\omega t}\langle \exp\{-\mathrm{i}\boldsymbol{\kappa}\,.\,\hat{\mathbf{R}}(0)\}\exp\{\mathrm{i}\boldsymbol{\kappa}\,.\,\hat{\mathbf{R}}(t)\}\rangle$$

$$= \frac{Nk'}{k}\, b^2 \frac{1}{2\pi\hbar}\int_{-\infty}^{\infty} \mathrm{d}t\, \mathrm{e}^{-\mathrm{i}\omega t} \exp\left\{\frac{-\kappa^2}{2M\beta}(t^2 - \mathrm{i}\hbar t\beta)\right\}$$

$$= N\frac{k'}{k}\, b^2\left(\frac{\beta M}{2\pi\hbar^2\kappa^2}\right)^{\frac{1}{2}} \exp\left\{-\frac{\beta M}{2\hbar^2\kappa^2}\left(\hbar\omega - \frac{\hbar^2\kappa^2}{2M}\right)^2\right\}. \tag{12.23}$$

The total cross-section can be expressed in terms of the probability function

$$\Phi(x) = (2/\sqrt{\pi})\int_0^x \mathrm{e}^{-t^2}\,\mathrm{d}t.$$

It is

$$\sigma = \frac{4\pi N|b|^2}{(1+\gamma)^2}\left\{\frac{\mathrm{e}^{-\alpha^2}}{\alpha\sqrt{\pi}} + \left(1+\frac{1}{2\alpha^2}\right)\Phi(\alpha)\right\}, \tag{12.24}$$

where $\alpha^2 = \gamma^{-1}\beta E = (M/m)\beta E$. This result was derived also in the example discussed in § 3.6, eqn (3.102).

In the limit $\alpha \ll 1$, which occurs for high temperatures or a very small incident neutron energy E, the total cross-section (12.24) becomes

$$\sigma \simeq \frac{4\pi N|b|^2}{(1+\gamma)^2}\frac{2}{\alpha\sqrt{\pi}}. \tag{12.25}$$

The cross-section diverges as α tends to zero because zero-energy neutrons are struck by moving molecules, and thus scattered even though there is no neutron flux. When the incident energy is very large, or alternatively when the temperature of the gas is very low, (12.24) becomes

$$\sigma \simeq \frac{4\pi N|b|^2}{(1+\gamma)^2}\left(1+\frac{1}{2\alpha^2}\right). \tag{12.26}$$

† The Fermi–Dirac and Bose–Einstein gases have been considered by L. Van Hove, *Phys. Rev.* **95**, 249 (1954).

12.4. Diatomic gas—partial differential cross-section

The partial differential cross-section (in the laboratory frame of reference) for the scattering from N identical diatomic molecules can be written in terms of a correlation function as

$$\frac{d^2\sigma}{d\Omega dE'} = N\frac{k'}{k}\frac{1}{2\pi\hbar}\int_{-\infty}^{\infty} dt\, e^{-i\omega t} \sum_{\xi=1,2} \overline{\langle \hat{b}\exp\{-i\boldsymbol{\varkappa}.\hat{\mathbf{R}}_{l\xi}(0)\}\hat{b}\exp\{i\boldsymbol{\varkappa}.\hat{\mathbf{R}}_{l\xi}(t)\}\rangle}, \tag{12.27}$$

where
$$\mathbf{R}_{l\xi} = \mathbf{l}+(\pm)^{\xi}\mathbf{r}/2$$

and the bar indicates an average over the orientations of the incident neutrons. This form of the cross-section follows immediately from Chapter 3. In (12.27) we have included only scattering from the same nuclei, the so-called direct scattering contribution (see below for a discussion of 'inner' and 'outer' scattering). If we now make the assumptions that the translational motion can be separated from the rotational and vibration motions, which are themselves taken to be independent of one another, and also that the molecule is initially in its vibrational ground state, then (12.27) becomes

$$\frac{d^2\sigma}{d\Omega dE'} = \frac{Nk'}{k}\frac{1}{2\pi\hbar}\int_{-\infty}^{\infty} dt\, e^{-i\omega t}\langle\exp\{-i\boldsymbol{\varkappa}.\hat{\mathbf{l}}(0)\}\exp\{i\boldsymbol{\varkappa}.\hat{\mathbf{l}}(t)\}\rangle\times$$
$$\times\left\{\sum_{\xi=1,2}\sum_{K} p_K \frac{1}{2K+1}\sum_{M_K}\sum_{K',M_{K'},\nu'} \exp\{it(E_{K'}-E_K)/\hbar\}\exp(i\omega_e \nu' t)\times\right.$$
$$\left.\times\overline{|\langle K'M_{K'}\nu'|\hat{b}\exp[(\pm)^{\xi}\tfrac{1}{2}i\boldsymbol{\varkappa}.\hat{\mathbf{r}}(0)]|KM_K\nu=0\rangle|^2}\right\}, \tag{12.28}$$

where $E_K = B_e K(K+1)$ and $p_K = \exp(-\beta E_K)\Big/\sum_K \exp(-\beta E_K)$. Let us first consider the calculation of the term in curly brackets, which we denote by $F(t)$. From our discussion in § 12.2, we must distinguish between the two cases in which $\Delta K = K'-K$ is an even integer and when it is an odd integer. In the latter case the total nuclear spin must change by one unit, whereas when ΔK is even, I remains unchanged. For unpolarized neutrons we therefore have

ΔK even:

$$F(t) = \{4A^2+B^2\tfrac{1}{4}I(I+1)\}\times$$
$$\times\sum_{K} p_K\frac{1}{2K+1}\sum_{M_K}\sum_{K',M_{K'},\nu'} \exp\{it(E_{K'}-E_K)/\hbar+i\omega_e\nu' t\}\times$$
$$\times|\langle K'M_{K'}\nu'|\cos\tfrac{1}{2}\boldsymbol{\varkappa}.\hat{\mathbf{r}}(0)|KM_K\nu=0\rangle|^2; \tag{12.29 a}$$

ΔK odd:

$$F(t) = \tfrac{1}{4}B^2\{4i(i+1)-I(I+1)\}\times$$
$$\times \sum_K p_K \frac{1}{2K+1} \sum_{M_K} \sum_{K',M_{K'},\nu'} \exp\{it(E_{K'}-E_K)/\hbar+\mathrm{i}\omega_\mathrm{e}\nu't\}\times$$
$$|\langle K'M_{K'}\nu'|\sin\tfrac{1}{2}\boldsymbol{\kappa}.\hat{\mathbf{r}}(0)|KM_K\nu=0\rangle|^2. \quad (12.29\,\mathrm{b})$$

We note that in (12.29 b) I is the total nuclear spin of the initial state of the nuclei.

To calculate the vibrational contribution to the matrix elements in F, let

$$\hat{\mathbf{r}}(0) = \mathbf{d}+\hat{\mathbf{u}}(0), \quad (12.30)$$

where $|\mathbf{d}| = r_\mathrm{e}$ and $\mathbf{u}$ is the displacement of the nuclei from their equilibrium separation. The motion of the two nuclei relative to each other is equivalent to a single, harmonically bound particle of mass $M/2$. Hence, from eqn (12.7),

$$\hat{u}(0) = \left(\frac{\hbar}{M\omega_\mathrm{e}}\right)^{\frac{1}{2}}(\hat{a}+\hat{a}^+)$$

and we require the matrix elements

$$\left\langle \nu'\middle|\exp\left\{\pm\frac{\mathrm{i}\kappa}{2}\eta\left(\frac{\hbar}{M\omega_\mathrm{e}}\right)^{\frac{1}{2}}(\hat{a}+\hat{a}^+)\right\}\middle|\nu=0\right\rangle, \quad (12.31)$$

where η is the cosine of the angle between $\boldsymbol{\kappa}$ and $\mathbf{r}$. Since

$$\exp\hat{A}\exp\hat{B} = \exp(\hat{A}+\hat{B}+\tfrac{1}{2}[\hat{A},\hat{B}]), \quad (12.32)$$

if $[\hat{A},\hat{B}]$ is a c-number, and

$$\exp(\pm\alpha\hat{a}^+)|\nu\rangle = \sum_{h=0}^{\infty}\left\{\frac{(\nu+h)!}{\nu!}\right\}^{\frac{1}{2}}\frac{(\pm\alpha)^h}{h!}|\nu+h\rangle,$$

we have

$$\langle\nu'|\exp\{\pm\alpha(\hat{a}+\hat{a}^+)\}|\nu=0\rangle$$
$$= \mathrm{e}^{-\frac{1}{2}\alpha^2}\langle\nu'|\exp(\pm\alpha\hat{a})\exp(\pm\alpha\hat{a}^+)\,|\nu=0\rangle$$
$$= \mathrm{e}^{-\frac{1}{2}\alpha^2}\left(\frac{1}{\nu'!}\right)^{\frac{1}{2}}\sum_{h=0}^{\infty}\frac{(\pm\alpha)^{\nu'+2h}}{h!} = \frac{(\pm\alpha)^{\nu'}}{(\nu'!)^{\frac{1}{2}}}\mathrm{e}^{\frac{1}{2}\alpha^2}.$$

So that, with $\alpha = \frac{1}{2}\mathrm{i}\kappa\eta(\hbar/\omega_\mathrm{e}M)^{\frac{1}{2}}$,

$$|\langle K'M_{K'}\nu'|\cos\tfrac{1}{2}\boldsymbol{\kappa}.\hat{\mathbf{r}}(0)|KM_K\nu=0\rangle|^2$$
$$= \frac{1}{\nu'!}|\langle K'M_{K'}|\exp(-\kappa^2\eta^2\hbar/8\omega_\mathrm{e}M)\times$$
$$\times\tfrac{1}{2}\{\exp(\tfrac{1}{2}\mathrm{i}\kappa\eta r_\mathrm{e})\alpha^{\nu'}+\exp(-\tfrac{1}{2}\mathrm{i}\kappa\eta r_\mathrm{e})(-\alpha)^{\nu'}\}|KM_K\rangle|^2$$
$$= \frac{1}{\nu'!}\left(\frac{\kappa^2\hbar}{4\omega_\mathrm{e}M}\right)^{\nu'}\left|\left\langle K'M_{K'}\middle|\eta^{\nu'}\exp\left(-\frac{\kappa^2\eta^2\hbar}{8M\omega_\mathrm{e}}+\frac{\mathrm{i}}{2}\kappa\eta r_\mathrm{e}\right)\middle|KM_K\right\rangle\right|^2. \quad (12.33)$$

The last line follows because ΔK is an even integer. The result for the corresponding matrix element of $\sin\frac{1}{2}\boldsymbol{\kappa}.\hat{\mathbf{r}}(0)$ is the same.

The rotational wave functions of a diatomic molecule are spherical harmonics $Y^K_{M_K}$. Thus

$$\left\langle K'M_{K'}\middle|\eta^{\nu'}\exp\left(-\frac{\kappa^2\eta^2\hbar}{8M\omega_e}+\frac{i}{2}\kappa\eta r_e\right)\middle|KM_K\right\rangle$$

$$=\int_0^{2\pi}d\phi\int_{-1}^{1}d\eta\; Y^{*K'}_{M_{K'}}\eta^{\nu'}\exp\left(-\frac{\kappa^2\eta^2\hbar}{8M\omega_e}+\frac{i}{2}\kappa\eta r_e\right)Y^K_{M_K}$$

$$=(-)^{M_{K'}}\sum_{l,m}\left\{\frac{(2K+1)(2K'+1)(2l+1)}{4\pi}\right\}^{\frac{1}{2}}\begin{pmatrix}K' & K & l\\ -M_{K'} & M_K & m\end{pmatrix}\times$$

$$\times\begin{pmatrix}K' & K & l\\ 0 & 0 & 0\end{pmatrix}\int_0^{2\pi}d\phi\int_{-1}^{1}d\eta\; Y^{*l}_{m}\eta^{\nu'}\exp\left(-\frac{\kappa^2\eta^2\hbar}{8M\omega_e}+\frac{i}{2}\kappa\eta r_e\right)$$

$$=(-)^{M_{K'}}\{(2K'+1)(2K+1)\}^{\frac{1}{2}}\sum_l(2l+1)\times$$

$$\times\begin{pmatrix}K' & K & l\\ -M_{K'} & M_K & 0\end{pmatrix}\begin{pmatrix}K' & K & l\\ 0 & 0 & 0\end{pmatrix}\mathscr{A}_{l\nu'}, \tag{12.34}$$

where
$$\mathscr{A}_{l\nu'}=\frac{1}{2}\int_{-1}^{1}d\eta\; P_l(\eta)\eta^{\nu'}\exp\left(-\frac{\kappa^2\eta^2\hbar}{8M\omega_e}+\frac{i}{2}\kappa\eta r_e\right) \tag{12.35}$$

and P_l is the Legendre polynomial of order l. In deriving (12.34) we have used an expansion of the product of two spherical harmonics in terms of $3j$ symbols,† namely

$$Y^{*l_1}_{m_1}Y^{l_2}_{m_2}=(-)^{m_1}\sum_{l,m}\left\{\frac{(2l_1+1)(2l_2+1)(2l+1)}{4\pi}\right\}^{\frac{1}{2}}\times$$

$$\times\begin{pmatrix}l_1 & l_2 & l\\ -m_1 & m_2 & m\end{pmatrix}Y^{*l}_{m}\begin{pmatrix}l_1 & l_2 & l\\ 0 & 0 & 0\end{pmatrix}$$

and the second $3j$ symbol is non-zero only if l_1+l_2+l is an even integer. The integral over ϕ in the second line is non-zero if $m=0$ and

$$Y^l_0=\left(\frac{2l+1}{4\pi}\right)^{\frac{1}{2}}P_l.$$

† We remind the reader that our definitions follow the convention adopted by Edmonds; all the relationships used in this and subsequent chapters are contained in this book. § 6.3 (a) contains a summary of $3j$ symmetry properties.

From (12.34),

$$\frac{1}{2K+1}\sum_{M_K,M_{K'}}\left|\left\langle K'M_{K'}\middle|\eta^{\nu'}\exp\left(-\frac{\eta^2\kappa^2\hbar}{8M\omega_{\mathrm{e}}}+\frac{\mathrm{i}}{2}\kappa\eta r_{\mathrm{e}}\right)\middle|KM_K\right\rangle\right|^2$$

$$= (2K'+1)\sum_{M_K,M_{K'}}\sum_{l,l'}(2l+1)(2l'+1)\begin{pmatrix}K' & K & l\\ -M_{K'} & M_K & 0\end{pmatrix}\times$$

$$\times\begin{pmatrix}K' & K & l\\ 0 & 0 & 0\end{pmatrix}\begin{pmatrix}K' & K & l'\\ -M_{K'} & M_K & 0\end{pmatrix}\begin{pmatrix}K' & K & l'\\ 0 & 0 & 0\end{pmatrix}\mathscr{A}_{l\nu'}\mathscr{A}^*_{l'\nu'}$$

$$= (2K'+1)\sum_{l=|K'-K|}^{K'+K}(2l+1)\begin{pmatrix}K' & K & l\\ 0 & 0 & 0\end{pmatrix}^2|\mathscr{A}_{l\nu'}|^2 \tag{12.36}$$

because of the orthogonality relation

$$\sum_{m_1,m_2}\begin{pmatrix}j_1 & j_2 & j_3\\ m_1 & m_2 & m_3\end{pmatrix}\begin{pmatrix}j_1 & j_2 & j_3'\\ m_1 & m_2 & m_3'\end{pmatrix}$$

$$= \frac{1}{2j_3+1}\delta_{j_3,j_3'}\delta_{m_3,m_3'}\Delta(j_1 j_2 j_3), \tag{12.37}$$

where $\Delta(j_1 j_2 j_3) = 1$ if j_1, j_2, j_3 satisfy the triangular condition, and is zero otherwise. This completes the calculation of $F(t)$.

The contribution to the cross-section (12.28) from the translation motion is, from the second line of (12.23),

$$\exp\left\{-\frac{\kappa^2}{4M\beta}(t^2-\mathrm{i}\hbar t\beta)\right\},$$

and the resulting time integral in (12.28) from this term and $F(t)$ is of the same form as for the scattering from free nuclei. The final expressions for the partial differential cross-sections for the two cases in which ΔK is an even integer ($\Delta I = 0$) and when ΔK is an odd integer ($\Delta I = \pm 1$) can be written

$$\frac{\mathrm{d}^2\sigma}{\mathrm{d}\Omega\mathrm{d}E'} = \begin{Bmatrix}[4A^2+B^2\tfrac{1}{4}I(I+1)], \ \Delta K \text{ even}\\ \tfrac{1}{4}B^2[4i(i+1)-I(I+1)], \ \Delta K \text{ odd}\end{Bmatrix}N\frac{k'}{k}\left(\frac{\beta M}{\pi\hbar^2\kappa^2}\right)^{\frac{1}{2}}\times$$

$$\times\sum_{K',K}p_K(2K'+1)\sum_{\nu'}\left(\frac{\hbar\kappa^2}{4M\omega_{\mathrm{e}}}\right)^{\nu'}\frac{1}{\nu'!}\times$$

$$\times\exp\left\{-\frac{\beta M}{\hbar^2\kappa^2}\left(E_{K'}-E_K+\nu'\hbar\omega_{\mathrm{e}}+\frac{\hbar^2\kappa^2}{4M}-\hbar\omega\right)^2\right\}\times$$

$$\times\sum_{l=|K'-K|}^{K'+K}(2l+1)\begin{pmatrix}K' & K & l\\ 0 & 0 & 0\end{pmatrix}^2|\mathscr{A}_{l\nu'}|^2. \tag{12.38}$$

If, as may well be the case, the incident neutron energy is less than $\hbar\omega_e$, the vibrational states need not be considered; i.e. we can put $\nu' = 0$ in (12.38) and use

$$\mathscr{A}_{0l} = \tfrac{1}{2}\int_{-1}^{1} \mathrm{d}\eta\, \exp(\tfrac{1}{2}\mathrm{i}\kappa r_e\,\eta)P_l(\eta) = \mathrm{i}^l j_l(\tfrac{1}{2}\kappa r_e) \tag{12.39 a}$$

$$\simeq \mathrm{i}^l\,\frac{(\kappa r_e/2)^l}{(2l+1)!!}, \tag{12.39 b}$$

if in addition $\frac{1}{2}\kappa r_e \ll 1$. In the latter case only the first few terms in the summation over l need be retained. The $l = 0$ term is non-zero only if $K+K'$ is an even integer, which requires ΔK to be even, cf. eqn (6.23).

As a particular example of (12.38) let us consider, as in § 12.2, the scattering from hydrogen. Here $\hbar\omega_e = 0{\cdot}54$ eV so that we can usually ignore the vibrational levels. For the scattering from para-hydrogen ($I = 0$) we then have

$$\frac{\mathrm{d}^2\sigma}{\mathrm{d}\Omega\mathrm{d}E'} = N\frac{k'}{k}\left(\frac{\beta m}{\pi\hbar^2\kappa^2}\right)^{\frac{1}{2}} \sum_{K=0,2,4,\ldots} p_K\Bigg[(3a^{(+)}+a^{(-)})^2 \sum_{K'=0,2,4,\ldots}(2K'+1)\times$$
$$\times\exp\left\{-\frac{\beta m}{\hbar^2\kappa^2}\left(E_{K'}-E_K+\frac{\hbar^2\kappa^2}{4m}-\hbar\omega\right)^2\right\}\times$$
$$\times\sum_{l=|K'-K|}^{K'+K}(2l+1)\begin{pmatrix}K' & K & l\\ 0 & 0 & 0\end{pmatrix}^2 j_l^2(\kappa r_e/2)+$$
$$+3(a^{(+)}-a^{(-)})^2\sum_{K'=1,3,5,\ldots}(2K'+1)\exp\left\{-\frac{\beta m}{\hbar^2\kappa^2}\left(E_{K'}-E_K+\frac{\hbar^2\kappa^2}{4m}-\hbar\omega\right)^2\right\}\times$$
$$\times\sum_{l=|K'-K|}^{K'+K}(2l+1)\begin{pmatrix}K' & K & l\\ 0 & 0 & 0\end{pmatrix}^2 j_l^2(\kappa r_e/2)\Bigg], \tag{12.40}$$

and for the scattering from ortho-hydrogen ($I = 1$),

$$\frac{\mathrm{d}^2\sigma}{\mathrm{d}\Omega\mathrm{d}E'} = N\frac{k'}{k}\left(\frac{\beta m}{\pi\hbar^2\kappa^2}\right)^{\frac{1}{2}}\times$$
$$\times\sum_{K=1,3,5,\ldots} p_K\Bigg[\{(3a^{(+)}+a^{(-)})^2+2(a^{(+)}-a^{(-)})^2\}\sum_{K'=1,3,5,\ldots}(2K'+1)\times$$
$$\times\exp\left\{-\frac{\beta m}{\hbar^2\kappa^2}\left(E_{K'}-E_K+\frac{\hbar^2\kappa^2}{4m}-\hbar\omega\right)^2\right\}\times$$
$$\times\sum_{l=|K'-K|}^{K'+K}(2l+1)\begin{pmatrix}K' & K & l\\ 0 & 0 & 0\end{pmatrix}^2 j_l^2(\kappa r_e/2)+$$
$$+(a^{(+)}-a^{(-)})^2\sum_{K'=0,2,4,\ldots}(2K'+1)\exp\left\{-\frac{\beta m}{\hbar^2\kappa^2}\left(E_{K'}-E_K+\frac{\hbar^2\kappa^2}{4m}-\hbar\omega\right)^2\right\}\times$$
$$\times\sum_{l=|K'-K|}^{K'+K}(2l+1)\begin{pmatrix}K' & K & l\\ 0 & 0 & 0\end{pmatrix}^2 j_l^2(\kappa r_e/2)\Bigg]. \tag{12.41}$$

Fig. 12.3 shows the partial differential cross-section calculated from eqn (12.38) for 1·0 eV incident neutrons scattered from ortho- and para-hydrogen. The structure due to the rotational levels is clearly visible superimposed on the contributions from the $\nu' = 0$ and $\nu' = 1$ vibrational contributions. The structure is more pronounced for para-hydrogen than ortho-hydrogen because in the former case only the odd

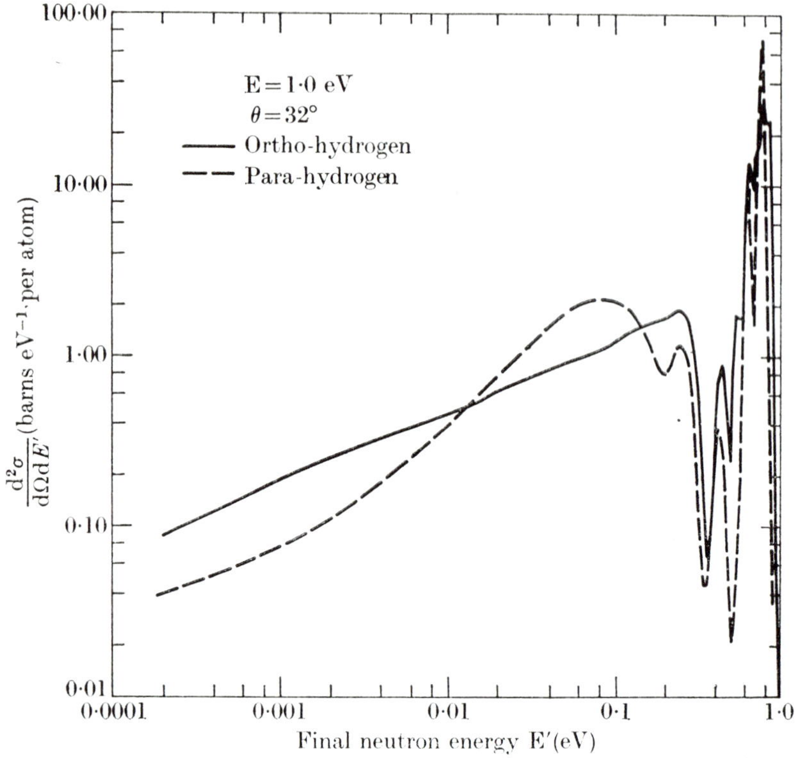

FIG. 12.3. Partial differential cross-section for the scattering of $E = 1{\cdot}0$ eV neutrons at $\theta = 32°$ from ortho- and para-hydrogen at 20·4 K (after J. A. Young and J. U. Koppel, *Nukleonik* **8**, 40 (1966)).

rotational levels make a significant contribution, whereas all levels contribute to the ortho-hydrogen cross-section. Only the $\nu' = 0$ vibrational contribution is involved at much lower incident neutron energies, as is clear from Fig. 12.4 where $E = 0{\cdot}22$ eV. A comparison of Figs. 12.4 (a) and 12.4 (b) shows that the structure in the cross-section is much less pronounced at higher scattering angles, which is a consequence of the larger recoil energy involved at higher scattering angles. Squires and Stuart† have measured the total cross-sections for para- and ortho-

† G. L. Squires and A. T. Stewart, *Proc. R. Soc.* **A230**, 19 (1955).

hydrogen at 20·4 K and their results are shown in Fig. 12.5 together with the total cross-sections calculated on the basis of eqn (12.34). Note the rapid increase in the total cross-section for para-hydrogen as E approaches 0·02 eV, which is the energy required in the laboratory frame

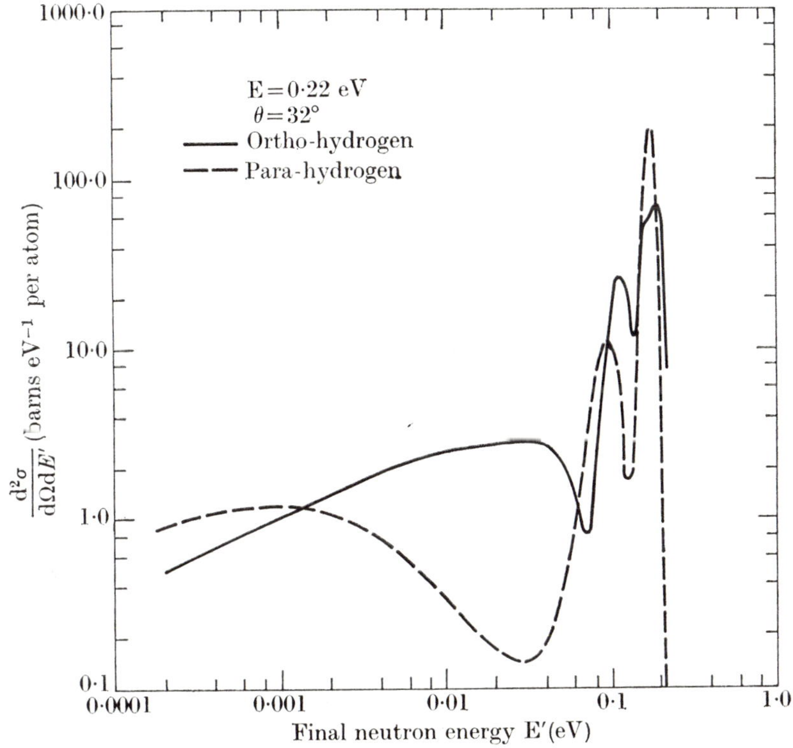

FIG. 12.4 (a). Partial differential cross-section for the scattering from ortho- and para-hydrogen at 20·4 K; including vibrational motion. The incident neutron energy is 0·22 eV and $\theta = 32°$ (after Young and Koppel).

of reference to cause the $K = 0$ to $K = 1$ rotational transition from para-hydrogen to ortho-hydrogen.

12.5. General expression for the partial differential cross-section

The cross-section given by eqn (12.27) is a particular case of the general expression for the cross-section for the scattering from N identical molecules,

$$\frac{d^2\sigma}{d\Omega dE'} = \frac{k'}{k}\frac{1}{2\pi\hbar}\int_{-\infty}^{\infty} dt\, e^{-i\omega t} \sum_{l,l'} \sum_{v,v'} \overline{\langle \hat{b}_{lv} \exp\{-i\boldsymbol{\kappa}.\hat{\mathbf{R}}_{lv}(0)\}\hat{b}_{l'v'} \exp\{i\boldsymbol{\kappa}.\hat{\mathbf{R}}_{l'v'}(t)\}\rangle}, \tag{12.42}$$

where $\mathbf{R}_{lv}$ is the position vector of the vth nucleus in the lth molecule. As usual, the angular brackets in (12.42) denote a thermal average. The cross-section must also be averaged over the orientations of the incident neutrons and also over the orientations of the molecule (the molecules do not, clearly, have any preferred orientation in space) and these are denoted collectively by a bar over the correlation function in (12.42).

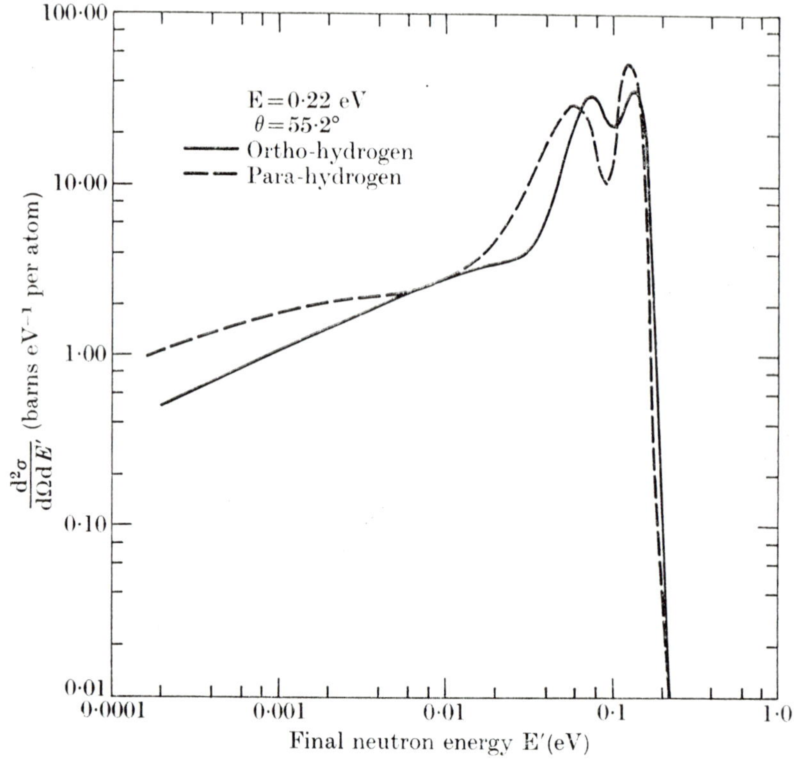

FIG. 12.4 (b). Partial differential cross-section for the scattering from ortho- and para-hydrogen at 20·4 K, including vibrational motion. The incident neutron energy is 0·22 eV and $\theta = 55°$ (after Young and Koppel).

In our discussion of the diatomic molecule we found that the nuclear spin and coordinates were not independent. The same is true, of course, in (12.42) and strictly speaking the wave functions used to evaluate the matrix elements must be symmetrized, i.e. for polyatomic molecules that contain two or more identical nuclei their spatial wave functions must be either symmetric or antisymmetric with respect to the interchange of two of the identical nuclei, depending on whether these have integral or half-integral spins. However, we know on quite general grounds that the purely quantum mechanical effect of spin correlation is negligible at high temperatures and for nuclei of mass $M \gg m$; we shall, therefore, assume that the matrix elements in (12.42) can be evaluated with unsymmetrized wave functions.†

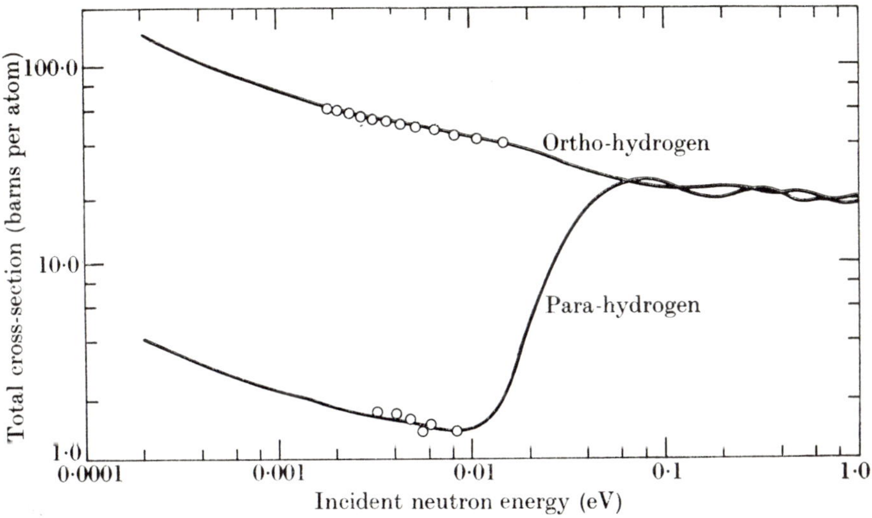

FIG. 12.5. Total cross-sections for para- and ortho-hydrogen at 20·4 K (calculated cross-section after Young and Koppel).

With this approximation the cross-section can be averaged over the orientations of the unpolarized neutrons, i.e.

$$\overline{\hat{b}_{lv}\hat{b}_{l'v'}} = A_{lv}A_{l'v'} + \delta_{l,l'}\delta_{v,v'}B^2_{lv}\tfrac{1}{4}i_{lv}(i_{lv}+1), \tag{12.43}$$

so that the neutron cross-section can now be written in the form

$$\frac{\mathrm{d}^2\sigma}{\mathrm{d}\Omega\mathrm{d}E'} = \frac{k'}{k}\frac{1}{2\pi\hbar}\int_{-\infty}^{\infty}\mathrm{d}t\,\mathrm{e}^{-\mathrm{i}\omega t}\Big[\sum_{l,v}\{A^2_{lv} + B^2_{lv}\tfrac{1}{4}i_{lv}(i_{lv}+1)\}\overline{\langle\Phi(lv;lv)\rangle} +$$
$$+\sum_{l}\sum_{v,v'}{}' A_{lv}A_{lv'}\overline{\langle\Phi(lv;lv')\rangle} + \sum_{l,l'}{}'\sum_{v}A_{lv}A_{l'v}\overline{\langle\Phi(lv;l'v)\rangle}\Big]. \tag{12.44}$$

The primes on the summation signs mean that the terms $l = l'$ or $v = v'$ are omitted from the sum, and we have defined

$$\Phi(lv;l'v') = \exp\{-\mathrm{i}\boldsymbol{\kappa}.\hat{\mathbf{R}}_{lv}(0)\}\exp\{\mathrm{i}\boldsymbol{\kappa}.\hat{\mathbf{R}}_{l'v'}(t)\}. \tag{12.45}$$

The bar over the correlation functions in (12.44) now denotes the average

† A detailed analysis of the effect of nuclear spin correlations on the cross-section has been made by S. K. Sinha and G. Venkataraman, *Phys. Rev.* **149**, 1 (1966). For methane they find that at 300 K nuclear spin correlations alter the cross-section by less than 0·7 per cent.

over the orientations of the molecules. The three terms in (12.44) are usually called the 'direct', 'inner', and 'outer' scattering terms, respectively, for obvious reasons. Because we have assumed that the molecules are non-interacting, the 'outer' term is independent of time and therefore gives purely elastic scattering.

To calculate the contributions to (12.45) from the translational, vibrational, and rotational motions of the nuclei, which we take to be independent of each other, we write

$$\hat{\mathbf{R}}_{lv}(t) = \hat{\mathbf{l}}(t)+\hat{\mathbf{d}}_v(t)+\hat{\mathbf{u}}_v(t), \tag{12.46}$$

where $\mathbf{d}_v$ denotes equilibrium positions of the nuclei relative to the centre of mass of the molecule and $\mathbf{u}_v$ are their displacements from the equilibrium configuration. The total Hamiltonian of the target system is then the sum of N single-molecule Hamiltonians, each of which we take to be

$$\mathscr{H}_{\mathrm{t}}+\mathscr{H}_{\mathrm{r}}+\mathscr{H}_{\mathrm{v}}, \tag{12.47}$$

where $\mathscr{H}_{\mathrm{t}}$ is the Hamiltonian of the translational motion of the centre of mass, and $\mathscr{H}_{\mathrm{r}}$ and $\mathscr{H}_{\mathrm{v}}$ the Hamiltonians that describe the rotational and vibrational motions of the nuclei. Obviously $\hat{\mathbf{l}}$, $\hat{\mathbf{d}}_v$, and $\hat{\mathbf{u}}_v$ do not commute with $\mathscr{H}_{\mathrm{t}}$, $\mathscr{H}_{\mathrm{r}}$, and $\mathscr{H}_{\mathrm{v}}$, respectively, but in addition $\hat{\mathbf{u}}_v$ does not commute with $\mathscr{H}_{\mathrm{r}}$ because the vectors $\mathbf{u}_v$ rotate with the molecule.

With (12.46) we can write

$$\langle\Phi(lv;l'v')\rangle$$
$$= \langle\exp\{-\mathrm{i}\boldsymbol{\kappa}.\hat{\mathbf{l}}(0)\}\exp\{\mathrm{i}\boldsymbol{\kappa}.\hat{\mathbf{l}}'(t)\}\rangle\overline{\langle\exp\{-\mathrm{i}\boldsymbol{\kappa}.\hat{\mathbf{d}}_v(0)\}\hat{O}_{vv'}(t)\exp\{\mathrm{i}\boldsymbol{\kappa}.\hat{\mathbf{d}}_{v'}(t)\}\rangle}, \tag{12.48 a}$$

where we have defined

$$\hat{O}_{vv'}(t) = \exp\{-\mathrm{i}\boldsymbol{\kappa}.\hat{\mathbf{u}}_v(0)\}\exp\{\mathrm{i}\boldsymbol{\kappa}.\hat{\mathbf{u}}_{v'}(t)\}. \tag{12.48 b}$$

12.6. Gas of isotropic harmonic oscillators

If we consider a single nucleus bound harmonically in a molecule of mass M_{mol}, then

$$\mathscr{H}_{\mathrm{t}}+\mathscr{H}_{\mathrm{v}} = \tfrac{1}{2}M_{\mathrm{mol}}(\dot{\mathbf{l}})^2+\tfrac{1}{2}M(\dot{\mathbf{u}})^2+\tfrac{1}{2}M\omega_0^2\,\mathbf{u}^2 \tag{12.49}$$

and

$$\langle\Phi(lv;l'v')\rangle$$
$$= \delta_{v,v'}\langle\exp\{-\mathrm{i}\boldsymbol{\kappa}.\hat{\mathbf{l}}(0)\}\exp\{\mathrm{i}\boldsymbol{\kappa}.\hat{\mathbf{l}}'(t)\}\rangle\langle\exp\{-\mathrm{i}\boldsymbol{\kappa}.\hat{\mathbf{u}}_v(0)\}\exp\{\mathrm{i}\boldsymbol{\kappa}.\hat{\mathbf{u}}_v(t)\}\rangle. \tag{12.50}$$

From (12.23) we have immediately that

$$\langle\exp\{-\mathrm{i}\boldsymbol{\kappa}.\hat{\mathbf{l}}(0)\}\exp\{\mathrm{i}\boldsymbol{\kappa}.\hat{\mathbf{l}}'(t)\}\rangle$$
$$= \exp[-\mathrm{i}\boldsymbol{\kappa}.\{\mathbf{l}(0)-\mathbf{l}'(0)\}]\exp\left\{\frac{-\kappa^2}{2\beta M_{\mathrm{mol}}}(t^2-\mathrm{i}\hbar t\beta)\right\}. \tag{12.51}$$

To evaluate the term involving the displacements $\mathbf{u}$ we expand $\hat{\mathbf{u}}(t)$ in terms of Bose operators $\hat{a}$ and $\hat{a}^+$ in analogy with the expansion used in the introduction, eqn (12.7). Thus

$$\hat{u}_v^{(\alpha)}(t) = \left(\frac{\hbar}{2M\omega_0}\right)^{\frac{1}{2}} (\hat{a}\,\mathrm{e}^{-\mathrm{i}\omega_0 t} + \hat{a}^+\,\mathrm{e}^{\mathrm{i}\omega_0 t}). \tag{12.52}$$

Since the required thermal average can be written as the product of the thermal average of the x, y, and z components, all of which are identical in form, we need consider only one such term and omit the superscript α on $\hat{u}_v^{(\alpha)}$ and $\kappa^{(\alpha)}$.

On applying the identity (12.32) to $\hat{u}_v(0)$ and $\hat{u}_v(t)$,

$$\begin{aligned}
&\langle \exp\{-\mathrm{i}\kappa\hat{u}_v(0)\}\exp\{\mathrm{i}\kappa\hat{u}_v(t)\}\rangle \\
&\quad = \exp\{\tfrac{1}{2}\kappa^2[\hat{u}_v(0), \hat{u}_v(t)]\}\langle \exp[-\mathrm{i}\kappa\{\hat{u}_v(0)-\hat{u}_v(t)\}]\rangle \\
&\quad = \exp\{\tfrac{1}{2}\kappa^2[\hat{u}_v(0), \hat{u}_v(t)]\}\exp[-\tfrac{1}{2}\kappa^2\langle\{\hat{u}_v(0)-\hat{u}_v(t)\}^2\rangle] \\
&\quad = \exp[-\kappa^2\{\langle\hat{u}_v^2\rangle - \langle\hat{u}_v(0)\hat{u}_v(t)\rangle\}].
\end{aligned} \tag{12.53}$$

In obtaining the second line of (12.53) we have made use of Bloch's theorem for the average of an exponential whose argument is a function of harmonic oscillator coordinates (cf. § 3.7). Now

$$\begin{aligned}
\langle\hat{u}(0)\hat{u}(t)\rangle &= \frac{\hbar}{2M\omega_0}\langle\hat{a}\hat{a}^+\,\mathrm{e}^{\mathrm{i}\omega_0 t} + \hat{a}^+\hat{a}\,\mathrm{e}^{-\mathrm{i}\omega_0 t}\rangle \\
&= \frac{\hbar}{2M\omega_0}\{(1+\langle\hat{a}^+\hat{a}\rangle)\mathrm{e}^{\mathrm{i}\omega_0 t} + \langle\hat{a}^+\hat{a}\rangle\mathrm{e}^{-\mathrm{i}\omega_0 t}\},
\end{aligned} \tag{12.54 a}$$

where

$$\langle\hat{a}^+\hat{a}\rangle = \{\exp(\hbar\omega_0\beta)-1\}^{-1}. \tag{12.54 b}$$

Thus

$$\begin{aligned}
&\exp[-\kappa^2\{\langle\hat{u}_v^2\rangle - \langle\hat{u}_v(0)\hat{u}_v(t)\rangle\}] \\
&= \exp\left[-\frac{\hbar\kappa^2}{2M\omega_0}\{2\langle\hat{a}^+\hat{a}\rangle+1\}\right]\exp\left[\frac{\hbar\kappa^2}{2M\omega_0}\{\mathrm{e}^{\mathrm{i}\omega_0 t}(\langle\hat{a}^+\hat{a}\rangle+1)+\mathrm{e}^{-\mathrm{i}\omega_0 t}\langle\hat{a}^+\hat{a}\rangle\}\right] \\
&= \exp\left\{-\frac{\hbar\kappa^2}{2M\omega_0}\coth(\tfrac{1}{2}\hbar\omega_0\beta)\right\}\sum_{n=-\infty}^{\infty}\exp(\tfrac{1}{2}n\hbar\omega_0\beta + in\omega_0 t)\times \\
&\qquad\qquad \times I_n\left\{\frac{\hbar\kappa^2}{2M\omega_0}\operatorname{cosech}(\tfrac{1}{2}\hbar\omega_0\beta)\right\}.
\end{aligned} \tag{12.55}$$

In (12.55) we have introduced the expansion

$$\exp\{\tfrac{1}{2}y(t+t^{-1})\} = \sum_{n=-\infty}^{\infty} t^n I_n(y),$$

where $I_n(y)$ is the modified Bessel function of the first kind.

If we now substitute (12.51) and (12.55) in (12.44), we see that the 'outer' term contains the function

$$\sum_{l,l'}{}' \exp[\mathrm{i}\boldsymbol{\kappa}.(\mathbf{l}'-\mathbf{l})]. \tag{12.56}$$

This can be approximately evaluated by assuming the neighbours of any given molecule to be smeared over a sphere of radius p; (12.56) can then be represented by

$$Nj_0(\kappa p). \tag{12.57}$$

It now remains only to perform the straightforward time integral in (12.42) to obtain the partial differential cross-section for the scattering from N molecules each containing one harmonically bound scattering nucleus of spin i and mass M,

$$\frac{\mathrm{d}^2\sigma}{\mathrm{d}\Omega\mathrm{d}E'} = N\{A^2+\tfrac{1}{4}B^2i(i+1)+A^2j_0(\kappa p)\}\frac{k'}{k}\exp\left\{-\frac{\hbar\kappa^2}{2M\omega_0}\coth(\tfrac{1}{2}\hbar\omega_0\beta)\right\}\times$$
$$\times\left(\frac{\beta M_{\mathrm{mol}}}{2\pi\hbar^2\kappa^2}\right)^{\frac{1}{2}}\sum_{n=-\infty}^{\infty} I_n\left\{\frac{\hbar\kappa^2}{2M\omega_0}\operatorname{cosech}(\tfrac{1}{2}\hbar\omega_0\beta)\right\}\times$$
$$\times\exp\left\{-\frac{\beta M_{\mathrm{mol}}}{2\hbar^2\kappa^2}\left(\hbar\omega-\frac{\hbar^2\kappa^2}{2M_{\mathrm{mol}}}-\hbar\omega_0 n\right)^2+\tfrac{1}{2}n\hbar\omega_0\beta\right\}. \tag{12.58}$$

The integer n is the number of units of energy $\hbar\omega_0$ lost ($n > 0$) or gained ($n < 0$) by the neutron in a given term in the sum and

$$\exp(\tfrac{1}{2}n\hbar\omega_0\beta)I_n\left\{\frac{\hbar\kappa^2}{2M\omega_0}\operatorname{cosech}(\tfrac{1}{2}\hbar\omega_0\beta)\right\}$$

is the statistical weight associated with the nth order process. The processes in which the neutron loses energy are more strongly weighted than those in which it gains energy ($I_n(x) = I_{-n}(x)$), as one would expect, and in the limit $\beta\to\infty$ only those terms for which $n \geqslant 0$ remain. If the nucleus is bound very tightly to the molecule ω_0 is large. As ω_0 increases the only non-zero term is that for $n = 0$, i.e. the neutron does not have sufficient energy to excite any of the harmonic oscillator energy levels.

We now calculate the total neutron cross-section at absolute zero. It is not difficult to show that in the limit $\beta\to\infty$, (12.58) reduces to

$$\frac{\mathrm{d}^2\sigma}{\mathrm{d}\Omega\mathrm{d}E'} = N\{A^2+\tfrac{1}{4}B^2i(i+1)+A^2j_0(\kappa p)\}\exp(-\hbar\kappa^2/2M\omega_0)\times$$
$$\times\frac{k'}{k}\sum_{n=0}^{\infty}\frac{1}{n!}\left(\frac{\hbar\kappa^2}{2M\omega_0}\right)^n\delta\left(\hbar\omega-\hbar n\omega_0-\frac{\hbar^2\kappa^2}{2M_{\mathrm{mol}}}\right). \tag{12.59}$$

If we drop the 'outer' scattering term, and integrate (12.59) over angles, we obtain

$$\frac{\mathrm{d}\sigma}{\mathrm{d}E'} = N\{A^2+\tfrac{1}{4}B^2i(i+1)\}\frac{2\pi M_{\text{mol}}}{\hbar^2k^2}\sum_{n=0}^{\infty}\frac{1}{n!}\left(\frac{\omega-n\omega_0}{\gamma\omega_0}\right)^n \times$$

$$\times \exp\left\{-\frac{1}{\gamma\omega_0}(\omega-n\omega_0)\right\}, \quad (12.60)$$

where $\gamma = M/M_{\text{mol}}$, and the energy satisfies the condition

$$\frac{\hbar^2}{2M_{\text{mol}}}(k-k')^2 \leqslant \hbar\omega - n\hbar\omega_0 \leqslant \frac{\hbar^2}{2M_{\text{mol}}}(k+k')^2. \quad (12.61)$$

The integration of (12.60) over the final energy of the neutron, subject to the condition given by (12.61), is very similar to that for the scattering of a neutron by a single free nucleus, which we discussed in § 1.6. By an argument analogous to that given in § 1.6, we find that (12.61) is equivalent to

$$\frac{-\gamma'k+\surd\{k^2-(2mn\omega_0/\hbar)(1+\gamma')\}}{1+\gamma'} \leqslant k' \leqslant \frac{\gamma'k+\surd\{k^2-(2mn\omega_0/\hbar)(1+\gamma')\}}{1+\gamma'}, \quad (12.62)$$

where $\gamma' = m/M_{\text{mol}}$.

Using (12.62) it is straightforward to obtain the total cross-section, which can be written as

$$\sigma = N\sigma_{\text{b}}\left(\frac{\hbar\omega_0}{4E}\right)\left(\frac{M}{m}\right)\sum_n \frac{1}{n!}\int_{x_-(n)}^{x_+(n)} \mathrm{d}x\, x^n\, \mathrm{e}^{-x}, \quad (12.63)$$

where σ_{b} is the cross-section for a single bound nucleus, i.e.

$$\sigma_{\text{b}} = 4\pi\{A^2+B^2\tfrac{1}{4}i(i+1)\},$$

and the upper limit of the summation is that value of n for which $E = n\hbar\omega_0(1+\gamma')$. The limits of integration in (12.63) are

$$x_{\pm}(n) = \frac{m}{M}\frac{1}{\hbar\omega_0(1+\gamma')^2}[E^{\frac{1}{2}} \pm \{E-n\hbar\omega_0(1+\gamma')\}^{\frac{1}{2}}]^2 \quad (12.64)$$

and $E = \hbar^2k^2/2m$ is the energy of the incident neutron. When $m = M$ and $\gamma' = 0$, which corresponds to a proton bound harmonically to a fixed point, the cross-section (12.63) reduces to the result first derived by Fermi.

Consider the $n = 0$ term. The integral in (12.63) is easily performed, giving a total neutron cross-section

$$\sigma = N\sigma_{\text{b}}\frac{\hbar\omega_0}{4E}\left(\frac{M}{m}\right)\left[1-\exp\left\{\frac{-4mE}{M\hbar\omega_0(1+\gamma')^2}\right\}\right]. \quad (12.65)$$

If $E/\hbar\omega_0 \ll 1$ this reduces to

$$\sigma \sim \frac{N\sigma_b}{(1+\gamma')^2}, \tag{12.66}$$

which is the total cross-section for the scattering from free nuclei of mass M_{mol}. Since the limit $E/\hbar\omega_0 \ll 1$ corresponds to nuclei that are very tightly bound to their corresponding molecules we should expect the mass of the molecules to enter the cross-section formula rather than that of the scattering nuclei.

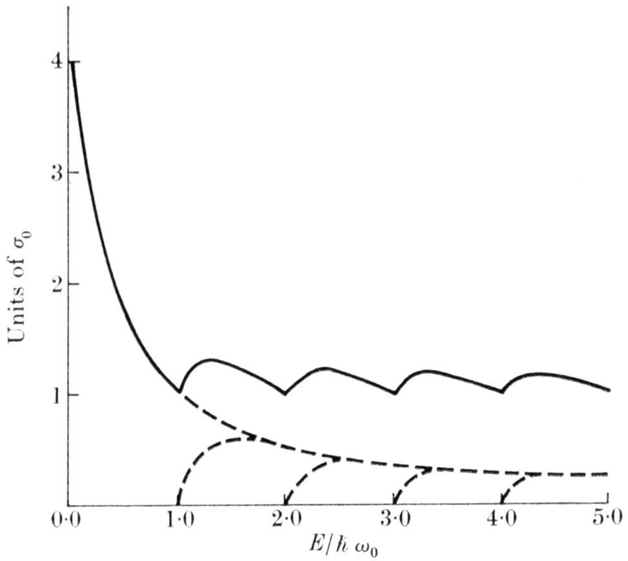

FIG. 12.6. Cross-section for a proton bound harmonically in a crystal (in units of the free single-nucleus cross-section $\sigma_0 = \sigma_b/4$) as a function of incident neutron energy E. The broken lines are the individual contributions from the terms in the summation.

In the opposite limit of $E \gg \hbar\omega_0$, the total cross-section has the asymptotic form

$$\sigma = \frac{N\sigma_b}{4}\frac{\hbar\omega_0}{E}\left(\frac{M}{m}\right)\sum_{n=0}^{E/\hbar\omega_0(1+\gamma')}\frac{1}{n!}\int_0^\infty \mathrm{d}x\, x^n\, \mathrm{e}^{-x}$$

$$= \frac{N\sigma_b}{4}\left(\frac{M}{m}\right)\left(\frac{1}{1+\gamma'}\right). \tag{12.67}$$

Fig. 12.6 shows the cross-section for a proton ($M = m$) bound harmonically in a crystal ($M_{mol} = \infty$), in units of $\sigma_b/4$, as a function of the incident energy E. The broken lines in Fig. 12.6 are the individual contributions from the terms in the summation. This cross-section corresponds to that expected for hydrogen atoms harmonically bound in crystals, the so-called Einstein model in solid state theory, and is realized in practice by very few solids. Whittemore and McReynolds measured the total neutron cross-section for the scattering by the hydrogen atoms in zirconium hydride and, as is seen from Fig. 12.7, found that it was almost an 'ideal Einstein crystal'.

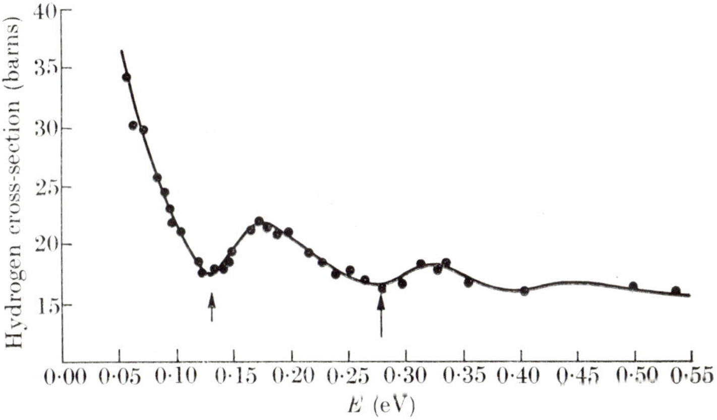

FIG. 12.7. Total cross-section for the scattering per hydrogen atom in zirconium hydride. The arrows indicate the excitation of the harmonic oscillator levels of the hydrogen in units of $\hbar\omega_0 = 0{\cdot}137$ eV (Whittemore and McReynolds, *Phys. Rev.* **113,** 806 (1959)).

12.7. Polyatomic gas

The calculation of the neutron cross-section for the scattering by a gas of polyatomic molecules is essentially the calculation of the second term in (12.48 a), the term that contains the dynamics of the rotational and vibrational motions of the nuclei. To do this we introduce a number of approximations. Though some of these are initially made more or less to make the calculation tractable, a detailed comparison of theory and experiment has shown that, for the two gases ammonia and methane, the approximations are very good ones.

Because the energy of the first vibrational level is in general much larger than the separation of the rotational levels and also larger than the energy associated with thermal neutrons (for methane the lowest vibrational energy level is at 0·168 eV) it is clear from the outset that the cross-section for the scattering of thermal neutrons will depend strongly on the rotational states of the nuclei and not on their vibrational states. It is usually sufficient, in fact, to assume that the nuclei are in their vibrational ground state.

Let us assume that we can separate the rotational and vibrational contributions and hence write

$$\langle \exp\{-i\boldsymbol{\varkappa}.\hat{\mathbf{d}}_v(0)\}\hat{O}_{vv'}(t)\exp\{i\boldsymbol{\varkappa}.\hat{\mathbf{d}}_{v'}(t)\}\rangle$$
$$\simeq \langle \exp\{-i\boldsymbol{\varkappa}.\hat{\mathbf{d}}_v(0)\}\exp\{i\boldsymbol{\varkappa}.\hat{\mathbf{d}}_{v'}(t)\}\rangle\langle \hat{O}_{vv'}(t)\rangle. \quad (12.68)$$

We first calculate the rotational term in (12.68), restricting ourselves to spherical and symmetric-top molecules. For a symmetric molecule

$$\langle \exp\{-i\boldsymbol{\varkappa}.\hat{\mathbf{d}}_v(0)\}\exp\{i\boldsymbol{\varkappa}.\hat{\mathbf{d}}_{v'}(t)\rangle$$
$$= \sum_{JK} P_{JK}\frac{1}{2J+1}\sum_{M}\langle JMK|\exp\{-i\boldsymbol{\varkappa}.\hat{\mathbf{d}}_v(0)\}\exp\{i\boldsymbol{\varkappa}.\hat{\mathbf{d}}_{v'}(t)\}|JMK\rangle, \quad (12.69)$$

where $|JMK\rangle$ are the normalized eigenfunctions of the symmetric-top molecule, eqn (12.14), and

$$P_{JK} = \frac{\exp(-\beta E_{JK})}{\sum\exp(-\beta E_{JK})} \quad (12.70)$$

is the Boltzmann factor.† We have taken account in (12.69) of the $(2J+1)$-fold degeneracy of the energy levels E_{JK}, eqn (12.13), with respect to the quantum number M (this degeneracy is merely a consequence of the molecule having no preferred orientation). If we introduce a complete set of eigenfunctions $|J'M'K'\rangle$ in (12.69) then we require to calculate the matrix elements

$$\langle J'M'K'|\exp\{i\boldsymbol{\varkappa}.\hat{\mathbf{d}}_v(t)\}|JMK\rangle$$
$$= \exp\{it(E_{J'K'}-E_{JK})/\hbar\}\langle J'M'K'|\exp(i\boldsymbol{\varkappa}.\mathbf{d}_v)|JMK\rangle. \quad (12.71)$$

If $\tilde{\mathbf{d}}_v'$ and $\tilde{\boldsymbol{\varkappa}}'$ are unit vectors along $\mathbf{d}_v$ and $\boldsymbol{\varkappa}$ referred to the rotating coordinate frame of reference in the molecule, then the exponential factor has the standard expansion

$$\exp(i\boldsymbol{\varkappa}.\mathbf{d}_v) = 4\pi\sum_{l=0}^{\infty}\sum_{n=-l}^{l} i^l j_l(\kappa d_v)Y_n^{*l}(\tilde{\mathbf{d}}_v')Y_n^l(\tilde{\boldsymbol{\varkappa}}'). \quad (12.72)$$

However, the scattering vector $\boldsymbol{\varkappa}$ is required referred to the laboratory frame of reference and not the frame of reference rotating with the molecule. If the rotation of the axes fixed in the molecule with respect to those in the laboratory is specified by the Euler angles $\{\alpha\beta\gamma\}$, then

$$Y_n^l(\tilde{\boldsymbol{\varkappa}}') = \sum_{n'=-l}^{l} Y_{n'}^l(\tilde{\boldsymbol{\varkappa}})\mathscr{D}_{n'n}^{(l)}(\alpha\beta\gamma), \quad (12.73)$$

† The following method of calculating (12.69) was developed independently by A. Rahman, *J. nucl. Energy*, **13**, 128 (1961), and S. Yip, thesis, University of Michigan (1963).

where $\tilde{\boldsymbol{\varkappa}}$ is the corresponding unit vector in the laboratory frame of reference. Thus, with (12.72) and (12.73),

$$\langle J'M'K'|\exp(\mathrm{i}\boldsymbol{\varkappa}.\mathbf{d}_v)|JMK\rangle = \int \mathrm{d}\Omega\left(\frac{2J'+1}{8\pi^2}\right)^{\frac{1}{2}}\{\mathscr{D}^{(J')}_{M'K'}(\alpha\beta\gamma)\}^*\times$$

$$\times 4\pi\sum_{l=0}^{\infty}\sum_{n=-l}^{l}\mathrm{i}^l j_l(\kappa d_v)Y^{*l}_n(\tilde{\mathbf{d}}'_v)\sum_{n'=-l}^{l}Y^l_{n'}(\tilde{\boldsymbol{\varkappa}})\mathscr{D}^{(l)}_{n'n}(\alpha\beta\gamma)\left(\frac{2J+1}{8\pi^2}\right)^{\frac{1}{2}}\mathscr{D}^{(J)}_{MK}(\alpha\beta\gamma)$$

$$= (-1)^{M'-K'}\{(2J+1)(2J'+1)\}^{\frac{1}{2}}4\pi\sum_{lnn'}\mathrm{i}^l j_l(\kappa d_v)Y^{*l}_n(\tilde{\mathbf{d}}'_v)Y^l_{n'}(\tilde{\boldsymbol{\varkappa}})\times$$

$$\times\begin{pmatrix}J & J' & l\\ M & -M' & n'\end{pmatrix}\begin{pmatrix}J & J' & l\\ K & -K' & n\end{pmatrix}. \quad (12.74)$$

In obtaining the second line we have used the relationship

$$\{\mathscr{D}^{(J)}_{MK}(\alpha\beta\gamma)\}^* = (-1)^{M-K}\mathscr{D}^{(J)}_{-M,-K}(\alpha\beta\gamma)$$

together with the following identity for the integral of a triple product of rotation matrices in terms of $3j$ symbols,

$$\frac{1}{8\pi^2}\int \mathrm{d}\Omega\,\mathscr{D}^{(j_1)}_{m'_1m_1}(\alpha\beta\gamma)\mathscr{D}^{(j_2)}_{m'_2m_2}(\alpha\beta\gamma)\mathscr{D}^{(j_3)}_{m'_3m_3}(\alpha\beta\gamma)$$

$$= \begin{pmatrix}j_1 & j_2 & j_3\\ m'_1 & m'_2 & m'_3\end{pmatrix}\begin{pmatrix}j_1 & j_2 & j_3\\ m_1 & m_2 & m_3\end{pmatrix}.$$

We can choose the z axis to coincide with the direction defined by $\tilde{\boldsymbol{\varkappa}}$, i.e. we can put

$$Y^l_n(\tilde{\boldsymbol{\varkappa}}) = \left(\frac{2l+1}{4\pi}\right)^{\frac{1}{2}}. \quad (12.75)$$

Thus, in conjunction with (12.74),

$$\frac{1}{2J+1}\sum_M\sum_{J'M'K'}\langle JMK|\exp\{-\mathrm{i}\boldsymbol{\varkappa}.\hat{\mathbf{d}}_v(0)\}|J'M'K'\rangle\times$$

$$\times\langle J'M'K'|\exp\{\mathrm{i}\boldsymbol{\varkappa}.\hat{\mathbf{d}}_{v'}(t)|JMK\rangle$$

$$= \frac{1}{2J+1}\sum_M\sum_{J'M'K'}\exp\{\mathrm{i}t(E_{J'K'}-E_{JK})/\hbar\}4\pi(2J+1)(2J'+1)\times$$

$$\times\sum_{lnn'}\sum_{l_1n_1n'_1}\{(2l+1)(2l_1+1)\}^{\frac{1}{2}}\mathrm{i}^l(-\mathrm{i})^{l_1}j_l(\kappa d_v)j_{l_1}(\kappa d_{v'})Y^{*l}_n(\tilde{\mathbf{d}}'_v)Y^{l_1}_{n_1}(\tilde{\mathbf{d}}'_{v'})\times$$

$$\times\begin{pmatrix}J & J' & l\\ M & -M' & n'\end{pmatrix}\begin{pmatrix}J & J' & l\\ K & -K' & n\end{pmatrix}\begin{pmatrix}J & J' & l_1\\ M & -M' & n'_1\end{pmatrix}\begin{pmatrix}J & J' & l_1\\ K & -K' & n_1\end{pmatrix}.$$

Because of the orthogonality relation for the $3j$ symbols given by eqn (12.37), this is

$$\sum_{J'K'}\exp\{\mathrm{i}t(E_{J'K'}-E_{JK})/\hbar\}4\pi(2J'+1)\sum_{l=|J-J'|}^{J+J'}j_l(\kappa d_v)j_l(\kappa d_{v'})\times$$

$$\times Y^{*l}_n(\tilde{\mathbf{d}}'_v)Y^l_n(\tilde{\mathbf{d}}'_{v'})\begin{pmatrix}J & J' & l\\ K & -K' & n\end{pmatrix}^2$$

$$= \sum_{J'K'}\exp\{\mathrm{i}t(E_{J'K'}-E_{JK})/\hbar\}\mathscr{L}_{vv'}(JJ';KK'). \quad (12.76)$$

We have therefore shown that, for a symmetric-top molecule,

$$\langle\exp\{-\mathrm{i}\boldsymbol{\kappa}.\hat{\mathbf{d}}_v(0)\}\exp\{\mathrm{i}\boldsymbol{\kappa}.\hat{\mathbf{d}}_{v'}(t)\}\rangle = \sum_{JJ'}\sum_{KK'} P_{JK}\exp\{\mathrm{i}t(E_{J'K'}-E_{JK})/\hbar\}\mathscr{L}_{vv'}(JJ';KK'), \quad (12.77)$$

where $\mathscr{L}$ is defined by eqn (12.76). The corresponding expression for the spherical-top molecule is readily obtained from eqn (12.77), for in this case the eigenstates are independent of the quantum number K (eqn (12.11)) so that

$$P_J = \frac{\exp(-\beta E_J)}{\sum_J (2J+1)\exp(-\beta E_J)} \quad (12.78\,\mathrm{a})$$

and

$$\sum_{KK'}\mathscr{L}_{vv'}(JJ';KK') = (2J'+1)\sum_{l=|J-J'|}^{J+J'} j_l(\kappa d_v)j_l(\kappa d_{v'})P_l(\cos\theta_{vv'}), \quad (12.78\,\mathrm{b})$$

where P_l is the Legendre polynomial of order l and $\theta_{vv'}$ is the angle between $\tilde{\mathbf{d}}'_v$ and $\tilde{\mathbf{d}}'_{v'}$.

The actual calculation of $\mathscr{L}$ for a given molecule is, clearly, a laborious task. In the limit $\kappa d_v \ll 1$, however, which corresponds to very low-energy incident neutrons, only the first few terms in the summation over l need be taken on account of the well-known expansion of spherical Bessel functions,

$$j_l(x) = \frac{x^l}{(2l+1)!!}\left\{1-\frac{x^2}{2(2l+3)}+\ldots\right\}.$$

The contribution to (12.68) from the vibrational motion of the nuclei can be calculated by essentially the same method as that adopted in § 12.6. From eqn (12.7) we can write

$$\hat{\mathbf{u}}_v(t) = \sum_\lambda \mathbf{c}_v^{(\lambda)}\hat{Q}^{(\lambda)}(t), \quad (12.79)$$

where $\hat{Q}^{(\lambda)}(t) = (\hbar/2\omega_\lambda)^{\frac{1}{2}}(\hat{a}_\lambda+\hat{a}^{+}_{-\lambda})$, so that

$$\exp\{\mathrm{i}\boldsymbol{\kappa}.\hat{\mathbf{u}}_v(t)\} = \prod_\lambda \exp\{\mathrm{i}\boldsymbol{\kappa}.\mathbf{c}_v^{(\lambda)}\hat{Q}^{(\lambda)}(t)\}. \quad (12.80)$$

The formal development to obtain $\langle\hat{O}_{vv'}\rangle$ now parallels that given in § 12.6, the only difference being that there is a product over all allowed eigenfrequencies ω_λ; we find

$$\langle\hat{O}_{vv'}\rangle = \prod_\lambda \exp[-\{(\boldsymbol{\kappa}.\mathbf{c}_v^{(\lambda)})^2+(\boldsymbol{\kappa}.\mathbf{c}_{v'}^{(\lambda)})^2\}(\hbar/4\omega_\lambda)\coth(\tfrac{1}{2}\hbar\omega_\lambda\beta)]\times$$

$$\times\sum_{n=-\infty}^{\infty}\exp(\mathrm{i}n\omega_\lambda t+\tfrac{1}{2}n\hbar\omega_\lambda\beta)I_n\{(\hbar/2\omega_\lambda)(\boldsymbol{\kappa}.\mathbf{c}_v^{(\lambda)})(\boldsymbol{\kappa}.\mathbf{c}_{v'}^{(\lambda)})\mathrm{cosech}(\tfrac{1}{2}\hbar\omega_\lambda\beta)\}. \quad (12.81)$$

As we have already mentioned, the molecule may usually be regarded as being in its vibrational ground state. We will, in addition, introduce an approximate method of averaging the vibrational contribution over the orientations of the molecule, namely

$$\overline{\langle \hat{O}_{vv'} \rangle} \simeq \exp\Big[-\sum_{\lambda} (\hbar/4\omega_\lambda)\overline{\{(\boldsymbol{\kappa}\,.\,\mathbf{c}_v^{(\lambda)})^2+(\boldsymbol{\kappa}\,.\,\mathbf{c}_{v'}^{(\lambda)})^2\}}\Big] = \exp(-\kappa^2\gamma_{vv'}), \tag{12.82 a}$$

where
$$\gamma_{vv'} = \sum_{\lambda} (\hbar/12\omega_\lambda)\{(\mathbf{c}_v^{(\lambda)})^2+(\mathbf{c}_{v'}^{(\lambda)})^2\}. \tag{12.82 b}$$

Within the various approximations that have been introduced, the neutron cross-section for the scattering from symmetric-top molecules is, from (12.44), (12.48 a), (12.68), (12.77), and (12.82),

$$\frac{\mathrm{d}^2\sigma}{\mathrm{d}\Omega\mathrm{d}E'} = \frac{Nk'}{k}\left(\frac{\beta M_{\mathrm{mol}}}{2\pi\hbar^2\kappa^2}\right)^{\frac{1}{2}} \sum_{v,v'} (A_v A_{v'}+\tfrac{1}{4}i_v(i_v+1)\delta_{vv'}B_v^2)\times$$
$$\times\exp(-\kappa^2\gamma_{vv'})\sum_{JJ'}\sum_{KK'} P_{JK}\mathscr{L}_{vv'}(JJ';KK')\times$$
$$\times\exp\left\{-\frac{\beta M_{\mathrm{mol}}}{2\hbar^2\kappa^2}\left(E_{J'K'}-E_{JK}\quad\hbar\omega+\frac{\hbar^2\kappa^2}{2M_{\mathrm{mol}}}\right)^2\right\}, \tag{12.83}$$

where we have dropped the 'outer' scattering term in (12.44). The corresponding cross-section for the scattering from spherical molecules has an analogous form to (12.83), it being only necessary to take account of eqns (12.78 a) and (12.78 b). These expressions for the cross-sections were first used by Grifting for spherical molecules and by Venkataraman, Rao, and Dasannacharya for symmetric molecules.†

A detailed experimental study of the partial differential cross-section for the scattering of cold ($\lambda = 4{\cdot}2$ Å) neutrons by the two gases methane and ammonia has been made at room temperature by Webb.‡ His results are shown in Figs. 12.8 and 12.9 together with the cross-sections evaluated from eqn (12.83). At small scattering angles there are two distinct contributions to the cross-section for both gases, a quasi-elastic peak near the incident neutron energy and an inelastic peak corresponding to a neutron energy gain of 0·014 eV. For higher values of the scattering angle the two peaks merge into a single inelastic peak, much as one would expect. The structure that is exhibited at small scattering angles is seen to be well reproduced by eqn (12.83) and the agreement at large scattering angles is very good.

† G. W. Grifting, *Phys. Rev.* **124**, 1489 (1961); G. Venkataraman, K. R. Rao, and B. A. Dasannacharya, *Proc. phys. Soc.* **89**, 379 (1966).

‡ F. J. Webb, *Proc. phys. Soc.* **92**, 912 (1967).

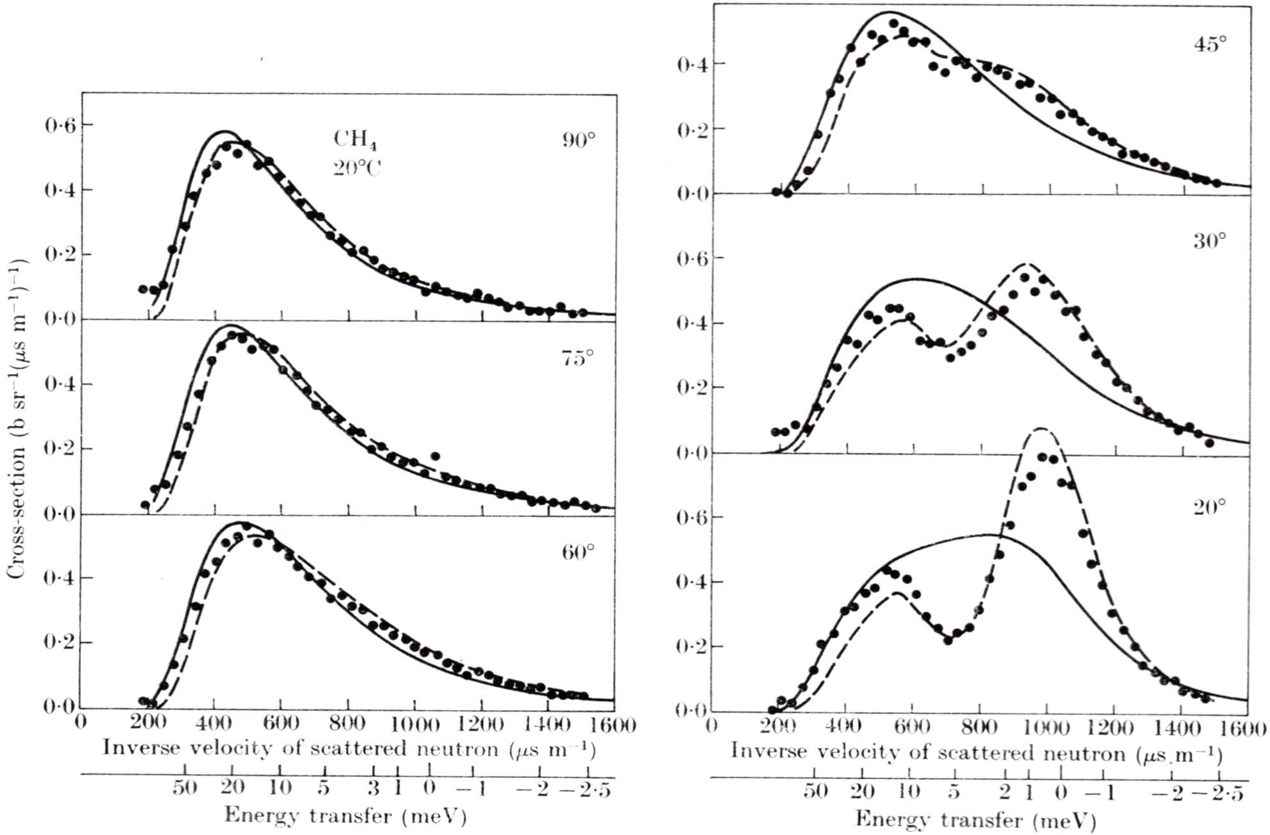

Fig. 12.8. The partial differential cross-section of methane at 20 °C for 4·2 Å neutrons. The experimental results are represented by dots; the cross-section as calculated from (12.83) is represented by a broken line and the Krieger–Nelkin approximation, eqn (12.88) by a solid line (after Webb). Only the incoherent proton scattering is included in the numerical calculations.

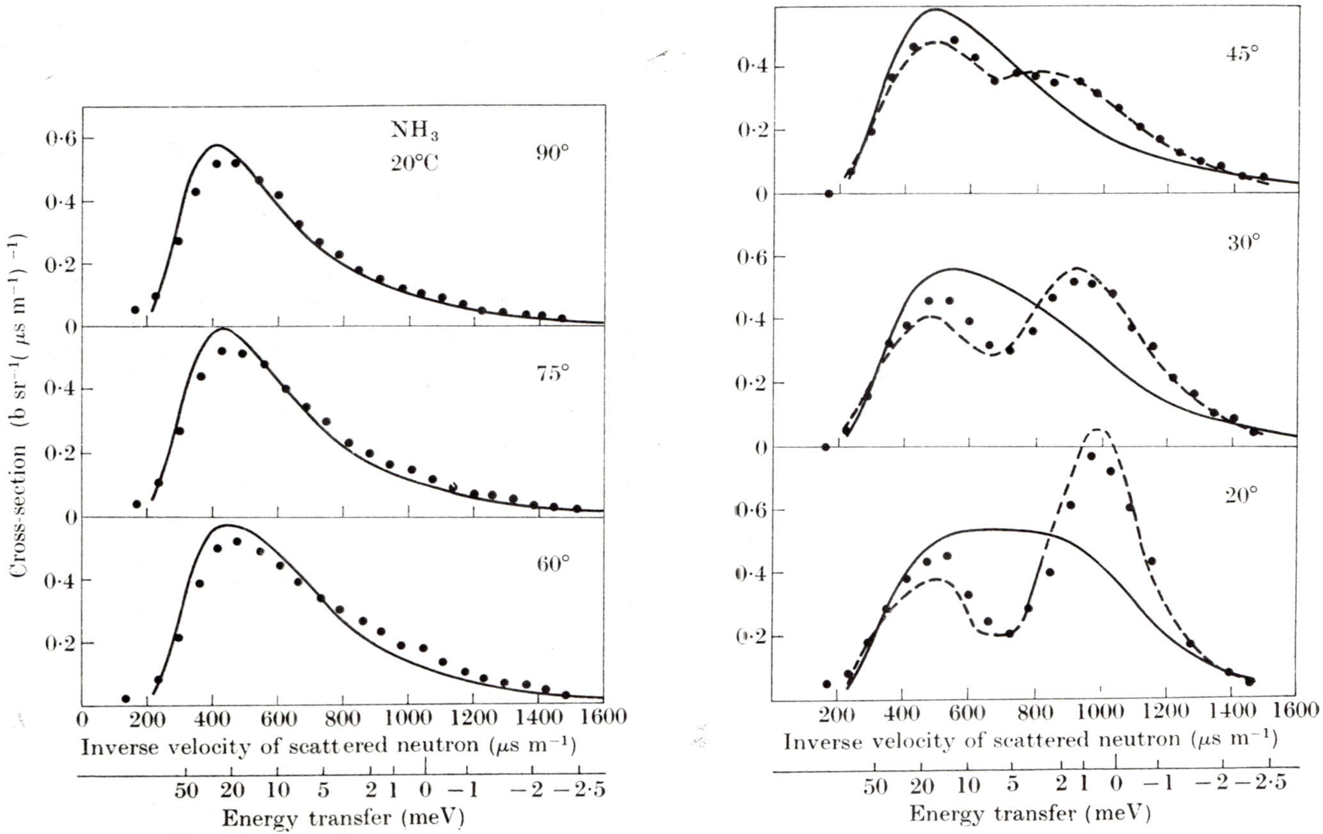

FIG. 12.9. The partial differential neutron cross-section for the scattering of 4·2 Å neutrons by ammonia at 20 °C. The experimental points are represented by dots. The cross-section as calculated from eqn (12.83) is represented by a broken line and the Krieger–Nelkin approximation, eqn (12.88), by a solid line (after Webb). Only the incoherent proton scattering is included in the numerical calculations.

The agreement between the measured cross-section and that calculated from (12.83) (only the incoherent proton scattering contribution has been included) is not, however, completely without fault. At small scattering angles there is a discrepancy in the height of the rotational peak relative to the translational peak, the former being some 15 per cent larger than that calculated from (12.83). Such a disagreement is, perhaps, quite small in view of the approximations made *en route* to establishing eqn (12.83), namely; the separation of the translational, vibrational, and rotational motions, the complete neglect of nuclear spin correlations, the separate, approximate averaging of the vibrational contribution over the orientations of the molecule and, in the calculations shown in Figs. 12.8 and 12.9, the neglect of the small coherent cross-section of the proton and the scattering from the carbon and nitrogen nuclei. All but the separation of motions into the product of three separate contributions are known with certainty to be small in the present case. Webb suggests that the coupling of the rotational and translation motions may be the prime cause of the observed disagreement, the vibrational states being, of course, too high in energy to be effective.

12.8. Quasi-classical approximation

Historically, the detailed treatment given in the previous section of the rotational contribution to the approximate cross-section was undertaken when the much simpler theories based on the mass tensor concept were found to be inadequate in explaining the structure of the measured partial differential cross-sections. There are situations, however, in which these simpler theories, which are quasi-classical approximations, remain of value, though care must be taken to ensure that they are not used outside their domain of validity. Because the use of the mass tensor approximation and its various extensions is limited, and because they have been treated in some detail elsewhere,† our discussion of them will be restricted to writing down the appropriate formulae, with plausible supporting arguments, and discussing the conditions under which they can legitimately be employed. In particular we give the total cross-section as obtained by the approximations used by Krieger and Nelkin.‡

The approximation to the cross-section that we discuss in this section is valid when the incident neutron energy is large compared to the level spacing of the rotational states and when many of these levels are thermally excited. Under these conditions it is not necessary to consider the contribution to the scattering from individual levels but only the

† D. E. Parks *et al.*, *Slow neutron scattering and thermalisation*, §§ 2–7.4, Benjamin (1970).
‡ T. J. Krieger and M. S. Nelkin, *Phys. Rev.* **106**, 290 (1957).

average of a small number; this is what we mean by a quasi-classical treatment of the scattering process. Mathematically this means that the operator nature of the variables associated with the rotational motion is ignored and the thermal averaging is carried through in a straightforward classical manner. Krieger and Nelkin found that such a procedure gave the result

$$\langle\exp\{-i\boldsymbol{\kappa}.\hat{\mathbf{l}}(0)\}\exp\{i\boldsymbol{\kappa}.\hat{\mathbf{l}}(t)\}\rangle\langle\exp\{-i\boldsymbol{\kappa}.\hat{\mathbf{d}}_v(0)\}\exp\{i\boldsymbol{\kappa}.\hat{\mathbf{d}}_{v'}(t)\}\rangle$$
$$\simeq \exp[-i\boldsymbol{\kappa}.\{\hat{\mathbf{d}}_v(0)-\hat{\mathbf{d}}_{v'}(0)\}]\exp\left\{-\frac{1}{2\beta}\boldsymbol{\kappa}.\boldsymbol{M}_v^{-1}.\boldsymbol{\kappa}(t^2-i\hbar t\beta)\right\}, \quad (12.84)$$

where $\boldsymbol{M}_v^{-1}$ is the *mass tensor* introduced by Sachs and Teller. Its components are

$$\{M_v^{-1}\}_{ii} = \frac{d_{vj}^2}{I_{vk}}+\frac{d_{vk}^2}{I_{vj}}+\frac{1}{M_{\text{mol}}}$$

and

$$\{M_v^{-1}\}_{ij} = -\frac{d_{vi}d_{vj}}{I_{vk}} \quad (i\neq j), \quad (12.85)$$

where i, j, k are referred to the directions of the principal axes of inertia and d_ξ, d_η, d_ζ are the coordinates of the vth nucleus under consideration with respect to the centre of mass of the molecule.

If we assume, as above, that the molecule can be regarded to be in its vibrational ground state, which requires, in particular, the incident neutron energy to be less than the energy needed to excite the first vibrational level, the cross-section is proportional to the integral over the time t of the product of the vibrational term

$$\exp\left[-\sum_\lambda \frac{\hbar}{4\omega_\lambda}\{(\boldsymbol{\kappa}.\mathbf{c}_v^{(\lambda)})^2+(\boldsymbol{\kappa}.\mathbf{c}_{v'}^{(\lambda)})^2\}\right] \quad (12.86)$$

with the two factors given by the right-hand side of (12.84). To average this product over the orientations of the molecule, we replace it by the average of the products and take the average of the arguments of the exponential in (12.86) and the second factor on the right-hand side of (12.84). The vibrational contribution is then identical with that given in the previous section, eqn. (12.82), while the translational and rotational motions contribute

$$\overline{\exp\{-i\boldsymbol{\kappa}.(\mathbf{d}_v-\mathbf{d}_{v'})\}\exp\left\{-\frac{1}{2\beta}\boldsymbol{\kappa}.\boldsymbol{M}_v^{-1}.\boldsymbol{\kappa}(t^2-i\hbar t\beta)\right\}}$$
$$\simeq \overline{\exp\{-i\boldsymbol{\kappa}.(\mathbf{d}_v-\mathbf{d}_{v'})\}}\exp\left\{-\frac{1}{2\beta}\overline{\boldsymbol{\kappa}.\boldsymbol{M}_v^{-1}.\boldsymbol{\kappa}}(t^2-i\hbar t\beta)\right\}$$
$$= j_0(\kappa|\mathbf{d}_v-\mathbf{d}_{v'}|)\exp\left\{-\frac{\kappa^2}{2\beta M_v^{(0)}}(t^2-i\hbar t\beta)\right\}, \quad (12.87)$$

where $3/M_v^{(0)} = \operatorname{Tr} M_v^{-1}$.

With these approximations the neutron cross-section is

$$\frac{d^2\sigma}{d\Omega dE'} = N\frac{k'}{k}\sum_{v,v'}(A_v A_{v'} + \tfrac{1}{4}i_v(i_v+1)\delta_{v,v'}B_v^2)\times$$

$$\times\frac{1}{2\pi\hbar}\int_{-\infty}^{\infty} dt\,\exp(-i\omega t)\exp(-\kappa^2\gamma_{vv'})j_0(\kappa|\mathbf{d}_v-\mathbf{d}_{v'}|)\exp\left\{-\frac{\kappa^2}{2\beta M_v^{(0)}}(t^2-i\hbar t\beta)\right\}$$

$$= N\frac{k'}{k}\sum_{v,v'}\{A_v A_{v'}+\tfrac{1}{4}i_v(i_v+1)\delta_{v,v'}B_v^2\}j_0(\kappa|\mathbf{d}_v-\mathbf{d}_{v'}|)\times$$

$$\times\exp(-\kappa^2\gamma_{vv'})\left(\frac{\beta M_v^{(0)}}{2\pi\hbar^2\kappa^2}\right)^{\frac{1}{2}}\exp\left\{-\frac{\beta M_v^{(0)}}{2\hbar^2\kappa^2}\left(\hbar\omega-\frac{\hbar^2\kappa^2}{2M_v^{(0)}}\right)^2\right\}. \quad (12.88)$$

As a function of the scattering angle, this expression will be most reliable at large scattering angles because here the large recoil energy means that an envelope of rotational states is involved.

The Krieger–Nelkin approximation to the partial differential cross-section, eqn (12.88), is shown for methane and ammonia in Figs. 12.8 and 12.9 (again only the incoherent proton scattering has been included). At small scattering angles the approximation is clearly inadequate, while at larger scattering angles the agreement with Webb's experimental results is tolerable, though not as good as that with the cross-section calculated from eqn (12.83). This is equivalent to saying that, at small scattering angles and for slow incident neutrons, the rotational motion gives a distinct contribution to the cross-section, as expected, which the Krieger–Nelkin approximation is by its very nature incapable of reproducing. This point has been well illustrated in calculations performed by Laurie of cross-sections for linear molecules.†

The total direct scattering cross-section in the Krieger–Nelkin approximation is

$$\sigma_{(v)} = \{A_v^2+\tfrac{1}{4}i_v(i_v+1)B_v^2\}\frac{N\pi}{2m\gamma_{vv}E}\times$$

$$\times[\Phi\{\sqrt{(M_v^{(0)}\beta E/m)}\}-(1-c)^{\frac{1}{2}}\exp(-M_v^{(0)}\beta Ec/m)\Phi\{\sqrt{(M_v^{(0)}\beta E(1-c)/m)}\}], \quad (12.89)$$

where

$$c = 1+\frac{\beta(1+M_v^{(0)}/m)^2}{8\gamma_{vv}M_v^{(0)}}.$$

Because the total cross-section is insensitive to the contribution of individual dynamical states of the nuclei, this expression should give a good representation of the total cross-section, provided that the energy of the incident neutrons is large compared to the separation of the

† N. A. Laurie, *J. chem. Phys.* **46**, 352 (1967).

rotational levels. Fig. 12.10 shows the total cross-section for the scattering from methane as measured by Melkonian† and that given by eqn (12.89). The agreement is very good at high energies and not unreasonable at the smallest incident energy considered (we recall that the rotational constant for methane is small, $\sim$ 0·0006 eV).

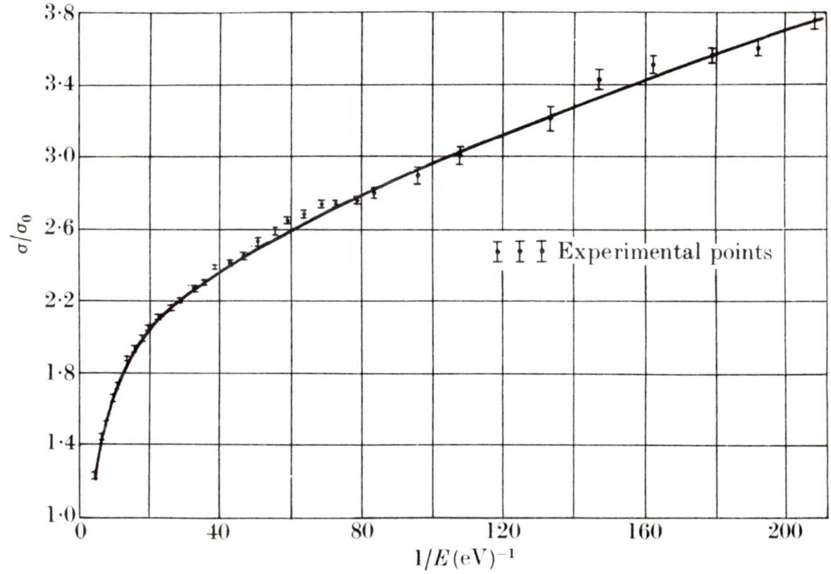

FIG. 12.10. The total neutron cross-section of methane at room temperature. The free-proton cross-section $\sigma_0 = 20{\cdot}36$ b. The experimental points are those of Melkonian and the solid curve is calculated from eqn (12.69) taking account of only the incoherent proton scattering (after Krieger and Nelkin).

Because the Krieger–Nelkin approximation to the cross-section is much simpler than that in which the rotational states are considered explicitly, and also because it can easily be applied to any polyatomic molecule, there have been several attempts to extend its domain of validity and a number of investigations to establish more precisely when it can be expected to be reliable.‡ In general, the mass tensor approximation to the partial differential cross-section is good when

$$\frac{m}{M^{(0)}}\{E+E'-2\surd(EE')\cos\theta\}$$

† E. Melkonian, *Phys. Rev.* **76,** 1744 (1949).

‡ I. P. Dzyub, *Soviet Phys. JETP,* **22,** 347 (1966); H. C. Volkin, *Phys. Rev.* **113,** 866 (1959), **117,** 1029 (1960); G. Kosaly and G. Solt, *Phys. Lett.* **6,** 51 (1963), **13,** 223 (1964), *Physica 's Grav.* **32,** 16 (1966); G. Summerfeld and P. Zweifel, *Phys. Rev.* **159,** 1 (1967); V. F. Sears, *Can. J. Phys.* **45,** 237 (1967).

is large compared to the rotational constant (Kosaly and Solt); thus it will be good when the energy gain is high ($E' \gg E$), and this situation is associated with temperatures at which a large number of rotational levels are populated. Furthermore, according to this criterion, the mass tensor approximation should be more reliable at large scattering angles. The validity of the above conditions is borne out by the evidence contained in Figs. 12.8, 12.9, and 12.10.

13

CRITICAL MAGNETIC SCATTERING

SEVERAL topics discussed in this book are not at present anywhere near being well understood, though their gross features may be. This is particularly true of critical magnetic scattering. There is no really adequate theory of critical scattering, but various aspects of it have received much attention and are understood. It is a field of intense effort, both theoretical and experimental, and to give a detailed account of either would be outside the scope of this book. It would also be inappropriate, because the field is moving so rapidly. Several reviews have been written on static critical phenomena.

In this chapter we aim (a) to set up a rigorous mathematical theory to describe critical magnetic scattering and (b) to examine the consequences of the theory within the *molecular-field* (or mean-field) approximation. Though this approximation is altogether inadequate to yield detailed information, it serves to give the gross overall picture observed in recent experiments on magnetic salts. Since all our theoretical discussions are based on the Heisenberg exchange Hamiltonian we shall not consider any experiments on magnetic metals.

13.1. Introduction

Throughout this chapter we shall be concerned with Bravais lattices and Heisenberg magnets. For such systems the partial differential cross-section is (cf. Chapter 8)

$$\frac{d^2\sigma}{d\Omega dE'} = \left(\frac{\gamma e^2}{m_e c^2}\right)^2 \frac{k'}{k} |\tfrac{1}{2}gF(\boldsymbol{\varkappa})|^2 \sum_{\alpha} (1-\tilde{\kappa}_{\alpha}^2) S^{\alpha}(\boldsymbol{\varkappa}, \omega). \tag{13.1}$$

Here

$$S^{\alpha}(\boldsymbol{\varkappa}, \omega) = \frac{1}{2\pi\hbar} \int_{-\infty}^{\infty} dt \exp(-i\omega t) \sum_{ll'} \exp\{i\boldsymbol{\varkappa}.(\mathbf{l}'-\mathbf{l})\}\langle \hat{S}_{l}^{\alpha}(0)\hat{S}_{l'}(t)\rangle$$

$$= \frac{1}{2\pi\hbar} \int_{-\infty}^{\infty} dt \exp(-i\omega t)\langle \hat{S}_{\boldsymbol{\varkappa}}^{\alpha}(0)\hat{S}_{-\boldsymbol{\varkappa}}^{\alpha}(t)\rangle, \tag{13.2}$$

where, as usual, $$\hat{S}_{\mathbf{q}}^{\alpha} = \sum_{l} \exp(-i\mathbf{q}.\mathbf{l})\hat{S}_{l}^{\alpha} \tag{13.3}$$

and $\alpha = x$, y, or z.

It is a simple matter to demonstrate that near the critical point there is a pronounced increase in the magnetic scattering.

If we denote $$\Delta\hat{S}^{\alpha}_{\mathbf{q}} = \sum_{l} \exp(-\mathrm{i}\mathbf{q}\,.\,\mathbf{l})\{\hat{S}^{\alpha}_{l} - \langle\hat{S}^{\alpha}_{l}\rangle\}, \tag{13.4}$$

then, for a ferromagnet ($\langle\hat{S}^{\alpha}_{l}\rangle \equiv \langle\hat{S}^{\alpha}\rangle$),

$$S^{\alpha}(\boldsymbol{\varkappa}, \omega) = N^2\delta_{\boldsymbol{\varkappa},0}\langle\hat{S}^{\alpha}\rangle^2\delta(\hbar\omega) + \frac{1}{2\pi\hbar}\int_{-\infty}^{\infty} \mathrm{d}t\, \exp(-\mathrm{i}\omega t)\langle\Delta\hat{S}^{\alpha}_{\boldsymbol{\varkappa}}(0)\Delta\hat{S}^{\alpha}_{-\boldsymbol{\varkappa}}(t)\rangle. \tag{13.5}$$

The first term on the right-hand side of (13.5) corresponds to magnetic Bragg scattering and is of no interest to us here (it is very small near T_c because $\langle\hat{S}^{\alpha}\rangle \to 0$ at T_c). If the spin fluctuations appear almost static to the incident neutrons the inelasticity of the scattering is small.

Assuming for the present that this is a valid assumption, the cross-section is determined by
$$\langle\Delta\hat{S}^{\alpha}_{\boldsymbol{\varkappa}}(0)\Delta\hat{S}^{\alpha}_{-\boldsymbol{\varkappa}}(0)\rangle. \tag{13.6}$$

Now the static susceptibility χ^{α} is related to the fluctuations in the magnetization $M^{\alpha} = -g\mu_{\mathrm{B}} \sum \langle\hat{S}^{\alpha}_{l}\rangle$ through
$$\chi^{\alpha} = \frac{\beta}{N}\langle(\Delta M^{\alpha})^2\rangle. \tag{13.7}$$

Thus it follows that at $\boldsymbol{\varkappa} = 0$ (13.6) becomes
$$\frac{N}{\beta}(g\mu_{\mathrm{B}})^{-2}\chi^{\alpha}. \tag{13.8}$$

Near the critical point the fluctuations in the magnetization become very large and in the limit $T \to T_c$, χ^{α} actually diverges. The corresponding pronounced increase in scattering is, of course, seen at all Bragg peaks, $\boldsymbol{\varkappa} = \boldsymbol{\tau}$.

In the remainder of this chapter we shall be concerned with removing the two restrictions made above in arriving at (13.8), i.e. we look at the wave-vector dependence of the diffuse scattering about the Bragg positions and also its energy dependence. To pursue both these investigations it is convenient to express the cross-section in terms of the relaxation function $\mathscr{R}^{\alpha}_{\boldsymbol{\varkappa}}(t)$, viz.

$$S^{\alpha}(\boldsymbol{\varkappa}, \omega) = \frac{\omega\exp(\hbar\omega\beta)}{\exp(\hbar\omega\beta)-1}\mathscr{R}^{\alpha}_{\boldsymbol{\varkappa}}(\omega) + \delta(\hbar\omega)\langle\hat{S}^{\alpha}_{\boldsymbol{\varkappa}}\rangle\langle\hat{S}^{\alpha}_{-\boldsymbol{\varkappa}}\rangle, \tag{13.9}$$

where $$\mathscr{R}^{\alpha}_{\boldsymbol{\varkappa}}(\omega) = \frac{1}{2\pi}\int_{-\infty}^{\infty} \mathrm{d}t\, \exp(\mathrm{i}\omega t)\mathscr{R}^{\alpha}_{\boldsymbol{\varkappa}}(t). \tag{13.10}$$

From (13.9) and (13.10) it follows that

$$\int\limits_{-\infty}^{\infty} \mathrm{d}\omega \left\{\frac{1-\exp(-\hbar\omega\beta)}{\omega}\right\}\{S^{\alpha}(\boldsymbol{\kappa},\omega)-S^{\alpha}_{\mathrm{Bragg}}(\boldsymbol{\kappa},\omega)\} = \mathscr{R}^{\alpha}_{\boldsymbol{\kappa}}(t=0). \quad (13.11)$$

But, from § 8.3,

$$\mathscr{R}^{\alpha}_{\boldsymbol{\kappa}}(t=0) = \int\limits_{0}^{\beta} \mathrm{d}\lambda\, \langle \hat{S}^{\alpha}_{\boldsymbol{\kappa}}(-\mathrm{i}\hbar\lambda)\hat{S}^{\alpha}_{-\boldsymbol{\kappa}}(0)\rangle - \beta\langle \hat{S}^{\alpha}_{\boldsymbol{\kappa}}\rangle\langle \hat{S}^{\alpha}_{-\boldsymbol{\kappa}}\rangle = \frac{N}{(g\mu_{\mathrm{B}})^2}\chi^{\alpha}_{\boldsymbol{\kappa}}, \quad (13.12)$$

where $\chi^{\alpha}_{\boldsymbol{\kappa}}$ is the *isothermal wave-vector dependent susceptibility*; $\chi^{\alpha}_{\boldsymbol{\kappa}}$ gives the response of a magnetic system to an external magnetic field applied in the direction α with a spatial dependence $\cos(\boldsymbol{\kappa}.\mathbf{l})$. Note that in the limit $\boldsymbol{\kappa} = 0$ when the total spin is a constant of motion

$$\hat{S}^{\alpha}_{\boldsymbol{\kappa}}(t) \to \hat{S}^{\alpha}_{\mathbf{0}}(t) = \hat{S}^{\alpha}_{\mathbf{0}}(0),$$

(13.12) becomes

$$\int\limits_{0}^{\beta} \mathrm{d}\lambda\, \langle \hat{S}^{\alpha}_{\mathbf{0}}(0)\hat{S}^{\alpha}_{\mathbf{0}}(0)\rangle - \beta\langle \hat{S}^{\alpha}_{\mathbf{0}}\rangle\langle \hat{S}^{\alpha}_{\mathbf{0}}\rangle = \beta\Big\{\Big\langle\Big(\sum_{l} \hat{S}^{\alpha}_{l}\Big)^2\Big\rangle - \Big\langle\sum_{l} \hat{S}^{\alpha}_{l}\Big\rangle^2\Big\}$$

$$= \frac{N}{(g\mu_{\mathrm{B}})^2}\chi^{\alpha},$$

i.e. $\chi^{\alpha}_{\boldsymbol{\kappa}=0} \equiv \chi^{\alpha}$, as required.

Using (13.12) in (13.11) and recalling that the susceptibility of an isolated ion is

$$\chi_0 = \tfrac{1}{3}(g\mu_{\mathrm{B}})^2 S(S+1)\beta, \quad (13.13)$$

we have the result

$$\hbar\int\limits_{-\infty}^{\infty} \mathrm{d}\omega \left\{\frac{1-\exp(-\hbar\omega\beta)}{\hbar\omega\beta}\right\}\{S^{\alpha}(\boldsymbol{\kappa},\omega)-S^{\alpha}_{\mathrm{Bragg}}(\boldsymbol{\kappa},\omega)\} = N\tfrac{1}{3}S(S+1)(\chi^{\alpha}_{\boldsymbol{\kappa}}/\chi_0). \quad (13.14)$$

If the inelasticity of the scattering is small we can replace

$$\{1-\exp(-\hbar\omega\beta)\}/\hbar\omega\beta$$

in the integrand by unity (often called the *quasi-static approximation*). This result represents the generalization of the previous discussion to non-zero $\boldsymbol{\kappa}$, the corresponding cross-section now being proportional to $\chi^{\alpha}_{\boldsymbol{\kappa}}$ in place of χ^{α}.

In discussing the energy dependence of the scattering we choose to define the *spectral weight function*

$$F^{\alpha}(\boldsymbol{\kappa},\omega) = \frac{1}{2\pi}\int\limits_{-\infty}^{\infty} \mathrm{d}t\,\exp(-\mathrm{i}\omega t)\{\mathscr{R}^{\alpha}_{\boldsymbol{\kappa}}(t)/\mathscr{R}^{\alpha}_{\boldsymbol{\kappa}}(t=0)\}, \quad (13.15)$$

which possesses the obvious property

$$\int_{-\infty}^{\infty} \mathrm{d}\omega \; F^{\alpha}(\boldsymbol{\varkappa}, \omega) = 1. \tag{13.16}$$

On combining (13.12) and (13.15) we obtain

$$S^{\alpha}(\boldsymbol{\varkappa}, \omega) - S^{\alpha}_{\text{Bragg}}(\boldsymbol{\varkappa}, \omega) = \frac{N}{(g\mu_{\text{B}})^2}\chi^{\alpha}_{\boldsymbol{\varkappa}} \frac{\omega \exp(\hbar\omega\beta)}{\exp(\hbar\omega\beta)-1} F^{\alpha}(\boldsymbol{\varkappa}, \omega). \tag{13.17}$$

Having decomposed $S^{\alpha}(\boldsymbol{\varkappa}, \omega)$ in this form it is convenient to discuss separately the structure of $\chi^{\alpha}_{\boldsymbol{\varkappa}}$ and $F^{\alpha}(\boldsymbol{\varkappa}, \omega)$.

13.2. Wavelength-dependent susceptibility $\chi^{\alpha}_{\boldsymbol{\varkappa}}$

The general features of $\chi^{\alpha}_{\boldsymbol{\varkappa}}$ in the vicinity of the critical temperature T_{c} can be derived from molecular-field theory. This approximation cannot be expected to yield detailed information, but it is adequate to give an overall picture. To go beyond this level of discussion is extremely difficult and we shall do no more than indicate the various results that have been derived. First we discuss the molecular-field approximation as applied to the Heisenberg ferromagnet.

If the axis of quantization is taken to coincide with the z axis, the molecular-field approximation consists of replacing the Hamiltonian

$$\mathscr{H} = -\sum_{m,n} J(\mathbf{n})\hat{\mathbf{S}}_m \cdot \hat{\mathbf{S}}_{m+n} \tag{13.18}$$

by a set of single-ion Hamiltonians

$$\mathscr{H}_m = -2\sum_{n} J(\mathbf{n})\hat{S}^z_m \langle \hat{S}^z_{m+n}\rangle. \tag{13.19}$$

It follows that the effective magnetic field on $\hat{S}_m$ is $2\sum_{n} J(\mathbf{n})\langle \hat{S}^z_{m+n}\rangle/g\mu_{\text{B}}$ and a simple calculation (compare, for example, the derivation of (5.39)) gives

$$\langle \hat{S}^z_m\rangle = SB_S\Big[\beta \sum_{n} J(\mathbf{n})\langle \hat{S}^z_{m+n}\rangle\Big] \tag{13.20}$$

or

$$\sigma = \frac{\langle \hat{S}^z\rangle}{S} = B_S\Big[S\beta\sigma \sum_{n} J(\mathbf{n})\Big], \tag{13.21}$$

where $B_S[x]$ is the Brillouin function defined by

$$B_S[x] = \left(\frac{2S+1}{2S}\right)\coth\{x(2S+1)\} - \frac{1}{2S}\coth x. \tag{13.22}$$

Let us determine the behaviour of σ near the critical temperature. The small argument expansion of $B_S[x]$ is

$$B_S[x] = \tfrac{2}{3}(S+1)\left\{x - \frac{2x^3}{15}(2S^2+2S+1)\right\}. \tag{13.23}$$

Hence in (13.21) either $\sigma = 0$ or

$$1 = \tfrac{2}{3}(S+1)[S\beta\mathscr{J}(0) - \sigma^2\tfrac{2}{15}(2S^2+2S+1)\{S\beta\mathscr{J}(0)\}^3],$$

where

$$\mathscr{J}(\mathbf{q}) = \sum_n J(\mathbf{n})\exp(i\mathbf{q}\,.\,\mathbf{n}). \tag{13.24}$$

Defining

$$\beta_c = \frac{3}{2S(S+1)\mathscr{J}(0)} = (k_B T_c)^{-1}, \tag{13.25}$$

we find, near T_c,

$$\sigma^2 = \frac{10(S+1)^2}{3(2S^2+2S+1)}\left(\frac{T_c - T}{T_c}\right). \tag{13.26}$$

Thus in the molecular-field approximation σ tends to zero as T tends to T_c like $(T_c - T)^\beta$ where $\beta = \frac{1}{2}$. Both experiment and more sophisticated theories indicate† $\beta \simeq \frac{1}{3}$. The molecular-field result (13.26) predicts that magnetic Bragg peaks decrease linearly with temperature as $T \to T_c$.

We turn now to the calculation of $\chi^\alpha_{\varkappa}$.

Above the critical temperature the system is isotropic and $\langle\hat{S}^\alpha\rangle = 0$. If we impose a magnetic field H^α_m on the system the Hamiltonian (13.18) becomes

$$\begin{aligned}\mathscr{H} &= -2\sum_{m,n} J(\mathbf{n})\hat{S}^\alpha_m\langle\hat{S}^\alpha_{m+n}\rangle + g\mu_B\sum_m \hat{S}^\alpha_m H^\alpha_m \\ &= \sum_m \Big\{g\mu_B H^\alpha_m - 2\sum_n J(\mathbf{n})\langle\hat{S}^\alpha_{m+n}\rangle\Big\}\hat{S}^\alpha_m.\end{aligned} \tag{13.27}$$

Thus, for $T > T_c$,

$$-g\mu_B\langle\hat{S}^\alpha_m\rangle = \chi_0\left\{H^\alpha_m - \left(\frac{2}{g\mu_B}\right)\sum_n J(\mathbf{n})\langle\hat{S}^\alpha_{m+n}\rangle\right\},$$

and hence

$$M^\alpha_{\varkappa} = \chi_0\left\{H^\alpha_{\varkappa} + \frac{2}{(g\mu_B)^2}\mathscr{J}(\boldsymbol{\varkappa})M^\alpha_{\varkappa}\right\},$$

or

$$\chi^\alpha_{\varkappa} = \frac{\chi_0}{1 - \{2/(g\mu_B)^2\}\chi_0\mathscr{J}(\boldsymbol{\varkappa})}. \tag{13.28}$$

Since

$$\frac{2}{(g\mu_B)^2}\chi_0\mathscr{J}(0) = \frac{T_c}{T}$$

(13.28) can be written in the form

$$\chi^\alpha_{\varkappa} = \frac{\chi_c}{(T-T_c)/T_c + \{1 - \mathscr{J}(\boldsymbol{\varkappa})/\mathscr{J}(0)\}}, \tag{13.29}$$

where

$$\chi_c = \frac{T}{T_c}\chi_0.$$

From (13.29) we easily recover the well-known result for the molecular-field susceptibility above the critical temperature. Other theories

† For a review of static critical phenomena see, for example, M. E. Fisher, *Rep. Prog. Phys.* **30**, 615 (1967).

indicate that $\chi^{\alpha}_{\varkappa=0}$ diverges at T_c like $(T_c-T)^{-\gamma}$, $\gamma \sim 1{\cdot}2$ to $1{\cdot}4$,† whereas molecular-field theory gives $\gamma = 1$.

Further, on writing $\boldsymbol{\varkappa} = \boldsymbol{\tau}+\mathbf{q}$ and making use of the property $J(\mathbf{n}) = J(-\mathbf{n})$,

$$\mathscr{J}(\boldsymbol{\varkappa}) = \sum_{n} \exp(i\boldsymbol{\varkappa}\,.\,\mathbf{n})J(\mathbf{n}) = \sum_{n} \exp(i\mathbf{q}\,.\,\mathbf{n})J(\mathbf{n}) \simeq \sum_{n} \{1+\tfrac{1}{2}(i\mathbf{q}\,.\,\mathbf{n})^2\}J(\mathbf{n})$$

for small $|\mathbf{q}|$. For cubic crystals we therefore have

$$\mathscr{J}(\boldsymbol{\varkappa}) \simeq \mathscr{J}(0)-\tfrac{1}{6}q^2J^{(2)}, \tag{13.30}$$

where

$$J^{(2)} = \sum_{n} n^2J(\mathbf{n}). \tag{13.31}$$

Combining (13.29) and (13.30) $\chi^{\alpha}_{\varkappa}$ in the vicinity of a Bragg setting has the form

$$\chi^{\alpha}_{\varkappa} = \frac{\chi_c}{r_1^2(q^2+\kappa_1^2)} \quad (\boldsymbol{\varkappa} = \boldsymbol{\tau}+\mathbf{q}), \tag{13.32}$$

where

$$r_1^2 = \tfrac{1}{6}\{J^{(2)}/\mathscr{J}(0)\}, \tag{13.33}$$

$$\kappa_1^2 = \left(\frac{T-T_c}{T_c}\right)r_1^{-2}. \tag{13.34}$$

To see the significance of κ_1, note for the quasi-static approximation,

$$\langle\Delta\hat{S}^{\alpha}_m(0)\Delta\hat{S}^{\alpha}_n(0)\rangle = \frac{1}{N^2}\sum_{\mathbf{q}} \exp\{i\mathbf{q}\,.\,(\mathbf{m}-\mathbf{n})\}\langle\Delta\hat{S}^{\alpha}_{\mathbf{q}}(0)\Delta\hat{S}^{\alpha}_{-\mathbf{q}}(0)\rangle$$

$$\simeq \frac{1}{N^2}\sum_{\mathbf{q}} \exp\{i\mathbf{q}\,.\,(\mathbf{m}-\mathbf{n})\}N\tfrac{1}{3}S(S+1)(\chi^{\alpha}_{\mathbf{q}}/\chi_0). \tag{13.35}$$

Using the result (13.32) for $\chi^{\alpha}_{\mathbf{q}}$ and replacing the summation over $\mathbf{q}$ by an integration (13.35) gives

$$\langle\Delta\hat{S}^{\alpha}_m(0)\Delta\hat{S}^{\alpha}_n(0)\rangle = \frac{v_0}{12\pi}S(S+1)\frac{T}{T_c}\frac{1}{r_1^2}\frac{\exp(-\kappa_1 r)}{r}, \tag{13.36}$$

where $r = |\mathbf{m}-\mathbf{n}|$. In view of the fact that (13.32) was derived for small $|\mathbf{q}|$, (13.36) gives the asymptotic form of $\langle\Delta\hat{S}^{\alpha}_m(0)\Delta\hat{S}^{\alpha}_n(0)\rangle$, as $|\mathbf{m}-\mathbf{n}| \to \infty$. The result tell us that κ_1 is an inverse correlation range; as $T \to T_c$, $\kappa_1 \to 0$ and the correlation between the spins is extremely long ranged, falling off as the inverse of their separation.

For $T < T_c$ we take the z axis to coincide with the axis of quantization. The calculation of $\chi^{z}_{\varkappa}$ proceeds as follows. From (13.27)

$$\sigma_m = B_S\Big[\tfrac{1}{2}\beta\Big\{2S\sum_{n} J(\mathbf{n})\sigma_{m+n}-g\mu_B H^z_m\Big\}\Big]. \tag{13.37}$$

We write

$$\sigma_m = \sigma-\lambda_m, \tag{13.38}$$

† Ibid.

where λ_m is the change in σ_m from its equilibrium value σ due to H_m, and satisfies $\sigma \gg \lambda_m$. By expanding the Brillouin function about σ, which satisfies eqn (13.21), the λ_m are found to be given by

$$2S\lambda_m = \tfrac{1}{2}\beta\Big\{2S\sum_n J(\mathbf{n})\lambda_{m+n}+g\mu_B H_m^z\Big\}\times$$
$$\times[\operatorname{cosech}^2\{\beta S\sigma\mathscr{J}(0)\}-(2S+1)^2\operatorname{cosech}^2\{(2S+1)\beta S\sigma\mathscr{J}(0)\}]. \quad (13.39)$$

Because we are particularly interested in temperatures just below T_c the function in the second bracket on the right-hand side of (13.39) can be expanded in σ; to order σ^2 it has the value

$$\tfrac{4}{3}S(S+1)-\tfrac{8}{15}S(S+1)(2S^2+2S+1)\{\beta S\sigma\mathscr{J}(0)\}^2. \quad (13.40)$$

Now $$\{\beta S\sigma\mathscr{J}(0)\}^2 = \tfrac{15}{2}(2S^2+2S+1)^{-1}\left(\frac{T_c-T}{T_c}\right), \quad (13.41)$$

so (13.40) reduces to $$\tfrac{4}{3}S(S+1)\left(\frac{3T}{T_c}-2\right). \quad (13.42)$$

Thus

$$S\sum_m \exp(-i\mathbf{q}.\mathbf{m})\lambda_m = \frac{(g\mu_B)^{-1}\chi_c H_{\mathbf{q}}^z}{(T/T_c)-\{\mathscr{J}(\mathbf{q})/\mathscr{J}(0)\}(3T/T_c-2)}, \quad (13.43)$$

which on making use of (13.30) yields the result

$$\chi_{\boldsymbol{\kappa}}^z = \frac{g\mu_B S}{H_{\boldsymbol{\kappa}}^z}\sum_m \exp(-i\boldsymbol{\kappa}.\mathbf{m})\lambda_m \simeq \frac{\chi_c}{r_1^2\{(\kappa_1^z)^2+q^2\}} \quad (\boldsymbol{\kappa} = \boldsymbol{\tau}+\mathbf{q}), \quad (13.44)$$

with $$(\kappa_1^z)^2 = \frac{12\mathscr{J}(0)}{J^{(2)}}\left(\frac{T_c-T}{T_c}\right) \quad (T < T_c), \quad (13.45)$$

and r_1^2 is given by eqn (13.33). On comparing the expression (13.45) with (13.34) we see that both above and below T_c the inverse correlation range κ_1 vanishes at the critical point as $|T_c-T|^{\frac{1}{2}}$, their absolute magnitudes for a given $|T_c-T|$ differing only by a factor of 2. Notice that $\chi_{\boldsymbol{\kappa}}^\alpha$ given by (13.32) and (13.44) diverge at $T = T_c$ only when $\mathbf{q} = 0$.

Finally, we require to calculate $\chi_{\boldsymbol{\kappa}}^x = \chi_{\boldsymbol{\kappa}}^y$ below the critical temperature. The magnetic field is now applied parallel to the x axis, say, so that the operative effective fields seen by the atomic spin at $\mathbf{m}$ are

$$-\frac{2}{g\mu_B}\sum_n J(\mathbf{m}-\mathbf{n})\langle\hat{S}_n^z\rangle \quad \text{along the } z \text{ axis}$$

and $$-\frac{2}{g\mu_B}\sum_n J(\mathbf{m}-\mathbf{n})\langle\hat{S}_n^x\rangle+H_m^x \quad \text{along the } x \text{ axis.}$$

The value of $\langle\hat{S}_m\rangle$ is parallel to the resultant of these two fields so that

$$\langle\hat{S}_m^x\rangle = \langle\hat{S}_m^z\rangle\frac{\left\{g\mu_B H_m^x - 2\sum_n J(\mathbf{m}-\mathbf{n})\langle\hat{S}_n^x\rangle\right\}}{-2\sum_n J(\mathbf{m}-\mathbf{n})\langle\hat{S}_n^z\rangle}.$$

Proceeding as above, we find

$$\chi_{\varkappa}^x = \chi_{\varkappa}^y = \frac{(g\mu_B)^2}{2\{\mathscr{J}(0)-\mathscr{J}(\varkappa)\}}. \tag{13.46}$$

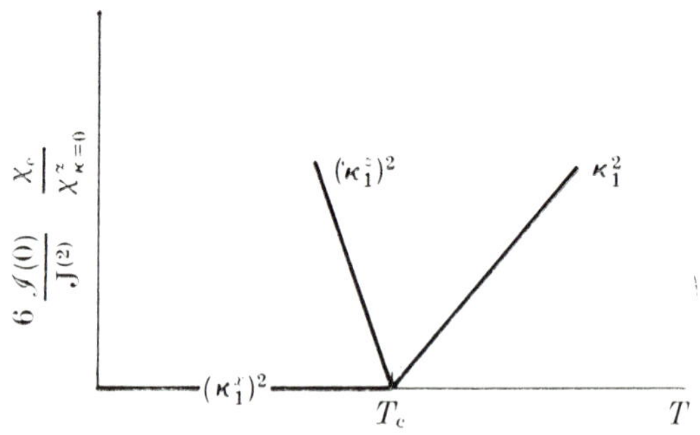

FIG. 13.1

Again expanding about $\boldsymbol{\tau} = \boldsymbol{\varkappa}-\mathbf{q}$,

$$\chi_{\varkappa}^x \simeq \frac{\chi_c}{r_1^2 q^2} \quad (T < T_c), \tag{13.47}$$

with r_1^2 given by (13.33). From (13.46) we conclude that for $T < T_c$, κ_1^x is zero. Fig. 13.1 contrasts the behaviour of the various inverse correlation ranges for T both above and below T_c. This simple molecular-field theory is easily extended to antiferromagnets.

For a simple two-sublattice antiferromagnet we can always find a vector $\mathbf{w}$ such that $\exp(i\mathbf{l}.\mathbf{w})$ has the value $+1$ on one sublattice and -1 on the other:

$$\text{simple cubic antiferromagnet of side } a,\ \mathbf{w} = \frac{\pi}{a}(1,1,1)$$

$$\text{body-centred} \quad ,, \quad ,, \quad ,, \quad a,\ \mathbf{w} = \frac{2\pi}{a}(1,0,0). \tag{13.48}$$

Hence the magnetic moments (in the molecular-field approximation) are simply

$$\langle\hat{S}_l^z\rangle = \exp(i\mathbf{w}.\mathbf{l})\langle\hat{S}^z\rangle \tag{13.49}$$

with $\langle\hat{S}^z\rangle$ given by eqn (13.21). This means that the wavelength-dependent susceptibility of an antiferromagnet above the critical temperature is of the same form as eqn (13.29) for the ferromagnet except that $\mathscr{J}(0)$ is replaced by $\mathscr{J}(\mathbf{w})$, the maximum value of $\mathscr{J}(\mathbf{q})$, and the critical temperature T_N is defined as

$$\beta_N = \frac{3}{2S(S+1)\mathscr{J}(\mathbf{w})} = (k_B T_N)^{-1}, \tag{13.50}$$

i.e.
$$\chi^{\alpha}_{\varkappa} = \frac{\chi_N}{\{(T-T_N)/T_N\}+\{1-\mathscr{J}(\varkappa)/\mathscr{J}(\mathbf{w})\}} \quad (T > T_N). \tag{13.51}$$

Below T_N,

$$\chi^{z}_{\varkappa} = \frac{\chi_N}{(T/T_N)-\{\mathscr{J}(\varkappa)/\mathscr{J}(\mathbf{w})\}(3T/T_N-2)} \simeq \frac{\chi_N}{r_1^2\{(\kappa_1^z)^2+q^2\}} \quad (\varkappa = \boldsymbol{\tau}+\mathbf{q}), \tag{13.52}$$

where
$$(\kappa_1^z)^2 = \frac{12\mathscr{J}(\mathbf{w})}{J^{(2)}}\left(\frac{T_N-T}{T_N}\right) \tag{13.53}$$

and
$$r_1^2 = \frac{1}{6}\frac{J^{(2)}}{\mathscr{J}(\mathbf{w})}. \tag{13.54}$$

In many instances the exchange parameter is not isotropic and the appropriate exchange Hamiltonian becomes

$$-\sum_{m,n} J(\mathbf{n})\hat{\mathbf{S}}_m.\hat{\mathbf{S}}_{m+n}+\hat{\mathbf{S}}_m.\mathbf{D}(\mathbf{m}-\mathbf{n}).\hat{\mathbf{S}}_n, \tag{13.55}$$

where, for example, $\mathbf{D}(\mathbf{m}-\mathbf{n})$ might represent dipole–dipole forces between the atoms.

Define
$$\mathscr{D}(\mathbf{q}) = \sum_n \exp(i\mathbf{q}.\mathbf{n})\mathbf{D}(\mathbf{n}); \tag{13.56}$$

then the spin ordering is determined by the tensor quantities

$$\mathscr{J}(\mathbf{q})\delta_{\alpha\beta}+\mathscr{D}_{\alpha\beta}(\mathbf{q}).$$

We assume the principal axes of this tensor coincide with the crystal axes. Then for a ferromagnetic system one of these principal values, say along the α axis, has a maximum at $\mathbf{q} = 0$. The critical temperature becomes

$$T_\alpha = \frac{2S(S+1)}{3k_B}\{\mathscr{J}(0)+\mathscr{D}_\alpha(0)\}. \tag{13.57}$$

Further, (13.29) is replaced by

$$\chi^{\alpha}_{\varkappa} = \frac{\chi_\alpha}{\{(T-T_\alpha)/T_\alpha\}+1-[\{\mathscr{J}(\varkappa)+\mathscr{D}_\alpha(\varkappa)\}/\{\mathscr{J}(0)+\mathscr{D}_\alpha(0)\}]} \tag{13.58}$$

for $T > T_\alpha$. The corresponding expressions for an antiferromagnet are obtained from (13.58) by replacing $\mathscr{J}(0)$ and $\mathscr{D}_\alpha(0)$ by $\mathscr{J}(\mathbf{w})$ and $\mathscr{D}_\alpha(\mathbf{w})$, respectively.

Below the critical temperature the wavelength-dependent susceptibilities are

$$\chi^z_{\boldsymbol{\kappa}} = \frac{\chi_c}{2\{(T-T_c)/T_c\}+1-[\{\mathscr{J}(\boldsymbol{\kappa})+\mathscr{D}_z(\boldsymbol{\kappa})\}/\{\mathscr{J}(0)+\mathscr{D}_z(0)\}]} \tag{13.59}$$

and

$$\chi^x_{\boldsymbol{\kappa}} = \frac{\chi_x}{\{(T_c-T_x)/T_x\}+1-[\{\mathscr{J}(\boldsymbol{\kappa})+\mathscr{D}_x(\boldsymbol{\kappa})\}/\{\mathscr{J}(0)+\mathscr{D}_x(0)\}]}, \tag{13.60}$$

etc.

Again, the corresponding expressions for an antiferromagnet are obtained from (13.59) and (13.60) by replacing $\mathscr{J}(0)$ and $\mathscr{D}_\alpha(0)$ by $\mathscr{J}(\mathbf{w})$ and $\mathscr{D}_\alpha(\mathbf{w})$, respectively.

13.3. Spectral weight function

(a) General features

There is not at the present time a single unified theory of the dynamics of Heisenberg magnets that adequately covers the entire temperature range. At temperatures well below the critical temperature we have spin-wave theory and the infinite-temperature limit is well understood. A consequence of this state of affairs is that we cannot present a really adequate theory for the evaluation of the spectral weight function $F^\alpha(\boldsymbol{\kappa}, \omega)$ particularly for temperatures close to the critical temperature. However, we can gain considerable insight into its functional form in various ranges of temperature and wave vector by evaluating its moments, namely, for $n \geqslant 1$,

$$\overline{\omega^{2n}} = \int_{-\infty}^{\infty} d\omega\, \omega^{2n} F^\alpha(\boldsymbol{\kappa}, \omega) = \frac{i(-1)^{n+1}}{\hbar \mathscr{R}^\alpha_{\boldsymbol{\kappa}}(t=0)} \frac{\partial^{2n-1}}{\partial t^{2n-1}} \langle[\hat{S}^\alpha_{\boldsymbol{\kappa}}(t), \hat{S}^\alpha_{-\boldsymbol{\kappa}}(0)]\rangle\bigg|_{t=0}. \tag{13.61}$$

The proof of (13.61) is as follows. From the definition of $F^\alpha(\boldsymbol{\kappa}, \omega)$, eqn (13.15), it follows that

$$\int_{-\infty}^{\infty} d\omega\, \omega^n F^\alpha(\boldsymbol{\kappa}, \omega) = i^{-n} \frac{\partial^n}{\partial t^n} \{\mathscr{R}^\alpha_{\boldsymbol{\kappa}}(t)/\mathscr{R}^\alpha_{\boldsymbol{\kappa}}(t=0)\}\bigg|_{t=0}. \tag{13.62}$$

Since $\mathscr{R}^\alpha_{\boldsymbol{\kappa}}(t)$ is an even function of t the right-hand side of (13.62) is non-zero only if n is an even integer. Further (cf. Chapter 8 and Appendix B),

$$\frac{\partial}{\partial t} \mathscr{R}^\alpha_{\boldsymbol{\kappa}}(t) = -\frac{N}{(g\mu_B)^2} \phi^\alpha_{\boldsymbol{\kappa}}(t) = -\frac{i}{\hbar} \langle[\hat{S}^\alpha_{\boldsymbol{\kappa}}(t), \hat{S}^\alpha_{-\boldsymbol{\kappa}}(0)]\rangle \tag{13.63}$$

from which (13.62) immediately follows.

With regard to eqn (13.62) for the moments $\overline{\omega^{2n}}$ note that

$$\mathscr{R}^\alpha_{\boldsymbol{\kappa}}(t=0) \propto \chi^\alpha_{\boldsymbol{\kappa}} \quad \text{(cf. eqn (13.12))},$$

and for magnets with an isotropic exchange the numerator on the right-hand side is proportional to κ^2 for small scattering vectors. In general we shall find that $\overline{\omega^4} \simeq 3(\overline{\omega^2})^2$, which implies that the spectral weight function is approximately Gaussian. This point is further amplified below. Two exceptions to the Gaussian form occur for (a) very small values of $|\boldsymbol{\varkappa}|$ at any value of T and (b) at $T \simeq T_c$ with $\boldsymbol{\varkappa}$ near to an ordering vector. In both cases $\overline{\omega^4} \gg (\overline{\omega^2})^2$ and hence $F^\alpha(\boldsymbol{\varkappa}, \omega)$ more closely resembles a Lorentzian function than a Gaussian. (Note that the spectral weight function is periodic in $\boldsymbol{\varkappa}$.) Both $\overline{\omega^2}$ and $\overline{\omega^4}$ are proportional to κ^2 and hence in case (a) $\overline{\omega^4} \gg (\overline{\omega^2})^2$, while for (b) $\overline{\omega^4} \sim \chi_{\boldsymbol{\varkappa}}^{-1}$ and $\chi_{\boldsymbol{\varkappa}}$ is close to a maximum.

It is useful to note the high-temperature limit of eqn (13.61). For small β

$$\mathscr{R}^\alpha_{\boldsymbol{\varkappa}}(t) \simeq \beta\{\langle \hat{S}^\alpha_{\boldsymbol{\varkappa}}(0)\hat{S}^\alpha_{-\boldsymbol{\varkappa}}(t)\rangle - \langle \hat{S}^\alpha_{\boldsymbol{\varkappa}}(0)\rangle\langle \hat{S}^\alpha_{-\boldsymbol{\varkappa}}(0)\rangle\} = \beta\langle \hat{S}^\alpha_{\boldsymbol{\varkappa}}(0)\hat{S}^\alpha_{-\boldsymbol{\varkappa}}(t)\rangle, \tag{13.64}$$

where the last line follows because in the paramagnetic region $\langle \hat{S}^\alpha_{\boldsymbol{\varkappa}}(0)\rangle = 0$. Utilizing (13.64) we obtain

$$\lim_{\beta\to 0} \int_{-\infty}^{\infty} \mathrm{d}\omega\, \omega^{2n} F^\alpha(\boldsymbol{\varkappa}, \omega) = \frac{(-1)^n}{\langle \hat{S}^\alpha_{\boldsymbol{\varkappa}}(0)\hat{S}^\alpha_{-\boldsymbol{\varkappa}}(0)\rangle} \frac{\partial^{2n}}{\partial t^{2n}} \langle \hat{S}^\alpha_{\boldsymbol{\varkappa}}(0)\hat{S}^\alpha_{-\boldsymbol{\varkappa}}(t)\rangle\Big|_{t=0} \tag{13.65}$$

and the right-hand side is asymptotically independent of temperature.

(b) Evaluation of moments at high temperatures

We consider the calculation of the first few moments of $F^\alpha(\boldsymbol{\varkappa}, \omega)$ for an isotropic ferromagnet and begin by considering $\chi^\alpha_{\boldsymbol{\varkappa}}$.

In the limit $T \to \infty$ there is no correlation between spins and hence it follows that

$$\begin{aligned}\lim_{\beta\to 0} \langle \hat{S}^\alpha_l(0)\hat{S}^\beta_{l'}(0)\rangle &= \delta_{\alpha,\beta}\delta_{l,l'}\langle \hat{S}^\alpha(0)\hat{S}^\alpha(0)\rangle \\ &= \delta_{\alpha,\beta}\delta_{l,l'}\tfrac{1}{3}S(S+1).\end{aligned} \tag{13.66}$$

To calculate the first temperature correction to this result we need to evaluate

$$\langle \hat{S}^\alpha_l(0)\hat{S}^\beta_{l'}(0)\rangle = \mathrm{Tr}\{\exp(-\beta\mathscr{H})\hat{S}^\alpha_l\, \hat{S}^\beta_{l'}\}/\mathrm{Tr}\{\exp(-\beta\mathscr{H})\} \tag{13.67}$$

to first order in β. Clearly

$$\mathrm{Tr}\exp(-\beta\mathscr{H}) \simeq \mathrm{Tr}(1-\beta\mathscr{H}) = 1$$

in virtue of the result (13.66) and the fact that $J(0) = 0$. Hence to order β, (13.67) becomes

$$\langle \hat{S}^\alpha_l(0)\hat{S}^\beta_{l'}(0)\rangle = \mathrm{Tr}(\hat{S}^\alpha_l\, \hat{S}^\beta_{l'} - \beta\mathscr{H}\, \hat{S}^\alpha_l\, \hat{S}^\beta_{l'}). \tag{13.68}$$

Now $$\operatorname{Tr} \mathscr{H} \hat{S}_l^\alpha \hat{S}_{l'}^\beta = -\sum_{m,n} J(\mathbf{m}-\mathbf{n}) \sum_\nu \operatorname{Tr} \hat{S}_m^\nu \hat{S}_n^\nu \hat{S}_l^\alpha \hat{S}_{l'}^\beta$$

and

$$\begin{aligned}\operatorname{Tr} \hat{S}_m^z \hat{S}_n^z \hat{S}_l^\alpha \hat{S}_{l'}^\beta &= \langle \hat{S}^z \hat{S}^\alpha \rangle \langle \hat{S}^z \hat{S}^\beta \rangle (\delta_{m,l}\delta_{n,l'} + \delta_{m,l'}\delta_{n,l}) \\ &= \delta_{\alpha,z}\delta_{\alpha,\beta}\{\tfrac{1}{3}S(S+1)\}^2(\delta_{m,l}\delta_{n,l'} + \delta_{m,l'}\delta_{n,l}).\end{aligned} \tag{13.69}$$

Hence (13.68) reads

$$\delta_{\alpha,\beta}\tfrac{1}{3}S(S+1)\Big\{\delta_{l,l'} + \beta\tfrac{1}{3}S(S+1)\sum_{m,n} J(\mathbf{m}-\mathbf{n}) \sum_\nu \delta_{\alpha,\nu}(\delta_{m,l}\delta_{n,l'} + \delta_{m,l'}\delta_{n,l})\Big\},$$

i.e. $$\langle \hat{S}_l^\alpha(0)\hat{S}_l^\beta(0)\rangle = \delta_{\alpha,\beta}\tfrac{1}{3}S(S+1). \tag{13.70 a}$$

and $$\langle \hat{S}_l^\alpha(0)\hat{S}_{l'}^\beta(0)\rangle = \delta_{\alpha,\beta}\, 2\beta\{\tfrac{1}{3}S(S+1)\}^2 J(\mathbf{l}-\mathbf{l}'), \tag{13.70 b}$$

to first order in β.

Now from (13.64)

$$\begin{aligned}\mathscr{R}_\varkappa^\alpha(t=0) &\simeq \beta\langle \hat{S}_\varkappa^\alpha(0)\hat{S}_{-\varkappa}^\alpha(0)\rangle \\ &= \beta\sum_{l,l'} \exp\{i\boldsymbol{\varkappa}.(\mathbf{l}'-\mathbf{l})\}\langle \hat{S}_l^\alpha(0)\hat{S}_{l'}^\alpha(0)\rangle \\ &\simeq \beta N\tfrac{1}{3}S(S+1)\{1+\tfrac{2}{3}\beta S(S+1)\mathscr{J}(\boldsymbol{\varkappa})\},\end{aligned} \tag{13.71}$$

i.e. $\mathscr{R}_\varkappa^\alpha(t=0)$ tends to zero as $\beta \to 0$ like

$$N\beta\tfrac{1}{3}S(S+1).$$

Further $$\chi_\varkappa^\alpha = \frac{(g\mu_B)^2}{N}\mathscr{R}_\varkappa^\alpha(t=0) \Rightarrow \chi_0.$$

If we retain the term in β^2 in (13.71), then

$$\chi_\varkappa^\alpha = \chi_0\left\{1+\frac{2}{(g\mu_B)^2}\chi_0\mathscr{J}(\boldsymbol{\varkappa})\right\}, \tag{13.72}$$

which are seen to be the first two terms in the expansion of the molecular-field result for $\chi_\varkappa^\alpha$, eqn (13.28). Thus the molecular-field approximation for the susceptibility becomes exact in the limit $\beta \to 0$; this is not surprising since there is no short-range order between the spins at $T = \infty$.

Turning now to the calculation of $\overline{\omega^2}$, we require

$$\begin{aligned}\overline{\omega^2} &= \frac{i}{\hbar}\frac{1}{\mathscr{R}_\varkappa^\alpha(t=0)}\frac{\partial}{\partial t}\langle[\hat{S}_\varkappa^\alpha(t), \hat{S}_{-\varkappa}^\alpha(0)]\rangle\Big|_{t=0} \\ &= \frac{1}{\hbar^2}\frac{1}{\mathscr{R}_\varkappa^\alpha(t=0)}\sum_{l,l'}\exp\{i\boldsymbol{\varkappa}.(\mathbf{l}'-\mathbf{l})\}\langle[[\hat{S}_l^\alpha, \mathscr{H}], \hat{S}_{l'}^\alpha]\rangle.\end{aligned} \tag{13.73}$$

Now $$[\hat{S}_l^x, \mathscr{H}] = -2i\sum_m J(\mathbf{l}-\mathbf{m})(\hat{S}_l^z\hat{S}_m^y - \hat{S}_l^y\hat{S}_m^z) \tag{13.74}$$

and

$$\begin{aligned}&\sum_{l,l'}\exp\{i\boldsymbol{\varkappa}.(\mathbf{l}'-\mathbf{l})\}[[\hat{S}_l^x, \mathscr{H}], \hat{S}_{l'}^x] \\ &\quad = 2\sum_{l,m} J(\mathbf{l}-\mathbf{m})(\hat{S}_l^z\hat{S}_m^z + \hat{S}_l^y\hat{S}_m^y)[1-\exp\{i\boldsymbol{\varkappa}.(\mathbf{l}-\mathbf{m})\}].\end{aligned} \tag{13.75}$$

Assembling these results,

$$\overline{\omega^2} = \frac{4N}{\hbar^2 \mathscr{R}^{\alpha}_{\varkappa}(t=0)} \sum_m J(\mathbf{m})(1-\cos\boldsymbol{\varkappa}.\mathbf{m})\langle \hat{S}^z_0 \hat{S}^z_m\rangle. \tag{13.76}$$

Evaluating this expression in the high-temperature limit merely requires the use of (13.70 b) and (13.71):

$$\overline{\omega^2} = \frac{8S(S+1)}{3\hbar^2} \sum_m J^2(\mathbf{m})(1-\cos\boldsymbol{\varkappa}.\mathbf{m}). \tag{13.77}$$

For nearest-neighbour exchange J,

$$\overline{\omega^2} = \frac{4J(S+1)}{3\hbar^2}\,\omega_{\varkappa}, \tag{13.78}$$

where $\omega_{\varkappa}$ is the dispersion relation for linear spin waves (cf. eqn (9.16)).
Also, writing $\boldsymbol{\varkappa} = \boldsymbol{\tau}+\mathbf{q}$ we obtain for small $\mathbf{q}$

$$\overline{\omega^2} = \frac{4S(S+1)}{9\hbar^2}\, J^{(2)} J q^2. \tag{13.79}$$

In some instances the appropriate spin Hamiltonian is of the form

$$\mathscr{H} = \mathscr{H}_{\mathrm{E}} + \mathscr{H}_{\mathrm{A}}, \tag{13.80}$$

where $\mathscr{H}_{\mathrm{E}}$ is the exchange Hamiltonian considered above and $\mathscr{H}_{\mathrm{A}}$ describes single ion anisotropy effects (cf. § 9.5). For this more general Hamiltonian†

$$\overline{\omega^2} = (\overline{\omega^2})_{\mathrm{E}} + (\overline{\omega^2})_{\mathrm{A}} \quad (\beta = 0), \tag{13.81}$$

where $(\overline{\omega^2})_{\mathrm{E}}$ is the second moment calculated with respect to $\mathscr{H}_{\mathrm{E}}$, and $(\overline{\omega^2})_{\mathrm{A}}$ the second moment evaluated with respect to $\mathscr{H}_{\mathrm{A}}$. The identity (13.81) is correct only in the limit of infinite temperatures.

The evaluation of the fourth moment of $F^{\alpha}(\boldsymbol{\varkappa}, \omega)$ requires considerable effort. For a pure exchange Hamiltonian the result is‡

$$\begin{aligned}\hbar^4\overline{\omega^4} = 32\left\{\frac{S(S+1)}{3}\right\}^2 \Big[-\sum_m J^4(\mathbf{m})\{1-\cos\boldsymbol{\varkappa}.\mathbf{m}\}\left\{4+\frac{3}{2S(S+1)}\right\} + \\ + \Big[\sum_m J^2(\mathbf{m})\{1-\cos\boldsymbol{\varkappa}.\mathbf{m}\}\Big]\Big[\sum_m J^2(\mathbf{m})\{7-3\cos\boldsymbol{\varkappa}.\mathbf{m}\}\Big] - \\ - 2\sum_{m,n} J(\mathbf{m})J(\mathbf{n})J(\mathbf{m}-\mathbf{n})\{1-\exp(\mathrm{i}\boldsymbol{\varkappa}.\mathbf{n})\}[J(\mathbf{n})\exp(-\mathrm{i}\boldsymbol{\varkappa}.\mathbf{m}) + \\ + J(\mathbf{m}-\mathbf{n})\{1-\exp(-\mathrm{i}\boldsymbol{\varkappa}.\mathbf{m})\}]\Big].\end{aligned} \tag{13.82}$$

It can be shown from this expression that $\overline{\omega^4}$ is proportional to q^2 for small $\mathbf{q}$, as noted previously.

† N. W. Dalton, private communication.
‡ M. F. Collins and W. Marshall, *Proc. phys. Soc.* **92**, 390 (1967).

In experiments with polycrystals the cross-section is averaged over all crystal orientations. The second and fourth moments in this instance are

$$\hbar^2\overline{\omega^2} = \tfrac{8}{3}S(S+1)\sum_m J^2(\mathbf{m})\left\{1-\frac{\sin(\kappa m)}{(\kappa m)}\right\} \tag{13.83 a}$$

and

$$\hbar^4\overline{\omega^4} = 32\left\{\frac{S(S+1)}{3}\right\}^2\left[\!\!\left[-\sum_m J^4(\mathbf{m})\left\{1-\frac{\sin(\kappa m)}{(\kappa m)}\right\}\left\{4+\frac{3}{2S(S+1)}\right\}+\right.\right.$$

$$+\sum_{m,m'} J(\mathbf{m})J(\mathbf{m}')\left(J(\mathbf{m})J(\mathbf{m}')\left\{7-10\frac{\sin(\kappa m')}{(\kappa m')}+3\frac{\sin(\kappa|\mathbf{m}-\mathbf{m}'|)}{\kappa|\mathbf{m}-\mathbf{m}'|}\right\}-\right.$$

$$-2J(\mathbf{m}-\mathbf{m}')\left[J(\mathbf{m}')\left\{\frac{\sin(\kappa m)}{(\kappa m)}-\frac{\sin(\kappa|\mathbf{m}-\mathbf{m}'|)}{\kappa|\mathbf{m}-\mathbf{m}'|}\right\}+\right.$$

$$\left.\left.\left.\left.+J(\mathbf{m}-\mathbf{m}')\left\{1-2\frac{\sin(\kappa m)}{(\kappa m)}+\frac{\sin(\kappa|\mathbf{m}-\mathbf{m}'|)}{\kappa|\mathbf{m}-\mathbf{m}'|}\right\}\right]\right)\right]\!\!\right]. \tag{13.83 b}$$

In discussing the implications of the above expressions for the second and fourth moments it is convenient to introduce the parameter

$$\alpha = \frac{\overline{\omega^4}}{3(\overline{\omega^2})^2}-1. \tag{13.84}$$

If the spectral weight function is a Gaussian function of ω, $\alpha = 0$, while for a Lorentzian function it is infinite. α is shown in Fig. 13.2 for $\boldsymbol{\varkappa}$ along the three symmetry axis of a simple cubic lattice with nearest-neighbour exchange coupling and $S = 5/2$. It is evident from the behaviour of α that at high temperatures the spectral weight function is approximately Gaussian for large values of the scattering vector but deviates sharply from this form for small $\boldsymbol{\varkappa}$ where it is more closely approximated by a Lorentzian function. This is in agreement with our previous general observations regarding the ω dependence of $F^{\alpha}(\boldsymbol{\varkappa}, \omega)$.

It is quite apparent from the simple example considered above that a major part of the calculation of the moments of $F^{\alpha}(\boldsymbol{\varkappa}, \omega)$ is the evaluation of traces of products of spin operators. This is particularly true when single-ion anisotropy terms are included in the Hamiltonian. However, it can be shown that†

$$\mathrm{Tr}(\hat{S}^{\nu_1}\hat{S}^{\nu_2}\ldots\hat{S}^{\nu_n}) = (2S+1)\sum_{p=1}^{n} b_p X^p, \tag{13.85}$$

† N. W. Dalton, *Phys. Lett.* **28A**, 172 (1968).

where $\hat{S}^{\nu} = \hat{S}^{z}, \hat{S}^{+}, \hat{S}^{-}$ according to $\nu = 0$, $+$ and $-$, respectively, and $X = S(S+1)$.

Table 13.1 contains the coefficients b_p (scaled by a certain factor) for $1 \leqslant p \leqslant 8$ as calculated by Dalton and Rimmer.† The rules for evaluating a particular trace are as follows.

(1) Transfer leading $\hat{S}^{z}$'s to end of product until sequence begins with either $\hat{S}^{+}$ or $\hat{S}^{-}$ (recall that the trace is invariant under cyclic permutation of operators).

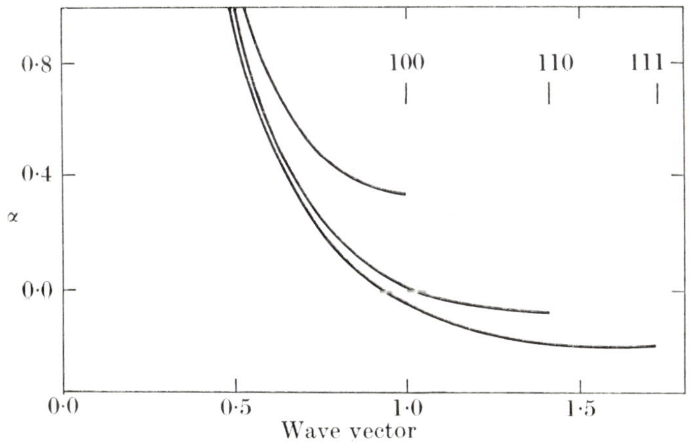

FIG. 13.2. Variation of α, eqn (13.84), with wave vector along the principal symmetry axes of a simple cubic lattice. α is a measure of the departure of the shape of the distribution from Gaussian. Nearest-neighbour exchange interactions are assumed.

(2) If sequence commences with $\hat{S}^{+}$ interchange with an $\hat{S}^{-}$. This operation multiplies the trace by $(-)^n$.

(3) Construct the n-digit integer (the *identifier*) from the sequence by means of the correspondence

$$\hat{S}^{z} \equiv 0, \qquad \hat{S}^{-} \equiv 1, \qquad \hat{S}^{+} \equiv 2. \tag{13.86}$$

The traces are compiled in increasing length of sequence n. For sequences of given length the identifiers are listed in increasing order of numerical value in column 2 of the table. Column 3 contains the representation of the operator sequence. The listed coefficients are scaled by a factor whose magnitude depends on n: the values of b_p are obtained by dividing the listed numbers by the integers in Table 13.2.

† N. W. Dalton and D. E. Rimmer, Atomic Energy Research Establishment, Report No. A.E.R.E.-R. 5958.

TABLE 13.1

		Power of X	(1)	2-order
1	00	(00)	1	
2	12	$(-+)$	2	

		Power of X	(1)	3-order
1	000	(000)	0	
2	102	$(-0+)$	1	
3	120	$(-+0)$	-1	

		Power of X	(1)	(2)	4-order
1	0000	(0000)	-1	3	
2	1002	$(-00+)$	1	2	
3	1020	$(-0+0)$	-4	2	
4	1122	$(--++)$	-6	8	
5	1200	$(-+00)$	1	2	
6	1212	$(-+-+)$	4	8	
7	1221	$(-++-)$	-6	8	

		Power of X	(1)	(2)	5-order
1	00000	(00000)	0	0	
2	10002	$(-000+)$	-1	3	
3	10020	$(-00+0)$	-2	1	
4	10122	$(-0-++)$	0	0	
5	10200	$(-0+00)$	2	-1	
6	10212	$(-0+-+)$	2	4	
7	10221	$(-0++-)$	-6	8	
8	11022	$(--0++)$	-6	8	
9	11202	$(--+0+)$	0	0	
10	11220	$(--++0)$	6	-8	
11	12000	$(-+000)$	1	-3	
12	12012	$(-+0-+)$	-2	-4	
13	12021	$(-+0+-)$	0	0	
14	12102	$(-+-0+)$	2	4	
15	12120	$(-+-+0)$	-2	-4	
16	12201	$(-++0-)$	6	-8	
17	12210	$(-++-0)$	0	0	

		Power of X	(1)	(2)	(3)	6-order
1	000000	(000000)	5	-15	15	
2	100002	$(-0000+)$	-5	8	6	
3	100020	$(-000+0)$	2	-13	6	
4	100122	$(-00-++)$	12	-22	8	
5	100200	$(-00+00)$	16	-20	6	
6	100212	$(-00+-+)$	-2	20	8	
7	100221	$(-00++-)$	-30	34	8	
8	101022	$(-0-0++)$	12	-22	8	
9	101202	$(-0-+0+)$	12	-22	8	
10	101220	$(-0-++0)$	12	-22	8	
11	102000	$(-0+000)$	2	-13	6	
12	102012	$(-0+0-+)$	-16	-8	8	
13	102021	$(-0+0+-)$	12	-22	8	
14	102102	$(-0+-0+)$	-2	20	8	
15	102120	$(-0+-+0)$	-16	-8	8	

TABLE 13.1 (*cont.*)

		Power of X	(1)	(2)	(3)	6-order
16	102201	(−0++0−)	54	−78	8	
17	102210	(−0++−0)	12	−22	8	
18	110022	(−−00++)	−30	34	8	
19	110202	(−−0+0+)	12	−22	8	
20	110220	(−−0++0)	54	−78	8	
21	111222	(−−−+++)	72	−132	48	
22	112002	(−−+00+)	12	−22	8	
23	112020	(−−+0+0)	12	−22	8	
24	112122	(−−+−++)	−12	−20	48	
25	112200	(−−++00)	−30	34	8	
26	112212	(−−++−+)	−12	−20	48	
27	112221	(−−+++−)	72	−132	48	
28	120000	(−+0000)	−5	8	6	
29	120012	(−+00−+)	−2	20	8	
30	120021	(−+00+−)	12	−22	8	
31	120102	(−+0−0+)	−16	−8	8	
32	120120	(−+0−+0)	−2	20	8	
33	120201	(−+0+0−)	12	−22	8	
34	120210	(−+0+−0)	12	−22	8	
35	121002	(−+−00+)	−2	20	8	
36	121020	(−+−0+0)	−16	−8	8	
37	121122	(−+−−++)	−12	−20	48	
38	121200	(−+−+00)	−2	20	8	
39	121212	(−+−+−+)	16	36	48	
40	121221	(−+−++−)	−12	−20	48	
41	122001	(−++00−)	−30	34	8	
42	122010	(−++0−0)	12	−22	8	
43	122100	(−++−00)	12	−22		
44	122112	(−++−−+)	−12	−20	48	
45	122121	(−++−+−)	−12	−20	48	
46	122211	(−+++−−)	72	−132	48	

		Power of X	(1)	(2)	(3)	7-order
1	0000000	(0000000)	0	0	0	
2	1000002	(−00000+)	5	−15	15	
3	1000020	(−0000+0)	10	−23	9	
4	1000122	(−000−++)	0	0	0	
5	1000200	(−000+00)	8	−10	3	
6	1000212	(−000+−+)	−10	16	12	
7	1000221	(−000++−)	−6	−10	24	
8	1001022	(−00−0++)	12	−22	8	
9	1001202	(−00−+0+)	0	0	0	
10	1001220	(−00−++0)	−12	22	−8	
11	1002000	(−00+000)	−8	10	−3	
12	1002012	(−00+0−+)	−8	−4	4	
13	1002021	(−00+0+−)	24	−44	16	
14	1002102	(−00+−0+)	−10	16	12	
15	1002120	(−00+−+0)	−8	−4	4	
16	1002201	(−00++0−)	54	−78	8	
17	1002210	(−00++−0)	24	−44	16	
18	1010022	(−0−00++)	24	−44	16	
19	1010202	(−0−0+0+)	12	−22	8	
20	1010220	(−0−0++0)	0	0	0	

TABLE 13.1 (*cont.*)

		Power of X	(1)	(2)	(3)	7-order
21	1011222	(−0− −+++)	−36	66	−24	
22	1012002	(−0−+00+)	0	0	0	
23	1012020	(−0−+0+0)	−12	22	−8	
24	1012122	(−0−+−++)	−12	22	−8	
25	1012200	(−0−++00)	−24	44	−16	
26	1012212	(−0−++−+)	12	−22	8	
27	1012221	(−0−+++−)	36	−66	24	
28	1020000	(−0+0000)	−10	23	−9	
29	1020012	(−0+00−+)	8	4	−4	
30	1020021	(−0+00+−)	12	−22	8	
31	1020102	(−0+0−0+)	−8	−4	4	
32	1020120	(−0+0−+0)	8	4	−4	
33	1020201	(−0+0+0−)	0	0	0	
34	1020210	(−0+0+−0)	12	−22	8	
35	1021002	(−0+−00+)	−10	16	12	
36	1021020	(−0+−0+0)	−8	−4	4	
37	1021122	(−0+−−++)	12	−22	8	
38	1021200	(−0+−+00)	8	4	−4	
39	1021212	(−0+−+−+)	8	18	24	
40	1021221	(−0+−++−)	−24	2	40	
41	1022001	(−0++00−)	−54	78	−8	
42	1022010	(−0++0−0)	0	0	0	
43	1022100	(−0++−00)	12	−22	8	
44	1022112	(−0++−−+)	−24	2	40	
45	1022121	(−0++−+−)	0	−42	56	
46	1022211	(−0+++−−)	108	−198	72	
47	1100022	(− −000++)	−6	−10	24	
48	1100202	(− −00+0+)	24	−44	16	
49	1100220	(− −00++0)	54	−78	8	
50	1101222	(− −0−+++)	36	−66	24	
51	1102002	(− −0+00+)	12	−22	8	
52	1102020	(− −0+0+0)	0	0	0	
53	1102122	(− −0+−++)	−24	2	40	
54	1102200	(− −0++00)	−54	78	−8	
55	1102212	(− −0++−+)	0	−42	56	
56	1102221	(− −0+++−)	108	−198	72	
57	1110222	(− − −0+++)	108	−198	72	
58	1112022	(− − −+0++)	36	−66	24	
59	1112202	(− − −++0+)	−36	66	−24	
60	1112220	(− − −+++0)	−108	198	−72	
61	1120002	(− −+000+)	0	0	0	
62	1120020	(− −+00+0)	−12	22	−8	
63	1120122	(− −+0−++)	−12	22	−8	
64	1120200	(− −+0+00)	−24	44	−16	
65	1120212	(− −+0+−+)	12	−22	8	
66	1120221	(− −+0++−)	36	−66	24	
67	1121022	(− −+−0++)	−24	2	40	
68	1121202	(− −+−+0+)	−12	22	−8	
69	1121220	(− −+−++0)	0	42	−56	
70	1122000	(− −++000)	6	10	−24	

TABLE 13.1 (*cont.*)

		Power of X	(1)	(2)	(3)	7-order
71	1122012	(−−++0−+)	24	−2	−40	
72	1122021	(−−++0+−)	−36	66	−24	
73	1122102	(−−++−0+)	12	−22	8	
74	1122120	(−−++−+0)	24	−2	−40	
75	1122201	(−−+++0−)	−108	198	−72	
76	1122210	(−−+++−0)	−36	66	−24	
77	1200000	(−+00000)	−5	15	−15	
78	1200012	(−+000−+)	10	−16	−12	
79	1200021	(−+000+−)	0	0	0	
80	1200102	(−+00−0+)	8	4	−4	
81	1200120	(−+00−+0)	10	−16	−12	
82	1200201	(−+00+0−)	−12	22	−8	
83	1200210	(−+00+−0)	0	0	0	
84	1201002	(−+0−00+)	−8	−4	4	
85	1201020	(−+0−0+0)	8	4	−4	
86	1201122	(−+0−−++)	24	−2	−40	
87	1201200	(−+0−+00)	10	−16	−12	
88	1201212	(−+0−+−+)	−8	−18	−24	
89	1201221	(−+0−++−)	−12	22	−8	
90	1202001	(−+0+00−)	−24	44	−16	
91	1202010	(−+0+0−0)	−12	22	−8	
92	1202100	(−+0+−00)	0	0	0	
93	1202112	(−+0+−−+)	−12	22	−8	
94	1202121	(−+0+−+−)	12	−22	8	
95	1202211	(−+0++−−)	36	−66	24	
96	1210002	(−+−000+)	−10	16	12	
97	1210020	(−+−00+0)	−8	−4	4	
98	1210122	(−+−0−++)	12	−22	8	
99	1210200	(−+−0+00)	8	4	−4	
100	1210212	(−+−0+−+)	8	18	24	
101	1210221	(−+−0++−)	−24	2	40	
102	1211022	(−+−−0++)	0	−42	56	
103	1211202	(−+−−+0+)	12	−22	8	
104	1211220	(−+−−++0)	24	−2	−40	
105	1212000	(−+−+000)	10	−16	−12	
106	1212012	(−+−+0−+)	−8	−18	−24	
107	1212021	(−+−+0+−)	−12	22	−8	
108	1212102	(−+−+−0+)	8	18	24	
109	1212120	(−+−+−+0)	−8	−18	−24	
110	1212201	(−+−++0−)	0	42	−56	
111	1212210	(−+−++−0)	−12	22	−8	
112	1220001	(−++000−)	6	10	−24	
113	1220010	(−++00−0)	−24	44	−16	
114	1220100	(−++0−00)	−12	22	−8	
115	1220112	(−++0−−+)	0	42	−56	
116	1220121	(−++0−+−)	24	−2	−40	
117	1220211	(−++0+−−)	−36	66	−24	
118	1221000	(−++−000)	0	0	0	
119	1221012	(−++−0−+)	−12	22	−8	
120	1221021	(−++−0+−)	12	−22	8	

TABLE 13.1 (*cont.*)

		Power of X	(1)	(2)	(3)	7-order
121	1221102	(−++−−0+)	−24	2	40	
122	1221120	(−++−−+0)	−12	22	−8	
123	1221201	(−++−+0−)	24	−2	−40	
124	1221210	(−++−+−0)	12	−22	8	
125	1222011	(−+++0−−)	−108	198	−72	
126	1222101	(−+++−0−)	−36	66	−24	
127	1222110	(−+++−−0)	36	−66	24	

		Power of X	(1)	(2)	(3)	(4)	8-order
1	00000000	(00000000)	−21	63	−70	35	
2	10000002	(−000000+)	21	−48	25	10	
3	10000020	(−00000+0)	6	−3	−20	10	
4	10000122	(−0000−++)	−36	78	−46	8	
5	10000200	(−0000+00)	−24	66	−47	10	
6	10000212	(−0000+−+)	−6	−12	44	8	
7	10000221	(−0000++−)	54	−150	98	8	
8	10001022	(−000−0++)	−36	78	−46	8	
9	10001202	(−000−+0+)	−36	78	−46	8	
10	10001220	(−000−++0)	−36	78	−46	8	
11	10002000	(−000+000)	−48	96	−56	10	
12	10002012	(−000+0−+)	24	−60	8	8	
13	10002021	(−000+0+−)	72	−120	26	8	
14	10002102	(−000+−0+)	−6	−12	44	8	
15	10002120	(−000+−+0)	24	−60	8	8	
16	10002201	(−000++0−)	90	−90	−46	8	
17	10002210	(−000++−0)	72	−120	26	8	
18	10010022	(−00−00++)	0	12	−22	8	
19	10010202	(−00−0+0+)	−36	78	−46	8	
20	10010220	(−00−0++0)	−72	144	−70	8	
21	10011222	(−00−−+++)	−36	90	−68	16	
22	10012002	(−00−+00+)	−36	78	−46	8	
23	10012020	(−00−+0+0)	−36	78	−46	8	
24	10012122	(−00−+−++)	36	−42	−20	16	
25	10012200	(−00−++00)	0	12	−22	8	
26	10012212	(−00−++−+)	36	−42	−20	16	
27	10012221	(−00−+++−)	−36	90	−68	16	
28	10020000	(−00+0000)	−24	66	−47	10	
29	10020012	(−00+00−+)	48	−48	−4	8	
30	10020021	(−00+00+−)	0	12	−22	8	
31	10020102	(−00+0−0+)	24	−60	8	8	
32	10020120	(−00+0−+0)	48	−48	−4	8	
33	10020201	(−00+0+0−)	−72	144	−70	8	
34	10020210	(−00+0+−0)	0	12	−22	8	
35	10021002	(−00+−00+)	−6	−12	44	8	
36	10021020	(−00+−0+0)	24	−60	8	8	
37	10021122	(−00+−−++)	36	−42	−20	16	
38	10021200	(−00+−+00)	48	−48	−4	8	
39	10021212	(−00+−+−+)	−24	54	52	16	
40	10021221	(−00+−++−)	−72	30	76	16	
41	10022001	(−00++00−)	−234	378	−94	8	
42	10022010	(−00++0−0)	−72	144	−70	8	
43	10022100	(−00++−00)	0	12	−22	8	
44	10022112	(−00++−−+)	−72	30	76	16	
45	10022121	(−00++−+−)	72	−234	172	16	

TABLE 13.1 (*cont.*)

		Power of X	(1)	(2)	(3)	(4)	8-order
46	10022211	(−00+++−−)	396	−702	220	16	
47	10100022	(−0−000++)	72	−120	26	8	
48	10100202	(−0−00+0+)	0	12	−22	8	
49	10100220	(−0−00++0)	−72	144	−70	8	
50	10101222	(−0−0−+++)	−144	288	−140	16	
51	10102002	(−0−0+00+)	−36	78	−46	8	
52	10102020	(−0−0+0+0)	−72	144	−70	8	
53	10102122	(−0−0+−++)	0	24	−44	16	
54	10102200	(−0−0++00)	−72	144	−70	8	
55	10102212	(−0−0++−+)	72	−108	4	16	
56	10102221	(−0−0+++−)	72	−108	4	16	
57	10110222	(−0−−0+++)	−252	486	−212	16	
58	10112022	(−0−−+0++)	−144	288	−140	16	
59	10112202	(−0−−++0+)	−36	90	−68	16	
60	10112220	(−0−−+++0)	72	−108	4	16	
61	10120002	(−0−+000+)	−36	78	−46	8	
62	10120020	(−0−+00+0)	−36	78	−46	8	
63	10120122	(−0−+0−++)	36	−42	−20	16	
64	10120200	(−0−+0+00)	0	12	−22	8	
65	10120212	(−0−+0+−+)	36	−42	−20	16	
66	10120221	(−0−+0++−)	−36	90	−68	16	
67	10121022	(−0−+−0++)	0	24	−44	16	
68	10121202	(−0−+−+0+)	36	−42	−20	16	
69	10121220	(−0−+−++0)	72	−108	4	16	
70	10122000	(−0−++000)	72	−120	26	8	
71	10122012	(−0−++0−+)	0	24	−44	16	
72	10122021	(−0−++0+−)	−144	288	−140	16	
73	10122102	(−0−++−0+)	36	−42	−20	16	
74	10122120	(−0−++−+0)	0	24	−44	16	
75	10122201	(−0−+++0−)	−252	486	−212	16	
76	10122210	(−0−+++−0)	−144	288	−140	16	
77	10200000	(−0+00000)	6	−3	−20	10	
78	10200012	(−0+000−+)	24	−60	8	8	
79	10200021	(−0+000+−)	−36	78	−46	8	
80	10200102	(−0+00−0+)	48	−48	−4	8	
81	10200120	(−0+00−+0)	24	−60	8	8	
82	10200201	(−0+00+0−)	−72	144	−70	8	
83	10200210	(−0+00+−0)	−36	78	−46	8	
84	10201002	(−0+0−00+)	24	−60	8	8	
85	10201020	(−0+0−0+0)	48	−48	−4	8	
86	10201122	(−0+0−−++)	0	24	−44	16	
87	10201200	(−0+0−+00)	24	−60	8	8	
88	10201212	(−0+0−+−+)	−48	0	−20	16	
89	10201221	(−0+0−++−)	0	24	−44	16	
90	10202001	(−0+0+00−)	−72	144	−70	8	
91	10202010	(−0+0+0−0)	−72	144	−70	8	
92	10202100	(−0+0+−00)	−36	78	−46	8	
93	10202112	(−0+0+−−+)	0	24	−44	16	
94	10202121	(−0+0+−+−)	72	−108	4	16	
95	10202211	(−0+0++−−)	72	−108	4	16	

TABLE 13.1 (*cont.*)

		Power of X	(1)	(2)	(3)	(4)	8-order
96	10210002	(−0+−000+)	−6	−12	44	8	
97	10210020	(−0+−00+0)	24	−60	8	8	
98	10210122	(−0+−0−++)	36	−42	−20	16	
99	10210200	(−0+−0+00)	48	−48	−4	8	
100	10210212	(−0+−0+−+)	−24	54	52	16	
101	10210221	(−0+−0++−)	−72	30	76	16	
102	10211022	(−0+−−0++)	72	−108	4	16	
103	10211202	(−0+−−+0+)	36	−42	−20	16	
104	10211220	(−0+−−++0)	0	24	−44	16	
105	10212000	(−0+−+000)	24	−60	8	8	
106	10212012	(−0+−+0−+)	−48	0	−20	16	
107	10212021	(−0+−+0+−)	0	24	−44	16	
108	10212102	(−0+−+−0+)	−24	54	52	16	
109	10212120	(−0+−+−+0)	−48	0	−20	16	
110	10212201	(−0+−++0−)	72	18	−164	16	
111	10212210	(−0+−++−0)	0	24	−44	16	
112	10220001	(−0++000−)	90	−90	−46	8	
113	10220010	(−0++00−0)	−72	144	−70	8	
114	10220100	(−0++0−00)	−72	144	−70	8	
115	10220112	(−0++0−−+)	72	18	−164	16	
116	10220121	(−0++0−+−)	72	18	−164	16	
117	10220211	(−0++0+−−)	−252	486	−212	16	
118	10221000	(−0++−000)	−36	78	−46	8	
119	10221012	(−0++−0−+)	0	24	−44	16	
120	10221021	(−0++−0+−)	72	−108	4	16	
121	10221102	(−0++−−0+)	−72	30	76	16	
122	10221120	(−0++−−+0)	0	24	−44	16	
123	10221201	(−0++−+0−)	72	18	−164	16	
124	10221210	(−0++−+−0)	72	−108	4	16	
125	10222011	(−0+++0−−)	−576	1080	−428	16	
126	10222101	(−0+++−0−)	−252	486	−212	16	
127	10222110	(−0+++−−0)	72	−108	4	16	
128	11000022	(−−0000++)	54	−150	98	8	
129	11000202	(−−000+0+)	72	−120	26	8	
130	11000220	(−−000++0)	90	−90	−46	8	
131	11001222	(−−00−+++)	−36	90	−68	16	
132	11002002	(−−00+00+)	0	12	−22	8	
133	11002020	(−−00+0+0)	−72	144	−70	8	
134	11002122	(−−00+−++)	−72	30	76	16	
135	11002200	(−−00++00)	−234	378	−94	8	
136	11002212	(−−00++−+)	72	−234	172	16	
137	11002221	(−−00+++−)	396	−702	220	16	
138	11010222	(−−0−0+++)	72	−108	4	16	
139	11012022	(−−0−+0++)	−36	90	−68	16	
140	11012202	(−−0−++0+)	−144	288	−140	16	
141	11012220	(−−0−+++0)	−252	486	−212	16	
142	11020002	(−−0+000+)	−36	78	−46	8	
143	11020020	(−−0+00+0)	−72	144	−70	8	
144	11020122	(−−0+0−++)	0	24	−44	16	
145	11020200	(−−0+0+00)	−72	144	−70	8	

TABLE 13.1 (*cont.*)

		Power of X	(1)	(2)	(3)	(4)	8-order
146	11020212	(−−0+0+−+)	72	−108	4	16	
147	11020221	(−−0+0++−)	72	−108	4	16	
148	11021022	(−−0+−0++)	−72	30	76	16	
149	11021202	(−−0+−+0+)	0	24	−44	16	
150	11021220	(−−0+−++0)	72	18	−164	16	
151	11022000	(−−0++000)	90	−90	−46	8	
152	11022012	(−−0++0−+)	72	18	−164	16	
153	11022021	(−−0++0+−)	−252	486	−212	16	
154	11022102	(−−0++−0+)	72	−108	4	16	
155	11022120	(−−0++−+0)	72	18	−164	16	
156	11022201	(−−0+++0−)	−576	1080	−428	16	
157	11022210	(−−0+++−0)	−252	486	−212	16	
158	11100222	(−−−00+++)	396	−702	220	16	
155	11102022	(−−−0+0++)	72	−108	4	16	
160	11102202	(−−−0++0+)	−252	486	−212	16	
161	11102220	(−−−0+++0)	−576	1080	−428	16	
162	11112222	(−−−−++++)	−720	1512	−832	128	
163	11120022	(−−−+00++)	−36	90	− 68	16	
164	11120202	(−−−+0+0+)	−144	288	−140	16	
165	11120220	(−−−+0++0)	−252	486	−212	16	
166	11121222	(−−−+−+++)	−72	324	−400	128	
167	11122002	(−−−++00+)	−36	90	−68	16	
168	11122020	(−−−++0+0)	72	−108	4	16	
169	11122122	(−−−++−++)	144	−72	−256	128	
170	11122200	(−−−+++00)	396	−702	220	16	
171	11122212	(−−−+++−+)	−72	324	−400	128	
172	11122221	(−−−++++−)	−720	1512	−832	128	
173	11200002	(−−+0000+)	−36	78	−46	8	
174	11200020	(−−+000+0)	−36	78	−46	8	
175	11200122	(−−+00−++)	36	−42	−20	16	
176	11200200	(−−+00+00)	0	12	−22	8	
177	11200212	(−−+00+−+)	36	−42	−20	16	
178	11200221	(−−+00++−)	−36	90	−68	16	
179	11201022	(−−+0−0++)	0	24	−44	16	
180	11201202	(−−+0−+0+)	36	−42	−20	16	
181	11201220	(−−+0−++0)	72	−108	4	16	
182	11202000	(−−+0+000)	72	−120	26	8	
183	11202012	(−−+0+0−+)	0	24	−44	16	
184	11202021	(−−+0+0+−)	−144	288	−140	16	
185	11202102	(−−+0+−0+)	36	−42	−20	16	
186	11202120	(−−+0+−+0)	0	24	−44	16	
187	11202201	(−−+0++0−)	−252	486	−212	16	
188	11202210	(−−+0++−0)	−144	288	−140	16	
189	11210022	(−−+−00++)	−72	30	76	16	
190	11210202	(−−+−0+0+)	0	24	−44	16	
191	11210220	(−−+−0++0)	72	18	−164	16	
192	11211222	(−−+−−+++)	144	−72	−256	128	
193	11212002	(−−+−+00+)	36	−42	−20	16	
194	11212020	(−−+−+0+0)	72	−108	4	16	
195	11212122	(−−+−+−++)	0	−60	−16	128	

TABLE 13.1 (*cont.*)

		Power of X	(1)	(2)	(3)	(4)	8-order
196	11212200	(−−+−++00)	72	−234	172	16	
197	11212212	(−−+−++−+)	−72	72	−64	128	
198	11212221	(−−+−+++−)	−72	324	−400	128	
199	11220000	(−−++0000)	54	−150	98	8	
200	11220012	(−−++00−+)	−72	30	76	16	
201	11220021	(−−++00+−)	−36	90	−68	16	
202	11220102	(−−++0−0+)	0	24	−44	16	
203	11220120	(−−++0−+0)	−72	30	76	16	
204	11220201	(−−++0+0−)	72	−108	4	16	
205	11220210	(−−++0+−0)	−36	90	−68	16	
206	11221002	(−−++−00+)	36	−42	−20	16	
207	11221020	(−−++−0+0)	0	24	−44	16	
208	11221122	(−−++−−++)	−72	72	−64	128	
209	11221200	(−−++−+00)	−72	30	76	16	
210	11221212	(−−++−+−+)	0	−60	−16	128	
211	11221221	(−−++−++−)	144	−72	−256	128	
212	11222001	(−−+++00−)	396	−702	220	16	
213	11222010	(−−+++0−0)	72	−108	4	16	
214	11222100	(−−+++−00)	−36	90	−68	16	
215	11222112	(−−+++−−+)	144	−72	−256	128	
216	11222121	(−−+++−+−)	−72	324	−400	128	
217	11222211	(−−++++−−)	−720	1512	−832	128	
218	12000000	(−+000000)	21	−48	25	10	
219	12000012	(−+0000−+)	−6	−12	44	8	
220	12000021	(−+0000+−)	−36	78	−46	8	
221	12000102	(−+000−0+)	24	−60	8	8	
222	12000120	(−+000−+0)	−6	−12	44	8	
223	12000201	(−+000+0−)	−36	78	−46	8	
224	12000210	(−+000+−0)	−36	78	−46	8	
225	12001002	(−+00−00+)	48	−48	−4	8	
226	12001020	(−+00−0+0)	24	−60	8	8	
227	12001122	(−+00−−++)	−72	30	76	16	
228	12001200	(−+00−+00)	−6	−12	44	8	
229	12001212	(−+00−+−+)	−24	54	52	16	
230	12001221	(−+00−++−)	36	−42	−20	16	
231	12002001	(−+00+00−)	0	12	−22	8	
232	12002010	(−+00+0−0)	−36	78	−46	8	
233	12002100	(−+00+−00)	−36	78	−46	8	
234	12002112	(−+00+−−+)	36	−42	−20	16	
235	12002121	(−+00+−+−)	36	−42	−20	16	
236	12002211	(−+00++−−)	−36	90	−68	16	
237	12010002	(−+0−000+)	24	−60	8	8	
238	12010020	(−+0−00+0)	48	−48	−4	8	
239	12010122	(−+0−0−++)	0	24	−44	16	
240	12010200	(−+0−0+00)	24	−60	8	8	
241	12010212	(−+0−0+−+)	−48	0	−20	16	
242	12010221	(−+0−0++−)	0	24	−44	16	
243	12011022	(−+0−−0++)	72	18	−164	16	
244	12011202	(−+0−−+0+)	0	24	−44	16	
245	12011220	(−+0−−++0)	−72	30	76	16	

TABLE 13.1 (*cont.*)

		Power of X	(1)	(2)	(3)	(4)	8-order
246	12012000	(−+0−+000)	−6	−12	44	8	
247	12012012	(−+0−+0−+)	−24	54	52	16	
248	12012021	(−+0−+0+−)	36	−42	−20	16	
249	12012102	(−+0−+−0+)	−48	0	−20	16	
250	12012120	(−+0−+−+0)	−24	54	52	16	
251	12012201	(−+0−++0−)	72	−108	4	16	
252	12012210	(−+0−++−0)	36	−42	−20	16	
253	12020001	(−+0+000−)	72	−120	26	8	
254	12020010	(−+0+00−0)	0	12	−22	8	
255	12020100	(−+0+0−00)	−36	78	−46	8	
256	12020112	(−+0+0−−+)	72	−108	4	16	
257	12020121	(−+0+0−+−)	0	24	−44	16	
258	12020211	(−+0+0+−−)	−144	288	−140	16	
259	12021000	(−+0+−000)	−36	78	−46	8	
260	12021012	(−+0+−0−+)	36	−42	−20	16	
261	12021021	(−+0+−0+−)	36	−42	−20	16	
262	12021102	(−+0+−−0+)	0	24	−44	16	
263	12021120	(−+0+−−+0)	36	−42	−20	16	
264	12021201	(−+0+−+0−)	0	24	−44	16	
265	12021210	(−+0+−+−0)	36	−42	−20	16	
266	12022011	(−+0++0−−)	−252	486	−212	16	
267	12022101	(−+0++−0−)	−144	288	−140	16	
268	12022110	(−+0++−−0)	−36	90	−68	16	
269	12100002	(−+−0000+)	−6	−12	44	8	
270	12100020	(−+−000+0)	24	−60	8	8	
271	12100122	(−+−00−++)	36	−42	−20	16	
272	12100200	(−+−00+00)	48	−48	−4	8	
273	12100212	(−+−00+−+)	−24	54	52	16	
274	12100221	(−+−00++−)	−72	30	76	16	
275	12101022	(−+−0−0++)	72	−108	4	16	
276	12101202	(−+−0−+0+)	36	−42	−20	16	
277	12101220	(−+−0−++0)	0	24	−44	16	
278	12102000	(−+−0+000)	24	−60	8	8	
279	12102012	(−+−0+0−+)	−48	0	−20	16	
280	12102021	(−+−0+0+−)	0	24	−44	16	
281	12102102	(−+−0+−0+)	−24	54	52	16	
282	12102120	(−+−0+−+0)	−48	0	−20	16	
283	12102201	(−+−0++0−)	72	18	−164	16	
284	12102210	(−+−0++−0)	0	24	−44	16	
285	12110022	(−+−−00++)	72	−234	172	16	
286	12110202	(−+−−0+0+)	72	−108	4	16	
287	12110220	(−+−−0++0)	72	18	−164	16	
288	12111222	(−+−−−+++)	−72	324	−400	128	
289	12112002	(−+−−+00+)	36	−42	−20	16	
290	12112020	(−+−−+0+0)	0	24	−44	16	
291	12112122	(−+−−+−++)	−72	72	−64	128	
292	12112200	(−+−−++00)	−72	30	76	16	
293	12112212	(−+−−++−+)	0	−60	−16	128	
294	12112221	(−+−−+++−)	144	−72	−256	128	
295	12120000	(−+−+0000)	−6	−12	44	8	

TABLE 13.1 (*cont.*)

		Power of X	(1)	(2)	(3)	(4)	8-order
296	12120012	(−+−+00−+)	−24	54	52	16	
297	12120021	(−+−+00+−)	36	−42	−20	16	
298	12120102	(−+−+0−0+)	−48	0	−20	16	
299	12120120	(−+−+0−+0)	−24	54	52	16	
300	12120201	(−+−+0+0−)	72	−108	4	16	
301	12120210	(−+−+0+−0)	36	−42	−20	16	
302	12121002	(−+−+−00+)	−24	54	52	16	
303	12121020	(−+−+−0+0)	−48	0	−20	16	
304	12121122	(−+−+−−++)	0	−60	−16	128	
305	12121200	(−+−+−+00)	−24	54	52	16	
306	12121212	(−+−+−+−+)	48	48	128	128	
307	12121221	(−+−+−++−)	0	−60	−16	128	
308	12122001	(−+−++00−)	72	−234	172	16	
309	12122010	(−+−++0−0)	72	−108	4	16	
310	12122100	(−+−++−00)	36	−42	−20	16	
311	12122112	(−+−++−−+)	0	−60	−16	128	
312	12122121	(−+−++−+−)	−72	72	−64	128	
313	12122211	(−+−+++−−)	−72	324	−400	128	
314	12200001	(−++0000−)	54	−150	98	8	
315	12200010	(−++000−0)	72	−120	26	8	
316	12200100	(−++00−00)	0	12	−22	8	
317	12200112	(−++00−−+)	72	−234	172	16	
318	12200121	(−++00−+−)	−72	30	76	16	
319	12200211	(−++00+−−)	−36	90	−68	16	
320	12201000	(−++0−000)	−36	78	−46	8	
321	12201012	(−++0−0−+)	72	−108	4	16	
322	12201021	(−++0−0+−)	0	24	−44	16	
323	12201102	(−++0−−0+)	72	18	−164	16	
324	12201120	(−++0−−+0)	72	−108	4	16	
325	12201201	(−++0−+0−)	−72	30	76	16	
326	12201210	(−++0−+−0)	0	24	−44	16	
327	12202011	(−++0+0−−)	72	−108	4	16	
328	12202101	(−++0+−0−)	−36	90	−68	16	
329	12202110	(−++0+−−0)	−144	288	−140	16	
330	12210000	(−++−0000)	−36	78	−46	8	
331	12210012	(−++−00−+)	36	−42	−20	16	
332	12210021	(−++−00+−)	36	−42	−20	16	
333	12210102	(−++−0−0+)	0	24	−44	16	
334	12210120	(−++−0−+0)	36	−42	−20	16	
335	12210201	(−++−0+0−)	0	24	−44	16	
336	12210210	(−++−0+−0)	36	−42	−20	16	
337	12211002	(−++−−00+)	−72	30	76	16	
338	12211020	(−++−−0+0)	0	24	−44	16	
339	12211122	(−++−−−++)	144	−72	−256	128	
340	12211200	(−++−−+00)	36	−42	−20	16	
341	12211212	(−++−−+−+)	0	−60	−16	128	
342	12211221	(−++−−++−)	−72	72	−64	128	
343	12212001	(−++−+00−)	−72	30	76	16	
344	12212010	(−++−+0−0)	0	24	−44	16	
345	12212100	(−++−+−00)	36	−42	−20	16	

TABLE 13.1 (*cont.*)

		Power of X	(1)	(2)	(3)	(4)	8-order
346	12212112	(−++−+−−+)	−72	72	−64	128	
347	12212121	(−++−+−+−)	0	−60	−16	128	
348	12212211	(−++−++−−)	144	−72	−256	128	
349	12220011	(−+++00−−)	396	−702	220	16	
350	12220101	(−+++0−0−)	72	−108	4	16	
351	12220110	(−+++0−−0)	−252	486	−212	16	
352	12221001	(−++++−00−)	−36	90	−68	16	
353	12221010	(−++++−0−0)	−144	288	−140	16	
354	12221100	(−++++−−00)	−36	90	−68	16	
355	12221112	(−+++−−−+)	−72	324	−400	128	
356	12221121	(−+++−−+−)	144	−72	−256	128	
357	12221211	(−+++−+−−)	−72	324	−400	128	
358	12222111	(−++++−−−)	−720	1512	−832	128	

TABLE 13.2

n	Dividing integers
1	1
2, 3	3
4, 5	15
6, 7	105
8	315

As an example, consider

$$\mathrm{Tr}(\hat{S}^z\hat{S}^z\hat{S}^-\hat{S}^+\hat{S}^-\hat{S}^z\hat{S}^+).$$

First transfer leading $\hat{S}^z$ operator to get

$$\mathrm{Tr}(\hat{S}^-\hat{S}^+\hat{S}^-\hat{S}^z\hat{S}^+\hat{S}^z\hat{S}^z)$$

and then construct the identifier, i.e. 1210200. This entry is number 99 in the $n = 7$ section. The symbol is (—+—0+0 0) and the tabulation reads

$$8 \quad 4 \quad -4.$$

Referring to Table 13.2 for the scale factor we have, finally,

$$\mathrm{Tr}(\hat{S}^z\hat{S}^z\hat{S}^-\hat{S}^+\hat{S}^-\hat{S}^z\hat{S}^+) = \frac{(2S+1)}{105}(8X+4X^2-4X^3), \qquad (13.87)$$

with $X = S(S+1)$.

(*c*) *High temperatures: large* $\mathbf{q} = \boldsymbol{\varkappa}-\boldsymbol{\tau}$

We have found in the foregoing discussion of the second and fourth moments of the spectral weight function $F^{\alpha}(\boldsymbol{\varkappa}, \omega)$ that the latter resembles a Gaussian function of ω for large values of the scattering vector. By large values of the scattering vector we mean values of $\mathbf{q} = \boldsymbol{\varkappa}-\boldsymbol{\tau}$

that are close to the values of **q** that define the first Brillouin zone. In this section we consider improvements to the Gaussian approximation for which

$$F(\boldsymbol{\kappa},\omega) \simeq F_G(\boldsymbol{\kappa},\omega) = (2\pi\overline{\omega^2})^{-\frac{1}{2}}\exp\left(-\frac{\omega^2}{2\overline{\omega^2}}\right). \tag{13.88}$$

In order to include information on the fourth moment in the expression for $F(\boldsymbol{\kappa},\omega)$ Collins and Marshall† suggested the use of the Gram–Charlier series

$$F(\boldsymbol{\kappa},\omega) = F_G(\boldsymbol{\kappa},\omega)\{1+C_2H_2(x)+C_4H_4(x)+\ldots\} \tag{13.89}$$

where $H_n(x)$ are Tchebychev–Hermite polynomials and

$$x = \frac{\omega}{\sqrt{(\overline{\omega^2})}}. \tag{13.90}$$

The first two polynomials in (13.89) are

$$H_2(x) = x^2-1, \qquad H_4(x) = x^4-6x^2+3. \tag{13.91}$$

The idea behind the particular expansion (13.89) arises from the orthogonality condition

$$\int\limits_{-\infty}^{\infty} H_n(x)H_m(x)F_G(\boldsymbol{\kappa},\omega)\,d\omega = n!\,\delta_{m,n}, \tag{13.92}$$

which means that the addition of a further term, $n+1$ say, leaves the first n moments unchanged. From (13.92),

$$C_n = \frac{1}{n!}\int\limits_{-\infty}^{\infty} d\omega\, H_n(x)F(\boldsymbol{\kappa},\omega), \tag{13.93}$$

so that

$$C_2 = \tfrac{1}{2}\int\limits_{-\infty}^{\infty} d\omega\left(\frac{\omega^2}{\overline{\omega^2}}-1\right)F(\boldsymbol{\kappa},\omega) = 0 \tag{13.94}$$

and

$$C_4 = \frac{1}{4!}\int\limits_{-\infty}^{\infty} d\omega\left\{\frac{\omega^4}{(\overline{\omega^2})^2}-\frac{6\omega^2}{\overline{\omega^2}}+3\right\}F(\boldsymbol{\kappa},\omega) = \tfrac{1}{8}\alpha. \tag{13.95}$$

Hence

$$F(\boldsymbol{\kappa},\omega) \simeq (2\pi\overline{\omega^2})^{-\frac{1}{2}}\exp\left(-\frac{\omega^2}{2\overline{\omega^2}}\right)\left\{1+\frac{\alpha}{8}H_4\left(\frac{\omega}{\sqrt{(\overline{\omega^2})}}\right)\right\}. \tag{13.96}$$

Fig. 13.3 illustrates the Gaussian and Lorentzian form for $F(\boldsymbol{\kappa},\omega)$ and also eqn (13.96) with $\alpha = 0{\cdot}2$ and $0{\cdot}6$.

For positive values of the parameter α, $F(\boldsymbol{\kappa},\omega)$ is intermediate between the Gaussian and Lorentzian shapes, while for negative α it becomes more square-shaped with a sharply cut-off tail.

† M. Collins and W. Marshall, *Proc. phys. Soc.* **92**, 390 (1967).

In order to gain a measure of the validity of expression (13.96) for $F(\boldsymbol{\kappa}, \omega)$ Collins and Marshall compared it with the *exact* result for a spin 5/2 Ising ferromagnet. With a value of $-0{\cdot}071$ for α it was found that (13.96) fitted the exact result to better than 1 per cent everywhere. This encourages belief in the adequacy of (13.96) but does not imply that it will be as good for exchange coupled systems.

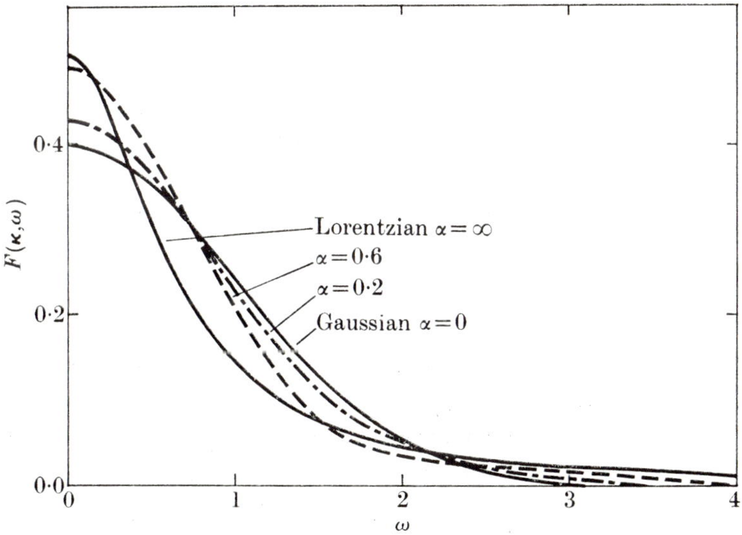

FIG. 13.3. Comparison of various approximate forms for the shape of the paramagnetic scattering function. The forms plotted are Gaussian and Lorentzian distributions and two distributions based on the Gram–Charlier series with the parameter α equal to 0·2 and 0·6. α is zero for a Gaussian shape and infinite for a Lorentzian.

(*d*) *High temperatures: small* $\mathbf{q} = \boldsymbol{\kappa} - \boldsymbol{\tau}$

Since both $\overline{\omega^2}$ and $\overline{\omega^4}$ vary like q^2 for small $\mathbf{q}$ the parameter α varies like q^{-2}. Thus $F(\boldsymbol{\kappa}, \omega)$ can in the limit of small $\mathbf{q}$ be represented by a cut-off Lorentzian, viz.

$$\begin{aligned} F(\boldsymbol{\kappa}, \omega) &= \frac{1}{\pi}\frac{\Gamma_{\kappa}}{\omega^2+\Gamma_{\kappa}^2} \qquad (|\omega| < \omega_c) \\ &= 0 \qquad\qquad\qquad (|\omega| > \omega_c), \end{aligned} \tag{13.97}$$

where

$$\Gamma_{\kappa} = \frac{\pi}{2\sqrt{3}}\left\{\frac{(\overline{\omega^2})^3}{\overline{\omega^4}}\right\}^{\frac{1}{2}}$$

and

$$\omega_c^2 = \frac{3\overline{\omega^4}}{\overline{\omega^2}} \tag{13.98}$$

are chosen to give the second and fourth moments correctly. Specializing to a ferromagnet with only nearest-neighbour exchange coupling J,

$$\overline{\omega^2} = \left(\frac{2J}{\hbar}\right)^2 \tfrac{2}{3} r S(S+1)(1-\gamma_{\varkappa}) \tag{13.99}$$

and

$$\overline{\omega^4} = \left(\frac{2J}{\hbar}\right)^4 \tfrac{4}{9} S^2(S+1)^2 r(1-\gamma_{\varkappa}) \left\{\frac{7r}{2} - \frac{3r}{2}\gamma_{\varkappa} - 2 - \frac{3}{4S(S+1)}\right\}, \tag{13.100}$$

from which we have, to leading order in $|\varkappa|^2$,

$$\Gamma_{\varkappa} = \frac{J\pi}{3\hbar} r(1-\gamma_{\varkappa}) \left\{\frac{S(S+1)}{r-1-\{3/8S(S+1)\}}\right\}^{\frac{1}{2}} \tag{13.101}$$

and

$$\omega_{\mathrm{c}} = \frac{4J}{\hbar}\left[S(S+1)\left\{r-1-\frac{3}{8S(S+1)}\right\}\right]^{\frac{1}{2}}. \tag{13.102}$$

(*e*) *Temperature dependence of* $F^{\alpha}(\varkappa, \omega)$ *close to the critical point*

If we go back to the definition of the moments $\overline{\omega^{2n}}$ given by eqn (13.61) we deduce that their temperature dependence close to the critical point is essentially governed by the wavelength-dependent susceptibility that appears in the denominator. The short-range correlation function in the numerator of (13.61) can only have a weak singularity (probably logarithmic) whereas the wavelength susceptibility $\chi_{\varkappa}^{\alpha}$ contains long-range correlation functions and is strongly temperature dependent. In view of this we estimate the moments just above T_{c} by

$$\overline{\omega^2} = \overline{\omega_{\infty}^2}\,\frac{\chi_0}{\chi_{\varkappa}^{\alpha}} \quad \text{and} \quad \overline{\omega^4} = \overline{\omega_{\infty}^4}\,\frac{\chi_0}{\chi_{\varkappa}^{\alpha}}, \tag{13.103}$$

where the suffix ∞ denotes that the moments are evaluated at infinite temperatures. If $\varkappa$ is close to an ordering vector then with $T \simeq T_{\mathrm{c}}$ $1/\chi_{\varkappa}^{\alpha}$ is small, and in consequence $\overline{\omega^4} \gg (\overline{\omega^2})^2$. Hence in this instance the spectral weight function is approximately a (cut-off) Lorentzian function of ω, with a width

$$\Gamma_{\varkappa}^{\alpha} = \frac{\pi}{2\sqrt{3}}\left\{\frac{(\overline{\omega_{\infty}^2})^3}{\overline{\omega_{\infty}^4}}\right\}^{\frac{1}{2}} \frac{\chi_0}{\chi_{\varkappa}^{\alpha}}. \tag{13.104}$$

For a ferromagnet with small $\mathbf{q} = \varkappa - \boldsymbol{\tau}$ we obtain the result

$$\Gamma_{\varkappa}^{\alpha} = \Lambda q^2 \frac{T_{\mathrm{c}}}{T} r_1^2(\kappa_1^2 + q^2), \tag{13.105}$$

where we have made use of the two results that the moments are proportional to q^2 and eqn (13.32) for $\chi_{\varkappa}^{\alpha}$.

In the case of a simple antiferromagnet we can replace the ratio of the infinite temperature moments in (13.104) by a constant because, as is illustrated in Fig. 13.2, they are almost independent of the scattering vector for $\boldsymbol{\varkappa}$ in the vicinity of $\mathbf{w}$, the superlattice vector. Thus, in this case we write

$$\Gamma^{\alpha}_{\boldsymbol{\varkappa}} \propto \kappa_1^2+q^2. \tag{13.106}$$

Note that these forms for the width predict that it should decrease as $T \to T_c$ like $(T-T_c)$, i.e. we should expect to see the inelasticity of the scattering decrease. This latter effect is usually referred to as thermodynamic slowing down of the spin fluctuations. It is emphasized that by small q we mean $q^2 \ll \kappa_1^2$, a condition that becomes increasingly difficult to fulfil as $T \to T_c$.

13.4. Experimental results for antiferromagnets

To facilitate the comparison of the mean-field results described above with experimental results, it is convenient to introduce critical indices† β, γ, ν for the power-law dependence on $\epsilon = (T-T_N)/T_N$ of the magnetization, susceptibility, and inverse correlation range, respectively. The mean-field values of the critical indices are summarized in Table 13.3.

Some experimental values of the indices for the Heisenberg antiferromagnets MnF_2, $RbMnF_3$, and $KMnF_3$, and dysprosium aluminium garnet (DAG), which approximates to an Ising antiferromagnet, are also given in Table 13.3. These are seen to show overall a marked discrepancy with the mean-field indices.

The wave-vector dependent susceptibility can be measured directly in the quasi-elastic approximation (cf. § 13.1). Schulhof *et al.*‡ verified the validity of the quasi-elastic approximation for MnF_2 with $0{\cdot}04 \leqslant T-T_N \lesssim 8$ K and incident energies E in the range

$$56 \text{ meV} \leqslant E \leqslant 134 \text{ meV},$$

and we now consider their results for the wave-vector dependent susceptibility.

Because of the uniaxial anisotropy in MnF_2 it is necessary to assign to it two Néel temperatures, one for the longitudinal properties and another for the transverse. These are determined by (cf. eqn (13.57))

$$T_\alpha = \frac{2S(S+1)}{3k_B}\{\mathscr{J}(\mathbf{w})+\mathscr{D}_\alpha(\mathbf{w})\}. \tag{13.107}$$

The true Néel temperature is that at which $\chi^{\alpha}_{\boldsymbol{\varkappa}}$ for $\boldsymbol{\varkappa} = \mathbf{w}$ first diverges as T decreases from infinity. For MnF_2 the longitudinal transition

† M. E. Fisher, *Rep. Prog. Phys.* **30**, 615 (1967).

‡ M. P. Schulhof, P. Heller, R. Nathans, and A. Linz, private communication.

TABLE 13.3

Critical indices $\epsilon = (T - T_N)/T_N$

Physical quantity	Behaviour of quantity		Mean-field index	Experimental indices
$\langle \hat{S}^z \rangle$	$\langle \hat{S}^z \rangle = 0$ $\langle \hat{S}^z \rangle \sim \lvert\epsilon\rvert^{\beta}$	$\epsilon > 0$ $\epsilon < 0$	$\beta = \frac{1}{2}$	$\beta = 0{\cdot}335 \pm 0{\cdot}01$; MnF_2 Heller and Benedek[1] $\beta = 0{\cdot}33$; $KMnF_3$ Cooper and Nathans[2]
χ	$\sim \epsilon^{-\gamma}$ $\sim \lvert\epsilon\rvert^{-\gamma'}$	$\epsilon > 0$ $\epsilon < 0$	$\gamma = \gamma' = 1$	$\gamma = 1{\cdot}35 \pm 0{\cdot}01$; χ_L MnF_2 Dietrich[3] $\gamma = 1{\cdot}27 \pm 0{\cdot}02$, $\gamma' = 1{\cdot}32 \pm 0{\cdot}06$ } χ_L MnF_2 Schulhof *et al.*[4] $\gamma = 1{\cdot}16 \pm 0{\cdot}04$; DAG Norvell *et al.*[5]
κ_1	$\sim \epsilon^{\nu}$ $\sim \lvert\epsilon\rvert^{\nu'}$	$\epsilon > 0$ $\epsilon < 0$	$\nu = \nu' = \frac{1}{2}$	$\nu = 0{\cdot}68 \pm 0{\cdot}03$; κ_{1L} MnF_2 Dietrich[3] $\nu = 0{\cdot}634 \pm 0{\cdot}02$, $\nu' = 0{\cdot}56 \pm 0{\cdot}05$ } κ_{1L} MnF_2 Schulhof *et al.*[4] $\nu = 0{\cdot}61 \pm 0{\cdot}02$; DAG Norvell *et al.*[5]
Spin correlation (cf. eqn (13.108))	$\sim r^{-(1+\eta)}$		$\eta = 0$	$\eta = 0{\cdot}05 \pm 0{\cdot}02$; MnF_2 Schulhof *et al.*[4] $\eta = 0{\cdot}12 \pm 0{\cdot}10$; DAG Norvell *et al.*[5] $\eta = 0{\cdot}067$; $RbMnF_3$ Corliss *et al.*[6]

(1) P. Heller and G. B. Benedek, *Phys. Rev. Lett.* **8,** 428 (1962); (2) M. J. Cooper and R. Nathans, *J. appl. Phys.* **37,** 1041 (1966); (3) O. W. Dietrich, *J. Phys.* C, **2,** 2022 (1969); (4) M. P. Schulhof, P. Heller, R. Nathans, and A. Linz, private communication; (5) J. C. Norvell *et al., Phys. Rev.* **186,** 567 (1969); (6) L. M. Corliss *et al., J. appl. Phys.* **40,** 1278 (1969).

determines the true Néel temperature $T_N = 67{\cdot}3$ K (Moriya† has estimated that the transverse Néel temperature $T_{\perp}$ lies 1·36 K below T_N). Because it is the longitudinal component of the susceptibility that diverges at T_N we expect to see thermodynamic slowing down only in the corresponding $\Gamma_{\boldsymbol{\kappa}}$ (see below). The longitudinal and transverse components of the cross-section can be isolated by utilizing the different orientation factors associated with either one. The transverse component has the orientation factor $1+\tilde{\kappa}_z^2$ (assuming the z axis to coincide with the axis of quantization), the longitudinal the factor $1-\tilde{\kappa}_z^2$; hence for $\boldsymbol{\kappa}$ parallel to the z axis only the transverse component is observed whereas for $\boldsymbol{\kappa}$ perpendicular to the z axis both are observed.

Fisher has argued that above T_N the mean-field form of the spin correlation function, eqn (13.36), should be replaced by

$$\langle \Delta\hat{S}_m \Delta\hat{S}_n \rangle \propto \exp(-\kappa_1 r)/r^{1+\eta} \quad (r = |\mathbf{m}-\mathbf{n}|), \tag{13.108}$$

and in consequence ($\boldsymbol{\kappa} = \mathbf{q}-\mathbf{w}$)

$$\chi_{\mathbf{q}} \propto \left(\frac{1}{\kappa_1^2+q^2}\right)^{1-\eta/2}. \tag{13.109}$$

In view of the above discussion of the properties of MnF_2 and eqn (13.109) for $\chi_{\mathbf{q}}$, the partial cross-sections for measurements around (100) and (001) were taken by Schulhof *et al.* to be of the form

$$\frac{d\sigma}{d\Omega} \propto \left(\frac{B_{\perp}(T)}{\kappa_{1\perp}^2+q^{*2}}\right)^{1-\eta/2} \quad ((001)\text{ reflection}), \tag{13.110}$$

and

$$\frac{d\sigma}{d\Omega} \propto \left(\frac{B_{\mathrm{L}}(T)}{\kappa_{1\mathrm{L}}^2+q^{*2}}\right)^{1-\eta/2} + \left(\frac{B'_{\perp}(T)}{\kappa_{1\perp}^2+q^{*2}}\right)^{1-\eta/2} \quad ((100)\text{ reflection}). \tag{13.111}$$

Here

$$q^{*2} = q_a^2+q_{a'}^2+\left(\frac{c}{a}\right)^2 q_c^2. \tag{13.112}$$

Using these forms for $d\sigma/d\Omega$ they obtained the following results for the critical indices (cf. Table 13.3):

$$\text{Longitudinal} \begin{cases} \chi_{\mathbf{q}=0} \sim \epsilon^{-\gamma},\ \gamma = 1{\cdot}27 \pm 0{\cdot}02 \\ \nu = 0{\cdot}634 \pm 0{\cdot}02 \end{cases}$$

$$\text{Transverse} \begin{cases} \chi_{\mathbf{q}=0} \sim \epsilon^{-\gamma},\ \gamma = 1{\cdot}47 \pm 0{\cdot}1,\ T_{\perp} = 64{\cdot}40 \pm 0{\cdot}5\,\mathrm{K} \\ \nu = 0{\cdot}63 \pm 0{\cdot}08,\ T_{\perp} = 64{\cdot}9 \pm 0{\cdot}7\,\mathrm{K}. \end{cases} \tag{13.113}$$

(Note that the values of $T_{\perp}$ differ considerably from Moriya's estimate. Since the latter was obtained from mean-field theory this difference is perhaps not surprising.)

† T. Moriya, *Prog. theor. Phys. Osaka*, **28**, 371 (1962).

A best fit value of $\eta = 0{\cdot}05 \pm 0{\cdot}02$ was obtained with the longitudinal measurements closest to T_N.

These values of γ, ν, and η differ markedly from the mean-field values, but γ_L, ν_L, and η are close to values obtained by other, better theories.†

Turning now to dynamical effects near T_N, several experiments have confirmed the effect of thermodynamic slowing down of the spin fluctuations but disagree quantitatively with the mean-field results derived in § 13.3 (*e*). Two other important features revealed by these experiments are: (a) for $T < T_N$ but close to the critical point there is a central peak in the longitudinal susceptibility, and (b) the spin-wave peaks persist above T_N and finally merge into the central peak as T is further increased.

To illustrate some of these features we again consider the measurements on MnF_2 by Schulhof *et al.* The partial differential cross-section is taken to have the following forms for measurements about (001) and (100):

$$\left(\frac{d^2\sigma}{d\Omega dE'}\right)_{\perp} \propto \left(\frac{1}{\kappa_{1\perp}^2+q^{*2}}\right)^{1-\eta/2}\left\{\frac{\Gamma_{\perp}}{\Gamma_{\perp}^2+(\omega-\omega_0)^2}+\frac{\Gamma_{\perp}}{\Gamma_{\perp}^2+(\omega+\omega_0)^2}\right\} \qquad (13.114)$$

and
$$\left(\frac{d^2\sigma}{d\Omega dE'}\right)_{L} \propto \left(\frac{1}{\kappa_{1L}^2+q^{*2}}\right)^{1-\eta/2}\left\{\frac{\Gamma_{L}}{\Gamma_{L}^2+\omega^2}\right\}+\left(\frac{d^2\sigma}{d\Omega dE'}\right)_{\perp}. \qquad (13.115)$$

The measured cross-sections are illustrated in Figs. 13.4 and 13.5. These show features (a) and (b) discussed above.

The longitudinal width Γ_L is shown in Fig. 13.6 as a function of temperature for various values of q^*. For small values of q^*, Γ_L tends to zero as $T \to T_N$ from above; this behaviour is very sensitive to the value of the wave vector. In contrast to the prediction of mean-field theory, Γ_L does not for $T < T_N$ increase with decreasing T. A similar conclusion for $T > T_N$ was reached by Nathans *et al.*‡ from their measurements on $RbMnF_3$, which unlike MnF_2 is cubic and shows no single-ion anisotropy.

A recent theoretical attempt to correlate the various critical indices and predict the behaviour of Γ which has received considerable support from neutron measurements is based on so-called scaling assumptions.§ While it is not appropriate to discuss scaling theories in detail it is worth while to sketch the basic assumptions made and indicate the present degree of support for them.

† Cf. M. E. Fisher, loc. cit.

‡ R. Nathans *et al.*, *J. appl. Phys.* **39**, 1237 (1968).

§ B. I. Halperin and P. C. Hohenberg, *Phys. Rev.* **177**, 952 (1969). See also the discussion in H. E. Stanley, *Introduction to phase transitions and critical phenomena*, § 13, Clarendon Press (1971).

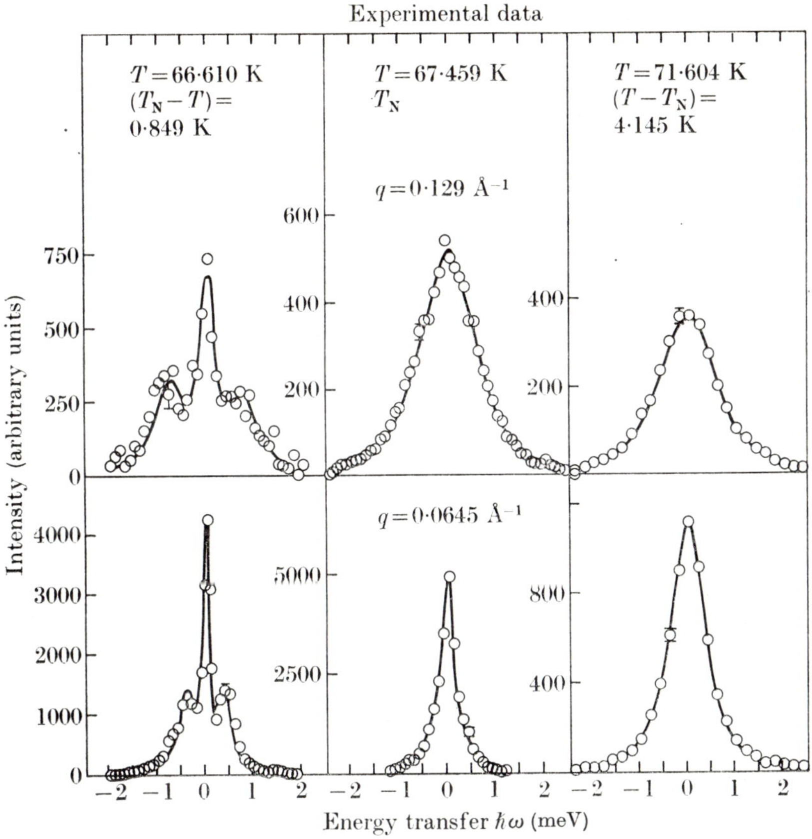

FIG. 13.4. The cross-section measured about (100) is shown as a function of energy transfer and for various values of temperature and wave vector. Both the transverse and longitudinal fluctuations contribute here and the central peak below T_N is to be noted.

We define $C^{\alpha}(\boldsymbol{\kappa}, \omega)$ as

$$C^{\alpha}(\boldsymbol{\kappa}, \omega) = \frac{1}{2\hbar}\int_{-\infty}^{\infty} dt\, e^{i\omega t}\langle \Delta\hat{S}^{\alpha}_{\boldsymbol{\kappa}}(t)\Delta\hat{S}^{\alpha}_{-\boldsymbol{\kappa}}(0)+\Delta\hat{S}^{\alpha}_{-\boldsymbol{\kappa}}(0)\Delta\hat{S}^{\alpha}_{\boldsymbol{\kappa}}(t)\rangle, \quad (13.116)$$

which is related to $S^{\alpha}(\boldsymbol{\kappa}, \omega)$ by

$$C^{\alpha}(\boldsymbol{\kappa}, \omega) = \pi\{1+\exp(-\hbar\omega\beta)\}S^{\alpha}(\boldsymbol{\kappa}, \omega). \quad (13.117)$$

Now,
$$\chi^{\alpha} \propto \lim_{\boldsymbol{\kappa}\to 0} \beta \int_{-\infty}^{\infty} d\omega\; C(\boldsymbol{\kappa}, \omega)$$

diverges at the critical point like $\epsilon^{-\gamma}$. For $\boldsymbol{\kappa} \neq 0$, $\int d\omega\; C(\boldsymbol{\kappa}, \omega)$ is a continuous function of T. In the mean-field calculation we found that the temperature dependence entered $\chi_{\varkappa}$ through the inverse correlation range κ_1, which vanishes at the critical point like ϵ^{ν}.

Static scaling assumes that κ_1 contains the most important effects of critical fluctuations on $\int d\omega\, C^{\alpha}(\boldsymbol{\varkappa}, \omega)$, and that this function diverges at

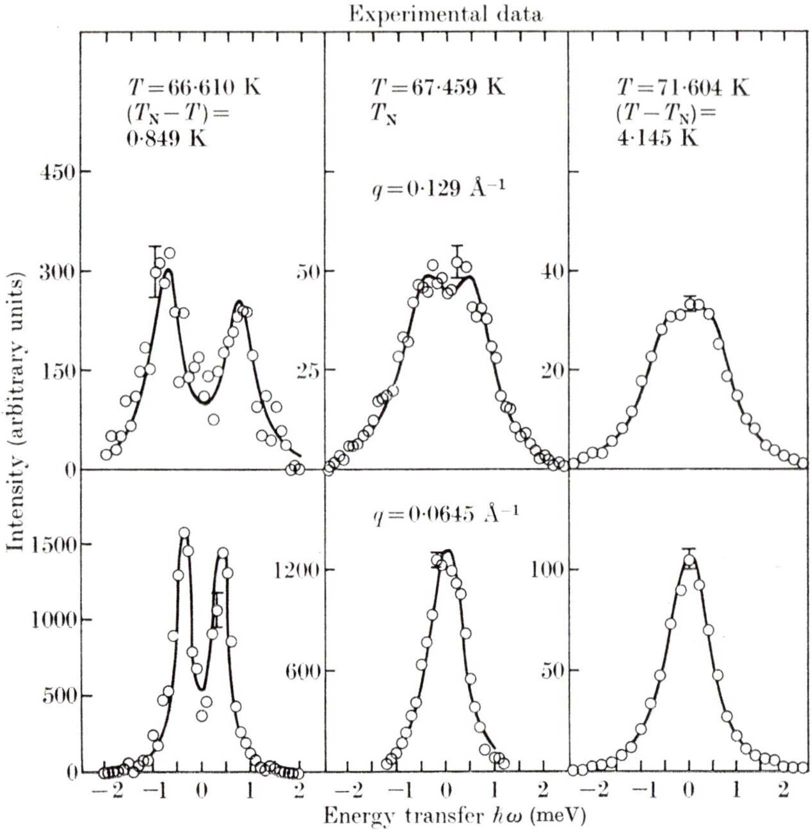

FIG. 13.5. The cross-section measured about (001) is shown as a function of energy transfer and for various values of temperature and wave vector. Only the transverse fluctuations contribute to the cross-section. For the larger of the two values of the wave vector two distinct spin-wave peaks exist at T_N and are still in evidence at $T_N + 4{\cdot}145$ K.

$\boldsymbol{\varkappa} = 0, \kappa_1 = 0$ but is continuous for all other values of $\boldsymbol{\varkappa}$ and κ_1. Furthermore, the function is assumed to be determined by its limiting behaviour in the hydrodynamic regime ($q \ll \kappa_1$; $T > T_N$, $T < T_N$) and moreover a *single* function describes the correlation in the *whole* $\boldsymbol{\varkappa}$, κ_1 plane. $\int d\omega\, C^{\alpha}(\boldsymbol{\varkappa}, \omega)$ is accordingly a homogeneous function of $|\boldsymbol{\varkappa}|$ and κ_1.

As a consequence of these assumptions, the relation

$$\gamma = \nu(2 - \eta) \tag{13.118}$$

is predicted between the critical indices. The longitudinal indices of Schulhof *et al.* (13.113) are seen to be consistent with (13.118) for $\eta = 0{\cdot}047$.

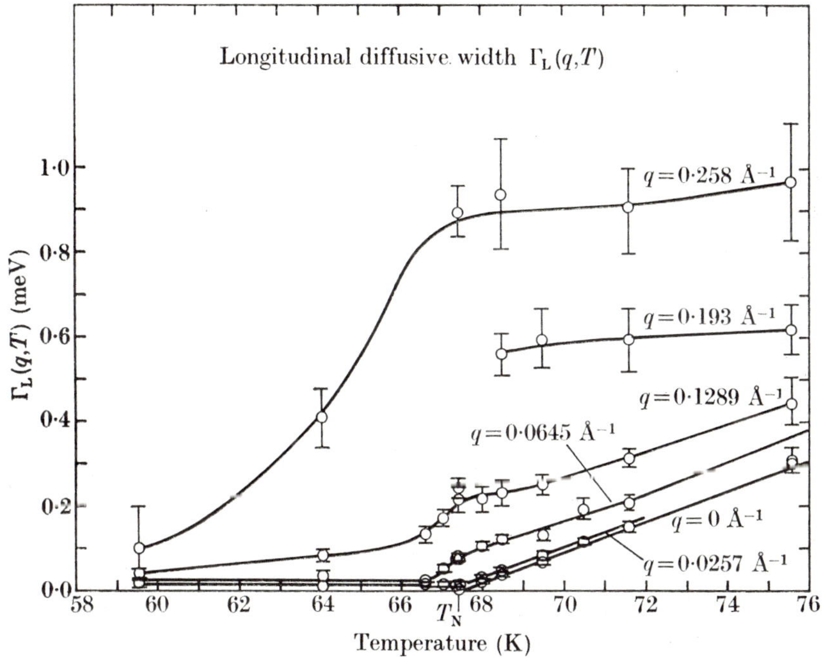

FIG. 13.6. The longitudinal width Γ_L for MnF_2 is shown as a function of temperature for various values of wave vector. For $q^* = 0$ the width is seen to go to zero, i.e. critical slowing down is observed.

To formulate dynamic scaling, $C^{\alpha}(\boldsymbol{\kappa}, \omega)$ is expressed in the form

$$C^{\alpha}(\boldsymbol{\kappa}, \omega) \propto \omega_{\kappa_1}^{-1}(\boldsymbol{\kappa}) f(\omega/\omega_{\kappa_1}(\boldsymbol{\kappa})), \tag{13.119}$$

where

$$\int_{-\infty}^{\infty} \mathrm{d}x\, f(x) = 1$$

and $\omega_{\kappa_1}(\boldsymbol{\kappa})$, the characteristic frequency, is determined by

$$\int_{-1}^{1} \mathrm{d}x\, f(x) = \tfrac{1}{2}. \tag{13.120}$$

If $C^{\alpha}(\boldsymbol{\kappa}, \omega)$ is a Lorentzian of width Γ centred about $\omega = 0$, then $\omega_{\kappa_1}(\boldsymbol{\kappa}) = \Gamma$. It is assumed that $\omega_{\kappa_1}(\boldsymbol{\kappa})$ is a homogeneous function of $\boldsymbol{\kappa}$ and κ_1 and that f depends not on κ_1 and $|\boldsymbol{\kappa}|$ separately but only on their product.

For an isotropic Heisenberg antiferromagnet, dynamic scaling leads to the predictions

$$\left.\begin{aligned} &\Gamma(q, T_{\mathrm{N}}) \sim q^{L} \\ \text{and} \qquad &\Gamma(0, T > T_{\mathrm{N}}) \sim \kappa_1^{L} + \text{const.} \end{aligned}\right\} \quad (L = \tfrac{3}{2}). \tag{13.121}$$

These predictions have been tested for $RbMnF_3$ by Lau *et al.*† At $T = T_{\mathrm{N}}$, where κ_1 is zero, they assume the partial differential cross-section to be of the form (13.115) (since $RbMnF_3$ is cubic and possesses no measurable anisotropy the transverse and longitudinal contributions cannot be separated as in the case of MnF_2), with Γ_{L}, $\Gamma_{\perp}$, and ω_0 all proportional to q^L. This choice of functional dependence ensures that $C^{\alpha}(\boldsymbol{\kappa}, \omega)$ satisfies the dynamic scaling assumption along $\kappa_1 = 0$. Using $\eta = 0{\cdot}067$, $0{\cdot}05\ \text{Å}^{-1} \leqslant q \leqslant 0{\cdot}25\ \text{Å}^{-1}$ and an incident neutron energy of 13 meV, a good fit to their measurements was obtained with $L = 1{\cdot}4 \pm 0{\cdot}2$, in agreement with the first dynamic scaling prediction.

To test the prediction $\Gamma(0, T > T_{\mathrm{N}}) \sim \kappa_1^L$ in the hydrodynamic region Lau *et al.* assumed a partial differential cross-section of the form

$$\frac{\mathrm{d}^2\sigma}{\mathrm{d}\Omega \mathrm{d}E'} \sim \left(\frac{1}{\kappa_1^2 + q^2}\right)^{1-\eta/2} \frac{\Gamma}{\Gamma^2 + \omega^2}. \tag{13.122}$$

The exponent L was found to agree well, within the estimated error of $\pm 0{\cdot}2$, with the predicted value of $1{\cdot}5$.

The experiments on $RbMnF_3$ by Lau *et al.* and other critical measurements give good support to the scaling assumptions. A first principles theory of critical phenomena is yet to come‡ but it would seem that such a theory must be in accord with scaling theories.

† H. Y. Lau, L. M. Corliss, A. Delapalme, J. M. Hastings, R. Nathans, and A. Tucciarone, *Phys. Rev. Lett.* **23**, 1225 (1969).

‡ A. A. Migdal, *Zh. éksp. teor. Fiz.* **55**, 1964 (1968) has proposed a microscopic argument to substantiate static scaling.

14

MAGNETIC DIFFUSE ELASTIC SCATTERING FROM ALLOYS

In the last decade a great interest has developed in the effects of substitutional impurities on the properties of solids, with experimental investigations often leading to a better understanding of the physical properties of the unperturbed host. The scattering of thermal neutrons is a very sensitive probe to the changes in the host system and the states of impurity atoms, and in this and the subsequent chapter we examine two different aspects of their use, first the diffuse elastic scattering from mixed magnetic systems (Chapter 14) and secondly the inelastic scattering from ferro- and antiferromagnetics and crystals containing substitutional mass impurities (Chapter 15).

The diffuse elastic scattering of long-wavelength neutrons by dilute binary alloys has been used, particularly by G. G. Low† and his collaborators, to gain information on the electronic structure of transition metals. The interaction of unpaired electron spins with the magnetic moment of the neutrons gives rise to the diffuse elastic scattering of slow neutrons, the intensity distribution of which depends, for dilute alloys, on the Fourier transform of the magnetic moment disturbance around the impurity sites (§ 3). Since the unpaired electron spins in transition metals correspond to outer electrons intimately concerned in the alloying process, a knowledge of their spatial distribution gives valuable and direct information about the electronic structure of the alloy. Experiments have also been made with concentrated binary alloys but are less easy to interpret (§ 14.5).

We first of all discuss some general properties of diffuse elastic neutron cross-sections and then go on to examine the structure factor for various models of the binary alloy that are suitable for different concentrations of impurity atoms.

14.1. General formulae‡

We are concerned here with the structure of the diffuse elastic neutron cross-section for various models of binary ferromagnetic and antiferro-

† For a general survey see *Adv. Phys.* **18**, 371 (1869).

‡ W. Marshall, *J. Phys.* C, **1**, 88 (1968).

magnetic systems; the concentration of atoms of type 1 is c and that of type 2 atoms is $1-c$.

From Chapter 5 the magnetic cross-section is

$$\frac{\mathrm{d}^2\sigma}{\mathrm{d}\Omega\mathrm{d}E'} = \frac{N}{4}\left(\frac{\gamma e^2}{m_e c^2}\right)^2 \frac{k'}{k} \sum_{\alpha\beta} (\delta_{\alpha\beta}-\tilde{\kappa}_\alpha \tilde{\kappa}_\beta)\times$$

$$\times \frac{1}{2\pi\hbar N} \sum_{m,n} \int_{-\infty}^{\infty} \mathrm{d}t \exp\{\mathrm{i}\boldsymbol{\kappa}.(\mathbf{m}-\mathbf{n})-\mathrm{i}\omega t\}\times$$

$$\times \langle g_n F_n(\boldsymbol{\kappa})\hat{S}_n^\alpha(0) g_m F_m(\boldsymbol{\kappa})\hat{S}_m^\beta(t)\rangle,$$

where $F_m(\boldsymbol{\kappa})$ is the form factor of the atom at site $\mathbf{m}$ and g_m the gyromagnetic ratio appropriate to the atom at $\mathbf{m}$.

The strictly elastic part is that which is proportional to $\delta(\omega)$, i.e.

$$\frac{\mathrm{d}\sigma}{\mathrm{d}\Omega} = \frac{N}{4}\left(\frac{\gamma e^2}{m_e c^2}\right)^2 \sum_{\alpha\beta} (\delta_{\alpha\beta}-\tilde{\kappa}_\alpha \tilde{\kappa}_\beta) \frac{1}{N} \sum_{m,n} \exp\{\mathrm{i}\boldsymbol{\kappa}.(\mathbf{m}-\mathbf{n})\}\times$$

$$\times g_n F_n \langle \hat{S}_n^\alpha \rangle g_m F_m \langle \hat{S}_m^\beta \rangle. \qquad (14.1)$$

We notice that $g_n\langle \hat{S}_n^\beta\rangle$ is the time-averaged magnetic moment of the atom on the site $\mathbf{n}$ in the direction β. We shall consider only those systems which are collinear, i.e. which have these time-averaged moments parallel to a given direction, which we shall take as the z axis. Hence (14.1) becomes

$$\frac{\mathrm{d}\sigma}{\mathrm{d}\Omega} = \frac{N}{4}\left(\frac{\gamma e^2}{m_e c^2}\right)^2 (1-\tilde{\kappa}_z^2) \frac{1}{N} \sum_{m,n} \exp\{\mathrm{i}\boldsymbol{\kappa}.(\mathbf{m}-\mathbf{n})\} F_n \mu_n F_m \mu_m, \qquad (14.2)$$

where μ_m is the time-averaged magnetic moment of the atom at the site $\mathbf{m}$.

We now recognize that the cross-section depends upon the precise distribution of moments μ_m. Because these are randomly distributed to a large extent, it follows that the cross-section is not exactly the same from sample to sample. However, by analogy to similar problems in statistical mechanics, we know that the variation in cross-section is negligible for any sample of reasonable size. It follows that we really need the average of (14.2) over an ensemble. Hence

$$\frac{\mathrm{d}\sigma}{\mathrm{d}\Omega} = \frac{N}{4}\left(\frac{\gamma e^2}{m_e c^2}\right)^2 (1-\tilde{\kappa}_z^2) \frac{1}{N} \sum_{m,n} \exp\{\mathrm{i}\boldsymbol{\kappa}.(\mathbf{m}-\mathbf{n})\} \overline{F_n \mu_n F_m \mu_m},$$

where the bar indicates the average over a grand canonical ensemble.

For a ferromagnetic alloy we can separate out the Bragg scattering (see Chapter 7) as

$$\left(\frac{d\sigma}{d\Omega}\right)_{\text{Bragg}} = \frac{N}{4}\left(\frac{\gamma e^2}{m_e c^2}\right)^2 (1-\tilde{\kappa}_z^2)\{\overline{\mu F(\boldsymbol{\kappa})}\}^2 \frac{(2\pi)^3}{v_0} \sum_{\boldsymbol{\tau}} \delta(\boldsymbol{\kappa}-\boldsymbol{\tau}), \quad (14.3)$$

where v_0 is the volume of a unit cell, $\boldsymbol{\tau}$ a reciprocal lattice vector, and

$$\overline{\mu F(\boldsymbol{\kappa})} = \frac{1}{N}\sum_m \mu_m F_m(\boldsymbol{\kappa}). \quad (14.4)$$

The remaining scattering, called the *diffuse elastic scattering*, is given by

$$\left(\frac{d\sigma}{d\Omega}\right)_{\text{diffuse}} = \frac{N}{4}\left(\frac{\gamma e^2}{m_e c^2}\right)^2 (1-\tilde{\kappa}_z^2)\Xi(\boldsymbol{\kappa}). \quad (14.5)$$

In (14.5)

$$\Xi(\boldsymbol{\kappa}) = \frac{1}{N}\sum_{m,n} \exp\{i\boldsymbol{\kappa}.(\mathbf{m}-\mathbf{n})\}\{\overline{F_n\mu_n F_m\mu_m} - \overline{F_n\mu_n}\,\overline{F_m\mu_m}\}$$

$$\equiv \frac{1}{N}\sum_{m,n} \exp\{i\boldsymbol{\kappa}.(\mathbf{m}-\mathbf{n})\}\overline{(F_n\mu_n - \overline{F_n\mu_n})(F_m\mu_m - \overline{F_m\mu_m})}, \quad (14.6)$$

and from the second line it is clear that the diffuse magnetic scattering involves the *spatial* fluctuations of the moments about their mean values.

In an experiment the diffuse magnetic scattering can be resolved from the total scattering by making use of the facts that (i) it is strictly elastic, (ii) it is proportional to $1-\tilde{\kappa}_z^2$, and (iii) it is distinct from Bragg scattering, which appears only in very sharp peaks when rigid geometrical conditions are satisfied.

The total magnetic moment of the target system is

$$M = \sum_m \mu_m, \quad (14.7)$$

so that, since the form factors are, by definition, unity in the forward direction $\boldsymbol{\kappa} = 0$,

$$\Xi(0) = \frac{1}{N}\{\overline{M^2} - (\overline{M})^2\} = \frac{1}{N}\overline{(\delta M)^2}, \quad (14.8)$$

where
$$\delta M = M - \overline{M}.$$

In other words, $\Xi(0)$ is proportional to the mean-square spatial fluctuation of the total magnetic moment M.

The total magnetic moment varies primarily because of concentration fluctuations, and we shall consider these as follows. We suppose that the sample contains $N_1 = cN$ atoms of type 1 and $N_2 = (1-c)N$ atoms of type 2. Each type of atom has a mean magnetic moment, μ_1 and μ_2, respectively. The mean moment μ, as observed in a macroscopic determination of the magnetization, is, clearly,

$$\mu = c\mu_1+(1-c)\mu_2, \tag{14.9}$$

and we can assume μ to be known for all concentrations. In principle both μ_1 and μ_2 also vary with concentration and therefore cannot be determined from magnetization measurements alone. For the variation of the total magnetic moment with concentration we note that

$$\delta M = \frac{\partial M}{\partial N_1}\delta N_1 = \frac{\partial \mu}{\partial c}\delta N_1, \tag{14.10}$$

so, from (14.8),

$$\Xi(0) = \left(\frac{\partial \mu}{\partial c}\right)^2 \frac{1}{N}\overline{\delta N_1^2}. \tag{14.11}$$

Let the function p_n be defined such that

$p_n = 1$, if an atom of type 1 is on the site n,

$p_n = 0$, if an atom of type 2 is on the site n.

Then

$$\overline{p_m} = c \tag{14.12}$$

and

$$N_1 = \sum_m p_m. \tag{14.13}$$

Hence

$$\overline{N_1^2} = \sum_{m,n} \overline{p_m p_n} = c^2N^2+c(1-c)N, \tag{14.14}$$

since

$$\overline{p_m p_n} = c^2+c(1-c)\delta_{m,n}. \tag{14.15}$$

From (14.14) we now have

$$\overline{\delta N_1^2} = \overline{N_1^2}-(\overline{N_1})^2 = c(1-c)N, \tag{14.16}$$

which together with (14.10) gives the result

$$\Xi(0) = c(1-c)\left(\frac{\partial \mu}{\partial c}\right)^2. \tag{14.17}$$

This is an important relationship because it tells us that, whatever the concentration, the scattering in the forward direction is related to the rate of change of magnetization with concentration and the latter can be determined by straightforward magnetization measurements to give a useful check on the neutron results.

In most alloys (14.17) is completely satisfactory, and it is rigorously valid for small concentrations, but in general the magnetization can vary through parameters other than the concentration. It is therefore worth while to consider the effect of short-range order. We suppose that a sample, with a fixed concentration, has a magnetization that varies with the heat treatment it has received in the past. It follows that the magnetization depends upon the short-range order which is established by the particular heat treatment. To discuss this we must describe short-range order parameters as follows.

Let the number of atoms, each of type 1, separated by a distance $\mathbf{R}$ be $N_{11}(\mathbf{R})$, then by definition

$$N_{11}(\mathbf{R}) = \sum_m p_m p_{m+R}, \tag{14.18}$$

which we write

$$N_{11}(\mathbf{R}) = \{c^2+s(\mathbf{R})\}N \quad (\mathbf{R} \neq 0). \tag{14.19}$$

Eqn (14.19) defines the short-range order parameter $s(\mathbf{R})$. We now examine the effect of fluctuations of $s(\mathbf{R})$ about the mean value of zero; thus

$$\overline{N_{11}(\mathbf{R})} = c^2N \tag{14.20 a}$$

and

$$\{\delta N_{11}(\mathbf{R})\}_c = N\delta s(\mathbf{R}) = \delta N_{11}(\mathbf{R})-2Nc\,\delta c, \tag{14.20 b}$$

where the subscript c means that the variation is at constant concentration. In place of the relationship given by eqn (14.10) we now have

$$\begin{aligned}\delta M &= \frac{\partial M}{\partial N_1}\delta N_1+\sum_R \frac{\partial M}{\partial N_{11}(\mathbf{R})}\{\delta N_{11}(\mathbf{R})\}_c \\ &= \frac{\partial \mu}{\partial c}\delta N_1+\sum_R \frac{\partial \mu}{\partial s(\mathbf{R})}\{\delta N_{11}(\mathbf{R})\}_c,\end{aligned} \tag{14.21}$$

where in the second term it is understood that the partial derivatives are taken at constant concentration.

Hence

$$\begin{aligned}\overline{\delta M^2} = \left(\frac{\partial \mu}{\partial c}\right)^2\overline{\delta N_1^2}+2\left(\frac{\partial \mu}{\partial c}\right)\sum_R \frac{\partial \mu}{\partial s(\mathbf{R})}\overline{\delta N_1\{\delta N_{11}(\mathbf{R})\}_c}+ \\ +\sum_{R,R'} \frac{\partial \mu}{\partial s(\mathbf{R})}\frac{\partial \mu}{\partial s(\mathbf{R}')}\overline{\{\delta N_{11}(\mathbf{R})\}_c\{\delta N_{11}(\mathbf{R}')\}_c}.\end{aligned} \tag{14.22}$$

To reduce this expression to a more convenient form we make use of the relationships

$$\delta N_1 = \sum_m p_m - cN \tag{14.23 a}$$

and

$$\delta N_{11}(\mathbf{R}) = \sum_m p_m p_{m+R} - c^2N, \tag{14.23 b}$$

which follow directly from (14.13) and (14.18), respectively, to write, following eqn (14.20 b),

$$\begin{aligned}\{\delta N_{11}(\mathbf{R})\}_c &= \delta N_{11}(\mathbf{R})-2c\delta N_1 \\ &= \sum_m p_m p_{m+R}-2c\sum_m p_m+c^2N.\end{aligned} \tag{14.24}$$

From the product of (14.23 a) and (14.24),

$$\overline{\delta N_1\{\delta N_{11}(\mathbf{R})\}_c} = \sum_{m,n}\overline{p_n p_m p_{m+R}} - 2c\sum_{m,n}\overline{p_n p_m} +$$

$$+c^2N\sum_n \overline{p_n} - cN\Big\{\sum_m \overline{p_m p_{m+R}} - 2c\sum_m \overline{p_m} + c^2N\Big\} = 0. \qquad (14.25\,\text{a})$$

In obtaining this result the identity

$$\overline{p_l p_m p_n} = c^3 + c^2(1-c)(\delta_{l,m}+\delta_{m,n}+\delta_{l,n}) + c(1-c)(1-2c)\delta_{l,m}\,\delta_{m,n} \qquad (14.25\,\text{b})$$

has been used in conjunction with the condition $\mathbf{R} \neq 0$. Also, from (14.24),

$$\overline{\{\delta N_{11}(\mathbf{R})\}_c\{\delta N_{11}(\mathbf{R}')\}_c}$$

$$= \overline{\Big(\sum_m p_m p_{m+R} - 2c\sum_m p_m + c^2N\Big)\Big(\sum_n p_n p_{n+R'} - 2c\sum_n p_n + c^2N\Big)}$$

$$= \sum_{m,n}\overline{p_m p_{m+R} p_n p_{n+R'}} - 2c\sum_{m,n}\overline{p_m p_n p_{n+R'}} -$$

$$-2c\sum_{mn}\overline{p_m p_{m+R} p_n} + 4c^2\sum_{mn}\overline{p_m p_n} - N^2c^4.$$

Using (14.25 b) and also the identity, for $\mathbf{R} \neq 0$ and $\mathbf{R}' \neq 0$,

$$\overline{p_m p_{m+R} p_n p_{n+R'}}$$

$$= c^4 + c^3(1-c)\{\delta_{m+R-n} + \delta_{m-n-R'} + \delta_{m+R-n-R'} + \delta_{m-n}\} +$$

$$+c^2(1-c)^2\{\delta_{m+R-n}\,\delta_{R+R'} + \delta_{m-n}\,\delta_{R-R'}\},$$

this can be evaluated to give

$$\overline{\{\delta N_{11}(\mathbf{R})\}_c\{\delta N_{11}(\mathbf{R}')\}_c} = c^2(1-c)^2N(\delta_{R,R'}+\delta_{R,-R'}). \qquad (14.26)$$

If we now put (14.16), (14.25 a), and (14.26) into eqn (14.22) for $\overline{\delta M^2}$, then from the general relationship between $\Xi(0)$ and $\overline{\delta M^2}$ given by eqn (14.8) we obtain finally

$$\Xi(0) = c(1-c)\left(\frac{\partial\mu}{\partial c}\right)^2 + 2c^2(1-c)^2\sum_R\left\{\frac{\partial\mu}{\partial s(\mathbf{R})}\right\}^2. \qquad (14.27)$$

This formula is a generalization of eqn (14.17). The correction term is negligible at either end of the concentration range because of the factor $c^2(1-c)^2$ and is, therefore, unimportant for dilute alloys. In general, there are an infinite number of short-range order parameters $s(\mathbf{R})$. It follows, therefore, that in order to estimate the second term in (14.27) it is necessary to subject the sample to an infinite set of heat treatments, in each case measuring the total magnetization and short-range order parameters. Fortunately in many cases it ought to be a reasonable approximation to assume that the magnetization changes only with a single parameter, the nearest-neighbour parameter, for instance. We define this parameter as just s, and the number of nearest-neighbour pairs in the sample is $\frac{1}{2}Nr$, where r is the number of nearest neighbours. The number of nearest-neighbour pairs which happen to be both of type 1 is $\frac{1}{2}Nr(c^2+s)$ and then (14.27) becomes

$$\Xi(0) \simeq c(1-c)\left(\frac{\partial\mu}{\partial c}\right)^2+\frac{2c^2}{r}(1-c)^2\left(\frac{\partial\mu}{\partial s}\right)^2. \tag{14.28}$$

This formula is more tractable than (14.27). To estimate the second term in (14.28) we need measurements only on two samples: the first quenched rapidly to ensure it has no short-range order; the second cooled to have a small residual short-range order s, measurable through X-ray determinations of the chemical order. Magnetization measurements on both then give an estimate of $\partial\mu/\partial s$.

The second term of (14.28) is usually small for the following reasons.

(a) At either end of the concentration range the factor $c^2(1-c)^2$ is small.

(b) The factor r^{-1} is of the order 0·1.

(c) In most cases the dependence of the magnetization on short-range order is small.

For these reasons it is usually satisfactory to retain only the first term in (14.28). But in special situations the second term in (14.28) can be important. For instance, in Fe–Co alloys the magnetization curve passes through a maximum at 40% Co. At this point $\partial\mu/\partial c$ is zero and the second term in (14.28) is the entire measure of the elastic diffuse scattering in the forward direction. A second example is given by the nickel-rich Ni–Mn alloys where the magnetization is extremely sensitive to short-range order, i.e. to the number of nearest-neighbour Mn–Mn pairs.

14.2. Shull–Wilkinson model

The simplest model of an alloy is that in which any fluctuation in the moments of either type 1 or type 2 atoms is completely neglected. In other words, all atoms of type 1 are taken to have a moment μ_1 independent of what their environment might be; similarly all atoms of type 2 are taken to have a moment μ_2. Hence, in general,

$$\mu_m = \mu_2+(\mu_1-\mu_2)p_m, \tag{14.29}$$

from which (14.9) immediately follows.

Furthermore,

$$F_m\mu_m = F_2\mu_2+(F_1\mu_1-F_2\mu_2)p_m, \tag{14.30}$$

and hence

$$\overline{F_m\mu_m F_n\mu_n} = (F_2\mu_2)^2+F_2\mu_2(F_1\mu_1-F_2\mu_2)(\bar{p}_m+\bar{p}_n)+ \\ +(F_1\mu_1-F_2\mu_2)^2\overline{p_m p_n} \\ = (F_2\mu_2)^2+F_2\mu_2(F_1\mu_1-F_2\mu_2)2c+ \\ +(F_1\mu_1-F_2\mu_2)^2c^2+(F_1\mu_1-F_2\mu_2)^2c(1-c)\delta_{m,n} \\ = \{cF_1\mu_1+(1-c)F_2\mu_2\}^2+ \\ +(F_1\mu_1-F_2\mu_2)^2c(1-c)\delta_{m,n}, \quad (14.31)$$

where use has been made of the identity (14.15). From the definition of $\Xi(\boldsymbol{\kappa})$, eqn (14.6), we obtain, with the result (14.31),

$$\Xi(\boldsymbol{\kappa}) = c(1-c)\{F_1(\boldsymbol{\kappa})\mu_1-F_2(\boldsymbol{\kappa})\mu_2\}^2. \quad (14.32)$$

This is the result used by Shull and Wilkinson† in their early experiments. In the forward direction, and within the context of this simple model, (14.32) is consistent with (14.17).

In experiments of the Shull–Wilkinson type the major interest is in the extrapolation of the observed cross-sections to the forward direction. Eqn (14.32) shows that in this limit we have a result for $(\mu_1-\mu_2)^2$. This, coupled with a magnetization measurement of μ, serves to give μ_1 and μ_2 separately.

Fig. 14.1 shows the diffuse elastic cross-section measured by Shull and Wilkinson for $\underline{\text{Fe}}$Cr. Drawn through the various sets of data is a calculated form factor for metallic iron, appropriately scaled, the intercepts of the form factors at zero scattering angle being used to calculate the magnetic moment data given in Table 14.1.

There are two possible signs of $\mu_1-\mu_2$ as determined from the neutron cross-section, as is clear from eqn (14.32). Usually this ambiguity is removed by taking into account some other independent experimental information. Collins and Forsyth‡ used polarized neutrons to determine the sign of $\mu_1-\mu_2$. From eqn (10.95) it is clear that there is an additional term in the cross-section

$$2\left(\frac{\gamma e^2}{2m_e c^2}\right)\mathbf{P}.(\tilde{\mathbf{z}}-\tilde{\kappa}_z\tilde{\boldsymbol{\kappa}})c(1-c)(b_1-b_2)\{F_1(\boldsymbol{\kappa})\mu_1-F_2(\boldsymbol{\kappa})\mu_2\}.$$

Here $\mathbf{P}$ is the polarization vector, and b_1 and b_2 are the coherent nuclear scattering lengths of the type 1 and type 2 atoms respectively. (The Debye–Waller factor of each type of atom is assumed equal to unity.) Implicit in this expression is the assumption that the different atoms

† C. G. Shull and M. K. Wilkinson, *Phys. Rev.* **97**, 304 (1955).
‡ M. F. Collins and J. B. Forsyth, *Phil. Mag.* **8**, 401 (1963).

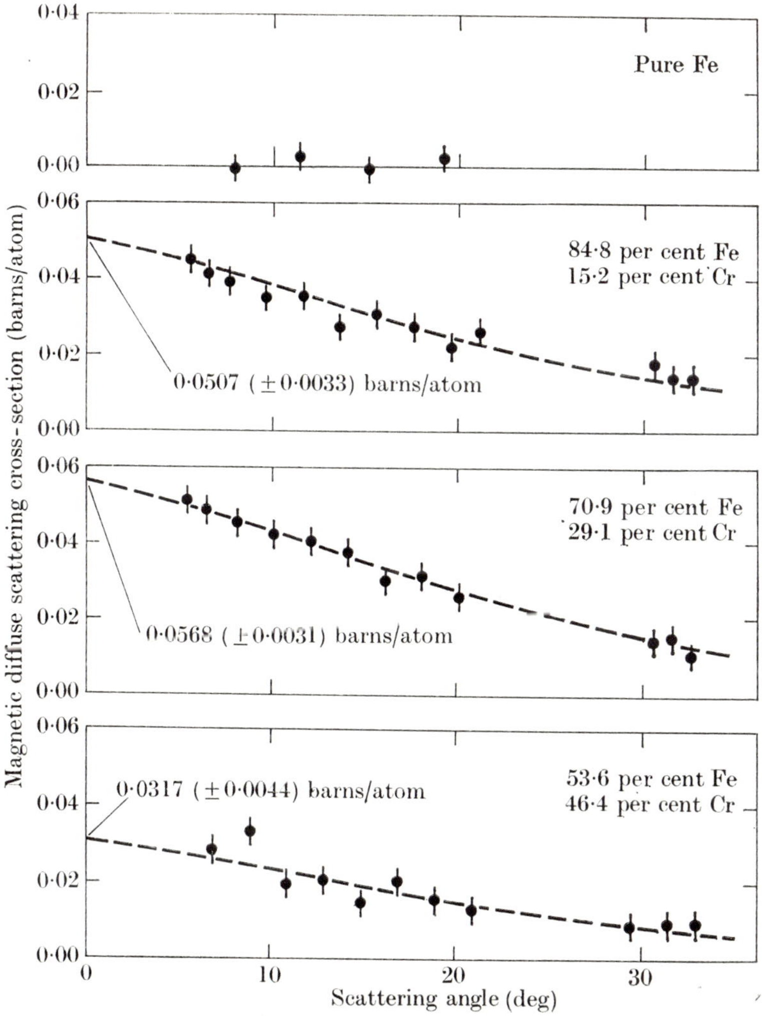

FIG. 14.1. Ferromagnetic diffuse scattering obtained for a series of disordered Fe–Cr alloys.

have distinct scattering lengths, as we assume them to have distinct magnetic moments. If the experimental arrangement is that the neutron beam is polarized parallel or antiparallel to the magnetization of the sample ($\mathbf{P}.\mathbf{z} = \pm 1$) and $\tilde{\boldsymbol{\kappa}}$ lies in a plane normal to this direction, then the above term has the two values

$$\pm 2\left(\frac{\gamma e^2}{2m_e c^2}\right)c(1-c)(b_1-b_2)(\mu_1-\mu_2)$$

TABLE 14.1

Magnetic moment data for Fe–Cr *disordered alloys as obtained from neutron scattering and magnetization experiments*

	$\mu_1-\mu_2$ (μ_B)	$\bar{\mu}$ (μ_B)
Fe	0	2·22
84·8 Fe–15·2 Cr	2·85±0·10	1·81
70·9 Fe–29·1 Cr	2·38±0·07	1·44
53·6 Fe–46·4 Cr	1·62±0·11	1·00

in the forward direction, according to the two directions of **P**. The magnitude of this term may be increased by the use of a particular isotope of either type of atom (or both) chosen to give the maximum value of b_1-b_2.

14.3. Dilute alloys: theory

When the concentration c is small the immediate environment of each type 1 atom is the same and, therefore, one can assume all atoms of type 1 to have the same moment μ_1. However, each type 1 atom will affect the magnitude of the moment of the type 2 atoms that surround it. To allow for this, we define $\phi(\mathbf{R})$ to be the increase in the z component of the magnetic moment of a type 2 atom at a distance **R** from a given type 1 atom, due to the presence of this atom. Thus, we now have

$$\mu_m = \mu_2+(\mu_1-\mu_2)p_m+\sum_R \phi(\mathbf{R})p_{m+R} \tag{14.33 a}$$

and

$$F_m\mu_m = F_2\mu_2+(F_1\mu_1-F_2\mu_2)p_m+F_2\sum_R \phi(\mathbf{R})p_{m+R}. \tag{14.33 b}$$

By definition $\phi(0)$ is zero and because any disturbance must vanish at large enough distances from the type 1 atom, $\phi(\mathbf{R})$ approaches zero as $|\mathbf{R}|$ becomes large. It is convenient to define the spatial Fourier transform

$$\Phi(\boldsymbol{\kappa}) = \sum_R \exp(i\boldsymbol{\kappa}.\mathbf{R})\phi(\mathbf{R}). \tag{14.34}$$

We shall assume the lattice to be such that every atom is a centre of symmetry so that $\phi(\mathbf{R}) = \phi(-\mathbf{R})$. $\Phi(\boldsymbol{\kappa})$ is then real and $\Phi(\boldsymbol{\kappa}) = \Phi(-\boldsymbol{\kappa})$. From (14.33 a)

$$\mu = \mu_2+(\mu_1-\mu_2)c+\Phi(0)c, \tag{14.35 a}$$

and hence

$$\frac{\partial\mu}{\partial c} = \mu_1-\mu_2+\Phi(0). \tag{14.35 b}$$

From (14.33 b) we calculate

$$\overline{F_m\mu_m F_n\mu_n} = [F_2\mu_2+c\{F_1\mu_1-F_2\mu_2+F_2\Phi(0)\}]^2+$$
$$+c(1-c)\{(F_1\mu_1-F_2\mu_2)^2\delta_{m,n}+$$
$$+2F_2(F_1\mu_1-F_2\mu_2)\phi(\mathbf{m}-\mathbf{n})+$$
$$+F_2^2\sum_R\phi(\mathbf{R})\phi(\mathbf{m}+\mathbf{R}-\mathbf{n})\}$$
$$= (\overline{F\mu})^2+c(1-c)\frac{1}{N}\sum_{\boldsymbol{\varkappa}}\exp\{-i\boldsymbol{\varkappa}.(\mathbf{m}-\mathbf{n})\}\times$$
$$\times\{F_1\mu_1-F_2\mu_2+F_2\Phi(\boldsymbol{\varkappa})\}^2. \quad (14.36)$$

If we substitute (14.36) into eqn (14.6), then $\Xi(\boldsymbol{\varkappa})$ for a dilute ferromagnetic alloy is given by

$$\Xi(\boldsymbol{\varkappa}) = c(1-c)\{F_1(\boldsymbol{\varkappa})\mu_1-F_2(\boldsymbol{\varkappa})\mu_2+F_2(\boldsymbol{\varkappa})\Phi(\boldsymbol{\varkappa})\}^2. \quad (14.37\,\text{a})$$

This formula is consistent with (14.17) in the forward direction.

To illustrate the behaviour of the diffuse scattering from a dilute ferromagnetic alloy let us assume that the magnetic moment disturbance is significant only for the type 2 atoms in the first-neighbour shell of a given type 1 atom and that at this distance ($R = \rho$) it has the value ϕ_1. If in addition $F_1 \simeq F_2 = F$ for all values of $\boldsymbol{\varkappa}$, then for a polycrystalline sample we can expect

$$\Xi(\boldsymbol{\varkappa}) \simeq c(1-c)F^2(\boldsymbol{\varkappa})\left\{\mu_1-\mu_2+r\phi_1\frac{\sin(\kappa\rho)}{(\kappa\rho)}\right\}^2, \quad (14.37\,\text{b})$$

where r is the number of nearest-neighbour atoms. If both $\mu_1-\mu_2$ and ϕ_1 are positive then the quantity in brackets in (14.37 b) has the value $\mu_1-\mu_2+r\phi_1$, i.e. $\partial\mu/\partial c$, at $\boldsymbol{\varkappa} = 0$ and falls rapidly to a value oscillating about $\mu_1-\mu_2$ as $\boldsymbol{\varkappa}$ increases. The forward peak, as a function of $\boldsymbol{\varkappa}$, has a width of about π/ρ and thus is well inside the Bragg peaks of most simple crystals. If on the other hand $\mu_1-\mu_2$ and ϕ_1 have opposite signs then there is a sharp dip in the forward direction.

In general we can draw the following conclusions about the Ξ for a dilute ferromagnetic alloy:

(a) Near the forward direction there will be either a peak or a dip of width about π/R_0, where R_0 characterizes the range of the disturbance of the magnetic moments of the type 2 atoms.

(b) If $\Xi(\boldsymbol{\varkappa})$ displays a maximum other than at $\kappa = 0$ then the disturbance $\phi(\mathbf{R})$, taken together with that on the central atom, $\mu_1-\mu_2$, cannot be of a uniform sign.

(c) At very large values of $\boldsymbol{\kappa}$ the oscillations of $\Phi(\boldsymbol{\kappa})$ are relatively unimportant and, to a good approximation, we obtain the Shull–Wilkinson result of § 14.2.

(d) The behaviour of $\Phi(\boldsymbol{\kappa})$ is repeated in every reciprocal lattice cell. Therefore, around each reciprocal lattice vector the same behaviour is observed, apart from the changes in the form factors F_1 and F_2. Moreover, for polycrystalline samples only the behaviour near the forward direction is easy to observe.

(e) In principle, it is possible to make a complete measurement of $\Phi(\boldsymbol{\kappa})$ and hence obtain $\phi(\mathbf{R})$ exactly, but there are limitations set on such a programme by the experimental techniques (see below).

(f) As the temperature of the ferromagnetic alloy is raised towards its Curie point, the elastic diffuse scattering increases rapidly in magnitude because there is an increase in the spatial range of the disturbance on the host. In fact there are two competing processes involved: first, the increase in the range of the disturbance, and secondly, the decrease in the magnitude of individual changes in moment of the type 2 atoms, but the former process dominates as the Curie temperature is approached so that in this limit $\sum_R \phi(\mathbf{R})$ tends to infinity.†

14.4. Dilute alloys: experimental results

In all the examples discussed here the alloy sample is polycrystalline. We shall assume that we can replace the average of $\Xi(\boldsymbol{\kappa})$ over all orientations of $\mathbf{R}$ relative to $\boldsymbol{\kappa}$ by the square of the average of

$$F_1(\boldsymbol{\kappa})\mu_1 - F_2(\boldsymbol{\kappa})\mu_2 + F_2(\boldsymbol{\kappa})\Phi(\boldsymbol{\kappa})$$

(apart from the factor $c(1-c)$, that is). This procedure would be exact if the symmetry of an atomic site was spherical and is likely to be good for the body-centred and face-centred cubic host lattices with which we shall be concerned.

For large values of $\kappa = 4\pi\sin(\frac{1}{2}\theta)/\lambda$, where by large in the present context we mean values of $\kappa \simeq 1{\cdot}5\ \text{Å}^{-1}$, we can expect Φ to be small whilst the form factors will be close to unity. (By definition,

$$F(\boldsymbol{\kappa} = 0) = 1.)$$

Thus by examining the asymptotic form of the scattered intensity the magnitude of the difference in the magnetic moment associated with the impurity atom from that of the unperturbed host atoms can be obtained. The sign of this difference has to be obtained by some other, independent, method.

† This is further discussed by S. W. Lovesey and W. Marshall, *Proc. phys. Soc.* **89**, 613 (1967) and S. W. Lovesey, ibid. **89**, 625 and 893.

The experimental problems involved in measuring the diffuse elastic scattering by dilute ferromagnetic alloys have been discussed in detail by Low.† The scattering of interest can be as much as two orders of magnitude smaller than other scattering coming, for example, from

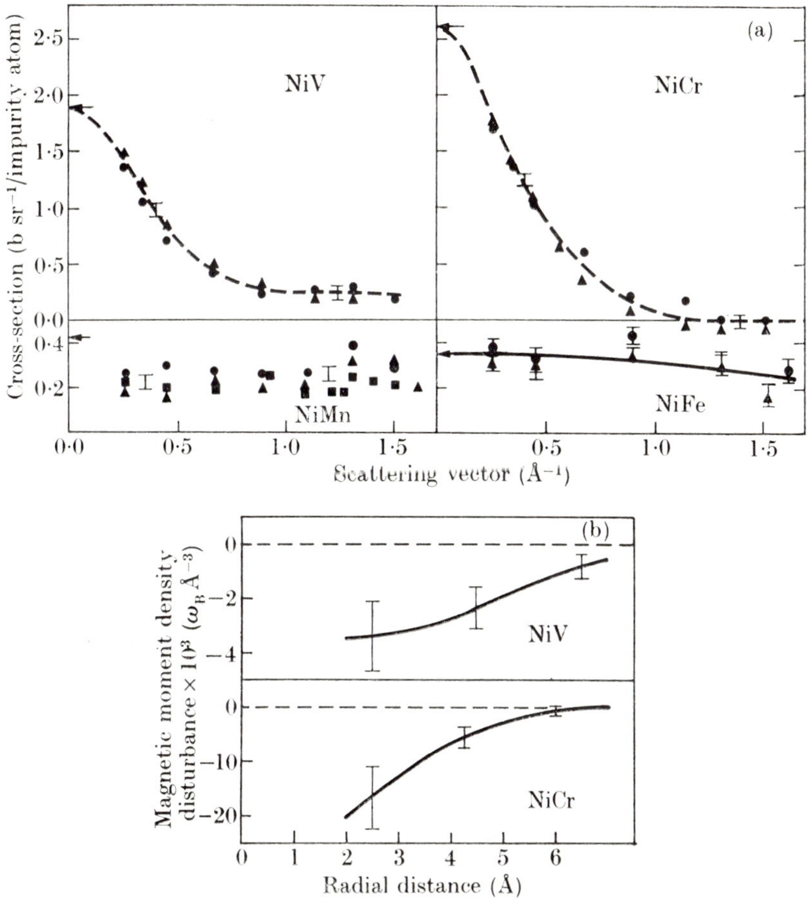

FIG. 14.2. Dilute Ni-based alloys.

nuclear incoherent cross-sections. For a ferromagnetic system a separation of the magnetic diffuse scattering can be effected by changing the direction of the magnetization of the sample with respect to $\boldsymbol{\kappa}$. The orientation factor $1-\tilde{\kappa}_z^2$ can be made zero or unity according to whether the sample magnetization is made parallel or perpendicular to the scattering vector $\boldsymbol{\kappa}$ by the application of an external magnetic field.

† G. G. Low, *Inelastic scattering of neutrons from solids and liquids*, vol. 1, pp. 413–31. I.A.E.A. (1965).

If this technique is used care must be taken to avoid the effects of multiple Bragg scattering. This scattering is diffuse in character and has a complicated dependence on both $\varkappa$ and the direction of magnetization. However, multiple Bragg scattering effects can be avoided

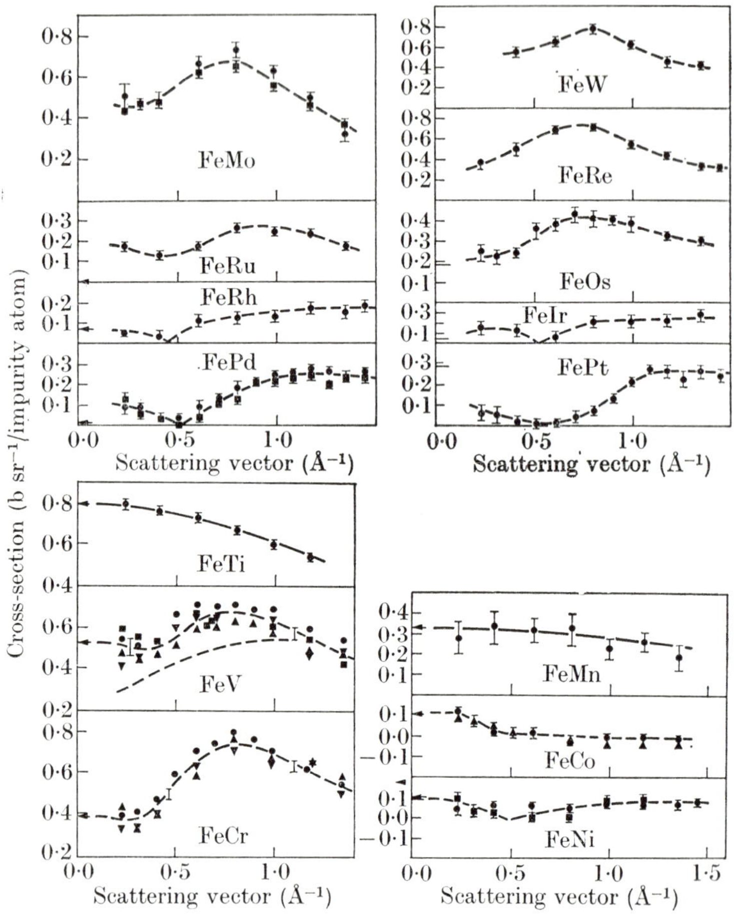

Fig. 14.3. Dilute Fe-based alloys.

by either using neutrons of a sufficiently long wavelength so that Bragg scattering cannot occur (λ greater than twice the largest interplanar spacing in the crystal), or alternatively by using a single crystal specimen.

As illustrations of the observed behaviour of the diffuse elastic cross-section for dilute binary ferromagnetic alloys Figs. 14.2 and 14.3 show some cross-sections measured by Low and Collins† for dilute Ni- and Fe-based alloys.

† M. F. Collins and G. G. Low, *Proc. phys. Soc.* **86**, 535 (1965).

The scattering intensity from $\underline{\mathrm{Ni}}$Fe is relatively independent of angle and the measurements agree well with a form factor appropriate to $3d$ electrons, the intercept of which at $\boldsymbol{\varkappa} = 0$ has been adjusted so as to be in agreement with the experimental value of $\partial\mu/\partial c = 2{\cdot}2\,\mu_{\mathrm{B}}$. From these facts we conclude that in this case the magnetic moment disturbance is confined to the type 1 atom sites, i.e. the presence of iron atoms does not influence the moments associated with the nickel atoms. Furthermore, the moment associated with the iron atoms can be obtained from a best fit of a $3d$ form factor: Low and Collins find that $\mu_1 = (2{\cdot}8 \pm 0{\cdot}2)\,\mu_{\mathrm{B}}$. The cross-section for the alloy $\underline{\mathrm{Ni}}$Mn is much the same as that for $\underline{\mathrm{Ni}}$Fe. In contrast to these two alloys, the cross-sections for $\underline{\mathrm{Ni}}$Cr and $\underline{\mathrm{Ni}}$V alloys show a very rapid decrease of intensity with increasing $\boldsymbol{\varkappa}$, indicating that there is a magnetic moment defect centred on the Cr or V atoms with a large spatial extent into the Ni host. From the intensity at large scattering angles, where $\Phi(\boldsymbol{\varkappa})$ can be neglected, a measure is obtained of the amount by which the moments localized on the impurity atoms themselves differ from the moment of the undisturbed host. In the case of $\underline{\mathrm{Ni}}$Cr the difference of the impurity atom moment from $\mu_2 = 0{\cdot}6\,\mu_{\mathrm{B}}$ is roughly zero while that for the alloy $\underline{\mathrm{Ni}}$V is approximately $(2{\cdot}0 \pm 0{\cdot}4)\,\mu_{\mathrm{B}}$, assuming the moment on vanadium is negative. In both these cases it therefore follows that the greater part of the diminution observed in the saturation magnetization for these alloys ($\partial\mu/\partial c = -6{\cdot}0\,\mu_{\mathrm{B}}$ for $\underline{\mathrm{Ni}}$Cr and $-5{\cdot}1\,\mu_{\mathrm{B}}$ for $\underline{\mathrm{Ni}}$V) results from a loss of moment on the nickel atoms in the vicinity of the impurities. This behaviour is clearly shown by the values of the magnetic moment disturbance around V and Cr shown in Fig. 14.2 (b). These values are obtained by Fourier transformation of the experimental results.

The experimental results of Collins and Low for Fe-based alloys are shown in Fig. 14.3. An examination of the cross-sections show that there are three distinct types of dependence on the scattering vector:

(i) The scattering decreases monotonically and corresponds more or less to an atomic $3d$ form factor. $\underline{\mathrm{Fe}}$Ti and $\underline{\mathrm{Fe}}$Mn are of this type.

(ii) The cross-section is non-zero everywhere and passes through a maximum at about $\kappa = 0{\cdot}75$ Å^{-1}. The cross-section at this point is greater than in the forward direction. The V, Cr, Mo, Ru, W, Re, and Os alloys give cross-sections of this type.

(iii) The cross-section is zero at about $\kappa = 0{\cdot}5$ Å^{-1}.

Type (i) corresponds closely to the behaviour already discussed in connection with $\underline{\mathrm{Ni}}$Fe and $\underline{\mathrm{Ni}}$Mn alloys, namely the defect is strongly localized and the cross-section follows approximately a $3d$ form factor. Again the $3d$ moments at the impurity sites can be found from the intensity at large κ. The values are given in Table 14.2, where the values quoted for the various alloys are calculated on the basis that the sign of the difference in the magnetic moments of the impurity atoms and the unperturbed host atoms is negative, i.e. $\mu_1 - \mu_2 < 0$.

TABLE 14.2

Magnetic moments localized in d orbitals on substitutional impurities in iron (after Collins and Low)

Ti	V	Cr	Mn		Co	Ni
−0·7 ±0·3	−0·4 ±0·4	−0·7 ±0·4	0 ±0·2		2·1 ±0·5	0·9 ±0·15
		Mo		Ru	Rh	Pd
		−0·1 ±0·6		0·9 ±0·5	0·5 ±0·3	0·1 ±0·2
		W	Re	Os	Ir	Pt
		−0·7 ±0·6	−0·3 ±0·6	0 ±0·5	0 ±0·5	0 ±0·3

The scattering from the type (ii) alloys illustrates conclusion (a) of the previous section. For $\partial\mu/\partial c$ is negative for all alloys of type (ii) and the scattering intensity does not pass through zero, so the amplitude is negative for all κ, indicating that the moment of the impurity atom is less than that of the unperturbed Fe atoms. On the other hand, the scattering from type (iii) alloys illustrates conclusion (b), because for these $\partial\mu/\partial c$ is always positive and there is clear evidence that the scattering amplitude for $\kappa \gtrsim 0{\cdot}5$ Å^{-1} is of the opposite sign to this. Hence in this case also the magnetic moment of the impurity must be less than that of the unperturbed Fe atoms.

The magnetic moment disturbance on the Fe atoms is found to be of two types, one characteristic of alloys where the impurities lie to the right of iron in the periodic table and one for alloys where the impurities lie to the left or under iron. Sketches of these patterns are shown in Fig. 14.4. At present the correct explanation of this behaviour is not entirely certain.†

The discussion given above is intended primarily to illustrate the types of behaviour of the diffuse elastic neutron cross-section discussed

† See the discussions by I. A. Campbell and A. A. Gomes, *Proc. phys. Soc.* **91,** 319 (1967) and G. G. Low, ibid. **92,** 938 (1967).

in § 14.3, but it also clearly illustrates the wealth of information that can be obtained about the magnetic structure of binary ferromagnetic alloys by this method. Many other alloys besides those discussed above have been examined. For further details and references the reader is referred to the review by G. Low.†

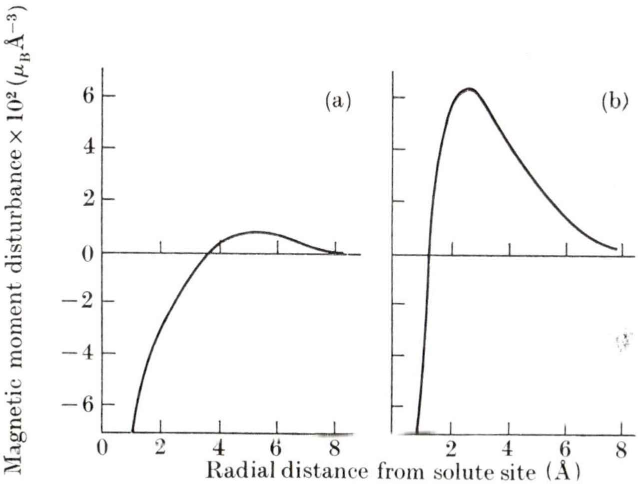

FIG. 14.4. Magnetic moment disturbance on Fe atoms.

14.5. Concentrated alloys

In order to extend the above theory to this, more general, case we make two assumptions. First, we assume that the mean short-range order parameter is zero; secondly, we assume that, within the range of environments allowed by statistical fluctuations, the disturbance in the magnetic moment is both linear and additive. This second assumption is best explained through the following simplified example.

Let us assume that the lattice is face-centred cubic and that each atom influences only its nearest neighbours. We further suppose that the concentration is $\frac{1}{3}$. Then, on average, each atom will have four nearest neighbours of type 1 and eight nearest neighbours of type 2. However, environments differing from this one have a quite reasonable probability; indeed the probability of n neighbours of type 1 and $12-n$ neighbours of type 2 is

$$P(n) = \frac{12!}{n!\,(12-n)!}\,c^n(1-c)^{12-n}.$$

† G. G. Low, *Z. angew. Phys.* **24**, 254 (1968).

Roughly speaking this means that only values of n ranging from 1 to 7 have appreciable probabilities distributed as follows:

$$\begin{array}{ll} P(1) = 0{\cdot}05 & P(5) = 0{\cdot}20 \\ P(2) = 0{\cdot}13 & P(6) = 0{\cdot}12 \\ P(3) = 0{\cdot}22 & P(7) = 0{\cdot}04 \\ P(4) = 0{\cdot}24 & \end{array} \tag{14.38}$$

If an atom has, say, three neighbours of type 1, we assume it has the same magnetic moment no matter which particular positions those three might take in the twelve possible positions in the nearest-neighbour shell. This means that the magnetic moment is only a function of n, and we further assume that, within the width of a probability distribution like (14.38), this variation with n is linear. In our particular example, because the total probability is dominated by the values for $n = 3$, 4, and 5, producing magnetic moments $\mu(3)$, $\mu(4)$, and $\mu(5)$, respectively, this assumption can roughly be summarized as $\mu(5)-\mu(4) = \mu(4)-\mu(3)$.

The assumption we actually make is more general than illustrated in the foregoing example in the sense that, in principle, each atom is allowed to be influenced by more than just the nearest neighbours (although the disturbance must, certainly, fall off at large distances). The mathematical expression of these assumptions is as follows. If an atom of type 2 is at $\mathbf{m}$, it has a moment

$$\mu_m = \mu_2 + \sum_R \phi(\mathbf{R})(p_{m+R}-c), \tag{14.39 a}$$

whereas if an atom of type 1 is at $\mathbf{m}$, it has a moment

$$\mu_m = \mu_1 + \sum_R \chi(\mathbf{R})(p_{m+R}-c), \tag{14.39 b}$$

where $\phi(\mathbf{R})$ and $\chi(\mathbf{R})$ are the disturbances produced by fluctuations in the number of type 1 atoms at distance $\mathbf{R}$. By definition,

$$\phi(0) = \chi(0) = 0 \tag{14.40}$$

and μ_1 and μ_2 are the mean moments for atoms of type 1 and 2 respectively. The two formulae (14.39 a) and (14.39 b) can be combined to give for the moment at a general site $\mathbf{m}$,

$$\mu_m = \mu_2 + (\mu_1-\mu_2)p_m + \sum_R \phi(\mathbf{R})(p_{m+R}-c) + \sum_R p_m(p_{m+R}-c)\{\chi(\mathbf{R})-\phi(\mathbf{R})\} \tag{14.41}$$

and

$$F_m \mu_m = F_2\mu_2 + (F_1\mu_1 - F_2\mu_2)p_m + F_2 \sum_R \phi(\mathbf{R})(p_{m+R}-c) + \sum_R p_m(p_{m+R}-c)\{F_1\chi(\mathbf{R}) - F_2\phi(\mathbf{R})\}. \tag{14.42}$$

We define $\Phi(\varkappa)$ by (14.34) and, in addition,

$$X(\varkappa) = \sum_R \exp(i\varkappa.\mathbf{R})\chi(\mathbf{R}), \tag{14.43}$$

$$\psi(\mathbf{R}) = \chi(\mathbf{R})-\phi(\mathbf{R}), \tag{14.44}$$

$$\Psi(\varkappa) = \sum_R \exp(i\varkappa.\mathbf{R})\psi(\mathbf{R}) = X(\varkappa)-\Phi(\varkappa), \tag{14.45}$$

$$\Upsilon(\varkappa) = \sum_R \exp(i\varkappa.\mathbf{R})\psi^2(\mathbf{R}), \tag{14.46}$$

$$\begin{aligned}\Psi(\varkappa, \mathbf{Q}) &= F_1(\varkappa)\Psi(\mathbf{Q})-F_2(\varkappa)\Phi(\mathbf{Q})\\ &= \sum_R \exp(i\mathbf{Q}.\mathbf{R})\{F_1(\varkappa)\chi(\mathbf{R})-F_2(\varkappa)\phi(\mathbf{R})\},\end{aligned} \tag{14.47}$$

and finally

$$\Upsilon(\varkappa, \mathbf{Q}) = \sum_R \exp(i\mathbf{Q}.\mathbf{R})\{F_1(\varkappa)\chi(\mathbf{R})-F_2(\varkappa)\phi(\mathbf{R})\}^2. \tag{14.48}$$

For the model of the concentrated binary alloy defined above, (14.39) and (14.41) represent the full variation of the moments with concentration fluctuations about the mean value of c and with short-range order fluctuations about the mean value of zero. It follows that

$$\frac{\partial\mu_2}{\partial c} = \Phi(0), \qquad \frac{\partial\mu_1}{\partial c} = X(0), \tag{14.49}$$

$$\begin{aligned}\frac{\partial\mu}{\partial c} &= \mu_1-\mu_2+(1-c)\Phi(0)+cX(0)\\ &= \mu_1-\mu_2+\Phi(0)+c\Psi(0),\end{aligned} \tag{14.50}$$

$$\frac{\partial^2\mu}{\partial c^2} = 2\Psi(0), \tag{14.51}$$

$$\frac{\partial\mu}{\partial s(\mathbf{R})} = \chi(\mathbf{R})-\phi(\mathbf{R}) = \psi(\mathbf{R}). \tag{14.52}$$

Using the general formula (14.27) we then have

$$\Xi(0) = c(1-c)\{\mu_1-\mu_2+\Phi(0)+c\Psi(0)\}^2+2c^2(1-c)^2\Upsilon(0). \tag{14.53}$$

From (14.6) and (14.42) it can be shown that for a random alloy $\Xi(\varkappa)$ reduces in this case to

$$\begin{aligned}\Xi(\varkappa) = c(1-c)\{F_1(\varkappa)\mu_1-F_2(\varkappa)\mu_2+F_2(\varkappa)\Phi(\varkappa)+c\Psi(\varkappa,\varkappa)\}^2+\\ +c^2(1-c)^2\{\Upsilon(\varkappa,0)+\Upsilon(\varkappa,\varkappa)\}.\end{aligned} \tag{14.54}$$

This expression reduces to (14.53) in the limit $\varkappa = 0$, as it should, and is remarkably simple in form. In most cases the second term will be an order of magnitude smaller than the first and can be neglected.

A case of special importance is that in which the form factors are sufficiently alike to be replaced by a single factor $F(\boldsymbol{\varkappa})$; this will usually be a good approximation for alloys within a given period. For this case,

$$\left.\begin{aligned}\Psi(\boldsymbol{\varkappa},\mathbf{Q}) &= F(\boldsymbol{\varkappa})\Psi(\mathbf{Q})\\ \Upsilon(\boldsymbol{\varkappa},\mathbf{Q}) &= \{F(\boldsymbol{\varkappa})\}^2\Upsilon(\mathbf{Q})\end{aligned}\right\}\quad \text{when } F_1 \simeq F_2 = F(\boldsymbol{\varkappa}), \tag{14.55}$$

and the expression (14.54) for $\Xi(\boldsymbol{\varkappa})$ simplifies to

$$\Xi(\boldsymbol{\varkappa}) = c(1-c)F^2(\boldsymbol{\varkappa})\{\mu_1-\mu_2+\Phi(\boldsymbol{\varkappa})+c\Psi(\boldsymbol{\varkappa})\}^2+ \\ +c^2(1-c)^2F^2(\boldsymbol{\varkappa})\{\Upsilon(0)+\Upsilon(\boldsymbol{\varkappa})\}. \tag{14.56}$$

A measurement of $\Xi(\boldsymbol{\varkappa})$, assuming $F(\boldsymbol{\varkappa})$ is known, then gives μ_1 and μ_2 and the combination $\Phi(\boldsymbol{\varkappa})+c\Psi(\boldsymbol{\varkappa})$. There is no unambiguous method of determining $\Phi(\boldsymbol{\varkappa})$ and $\Psi(\boldsymbol{\varkappa})$ separately but, because both should vary smoothly with concentration, measurements at slightly different concentrations should be sufficient to determine them with fair accuracy. The consistency conditions represented by (14.9), (14.49), (14.50), and (14.51) are quite restrictive.

It is to be noted that, for this model, (14.51) and (14.52) give some relation between the variation with short-range order and the second derivative of the total magnetization with respect to concentration. This relationship can be made precise only with some further assumption, such as that used in deriving (14.28). In the notation of the present model, this assumption is that $\psi(\mathbf{R})$ is zero beyond the nearest neighbours, i.e.

$$\psi(\mathbf{R}) \simeq \psi \sum_{\rho} \delta(\mathbf{R}-\boldsymbol{\rho}), \tag{14.57}$$

where the sum is over the r nearest neighbours to the site $\mathbf{R}$. Then

$$\Upsilon(0) \simeq r\psi^2, \tag{14.58 a}$$

$$\frac{1}{2}\frac{\partial^2\mu}{\partial c^2} = \Psi(0) \simeq r\psi. \tag{14.58 b}$$

Hence, the second term of (14.53) becomes

$$\frac{1}{2r}c^2(1-c)^2\left(\frac{\partial^2\mu}{\partial c^2}\right)^2, \tag{14.59}$$

which is to be compared to the first term, which is

$$c(1-c)\left(\frac{\partial\mu}{\partial c}\right)^2. \tag{14.60}$$

Except for concentrations where $\partial\mu/\partial c$ happens to be near zero, it is usually satisfactory to neglect (14.59) compared with (14.60).

14.6. Short-range order

So far we have always assumed the alloy to be completely random but we now relax this condition and examine the effect of chemical order. We shall assume that the short-range order is small and described by parameters $s(\mathbf{R})$. We can retain both (14.41) and (14.42) and note that we need to evaluate products like $\overline{p_m p_n}$, $\overline{p_l p_m p_n}$, and $\overline{p_k p_l p_m p_n}$. Whereas the evaluation of these products was straightforward for the random alloy much more care is required when short-range order is present. For, by definition, we now have

$$\overline{p_m} = c$$

and

$$\overline{p_m p_n} = c^2+c(1-c)\delta_{m,n}+s(\mathbf{m}-\mathbf{n}),$$

and to calculate the average of any function we have to average over a canonical ensemble subject to these restrictions. Since the algebra is rather tedious we shall content ourselves only with the main results; namely,

$$\bar{\mu} = \overline{\mu_m} = (1-c)\mu_2+c\mu_1+\sum_R s(\mathbf{R})\psi(\mathbf{R}), \tag{14.61}$$

$$c\overline{\mu_1} = \overline{p_m \mu_m} = c\mu_1+\sum_R s(\mathbf{R})\chi(\mathbf{R}), \tag{14.62}$$

$$(1-c)\overline{\mu_2} = \overline{(1-p_m)\mu_m} = (1-c)\mu_2-\sum_R s(\mathbf{R})\phi(\mathbf{R}), \tag{14.63}$$

and

$$\overline{F_m \mu_m} = (1-c)F_2\mu_2+cF_1\mu_1+\sum_R s(\mathbf{R})\{F_1\chi(\mathbf{R})-F_2\phi(\mathbf{R})\}, \tag{14.64}$$

where μ_1 and μ_2 are the mean moments in the absence of short-range order. Furthermore, when $F_1 = F_2 = F(\boldsymbol{\varkappa})$ it is found that

$$\Xi(\boldsymbol{\varkappa})/F^2(\boldsymbol{\varkappa}) = \Big\{c(1-c)+\sum_R \exp(i\boldsymbol{\varkappa}.\mathbf{R})s(\mathbf{R})\Big\}\times$$

$$\times\left[\mu_1-\mu_2+\Phi(\boldsymbol{\varkappa})+c\Psi(\boldsymbol{\varkappa})+\frac{1-2c}{c(1-c)}\Big\{\sum_R s(\mathbf{R})\psi(\mathbf{R})\{1+\exp(i\boldsymbol{\varkappa}.\mathbf{R})\}\Big\}\right]^2+$$

$$+c^2(1-c)^2\{\Upsilon(0)+\Upsilon(\boldsymbol{\varkappa})\}+(1-2c)^2\sum_R s(\mathbf{R})\psi^2(\mathbf{R})\{1+\exp(i\boldsymbol{\varkappa}.\mathbf{R})\}+$$

$$+c(1-c)\frac{1}{N}\sum_Q\sum_R s(\mathbf{R})\exp(i\boldsymbol{\varkappa}.\mathbf{R})\{\Psi(\mathbf{Q})+\Psi(\boldsymbol{\varkappa}+\mathbf{Q})\}^2. \tag{14.65}$$

In many alloys only the first term of (14.65) is of importance. This formula allows, in principle, corrections to be made for small known amounts of short-range order but it is, clearly, preferable to use a random alloy wherever possible because then the relevant formulae, eqns (14.54) and (14.56), are relatively simple.

14.7. Antiferromagnets

Much of the theory given above can be extended to a simple antiferromagnetic alloy, i.e. an alloy in which the magnetic structure can be divided into two identical sublattices, the magnetic moments associated with one sublattice being aligned antiparallel to those on the other. Experimentally the difficulty involved in observing the magnetic elastic diffuse scattering from such systems is, however, much greater than for a ferromagnetic alloy. This stems from the fact that one cannot, as in the ferromagnetic case, apply an external magnetic field to orientate the magnetization of the sample parallel and then perpendicular to $\boldsymbol{\kappa}$, which enables the diffuse magnetic scattering to be identified. In principle, the diffuse magnetic scattering could be isolated by using a pure sample as a reference system, but the preparation of two samples identical in all respects except for a concentration c of impurities in one is difficult.

If $\mathbf{w}$ is the smallest reciprocal lattice vector of the superlattice, $\exp(\mathrm{i}\mathbf{w}.\mathbf{m})$ is $+1$ or -1 depending upon whether $\mathbf{m}$ belongs to one sublattice or the other. We define

$$\nu_m = \mu_m \exp(\mathrm{i}\mathbf{w}.\mathbf{m}), \tag{14.66}$$

where $$\bar{\mu}_m = 0$$

but $$\bar{\nu}_m = \nu,$$

ν being non-zero at temperatures below the Néel point of the alloy.

Substituting (14.66) into (14.2) gives

$$\frac{\mathrm{d}\sigma}{\mathrm{d}\Omega} = \frac{N}{4}\left(\frac{\gamma e^2}{m_\mathrm{e} c^2}\right)^2 (1-\tilde{\kappa}_z^2)\frac{1}{N}\sum_{m,n} \exp\{\mathrm{i}(\boldsymbol{\kappa}-\mathbf{w}).(\mathbf{m}-\mathbf{n})\}\overline{F_n \nu_n F_m \nu_m}.$$

We again separate off the Bragg scattering

$$\left(\frac{\mathrm{d}\sigma}{\mathrm{d}\Omega}\right)_{\mathrm{Bragg}} = \frac{N}{4}\left(\frac{\gamma e^2}{m_\mathrm{e} c^2}\right)^2 (1-\tilde{\kappa}_z^2)\{\overline{\nu F(\boldsymbol{\kappa})}\}^2 \frac{(2\pi)^3}{v_0}\sum_{\boldsymbol{\tau}} \delta(\boldsymbol{\kappa}-\mathbf{w}-\boldsymbol{\tau}), \tag{14.67}$$

where $$\overline{\nu F(\boldsymbol{\kappa})} = \frac{1}{N}\sum_m \mu_m F_m(\boldsymbol{\kappa})\exp(\mathrm{i}\mathbf{w}.\mathbf{m}).$$

The elastic diffuse scattering is then given by (14.5) but with

$$\Xi(\boldsymbol{\kappa}) = \frac{1}{N}\sum_{m,n} \exp\{\mathrm{i}(\boldsymbol{\kappa}-\mathbf{w}).(\mathbf{m}-\mathbf{n})\}\{\overline{F_n \nu_n F_m \nu_m} - (\overline{F_n \nu_n})(\overline{F_m \nu_m})\}. \tag{14.68}$$

We now seek general formulae for Ξ analogous to those derived for $\Xi(0)$ in § 14.1 for the ferromagnetic alloy. However, when the form factors $F_1(\boldsymbol{\kappa})$ and $F_2(\boldsymbol{\kappa})$ are not equal, it is not possible to derive formulae of equal generality to those of § 14.1. Thus throughout this section we shall assume that the difference between the form factors is negligible and write $F_1 = F_2 = F(\boldsymbol{\varkappa})$. Eqn (14.68) then simplifies to

$$\Xi(\boldsymbol{\varkappa}) = |F(\boldsymbol{\varkappa})|^2 \frac{1}{N} \sum_{m,n} \exp\{\mathrm{i}(\boldsymbol{\varkappa}-\mathbf{w}).(\mathbf{m}-\mathbf{n})\}\{\overline{\nu_n \nu_m} - \overline{\nu_n}\,\overline{\nu_m}\}. \tag{14.69}$$

The staggered magnetization M_s is defined by

$$M_s = \sum_m \nu_m = \sum_m \mu_m \exp(\mathrm{i}\mathbf{w}.\mathbf{m}), \tag{14.70}$$

and we notice the exact analogy to the results of § 14.1. For instance, the formula analogous to (14.17) is

$$\Xi(\mathbf{w}) = |F(\mathbf{w})|^2 c(1-c)\left(\frac{\partial \nu}{\partial c}\right)^2, \tag{14.71}$$

and the more general formula (14.27) is replaced by

$$\Xi(\mathbf{w}) = |F(\mathbf{w})|^2 \left[c(1-c)\left(\frac{\partial \nu}{\partial c}\right)^2 + 2c^2(1-c)^2 \sum_R \left\{\frac{\partial \nu}{\partial s(\mathbf{R})}\right\}^2\right]. \tag{14.72}$$

The Shull–Wilkinson model of an antiferromagnet, with both atoms of type 1 and type 2 distributed randomly, would be represented by the formula

$$\nu_m = \nu_2 + (\nu_1 - \nu_2)p_m, \tag{14.73}$$

where it is assumed that all atoms of type 1 have a moment of magnitude ν_1 and of a sign determined by the sublattice they belong to, and a similar assumption is made for atoms of type 2. Then

$$\nu = c\nu_1 + (1-c)\nu_2 \tag{14.74}$$

can be measured as a function of concentration from the intensity of the Bragg superlattice peaks. (14.69) and (14.73) then give the Shull–Wilkinson result analogous to (14.32), namely

$$\Xi(\boldsymbol{\varkappa}) = |F(\boldsymbol{\varkappa})|^2 c(1-c)(\nu_1 - \nu_2)^2. \tag{14.75}$$

We notice that this has a dependence on $\boldsymbol{\varkappa}$ only through the form factor $F(\boldsymbol{\varkappa})$ and must therefore display a maximum about the forward direction.

For a dilute antiferromagnetic alloy we must replace (14.73) by

$$\nu_m = \nu_2 + (\nu_1 - \nu_2)p_m + \sum_R \phi(\mathbf{R})p_{m+R}, \tag{14.76}$$

from which

$$\nu = c\nu_1 + (1-c)\nu_2 + c\Phi(0), \tag{14.77 a}$$

$$\frac{\partial \nu}{\partial c} = \nu_1 - \nu_2 + \Phi(0), \tag{14.77 b}$$

and, finally,

$$\Xi(\boldsymbol{\varkappa}) = c(1-c)|F(\boldsymbol{\varkappa})|^2\{\nu_1 - \nu_2 + \Phi(\boldsymbol{\varkappa}-\mathbf{w})\}^2. \tag{14.78}$$

When the effect of an atom of type 1 is to produce a considerable disturbance in the magnetic order of its environment, then $\Phi(0)$ is large and $\Xi(\boldsymbol{\varkappa})$ peaks near the *superlattice* positions given by $\mathbf{w}+\boldsymbol{\tau}$.

15

INELASTIC DEFECT SCATTERING

15.1. Introduction

THERE has been much theoretical work on the change in the continuous energy spectrum of elementary excitations in solids caused by the introduction of substitutional impurities. Inelastic neutron scattering offers a very sensitive probe for detecting these changes, much more so than measuring the change in some macroscopic quantity (specific heat, magnetization, etc.), because the neutrons sample a narrow band of eigenmodes of the composite system whereas the macroscopic quantities involve an integration over the entire excitation spectrum.

The perturbation produced by an impurity is characterized by its spatial extent and for sufficiently small perturbation and simple models the energy levels of the mixed system may be calculated exactly. The impurity and perturbed region of the host system form a complex with discrete energy levels that can lie within and outside the energy band of the host, virtual and localized states, respectively. The virtual levels decay into the energy band of the host because of their interaction with the host excitations, and can significantly alter the thermodynamic behaviour of the system from that of the pure host. Localized levels have a long lifetime and are in general of less importance to the thermodynamic properties of the system. For if they lie below the excitation spectrum they will be completely occupied and therefore ineffective in determining the thermodynamic properties of the system, while if they lie above the band they are again ineffective because of the large energy required to excite them.

In this chapter we consider first the modification of the one-phonon cross-section for harmonic crystals due to the introduction of a small number of impurities with a mass different from those of the host ions, and secondly the modification of the one-magnon cross-section of a Heisenberg ferromagnet due to impurities with both exchange interactions and spins that differ from those of the host system. For both these problems the most convenient method of calculation is to employ thermal, two-time Green functions. All the theory of these functions necessary for our present discussions is given in Appendix B.

15.2. Harmonic crystal with mass defect

We adopt the harmonic approximation throughout and restrict our attention to a Bravais lattice. The ions have position vectors

$$\mathbf{R}_l = \mathbf{l}+\mathbf{u}(l),$$

where $\mathbf{l}$ is the lattice position and $\mathbf{u}(l)$ the displacement from this position. The unperturbed Hamiltonian is (cf. § 4.1)

$$\mathscr{H}_0 = \sum_{l,\alpha} \frac{1}{2M}\{\hat{p}_\alpha(l)\}^2+\tfrac{1}{2}\sum_{l,l'}\sum_{\alpha,\beta}\hat{u}_\alpha(l)A_{\alpha\beta}(l,l')\hat{u}_\beta(l'), \qquad (15.1)$$

where $A_{\alpha\beta}(l,l')$ are the second-order atomic force constants of the crystal. If we remove the ions at the sites $l = s$ and replace them by ions of mass M' then the total Hamiltonian is

$$\mathscr{H} = \mathscr{H}_0+\mathscr{H}_1, \qquad (15.2)$$

where

$$\mathscr{H}_1 = \frac{1}{2}\left(\frac{1}{M'}-\frac{1}{M}\right)\sum_{s,\alpha}\{\hat{p}_\alpha(s)\}^2$$

$$= \frac{1}{2}\frac{\lambda}{M'}\sum_{s,\alpha}\{\hat{p}_\alpha(s)\}^2. \qquad (15.3)$$

We shall find later on that an important parameter is

$$\frac{1}{\lambda} = \frac{1}{1-M'/M}$$

and we note that if $M' > M, \lambda^{-1}$ is negative and ranges from $-\infty$ to zero whereas if $M' < M, \lambda^{-1}$ is positive and ranges from $+\infty$ to $+1$.

Let us now derive expressions for the one-phonon coherent and incoherent cross-sections that lend themselves most readily to our discussion. From Chapter 4 the coherent and incoherent inelastic neutron cross-sections can be written in the forms

$$\left(\frac{d^2\sigma}{d\Omega dE'}\right)_{\text{coh}}$$

$$= \frac{k'}{k}\frac{1}{2\pi\hbar}\int_{-\infty}^{\infty} dt\, e^{-i\omega t}\sum_{l,l'}\bar{b}_l\bar{b}_{l'}[\langle\exp\{-i\boldsymbol{\varkappa}.\hat{\mathbf{R}}_l(0)\}\exp\{i\boldsymbol{\varkappa}.\hat{\mathbf{R}}_{l'}(t)\}\rangle-1] \qquad (15.4)$$

and

$$\left(\frac{d^2\sigma}{d\Omega dE'}\right)_{\text{incoh}} = \frac{k'}{k}\frac{1}{2\pi\hbar}\int_{-\infty}^{\infty} dt\, e^{-i\omega t}\times$$

$$\times\sum_{l}(\overline{b_l^2}-\bar{b}_l^2)[\langle\exp\{-i\boldsymbol{\varkappa}.\hat{\mathbf{R}}_l(0)\}\exp\{i\boldsymbol{\varkappa}.\hat{\mathbf{R}}_l(t)\}\rangle-1], \qquad (15.5)$$

where $\bar{b}_l$ is the bound scattering length (assumed real) for the nucleus at the lth lattice site.

For a harmonic crystal

$$
\begin{aligned}
&\langle\exp\{-\mathrm{i}\boldsymbol{\varkappa}.\hat{\mathbf{u}}(l,0)\}\exp\{\mathrm{i}\boldsymbol{\varkappa}.\hat{\mathbf{u}}(l',t)\}\rangle\\
&\quad=\exp\{\tfrac{1}{2}[\boldsymbol{\varkappa}.\hat{\mathbf{u}}(l,0),\boldsymbol{\varkappa}.\hat{\mathbf{u}}(l',t)]\}\langle\exp\{-\mathrm{i}\boldsymbol{\varkappa}.\hat{\mathbf{u}}(l,0)+\mathrm{i}\boldsymbol{\varkappa}.\hat{\mathbf{u}}(l',t)\}\rangle\\
&\quad=\exp\{\tfrac{1}{2}[\boldsymbol{\varkappa}.\hat{\mathbf{u}}(l,0),\boldsymbol{\varkappa}.\hat{\mathbf{u}}(l',t)]\}\exp(-\tfrac{1}{2}\langle[\boldsymbol{\varkappa}.\{\hat{\mathbf{u}}(l,0)-\hat{\mathbf{u}}(l',t)\}]^2\rangle)\\
&\quad=\exp[-\tfrac{1}{2}\langle\{\boldsymbol{\varkappa}.\hat{\mathbf{u}}(l,0)\}^2\rangle]\exp[-\tfrac{1}{2}\langle\{\boldsymbol{\varkappa}.\hat{\mathbf{u}}(l',0)\}^2\rangle]\times\\
&\qquad\qquad\times\exp\{\langle\boldsymbol{\varkappa}.\hat{\mathbf{u}}(l,0)\boldsymbol{\varkappa}.\hat{\mathbf{u}}(l',t)\rangle\}, \qquad (15.6)
\end{aligned}
$$

where we have used the identities (3.53) and (3.83). If we now define

$$\mathscr{b}_l=\bar{b}_l\exp[-\tfrac{1}{2}\langle\{\boldsymbol{\varkappa}.\hat{\mathbf{u}}(l,0)\}^2\rangle] \qquad (15.7)$$

and retain only the first two terms in the expansion of the third exponential in the last line of (15.6), then the one-phonon cross-sections are

$$
\begin{aligned}
\left(\frac{\mathrm{d}^2\sigma}{\mathrm{d}\Omega\mathrm{d}E'}\right)^{(1)}_{\mathrm{coh}}=\frac{k'}{k}\frac{1}{2\pi\hbar}\int_{-\infty}^{\infty}\mathrm{d}t\,\mathrm{e}^{\mathrm{i}\omega t}\sum_{l,l'}\mathscr{b}_l\mathscr{b}_{l'}\exp\{-\mathrm{i}\boldsymbol{\varkappa}.(\mathbf{l}-\mathbf{l}')\}\times\\
\times\sum_{\alpha,\beta}\kappa_\alpha\kappa_\beta\langle\hat{u}_\alpha(l,t)\hat{u}_\beta(l',0)\rangle \qquad (15.8)
\end{aligned}
$$

and

$$\left(\frac{\mathrm{d}^2\sigma}{\mathrm{d}\Omega\mathrm{d}E'}\right)^{(1)}_{\mathrm{incoh}}=\frac{k'}{k}\frac{1}{2\pi\hbar}\int_{-\infty}^{\infty}\mathrm{d}t\,\mathrm{e}^{\mathrm{i}\omega t}\sum_{l}\alpha_l^2\sum_{\alpha,\beta}\kappa_\alpha\kappa_\beta\langle\hat{u}_\alpha(l,t)\hat{u}_\beta(l,0)\rangle, \qquad (15.9)$$

where
$$\alpha_l^2=(\overline{b_l^2}-\bar{b}_l^2)\exp[-\langle\{\boldsymbol{\varkappa}.\hat{\mathbf{u}}(l,0)\}^2\rangle].$$

In obtaining these expressions we have used the relationship

$$\langle\hat{u}_\alpha(l,0)\hat{u}_\beta(l',-t)\rangle=\langle\hat{u}_\alpha(l,t)\hat{u}_\beta(l',0)\rangle.$$

It is evident from eqns (15.8) and (15.9) that the evaluation of the neutron cross-sections for the mixed system has reduced to the calculation of the displacement–displacement correlation function

$$\langle\hat{u}_\alpha(l,t)\hat{u}_\beta(l',0)\rangle$$

for the system described by the Hamiltonian $\mathscr{H}$, eqns (15.2) and (15.3). To this end we introduce a retarded thermal Green function

$$G_{\alpha\beta}(l,l';t)=\frac{2\pi}{\hbar}\langle\langle\hat{u}_\alpha(l,t);\,\hat{u}_\beta(l',0)\rangle\rangle, \qquad (15.10)$$

where
$$\langle\langle\hat{u}_\alpha(l,t);\,\hat{u}_\beta(l',0)\rangle\rangle=-\mathrm{i}\theta(t)\langle[\hat{u}_\alpha(l,t),\hat{u}_\beta(l',0)]\rangle.$$

Then

$$\langle \hat{u}_\alpha(l,t)\hat{u}_\beta(l',0)\rangle$$

$$= \left(\frac{\hbar}{2\pi}\right)\int_{-\infty}^{\infty} \mathrm{d}\omega' \exp(-\mathrm{i}\omega' t + \hbar\omega'\beta)(-2)n(\omega')\,\mathrm{Im}\, G_{\alpha\beta}(l,l';\omega'), \quad (15.11)$$

where
$$n(\omega) = (\mathrm{e}^{\hbar\omega\beta}-1)^{-1} \quad (15.12)$$

and
$$G_{\alpha\beta}(l,l';\omega) = \frac{1}{2\pi}\int_{-\infty}^{\infty} \mathrm{d}t\; \mathrm{e}^{\mathrm{i}\omega t} G_{\alpha\beta}(l,l';t).$$

Eqn (15.11) follows from eqns (B.33 b) and (B.71) of Appendix B. (The coefficient Υ in eqn (B.71) is clearly zero for the present case.)

Thus
$$\frac{1}{2\pi\hbar}\int_{-\infty}^{\infty} \mathrm{d}t\; \mathrm{e}^{\mathrm{i}\omega t}\langle \hat{u}_\alpha(l,t)\hat{u}_\beta(l',0)\rangle$$

$$= \frac{1}{\hbar}\left(-\frac{\hbar}{\pi}\right)\int_{-\infty}^{\infty} \mathrm{d}\omega'\; \mathrm{e}^{\hbar\omega'\beta} n(\omega')\,\mathrm{Im}\, G_{\alpha\beta}(l,l';\omega') \times \frac{1}{2\pi}\int_{-\infty}^{\infty} \mathrm{d}t \exp(\mathrm{i}\omega t - \mathrm{i}\omega' t)$$

$$= -\frac{1}{\pi}\int_{-\infty}^{\infty} \mathrm{d}\omega'\; \mathrm{e}^{\hbar\omega'\beta} n(\omega')\,\mathrm{Im}\, G_{\alpha\beta}(l,l';\omega')\delta(\omega-\omega')$$

$$= -\frac{1}{\pi}\{1+n(\omega)\}\,\mathrm{Im}\, G_{\alpha\beta}(l,l';\omega)$$

and the coherent cross-section is then

$$\left(\frac{\mathrm{d}^2\sigma}{\mathrm{d}\Omega \mathrm{d}E'}\right)^{(1)}_{\mathrm{coh}}$$

$$= -\frac{1}{\pi}\frac{k'}{k}\sum_{l,l'}\sum_{\alpha,\beta} \kappa_\alpha \kappa_\beta \bar{b}_l \bar{b}_{l'} \exp\{-\mathrm{i}\boldsymbol{\kappa}.(\mathbf{l}-\mathbf{l}')\}\{1+n(\omega)\}\,\mathrm{Im}\, G_{\alpha\beta}(l,l';\omega).$$
$$(15.13)$$

Similarly for the incoherent cross-section.

If the impurity ions are allowed to have a scattering length that differs from that of the ions, then we need to calculate not only

$$\sum_{l,l'} \exp\{-\mathrm{i}\boldsymbol{\kappa}.(\mathbf{l}-\mathbf{l}')\} G_{\alpha\beta}(l,l';\omega)$$

but also the same expression with either one lattice vector restricted to defect sites or both lattice vectors restricted to just defect sites. To see this, let

$$\bar{b}_l = \bar{b}_\mathrm{h} + p_l(\bar{b}_\mathrm{d} - \bar{b}_\mathrm{h}),$$

where, as in the previous chapter, p_l is unity or zero according to whether l is a defect or host site. Then

$$\sum_{l,l'} b_l b_{l'} \exp\{-i\boldsymbol{\varkappa}.(\mathbf{l}-\mathbf{l}')\}G_{\alpha\beta}(l,l';\omega)$$
$$= b_h^2 \sum_{l,l'} \exp\{-i\boldsymbol{\varkappa}.(\mathbf{l}-\mathbf{l}')\}G_{\alpha\beta}(l,l';\omega)+$$
$$+2b_h(b_d-b_h)\sum_{l,l'} p_l \exp\{-i\boldsymbol{\varkappa}.(\mathbf{l}-\mathbf{l}')\}G_{\alpha\beta}(l,l';\omega)+$$
$$+(b_d-b_h)^2\sum_{l,l'} p_l p_{l'} \exp\{-i\boldsymbol{\varkappa}.(\mathbf{l}-\mathbf{l}')\}G_{\alpha\beta}(l,l';\omega). \quad (15.14)$$

First, we calculate the Green function of the unperturbed system; this is denoted by $P_{\alpha\beta}(l,l';\omega)$. From the equation of motion

$$\hbar\omega\langle\langle\hat{u}_\alpha(l);\hat{u}_\beta(l')\rangle\rangle_\omega = \frac{\hbar}{2\pi}\langle[\hat{u}_\alpha(l),\hat{u}_\beta(l')]\rangle+\langle\langle[\hat{u}_\alpha(l),\mathscr{H}_0];\hat{u}_\beta(l')\rangle\rangle_\omega \quad (15.15)$$

we obtain
$$\omega P_{\alpha\beta}(l,l';\omega) = \frac{i}{M}\left(\frac{2\pi}{\hbar}\right)\langle\langle\hat{p}_\alpha(l);\hat{u}_\beta(l')\rangle\rangle_\omega. \quad (15.16)$$

In deriving this equation we have used the commutation relations

$$[\hat{u}_\alpha(l),\hat{u}_\beta(l')] = 0, \qquad [\hat{p}_\alpha(l),\hat{p}_\beta(l')] = 0,$$

and
$$[\hat{u}_\alpha(l),\hat{p}_\beta(l')] = i\hbar\delta_{\alpha\beta}\delta_{ll'}. \quad (15.17)$$

Since

$$[\hat{p}_\gamma(l''),\mathscr{H}_0] = \tfrac{1}{2}\sum_{l,l'}\sum_{\alpha,\beta} A_{\alpha\beta}(l,l')[\hat{p}_\gamma(l''),\hat{u}_\alpha(l)\hat{u}_\beta(l')]$$
$$= -\frac{i\hbar}{2}\left\{\sum_{\beta,l'} A_{\gamma\beta}(l'',l')\hat{u}_\beta(l')+\sum_{\alpha,l} A_{\alpha\gamma}(l,l'')\hat{u}_\alpha(l)\right\}$$
$$= -i\hbar\sum_{\beta,l'} A_{\gamma\beta}(l'',l')\hat{u}_\beta(l'),$$

where in the last line we have used the identity

$$A_{\alpha\beta}(l,l') = A_{\beta\alpha}(l',l),$$

the equation of motion for the Green function on the right-hand side of (15.16) is

$$\hbar\omega\langle\langle\hat{p}_\alpha(l);\hat{u}_\beta(l')\rangle\rangle_\omega$$
$$= \frac{\hbar}{2\pi}(-i\hbar)\delta_{\alpha\beta}\delta_{ll'}-i\hbar\left(\frac{\hbar}{2\pi}\right)\sum_{\gamma,l''} A_{\alpha\gamma}(l,l'')P_{\gamma\beta}(l'',l';\omega). \quad (15.18)$$

If the result (15.18) is substituted into eqn (15.16) we have, finally,

$$M\omega^2 P_{\alpha\beta}(l,l';\omega) = \delta_{\alpha\beta}\delta_{l,l'}+\sum_{\gamma,l''} A_{\alpha\gamma}(l,l'')P_{\gamma\beta}(l'',l';\omega), \quad (15.19)$$

or
$$\sum_{\gamma,l''}\{M\omega^2\delta_{\alpha,\gamma}\delta_{l,l''}-A_{\alpha\gamma}(l,l'')\}P_{\gamma\beta}(l'',l';\omega) = \delta_{\alpha\beta}\delta_{ll'}. \quad (15.20)$$

To solve this equation we expand $P_{\alpha\beta}(l,l';\omega)$ in terms of the eigenvectors of the force-constant matrix $A_{\alpha\beta}(l,l'')$. These are discussed in detail in § 4.1. We recall that the polarization vectors $\boldsymbol{\sigma}^j(\mathbf{q})$ are real and are defined such that (cf. (4.10))

$$\frac{1}{M}\sum_{l'',\gamma} A_{\alpha\gamma}(l,l'')\exp(i\mathbf{q}.\mathbf{l}'')\sigma^j_\gamma(\mathbf{q}) = \omega_j^2(\mathbf{q})\exp(i\mathbf{q}.\mathbf{l})\sigma^j_\beta(\mathbf{q}), \quad (15.21)$$

with the normalization

$$\sum_j \sigma^j_\alpha(\mathbf{q})\sigma^j_\beta(\mathbf{q}) = \delta_{\alpha,\beta} \quad (15.22\,\mathrm{a})$$

and

$$\sum_\alpha \sigma^j_\alpha(\mathbf{q})\sigma^{j'}_\alpha(\mathbf{q}) = \delta_{j,j'}. \quad (15\,22\,\mathrm{b})$$

In (15.21), $\omega_j(\mathbf{q})$ is the frequency of the jth phonon branch.

From (15.21) it follows that

$$\sum_{l'',\gamma}\{M\omega^2\delta_{l,l''}\delta_{\alpha\gamma}-A_{\alpha\gamma}(l,l'')\}\exp(i\mathbf{q}.\mathbf{l}'')\sigma^j_\gamma(\mathbf{q})$$
$$= M\{\omega^2-\omega_j^2(\mathbf{q})\}\exp(i\mathbf{q}.\mathbf{l})\sigma^j_\alpha(\mathbf{q}), \quad (15.23)$$

so that, if

$$P_{\alpha\beta}(l,l';\omega) = \frac{1}{NM}\sum_{j,\mathbf{q}} \sigma^j_\alpha(\mathbf{q})\sigma^j_\beta(\mathbf{q})\exp\{i\mathbf{q}.(\mathbf{l}-\mathbf{l}')\}\mathscr{P}^j(\mathbf{q}) \quad (15.24)$$

then, from eqn (15.20), $\mathscr{P}^j(\mathbf{q})$ satisfies

$$\frac{1}{N}\sum_{j,\mathbf{q}}\{\omega^2-\omega_j^2(\mathbf{q})\}\sigma^j_\alpha(\mathbf{q})\sigma^j_\beta(\mathbf{q})\exp\{-i\mathbf{q}.(\mathbf{l}-\mathbf{l}')\}\mathscr{P}^j(\mathbf{q}) = \delta_{\alpha,\beta}\delta_{l,l'}. \quad (15.25)$$

From the orthogonality relation (15.22 a) it then follows that

$$\mathscr{P}^j(\mathbf{q}) = \{\omega^2-\omega_j^2(\mathbf{q})\}^{-1}. \quad (15.26)$$

According to the convention adopted in Appendix B we evaluate the imaginary part of the Green function by giving the energy a small, positive imaginary part and then take the limit as this goes to zero. Hence

$$\mathscr{P}^j(\mathbf{q}) = \lim_{\delta\to 0+}\frac{1}{(\omega+i\delta)^2-\omega_j^2(\mathbf{q})}$$
$$= \frac{1}{2\omega_j(\mathbf{q})}\lim_{\delta\to 0+}\left\{\frac{1}{\omega+i\delta-\omega_j(\mathbf{q})}-\frac{1}{\omega+i\delta+\omega_j(\mathbf{q})}\right\} \quad (15.27)$$

and

$$\mathrm{Im}\,\mathscr{P}^j(\mathbf{q}) = -\frac{\pi}{2\omega_j(\mathbf{q})}[\delta\{\omega-\omega_j(\mathbf{q})\}-\delta\{\omega+\omega_j(\mathbf{q})\}].$$

If we now combine this result with (15.24) and substitute into our expression for the one-phonon coherent cross-section, eqn (15.13), the latter becomes

$$\left(\frac{d^2\sigma}{d\Omega dE'}\right)^{(1)}_{\text{coh}} = -\left(\frac{1}{\pi}\right)\left(-\frac{\pi}{2}\right)\frac{k'}{k}\{1+n(\omega)\}e^{-2W}\bar{b}^2\times$$

$$\times\sum_{j,\mathbf{q}}\frac{1}{\omega_j(\mathbf{q})}\sum_{\alpha,\beta}\kappa_\alpha\kappa_\beta\sigma^j_\alpha(\mathbf{q})\sigma^j_\beta(\mathbf{q})[\delta\{\omega-\omega_j(\mathbf{q})\}-\delta\{\omega+\omega_j(\mathbf{q})\}]\times$$

$$\times\frac{1}{N}\left|\sum_l \exp\{i\mathbf{l}.(\mathbf{q}-\boldsymbol{\varkappa})\}\right|^2$$

$$= \frac{k'}{k}\frac{(2\pi)^3}{2v_0}e^{-2W}\bar{b}^2\sum_{j,\mathbf{q},\boldsymbol{\tau}}\omega_j^{-1}(\mathbf{q})|\{\boldsymbol{\sigma}^j(\mathbf{q}).\boldsymbol{\varkappa}\}|^2\times$$

$$\times\delta(\boldsymbol{\varkappa}-\mathbf{q}-\boldsymbol{\tau})[[1+n\{\omega_j(\mathbf{q})\}]\delta\{\omega-\omega_j(\mathbf{q})\}+n\{\omega_j(\mathbf{q})\}\delta\{\omega+\omega_j(\mathbf{q})\}], \tag{15.28}$$

which is in agreement with the result derived in § 4.3.

It is not difficult to show that the equation of motion of the Green function of the perturbed system, $G_{\alpha\beta}(l,l';\omega)$, is

$$\sum_{\gamma,l''}\{M\omega^2\delta_{\alpha\gamma}\delta_{l,l''}-A_{\alpha\gamma}(l,l'')\}G_{\gamma\beta}(l'',l';\omega)$$

$$= \delta_{\alpha\beta}\delta_{l,l'}+\sum_{\gamma,l''}V_{\alpha\gamma}(l,l'')G_{\gamma\beta}(l'',l';\omega), \tag{15.29}$$

where $$V_{\alpha\beta}(l,l') = M\lambda\omega^2\sum_s\delta_{\alpha\beta}\delta_{l,s}\delta_{s,l'}.$$

If we write the equation of motion for the unperturbed Green function in matrix form, so that (15.20) reads

$$(M\omega^2\mathbf{I}-\mathbf{A})\mathbf{P} = \mathbf{I}$$

or $$(M\omega^2\mathbf{I}-\mathbf{A}) = \mathbf{P}^{-1}$$

then eqn (15.29) can be written

$$\mathbf{P}^{-1}\mathbf{G} = \mathbf{I}+\mathbf{VG},$$

i.e. $$\mathbf{G} = \mathbf{P}+\mathbf{PVG}. \tag{15.30}$$

Eqn (15.30) is usually referred to as the Dyson equation for the Green function $\mathbf{G}$. We have immediately

$$\mathbf{G} = \frac{1}{\mathbf{I}-\mathbf{PV}}.\mathbf{P} \tag{15.31}$$

so that the poles of the Green function are determined by the condition

$$\operatorname{Re}\det|\mathbf{I}-\mathbf{PV}| = 0. \tag{15.32}$$

(a) Single impurity

We now examine eqns (15.30) and (15.32) in some detail for the particular case of a single impurity ion at $l = 0$, say. Eqn (15.30) reduces to

$$G_{\alpha\beta}(l,l';\omega) = P_{\alpha\beta}(l,l';\omega)+\lambda\omega^2 M\sum_{\gamma} P_{\alpha\gamma}(l,0;\omega)G_{\gamma\beta}(0,l';\omega). \tag{15.33}$$

For orthorhombic, tetragonal, and cubic Bravais crystals $P_{\alpha\beta}(l,l;\omega)$ is diagonal in α, β. Furthermore, for cubic crystals $P_{\alpha\alpha}(l,l;\omega)$ is independent of α. For these crystals it is convenient to define

$$\begin{aligned}\mathscr{P}(\omega) &= MP_{\alpha\alpha}(l,l;\omega)\\ &= \frac{1}{3N}\sum_{j,\mathbf{q}}\frac{1}{\omega^2-\omega_j^2(\mathbf{q})}.\end{aligned} \tag{15.34}$$

The factor $\frac{1}{3}$ arises because, for all α, $|\sigma_\alpha^j|^2 = \frac{1}{3}$ by virtue of eqn (15.22 a).

With the aid of these results we have immediately

$$G_{\alpha\beta}(0,l';\omega) = \frac{P_{\alpha\beta}(0,l';\omega)}{1-\lambda\omega^2\mathscr{P}(\omega)} \tag{15.35}$$

and thus

$$G_{\alpha\beta}(l,l';\omega) = P_{\alpha\beta}(l,l';\omega)+\frac{\lambda\omega^2 M}{1-\lambda\omega^2\mathscr{P}(\omega)}\sum_{\gamma} P_{\alpha\gamma}(l,0;\omega)P_{\gamma\beta}(0,l';\omega). \tag{15.36}$$

The condition for the existence of an impurity mode is, from eqn (15.32),

$$\mathrm{Re}\{1-\lambda\omega^2\mathscr{P}(\omega)\} = 0. \tag{15.37}$$

If $Z(\omega)$ is the density of states for the host lattice, normalized to unity,

$$\begin{aligned}\mathscr{P}(\omega) = \mathscr{P}'(\omega)+\mathrm{i}\mathscr{P}''(\omega) &= \lim_{\delta\to 0}\int_0^{\omega_{\mathrm{m}}}\mathrm{d}\omega'\,\frac{Z(\omega')}{(\omega+\mathrm{i}\delta)^2-\omega'^2}\\ &= \mathrm{P}\int_0^{\omega_{\mathrm{m}}}\frac{\mathrm{d}\omega' Z(\omega')}{\omega^2-\omega'^2}-\frac{\mathrm{i}\pi}{2\omega}Z(\omega),\end{aligned} \tag{15.38}$$

where $\mathrm{P}\int$ denotes the principal part integral and ω_{m} is the maximum frequency of the band.

It is clear that $\mathscr{P}''(\omega)$ can be non-zero only for $0 < \omega < \omega_{\mathrm{m}}$.

To gain some understanding of the solutions of eqn (15.37) let us employ the Debye model, for which $Z(\omega) = 3\omega^2/\omega_{\mathrm{D}}^3$. Then eqn (15.37) can be written

$$\lambda^{-1} = f(x_0) = -3x_0^2+\frac{3x_0^3}{2}\ln\left|\frac{1+x_0}{1-x_0}\right|, \tag{15.39}$$

where $x = \omega/\omega_D$. $f(x)$ is shown in Fig. 15.1 (a), and the values of x and M'/M that satisfy (15.39) are shown in Fig. 15.1 (b). Because

$$\lambda^{-1} = \frac{1}{1-M'/M}$$

can only have values in the range $(\infty, 1)$, and $(0, -\infty)$, there is a range $0 \leqslant \lambda^{-1} \leqslant 1$ in which there are no solutions of (15.39). For both $M'/M > 1$ and $M'/M < 1$ impurity modes can exist within the frequency spectrum of the host, the so-called virtual modes. These modes decay into the band because they interact with the eigenmodes of the host, the strength of the interaction being proportional to the density of states of the unperturbed host $Z(\omega)$. It therefore follows that the virtual modes are well defined only when they occur near the bottom of the frequency spectrum, i.e. for $M'/M \gg 1$. On the other hand, the so-called localized modes, which occur at frequencies greater than the maximum frequency of the host spectrum, exist for only light mass impurity ions, $M'/M < 1$. The wave functions associated with these states cannot propagate in the host, so that the disturbance caused by the impurity ion is well localized.

The use of the Debye spectrum to evaluate $\mathscr{P}'(\omega)$ gives the correct qualitative behaviour except for $\omega = \omega_D$, where it gives a result that diverges. In Fig. 15.2 we show the results of a more realistic calculation of $\omega^2\mathscr{P}'(\omega)$. Here we see that localized modes can exist only for M'/M less than some critical value; in this example the critical value is $M'/M = 0{\cdot}785$ corresponding to the high peak at about $\omega/\omega_m \approx 0{\cdot}9$.

The neutron cross-section for the single-impurity problem can easily be obtained from our result for $G_{\alpha\beta}(l, l'; \omega)$, eqn (15.36). We might then expect that for a small concentration of impurities c, the cross-section would be well represented by multiplying the result for the single impurity problem by Nc. This procedure is quite feasible when just the contributions from the localized modes need to be taken into account, but we cannot expect it to be so good when virtual modes are important, because these have wave functions that propagate in the host lattice. In the following section we will develop a theory to calculate G for a small, finite concentration of defects, but it is nevertheless worth while to carry through the calculation of the inelastic one-phonon cross-section for a single impurity when only localized modes need to be considered.

For a single impurity the function p_l in eqn (15.14) becomes a Kronecker delta function, $\delta_{l,0}$ so that to calculate the cross-section we need

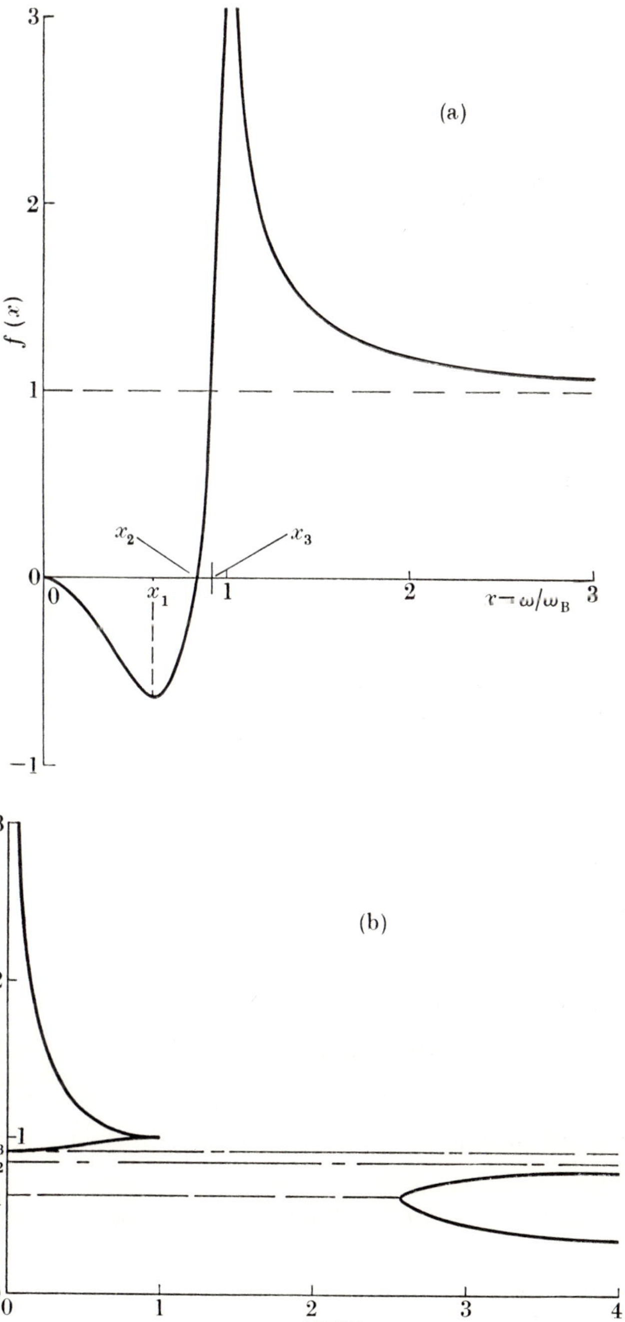

FIG. 15.1

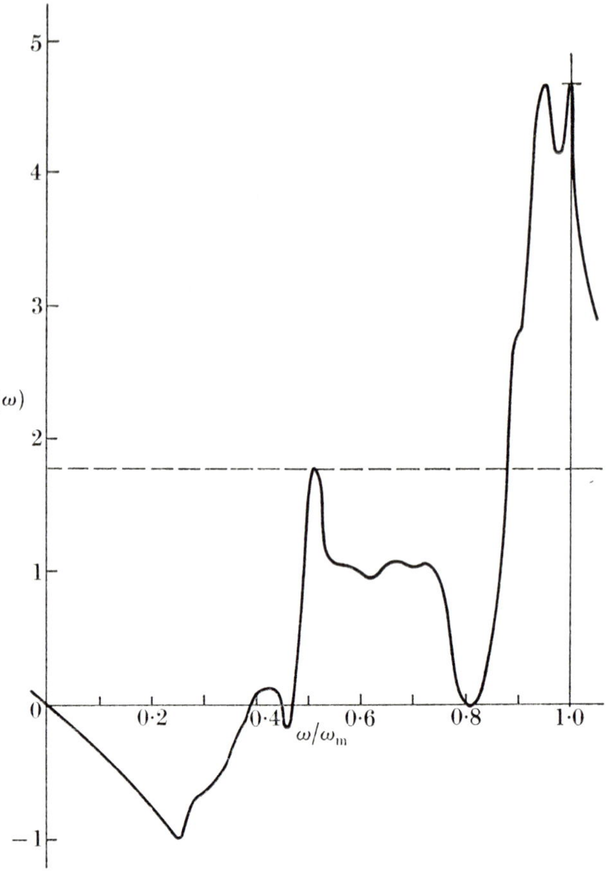

FIG. 15.2. $\omega^2\mathscr{P}'(\omega)$ as a function of ω/ω_m for a face-centred cubic crystal with nearest-neighbour, central-force interactions (G. F. Nardelli and N. Tettamanzi, *Phys. Rev.* **126,** 1283 (1962)).

$G_{\alpha\beta}(0,0;\omega)$ and $\sum_l \exp(\mathrm{i}\boldsymbol{\kappa}.\mathbf{l})G_{\alpha\beta}(0,l;\omega)$ in addition to

$$\sum_{l,l'} \exp\{-\mathrm{i}\boldsymbol{\kappa}.(\mathbf{l}-\mathbf{l}')\}G_{\alpha\beta}(l,l';\omega).$$

However, the first two functions are readily calculated from eqn (15.35) and the latter from eqn (15.36). We find ($\sigma^j_\alpha(\mathbf{q})$ is real for a cubic Bravais lattice)

$$\sum_{l,l'} \exp\{-\mathrm{i}\boldsymbol{\kappa}.(\mathbf{l}-\mathbf{l}')\}\{b_l b_{l'} G_{\alpha\beta}(l,l';\omega) - b_\mathrm{h}^2 P_{\alpha\beta}(l,l';\omega)\}$$
$$= \frac{1}{M}\sum_j \sigma^j_\alpha(\boldsymbol{\kappa})\sigma^j_\beta(\boldsymbol{\kappa})[b_\mathrm{h}^2\lambda\omega^2\{\mathscr{P}^j(\boldsymbol{\kappa})\}^2 +$$
$$+2b_\mathrm{h}(b_\mathrm{d}-b_\mathrm{h})\mathscr{P}^j(\boldsymbol{\kappa})+(b_\mathrm{d}-b_\mathrm{h})^2\mathscr{P}(\omega)]\{1-\lambda\omega^2\mathscr{P}(\omega)\}^{-1}. \quad (15.40)$$

The frequency of the localized mode ω_l ($> \omega_{\mathrm{m}}$) is given by

$$1-\lambda\omega_l^2\mathscr{P}(\omega_l) = 0, \tag{15.41}$$

and to evaluate the cross-section we need

$$\mathrm{Im}\,\frac{1}{1-\lambda\omega^2\mathscr{P}(\omega)}$$

as $\omega \to \omega_l$.

If $F^*(z) = F(z^*)$, $z = x+\mathrm{i}y$, and $F(x_0) = 0$, then

$$\lim_{y\to 0+}\mathrm{Im}\,\frac{1}{F(z)} = \frac{1}{2\mathrm{i}}\lim_{y\to 0+}\left\{\frac{1}{F(z)}-\frac{1}{F(z^*)}\right\}$$

$$= \frac{1}{2\mathrm{i}}\left\{\frac{\mathrm{d}F}{\mathrm{d}x}\bigg|_{x=x_0}\right\}^{-1}\lim_{y\to 0+}\left\{\frac{1}{x+\mathrm{i}y-x_0}-\frac{1}{x-\mathrm{i}y-x_0}\right\}$$

$$= \frac{1}{2\mathrm{i}}\left\{\frac{\mathrm{d}F}{\mathrm{d}x}\bigg|_{x=x_0}\right\}^{-1}\pi(-2\mathrm{i})\delta(x-x_0). \tag{15.42}$$

Now
$$\frac{\mathrm{d}}{\mathrm{d}\omega}\{1-\lambda\omega^2\mathscr{P}(\omega)\} = 2\lambda\omega\,B(\omega), \tag{15.43}$$

where
$$B(\omega) = \int_0^{\omega_{\mathrm{m}}}\frac{\mathrm{d}\omega'\,\omega'^2 Z(\omega')}{(\omega^2-\omega'^2)^2}\quad (\omega > \omega_{\mathrm{m}}). \tag{15.44}$$

Hence by combining (15.42) and (15.43), and noting that there are two solutions of (15.41), $\omega = \pm\omega_l$, we obtain

$$\mathrm{Im}\,\frac{1}{1-\lambda\omega_l^2\mathscr{P}(\omega_l)} = \frac{-\pi}{2\lambda\omega_l\,B(\omega_l)}\{\delta(\omega-\omega_l)-\delta(\omega+\omega_l)\}. \tag{15.45}$$

The imaginary part of eqn (15.40) for $\omega > \omega_{\mathrm{m}}$ is, therefore,

$$\frac{-\pi}{2M\lambda\omega_l\,B(\omega_l)}\{\delta(\omega-\omega_l)-\delta(\omega+\omega_l)\}\lambda\omega_l^2\sum_j\sigma_\alpha^j(\boldsymbol{\kappa})\sigma_\beta^j(\boldsymbol{\kappa})\left[b_{\mathrm{h}}\mathscr{P}^j(\boldsymbol{\kappa})+\frac{b_{\mathrm{d}}-b_{\mathrm{h}}}{\lambda\omega_l^2}\right]^2, \tag{15.46}$$

where we have made use of eqn (15.41) to eliminate $\mathscr{P}(\omega_l)$.

Thus the one-phonon coherent cross-section for the single-impurity case when the neutron energy change is greater than $\hbar\omega_{\mathrm{m}}$ is, from (15.13) and (15.46),

$$\left(\frac{\mathrm{d}\sigma^2}{\mathrm{d}\Omega\mathrm{d}E'}\right)_{\mathrm{coh}}^{(1)} = \frac{k'}{2kM}[\{1+n(\omega_l)\}\delta(\omega-\omega_l)+n(\omega_l)\delta(\omega+\omega_l)]\times$$

$$\times\frac{\omega_l}{B(\omega_l)}\sum_j\{\boldsymbol{\kappa}\cdot\boldsymbol{\sigma}^j(\boldsymbol{\kappa})\}^2\left\{\frac{b_{\mathrm{h}}}{\omega_l^2-\omega_j^2(\boldsymbol{\kappa})}+\frac{b_{\mathrm{d}}-b_{\mathrm{h}}}{\lambda\omega_l^2}\right\}^2. \tag{15.47}$$

If this expression is multiplied by the number of impurities Nc, we have an approximate expression for a small finite concentration of impurities first derived by Elliott and Maradudin.†

The cross-section given by eqn (15.47) has a simple form if the scattering vector $\boldsymbol{\varkappa}$ lies along one of the symmetry directions of the cubic lattice, i.e. along either the $(1,0,0)$, $(1,1,0)$, or $(1,1,1)$ directions. For lattice waves propagating along these directions are purely longitudinal and purely transverse, so that $\boldsymbol{\varkappa}.\boldsymbol{\sigma}^j(\boldsymbol{\varkappa}) = 0$ unless j refers to the longitudinal branch.

The calculation of the one-phonon incoherent cross-section, eqn (15.9), for the single-impurity case when $(\hbar^2/2m)(k^2-k'^2) > \hbar\omega_\mathrm{m}$, follows along similar lines to that given above for the coherent cross-section. If

$$\alpha_l^2 = \alpha_\mathrm{h}^2+(\alpha_\mathrm{d}^2-\alpha_\mathrm{h}^2)\delta_{l,0}$$

then

$$\left(\frac{\mathrm{d}\sigma^2}{\mathrm{d}\Omega\mathrm{d}E'}\right)^{(1)}_{\text{incoh}} = -\frac{1}{\pi}\frac{k'}{k}\sum_{\alpha,\beta}\kappa_\alpha\kappa_\beta\{1+n(\omega)\}\times$$
$$\times\Big\{\alpha_\mathrm{h}^2\sum_l \mathrm{Im}\,G_{\alpha\beta}(l,l;\omega)+(\alpha_\mathrm{h}^2-\alpha_\mathrm{d}^2)\mathrm{Im}\,G_{\alpha\beta}(0,0;\omega)\Big\}. \quad (15.48)$$

The term in curly brackets is

$$\delta_{\alpha\beta}\Bigg\{\alpha_\mathrm{h}^2\,\mathrm{Im}\left[\frac{\lambda\omega^2}{1-\lambda\omega^2\mathscr{P}(\omega)}\frac{1}{M}\frac{1}{3N}\sum_{j,\mathbf{q}}\{\mathscr{P}^j(\mathbf{q})\}^2\right]+$$
$$+(\alpha_\mathrm{d}^2-\alpha_\mathrm{h}^2)\mathrm{Im}\left[\frac{\mathscr{P}(\omega)/M}{1-\lambda\omega^2\mathscr{P}(\omega)}\right]\Bigg\}.$$

But
$$\frac{1}{3N}\sum_{j,\mathbf{q}}\{\mathscr{P}^j(\mathbf{q})\}^2 = -\frac{1}{2\omega}\frac{\mathrm{d}\mathscr{P}(\omega)}{\mathrm{d}\omega}$$

so that (15.48) is

$$\left(\frac{\mathrm{d}\sigma^2}{\mathrm{d}\Omega\mathrm{d}E'}\right)^{(1)}_{\text{incoh}} = \frac{k'}{2kM}\frac{\kappa^2}{B(\omega_l)}[\{1+n(\omega_l)\}\delta(\omega-\omega_l)+n(\omega_l)\delta(\omega+\omega_l)]\times$$
$$\times\left\{\frac{\alpha_\mathrm{d}^2-\alpha_\mathrm{h}^2}{\lambda^2\omega_l^3}-\frac{\alpha_\mathrm{h}^2}{2}\left(\frac{\mathrm{d}\mathscr{P}(\omega)}{\mathrm{d}\omega}\right)\Bigg|_{\omega=\omega_l}\right\}. \quad (15.49)$$

(b) *Low concentration theory*

The mass defects are assumed to be distributed randomly and of small, finite concentration c. In the following argument the Cartesian component labels α, β are redundant so that we shall omit them for ease of notation. (We also omit ω from the argument of the Green functions.) Also, we denote the site positions simply by 1, 2, 3, etc.

† R. J. Elliott and A. A. Maradudin, *Inelastic scattering of neutrons*, vol. 1, p. 231, Vienna I.A.E.A. (1965).

It is to be expected that the behaviour of the Green function $G(1,2)$ is mainly determined by whether or not either, or both, its site labels 1 and 2 refer to defect sites. The first site label of the Green function on the right-hand side of eqn (15.30) always refers to a defect site. We define two additional Green functions according to whether their first site label refers to defect or host sites: namely

$$G(1,2) = G^{\mathrm{d}}(1,2) \qquad \text{(defect at site 1)},$$
$$G(1,2) = G^{\mathrm{h}}(1,2) \qquad \text{(host atom at site 1)}. \tag{15.50}$$

Thus eqn (15.30) becomes

$$\mathbf{G} = \mathbf{P}+M\lambda\omega^2\mathbf{P}\mathbf{G}^{\mathrm{d}}. \tag{15.51}$$

The equations for G^{d} and G^{h} are

$$G^{\mathrm{d}}(1,2) = P(1,2)+M\lambda\omega^2\sum_3 P(1,3)G^{\mathrm{d}}(3,2) \tag{15.52 a}$$

and

$$G^{\mathrm{h}}(1,2) = P(1,2)+M\lambda\omega^2\sum_3 P(1,3)G^{\mathrm{d}}(3,2)-M\lambda\omega^2P(1,1)G^{\mathrm{d}}(1,2). \tag{15.52 b}$$

As in the previous section, we are concerned with cubic Bravais lattices; thus $\mathscr{P} = MP(1,1)$.

If we assume that G is completely determined by the type of atom at site 1 we can equate the two summations in eqns (15.52); this neglects the influence of the configuration of all the remaining defects in the system. This approximation gives immediately the following relationship between G^{d} and G^{h}:

$$\{1-\lambda\omega^2\mathscr{P}(\omega)\}G^{\mathrm{d}}(1,2) = G^{\mathrm{h}}(1,2). \tag{15.53}$$

Further

$$\bar{G}(1,2)= cG^{\mathrm{d}}(1,2)+(1-c)G^{\mathrm{h}}(1,2) \tag{15.54}$$

is the Green function averaged over all configurations of the defect atoms. (15.53) and (15.54) give

$$G^{\mathrm{d}}(1,2) = \{1-(1-c)\lambda\omega^2\mathscr{P}(\omega)\}^{-1}\bar{G}(1,2), \tag{15.55}$$

which together with (15.51) leads to

$$\bar{G}(1,2) = P(1,2)+\frac{M\lambda\omega^2}{1-(1-c)\lambda\omega^2\mathscr{P}(\omega)}\,c\sum_3 P(1,3)\bar{G}(3,2), \tag{15.56}$$

where the summation over defect sites has been replaced by c times the summation over all sites.†

The arguments leading to eqn (15.56) are rather intuitive and it is by no means clear exactly what approximations have been made in its

† The argument given to derive eqn (15.56) follows that of R. J. Elliott and D. W. Taylor, *Proc. R. Soc* **296A**, 161 (1967).

derivation. The same equation can be derived by iterating eqn (15.30), retaining all those terms that contain only one defect site label, which essentially means keeping all terms to order c. However, the possibility of more than one defect at a given site must be removed. The summations over sites are, therefore, split up into summations in which none of the summation indices are the same. The latter are simply averaged by multiplying by c^m, where m is the number of different summation indices involved. These restricted summations can then be regrouped into unrestricted summations and the complete iterative solution summed to give eqn (15.56).

In such an analysis, the scattering from all single defects is included together with that from one-, two-, and three-fold scattering from separate defect sites. For example the equation

$$\overline{\mathbf{G}}^{(1)} = \mathbf{P}+c\mathbf{PV}\frac{1}{\mathbf{I}-\mathbf{PV}}\mathbf{P} = \mathbf{P}+c\mathbf{PTP},$$

which introduces the T-matrix $\mathbf{T}$, accounts for all scattering from single defects, while in

$$\overline{\mathbf{G}}^{(2)} = \mathbf{P}+c\mathbf{PTP}+c^2\mathbf{PTPTP}$$

account has been taken of the scattering of the same wave by two separate defects. Clearly

$$\overline{\mathbf{G}} = \mathbf{P}+c\mathbf{PT}\overline{\mathbf{G}}.$$

This agrees with (15.56) to order c. The correction of excluding the possibility of more than one defect at a given site leads to $\mathbf{T}$ being replaced by the expression given in eqn (15.56).

If
$$T' = \frac{\lambda\omega^2}{1-(1-c)\lambda\omega^2\mathscr{P}(\omega)}, \tag{15.57}$$

then eqn (15.56) in full is

$$\bar{G}_{\alpha\beta}(l,l') = P_{\alpha\beta}(l,l')+cT'M\sum_{\gamma,l''}P_{\alpha\gamma}(l,l'')\bar{G}_{\gamma\beta}(l'',l'). \tag{15.58}$$

The configurational averaging has restored the translational symmetry, so that eqn (15.58) can be solved for $\bar{G}_{\alpha\beta}(l,l')$ by expressing the latter in terms of the eigenvectors of the force-constant matrix. Thus we define

$$\bar{G}_{\alpha\beta}(l,l') = \frac{1}{N}\sum_{\mathbf{q}}\exp(-i\mathbf{q}.\mathbf{l})\bar{G}_{\alpha\beta}(\mathbf{q})\exp(i\mathbf{q}.\mathbf{l}') \tag{15.59 a}$$

and
$$\bar{G}_{\alpha\beta}(\mathbf{q}) = \frac{1}{M}\sum_{j}\sigma_\alpha^j(\mathbf{q})\sigma_\beta^j(\mathbf{q})\bar{G}^j(\mathbf{q}). \tag{15.59 b}$$

From eqn (15.58)

$$\bar{G}^j(\mathbf{q}) = \mathscr{P}^j(\mathbf{q}) + cT'\mathscr{P}^j(\mathbf{q})\bar{G}^j(\mathbf{q})$$

or

$$\bar{G}^j(\mathbf{q}) = \frac{1}{\omega^2 - \omega_j^2(\mathbf{q}) - cT'}. \tag{15.60}$$

Thus T' plays the role of a self energy. The imaginary part of (15.60) will, if T' is small, be sharply peaked near $\omega_j(\mathbf{q})$ with a characteristic width and energy shift.

From eqn (15.51) we can obtain G^{d}. On performing the configuration average and Fourier transforming

$$G^{j,\mathrm{d}}(\mathbf{q}) = \frac{cT'}{\lambda\omega^2}\bar{G}^j(\mathbf{q}). \tag{15.61}$$

We also need G^{dd} to calculate the neutron cross-section. From eqn (15.30)

$$G_{\alpha\beta}(l,l') = P_{\alpha\beta}(l,l') + M\lambda\omega^2 \sum_{\gamma,s} P_{\alpha\gamma}(l,s)P_{\gamma\beta}(s,l') +$$
$$+ (M\lambda\omega^2)^2 \sum_{\gamma_1,\gamma_2} \sum_{s_1,s_2} P_{\alpha\gamma_1}(l,s_1)G_{\gamma_1\gamma_2}(s_1,s_2)P_{\gamma_2\beta}(l_2,l'). \tag{15.62}$$

If in the second term on the right-hand side of (15.62) we replace the restricted summation by c times a summation over all sites and then Fourier transform we obtain

$$\bar{G}^j(\mathbf{q}) = \mathscr{P}^j(\mathbf{q}) + c\lambda\omega^2\{\mathscr{P}^j(\mathbf{q})\}^2 + (\lambda\omega^2)^2\{\mathscr{P}^j(\mathbf{q})\}^2 G^{j,\mathrm{dd}}(\mathbf{q}).$$

Thus

$$G^{j,\mathrm{dd}}(\mathbf{q}) = \frac{\bar{G}^j(\mathbf{q}) - \mathscr{P}^j(\mathbf{q}) - c\lambda\omega^2\{\mathscr{P}^j(\mathbf{q})\}^2}{(\lambda\omega^2)^2\{\mathscr{P}^j(\mathbf{q})\}^2}$$
$$= c(1-c)\frac{T'}{\lambda\omega^2}\mathscr{P}(\omega) + \left(\frac{cT'}{\lambda\omega^2}\right)^2 \bar{G}^j(\mathbf{q}). \tag{15.63}$$

Now let us examine the analytic behaviour of $\bar{G}^j(\mathbf{q})$ as a function of ω. If

$$cT''(\omega) = \omega^2\{\Delta(\omega) + \mathrm{i}\Gamma(\omega)\} \tag{15.64}$$

then

$$\operatorname{Im}\bar{G}^j(\mathbf{q}) = \frac{\omega^2\Gamma(\omega)}{\{\omega^2 - \omega_j^2(\mathbf{q}) - \omega^2\Delta(\omega)\}^2 + \{\omega^2\Gamma(\omega)\}^2}. \tag{15.65}$$

Since $\Gamma(\omega)$ is proportional to the density of states $Z(\omega)$ in the unperturbed crystal there are additional contributions to (15.65) when, for $\omega_l > \omega_m$,

$$\omega_l^2 - \omega_j^2(\mathbf{q}) - \omega^2\Delta(\omega_l) = 0. \tag{15.66}$$

Eqn (15.66) is the expression analogous to the condition for the appearance of localized modes for the single-impurity problem, eqn (15.41). Since

$$\frac{\mathrm{d}}{\mathrm{d}\omega}\{\omega^2 - \omega_j^2(\mathbf{q}) - \omega^2\Delta(\omega)\}|_{\omega=\omega_l} = 2\omega_l\left[1 - \Delta(\omega_l)\left\{1 - \frac{\omega_l^2}{c}(1-c)\Delta(\omega_l)B(\omega_l)\right\}\right], \tag{15.67}$$

where $B(\omega)$ is defined by eqn (15.44),

$$\operatorname{Im}\bar{G}^j(\mathbf{q}) = \frac{-\pi}{2\omega_l[1-\Delta(\omega_l)\{1-(\omega_l^2/c)(1-c)\Delta(\omega_l)B(\omega_l)\}]}\times$$

$$\times\{\delta(\omega-\omega_l)-\delta(\omega+\omega_l)\}, \qquad (15.68\,\mathrm{a})$$

when $\omega > \omega_{\mathrm{m}}$. For very small c,

$$\operatorname{Im}\bar{G}^j(\mathbf{q}) = \frac{-c\omega_l\pi}{2\{\omega_l^2-\omega_j^2(\mathbf{q})\}^2B(\omega_l)}\{\delta(\omega-\omega_l)-\delta(\omega+\omega_l)\}$$

$$= \frac{c\pi}{4B(\omega_l)}\left(\frac{\mathrm{d}}{\mathrm{d}\omega}\mathscr{P}^j(\mathbf{q})\right)\bigg|_{\omega=\omega_l}\{\delta(\omega-\omega_l)-\delta(\omega+\omega_l)\}. \qquad (15.68\,\mathrm{b})$$

Where in the single-impurity problem the ratio M'/M is such as to give a localized mode, the present theory predicts an impurity band with a well-defined frequency for each j, $\mathbf{q}$, which spans the frequency of the isolated mode.†

With the aid of eqns (15.61) and (15.63),

$$\frac{1}{N}\sum_{l,l'}\ell_l\ell_{l'}\exp\{-\mathrm{i}\boldsymbol{\varkappa}.(\mathbf{l}-\mathbf{l}')\}\sum_{\alpha\beta}\kappa_\alpha\kappa_\beta G_{\alpha\beta}(l,l')$$

$$=\frac{1}{N^2}\sum_{l,l'}\sum_{\alpha,\beta}\sum_{j,\mathbf{q}}\exp\{-\mathrm{i}\boldsymbol{\varkappa}.(\mathbf{l}-\mathbf{l}')+\mathrm{i}\mathbf{q}.(\mathbf{l}-\mathbf{l}')\}\frac{\kappa_\alpha\kappa_\beta}{M}\sigma_\alpha^j(\mathbf{q})\sigma_\beta^j(\mathbf{q})\times$$

$$\times\{\ell_{\mathrm{h}}^2\bar{G}^j(\mathbf{q})+2\ell_{\mathrm{h}}(\ell_{\mathrm{d}}-\ell_{\mathrm{h}})G^{j,\mathrm{d}}(\mathbf{q})+(\ell_{\mathrm{d}}-\ell_{\mathrm{h}})^2G^{j,\mathrm{dd}}(\mathbf{q})\}$$

$$=\frac{1}{M}\sum_{\alpha,\beta}\sum_j\kappa_\alpha\kappa_\beta\sigma_\alpha^j(\boldsymbol{\varkappa})\sigma_\beta^j(\boldsymbol{\varkappa})\left[c(1-c)\mathscr{P}(\omega)\frac{T'}{\lambda\omega^2}(\ell_{\mathrm{d}}-\ell_{\mathrm{h}})^2+\right.$$

$$\left.+\bar{G}^j(\boldsymbol{\varkappa})\left\{\ell_{\mathrm{h}}+(\ell_{\mathrm{d}}-\ell_{\mathrm{h}})\left(\frac{cT'}{\lambda\omega^2}\right)\right\}^2\right].$$

Hence, the one-phonon coherent cross-section is

$$\left(\frac{\mathrm{d}\sigma^2}{\mathrm{d}\Omega\mathrm{d}E'}\right)^{(1)}_{\mathrm{coh}} = -\frac{N}{\pi M}\{1+n(\omega)\}\frac{k'}{k}\operatorname{Im}\left[\kappa^2c(1-c)\mathscr{P}(\omega)\left(\frac{T'}{\lambda\omega^2}\right)(\ell_{\mathrm{d}}-\ell_{\mathrm{h}})^2+\right.$$

$$\left.+\left\{\ell_{\mathrm{h}}+(\ell_{\mathrm{d}}-\ell_{\mathrm{h}})\left(\frac{cT'}{\lambda\omega^2}\right)\right\}^2\sum_j\{\boldsymbol{\varkappa}.\boldsymbol{\sigma}^j(\boldsymbol{\varkappa})\}^2\bar{G}^j(\boldsymbol{\varkappa})\right]. \qquad (15.69)$$

The imaginary part of the first term in the square brackets in eqn (15.69) can be reduced to

$$\frac{\kappa^2(\ell_{\mathrm{d}}-\ell_{\mathrm{h}})^2\Gamma(\omega)}{(\lambda\omega)^2}. \qquad (15.70)$$

† Ibid.

Similarly, the one-phonon incoherent cross-section is

$$\left(\frac{d\sigma^2}{d\Omega dE'}\right)^{(1)}_{\text{incoh}} = -\frac{N}{\pi M}\{1+n(\omega)\}\frac{k'}{k}\kappa^2\times$$

$$\times \operatorname{Im}\left[\alpha_{\mathrm{h}}^2\,\bar{G}(\omega)+(\alpha_{\mathrm{d}}^2-\alpha_{\mathrm{h}}^2)\left\{c(1-c)\mathscr{P}(\omega)\left(\frac{T'}{\lambda\omega^2}\right)+\left(\frac{cT'}{\lambda\omega^2}\right)^2\bar{G}(\omega)\right\}\right], \quad (15.71)$$

where
$$\bar{G}(\omega) = \frac{1}{3N}\sum_{j,\mathbf{q}}\bar{G}^j(\mathbf{q}). \quad (15.72)$$

An examination of (15.69) shows that the first term is a diffuse contribution arising solely from the impurities. The second term is a *modified coherent* contribution, and we shall refer to this part of the cross-section (15.69) as such. The first term could be more properly grouped with the terms coming from the incoherent cross-section, eqn (15.71).

The modified coherent cross-section has a peak shifted from that corresponding to the pure host by

$$\frac{\Delta\omega}{\omega} \simeq \tfrac{1}{2}\Delta\{\omega_j(\boldsymbol{\varkappa})\} \quad (15.73\,\text{a})$$

and a relative width
$$\frac{\gamma}{\omega} \simeq \tfrac{1}{2}\Gamma\{\omega_j(\boldsymbol{\varkappa})\}. \quad (15.73\,\text{b})$$

These are illustrated in Fig. 15.3 for 10% Au in Cu, a system studied by Svensson, Brockhouse, and Rowe.† Though the magnitude of the shift and relative width is in good agreement (Figs. 15.3 and 15.4), the experimental shift does not display the reduction in size and sign change in the resonance region. This disagreement could arise because of force-constant changes, but the concentration of 10% Au is probably outside the range of validity of the present theory. Møller and Mackintosh‡ found that the frequency shift for the system 3% W in Cr more closely resembles the theoretical predictions. The maximum width, however, occurs at a lower frequency relative to the zero in the shift than predicted by theory.

There are in fact two peaks in Im $\bar{G}^j(\boldsymbol{\varkappa})$, one due to the shifted $\omega_j(\boldsymbol{\varkappa})$ and one due to the resonance. For heavy mass defects, and a value of $\omega_j(\boldsymbol{\varkappa})$ in the region of the virtual resonance, both peaks are of the same order of magnitude and lie close together. The experimental resolution is not sufficient to reveal this structure and a single asymmetric peak is observed as shown in Fig. 15.4 (a). Eqn (15.73 a) gives a reasonable prescription for the position of the single peak relative to $\omega_j(\boldsymbol{\varkappa})$. For

† E. C. Svensson, B. N. Brockhouse, and J. M. Rowe, *Solid St. Communs.* **3**, 245 (1965).
‡ H. B. Møller and A. R. Mackintosh, *Phys. Rev. Lett.* **15**, 623 (1965).

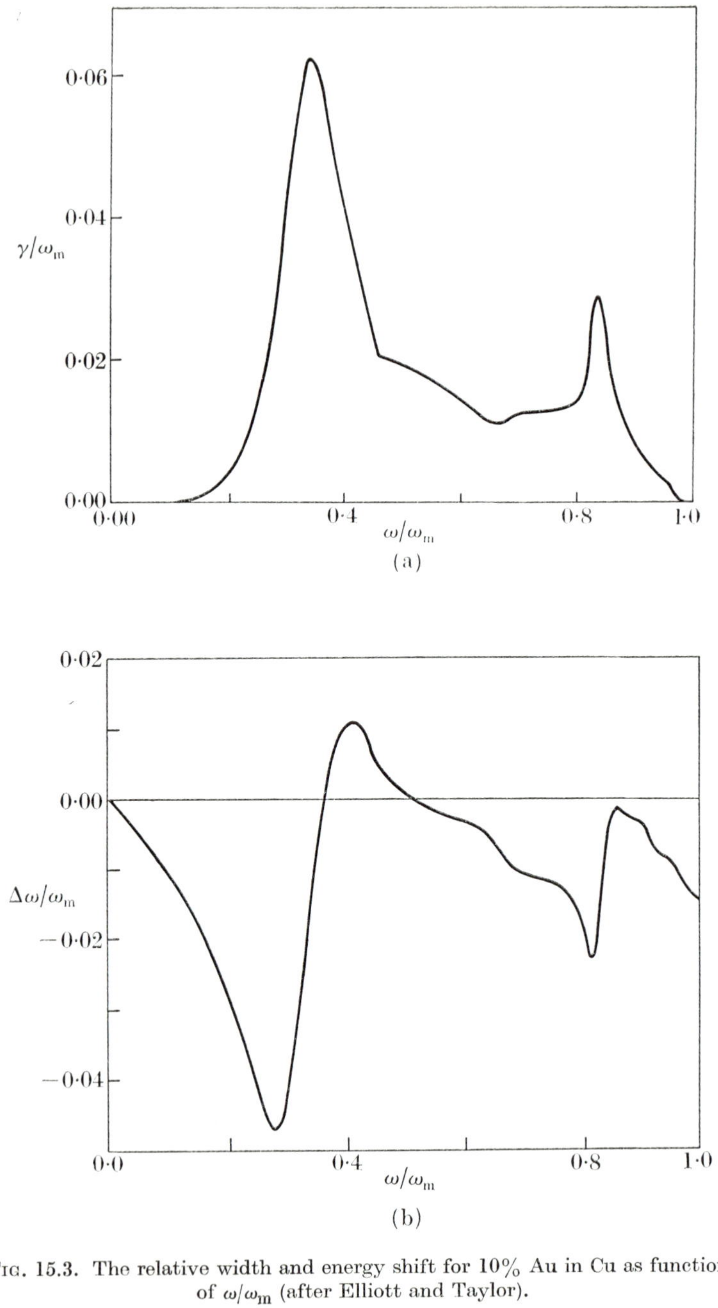

FIG. 15.3. The relative width and energy shift for 10% Au in Cu as functions of ω/ω_m (after Elliott and Taylor).

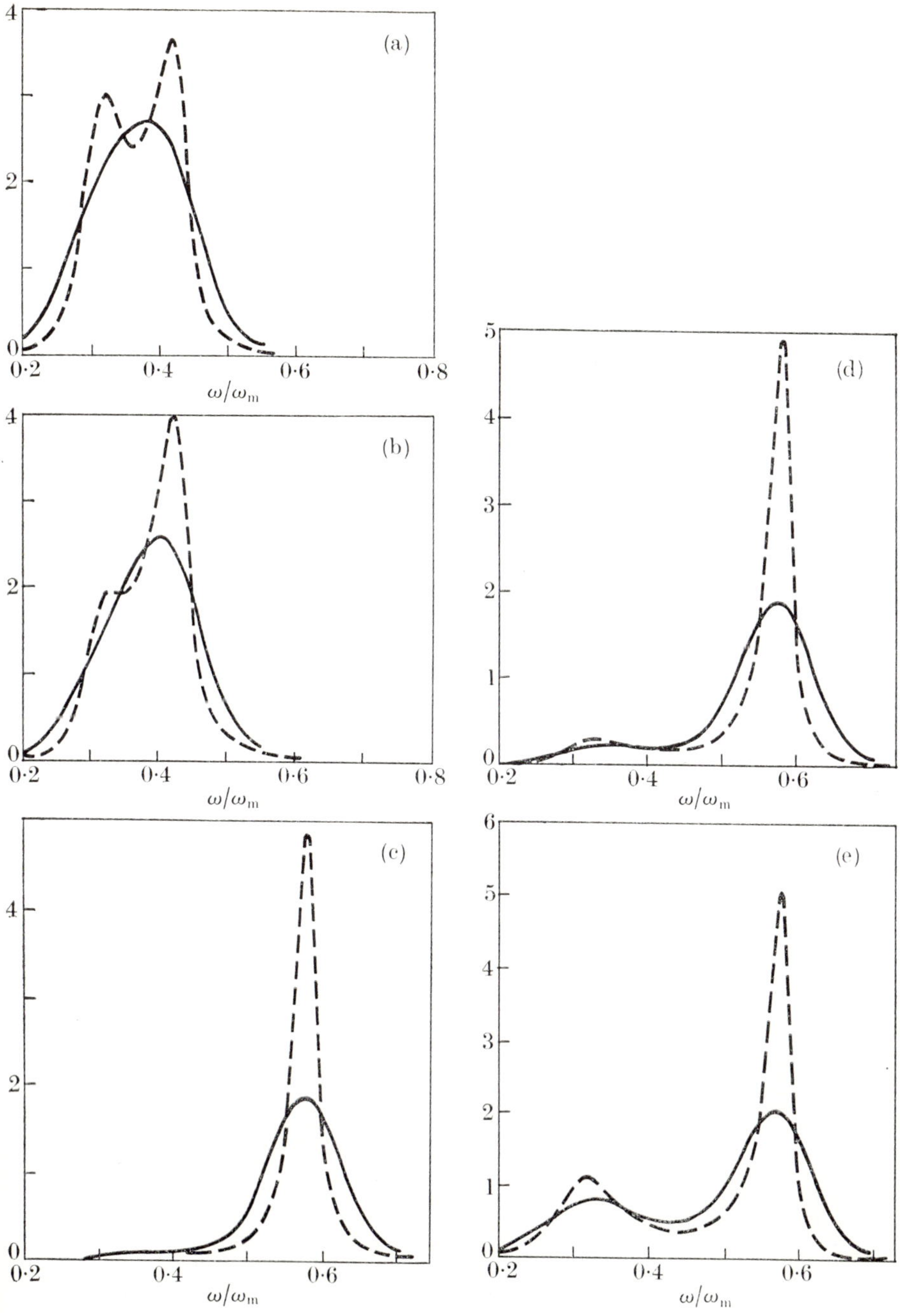

FIG. 15.4. The *modified coherent* cross-section for various values of $\omega_j(\varkappa)$ and $\delta b = (b_d - b_h)/b_h$ for 10% Au in Cu. The solid curves include the effect of resolution broadening, which is taken to have a full width at half maximum of $0{\cdot}095\omega_m$.

(a) $\omega_j(\varkappa) = 0{\cdot}4\omega_m$. Scattering lengths are those appropriate to Au and Cu.
(b) $\omega_j(\varkappa) = 0{\cdot}4\omega_m$; $\delta b = 1$.
(c) $\omega_j(\varkappa) = 0{\cdot}58\omega_m$; b_d is that for Au.
(d) $\omega_j(\varkappa) = 0{\cdot}58\omega_m$; $\delta b = 1$.
(e) $\omega_j(\varkappa) = 0{\cdot}58\omega_m$; $\delta b = 4$.

Scattering length of Cu is kept constant in (b)–(e).

$\omega_j(\boldsymbol{\varkappa})$ well away from the virtual resonance the two peaks are well separated but the resonance peak is too small to be observed.

In Fig. 15.4 (b)–(e) the modified coherent part of (15.69) is shown for various values of $\delta\ell = (\ell_{\mathrm{d}}-\ell_{\mathrm{h}})/\ell_{\mathrm{h}}$, using the parameters for 10% Au in Cu. The dominant effect of allowing $\delta\ell \neq 0$ is seen to increase or reduce the strength of the virtual resonance mode in $\bar{G}^j(\boldsymbol{\varkappa})$ dependent upon whether $|\ell_{\mathrm{d}}| \lessgtr |\ell_{\mathrm{h}}|$.

When the neutron energy change is greater than $\hbar\omega_{\mathrm{m}}$ the additional contributions to $\operatorname{Im}\bar{G}^j(\boldsymbol{\varkappa})$ given by eqn (15.68a) must be taken into account. For small c we can use (15.68b) (we recall that for $\omega > \omega_{\mathrm{m}}$ only $\bar{G}^j(\boldsymbol{\varkappa})$ can give a contribution to the cross-section (15.69)) and obtain

$$\left(\frac{\mathrm{d}\sigma^2}{\mathrm{d}\Omega \mathrm{d}E'}\right)^{(1)}_{\mathrm{coh}} = cN\frac{k'}{2kM}[\{1+n(\omega_l)\}\delta(\omega-\omega_l)+n(\omega_l)\delta(\omega+\omega_l)]\times$$
$$\times\frac{\omega_l}{B(\omega_l)}\sum_j\{\boldsymbol{\varkappa}\cdot\boldsymbol{\sigma}^j(\boldsymbol{\varkappa})\}^2\left\{\frac{\ell_{\mathrm{h}}}{\omega_l^2-\omega_j^2(\boldsymbol{\varkappa})}+\frac{\ell_{\mathrm{d}}-\ell_{\mathrm{h}}}{\lambda\omega_l^2}\right\}^2. \tag{15.74}$$

This has an identical form to the corresponding single defect cross-section, eqn (15.47), but here the ω_l is determined from eqn (15.66) instead of eqn (15.41).

Let us now examine the incoherent cross-section, eqn (15.71), for the same circumstances. In view of the second expression for $\operatorname{Im}\bar{G}^j$ given in eqn (15.68b),

$$\operatorname{Im}\bar{G}(\omega) = \frac{c\pi}{4B(\omega_l)}\{\delta(\omega-\omega_l)-\delta(\omega+\omega_l)\}\left(\frac{\mathrm{d}\mathscr{P}(\omega)}{\mathrm{d}\omega}\right)\bigg|_{\omega=\omega_l} \tag{15.75}$$

for $\omega > \omega_{\mathrm{m}}$.

The other contribution to (15.71) for $\omega > \omega_{\mathrm{m}}$ is

$$\operatorname{Im}\left(\frac{cT'}{\lambda\omega^2}\right)^2\bar{G}(\omega)$$
$$= -\frac{c\omega_l\pi}{2B(\omega_l)}\{\delta(\omega-\omega_l)-\delta(\omega+\omega_l)\}\frac{1}{3N}\sum_{j,\mathbf{q}}\left(\frac{\omega_l^2-\omega_j^2(\mathbf{q})}{\lambda\omega_l^2}\right)^2\left(\frac{1}{\omega_l^2-\omega_j^2(\mathbf{q})}\right)^2$$
$$= -\frac{c\pi}{2B(\omega_l)\lambda^2\omega_l^3}\{\delta(\omega-\omega_l)-\delta(\omega+\omega_l)\}. \tag{15.76}$$

On combining eqns (15.75) and (15.76) in the expression for the incoherent cross-section the latter becomes

$$\left(\frac{\mathrm{d}^2\sigma}{\mathrm{d}\Omega \mathrm{d}E'}\right)^{(1)}_{\mathrm{incoh}} = Nc\frac{k'\kappa^2}{2kMB(\omega_l)}[\{1+n(\omega_l)\}\delta(\omega-\omega_l)+n(\omega_l)\delta(\omega+\omega_l)]\times$$
$$\times\left\{\frac{\alpha_{\mathrm{d}}^2-\alpha_{\mathrm{h}}^2}{\lambda^2\omega_l^3}-\frac{\alpha_{\mathrm{h}}^2}{2}\left(\frac{\mathrm{d}\mathscr{P}(\omega)}{\mathrm{d}\omega}\right)\bigg|_{\omega=\omega_l}\right\}. \tag{15.77}$$

This expression has exactly the same form as the corresponding cross-section for the single defect problem, eqn. (15.49), as expected: as with the coherent cross-section, ω_l here satisfies eqn (15.66).

15.3. Spin-wave cross-section for a mixed magnetic system

(*a*) *Green function*

The model system we employ for our discussion is a Heisenberg ferromagnet with only a nearest-neighbour exchange coupling, J. The host is then described by the Hamiltonian

$$\mathscr{H}_0 = -J \sum_{m,l(m)} \hat{S}^z_m \hat{S}^z_l + \hat{S}^+_m \hat{S}^-_l. \tag{15.78}$$

In (15.78) the notation $l(m)$ means those r ions at sites l that are nearest neighbours to the site m. The spin operators satisfy the commutation relations

$$[\hat{S}^z_m, \hat{S}^\pm_n] = \pm\delta_{m,n}\hat{S}^\pm_m,$$
$$[\hat{S}^+_m, \hat{S}^-_n] = 2\delta_{m,n}\hat{S}^z_m. \tag{15.79}$$

The total exchange Hamiltonian for a single impurity at the site $m = s$, coupled to the host with an exchange coupling $\pm J'$, is

$$\mathscr{H} = \mathscr{H}_0 + \mathscr{H}_1, \tag{15.80}$$

where

$$\mathscr{H}_1 = 2J \sum_{l(s)} \mathbf{\hat{S}}_s . \mathbf{\hat{S}}_l \mp 2J' \sum_{l(s)} \mathbf{\hat{S}}'_s . \mathbf{\hat{S}}_l. \tag{15.81}$$

With the upper sign in (15.81) the ground state is, clearly, that in which the impurity spin is fully aligned with the host spins. This we refer to as the ferromagnetic impurity case. If the other possible sign of the impurity–host exchange is taken, then the fully aligned state is, of course, no longer the ground state. Instead we use the approximate ground state in which the impurity is aligned antiparallel to the host spins. We refer to this as the antiferromagnetic impurity.

The inelastic neutron cross-section for scattering from a magnetic system in which the orbital angular momentum of the ions is zero, or quenched, is in the absence of lattice vibrations expressible in the form (cf. Chapter 8)

$$\frac{\mathrm{d}^2\sigma}{\mathrm{d}\Omega\mathrm{d}E'} = \frac{1}{4}\left(\frac{\gamma e^2}{m_\mathrm{e} c^2}\right)^2 \frac{k'}{k}(1+\tilde{\kappa}_z^2)\frac{1}{2\pi\hbar}\int_{-\infty}^{\infty} \mathrm{d}t\, \mathrm{e}^{-\mathrm{i}\omega t} \times$$
$$\times \sum_{m,n} F_m(\boldsymbol{\kappa})F_n(\boldsymbol{\kappa})\exp\{-\mathrm{i}\boldsymbol{\kappa}.(\mathbf{m}-\mathbf{n})\}\langle \hat{S}^+_m(0)\hat{S}^-_n(t) + \hat{S}^-_m(0)\hat{S}^+_n(t)\rangle. \tag{15.82}$$

$F_m(\boldsymbol{\kappa})$ is the atomic form factor of the ions at the site m.

(15.82) can be separated into the sum of creation (+) and annihilation (−) parts corresponding, respectively, to the first and second correlation functions; i.e.

$$\frac{d^2\sigma}{d\Omega dE'} = \frac{d^2\sigma^{(+)}}{d\Omega dE'} + \frac{d^2\sigma^{(-)}}{d\Omega dE'}.$$

If we define a retarded Green function $G(m,n;\omega)$ by

$$G(m,n;\omega) = \langle\langle \hat{S}^+_m ; \hat{S}^-_n \rangle\rangle_\omega, \tag{15.83}$$

then

$$\frac{d^2\sigma^{(\pm)}}{d\Omega dE'} = \mp\frac{1}{2\hbar}\left(\frac{\gamma e^2}{m_e c^2}\right)^2 \frac{k'}{k}(1+\kappa_z^2)\{n(\omega)+1\}\times$$
$$\times \sum_{m,n} F_m(\boldsymbol{\varkappa})F_n(\boldsymbol{\varkappa})\exp\{\mp i\boldsymbol{\varkappa}.(\mathbf{m}-\mathbf{n})\}\text{Im}\, G(m,n;\pm\omega). \tag{15.84}$$

This expression follows immediately from the relation

$$\langle \hat{S}^-_n(0)\hat{S}^+_m(t)\rangle = -2\int_{-\infty}^{\infty} d\omega'\, n(\omega')e^{-i\omega' t}\,\text{Im}\, G(m,n;\omega').$$

Thus, as in the previous section, the calculation of the neutron cross-section has reduced essentially to the calculation of a Green function for the mixed system; $G(m,n;\omega)$ is determined by its equation of motion

$$\hbar\omega G(m,n;\omega) = \frac{\hbar}{2\pi}\langle[\hat{S}^+_m, \hat{S}^-_n]\rangle + \langle\langle[\hat{S}^+_m, \mathscr{H}]; \hat{S}^-_n\rangle\rangle_\omega. \tag{15.85}$$

The Green function on the right-hand side of eqn (15.85) is of a higher order than G, because the commutator $[\hat{S}^+, \mathscr{H}]$ contains $\hat{S}^z\hat{S}^+$, so that an approximation scheme must be introduced to reduce this Green function to an expression involving only G. If we make the replacement

$$\langle\langle \hat{S}^z_l \hat{S}^+_m; \hat{S}^-_n\rangle\rangle_\omega \to S_l\, G(m,n;\omega), \tag{15.86}$$

where S_l is the magnitude of the spin at the lth site (a c-number), then our approximation scheme is equivalent to the linear spin-wave approximation. An alternative way of regarding (15.86) is to replace the commutator $[\hat{S}^+, \hat{S}^-] = 2\hat{S}^z$ by $[\hat{S}^+, \hat{S}^-] = 2S$ (cf. § 9.2).

To illustrate the approximation scheme we calculate the Green function for the unperturbed host $P(m,n;\omega)$, i.e. the function that satisfies (15.85) with $\mathscr{H} = \mathscr{H}_0$. With the use of (15.86),

$$\hbar\omega P(m,n;\omega) = \frac{\hbar}{2\pi}(2S)\delta_{m,n} - 2JS\sum_{l(m)}\{P(l,n;\omega) - P(m,n;\omega)\}$$

or

$$(\mathscr{E}-1)P(m,n;\mathscr{E}) + \frac{1}{r}\sum_{l(m)} P(l,n;\mathscr{E}) = \frac{\hbar}{2\pi}\frac{2S}{2rJS}\delta_{m,n} = \alpha S\delta_{m,n}, \tag{15.87}$$

where the reduced energy $\mathscr{E}$ is defined by

$$\mathscr{E} = \frac{\hbar\omega}{2rJS}. \tag{15.88}$$

If
$$P(m,n;\mathscr{E}) = \frac{1}{N}\sum_{\mathbf{q}} P_{\mathbf{q}}(\mathscr{E})\exp\{-\mathrm{i}\mathbf{q}.(\mathbf{m}-\mathbf{n})\}, \tag{15.89 a}$$

then, from (15.87),
$$P_{\mathbf{q}}(\mathscr{E}) = \frac{\alpha S}{\mathscr{E}-1+\gamma_{\mathbf{q}}} = \frac{\alpha S}{\mathscr{E}-\mathscr{E}_{\mathbf{q}}}, \tag{15.89 b}$$

where
$$\gamma_{\mathbf{q}} = \frac{1}{r}\sum_{l(0)} \exp(\mathrm{i}\mathbf{q}.\mathbf{l}).$$

We recognize
$$\hbar\omega_{\mathbf{q}} = 2rJS\mathscr{E}_{\mathbf{q}} = 2rJS(1-\gamma_{\mathbf{q}}) \tag{15.90}$$

as the energy spectrum of linear spin waves. Also

$$\operatorname{Im} P_{\mathbf{q}}(\mathscr{E}) = \lim_{\delta\to 0+} (\mathscr{E}+\mathrm{i}\delta-\mathscr{E}_{\mathbf{q}})^{-1}\alpha S = -\pi\alpha S\delta(\mathscr{E}-\mathscr{E}_{\mathbf{q}}), \tag{15.91}$$

so that for the pure host the formula (15.84) for the cross-section reduces to

$$\frac{\mathrm{d}^2\sigma^{(\pm)}}{\mathrm{d}\Omega\mathrm{d}E'} = \mp\left(\frac{1}{2\hbar}\right)\left(-\pi\frac{\hbar}{2\pi}\frac{2S}{2rJS}\right)\left(\frac{\gamma e^2}{m_{\mathrm{e}}c^2}\right)^2\frac{k'}{k}(1+\tilde{\kappa}_z^2)F^2(\boldsymbol{\kappa})\times$$
$$\times\{n(\omega)+1\}\frac{1}{N}\sum_{\mathbf{q}}\left|\sum_m \exp\{\mathrm{i}\mathbf{m}.(\boldsymbol{\kappa}\pm\mathbf{q})\}\right|^2\delta(\mathscr{E}\mp\mathscr{E}_{\mathbf{q}})$$
$$= \frac{SF^2(\boldsymbol{\kappa})}{2}\left(\frac{\gamma e^2}{m_{\mathrm{e}}c^2}\right)^2\frac{k'}{k}(1+\tilde{\kappa}_z^2)\frac{(2\pi)^3}{v_0}\times$$
$$\times\sum_{\mathbf{q},\boldsymbol{\tau}}\{n(\omega_{\mathbf{q}})\pm\tfrac{1}{2}+\tfrac{1}{2}\}\delta(\boldsymbol{\kappa}\mp\mathbf{q}-\boldsymbol{\tau})\delta(\hbar\omega_{\mathbf{q}}\mp\hbar\omega) \tag{15.92}$$

in agreement with the expression derived in § 9.2 (*b*).

(*b*) *Ferromagnetic impurity*

Taking the upper sign in eqn (15.81) for $\mathscr{H}_1$ and using the decoupling scheme (15.86), the equation of motion for $G(m,n;\omega)$, eqn (15.85), becomes

$$(\mathscr{E}-1)G(m,n;\mathscr{E})+\frac{1}{r}\sum_{l(m)} G(l,n;\mathscr{E}) = \alpha\delta_{m,n}S_n+\sum_l V_{m,l}G(l,n;\mathscr{E}). \tag{15.93}$$

Here
$$S_n = S+\delta_{n,s}(S'-S) \tag{15.94}$$

and the perturbation matrix $V_{m,n}$, of dimension $(r+1)\times(r+1)$, is

$$V_{m,n} = \epsilon\delta_{m,s}\delta_{s,n}+\frac{1}{r}\sum_{l(s)}\{\rho\delta_{m,l}\delta_{l,n}-\rho\delta_{m,s}\delta_{l,n}-\epsilon\delta_{m,l}\delta_{s,n}\} \qquad (15.95)$$

$$\equiv r^{-1}\begin{pmatrix} r\epsilon & -\rho & -\rho & . & . & . & -\rho \\ -\epsilon & \rho & 0 & . & . & . & 0 \\ -\epsilon & 0 & \rho & . & . & . & 0 \\ -\epsilon & 0 & 0 & . & . & . & 0 \\ . & . & . & . & . & . & . \\ -\epsilon & 0 & 0 & . & . & . & \rho \end{pmatrix},$$

where the perturbation parameters ϵ and ρ are defined to be

$$\epsilon = \frac{J'}{J}-1 \quad\text{and}\quad \rho = \frac{J'S'}{JS}-1. \qquad (15.96)$$

(15.93) can be written in terms of P, which satisfies the same equation with $V = 0$; namely,

$$G(m,n;\mathscr{E}) = \alpha S_n\, P(m-n;\mathscr{E})+(PVG)_{m,n}. \qquad (15.97)$$

This equation for the spin operator Green function $G(m,n;\mathscr{E})$ is analogous to eqn (15.30) for the phonon problem. There are, however, two important differences between (15.30) and (15.97); the latter is not in the form of a Dyson equation because of the function S_n in the first term on the right-hand side and, unlike the perturbation matrix for the phonon problem, $V_{m,n}$ in (15.97) is not diagonal in m and n. This, of course, is a consequence of the fact that $\mathscr{H}$ couples nearest-neighbour pairs, and makes the problem of a finite number of magnetic impurities more difficult than for mass defects in a harmonic crystal.

Because the perturbation $V_{m,n}$ is not diagonal in m and n the algebra for this problem is more complicated than for the mass defect problem: also the perturbation appears in (15.97) not only in the second term on the right-hand side but also in the first term, because S_n has a different value on the impurity site from the value it has on the host. We shall, therefore, concern ourselves primarily with the single magnetic impurity problem, and content ourselves with a discussion of the many-impurity problem for a particular set of perturbation parameters that allow us to use analogues between the present problem and the mass-defect problem. For the present we discuss the solution of (15.97) and, in particular, the impurity energy modes.

The poles of $G(m,n;\mathscr{E})$ are determined by

$$\operatorname{Re}\det|I-PV| = 0. \tag{15.98a}$$

The matrix PV has the symmetry of the crystal lattice and the solution of eqn (15.98a) is greatly simplified by making use of this fact. For if the unitary matrix U is constructed from the vectors that transform as the irreducible representations of the point group of the crystal structure then $U^{+}PVU$ is diagonal and the criterion (15.98a) reads

$$\operatorname{Re}\det|I-PV| = \operatorname{Re}\det|U^{+}(I-PV)U|$$

$$= \prod_{\xi=1}^{r+1} \operatorname{Re} D_{\xi}(\mathscr{E}_{\xi}) = 0, \tag{15.98b}$$

where D_{ξ} are the elements of the diagonal matrix $U^{+}(I-PV)U$. These elements are usually labelled s, p, d, or f because the associated functions of the various irreducible representations of the point group O_{h}, to which they belong, are of the same form as the wave functions of the hydrogen atom. In Table 15.1 the irreducible representations of the O_{h} group are given for each of the three cubic lattices together with the degeneracies of the various factors in (15.98b).†

For all three lattices the s-like factor has exactly the same form

$$D_{\mathrm{s}}^{(+)}(\mathscr{E}) = 1+\epsilon\{1-\mathscr{E}P(\mathscr{E})\}+\rho\mathscr{E}R(\mathscr{E}), \tag{15.99}$$

where $$P(\mathscr{E}) \equiv P(0,0,0;\mathscr{E})$$

and $$R(\mathscr{E}) = 1-(\mathscr{E}-1)P(\mathscr{E}) = \frac{1}{r}\sum_{l(0)} P(l;\mathscr{E}). \tag{15.100}$$

TABLE 15.1

Lattice	Γ_1(s-like)	Γ_4'(p-like)	Γ_3(d-like)	Γ_5(d-like)	Γ_2'(f-like)	Γ_5'(f-like)
s.c.	2	3	2			
b.c.c.	2	3		3	1	
f.c.c.	2	3	2	3		3

The remaining factors are

$$\text{s.c.}\begin{cases} D_{\mathrm{p}}(\mathscr{E}) = \dfrac{1}{r}[P(0,0,0;\mathscr{E})-P(2,0,0;\mathscr{E})]-\rho^{-1} \\[2ex] D_{\mathrm{d}}(\mathscr{E}) = \dfrac{1}{r}[P(0,0,0;\mathscr{E})-2P(1,1,0;\mathscr{E})+P(2,0,0;\mathscr{E})]-\rho^{-1} \end{cases} \tag{15.101}$$

† Bethe's notation: see, for example, R. S. Knox and A. Gold, *Symmetry in the solid state*, W. A. Benjamin (1964).

$$\text{b.c.c.}\begin{cases} D_{\mathrm{p}}(\mathscr{E}) = \dfrac{1}{r}[P(0,0,0;\mathscr{E})+P(1,0,0;\mathscr{E})-P(1,1,0;\mathscr{E})- \\ \qquad\qquad -P(1,1,1;\mathscr{E})]-\rho^{-1} \\ D_{\mathrm{d}}(\mathscr{E}) = \dfrac{1}{r}[P(0,0,0;\mathscr{E})-P(1,0,0;\mathscr{E})-P(1,1,0;\mathscr{E})+ \\ \qquad\qquad +P(1,1,1;\mathscr{E})]-\rho^{-1} \\ D_{\mathrm{f}}(\mathscr{E}) = \dfrac{1}{r}[P(0,0,0;\mathscr{E})-3P(1,0,0;\mathscr{E})+3P(1,1,0;\mathscr{E})- \\ \qquad\qquad -P(1,1,1;\mathscr{E})]-\rho^{-1} \end{cases} \tag{15.102}$$

$$\text{f.c.c.}\begin{cases} D_{\mathrm{p}}(\mathscr{E}) = \dfrac{1}{r}[P(0,0,0;\mathscr{E})+2P(\tfrac{1}{2},\tfrac{1}{2},0;\mathscr{E})-P(1,1,0;\mathscr{E})- \\ \qquad\qquad -2P(1,\tfrac{1}{2},\tfrac{1}{2};\mathscr{E})]-\rho^{-1} \\ D_{\mathrm{d_1}}(\mathscr{E}) = \dfrac{1}{r}[P(0,0,0;\mathscr{E})-2P(\tfrac{1}{2},\tfrac{1}{2},0;\mathscr{E})+2P(1,0,0;\mathscr{E})+ \\ \qquad\qquad +P(1,1,0;\mathscr{E})-2P(1,\tfrac{1}{2},\tfrac{1}{2};\mathscr{E})]-\rho^{-1} \\ D_{\mathrm{d_2}}(\mathscr{E}) = \dfrac{1}{r}[P(0,0,0;\mathscr{E})-2P(1,0,0;\mathscr{E})+P(1,1,0;\mathscr{E})]-\rho^{-1} \\ \mathrm{D}_{\mathrm{f}}(\mathscr{E}) = \dfrac{1}{r}[P(0,0,0;\mathscr{E})-2P(\tfrac{1}{2},\tfrac{1}{2},0;\mathscr{E})-P(1,1,0;\mathscr{E})+ \\ \qquad\qquad +2P(1,\tfrac{1}{2},\tfrac{1}{2};\mathscr{E})]-\rho^{-1}. \end{cases} \tag{15.103}$$

In eqns (15.101), (15.102), and (15.103), $P(l;\mathscr{E})$ is the crystal Green function appropriate to the particular lattice and r is the number of nearest neighbours.

It can be shown that the spin-wave density of states $N(\mathscr{E})$ is given by†

$$N(\mathscr{E}) = -\frac{1}{\pi}\left[P''(\mathscr{E})+\frac{1}{\pi N}\,\mathrm{Im}\,\frac{\mathrm{d}}{\mathrm{d}\mathscr{E}}\left\{\ln\prod_{\xi} D_{\xi}(\mathscr{E})\right\}\right].$$

For the simple cubic lattice this is

$$N(\mathscr{E}) = N_0(\mathscr{E})-\frac{1}{\pi N}\,\mathrm{Im}\,\frac{\mathrm{d}}{\mathrm{d}\mathscr{E}}\,\{D_{\mathrm{s}}^{(+)}(\mathscr{E})D_{\mathrm{p}}^{3}(\mathscr{E})D_{\mathrm{d}}^{2}(\mathscr{E})\}.$$

From an argument similar to that leading to (15.42) this means that for $\mathscr{E} \sim \mathscr{E}_{\xi}$, where $\mathscr{E}_{\xi}$ is the solution of

$$\mathrm{Re}\,D_{\xi}(\mathscr{E}_{\xi}) = 0 \quad (\xi = \text{s, p, or d (s.c.)}),$$

there is a contribution to the density of states

$$\delta N_{\xi} \simeq \frac{1}{\pi N}\,\frac{\Gamma_{\xi}}{(\mathscr{E}-\mathscr{E}_{\xi})^2+(\Gamma_{\xi})^2},$$

† T. Wolfram and J. Callaway, *Phys. Rev.* **130**, 2207.

where $$\Gamma_\xi = \operatorname{Im} D_\xi(\mathscr{E}_\xi) \Big/ \left[\operatorname{Re}\left\{\frac{\mathrm{d}}{\mathrm{d}\mathscr{E}} D_\xi(\mathscr{E})\right\}\bigg|_{\mathscr{E}=\mathscr{E}_\xi}\right]$$

characterizes the width of the contribution to the density of states.

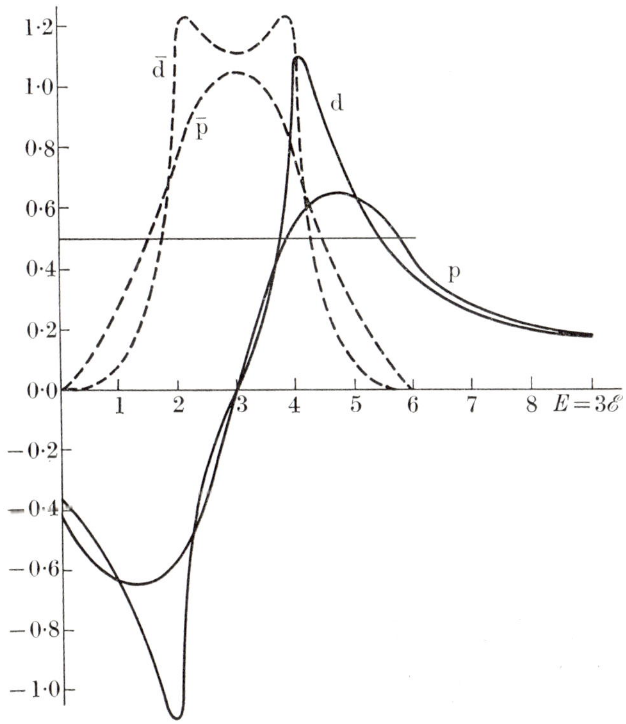

FIG. 15.5. The two solid lines are

$$\operatorname{Re}\frac{2}{r}\,[P(0,\,0,\,0\,;E)-P(2,\,0,\,0\,;E)] \quad \text{(p)}$$

and $$\operatorname{Re}\frac{2}{r}\,[P(0,\,0,\,0\,;E)-2P(1,\,1,\,0\,;E)+P(2,\,0,\,0\,;E) \quad \text{(d)}$$

as functions of $E = 3\mathscr{E}$. The intersection of these curves with $2/\rho$ gives the locations of the p and d type modes (the host spin-wave band runs from 0 to 6). The dashed lines, $\bar{\text{p}}$ and $\bar{\text{d}}$, are the imaginary parts of these two functions, from which the sign of Γ_ξ can be determined. For a true resonance state $\Gamma_\xi > 0$, so that these occur only near the top of the band.

Let us now examine the locations of the various impurity modes as a function of the parameters J'/J and $\sigma = S'/S$. Fig. 15.5 shows the two functions

$$\operatorname{Re}\frac{2}{r}\,[P(0,0,0;E)-P(2,0,0;E)] \qquad \text{(p)}$$

and $$\operatorname{Re}\frac{2}{r}\,[P(0,0,0;E)-2P(1,1,0;E)+P(2,0,0;E)] \quad \text{(d)}$$

for the simple cubic lattice. From eqn (15.101) it follows that if we equate $2/\rho$ to the first we obtain the location of the p-like modes, whereas the second gives the positions of the d-like modes. The parameter $2/\rho$ has values in the range $-\infty$ to -2 and 0 to $+\infty$; from Fig. 15.5 it follows that there are no p or d modes near the bottom of the spin-wave band, a characteristic feature of those modes in all the cubic lattices.

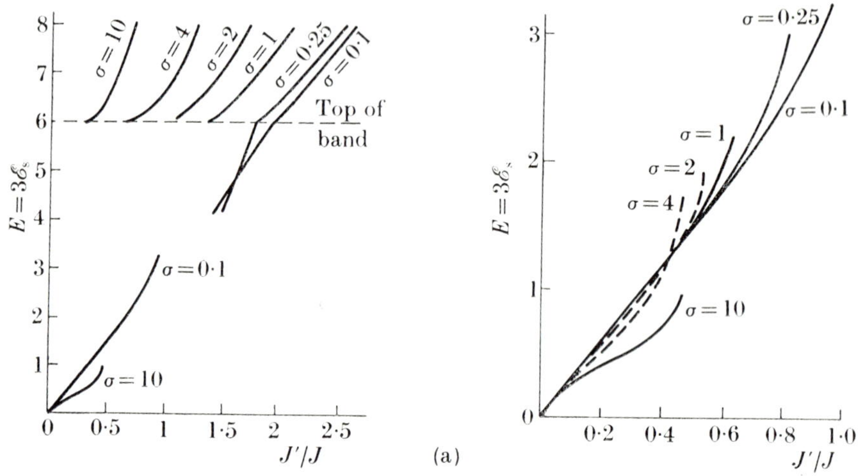

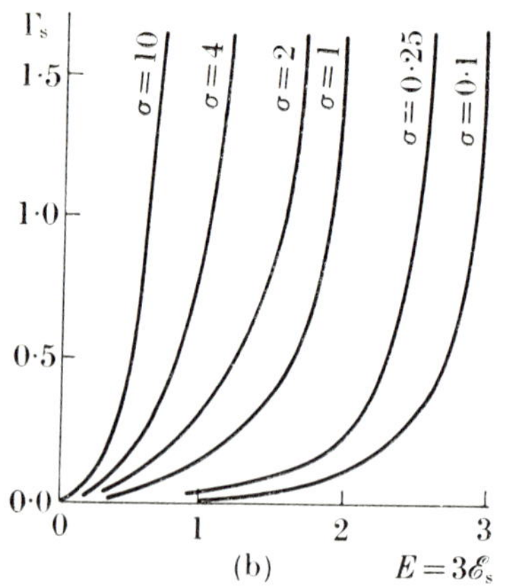

FIG. 15.6 (a). The energy of the s modes, $\mathscr{E}_s$, as a function of J'/J for various $\sigma = S'/S$ for the simple cubic lattice; i.e. the solutions of Re $D_s^{(+)}(\mathscr{E}_s) = 0$.

FIG. 15.6 (b). The widths of the virtual s modes at the bottom of the band for a simple cubic host.

(After Y. Izuymov and M. V. Medvedev, *Soviet Phys. JETP*, **22,** 1289 (1966).)

In Fig. 15.6 the energy and width of the s-mode resonances are shown for a simple cubic host. These figures clearly show that for small energies the dominant perturbation parameter is J'/J.

The energies of the virtual and localized modes for body-centred and face-centred cubic host lattices are given in Fig. 15.7.

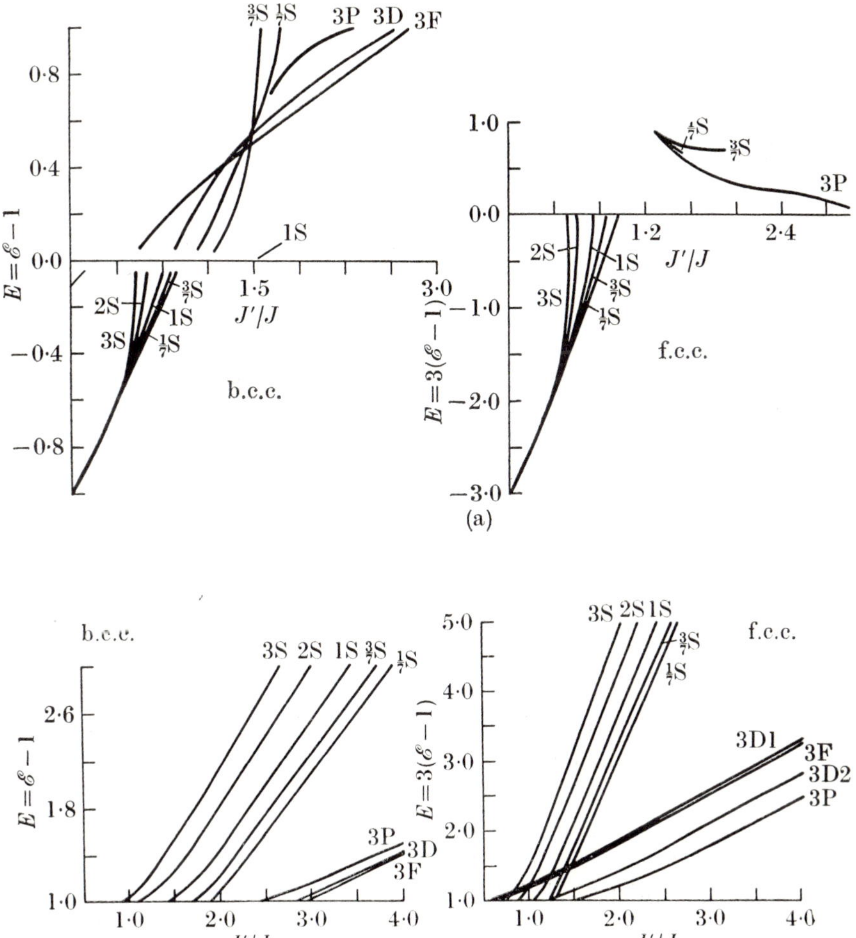

FIG. 15.7. The energies of the virtual (a) and localized (b) modes for body-centred and face-centred cubic host lattices as functions of J'/J for various $\sigma = S'/S$, given with each curve.

(S. Vashishta, *Proc. phys. Soc.* **91**, 372 (1967).)

Eqn (15.97) for $G(m, n; \mathscr{E})$ is, explicitly,

$$G(m, n; \mathscr{E})$$

$$= \alpha S_n P(m-n; \mathscr{E}) + P(m-s; \mathscr{E})\left\{\epsilon\mathscr{E} G(s, n; \mathscr{E}) - \frac{\rho}{r}\sum_{l(s)} G(l, n; \mathscr{E})\right\} -$$

$$-\epsilon\delta_{m,s} G(s, n; \mathscr{E}) + \frac{\rho}{r}\sum_{l(s)} P(m-l; \mathscr{E}) G(l, n; \mathscr{E}). \quad (15.104)$$

If we set $m = s$ and then m equal to a nearest neighbour to s, the two functions occurring on the right-hand side, $G(s,n;\mathscr{E})$ and $\sum G(l,n;\mathscr{E})$, can be solved for with the aid of the relation

$$\frac{1}{r}\sum_{l(m)} P(l-n;\mathscr{E}) = \delta_{m,n}-(\mathscr{E}-1)P(m-n;\mathscr{E}). \tag{15.105}$$

For instance

$$G(s,n;\mathscr{E}) = \alpha S_n \frac{[(1+\rho)P(s-n;\mathscr{E})+\rho\{1-\mathscr{E}P(\mathscr{E})\}\delta_{s,n}]}{D_s^{(+)}(\mathscr{E})}. \tag{15.106}$$

The r nearest-neighbour Green functions occurring in the last term of (15.104) are conveniently obtained by forming a matrix equation for them. If (g,f) denote the r nearest-neighbour positions to s, then from (15.104)

$$-\rho G(g,n;\mathscr{E}) = \sum_{f=1}^{r}(M^{-1})_{g,f}\times$$
$$\times\left[\alpha S_n P(f-n;\mathscr{E})+P(f-n;\mathscr{E})\left\{\epsilon\mathscr{E}G(s,n;\mathscr{E})-\frac{\rho}{r}\sum_{l(s)}G(l,n;\mathscr{E})\right\}\right]. \tag{15.107}$$

M is a symmetric matrix whose elements are functions of the crystal Green function P, and has the symmetry of the crystal lattice. The latter property means that $U^{+}MU$ is a diagonal matrix; hence M^{-1} is easily calculated. For a simple cubic host lattice

$$\sum_{g,f=1}^{r}(M^{-1})_{g,f}\exp(i\mathbf{q}.\mathbf{g}-i\mathbf{q}'.\mathbf{f})$$
$$= \gamma_{\mathbf{q}}\gamma_{\mathbf{q}'}\{-(\mathscr{E}-1)R(\mathscr{E})-\rho^{-1}\}^{-1}+\frac{3A^{(p)}(\mathbf{q},\mathbf{q}')}{D_p(\mathscr{E})}+\frac{2A^{(d)}(\mathbf{q},\mathbf{q}')}{D_d(\mathscr{E})}$$
$$= \gamma_{\mathbf{q}}\gamma_{\mathbf{q}'}\{-(\mathscr{E}-1)R(\mathscr{E})-\rho^{-1}\}^{-1}+H(\mathbf{q},\mathbf{q}';\mathscr{E}), \tag{15.108}$$

where

$$A^{(p)}(\mathbf{q},\mathbf{q}') = \tfrac{2}{3}(\sin u\sin u'+\sin v\sin v'+\sin w\sin w'),$$
$$A^{(d)}(\mathbf{q},\mathbf{q}') = \tfrac{1}{3}\{2(\cos u\cos u'+\cos v\cos v'+\cos w\cos w')-$$
$$-\cos u(\cos v'+\cos w')-\cos v(\cos u'+\cos w')-$$
$$-\cos w(\cos u'+\cos v')\}. \tag{15.109}$$

In (15.109), $u = aq_x$, $v = aq_y$, and $w = aq_z$, where a is the lattice spacing.

If (15.108) is used in (15.104) then the latter can be brought into the form

$$\frac{1}{\alpha S_n} G(m,n;\mathscr{E}) = P(m-n;\mathscr{E})+\mathscr{L}^{(+)}(m,n;\mathscr{E})-$$

$$-\frac{1}{N^2}\sum_{\mathbf{q},\mathbf{q}'} P_{\mathbf{q}}(\mathscr{E})P_{\mathbf{q}'}(\mathscr{E})\exp(-\mathrm{i}\mathbf{q}.\mathbf{m}+\mathrm{i}\mathbf{q}'.\mathbf{n})H(\mathbf{q},\mathbf{q}';\mathscr{E}), \quad (15.110\,\mathrm{a})$$

where
$$D_s^{(+)}(\mathscr{E})\mathscr{L}^{(+)}(m,n;\mathscr{E}) = \{\mathscr{E}P(m-s;\mathscr{E})-\delta_{m,s}\}\times$$
$$\times\{(\rho\mathscr{E}-\rho+\epsilon)P(s-n;\mathscr{E})-\rho\delta_{s,n}\}. \quad (15.110\,\mathrm{b})$$

The inelastic one-magnon cross-section eqn (15.84) contains

$$\sum F_m F_n\, G(m,n;\mathscr{E})\exp\{\mathrm{i}\boldsymbol{\varkappa}.(\mathbf{m}-\mathbf{n})\}.$$

For a single impurity at $\mathbf{m} = \mathbf{s}$

$$F_m(\boldsymbol{\varkappa}) = F(\boldsymbol{\varkappa})+\delta_{m,s}\{F'(\boldsymbol{\varkappa})-F(\boldsymbol{\varkappa})\},$$

where $F(\boldsymbol{\varkappa})$ is the form factor associated with the host ions and $F'(\boldsymbol{\varkappa})$ that of the impurity. It therefore follows that the above summation has exactly the same form as (15.14) with b_l replaced by F_m; the functions $G(s,s;\mathscr{E})$ and $\sum G(s,n;\mathscr{E})$ are easily calculated from (15.106). With some tedious algebra we obtain

$$\sum_{m,n} F_m(\boldsymbol{\varkappa})F_n(\boldsymbol{\varkappa})\exp\{\mathrm{i}\boldsymbol{\varkappa}.(\mathbf{m}-\mathbf{n})\}G(m,n;\mathscr{E})$$
$$= \alpha[SF^2\{NP_{\boldsymbol{\varkappa}}(\mathscr{E})+P_{\boldsymbol{\varkappa}}^2(\mathscr{E})T(\boldsymbol{\varkappa},\mathscr{E})\}+$$
$$+2\sigma SF(F'-F)\{(1+\epsilon)P_{\boldsymbol{\varkappa}}(\mathscr{E})+\rho R(\mathscr{E})-\epsilon P(\mathscr{E})\}/D_{\mathrm{s}}^{(+)}(\mathscr{E})+$$
$$+\sigma S(F'-F)^2\{P(\mathscr{E})+\rho R(\mathscr{E})\}/D_{\mathrm{s}}^{(+)}(\mathscr{E})], \quad (15.111)$$

where

$$T(\boldsymbol{\varkappa},\mathscr{E}) = \frac{1}{D_{\mathrm{s}}^{(+)}(\mathscr{E})}[(\rho\mathscr{E}-\rho+\epsilon)\{\mathscr{E}-2P_{\boldsymbol{\varkappa}}^{-1}(\mathscr{E})\}+$$
$$+P_{\boldsymbol{\varkappa}}^{-2}(\mathscr{E})\{\sigma\rho-(\sigma-1)P(\mathscr{E})(\rho\mathscr{E}-\rho+\epsilon)\}]-H(\boldsymbol{\varkappa},\boldsymbol{\varkappa};\mathscr{E}). \quad (15.112)$$

From (15.111) it is clear that the neutron cross-section will display peaks whenever the neutron energy change is such that for a given set of perturbations, the real parts of parameters $D_{\mathrm{s}}^{(+)}$, D_{p}, or D_{d} are zero, i.e. whenever the neutrons excite an impurity energy mode. If the mode lies outside the spin-wave band, a localized mode, then the cross-section behaves like a delta function in energy at the position of the impurity mode. If, on the other hand, a virtual mode is excited, then this will have a certain width associated with it, Γ_ξ. Because the width is proportional to the density of spin-wave states in the pure host only those virtual states near the bottom of the band will give a sharp peak in the cross-section. As we have already noted, only the s states exist near the bottom of the band so that these must dominate the behaviour of the cross-section when $\hbar^2(k^2-k'^2)/2m$ is less than the width of the spin-wave band, $4rJS$ (cf. Fig. 15.6 (b)).

The expression (15.111) contains a diffuse contribution whose energy dependence is determined by $D_{\mathrm{s}}^{(+)}(\mathscr{E})$, i.e. it depends only on the s mode. From symmetry conditions this is to be expected. The diffuse part of (15.111) is

$$\frac{S}{D_{\mathrm{s}}^{(+)}(\mathscr{E})}[\sigma F'^2\{\rho R(\mathscr{E})+P(\mathscr{E})\}-2\sigma FF'(1+\epsilon)P(\mathscr{E})+F^2(\rho\mathscr{E}+1+\epsilon)P(\mathscr{E})]. \tag{15.113}$$

When $F = F'$, $S = S'$, and $J = J'$ this term is zero. Also, when $F = F'$, $S = S'$ it reduces to the very simple result

$$\frac{\epsilon SF^2}{D_{\mathrm{s}}^{(+)}(\mathscr{E})}. \tag{15.114}$$

Now let us examine the contributions from the p and d modes, which are contained in $H(\boldsymbol{\varkappa},\boldsymbol{\varkappa};\mathscr{E})$. Because the impurity ion does not participate in these modes the form factor and spin associated with this term must be those of the host ions, as is seen to be the case. Also, these terms contribute only to the coherent part of the cross-section, reflecting as they do the symmetry of the crystal lattice, in this instance simple cubic. In contrast to this, the diffuse part of (15.111) has exactly the same form for all three cubic lattices. $H(\boldsymbol{\varkappa},\boldsymbol{\varkappa};\mathscr{E})$ is zero at the magnetic Bragg positions, i.e. when $\boldsymbol{\varkappa}$ coincides with a reciprocal lattice vector. For the Bragg peaks are a consequence of the translational symmetry of the crystal lattice and this is destroyed by the introduction of a magnetic impurity; $A^{(\mathrm{p})}$ and $A^{(\mathrm{d})}$ have zero amplitude at the Bragg positions.

The neutron cross-section for a finite concentration of magnetic impurities is well represented by the cross-section for a single impurity multiplied by Nc, provided the neutron energy change is such that only localized modes contribute to the cross-section. The cross-section for the single impurity is, in essence, the imaginary part of (15.111). Using the result (15.42), the approximate expression for the one-magnon cross-section outside the spin-wave band, i.e. for $(\hbar^2/2m)(k^2-k'^2) > 4rJS$, and Nc impurities, is, from eqns (15.84) and (15.111),

$$\frac{\mathrm{d}^2\sigma^{(\pm)}}{\mathrm{d}\Omega\mathrm{d}E'} = \frac{Nc}{4rJS}\left(\frac{\gamma e^2}{m_e c^2}\right)^2 \frac{k'}{k}(1+\tilde{\kappa}_z^2)\{n(\omega)\pm\tfrac{1}{2}+\tfrac{1}{2}\}\times$$
$$\times\Bigg[\delta(\mathscr{E}-\mathscr{E}_s)\frac{1}{\{\mathrm{d}D_s^{(+)}(\mathscr{E})/\mathrm{d}\mathscr{E}\}}[\![\,[(\rho\mathscr{E}-\rho+\epsilon)\{\mathscr{E}P_{\varkappa}(\mathscr{E})-2\}P_{\varkappa}(\mathscr{E})+$$
$$+\{\sigma\rho-(\sigma-1)(\rho\mathscr{E}-\rho+\epsilon)P(\mathscr{E})\}]SF^2(\boldsymbol{\varkappa})+$$
$$+2\sigma SF(\boldsymbol{\varkappa})\{F'(\boldsymbol{\varkappa})-F(\boldsymbol{\varkappa})\}\{(1+\epsilon)P_{\varkappa}(\mathscr{E})+\rho R(\mathscr{E})-\epsilon P(\mathscr{E})\}+$$
$$+\sigma S\{F'(\boldsymbol{\varkappa})-F(\boldsymbol{\varkappa})\}^2[P(\mathscr{E})+\rho R(\mathscr{E})]\,]\!]-$$
$$-SF^2(\boldsymbol{\varkappa})P_{\varkappa}^2(\mathscr{E})\Bigg\{\delta(\mathscr{E}-\mathscr{E}_p)\frac{1}{\{\mathrm{d}D_p(\mathscr{E})/\mathrm{d}\mathscr{E}\}}3A^{(p)}(\boldsymbol{\varkappa},\boldsymbol{\varkappa})+$$
$$+\delta(\mathscr{E}-\mathscr{E}_d)\frac{1}{\{\mathrm{d}D_d(\mathscr{E})/\mathrm{d}\mathscr{E}\}}2A^{(d)}(\boldsymbol{\varkappa},\boldsymbol{\varkappa})\Bigg\}\Bigg]. \quad (15.115)$$

In (15.115) we have retained the use of the reduced energy

$$\mathscr{E} = \hbar\omega/2rJS$$

for simplicity. The expression is for a simple cubic host lattice. The analogous result for the body-centred and face-centred host lattices has the same structure, the diffuse parts being identical in form for all three cubic lattices. For $F(\boldsymbol{\varkappa}) = F'(\boldsymbol{\varkappa})$ and $\sigma = S'/S = 1$, the coefficient of $\delta(\mathscr{E}-\mathscr{E}_s)$ in (15.115) reduces to

$$\left\{\frac{1}{\epsilon N^{-1}\Sigma_{\mathbf{q}}\,\mathscr{E}_{\mathbf{q}}^2/(\mathscr{E}-\mathscr{E}_{\mathbf{q}})^2}\right\}\frac{SF^2(\boldsymbol{\varkappa})\epsilon\mathscr{E}_{\varkappa}^2}{(\mathscr{E}-\mathscr{E}_{\varkappa})^2} = SF^2(\boldsymbol{\varkappa})P_{\varkappa}^2(\mathscr{E})\mathscr{E}_{\varkappa}^2\left\{\frac{1}{N}\sum_{\mathbf{q}}P_{\mathbf{q}}^2(\mathscr{E})\mathscr{E}_{\mathbf{q}}^2\right\}^{-1}. \quad (15.116)$$

We further note that

$$D_p(\mathscr{E}) = \rho^{-1}+\frac{1}{rN}\sum_{\mathbf{q}}P_{\mathbf{q}}(\mathscr{E})A^{(p)}(\mathbf{q},\mathbf{q})$$

and

$$D_d(\mathscr{E}) = \rho^{-1}+\frac{1}{rN}\sum_{\mathbf{q}}P_{\mathbf{q}}(\mathscr{E})A^{(d)}(\mathbf{q},\mathbf{q}). \quad (15.117\,\mathrm{a})$$

So

$$\frac{\mathrm{d}D_p(\mathscr{E})}{\mathrm{d}\mathscr{E}} = -\frac{1}{rN}\sum_{\mathbf{q}}P_{\mathbf{q}}^2(\mathscr{E})A^{(p)}(\mathbf{q},\mathbf{q})$$

and

$$\frac{\mathrm{d}D_d(\mathscr{E})}{\mathrm{d}\mathscr{E}} = -\frac{1}{rN}\sum_{\mathbf{q}}P_{\mathbf{q}}^2(\mathscr{E})A^{(d)}(\mathbf{q},\mathbf{q}). \quad (15.117\,\mathrm{b})$$

(c) *Ferromagnetic impurities—low concentration theory*

For the particular choice $\sigma = S'/S = 1$, eqn (15.97) reduces to a Dyson equation. Also, if $F(\boldsymbol{\varkappa}) = F'(\boldsymbol{\varkappa})$, which is consistent with $\sigma = 1$, the inelastic neutron cross-section for a small, finite concentration c of impurities, involves only $\sum_{m,n} \exp\{i\boldsymbol{\varkappa}.(\mathbf{m}-\mathbf{n})\}\bar{G}(m,n;\mathscr{E})$. The fact that for this choice of parameters G satisfies a Dyson equation means that we can use directly the results of the previous section on mass defects in harmonic crystals to calculate $\bar{G}$.

In § 15.3 (*b*) it was shown that the only virtual modes near the bottom of the spin-wave band are s-like and that these occur at energies which are virtually independent of S'/S. Hence, it is to be expected that the cross-section for the set of parameters $J'/J \neq 1$, $S'/S = 1$, and $F'(\boldsymbol{\varkappa}) = F(\boldsymbol{\varkappa})$ will be a good approximation to that in which $S'/S \neq 1$ for the case when the neutron energy change is small compared to $2rJS$.

From eqn (15.111)

$$\frac{1}{\alpha N}\sum_{m,n}\exp\{i\mathbf{q}.(\mathbf{m}-\mathbf{n})\}G(m,n;\mathscr{E}) = P_{\mathbf{q}}(\mathscr{E})+\frac{1}{N}P^2_{\mathbf{q}}(\mathscr{E})T(\mathbf{q},\mathscr{E}), \tag{15.118}$$

where, for $S'/S = 1$,

$$T(\mathbf{q},\mathscr{E}) = \frac{1}{D_{\mathrm{s}}^{(+)}(\mathscr{E})}\epsilon\mathscr{E}^2_{\mathbf{q}}-H(\mathbf{q},\mathbf{q};\mathscr{E}). \tag{15.119}$$

If we now regard $1/N$ as c, our discussion on p. 543 leads us to the result, to first order in c,

$$\bar{G}(\mathbf{q}) = \frac{\alpha P_{\mathbf{q}}(\mathscr{E})}{1-cP_{\mathbf{q}}(\mathscr{E})T(\mathbf{q},\mathscr{E})} = \frac{\alpha}{\mathscr{E}-\mathscr{E}_{\mathbf{q}}-cT(\mathbf{q},\mathscr{E})} \tag{15.120}$$

for the Fourier transform of the Green function of the system with a small number of randomly distributed magnetic impurities in a simple cubic host lattice. Eqn (15.120) is directly analogous to (15.60). If

$$cT(\mathbf{q},\mathscr{E}) = \delta(\mathbf{q},\mathscr{E})-i\gamma(\mathbf{q},\mathscr{E}) \tag{15.121}$$

then

$$\operatorname{Im}\bar{G}(\mathbf{q}) = \frac{-\gamma(\mathbf{q},\mathscr{E})\alpha}{\{\mathscr{E}-\mathscr{E}_{\mathbf{q}}-\delta(\mathbf{q},\mathscr{E})\}^2+\{\gamma(\mathbf{q},\mathscr{E})\}^2}. \tag{15.122}$$

For small $\mathbf{q}$ the width γ has the form†

$$\gamma(\mathbf{q},\omega) \simeq 2c\pi\left(\frac{2\epsilon}{1+\Lambda\epsilon}\right)^2 N(\omega_{\mathbf{q}})(2JSq^2a^2)^2, \tag{15.123}$$

where Λ is a constant that depends on the host lattice; for the simple cubic lattice, $\Lambda = 0{\cdot}21$. The energy spectrum of the spin waves can be approximately calculated for small $\mathbf{q}$ from the expression

$$\mathscr{E} \simeq \mathscr{E}_{\mathbf{q}}+\delta(\mathbf{q},\mathscr{E}_{\mathbf{q}}).$$

We find

$$\hbar\omega \simeq a^2q^2\,2JS\left(1+c\frac{2\epsilon}{1+\Lambda\epsilon}\right), \tag{15.124}$$

† J. Callaway, *Phys. Rev.* **132**, 2003 (1963).

in agreement with Murray,† who also finds the same result for body-centred and face-centred host lattices, for which $\Lambda = 0{\cdot}158$ and $0{\cdot}123$ respectively.

The one-magnon cross-section, for $\hbar\omega \ll 2rJS$ follows immediately from (15.122):

$$\frac{d^2\sigma^{(\pm)}}{d\Omega dE'} = \frac{N}{4\pi rJS}\left(\frac{\gamma e^2}{m_e c^2}\right)^2 \frac{k'}{k} SF^2(\boldsymbol{\varkappa})(1+\tilde{\kappa}_z^2)\{n(\omega)\pm\tfrac{1}{2}+\tfrac{1}{2}\}\times$$
$$\times\frac{\gamma(\boldsymbol{\varkappa},\mathscr{E})}{\{\mathscr{E}-\mathscr{E}_{\varkappa}-\delta(\boldsymbol{\varkappa},\mathscr{E})\}^2+\{\gamma(\boldsymbol{\varkappa},\mathscr{E})\}^2}, \quad (15.125)$$

where, as in previous formulae, we have retained the use of the reduced energy $\mathscr{E} = \hbar\omega/2rJS$.

(*d*) *Antiferromagnetic impurity*

The classical ground state of the system consisting of a single antiferromagnetic impurity in a Heisenberg ferromagnet is that in which the impurity spin is aligned antiparallel to those of the host. We shall seek spin-wave excitations from this ground state. As in the spin-wave theory of a two-sublattice Heisenberg antiferromagnet, the approximate nature of this ground state manifests itself in the appearance of a non-zero spin deviation at absolute zero.‡

We rotate the axes of the impurity through π about the x axis, which constitutes a canonical transformation, and use the decoupling scheme

$$\langle\langle \hat{S}_l^z \hat{S}_m^+; \hat{S}_n^-\rangle\rangle_\omega = S_l G(m,n;\omega),$$

where

$$S_l = S-\delta_{l,s}(S'+S). \quad (15.126)$$

From

$$[\hat{S}_m^+,\mathscr{H}_1] = -2J\sum_{l(s)}(\delta_{m,s}-\delta_{m,l})(\hat{S}_s^+\hat{S}_l^z-\hat{S}_s^z\hat{S}_l^+)+$$
$$+2J'\sum_{l(s)}\delta_{m,s}(\hat{S}_s^+\hat{S}_l^z+\hat{S}_s^z\hat{S}_l^-)+\delta_{m,l}(\hat{S}_s^z\hat{S}_l^++\hat{S}_s^-\hat{S}_l^z) \quad (15.127)$$

it is evident that the equation of motion for $G(m,n;\omega)$ contains a new Green function $\langle\langle \hat{S}_m^-; \hat{S}_n^-\rangle\rangle_\omega$. An equation of motion for this can be formed with the adjoint of eqn (15.127), and hence an equation for $G(m,n;\omega)$ obtained. The mathematical analysis then parallels that given in §15.3 (*b*).

† G. A. Murray, *Proc. phys. Soc.* **89**, 87 (1966).

‡ J. B. Parkinson, *Solid St. Communs.* **5**, 419 (1967). Y. Wang and H. B. Callen, *Phys. Rev.* **160**, 358 (1967).

The result is

$$G(m,n;\mathscr{E}) =$$

$$\alpha S_n\Big[P(m-n;\mathscr{E})+\frac{1}{D_s^{(-)}(-\mathscr{E})}[\mathscr{E}(2+\rho+\epsilon-\rho\mathscr{E})P(m-s;\mathscr{E})P(s-n;\mathscr{E})+$$
$$+(\rho\mathscr{E}-1-\epsilon)\{P(m-s;\mathscr{E})\delta_{n,s}+\delta_{m,s}P(s-n;\mathscr{E})\}-$$
$$-\delta_{m,s}\delta_{s,n}(\rho\mathscr{E}-1-\epsilon)P(\mathscr{E})]+\delta_{m,s}\delta_{s,n}\left\{\frac{P(-\mathscr{E})+\rho R(-\mathscr{E})}{D_s^{(-)}(\mathscr{E})}\right\}-$$
$$-\frac{1}{N^2}\sum_{\mathbf{q},\mathbf{q}'}P_{\mathbf{q}}(\mathscr{E})P_{\mathbf{q}'}(\mathscr{E})H(\mathbf{q},\mathbf{q}';\mathscr{E})\exp(-i\mathbf{q}.\mathbf{m}+i\mathbf{q}'.\mathbf{n})\Big], \quad (15.128)$$

where

$$D_s^{(-)}(\mathscr{E}) = \mathscr{E}\{P(-\mathscr{E})+\rho R(-\mathscr{E})\}+(1+\epsilon)\{1+\mathscr{E}P(-\mathscr{E})\}. \quad (15.129)$$

The contribution from the p and d modes is exactly the same as that for the ferromagnetic impurity, because the impurity spin does not participate in these modes. A major difference between the present case and the ferromagnetic impurity is the appearance of two s-like modes, one corresponding to the solutions of

$$\operatorname{Re} D_s^{(-)}(\mathscr{E}_0) = 0 \quad (\text{s}_0\text{-mode}), \quad (15.130)$$

and another corresponding to the solutions of

$$\operatorname{Re} D_s^{(-)}(-\mathscr{E}_1) = 0 \quad (\text{s}_1\text{-mode}). \quad (15.131)$$

There is a solution of first condition for all ϵ and ρ, as shown in Fig. 15.8. This mode, denoted by s_0, cannot decay into the spin wave energy band of the host, i.e. it is a localized mode. (Clearly, $\operatorname{Im} P(-\mathscr{E}) = 0$ for all $\mathscr{E} > 0$.) The s_1 mode, whose energy is determined by the condition (15.131) can be either virtual or localized, but the virtual modes occur only at the top of the spin-wave band. Another point to be noted is that in the limit $J'/J = S'/S = 1$ eqn (15.128) does not go over to the unperturbed Green function, because, even when the magnitudes of the exchange couplings and spins are equal, the spin at the site $m = s$ is still an impurity since it is oppositely aligned to the host spins.

From (15.128)

$$\alpha^{-1}\sum_{m,n}F_m F_n\, G(m,n;\mathscr{E})\exp\{i\boldsymbol{\varkappa}.(\mathbf{m}-\mathbf{n})\}$$
$$= NSF^2(\boldsymbol{\varkappa})P_{\boldsymbol{\varkappa}}(\mathscr{E})-\frac{S'\{F'(\boldsymbol{\varkappa})\}^2}{D_s^{(-)}(\mathscr{E})}\{P(-\mathscr{E})+\rho R(-\mathscr{E})\}+$$
$$+\frac{SF^2(\boldsymbol{\varkappa})}{D_s^{(-)}(-\mathscr{E})}\{\mathscr{E}(2+\rho+\epsilon-\rho\mathscr{E})P_{\boldsymbol{\varkappa}}^2(\mathscr{E})+2(\rho\mathscr{E}-1-\epsilon)P_{\boldsymbol{\varkappa}}(\mathscr{E})-$$
$$-(\rho\mathscr{E}-1-\epsilon)P(\mathscr{E})\}-SF^2(\boldsymbol{\varkappa})H(\boldsymbol{\varkappa},\boldsymbol{\varkappa};\mathscr{E}). \quad (15.132)$$

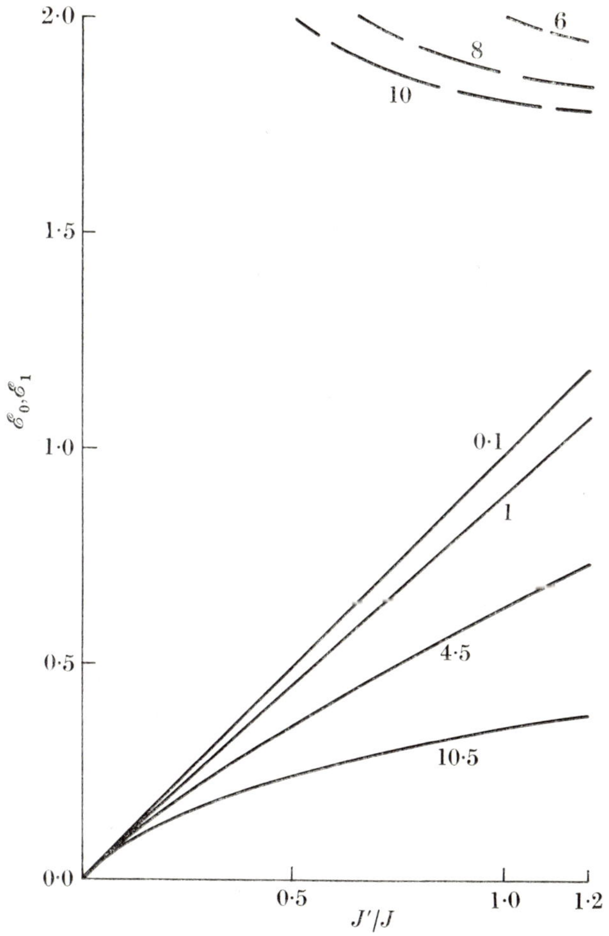

FIG. 15.8. The full curves are the energies of the s_0 mode of the antiferromagnetic impurity. They are given as functions of J'/J and the values of S'/S are given with each curve. The broken curves correspond to the energies of the virtual levels of the s_1 mode.

The first and last terms in this expression are identical to the first two terms in (15.111). The significant fact about the remaining two terms in (15.132) is that one can be attributed to the impurity and the other to the perturbed host spins. This is even more evident when it is noted that

$$G(s,s;\mathscr{E}) = -\frac{\alpha S'\{P(-\mathscr{E})+\rho R(-\mathscr{E})\}}{D_{\mathrm{s}}^{(-)}(\mathscr{E})}. \qquad (15.133)$$

To understand this result fully it is necessary to examine the physical nature of the s_0 and s_1 modes in more detail than we have hitherto.

The s_0 mode is strictly localized (the fact that $G(m, s; \mathscr{E}) \propto \delta_{m,s}$ is indicative of this) and corresponds to a state in which the impurity is precessing in a natural sense while the host spins, driven through the exchange coupling, precess in an unnatural sense. The precessional senses are reversed in the s_1 mode. It is therefore to be expected that the contributions to the neutron cross-section from the s_0 and s_1 modes should appear as coming from the impurity and perturbed host spins respectively, and that the former should give only a diffuse contribution because it cannot propagate in the host. Since neither the p and d modes nor the s_1 modes occur near the bottom of the spin-wave band the additional structure in the neutron cross-section due to the impurity for $\hbar\omega \ll 2rJS$ will be determined by the second term in eqn (15.132). For a small concentration c of impurities this term will contribute, to a very good approximation,

$$\frac{\mathrm{d}^2\sigma^{(\pm)}}{\mathrm{d}\Omega\mathrm{d}E'} = -\frac{Nc}{4rJS}\left(\frac{\gamma e^2}{m_{\mathrm{e}} c^2}\right)^2 (1+\tilde{\kappa}_z^2)\frac{k'}{k} S' F'^2(\boldsymbol{\varkappa})\{n(\omega_0)\pm\tfrac{1}{2}+\tfrac{1}{2}\}\times$$
$$\times\,\delta(\mathscr{E}-\mathscr{E}_0)\{P(-\mathscr{E}_0)+\rho R(-\mathscr{E}_0)\}\bigg/\left\{\frac{\mathrm{d}}{\mathrm{d}\mathscr{E}} D_{\mathrm{s}}^{(-)}(\mathscr{E})\right\}\bigg|_{\mathscr{E}=\mathscr{E}_0}, \qquad (15.134)$$

where $\mathscr{E} = \hbar\omega/2rJS$.

APPENDIX A

THIS appendix is devoted to (1) a more detailed derivation of the cross-section formula (1.10). In particular we justify the choice of phase for the scattering amplitude f, eqn (1.8 b), and clarify the relationship with the formal theory of scattering.† Also (2) we discuss the concept of a complex scattering length in relation to the absorption of thermal neutrons by nuclei.

A.1. The scattering amplitude f

First we discuss the scattering of a neutron with momentum $\mathbf{p} = \hbar\mathbf{k}$ by a target system consisting of a single particle moving in a potential well. This system is described by the Hamiltonian $\mathscr{H}_{\mathrm{T}}$. We choose an arbitrary origin close to the target system with respect to which the position vectors of the neutron and target particle are $\mathbf{r}$ and $\mathbf{R}$ respectively. If the interaction potential between the neutron and the target particle is $\hat{V}(\mathbf{r}, \mathbf{R})$ the Schrödinger equation for the system that consists of the incident neutron and the target system is

$$\left[\mathscr{H}_{\mathrm{T}}+\frac{1}{2m}\hat{\mathbf{p}}^2+\hat{V}(\mathbf{r},\mathbf{R})-\mathscr{E}\right]\Psi(\mathbf{r},\mathbf{R}) = 0. \tag{A.1}$$

In (A.1), $\mathscr{E}$ is the total energy, i.e. the sum of the energy of the incident neutron E and the initial energy of the target.

The total wave function Ψ can be expanded in terms of the complete set of eigenfunctions of the target Hamiltonian $\phi_\lambda(\mathbf{R})$, which satisfy

$$(\mathscr{H}_{\mathrm{T}}-E_\lambda)\phi_\lambda(\mathbf{R}) = 0.$$

If

$$\Psi(\mathbf{r},\mathbf{R}) = \sum_{\lambda'} A_{\lambda'}(\mathbf{r})\phi_{\lambda'}(\mathbf{R}), \tag{A.2}$$

the coefficients $A_{\lambda'}$ are found to satisfy the following inhomogeneous differential equation:

$$(\nabla^2+k_{\lambda'}^2)A_{\lambda'}(\mathbf{r}) = \frac{2m}{\hbar^2}\int \mathrm{d}\mathbf{R}\,\phi_{\lambda'}^*(\mathbf{R})\hat{V}(\mathbf{r},\mathbf{R})\Psi(\mathbf{r},\mathbf{R}), \tag{A.3}$$

in which

$$k_{\lambda'}^2 = \frac{2m}{\hbar^2}(\mathscr{E}-E_{\lambda'}).$$

Eqn (A.3) can be recast as an integral equation with the aid of the Green function $\mathscr{G}_\lambda(\mathbf{r}-\mathbf{r}')$, which satisfies (A.3) when the source term is replaced by $\delta(\mathbf{r}-\mathbf{r}')$. Explicitly,

$$\mathscr{G}_\lambda(\mathbf{r}-\mathbf{r}') = \frac{-1}{4\pi|\mathbf{r}-\mathbf{r}'|}\exp\{\mathrm{i}k_\lambda|\mathbf{r}-\mathbf{r}'|\},$$

and eqn (A.3) can be written

$$A_{\lambda'}(\mathbf{r}) = A_{\lambda'}^{(0)}(\mathbf{r})+\frac{2m}{\hbar^2}\int \mathrm{d}\mathbf{r}'\,\mathscr{G}_{\lambda'}(\mathbf{r}-\mathbf{r}')\int \mathrm{d}\mathbf{R}\,\phi_{\lambda'}^*(\mathbf{R})\hat{V}(\mathbf{r}',\mathbf{R})\Psi(\mathbf{r}',\mathbf{R}). \tag{A.4}$$

Here $A_{\lambda'}^{(0)}(\mathbf{r})$ is a solution of the homogeneous equation and is like $\exp(\mathrm{i}\mathbf{k}_{\lambda'}.\mathbf{r})$, i.e. like $\exp(\mathrm{i}k_{\lambda'}r\tilde{\mathbf{n}}.\tilde{\mathbf{n}}')$ where $\tilde{\mathbf{n}}$ is a unit vector along the direction of the incident plane wave and $\tilde{\mathbf{n}}'$ a unit vector along $\mathbf{r}$.

† A good introduction is found in A. Messiah, *Quantum mechanics*, vol. 2 (North-Holland).

Before we can proceed any further we must establish the boundary conditions on the coefficients A_λ. If the target is in the state ϕ_λ before the scattering process, then, clearly, A_λ must have an asymptotic form that represents the sum of the incident and scattered wave; thus A_λ must behave like

$$A_\lambda \sim \exp(ik_\lambda r\tilde{\mathbf{n}}.\tilde{\mathbf{n}}')+f_{\lambda\lambda}(\tilde{\mathbf{n}},\tilde{\mathbf{n}}')\frac{1}{r}\exp(ik_\lambda r)$$

for $r = |\mathbf{r}|$ much greater than the dimensions of the target. The functions $A_{\lambda'\neq\lambda}$ must represent scattered waves only, and so have the asymptotic form

$$A_{\lambda'\neq\lambda} \sim f_{\lambda\lambda'}(\tilde{\mathbf{n}},\tilde{\mathbf{n}}')\frac{1}{r}\exp(ik_{\lambda'} r).$$

From these expressions we deduce that $r^{-2}|f_{\lambda\lambda'}(\tilde{\mathbf{n}},\tilde{\mathbf{n}}')|^2$ is the number of neutrons per unit volume at a distance $\mathbf{r}$ from the target. Of these, the number crossing unit area per unit time is proportional to $k_{\lambda'}r^{-2}|f_{\lambda\lambda'}|^2$, whereas in the incident beam the number crossing unit area per unit time is proportional to k_λ. Hence, by definition,

$$\left(\frac{\mathrm{d}\sigma}{\mathrm{d}\Omega}\right)_{\lambda\lambda'} = \frac{k_{\lambda'}}{k_\lambda}|f_{\lambda\lambda'}(\tilde{\mathbf{n}},\tilde{\mathbf{n}}')|^2. \tag{A.5}$$

Thus, in view of this discussion, we require solutions of (A.4) for $|\mathbf{r}'| \ll r$. Since $|\mathbf{r}-\mathbf{r}'| \sim \mathbf{r}-\tilde{\mathbf{n}}'.\mathbf{r}'$ the scattering amplitude $f_{\lambda\lambda'}$ is

$$f_{\lambda\lambda'}(\tilde{\mathbf{n}},\tilde{\mathbf{n}}') = -\frac{m}{2\pi\hbar^2}\int \mathrm{d}\mathbf{r}'\exp(-ik_{\lambda'}\tilde{\mathbf{n}}'.\mathbf{r}')\int \mathrm{d}\mathbf{R}\,\phi^*_{\lambda'}(\mathbf{R})\hat{V}(\mathbf{r}',\mathbf{R})\Psi(\mathbf{r}',\mathbf{R}). \tag{A.6}$$

If the perturbation on the incident wave due to scattering is small then to a good approximation $\Psi(\mathbf{r},\mathbf{R})$ can be replaced by the initial wave function of the total system, i.e. $\Psi(\mathbf{r},\mathbf{R}) \simeq \phi_\lambda(\mathbf{R})\exp(ik_\lambda\tilde{\mathbf{n}}.\mathbf{r})$. This approximation is called the Born approximation. Inserting it in the above equation for $f_{\lambda\lambda'}(\tilde{\mathbf{n}},\tilde{\mathbf{n}}')$ we have

$$f_{\lambda\lambda'}(\tilde{\mathbf{n}},\tilde{\mathbf{n}}') = -\frac{m}{2\pi\hbar^2}\int \mathrm{d}\mathbf{r}'\exp(-ik_{\lambda'}\tilde{\mathbf{n}}'.\mathbf{r}')\int \mathrm{d}\mathbf{R}\,\phi^*_{\lambda'}(\mathbf{R})\hat{V}(\mathbf{r}',\mathbf{R})\phi_\lambda(\mathbf{R})\exp(ik_\lambda\tilde{\mathbf{n}}.\mathbf{r}'). \tag{A.7}$$

This expression verifies the choice of phase adopted in Chapter 1.

It is worth while to mention that in the formal theory of scattering† it is usual to define an operator $\hat{T}$, the T-matrix operator, such that when it operates on the initial state of the total system it is equal to $\hat{V}\Psi$. An equation for $\hat{T}$ is easily found which can, in principle, be solved to any order in the interaction potential $\hat{V}$. The first approximation is $\hat{T} \simeq \hat{V}$ and this leads immediately to eqn (A.7) for the scattering amplitude.

The expression given in (A.7) for $f_{\lambda\lambda'}$ has been derived for the case of a single target particle in a potential well, whereas we are for the most part concerned with scattering from an array of such target systems. The amplitude for the latter problem is obtained to an approximation consistent with that used to derive (A.7) if $\hat{V}$ is replaced by the sum of the potentials between the incident neutron and the individual target systems. Clearly this would be an invalid procedure if there was an appreciable probability that, before escaping to a large distance from the target, the incident neutron makes a collision with more than one of the individual target systems in the array. This probability will be very small if the amplitude scattered by each target system is small, which is the condition under which the use of the Born approximation is justified.

If we build the condition of conservation of energy into (A.5), as in Chapter 1,

† Ibid.

then the partial differential cross-section is, from (A.7),

$$\left(\frac{\mathrm{d}^2\sigma}{\mathrm{d}\Omega\,\mathrm{d}E'}\right)_{\lambda\lambda'} = \frac{k'}{k}\left(\frac{m}{2\pi\hbar^2}\right)^2 \left|\left\langle\lambda'\right| \int \mathrm{d}\mathbf{r}\exp(-\mathrm{i}\mathbf{k}'.\mathbf{r})\hat{V}\exp(\mathrm{i}\mathbf{k}.\mathbf{r})\left|\lambda\right\rangle\right|^2 \delta(\hbar\omega+E_\lambda-E_{\lambda'}) \tag{A.8}$$

in agreement with the result quoted in Chapter 1. Because of the delta function in (A.8) we are able to drop the suffix λ' on $k_{\lambda'}$ and simply denote the final energy of the neutron by $E' = \hbar^2k'^2/2m$.

A.2. Complex and energy-dependent scattering lengths

In the text we have for the most part assumed the scattering length b to be independent of the incident neutron energy $E = \hbar^2k^2/2m$. On the whole this is a valid assumption for thermal neutrons, but in some instances the scattering displays a very pronounced energy dependence as is evident from Fig. A.1, which shows the *total* cross-section for lutetium in the range $0{\cdot}01$ eV $\leqslant E \leqslant 1{\cdot}0$ eV. The peak in the cross-section at $E = 0{\cdot}14$ eV is due to resonant absorption of neutrons by the Lu nuclei. Our understanding of this and related processes is based upon the compound theory of nuclei introduced by Bohr in 1936.†

The scattering process we consider is more general than that discussed elsewhere, for we must allow the target nuclei to possess internal structure. Elastic scattering thus refers to processes in which the state of the scattered neutron is identical to that of the incident neutron. Inelastic scattering includes processes in which the neutron is scattered leaving the nucleus in an excited state or when it is captured by the nucleus. For thermal neutrons the dominant inelastic process is radiative capture mentioned above in which neutrons are captured by nuclei which subsequently emit γ-rays.

The ingoing neutron shares its energy among all the nucleons. For a large number of nucleons the probability that the additional energy is concentrated in any one is small. Furthermore any nucleon must overcome reflection at the surface of the nucleus before it can be emitted and for protons there is in addition to this the Coulomb potential barrier. Hence the compound system consisting of the incident neutron plus target nucleus exists for a long period of time and, in consequence, the decay of the compound nucleus is independent of its formation. Also, it follows from the uncertainty principle that the energy width of the state is very small, particularly for large Z nuclei, and hence neutron cross-sections can show resonance behaviour. Measurements show that in general both the width and spacing of resonances decrease as the mass number increases. Magic number nuclei form an exception to this trend. As we have already mentioned, the commonest reaction in the thermal energy range is radiative capture and in fact this is observed to be present to some extent for all nuclei.

It is not appropriate for us to give here anything bordering on a detailed discussion of resonance reactions. Our aim is to demonstrate the connection between a complex scattering amplitude and absorption of neutrons and the form of the elastic and reaction cross-sections in the vicinity of a resonance.

We represent the nuclei by a spherical scattering potential. The standard analysis of scattering by such a potential‡ in terms of partial waves of angular

† For a detailed analysis of neutron resonance reactions see, for example, J. E. Lynn, *Theory of neutron resonance reactions*, O.U.P. (1968).

‡ L. D. Landau and E. M. Lifshitz, *Quantum mechanics*, 2nd edn., Pergamon Press (1965).

momentum l gives the elastic σ_e and reaction σ_r cross-sections as the sum of the components

$$\sigma_e^{(l)} = \frac{\pi}{k^2}(2l+1)|1-\alpha_l|^2 \tag{A.9 a}$$

and

$$\sigma_r^{(l)} = \frac{\pi}{k^2}(2l+1)(1-|\alpha_l|^2), \tag{A.9 b}$$

where α_l is related to the (complex) phase shift η_l by

$$\alpha_l = \exp(2i\eta_l). \tag{A.10}$$

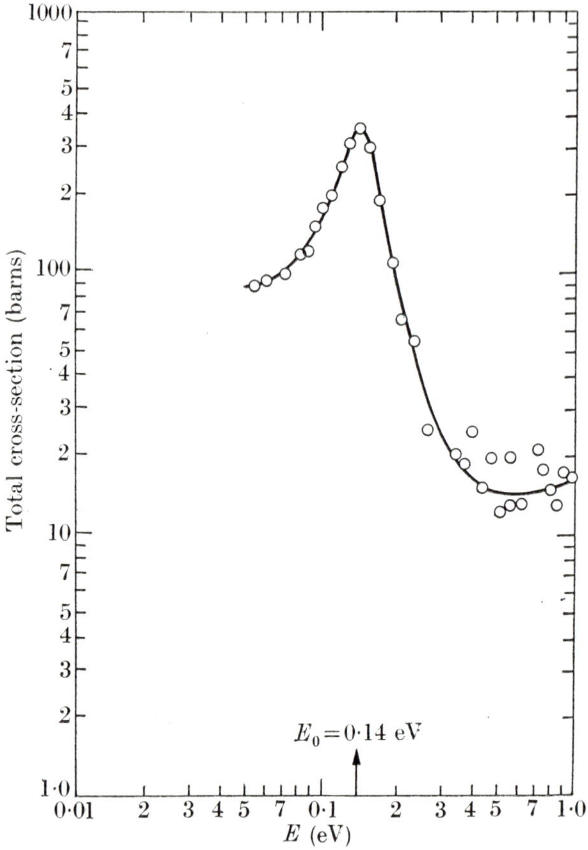

FIG. A.1. Total cross-section for lutetium. The parameters characterizing the resonance are:

$E_0 = 0{\cdot}142 \pm 0{\cdot}001$ eV, $\Gamma_n = 0{\cdot}093 \pm 0{\cdot}009$ meV,
$\Gamma_\gamma = 63 \pm 5$ meV,

The wavelength of the incident thermal neutrons is much greater than both the mean radius of the nucleus and the range of interaction of the nucleon–nucleon force. This permits us to restrict attention to the s-wave ($l = 0$) components of the partial wave expansion. For very small k

$$\alpha_0 \simeq 1+2ikf, \tag{A.11}$$

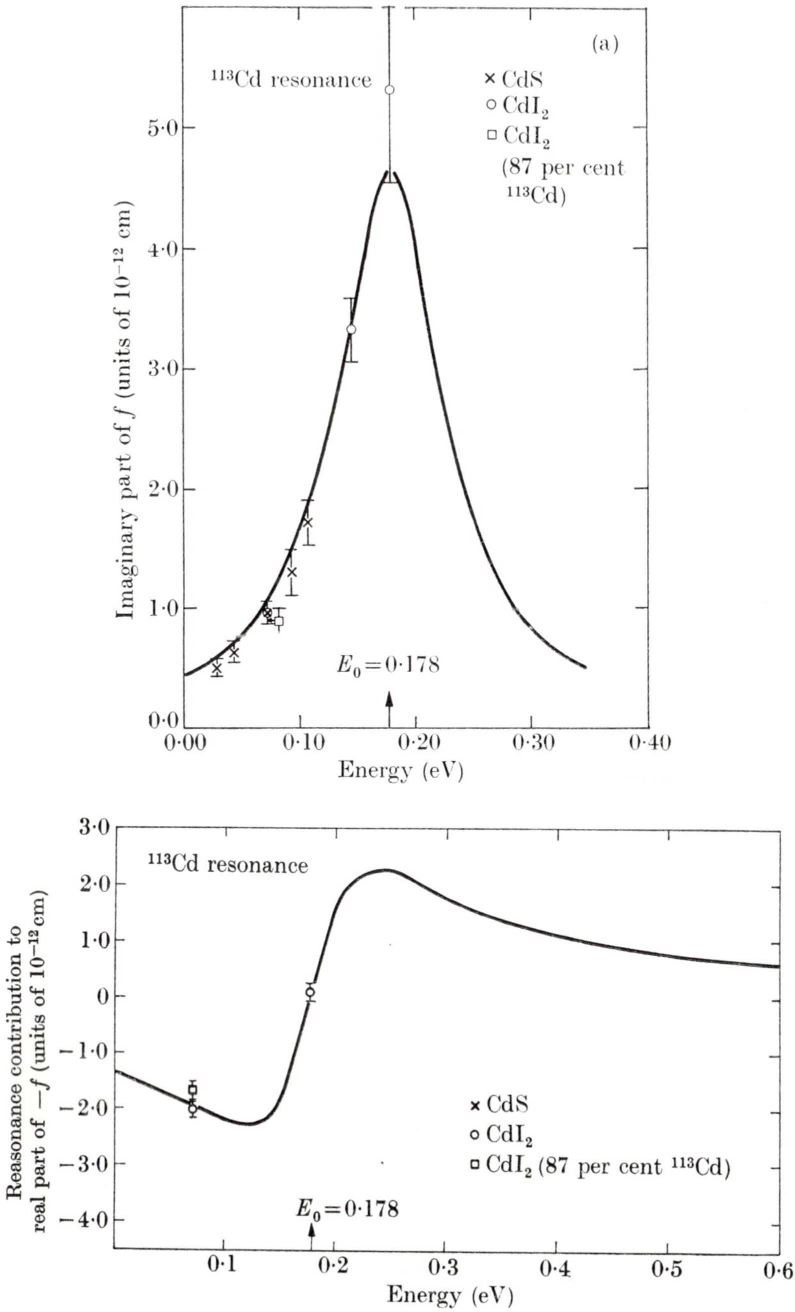

(b)

Fɪɢ. A.2. Energy dependence of the amplitude of neutron scattering from ^{113}Cd. (a) Imaginary component. (b) Real component.

where the scattering amplitude f is complex and independent of k. Thus as $k \to 0$

$$\sigma_e^{(0)} \simeq 4\pi |f|^2 \tag{A.12}$$

and

$$\sigma_r^{(0)} \simeq \frac{2\pi i}{k}(f^* - f) = \frac{4\pi}{k}\,\mathrm{Im}\,f. \tag{A.13}$$

Eqn (A.12) agrees with (1.13 b) if f is replaced by b. It is straightforward to show that the reaction cross-section (A.13) corresponds to absorption of the incident neutrons.

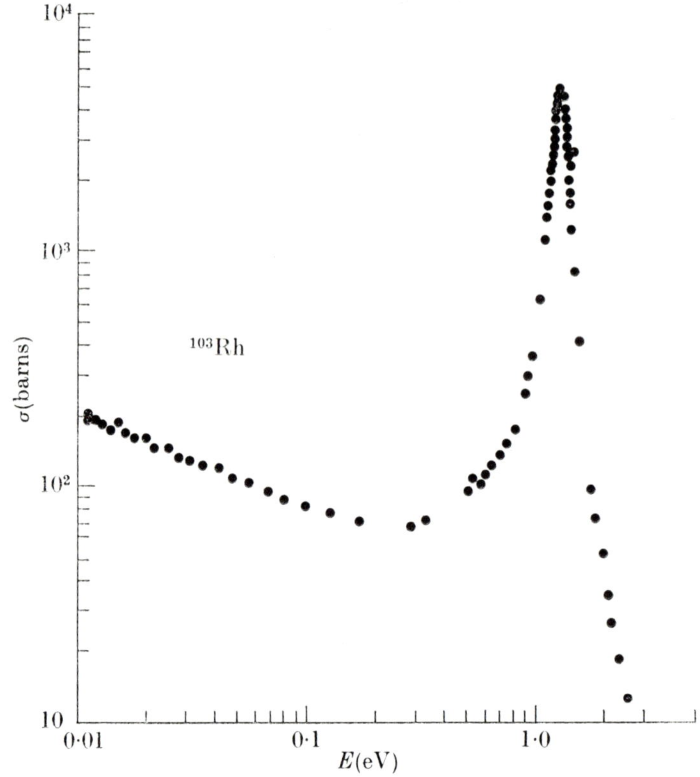

FIG. A.3. Capture cross-section of ^{103}Rh showing the '$1/v$' rise due to the resonance towards low energies.

The s-wave component of the net outward radial flux of neutrons from the spherical scattering potential is

$$\frac{\hbar}{4mr^2k}(|\alpha_0|^2 - 1) \tag{A.14}$$

and this has the value

$$-\frac{\hbar}{mr^2}\,\mathrm{Im}\,f \tag{A.15}$$

in the limit $k \to 0$. Thus the net outward flux is proportional to $\mathrm{Im}\,b$ and is negative, as it must be if absorption occurs.

In the presence of resonance reactions we can no longer take the scattering amplitude to be independent of k. We denote the energy at which the resonance occurs by E_0 and assume the nucleus to have a well-defined radius R. For zero-spin nuclei f is found to have the form

$$f = -R - \frac{\Gamma_n(E-E_0)}{2k\{(E-E_0)^2+\frac{1}{4}\Gamma^2\}} + \frac{\mathrm{i}\Gamma\Gamma_n}{4k\{(E-E_0)^2+\frac{1}{4}\Gamma^2\}}. \tag{A.16}$$

Here $\Gamma_n \propto \sqrt{E}$ is the neutron width and Γ the total width of the resonance. The first term on the right-hand side is simply the amplitude for scattering by an impenetrable sphere of radius R. Note that the two real terms can interfere with one another; their sum for $E < E_0$ may be positive. Fig. A.2 shows the resonance parts of the scattering amplitude for cadmium.† Here $E_0 = 0{\cdot}178$ eV, the radiative capture width $\Gamma_\gamma = 0{\cdot}113$ eV, and $(\Gamma_n/\Gamma_\gamma)_{E=E_0} = 0{\cdot}006$. The smooth curves represent the appropriate forms of the real and imaginary parts of the resonance contribution to the scattering amplitude evaluated with the above set of parameters.

The elastic scattering cross-section is given by substituting (A.16) into (A.9 a) to give

$$\sigma_{\mathrm{e}}^{(0)} = 4\pi R^2 + \frac{\pi\Gamma_n^2+4\pi kR\Gamma_n(E-E_0)}{k^2\{(E-E_0)^2+\frac{1}{4}\Gamma^2\}} \tag{A.17}$$

and near resonance the radiative capture cross-section has the form

$$\sigma_{\mathrm{r}}^{(0)} = \frac{\pi\Gamma_n\Gamma_\gamma}{k^2\{(E-E_0)^2+\frac{1}{4}\Gamma^2\}} \tag{A.18}$$

with $\Gamma = \Gamma_n+\Gamma_\gamma$. This is obtained by substituting (A.16) into (A.9 b).

From the latter and the knowledge that $\Gamma_n \propto k$, it follows that, at energies much below the resonance, $\sigma_{\mathrm{r}}^{(0)} \propto E^{-\frac{1}{2}}$, i.e. the capture cross-section increases as $k \to 0$ as the inverse of the neutron velocity (cf. A.13). This '$1/v$' rise in the capture cross-section is naturally most pronounced when there is a resonance at low energies and is well illustrated by ^{103}Rh shown in Fig. A.3.

† H. A. Smith, and S. W. Peterson, *J. Phys., Paris* **25**, 615 (1964).

APPENDIX B

GREEN FUNCTIONS AND LINEAR RESPONSE THEORY

In this appendix we discuss linear response theory and thermal two-time Green functions. Our objectives are twofold. First, to give a detailed account of the theory which underlies many of our discussions in the text of both nuclear and magnetic scattering. Secondly, to make contact with the various approaches involved in actual calculations. For instance, sometimes it is convenient to work directly in terms of the generalized susceptibility function (see, for example, § 9.2), while for other problems it proves convenient to use two-time Green functions (e.g. § 9.1 and Chapter 15) or relaxation functions (e.g. Chapter 13). As a consequence of these two objectives, particularly the second one, we are obliged to introduce many functions in the following sections, and for this reason alone the appendix may appear difficult to read. However, the theory is of importance and we believe it is essential to present a thorough foundation.†

B.1. The generalized susceptibility, $\chi_{AB}[\omega]$

We consider the behaviour of an isolated system initially in equilibrium, when a small time-dependent perturbation is allowed to act on it. At the time $t = -\infty$ the system is described by the Hamiltonian $\mathscr{H}_0$; at the later time t the total Hamiltonian is

$$\mathscr{H} = \mathscr{H}_0 - \mathscr{H}_1 = \mathscr{H}_0 - \hat{B}f(t). \tag{B.1}$$

The response of the system to the external perturbation $\mathscr{H}_1$ is observed in the change of a variable A, which itself does not depend explicitly on time. Quite generally we can write for a linear response,

$$\overline{A(t)} = \langle \hat{A} \rangle + \int_{-\infty}^{t} \mathrm{d}t'\, \phi_{AB}(t-t')f(t'), \tag{B.2}$$

where $\hat{A}$ is the operator corresponding to the variable A and

$$\langle \hat{A} \rangle = \mathrm{Tr}\,\hat{\rho}_0 \hat{A} = \frac{1}{Z}\,\mathrm{Tr}\exp(-\beta\mathscr{H}_0)\,\hat{A}. \tag{B.3}$$

$\phi_{AB}(t)$ is called a *response function*. It measures the change in A due to a perturbation localized at an earlier time $t = 0$, i.e. when $f(t) \propto \delta(t)$. Clearly $\phi_{AB}(t)$ is purely real for a real perturbation.

Any time-dependent perturbation can, by Fourier transformation, be represented as a set of monochromatic components whose time dependence is given through $\exp(-\mathrm{i}\omega t)$. In such cases $f(t) \propto \exp(-\mathrm{i}\omega t)$. When we derive an explicit expression for $\phi_{AB}(t)$ in terms of the operators $\hat{A}$ and $\hat{B}$ (cf. § B.3) we assume that $\mathscr{H}_1$ describes an adiabatic process, i.e. a process that at every instant of time leaves the system in equilibrium. The condition on $\mathscr{H}_1$ is fulfilled if we make $f(t)$ increase very slowly with increasing t, and this is achieved by replacing $\exp(-\mathrm{i}\omega t)$ by

† Several examples of the use of the theory are discussed by P. C. Martin, *Many-body physics*, Les Houches (1967), ed. C. de Witt and R. Balian (Gordon and Breach). See also G. V. Chester, *Rep. Prog. Phys.* **26**, 411 (1963).

$\exp(-i\omega t+\epsilon t)$, where ϵ is a small positive quantity. For in this instance the perturbation is zero for $t = -\infty$ and it can be made to increase as slowly as required through the choice of ϵ; we shall take the limit of $\epsilon \to 0^+$. Thus, taking $f(t) = f\exp(\epsilon t)\cos\omega t$, where f is a real constant, we obtain from (B.2)

$$\overline{A(t)} = \langle\hat{A}\rangle + f\,\mathrm{Re}\{\exp(-i\omega t)\chi_{AB}[\omega]\}, \tag{B.4}$$

where
$$\chi_{AB}[\omega] = \lim_{\epsilon\to 0^+}\int_0^\infty dt\,\phi_{AB}(t)\exp(i\omega t-\epsilon t) \tag{B.5}$$

defines the *generalized susceptibility function* $\chi_{AB}[\omega]$.

(Note that we define the one-sided Fourier transform in (B.5) by $\chi_{AB}[\omega]$. We reserve the notation of round brackets about the argument ω for a conventional Fourier transform.)

$\chi_{AB}[\omega]$ embodies the details of the dynamical properties of the system described by $\mathscr{H}_0$, since if $\hbar\omega$ coincides with the energy of an excitation in the system χ_{AB} must peak at this energy. Thus, in principle, the energy spectrum of the excitations can be obtained by examining the poles of the generalized susceptibility function.

Let us consider more closely the structure of the response function $\phi_{AB}(t)$. From both (B.2) and (B.5) it is observed that $\phi_{AB}(t)$ is required only for $t > 0$. This is simply a consequence of the fact that a response to a perturbation cannot be observed before the latter has been applied. It is therefore usual to call $\phi_{AB}(t)$ a *causal* function. Nonetheless it proves convenient in subsequent discussions to have $\phi_{AB}(t)$ defined for all values of t. This is most naturally achieved by using the explicit expression for $\phi_{AB}(t)$ derived in § B.3, namely

$$\phi_{AB}(t-t') = \frac{i}{\hbar}\langle[\hat{A}(t), \hat{B}(t')]\rangle. \tag{B.24 b}$$

B.2. Analytic properties of χ_{AB}

$\chi_{AB}[\omega]$ is, in general, a complex function, which we write

$$\chi_{AB} = \chi'_{AB} + i\chi''_{AB}. \tag{B.6}$$

Since $\phi_{AB}(t)$ must be real for a real external perturbation it follows from (B.5) that

$$\chi'[\omega] = \chi'[-\omega], \qquad \chi''[\omega] = -\chi''[-\omega]. \tag{B.7}$$

For the present discussion we omit the suffixes AB on both the generalized susceptibility and the response function.

χ' and χ'' satisfy certain relationships: in particular they are conjugate functions, which can be derived as follows. We define

$$\chi'(t) = \int_{-\infty}^{\infty} d\omega\,\chi'[\omega]\exp(-i\omega t), \tag{B.8}$$

which, from (B.5),

$$\begin{aligned} &= \int_{-\infty}^{\infty} d\omega\exp(-i\omega t)\int_0^\infty dt'\phi(t')\cos(\omega t') \\ &= \pi\phi(t), \quad \text{if } t > 0, \\ &= \pi\phi(-t), \quad \text{if } t < 0. \end{aligned} \tag{B.9}$$

Also,
$$\chi''(t) = \int_{-\infty}^{\infty} d\omega \exp(-i\omega t) \int_0^{\infty} dt' \phi(t')\sin(\omega t')$$
$$= -i\chi'(t)\operatorname{sgn} t. \tag{B.10}$$

Note that $\chi'(t)$ is purely real and that $\chi''(t)$ is purely imaginary.

The result (B.10) can be used to derive integral relationships between the real and imaginary parts of the generalized susceptibility function.

On utilizing the result
$$\operatorname{sgn} t = 2\theta(t) - 1, \tag{B.11}$$
where $\theta(t)$ is the unit step function, which we can express in the form
$$\theta(t) = \frac{i}{2\pi} \int_{-\infty}^{\infty} \frac{du \exp(-iut)}{u + i0^+} \tag{B.12}$$
and the integral identity
$$\int \frac{du}{\omega - u + i0^+} = P \int \frac{du}{\omega - u} - i\pi \int du\, \delta(\omega - u), \tag{B.13}$$
we obtain from (B.10),
$$\chi''[\omega] = \frac{1}{2\pi} \int_{-\infty}^{\infty} dt\, \{-i\chi'(t)\operatorname{sgn} t\}\exp(i\omega t)$$
$$= \frac{-i}{2\pi}\left\{-2\pi\chi'[\omega] + 2\left(\frac{i}{2\pi}\right) \int_{-\infty}^{\infty} \frac{du}{u + i0^+} \int_{-\infty}^{\infty} dt\, \chi'(t)\exp(i\omega t - iut)\right\}$$
$$= \frac{1}{\pi} P \int_{-\infty}^{\infty} \frac{du\, \chi'[u]}{\omega - u}. \tag{B.14 a}$$

Conversely,
$$\chi'[\omega] = -\frac{1}{\pi} P \int_{-\infty}^{\infty} \frac{du\, \chi''[u]}{\omega - u}. \tag{B.14 b}$$

Furthermore,
$$\int_{-\infty}^{\infty} d\omega\, |\chi'[\omega]|^2 = \frac{1}{2\pi} \int_{-\infty}^{\infty} dt\, |\chi'(t)|^2 = \int_{-\infty}^{\infty} d\omega\, |\chi''[\omega]|^2$$
$$= \frac{1}{2\pi} \int_{-\infty}^{\infty} dt\, |\chi''(t)|^2. \tag{B.15}$$

(B.14 a) and (B.14 b) are usually called the Kramers–Krönig relationships, and, as shown below, they can be regarded as a consequence of causality. If we formally regard $\chi[\omega]$ to be the limit of a function $\chi[z]$ of the complex variable $z = \omega + iy$, then (B.14 a) and (B.14 b) form one of the conditions that $\chi[z]$ should be analytic in the upper half plane. The remaining condition is that $\chi[z]$ should be square-integrable.† The derivation of the properties of $\chi[\omega]$ from arguments on the analytic properties of $\chi[z]$ is given, for example, by Landau and Lifshitz.‡

† E. C. Titchmarsh, *Introduction to the theory of Fourier integrals*, Clarendon Press, Oxford (1948).

‡ L. D. Landau and E. M. Lifshitz, *Statistical physics*, Pergamon Press (1959).

Eqns (B.9) and (B.10) together give

$$\chi(t) = \chi'(t)+\mathrm{i}\chi''(t) = 2\pi\phi(t)\theta(t), \tag{B.16}$$

that is to say, the time-dependent susceptibility function $\chi(t)$ is zero for $t < 0$; this is simply the condition of causality for the system. It follows from the definitions of $\chi'(t)$ and $\chi''(t)$ that (B.16) leads to

$$\chi[\omega] = \mathrm{i}\int_{-\infty}^{\infty} \frac{\mathrm{d}u\,\Phi(u)}{\omega-u+\mathrm{i}0^+}, \tag{B.17}$$

where

$$\Phi(\omega) = \frac{1}{2\pi}\int_{-\infty}^{\infty} \mathrm{d}t\,\phi(t)\exp(\mathrm{i}\omega t)$$

$$= \Phi'(\omega)+\mathrm{i}\Phi''(\omega) = \Phi^*(-\omega). \tag{B.18}$$

If $\phi(t)$ is an even function of t, then

$$\chi'[\omega] = \pi\Phi'(\omega) = \tfrac{1}{2}\int_{-\infty}^{\infty} \mathrm{d}t\,\phi(t)\cos(\omega t)$$

and

$$\Phi''(\omega) = 0, \tag{B.19}$$

while, if $\phi(t)$ is an odd function of t,

$$\chi''[\omega] = \pi\Phi''(\omega) = \tfrac{1}{2}\int_{-\infty}^{\infty} \mathrm{d}t\,\phi(t)\sin(\omega t)$$

and

$$\Phi'(\omega) = 0. \tag{B.20}$$

Also note that since

$$\phi(t) = \frac{1}{\pi}\int_{-\infty}^{\infty} \mathrm{d}\omega\exp(-\mathrm{i}\omega t)\chi'[\omega], \quad \text{if } \phi(t) \text{ is even in } t$$

and

$$\phi(t) = \frac{\mathrm{i}}{\pi}\int_{-\infty}^{\infty} \mathrm{d}\omega\exp(-\mathrm{i}\omega t)\chi''[\omega], \quad \text{if } \phi(t) \text{ is odd in } t,$$

it follows that

$$\frac{1}{\pi}\int_{-\infty}^{\infty} \mathrm{d}\omega\,\omega^{2n}\chi'[\omega] = (-)^n \frac{\partial^{2n}}{\partial t^{2n}}\phi(t)\bigg|_{t=0}, \quad \text{if } \phi(t) \text{ is even in } t \tag{B.21}$$

and

$$\frac{1}{\pi}\int_{-\infty}^{\infty} \mathrm{d}\omega\,\omega^{2n+1}\chi''[\omega] = (-)^n \frac{\partial^{2n+1}}{\partial t^{2n+1}}\phi(t)\bigg|_{t=0}, \quad \text{if } \phi(t) \text{ is odd in } t.$$

B.3. Calculation of the response function, $\phi_{AB}(t)$

The external perturbation $\mathscr{H}_1$ is defined such that its average value is very small compared to that of $\mathscr{H}_0$ and such that the density matrix $\hat{\rho}$ at time t is determined by

$$\mathrm{i}\hbar\dot{\hat{\rho}}(t) = [\mathscr{H},\hat{\rho}] \tag{B.22}$$

with the boundary condition $\hat{\rho}(-\infty) = \hat{\rho}_0$.

The condition that $\mathscr{H}_1$ must satisfy in order that the equation of motion (B.22) can be used to calculate $\hat{\rho}$ is that at every instant $-\infty < t$ the system is in equilibrium at time t. This (adiabatic) process is usually achieved by taking $\mathscr{H}_1$ to have an overriding time dependence of the form $\exp(\epsilon t)$, $\epsilon > 0$, for $t < 0$, but we can also admit a perturbation that is applied instantaneously, provided it is such that the system as a whole is in an equilibrium state immediately following its application.

With $\hat{\rho}(t) = \hat{\rho}_0 + \Delta\hat{\rho}(t)$, and neglecting second-order quantities, we obtain from (B.22)

$$i\hbar\Delta\dot{\hat{\rho}} = [-\mathscr{H}_1, \hat{\rho}_0] - [\Delta\hat{\rho}, \mathscr{H}_0].$$

If $\Delta\hat{\rho}(t) = \exp(-it\mathscr{H}_0/\hbar)\delta\hat{\rho}(t)\exp(it\mathscr{H}_0/\hbar)$, then $\delta\hat{\rho}(t)$ satisfies

$$i\hbar\,\delta\dot{\hat{\rho}} = \exp(it\mathscr{H}_0/\hbar)[-\mathscr{H}_1, \hat{\rho}_0]\exp(-it\mathscr{H}_0/\hbar),$$

and hence,
$$\Delta\hat{\rho}(t) = \frac{1}{i\hbar}\int_{-\infty}^{\infty} dt'\, f(t')[\hat{\rho}_0, \hat{B}(t'-t)]. \tag{B.23}$$

We then have, from (B.23),

$$\overline{A(t)} - \langle\hat{A}\rangle = \frac{1}{i\hbar}\int_{-\infty}^{t} dt' f(t')\mathrm{Tr}\{[\hat{\rho}_0, \hat{B}(0)]\hat{A}(t-t')\},$$

so that from the definition of $\phi_{AB}(t-t')$, eqn (B.2),

$$\phi_{AB}(t-t') = \frac{1}{i\hbar}\mathrm{Tr}\{[\hat{\rho}_0, \hat{B}(0)]\hat{A}(t-t')\} \tag{B.24 a}$$

$$= \frac{i}{\hbar}\langle[\hat{A}(t), \hat{B}(t')]\rangle. \tag{B.24 b}$$

With the aid of the identity

$$[\hat{\rho}_0, \hat{B}] = \hat{\rho}_0\int_0^{\beta} d\lambda\, \exp(\lambda\mathscr{H}_0)[\hat{B}, \mathscr{H}_0]\exp(-\lambda\mathscr{H}_0)$$

$$= (i\hbar)\hat{\rho}_0\int_0^{\beta} d\lambda\, \dot{\hat{B}}(-i\hbar\lambda),$$

we can express ϕ_{AB} in the form introduced by Kubo,†

$$\phi_{AB}(t) = \int_0^{\beta} d\lambda\, \langle\dot{\hat{B}}(-i\hbar\lambda)\hat{A}(t)\rangle. \tag{B.25}$$

B.4. The relaxation function, $R_{AB}(t)$

If $f(t) = f\exp(\epsilon t)$ for $t < 0$ and is zero for $t > 0$, then the response for $t > 0$ will relax according to

$$\bar{A} - \langle\hat{A}\rangle = fR_{AB}(t),$$

where
$$R_{AB}(t) = \lim_{\epsilon\to 0^+}\int_t^{\infty} dt'\,\phi_{AB}(t')\exp(-\epsilon t') \tag{B.26}$$

is called, following Kubo,‡ the *relaxation function*. With the aid of the obvious relationship

$$\dot{R}_{AB}(t) = -\phi_{AB}(t), \tag{B.27}$$

we obtain from (B.5)

$$\chi_{AB}[\omega] = R_{AB}(t=0) + i\omega\lim_{\epsilon\to 0^+}\int_0^{\infty} dt\, R_{AB}(t)\exp(i\omega t - \epsilon t). \tag{B.28}$$

† R. Kubo, *J. phys. Soc. Japan* **12**, 570 (1957).
‡ Ibid.

It is convenient to use eqn (B.27) to extend the definition of $R_{AB}(t)$ to negative times in a way appropriate to the extension of $\phi_{AB}(t)$ (cf. § B.1).

A very useful representation of $R_{AB}(t)$ can be derived with the aid of (B.25), namely

$$R_{AB}(t) = \int_0^\beta \mathrm{d}\lambda\, \langle \hat{B}(-\mathrm{i}\hbar\lambda)\hat{A}(t)\rangle - \frac{\beta}{Z}\sum_m \exp(-\beta E_m)A_{mm}B_{mm}. \tag{B.29}$$

In (B.29) the *exact* eigenfunctions of $\mathscr{H}_0$ are $\{\psi_m\}$ and its eigenvalues E_m, and

$$A_{mm} \equiv (\psi_m^*, \hat{A}\psi_m).$$

The second term on the right-hand side of (B.29) ensures that

$$\lim_{t\to\infty} R_{AB}(t) = 0. \tag{B.30}$$

To see this we express the first term of (B.29) in the representation in which $\mathscr{H}_0$ is diagonal,

$$\int_0^\beta \mathrm{d}\lambda\, \frac{1}{Z}\sum_{m,n} \exp\{\lambda(E_m-E_n)-\beta E_m\}B_{mn}A_{nm}\exp\{\mathrm{i}t(E_n-E_m)/\hbar\}. \tag{B.31}$$

In the limit $t\to\infty$ the terms $\exp\{\mathrm{i}t(E_n-E_m)/\hbar\}$, with $E_n \neq E_m$, oscillate rapidly in value. Thus only those terms for which $E_n = E_m$ give a non-zero contribution and this is equal in magnitude to the second term on the right-hand side of (B.29).

From the representation of $R_{AB}(t)$ given by eqn (B.29) we can show that

$$R_{AB}(t) = R_{BA}(-t). \tag{B.32}$$

To prove this relation, we need consider only the first term on the right-hand side of (B.29). Using the invariance of the trace under a cyclic permutation of the operators,

$$\begin{aligned}
&\int_0^\beta \mathrm{d}\lambda\, \frac{1}{Z}\mathrm{Tr}\{\exp(-\beta\mathscr{H}_0)\exp(\lambda\mathscr{H}_0)\hat{B}\exp(-\lambda\mathscr{H}_0)\hat{A}(t)\} \\
&\quad= \int_0^\beta \mathrm{d}\lambda\, \frac{1}{Z}\mathrm{Tr}[\exp(-\beta\mathscr{H}_0)\exp\{\mathscr{H}_0(\beta-\lambda)\}\hat{A}\exp\{-\mathscr{H}_0(\beta-\lambda)\}\hat{B}(-t)] \\
&\quad= \int_\beta^0 (-\mathrm{d}\lambda)\, \frac{1}{Z}\mathrm{Tr}\{\exp(-\beta\mathscr{H}_0)\exp(\lambda\mathscr{H}_0)\hat{A}\exp(-\lambda\mathscr{H}_0)\hat{B}(-t)\} \\
&\quad= \int_0^\beta \mathrm{d}\lambda\, \langle \hat{A}(-\mathrm{i}\hbar\lambda)\hat{B}(-t)\rangle,
\end{aligned}$$

from which (B.32) follows immediately.

It follows from (B.27) that if the response function $\phi_{AB}(t)$ is an odd function of t then $R_{AB}(t)$ is an even function of t and vice versa. Hence (B.32) tells us that when $\phi_{AB}(t)$ is an odd function of t, $R_{AB}(t) = R_{BA}(t) = R_{AB}(-t) = R_{BA}(-t)$.

B.5. Correlation functions and symmetry relationships

Consider two arbitrary operators $\hat{O}$ and $\hat{P}$ and define the *spectral functions* $G^{<}_{OP}(\omega)$ and $G^{>}_{OP}(\omega)$ such that

$$\langle\hat{P}(0)\hat{O}(t)\rangle = \int_{-\infty}^{\infty} \mathrm{d}\omega\, G^{<}_{OP}(\omega)\exp(-\mathrm{i}\omega t) \tag{B.33 a}$$

and

$$\langle\hat{O}(t)\hat{P}(0)\rangle = \int_{-\infty}^{\infty} \mathrm{d}\omega\, G^{>}_{OP}(\omega)\exp(-\mathrm{i}\omega t). \tag{B.33 b}$$

Since

$$\langle\hat{O}(t)\hat{P}(0)\rangle = \langle\hat{O}(0)\hat{P}(-t)\rangle = \langle\hat{P}(0)\hat{O}(t+\mathrm{i}\hbar\beta)\rangle, \tag{B.34}$$

it follows that

$$G^{>}_{OP}(\omega) = \exp(\hbar\omega\beta)G^{<}_{OP}(\omega). \tag{B.35}$$

The second identity in (B.34) is readily verified by using the invariance of the trace of a product of operators to their cyclic permutation.

It is also convenient to define a function $g_{OP}(t)$ as

$$g_{OP}(t) = \langle\hat{O}(t)\hat{P}(0)-\eta\hat{P}(0)\hat{O}(t)\rangle, \tag{B.36}$$

where $\eta = \pm 1$. The *spectral weight function* $g_{OP}(\omega)$ is then given by

$$g_{OP}(\omega) = \frac{1}{2\pi}\int_{-\infty}^{\infty} \mathrm{d}t\, g_{OP}(t)\exp(\mathrm{i}\omega t) = G^{<}_{OP}(\omega)\{\exp(\hbar\omega\beta)-\eta\}. \tag{B.37}$$

$g_{OP}(\omega)$ satisfies, in general, a chain of *sum rules*. From its definition,

$$\int_{-\infty}^{\infty} \mathrm{d}\omega\, g_{OP}(\omega) = \langle\hat{O}\hat{P}-\eta\hat{P}\hat{O}\rangle$$

$$\equiv \langle[\hat{O},\hat{P}]_{\pm}\rangle. \tag{B.38 a}$$

Furthermore,

$$\int_{-\infty}^{\infty} \mathrm{d}\omega\, \omega g_{OP}(\omega) = \mathrm{i}\frac{\partial}{\partial t}\int_{-\infty}^{\infty} \mathrm{d}\omega \exp(-\mathrm{i}\omega t)g_{OP}(\omega)\Big|_{t=0}$$

$$= \frac{1}{\hbar}\langle[[\hat{O},\mathscr{H}_0],\hat{P}]_{\pm}\rangle \tag{B.38 b}$$

and

$$\int_{-\infty}^{\infty} \mathrm{d}\omega\, \omega^2 g_{OP}(\omega) = \frac{1}{\hbar^2}\langle[[[\hat{O},\mathscr{H}_0],\mathscr{H}_0],\hat{P}]_{\pm}\rangle, \tag{B.38 c}$$

etc.

From the representation of the relaxation function given by eqn (B.29) we can readily derive its Fourier transform in terms of $G^{<}_{AB}(\omega)$. Since

$$\langle\hat{B}(-\mathrm{i}\hbar\lambda)\hat{A}(t)\rangle = \int_{-\infty}^{\infty} \mathrm{d}\omega\, G^{<}_{AB}(\omega)\exp(-\mathrm{i}\omega t+\lambda\hbar\omega),$$

it follows from (B.29) that

$$R_{AB}(t) = \int_{-\infty}^{\infty} d\omega \left\{\frac{\exp(\hbar\omega\beta)-1}{\hbar\omega}\right\}G^{<}_{AB}(\omega)\exp(-\mathrm{i}\omega t)-\beta\frac{1}{Z}\sum_m \exp(-\beta E_m)A_{mm}B_{mm}. \tag{B.39 a}$$

Thus

$$R_{AB}(\omega) = \frac{1}{2\pi}\int_{-\infty}^{\infty} \mathrm{d}t \exp(\mathrm{i}\omega t) R_{AB}(t)$$

$$= \left\{\frac{\exp(\hbar\omega\beta)-1}{\hbar\omega}\right\} G^{<}_{AB}(\omega) - \delta(\omega)\beta\frac{1}{Z}\sum_{m}\exp(-\beta E_m)A_{mm}B_{mm}, \quad \text{(B.39 b)}$$

or

$$\frac{\hbar\omega}{\exp(\hbar\omega\beta)-1} R_{AB}(\omega) = G^{<}_{AB}(\omega) - \delta(\omega)\frac{1}{Z}\sum_{m}\exp(-\beta E_m)A_{mm}B_{mm}. \quad \text{(B.40 a)}$$

Note that from (B.32) we have the general relationship

$$R_{AB}(\omega) = R_{BA}(-\omega) \quad \text{(B.40 b)}$$

and from (B.39 a) and (B.32)

$$R^{*}_{AB}(\omega) = R_{BA}(\omega). \quad \text{(B.40 c)}$$

Consider also the function

$$\lim_{\epsilon\to 0+}\int_{0}^{\infty}\mathrm{d}t \exp(\mathrm{i}\omega t - \epsilon t) R_{AB}(t)$$

that occurs on the right-hand side of (B.28). From the identity

$$\mathrm{Re}\lim_{\epsilon\to 0+}\int_{0}^{\infty}\mathrm{d}t\exp(\mathrm{i}xt-\epsilon t) = \lim_{\epsilon\to 0+}\frac{\epsilon}{x^2+\epsilon^2}$$

$$= \pi\delta(x)$$

we obtain from (B.39)

$$\mathrm{Re}\lim_{\epsilon\to 0+}\int_{0}^{\infty}\mathrm{d}t\exp(\mathrm{i}\omega t-\epsilon t)R_{AB}(t) = \pi R_{AB}(\omega).$$

Thus it follows from (B.28) that when $R_{AB}(t=0)$ is real, as is often the case (see below), and $\phi_{AB}(t)$ is odd in t,

$$\chi''_{AB}[\omega] = \pi\omega R_{AB}(\omega). \quad \text{(B.40 d)}$$

We now derive very general symmetry relations for $G^{<}_{OP}$ and $G^{>}_{OP}$. If, as in § B.4, $(\mathscr{H}_0 - E_m)\psi_m = 0$, then

$$[G^{<}_{OP}(\omega)]^{*} = \frac{\hbar}{Z}\sum_{mn}(P_{mn})^{*}(O_{nm})^{*}\exp(-\beta E_m)\delta(\hbar\omega+E_n-E_m)$$

$$= \frac{\hbar}{Z}\sum_{mn}(\hat{P}^{+})_{nm}(\hat{O}^{+})_{mn}\exp(-\beta E_m)\delta(\hbar\omega+E_n-E_m)$$

$$= G^{<}_{P^{+}O^{+}}(\omega), \quad \text{(B.41 a)}$$

and, also,

$$G^{<}_{OP}(-\omega) = \exp(\hbar\omega\beta)G^{<}_{PO}(\omega) = G^{>}_{PO}(\omega). \quad \text{(B.41 b)}$$

Thus, from (B.41 a),

$$\{G^{<}_{OP}(-\omega)\}^{*} = G^{>}_{O^{+}P^{+}}(\omega), \quad \text{(B.42 a)}$$

which could have alternatively been derived from the identity

$$\langle\hat{P}(0)\hat{O}(t)\rangle^{*} = \langle\hat{O}^{+}(t)\hat{P}^{+}(0)\rangle. \quad \text{(B.42 b)}$$

From (B.37) and (B.41) we obtain for the spectral weight function

$$g^*_{OP}(\omega) = g_{P^+O^+}(\omega) \tag{B.43}$$

and

$$g_{OP}(-\omega) = -\eta g_{PO}(\omega). \tag{B.44}$$

If $\hat{P} = \hat{O}^+$, then, from (B.41 a), it follows that $G^{\lessgtr}_{OP}$ and therefore $g_{OP}(\omega)$, through (B.43), is purely real.

Let us now consider what the above relationships imply for the operators $\hat{A}$ and $\hat{B}$, which are, by definition, Hermitian. From (B.41) we have ($\eta = 1$)

$$g^*_{AB}(t) = -g_{AB}(t), \tag{B.45}$$

so that $g_{AB}(t)$ is purely imaginary, and therefore, from (B.24 b) and (B.36), $\phi_{AB}(t)$ is purely real. (B.43) and (B.44) together give

$$g^*_{AB}(\omega) = -g_{AB}(-\omega) \tag{B.46}$$

and from (B.17),

$$\chi_{AB}[\omega] = -\frac{1}{\hbar}\int\limits_{-\infty}^{\infty} du\, \frac{g_{AB}(u)}{\omega - u + i0^+}, \tag{B.47}$$

so that (B.46) is equivalent to (B.7).

So far we have not imposed any restrictions on the form of the two Hermitian operators $\hat{A}$ and $\hat{B}$. Let $\hat{A}$ and $\hat{B}$ be unchanged under time reversal; the operators $\hat{A}$ and $\hat{B}$ are then purely real.† In addition we assume the body to have no magnetic structure and not to be subject to an external magnetic field, so the wave functions of the stationary states are also real. Hence

$$(\psi^*_n, \hat{A}\psi_m) = (\psi^*_n, \hat{A}^+\psi_m) = (\psi_m, \hat{A}^*\psi^*_n) = (\psi_m, \hat{A}\psi^*_n), \tag{B.48}$$

so

$$\langle \hat{B}(0)\hat{A}(t)\rangle = \langle \hat{A}(0)\hat{B}(t)\rangle \tag{B.49 a}$$

and

$$G^{\lessgtr}_{AB}(\omega) = G^{\lessgtr}_{BA}(\omega), \tag{B.49 b}$$

from which it follows that ($\eta = 1$)

$$\begin{aligned} g_{AB}(-t) &= \langle \hat{A}(0)\hat{B}(t) - \hat{B}(t)\hat{A}(0)\rangle \\ &= \langle \hat{B}(0)\hat{A}(t) - \hat{A}(t)\hat{B}(0)\rangle \\ &= -g_{AB}(t), \end{aligned} \tag{B.49 c}$$

or

$$\phi_{AB}(t) = -\phi_{AB}(-t). \tag{B.50 a}$$

Also, we know that when $\phi_{AB}(t)$ is an odd function of t, $R_{AB}(t)$ is an even function of t. Whence when (B.49) is valid we have from (B.32)

$$R_{AB}(t) = R_{BA}(t) \tag{B.50 b}$$

and

$$R_{AB}(\omega) = R_{AB}(-\omega), \tag{B.50 c}$$

with both $R_{AB}(t)$ and $R_{AB}(\omega)$ purely real.

In the presence of a magnetic field $\mathbf{H}$ the matrix elements of $\hat{A}$, say, satisfy

$$\{\hat{A}(\mathbf{H})\}_{mn} = \{\hat{A}(-\mathbf{H})\}_{nm};$$

so in place of (B.49 a) we have

$$\langle \hat{B}(0)\hat{A}(t)\rangle_{\mathbf{H}} = \langle \hat{A}(0)\hat{B}(t)\rangle_{-\mathbf{H}}, \tag{B.51}$$

and an analogous relationship holds for the transformation properties of the response function.

† See, for example, A. Messiah, *Quantum mechanics*, vol. 2, North-Holland (1962).

Finally, we assume that $\hat{A}$, say, changes sign under time reversal; then $\hat{A}$ must be purely imaginary.† Hence

$$(\psi_n^*, \hat{A}\psi_m) = (\psi_n^*, \hat{A}^+\psi_m) = (\psi_m, \hat{A}^*\psi_n^*) = -(\psi_m, \hat{A}\psi_n^*),$$

i.e.

$$A_{nm} = -A_{mn},$$

and it follows that

$$\langle \hat{B}(0)\hat{A}(t)\rangle = -\langle \hat{A}(0)\hat{B}(t)\rangle \tag{B.52 a}$$

and

$$G^{<}_{AB}(\omega) = -G^{<}_{BA}(\omega), \tag{B.52 b}$$

with the result that $\phi_{AB}(t)$ must be an even function of t.

If both $\hat{A}$ and $\hat{B}$ change sign under time reversal, then the results (B.49) and (B.50) apply.

The extension of (B.52 a) to the case where an external magnetic field is present is straightforward since, from (B.51), we have

$$\langle \hat{B}(0)\hat{A}(t)\rangle_{\mathbf{H}} = -\langle \hat{A}(0)\hat{B}(t)\rangle_{-\mathbf{H}}. \tag{B.52 c}$$

B.6. Fluctuation–dissipation theorem

This important theorem relates the fluctuation properties of the system in thermal equilibrium to the response of the system to an external perturbation.‡ From the preceding section it is easy to derive the relation

$$\pi R_{AB}(\omega)\hbar\omega = \tanh(\tfrac{1}{2}\hbar\omega\beta)\int_{-\infty}^{\infty} \mathrm{d}t \exp(\mathrm{i}\omega t)\langle\{\hat{A}(t)\hat{B}(0)\}\rangle, \tag{B.53}$$

where the integrand on the right-hand side contains the symmetrized correlation function

$$\langle\{\hat{A}(t)\hat{B}(0)\}\rangle = \tfrac{1}{2}\langle \hat{A}(t)\hat{B}(0)+\hat{B}(0)\hat{A}(t)\rangle. \tag{B.54}$$

One representation of the theorem is readily derived from (B.53) through (B.19) and (B.20). If ϕ_{AB} is an even function of t,

$$\langle\{\hat{A}(t)\hat{B}(0)\}\rangle = \frac{\hbar}{2\pi\mathrm{i}}\int_{-\infty}^{\infty} d\omega \exp(-\mathrm{i}\omega t)\coth(\tfrac{1}{2}\hbar\omega\beta)\,.\,\chi'_{AB}[\omega], \tag{B.55}$$

while if $\phi_{AB}(t)$ is an odd function of t,

$$\langle\{\hat{A}(t)\hat{B}(0)\}\rangle = \frac{\hbar}{2\pi}\int_{-\infty}^{\infty} d\omega \exp(-\mathrm{i}\omega t)\coth(\tfrac{1}{2}\hbar\omega\beta)\,.\,\chi''_{AB}[\omega]. \tag{B.56}$$

The fluctuation–dissipation theorem as expressed by (B.55) or (B.56) is abstract and therefore its physical content a little obscured. From (B.56) we can, in fact, show that the power spectrum of a fluctuating variable A, say, is related to the rate of energy dissipation of an external field coupled to the system by its corresponding operator $\hat{A}$. It is worth while to indicate how we can arrive at this result by a more physical argument than that used to derive the expressions given above.

For this discussion it is convenient to assume that $\langle\hat{A}\rangle$ is zero. Alternatively, we could employ an operator $\hat{\mathscr{A}}$ defined as

$$\hat{\mathscr{A}} = \hat{A}-\langle\hat{A}\rangle$$

so that $\langle(\hat{\mathscr{A}})^2\rangle$ is the mean square fluctuation of the variable A.

† Ibid.

‡ R. Kubo, *Rep. Prog. Phys.* **29**, 255 (1966).

As a result of the action of the external field, the system described by the Hamiltonian $\mathscr{H}_0$ must change its state and, in doing so, absorb energy from the field. The rate of absorption of energy from the field can be calculated as follows. If E is the mean energy of the complete system then, denoting the average of the total system by a bar,† and setting $\hat{B} \equiv \hat{A}$,

$$\frac{\mathrm{d}E}{\mathrm{d}t} = \overline{\frac{\partial \mathscr{H}}{\partial t}} = -\bar{A}\frac{\mathrm{d}f}{\mathrm{d}t}.$$

This is the rate at which the external field feeds energy into the system, i.e. the rate at which energy is dissipated from the external field, Q. We now put

$$f(t) = \tfrac{1}{2}\{f_0 \exp(-\mathrm{i}\omega t) + f_0^* \exp(\mathrm{i}\omega t)\},$$

and from (B.4) and (B.5) we have

$$\bar{A}(t) = \tfrac{1}{2} f_0\, \chi_{AA}[\omega]\exp(-\mathrm{i}\omega t) + \tfrac{1}{2} f_0^*\, \chi_{AA}^*[\omega]\exp(\mathrm{i}\omega t),$$

and hence

$$\begin{aligned} Q &= -\bar{A}(t)\frac{\mathrm{d}f}{\mathrm{d}t} \\ &= \frac{-\mathrm{i}\omega}{4}[f_0 f_0^*\{\chi_{AA}[\omega]-\chi_{AA}^*[\omega]\} - f_0^2\chi_{AA}[\omega]\exp(-2\mathrm{i}\omega t) - f_0^{*2}\chi_{AA}^*[\omega]\exp(2\mathrm{i}\omega t)]. \end{aligned}$$

The last two terms oscillate with time and give no mean dissipation rate. Hence only the first term survives to give the average dissipation rate as

$$Q = \tfrac{1}{2}\omega |f_0|^2 \chi_{AA}''[\omega]. \tag{B.57}$$

(B.57) shows that the energy dissipation Q is determined by the imaginary part of the generalized susceptibility.

If W_{mn} is the probability per unit time for the transition from the state m of $\mathscr{H}_0$ to the state n as a consequence of the external perturbation, then it follows that the energy dissipated per second is

$$Q_m = \sum_n W_{mn}(E_n - E_m).$$

The total rate of dissipation of energy is obtained from this by averaging over all the states m, i.e.

$$Q = \frac{1}{Z}\sum_{mn} \exp(-\beta E_m) W_{mn}(E_n - E_m). \tag{B.58}$$

From perturbation theory we know that

$$W_{mn} = \frac{\pi}{2\hbar}|f_0|^2 |A_{mn}|^2\{\delta(\hbar\omega + E_n - E_m) + \delta(\hbar\omega + E_m - E_n)\},$$

so that from § B.5

$$Q = \frac{\pi\omega}{2\hbar}|f_0|^2\{\exp(\hbar\omega\beta) - 1\}G_{AA}^{<}(\omega). \tag{B.59}$$

The *power spectrum* of A is defined by

$$\mathscr{G}_A(\omega) = \frac{1}{2\pi}\int_{-\infty}^{\infty} \mathrm{d}t\, \exp(\mathrm{i}\omega t)\langle\{\hat{A}(t)\hat{A}(0)\}\rangle = \tfrac{1}{2}\{\exp(\hbar\omega\beta)+1\}G_{AA}^{<}(\omega). \tag{B.60}$$

† See, for example, L. D. Landau and E. M. Lifshitz, *Statistical physics*, Pergamon Press (1956).

Thus, on combining (B.57) with (B.59) and (B.60), we have

$$\mathscr{G}_A(\omega) = \frac{\hbar}{2\pi}\chi''_{AA}[\omega]\coth(\tfrac{1}{2}\hbar\omega\beta), \tag{B.61}$$

which is the desired relationship.

It follows immediately from (B.61) that

$$\langle(\hat{A})^2\rangle = \frac{\hbar}{\pi}\int_0^\infty \mathrm{d}\omega\, \chi''_{AA}[\omega]\coth(\tfrac{1}{2}\hbar\omega\beta), \tag{B.62}$$

which relates the mean square fluctuation of $\hat{A}$ to the response of the system to an external field coupled to the system by $\hat{A}$. This could, as mentioned above, have been obtained directly from (B.56), since $\phi_{AA}(t)$ must be an odd function of t. However, the foregoing derivation of (B.62) demonstrates more clearly the physical ideas involved.

Furthermore, we see why the theorem arises in the study of the Born approximation to the differential scattering cross-section, since, for an interaction of the form considered above, the cross-section for the process in which an incident wave $\exp(\mathrm{i}\mathbf{k}\,.\,\mathbf{r})$ is scattered into the state $\exp(\mathrm{i}\mathbf{k}'\,.\,\mathbf{r})$, the target state changing from m to n, is proportional to

$$\frac{k'}{k}\int \mathrm{d}\mathbf{r}\exp(-\mathrm{i}\mathbf{k}'\,.\,\mathbf{r})W_{mn}\exp(\mathrm{i}\mathbf{k}\,.\,\mathbf{r}).$$

Therefore Q is proportional to the negative of the mean energy lost by the incident particle, i.e. the differential retardation.

B.7. Isothermal response function

We consider now the change in the average value of an observable A due to a small time-independent perturbation $\mathscr{H}_1$. As before we write the total Hamiltonian

$$\mathscr{H} = \mathscr{H}_0 - \mathscr{H}_1$$

and seek

$$\frac{1}{Z'}\operatorname{Tr}\exp(-\beta\mathscr{H})\hat{A},$$

where

$$Z' = \operatorname{Tr}\exp(-\beta\mathscr{H}).$$

A general solution to the problem can be obtained by defining $\hat{S}(\beta)$ by the equation

$$\exp\{-\beta(\mathscr{H}_0-\mathscr{H}_1)\} = \exp(-\beta\mathscr{H}_0)\hat{S}(\beta),$$

for then it satisfies

$$\frac{\partial\hat{S}}{\partial\beta} = \mathscr{H}_1(-\mathrm{i}\hbar\beta)\hat{S}(\beta),$$

with the boundary condition $\hat{S}(0) = 1$. In principle $\hat{S}(\beta)$ can be obtained to any order in the perturbation by iterating the integral equation

$$\hat{S}(\beta) = 1 + \int_0^\beta \mathrm{d}\lambda\, \mathscr{H}_1(-\mathrm{i}\hbar\lambda)\hat{S}(\lambda).$$

For our purposes, however, it is sufficient to take just the first order solution because we have specified that the perturbation is small. Thus

$$\hat{S}(\beta) = 1 + \int_0^\beta \mathrm{d}\lambda\, \mathscr{H}_1(-\mathrm{i}\hbar\lambda),$$

$$Z' = Z(1+\beta\langle\mathscr{H}_1\rangle),$$

and $$\frac{1}{Z'}\mathrm{Tr}\exp(-\beta\mathscr{H})\hat{A} = \left\{\langle\hat{A}\rangle+\int_0^\beta \mathrm{d}\lambda\,\langle\mathscr{H}_1(-\mathrm{i}\hbar\lambda)\hat{A}\rangle\right\}\{1+\beta\langle\mathscr{H}_1\rangle\}^{-1},$$

and so the change in the average value of A is

$$\int_0^\beta \mathrm{d}\lambda\,\langle\mathscr{H}_1(-\mathrm{i}\hbar\lambda)\hat{A}\rangle-\beta\langle\hat{A}\rangle\langle\mathscr{H}_1\rangle$$

to first order in the perturbation.

If we write $\mathscr{H}_1 = f\hat{B}$, where f is the relevant perturbation parameter, the ***isothermal susceptibility*** is, from the last expression,

$$\chi_{AB} = \int_0^\beta \mathrm{d}\lambda\,\langle\hat{B}(-\mathrm{i}\hbar\lambda)\hat{A}\rangle-\beta\langle\hat{A}\rangle\langle\hat{B}\rangle. \tag{B.63}$$

The static susceptibility $\chi_{AB}[0]$ is given from (B.28) and (B.29) by

$$\chi_{AB}[0] = \int_0^\beta \mathrm{d}\lambda\,\langle\hat{B}(-\mathrm{i}\hbar\lambda)\hat{A}\rangle-\frac{\beta}{Z}\sum_m \exp(-\beta E_m)A_{mm}B_{mm}. \tag{B.64}$$

The two expressions (B.63) and (B.64) differ because, as pointed out in § B.3, $\chi_{AB}[\omega]$ is calculated under the assumption that the external perturbation behaves adiabatically, while χ_{AB} is an isothermal response, as is clear from its derivation. However, in some circumstances the difference between $\chi_{AB}[0]$ and χ_{AB} is entirely negligible. To see when this will arise consider the long-time behaviour of the correlation function $\langle\hat{B}(0)\hat{A}(t)\rangle$. We might expect for most systems as $t\to\infty$ that

$$\langle\hat{B}(0)\hat{A}(t)\rangle \to \langle\hat{B}\rangle\langle\hat{A}\rangle. \tag{B.65 a}$$

But, by the same argument that we used in § B.4,

$$\lim_{t\to\infty}\langle\hat{B}(0)\hat{A}(t)\rangle = \frac{1}{Z}\sum_m B_{mm}A_{mm}\exp(-\beta E_m), \tag{B.65 b}$$

which is, apart from the factor β, the second term on the right-hand side of (B.64). Whether or not (B.65 a) does hold for a given system must be investigated for that case, but it is found that in most systems of interest it is in fact valid.†

B.8. Thermal two-time Green functions

We define the advanced and retarded Green functions, $G^{(-)}$ and $G^{(+)}$ respectively, for two arbitrary operators $\hat{O}$ and $\hat{P}$, as (see B.36)‡

$$G^{(+)}_{OP}(t-t') = -\mathrm{i}\theta(t-t')g_{OP}(t-t') \equiv \langle\langle\hat{O}(t);\,\hat{P}(t')\rangle\rangle^{(+)} \tag{B.66 a}$$

and $$G^{(-)}_{OP}(t-t') = \mathrm{i}\theta(t'-t)g_{OP}(t-t') \equiv \langle\langle\hat{O}(t);\,\hat{P}(t')\rangle\rangle^{(-)}. \tag{B.66 b}$$

The boundary conditions for $G^{(\pm)}_{OP}$ are determined by taking the limit of zero argument, i.e. for $G^{(+)}_{OP}$, say,

$$\lim_{t\to 0+} G^{(+)}_{OP}(t) = -\mathrm{i}g_{OP}(t=0).$$

† R. Kubo, *J. phys. Soc. Japan* **12**, 570 (1957).

‡ See, for example, D. N. Zubarev, *Soviet Phys. Usp.* **3**, 320 (1960), J. Hubbard, Harwell lectures (T.P. 174).

With the help of the integral expression for $\theta(t)$, eqn (B.12), the Fourier transforms of the Green functions

$$G_{OP}^{(\pm)}(\omega) = \frac{1}{2\pi}\int_{-\infty}^{\infty} dt\, \exp(i\omega t)\, G_{OP}^{(\pm)}(t)$$

can be expressed as
$$G_{OP}^{(\pm)}(\omega) = \frac{1}{2\pi}\int_{-\infty}^{\infty} \frac{du\, g_{OP}(u)}{\omega - u \pm i0^{+}}. \tag{B.67}$$

By comparing (B.47) and (B.67) we see that if in the latter we take $\hat{O} = \hat{A}$, $\hat{P} = \hat{B}$, and $\eta = 1$, we have

$$\chi_{AB}[\omega] = -\frac{2\pi}{\hbar}\, G_{AB}^{(+)}(\omega). \tag{B.68}$$

In § B.2 we discussed the analytic properties of $\chi_{AB}[\omega]$ and concluded that it was the limit of a function $\chi_{AB}[z]$ which is analytic in the upper half plane. The relations we derived between the real and imaginary parts of $\chi_{AB}[\omega]$ are, in fact, also applicable to the Fourier transforms of the Green functions even though, in general, $g_{OP}(\omega)$ does not satisfy (B.18). To see this we argue from (B.67) that $G_{OP}^{(+)}(\omega)$ can be continued analytically in to the upper half plane, and $G_{OP}^{(-)}(\omega)$ into the lower half plane, with a cut along the real axis. Then, with the assumption that $G_{OP}^{(\pm)}(\omega)$ vanishes for $|\omega| = \infty$, Cauchy's integral theorem gives

$$\int_{-\infty}^{\infty} \frac{du\; G_{OP}^{(\pm)}(u)}{\omega - u \mp i0^{+}} = 0. \tag{B.69}$$

On separating the real and imaginary parts of (B.69) we obtain relations analogous to (B.14). If $G_{OP}^{(\pm)}(\omega)$ does not tend to zero as $|\omega| \to \infty$, but tends to some constant value, it is straightforward to generalize (B.69) to cover this case by merely applying Cauchy's integral theorem to the function formed from $G_{OP}^{(\pm)}(\omega)$ on subtracting off the constant value.

To calculate the correlation function $\langle \hat{P}(0)\hat{O}(t)\rangle$ from the Green functions we require to know $G_{OP}^{<}(\omega)$, say. From (B.67) we obtain

$$g_{OP}(\omega) = i\{G_{OP}^{(+)}(\omega) - G_{OP}^{(-)}(\omega)\} \equiv -2\,\mathrm{Im}\, G_{OP}^{(+)}(\omega). \tag{B.70}$$

Hence,

$$\begin{aligned} G_{OP}^{<}(\omega) &= i\{\exp(\hbar\omega\beta) - \eta\}^{-1}\{G_{OP}^{(+)}(\omega) - G_{OP}^{(-)}(\omega)\} + \tfrac{1}{2}\Upsilon(1+\eta)\delta(\omega) \\ &= -2\{\exp(\hbar\omega\beta) - \eta\}^{-1}\,\mathrm{Im}\, G_{OP}^{(+)}(\omega) + \tfrac{1}{2}\Upsilon(1+\eta)\delta(\omega). \end{aligned} \tag{B.71}$$

The zero-frequency term in (B.71) arises because for $\eta = 1$ the first term on the right-hand side of (B.71) is defined at $\omega = 0$ to within a δ-function, the coefficient of which we can choose to give the correct answer, taking account that we choose to define the integral of the first term by its principal part. Explicitly, the coefficient Υ is given by (B.65 b), on replacing $\hat{A}$ and $\hat{B}$ with $\hat{O}$ and $\hat{P}$, and hence our remarks there are equally applicable to the present problem.

The Green functions $G_{OP}^{(\pm)}(t)$ are calculated from their equations of motion. These are readily shown to have the form

$$i\hbar\frac{\partial}{\partial t} G_{OP}^{(\pm)}(t) = \hbar\delta(t) g_{OP}(t=0) + \langle\langle[\hat{O}(t), \mathscr{H}];\, \hat{P}(0)\rangle\rangle^{(\pm)}. \tag{B.72}$$

The Green function on the right-hand side of (B.72) is usually of a higher order than that ocurring on the left-hand side, i.e. it contains three or more operators instead of two. Thus (B.72) forms, in general, one of a chain of equations and some form of approximation scheme must be introduced by which the higher order Green function is expressed as a function of a lower order Green function. Alternatively one may make some approximation to the Hamiltonian that ensures that the right-hand side of (B.72) does not contain a higher order Green function.

In terms of $G^{(\pm)}_{OP}(\omega)$ the equation of motion (B.72) reads

$$\hbar\omega G^{(\pm)}_{OP}(\omega) = \frac{\hbar}{2\pi} g_{OP}(t=0) + G^{(\pm)}_{[O,\mathscr{H}_1],P}(\omega). \tag{B.73}$$

B.9. Spatially varying perturbations

Our discussion has so far been concerned with the time dependence of the response of a system. It is very easy to extend all our previous analysis to cover the response of a system described by $\mathscr{H}_0$ to a magnetic field $\mathbf{H}_\mathbf{l}(t)$, i.e. we take

$$\mathscr{H}_1 = -g\mu_\mathrm{B} \sum_\mathbf{l} \mathbf{H}_\mathbf{l}(t)\,.\,\hat{\mathbf{S}}_\mathbf{l}, \tag{B.74}$$

where $\hat{\mathbf{S}}_\mathbf{l}$ is the spin operator associated with the unpaired electrons at the Bravais lattice site $\mathbf{l}$.

If we define spin raising and lowering operators $\hat{S}^\pm$ by

$$\hat{S}^\pm = \hat{S}^x \pm \mathrm{i}\hat{S}^y, \tag{B.75}$$

and transverse components of the magnetic field

$$H^\pm = \tfrac{1}{2}(H^x \pm \mathrm{i}H^y), \tag{B.76}$$

then

$$\mathbf{H}\,.\,\hat{\mathbf{S}} \equiv \sum_\alpha H^\alpha \hat{S}^\alpha \tag{B.77 a}$$

$$= \sum_\nu H^\nu (\hat{S}^\nu)^+, \tag{B.77 b}$$

where we define ν (or μ) $= 0, +, -$ with $\nu = 0 \equiv z$, the indices α, β being retained to define cartesian components of a vector.

We seek the change in the magnetization $\mathbf{M}$ of the system described by $\mathscr{H}_0$. By definition

$$M^\alpha_\mathbf{l}(t) = -g\mu_\mathrm{B}\,\mathrm{Tr}\,\hat{\rho}(t)\hat{S}^\alpha_\mathbf{l}, \tag{B.78}$$

which, from § B.3, is given by (assuming that $\mathscr{H}_0$ describes a spatial uniform system)

$$M^\mathbf{l}_\alpha(t) = -g\mu_\mathrm{B}\,\mathrm{Tr}\,\hat{\rho}_0\,\hat{S}^\alpha_\mathbf{l} - g\mu_\mathrm{B}\,\mathrm{Tr}\,\Delta\hat{\rho}(t)\hat{S}^\alpha_\mathbf{l}$$

$$= -g\mu_\mathrm{B}\langle\hat{S}^\alpha\rangle + \sum_{\mathbf{l}'} \int_{-\infty}^{t} \mathrm{d}t' \sum_\beta \phi^{\alpha\beta}(\mathbf{l}-\mathbf{l}', t-t') H^\beta_{\mathbf{l}'}(t'), \tag{B.79}$$

where

$$\phi^{\alpha\beta}(\mathbf{l}-\mathbf{l}', t-t') = (g\mu_\mathrm{B})^2 \frac{\mathrm{i}}{\hbar} \langle[\hat{S}^\alpha_\mathbf{l}(t), \hat{S}^\beta_{\mathbf{l}'}(t')]\rangle. \tag{B.80}$$

We adopt the following convention for spatial Fourier transforms. If $\mathbf{A}_\mathbf{l}$ is an arbitrary vector function of the lattice site $\mathbf{l}$,

$$A^\alpha_\mathbf{l} = \frac{1}{N}\sum_\mathbf{q} \exp(\mathrm{i}\mathbf{q}\,.\,\mathbf{l})A^\alpha_\mathbf{q} \quad (\text{for } \alpha = x, y, \text{ or } z),$$

$$A^+_\mathbf{l} = \frac{1}{N}\sum_\mathbf{q} \exp(\mathrm{i}\mathbf{q}\,.\,\mathbf{l})A^+_\mathbf{q},$$

$$A^-_\mathbf{l} = \frac{1}{N}\sum_\mathbf{q} \exp(-\mathrm{i}\mathbf{q}\,.\,\mathbf{l})A^-_\mathbf{q}\,; \tag{B.81 a}$$

then

$$A^{\alpha}_{\mathbf{q}} = \sum_{\mathbf{l}} \exp(-\mathrm{i}\mathbf{q}\,.\,\mathbf{l})A^{\alpha}_{\mathbf{l}},$$

$$A^{+}_{\mathbf{q}} = \sum_{\mathbf{l}} \exp(-\mathrm{i}\mathbf{q}\,.\,\mathbf{l})A^{+}_{\mathbf{l}} = A^{x}_{\mathbf{q}}+\mathrm{i}A^{y}_{\mathbf{q}},$$

$$A^{-}_{\mathbf{q}} = \sum_{\mathbf{l}} \exp(\mathrm{i}\mathbf{q}\,.\,\mathbf{l})A^{-}_{\mathbf{l}} = (A^{+}_{\mathbf{q}})^{+} = A^{x}_{-\mathbf{q}}-\mathrm{i}A^{y}_{-\mathbf{q}}, \tag{B.81 b}$$

since

$$\sum_{\mathbf{l}} \exp(\mathrm{i}\mathbf{q}\,.\,\mathbf{l}) = N\delta_{\mathbf{q},0}$$

and

$$\sum_{\mathbf{q}} \exp(\mathrm{i}\mathbf{q}\,.\,\mathbf{l}) = N\delta_{\mathbf{l},0}. \tag{B.81 c}$$

On taking the spatial Fourier transform of eqn (B.79) we obtain

$$M^{\alpha}_{\mathbf{q}}(t) = -g\mu_{\mathrm{B}}N\delta_{\mathbf{q},0}\langle\hat{S}^{\alpha}\rangle + \int_{-\infty}^{t} \mathrm{d}t' \sum_{\beta} \phi^{\alpha\beta}_{\mathbf{q}}(t-t')H^{\beta}_{\mathbf{q}}(t'), \tag{B.82}$$

and from (B.80)

$$\phi^{\alpha\beta}_{\mathbf{q}}(t-t') = \frac{(g\mu_{\mathrm{B}})^2}{N}\frac{\mathrm{i}}{\hbar}\langle[\hat{S}^{\alpha}_{\mathbf{q}}(t),\, \hat{S}^{\beta}_{-\mathbf{q}}(t')]\rangle. \tag{B.83}$$

The calculation of the μth component of $\mathbf{M}$ follows exactly that leading to (B.79), namely

$$M^{\mu}_{\mathbf{l}}(t) = -g\mu_{\mathrm{B}}\langle\hat{S}^{\mu}\rangle + \sum_{\mathbf{l}'} \int_{-\infty}^{t} \mathrm{d}t' \sum_{\nu} \phi^{\mu\nu}(\mathbf{l}-\mathbf{l}',\, t-t')H^{\nu}_{\mathbf{l}'}(t'), \tag{B.84}$$

where

$$\phi^{\mu\nu}(\mathbf{l}-\mathbf{l}',\, t-t') = (g\mu_{\mathrm{B}})^2\frac{\mathrm{i}}{\hbar}\langle[\hat{S}^{\mu}_{\mathbf{l}}(t),\, \{\hat{S}^{\nu}_{\mathbf{l}'}(t')\}^{+}]\rangle. \tag{B.85}$$

If $\mathscr{H}_0$ commutes with $\hat{S}^{z}_{\mathrm{tot}} = \sum \hat{S}^{z}_{\mathbf{l}}$ then the only non-zero terms in the summation over ν in the second term of (B.84) are those for which $\mu = \nu$. For in this case the eigenfunctions of $\mathscr{H}_0$ are also eigenfunctions of $\hat{S}^{z}_{\mathrm{tot}}$ and thus sums over matrix elements of the type

$$\langle m|\hat{S}^{\mu}_{\mathbf{l}}|n\rangle\langle n|\{\hat{S}^{\nu}_{\mathbf{l}'}\}^{+}|m\rangle,$$

which occur in $\phi^{\mu\nu}$, are non-zero only if $\mu = \nu$. In view of this fact we define

$$\phi^{(0)}(\mathbf{l}-\mathbf{l}',t-t') = (g\mu_{\mathrm{B}})^2\frac{\mathrm{i}}{\hbar}\langle[\hat{S}^{z}_{\mathbf{l}}(t),\, \hat{S}^{z}_{\mathbf{l}'}(t')]\rangle, \tag{B.86}$$

$$\phi^{(+)}(\mathbf{l}-\mathbf{l}',t-t') = (g\mu_{\mathrm{B}})^2\frac{\mathrm{i}}{\hbar}\langle[\hat{S}^{+}_{\mathbf{l}}(t),\, \hat{S}^{-}_{\mathbf{l}'}(t')]\rangle, \tag{B.87 a}$$

and

$$\phi^{(-)}(\mathbf{l}-\mathbf{l}',t-t') = (g\mu_{\mathrm{B}})^2\frac{\mathrm{i}}{\hbar}\langle[\hat{S}^{-}_{\mathbf{l}}(t),\, \hat{S}^{+}_{\mathbf{l}'}(t')]\rangle = \{\phi^{(+)}(\mathbf{l}-\mathbf{l}',t-t')\}^{*}, \tag{B.87 b}$$

and (B.84) becomes

$$M^{\mu}_{\mathbf{l}}(t) = -g\mu_{\mathrm{B}}\langle\hat{S}^{\mu}\rangle + \sum_{\mathbf{l}'} \int_{-\infty}^{t} \mathrm{d}t'\,\phi^{\mu}(\mathbf{l}-\mathbf{l}',t-t')H^{\mu}_{\mathbf{l}'}. \tag{B.87 c}$$

We draw attention to the fact that the functions $\phi^{(\pm)}$ defined by eqns (B.87 a) and (B.87 b) are not response functions in the sense of § B.1 because they are defined in terms of non-hermitian operators. However, we can formally define corresponding susceptibility $\chi^{(\pm)}$ and relaxation functions $R^{(\pm)}$ through eqns (B.5) and (B.26), respectively, and the inter-relations between the various functions remain valid provided one takes account of the difference in the symmetry properties between these 'transverse' functions and those of §§ B.1–8.

The fundamental difference is that $\phi^{(\pm)}$ are not real, as is evident from (B.84), for they measure the response to a complex perturbation. From (B.87 b) we note that the spatial Fourier transforms of $\phi^{(\pm)}(\mathbf{l},t)$ satisfy

$$\phi_{\mathbf{q}}^{(+)}(t) = \sum_{\mathbf{l}} \exp(-\mathrm{i}\mathbf{q}\,.\,\mathbf{l})\phi^{(+)}(\mathbf{l},t) = \{\phi_{\mathbf{q}}^{(-)}(t)\}^*. \tag{B.88}$$

For the susceptibility functions $\chi_{\mathbf{q}}^{(\pm)}[\omega]$ we note the following symmetry properties. If we take the time dependence of $H_{\mathbf{l}}^{+}(t)$ to be $\exp(-\mathrm{i}\omega t)$, then

$$M_{\mathbf{l}}^{+}(t) = \exp(-\mathrm{i}\omega t)\sum_{\mathbf{l}'}\int_0^\infty \mathrm{d}t'\;\phi^{(+)}(\mathbf{l}-\mathbf{l}',t')\exp(\mathrm{i}\omega t')H_{\mathbf{l}'}^{+} \tag{B.89}$$

and the corresponding equation for $M_{\mathbf{l}}^{-}(t)$ is $[H_{\mathbf{l}}^{+}(t) = \{H_{\mathbf{l}}^{-}(t)\}^*]$

$$M_{\mathbf{l}}^{-}(t) = \exp(\mathrm{i}\omega t)\sum_{\mathbf{l}'}\int_0^\infty \mathrm{d}t'\;\phi^{(-)}(\mathbf{l}-\mathbf{l}',t')\exp(-\mathrm{i}\omega t')H_{\mathbf{l}'}^{-}. \tag{B.90}$$

Since we have $M_{\mathbf{l}}^{+}(t) = \{M_{\mathbf{l}}^{-}(t)\}^*$, it follows from (B.90) that the spatial Fourier transforms of $\chi^{(\pm)}$, namely

$$\chi_{\mathbf{q}}^{(\pm)}[\omega] = \sum_{\mathbf{l}} \exp(\mp\mathrm{i}\mathbf{q}\,.\,\mathbf{l})\int_0^\infty \mathrm{d}t\exp(\mathrm{i}\omega t)\phi^{(\pm)}(\mathbf{l},t) \tag{B.91}$$

satisfy

$$\chi_{\mathbf{q}}^{(+)}[\omega] = \{\chi_{\mathbf{q}}^{(-)}[-\omega]\}^*. \tag{B.92}$$

As the final topic in this section, consider the isothermal susceptibility corresponding to a time-independent perturbation,

$$\mathscr{H}_1 = -g\mu_{\mathrm{B}}\sum_{\mathbf{l}} \mathbf{H}_{\mathbf{l}}\,.\,\hat{\mathbf{S}}_{\mathbf{l}}. \tag{B.93}$$

From § B.7 it follows that

$$M_{\mathbf{l}}^{\alpha} = -g\mu_{\mathrm{B}}\langle\hat{S}^{\alpha}\rangle + \sum_{\mathbf{l}'}\sum_{\beta}\chi^{\alpha\beta}(\mathbf{l}-\mathbf{l}')H_{\mathbf{l}'}^{\beta}, \tag{B.94}$$

where

$$\chi^{\alpha\beta}(\mathbf{l}-\mathbf{l}') = (g\mu_{\mathrm{B}})^2\left\{\int_0^\beta \mathrm{d}\lambda\,\langle\hat{S}_{\mathbf{l}'}^{\beta}(-\mathrm{i}\hbar\lambda)\hat{S}_{\mathbf{l}}^{\alpha}(0)\rangle - \beta\langle\hat{S}_{\mathbf{l}'}^{\beta}\rangle\langle\hat{S}_{\mathbf{l}}^{\alpha}\rangle\right\}. \tag{B.95}$$

Taking a spatial Fourier transform of (B.94) we obtain

$$M_{\mathbf{q}}^{\alpha} = -g\mu_{\mathrm{B}}N\delta_{\mathbf{q},0}\langle\hat{S}^{\alpha}\rangle + \sum_{\beta}\chi_{\mathbf{q}}^{\alpha\beta}H_{\mathbf{q}}^{\beta}, \tag{B.96}$$

where

$$\chi_{\mathbf{q}}^{\alpha\beta} = \frac{(g\mu_{\mathrm{B}})^2}{N}\left\{\int_0^\beta \mathrm{d}\lambda\,\langle\hat{S}_{\mathbf{q}}^{\beta}(-\mathrm{i}\hbar\lambda)\hat{S}_{-\mathbf{q}}^{\alpha}(0)\rangle - \beta\langle\hat{S}_{\mathbf{q}}^{\beta}\rangle\langle\hat{S}_{-\mathbf{q}}^{\alpha}\rangle\right\}. \tag{B.97}$$

APPENDIX C

THE purpose of this appendix is to outline the first-order calculation of the phonon lifetime in a Bravais crystal due to anharmonic interactions. The effect of anharmonic interactions on the one-phonon neutron cross-section has been discussed on pages 91–2, but the actual form of the phonon self-energy was not specified. Here we show how the latter can be evaluated to first order by utilizing the same scheme as that adopted in calculating the ferromagnetic magnon lifetime (compare § 9.2 (*d*)). Because the calculation of the phonon self-energy parallels very closely that of the magnon self-energy, the algebra is not given in full.

The anharmonic contributions to the phonon Hamiltonian are derived by retaining cubic and higher-order terms in the Taylor expansion of the potential energy of the nuclei. In lowest order, i.e. in the Hartree–Fock approximation, these do not lead to a finite phonon lifetime, but give only a shift Δ in the phonon energies $\hbar\omega_j(\mathbf{q})$. The first-order approximation to the phonon lifetime is obtained in second order where

$$\omega_j(\mathbf{q}) \to \omega_j(\mathbf{q})+\Delta_j(\mathbf{q},\omega)+\mathrm{i}\Gamma_j(\mathbf{q},\omega). \qquad \text{(C.1)}$$

It follows that we must treat the cubic terms to second order and quartic terms to first order to be consistent. In fact, there is no contribution to the self-energy from the cubic terms in first order.

If cubic and quartic terms in the expansion of the potential are retained, the Hamiltonian is

$$\mathscr{H} = \sum_{j\mathbf{q}} \hbar\omega_j(\mathbf{q})\hat{a}_j^+(\mathbf{q})\hat{a}_j(\mathbf{q})+\sum_{\alpha,\beta,\gamma} V^{(3)}(\alpha,\beta,\gamma)\hat{A}(\alpha)\hat{A}(\beta)\hat{A}(\gamma)+$$
$$+\sum_{\alpha,\beta,\gamma,\delta} V^{(4)}(\alpha,\beta,\gamma,\delta)\hat{A}(\alpha)\hat{A}(\beta)\hat{A}(\gamma)\hat{A}(\delta), \qquad \text{(C.2)}$$

where for brevity we have denoted the polarization index and the wave vector collectively by α, β, γ, or δ in the anharmonic terms, and (compare eqn (4.22a))

$$\hat{A}(\alpha) = \hat{a}_j(\mathbf{q})+\hat{a}_j^+(-\mathbf{q}). \qquad \text{(C.3)}$$

The coefficients $V^{(3)}$ and $V^{(4)}$ are fully symmetric in their indices and satisfy the relations

$$\{V^{(3)}(\alpha,\beta,\gamma)\}^* = V^{(3)}(-\alpha,-\beta,-\gamma)$$

and

$$\{V^{(4)}(\alpha,\beta,\gamma,\delta)\}^* = V^{(4)}(-\alpha,-\beta,-\gamma,-\delta). \qquad \text{(C.4)}$$

In the equation of motion for $\hat{a}_j(\mathbf{q})$, say, the cubic term contributes products of operators and the quartic term contributes $\hat{a}\hat{a}\hat{a}^+$, $\hat{a}\hat{a}^+\hat{a}$, and $\hat{a}^+\hat{a}\hat{a}$, when combinations that are zero in the Hartree–Fock approximation are neglected. To proceed, we form the equations of motion for the products of operators and approximate the higher-order terms therein by their Hartree–Fock values. The quartic terms we treat in first order, and replace the corresponding triple products by their Hartree–Fock values. This scheme creates a linear equation for the operator $\hat{a}_j(\mathbf{q})$ and we obtain the desired result, namely,†

$$\Delta_j(\mathbf{q},\omega)+\mathrm{i}\Gamma_j(\mathbf{q},\omega) = 12\sum_{\alpha} V^{(4)}(-\alpha,q,\alpha,-q)\{2n_{\alpha}+1\}-$$

$$-18\sum_{\alpha,\beta}|V^{(2)}(-q,\beta,-\alpha)|^2\left\{\frac{n_{\alpha}+n_{\beta}+1}{\omega+\mathrm{i}0^{+}+\omega_{\alpha}+\omega_{\beta}}-\right.$$

$$\left.-\frac{n_{\alpha}+n_{\beta}+1}{\omega+\mathrm{i}0^{+}-\omega_{\alpha}-\omega_{\beta}}-2\frac{n_{\alpha}-n_{\beta}}{\omega+\mathrm{i}0^{+}+\omega_{\alpha}-\omega_{\beta}}\right\}. \quad \text{(C.5)}$$

The structure of the energy-dependent term, which gives Γ, is worth noting. First, it depends on the square of the cubic interaction constant, i.e. its contribution is second order. Secondly, it arises from processes involving the annihilation and creation of two phonons. $V^{(3)}$ and $V^{(4)}$ depend intimately on the structure of the phonon spectrum and hence we cannot obtain a meaningful analytic form for $\Gamma_j(\mathbf{q},\omega)$, valid in the region of small $\mathbf{q}$ and ω. Maradudin and Fein‡ derived the result (C.5) using diagrammatic perturbation theory, and evaluated Γ for an approximate model of phonons in lead.

† E. Balcar, *Acta Phys. austriaca* **31,** 300 (1970).
‡ A. A. Maradudin and A. E. Fein, *Phys. Rev.* **128,** 2589 (1962).

AUTHOR INDEX

SUBJECT INDEX